HANDBOOK OF

Chemical Technology and Pollution Control

(Third Edition)

HANDBOOK OF
Chemical Technology and Pollution Control
(Third Edition)

Martin B. Hocking
Department of Chemistry
University of Victoria
Victoria, British Columbia
Canada

AMSTERDAM • BOSTON • HEIDELBERG • LONDON
NEW YORK • OXFORD • PARIS • SAN DIEGO
SAN FRANCISCO • SINGAPORE • SYDNEY • TOKYO

Academic Press is an imprint of Elsevier

ELSEVIER

Academic Press is an imprint of Elsevier

525 B Street, Suite 1900, San Diego, CA 92101-4495, USA
30 Corporate Drive, Suite 400, Burlington, MA 01803, USA
Radarweg 29, PO Box 211, 1000 AE Amsterdam, The Netherlands
84 Theobald's Road, London WC1X 8RR, UK
Linacre House, Jordan Hill, Oxford OX2 8DP, UK
The Boulevard, Langford Lane, Kidlington, Oxford OX5 1GB, UK

Third edition 2005

Notice
No responsibility is assumed by the publisher for any injury and/or damage to persons
or property as a matter of products liability, negligence or otherwise, or from any use
or operation of any methods, products, instructions or ideas contained in the material
herein. Because of rapid advances in the medical sciences, in particular, independent
verification of diagnoses and drug dosages should be made

ISBN-13: 978-0-12-088796-5
ISBN-10: 0-12-088796-7

For information on all Academic Press
publications visit our website at books.elsevier.com

Printed and bound in U.S.A

05   06   07   08   09   10   10   9   8   7   6   5   4   3   2   1

Working together to grow
libraries in developing countries

www.elsevier.com | www.bookaid.org | www.sabre.org

ELSEVIER   BOOK AID
             International   Sabre Foundation

FOREWORD TO THE THIRD EDITION

The complexity and magnitude of the sustainable development challenges are huge. The issues have been on the agenda for decades. Most recently, at the World Summit on Sustainable Development (WSSD) in Johannesburg in 2002, all Member States of the United Nations agreed to work to fulfil the Johannesburg Plan of Implementation[1], on poverty alleviation and sustainable consumption and production. They confirmed that action must be an integration of efforts at local and global levels. To realise such objectives, the scientific and technical communities, as well as policy makers, are at the centre.

These urgent socio-economic and environmental challenges must be effectively addressed in a successful transition to sustainability. Sustainable development criteria need to be met for people in all countries. This will require a deeper understanding of the nature, character, and magnitude of the challenges, linked to combinations of existing or emerging technologies, resource management, and policy options. Researchers and professional groups will be key actors in the transition.

The third edition of the Handbook of Chemical Technology and Pollution Control provides excellent support in this context, in one of the areas of most concern in environment and health protection. It builds on the fundamental idea that environmental soundness has to be incorporated from the outset in industrial operations, not be restricted to add-on pieces of clean-up equipment. Senior level students, professionals, and consultants, will find a wealth

[1]Johannesburg Plan of Implementation, A/Conf.199/20.

of information and knowledge to tap here in their search for concrete and practical steps to help implement sustainable production systems.

The original concept of the book had its genesis during the author's 8 years work as a research scientist involved with process development and emission control with Dow Canada Inc. The content, structure, and presentation of the first edition, published in 1986, took shape substantially during 30 years of teaching this material to upper level students in a number of courses in industrial and environmental applied chemistry at the University of Victoria. The content was refined with use, and resulted in an enlarged and much improved 2nd edition in 1998. That edition incorporated a number of innovations including a selection of tested review questions.

The book builds on a unique combination of the author's teaching experience, research activities, and visits to most of the operating facilities described. The work also benefited from study leave experiences in Australia, Sweden, and the U.K. that provided opportunities for comparative process data collection, adding an international dimension to the content.

This 3rd edition is extensively rewritten to incorporate more recent material. It contains production data updated to year 2000, has additional figures and review questions, and includes hundreds of additional references to more current sources of information, selected for suitability for practitioners, instructors, and students alike.

It is my conviction that the Handbook of Chemical Technology and Pollution Control will be a standard textbook and reference volume world wide.

Lund, January 10, 2005

Thomas B Johansson
Professor and Director
International Institute for Industrial Environmental Economics (IIIEE)
Lund University, Lund, Sweden
thomas.b.johansson@iiiee.lu.se

CONTENTS

I Background and Technical Aspects

2 Air Quality Measurement and Effects of Pollution

3 Air Pollution Control Priorities and Methods

4 Water Quality Measurement

5 Raw Water Processing and Wastewater Treatment

6 Natural and Derived Sodium and Potassium Salts

7 Industrial Bases by Chemical Routes

8 Electrolytic Sodium Hydroxide, Chlorine, and Related Commodities

9 Sulfur and Sulfuric Acid

12 Aluminum and Compounds

13 Ore Enrichment and Smelting of Copper

14 Production of Iron and Steel

15 Production of Pulp and Paper

16 Fermentation and Other Microbiological Processes

17 Petroleum Production and Transport

18 Petroleum Refining

19 Petrochemicals

20 Condensation (Step-Growth) Polymer Theory

21 Commercial Polycondensation (Step-Growth) Polymers

22 Addition (Chain Reaction) Polymer Theory

23 Commercial Addition (Vinyl-Type) Polymers

PREFACE TO THE THIRD EDITION

The objectives of the second edition have been maintained and updated to 2005 in the current volume where users will find that one third of the references, now totaling more than 1300, are new to this edition. At the same time most of the in-depth *Further Reading* suggestions are new to this volume, and production data of some 30 tables has been updated. Thirty percent of the 175 *Review Questions* are new to this edition. All have been tested by students. The scope of this book has also been expanded by the addition of two new appendices. The first comprises a select list of references relating to soil pollution and remediation methodologies. The second covers an organized selection of web sites relevant to the topics covered in the book. All of these changes have been achieved in a volume which is only slightly larger than the second edition by summarizing less essential content, and by the deletion of a few outdated technologies with referral of readers to the second edition and other sources for details.

As with the earlier editions, I invite users of this book to offer their suggestions for improvement.

Martin B. Hocking
May 2005

PREFACE TO THE SECOND EDITION

The objectives which motivated the first edition, a unified treatment of the fields of industrial and environmental chemistry, have been maintained here. The result is intended to be of interest to senior students in applied chemistry, science, engineering, and environmental programs in universities and colleges, as well as to professionals and consultants employed in these fields.

This edition further develops, refines, and updates the earlier material by drawing on progress in these fields, and responds to comments from users of the first edition. Sections relating to air and water pollution assessment and theory have been expanded, chapters on petrochemical production and basic polymer theory and practice have been added, and the original material has been supplemented by new data. In addition review questions have now been added to each chapter. These will be primarily of interest to students but could be of conceptual value to all users.

The new edition has been assembled to make it easy to use on any or all of three levels. Basic principles and theory of each process are discussed initially, followed by more recent refinements and developments of each process, finally supplemented with material which relates to possible process losses and integral and end-of-pipe emission control measures. The user's interest can dictate the level of approach to the material in the book, from a survey of a selection of basic processes to an in-depth referral to one or more particular processes, as appropriate. Chemical reactions and quantitative assessment are emphasized throughout, using worked examples to aid understanding.

Extensive current and retrospective production and consumption data has been maintained and expanded from the first edition to give an idea of the scale and volume trends of particular processes, and an indication as to regional similarities and differences. This material also provides a basis for consideration of technological changes as these relate to changes in chemical processes. Specific mention should be made of the difficulties in providing recent information for Germany and the region encompassed by the former U.S.S.R. because of their political changes during this period.

The author would appreciate receiving any suggestions for improvement.

Martin B. Hocking

PREFACE TO THE FIRST EDITION

This text of applied chemistry considers the interface between chemistry and chemical engineering using examples of some of the important process industries. Integrated with this is a detailed consideration of measures which may be taken for avoidance or control of potential emissions. This new emphasis in applied chemistry has been developed through eight years of experience gained from working in industry in research, development and environmental control fields, plus twelve years of teaching here using this approach. It is aimed primarily towards science and engineering students as well as environmentalists and practising professionals with responsibilities or an interest in this interface.

By providing the appropriate process information back to back with emissions and control data, the potential for process fine-tuning is improved for both raw material efficiency and emission control objectives. This approach emphasizes integral process changes rather than add-on units for emission control. Add-on units do have a place when rapid action on an emission problem is required, or when control is not feasible by process integral changes alone. Fundamental process changes for emission containment are best conceived at the design stage. This approach to control should appeal to industrialists in particular since something more substantial than decreased emissions may be achieved.

This book provides a general source of information on the details of process chemistry and air and water pollution chemistry. Many references are cited in each area to provide easy access to additional background material. Article titles are given with the citation for any anonymous material to aid in retrieval and consultation. Sources of further information on the subject of each chapter, but generally not cited in the text, are also given in a short Relevant Bibliography list immediately following the text. Tradenames have been recognized by capitalization, when known. It would be appreciated if any unrecognized tradenames are brought to the author's attention.

Martin B. Hocking
From *Modern Chemical Technology*
and Emission Control

■ ACKNOWLEDGMENTS

ACKNOWLEDGMENTS TO THE THIRD EDITION

Again the improvements in this volume owe a lot to the collective use of the second edition by many classes of students, from their comments relating to content, to the reworking of *Review Questions* to clarify objectives. I thank Kristin Hoffmann and Kathleen Nelson who assisted with the retrieval of some difficult to locate reference material. Brett Boniface and Brandon Grieve-Heringa provided invaluable assistance combing web data bases for technological updates, and David Flater (NorskeCanada Pulp and Paper), Thor Hægh (Norsk Hydro ASA), Gary Kjersem (Shell Canada Ltd.), Nikolaos Korovessis (Hellenic Saltworks S.A., Athens), Bruce Peachy (New Paradigm Engineering Ltd.), and Kevin Taylor (Taylor Industrial Research) are thanked for providing personal insights. Last, but not least, I again most gratefully thank my wife Diana for handling all of the file changes necessitated by the updating of text, tables, and several new figures added to this volume.

Martin Hocking

ACKNOWLEDGMENTS TO THE SECOND EDITION

Students using the first edition are thanked for providing useful feedback to improve the presentation in a general way and for testing the concepts of most of the problems. My former and present students in polymer chemistry have read and reviewed in detail the material of the new chapters 19-23. They, several more casual readers, and Diana Hocking who read the whole manuscript, are thanked for their contributions. Elizabeth Small, Carol Jenkins, Susanne Reiser, and Diana Hocking completed most of the typing, and Devon Greenway provided bibliographic assistance. Ken Josephson and Ole Heggen executed the new graphics for preparation of the final manuscript.

ACKNOWLEDGMENTS TO THE FIRST EDITION

I am grateful to numerous contacts in industry and environmental laboratories who contributed technical information included in this book. I would particularly like to thank the following: B.R. Buchanan, Dow Chemical Inc.; W. Cary, Suncor; R.G.M. Cosgrove, Imperial Oil Enterprises Ltd.; F.G. Colladay, Morton Salt Co.; J.F.C. Dixon, Canadian Industries Ltd.; R.W. Ford, Dow Chemical Inc.; T. Gibson, B.C. Cement Co.; G.J. Gurnon, Alcan Smelters and Chemicals Ltd.; D. Hill, B.C. Forest Products; J.A. McCoubrey, Lambton Industrial Society; R.D. McInerney, Canadian Industries Ltd.; R.C. Merrett, Canoxy, Canadian Occidental Petroleum; S.E. Moschopedis, Alberta Research Council; J.C. Mueller, B.C. Research; J.A. Paquette, Kalium Chemicals; J.N. Pitts, Jr., Air Pollution Research Centre, University of California; J.R. Prough, Kamyr Inc.; J.G. Sanderson, MacMillan-Bloedel Ltd.; A.D. Shendrikar, The Oil Shale Corp.; J.G. Speight, Exxon; A. Stelzig, Environmental Protection Service; H.E. Worster, MacMillan-Bloedel Ltd. They have been credited wherever possible through references to their own recent publications.

I also thank all of the following individuals, each of whom read sections of the text in manuscript form, and C.G. Carlson who read all of it, for their valuable comments and suggestions:

R.D. Barer, Metallurgy Division, Defence Research Establishment Pacific
G. Bonser, Husky Oil Limited
R.A. Brown, formerly of Shell Canada
M.J.R. Clark, Environmental Chemistry, Waste Management Branch, B.C. Government
H. Dotti, Mission Hill Vineyards
M. Kotthuri, Meteorology Section, Waste Management Branch, B.C. Government
J. Leja, Department of Mining and Mineral Process Engineering, University of British Columbia
L.J. Macaulay, Labatt Breweries of B.C., Ltd.
D.J. MacLaurin, formerly of MacMillan-Bloedel Ltd.
R.N. O'Brien, Department of Chemistry, University of Victoria
M.E.D. Raymont, Sulphur Development Institute of Canada
W.G. Wallace, Alcan Smelters and Chemicals, Ltd.
R.F. Wilson, Dow Chemical Canada Inc.
M.D. Winning, Shell Canada Resources Ltd.

Without the support of the University of Victoria, the Department of Chemistry, and my family to work within, this book would never have been completed. I owe a debt of gratitude to the inexhaustible patience of my wife, who handled the whole of the initial inputting of the manuscript into the computer, corrected several drafts, and executed all of the original line drawings. Thanks also go to K. Hartman who did the photographic work, to B.J. Hiscock and L. J. Proctor, who unfailingly encouraged adoption of the computer for manuscript preparation, and to L.G. Charron and M. Cormack, who completed the final manuscript.

Some of the line drawings and one photograph are borrowed courtesy of other publishers and authors, as acknowledged with each of these illustrations. To all of these I extend my thanks.

It would be tempting to blame any final errors on computer programming glitches which may, occasionally, have been the case. It would be appreciated if errors from any source were brought to my attention.

1
BACKGROUND AND TECHNICAL ASPECTS

Take calculated risks. That is quite different from being rash.

—George S. Patton

1.1. IMPORTANT GENERAL CHARACTERISTICS

The business niche occupied by the chemical industry is of primary importance to the developed world in its ability to provide components of food, clothing, transportation, accommodation, and employment enjoyed by modern humanity. Most material goods are either chemical in origin or have involved one or more chemicals during the course of their manufacture. In some cases, the chemical interactions involved in the generation of final products are relatively simple ones. In others, such as the fabrication of some of the more complex petrochemicals and drugs, more complicated and lengthy procedures are involved. Also, most chemical processes use raw materials naturally occurring on or near the earth's crust to produce the commodities of interest.

Consider the sources of some of the common chemical raw materials and relate these to products that are accessible via one or two chemical transformations in a typical chemical complex. Starting with just a few simple components—air, water, salt (NaCl), and ethane—together with an external source of energy, quite a range of finished products is possible (Fig. 1.1). While it is unlikely that all of these will be produced at any one location, many will be, and all are based on commercially feasible processes [1]. Thus, a company which focuses on the electrolytic production of chlorine and sodium hydroxide from salt will often be sited on or near natural salt beds in order to provide a secure source of this raw material. A large source of freshwater, such as a river or a lake will generally be used for feedstock and cooling water

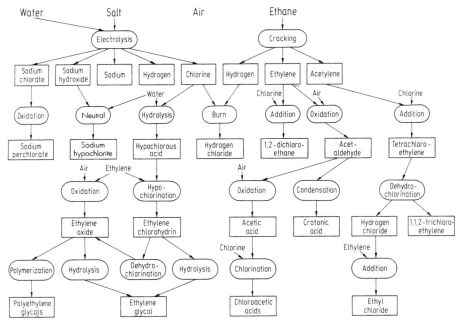

FIGURE 1.1 Flow sheet of a hypothetical though credible chemical complex based on only air, water, salt, and ethane raw materials. Ellipses represent processes, rectangles indicate products.

requirements. Quite often an oil refinery is one of a cluster of companies, which find it mutually advantageous to locate together. This can provide a supply of ethane, benzene, or other hydrocarbon feedstocks. In this manner all the simple raw material requirements of the complex can flow smoothly into the production of more than a dozen products for sale (Fig. 1.1).

A rapid rise in the numbers of chemicals produced commercially, and a steady growth in the uses and consumption of these chemicals historically (since the 1930–1940 period), has given the chemical industry a high growth rate relative to other industrial activities. In current dollars, the average annual growth rate in the U.S.A. was about 11% per year in the 1940s and just over 14% per year through the 1970s, seldom dropping below 6% in the intervening period. Plastics and basic organic chemicals have generally been the stronger performing sectors of the chemical industry. Basic inorganic chemicals production, a "mature" area of the industry, has shown slower growth. World chemical export growth has been strong too, having averaged just over a 17% annual growth rate during the 1968–1978 interval. However, growth rates based on current dollar values, such as these are, fail to recognize the salutary influence of inflation. Using a constant value dollar, and smoothing the values over a 10-year running average basis gives the maximum for the real growth rate of about 9% per year occurring in 1959, tapering down to about 1–3% per year by 1990. The slowing of the real growth rate in recent years may be because the chemical industry is gaining maturity. More recently, there may also have been a contribution from the global business recessions.

Most of the machinery and containment vessels required for chemical processing are expensive, in part because of the high degree of automation used by this industry. This means that the labor requirement is relatively low, based on the value of products. Put in another way, in the U.S., the investment in chemical plant per employee has amounted to about $30,000 per worker at the time when the average for all manufacturing stood at $14,000 per worker. In the U.K., this ratio of capital investment per employee in the chemical industry versus the investment by all manufacturing is very similar to the experience in the U.S.A. In 1963, these figures stood at 7,000 and 3,000 pounds, and in 1972, 17,000 and 7,000 pounds, respectively [2].

Yet another way of considering the relationship of investment to the number of employees is in terms of the "value added per employee." The value added is defined as the market price of a good minus the cost of raw materials required to produce that good [3]. It can be used as a measure of the worth of processing a chemical in terms of its new (usually greater) value after processing than before. When the gross increase in value of the products of a chemical complex is divided by the numbers of employees operating the complex, one arrives at a "value added per employee," one kind of productivity index. Using this index, the productivity of a worker in the chemical industry is at the high end of the range in comparison with the productivity of employees in all manufacturing within any particular country. There are also quite significant differences in relative productivity when the values added per employee in the chemical industry of countries are compared. In 1978 and 1999, the values added for the U.S.A. stood at $58,820 and $161,290/employee/year, as compared to values of $17,800 and $62,390 for Spain, the extremes of the range among the countries compared in Table 1.1. This comparison also

TABLE 1.1 Employment in Chemicals Production, and Value Added per Employee[a]

Country	Thousands employed		Value added per employee, US$		
	1978	1999	1968	1978	1999
Austria	61	29	–	19,670	78,520
Belgium	62.6	71	8,410	37,100	115,170
Canada	84.7	91	18,020	39,130	92,630
France	305	218	9,540	32,540	117,140
Germany (West)	548	522	11,350	46,220	76,860[b]
Italy	292	231	7,890	20,000	81,450
Japan	470	451	9,460	33,600	176,920
Netherlands	87.1	77	9,810	44,100	99,520
Norway	17.4	n/a	8,290	23,800	n/a
Spain	144.1	141	5,780	17,800	62,340
Sweden	39.7	n/a	12,180	36,500	n/a
U.K.	467	250	8,110	19,800	98,780
U.S.A.	1,088	1,037	24,760	58,820	161,290

[a]Data from Cairns [4] and OECD [5], and calculated for 1999 from the "OECD STAN database for industrial analysis"[6].
[b]Germany.

reflects the much higher investment per employee and the higher degree of automation generally used by American chemical companies versus their Spanish counterparts. However, the range of values given here is also dependent on a number of other factors such as scale and capacity usage rates, which have not yet been discussed. Relative positions may also change in 10- or 30-year spans, as shown by Canada and West Germany, from other factors.

The products of the chemical industry tend to have a high rate of obsolescence, because of the steady stream of better performing products being developed. During 1950s and 1960s, most of sales by chemical companies were from products developed in the preceding 15 years. Since the 1990s, the pace of product development has accelerated with Du Pont achieving 22–24% of sales from products developed in the last 5 years and setting its sights on 33% by 2005, and Kraton Polymers achieving 31% for the same period [7, 8]. Corning, in 2003, recorded 88% of products sold were developed in the previous 4 years [9]. To provide the steady stream of improved products required to maintain these records requires a substantial commitment to research for a company to keep up with its competition. This requirement also provides the incentive for a chemical company to employ chemists, engineers, biologists, and other professionals to help ensure the continuing discovery and development of new products for its success.

From 2.5 to 3.5% of the value of sales of U.S. chemical companies is spent on research and development activity, about the same proportion of sales as spent by all industry. The German companies tend to place a somewhat greater emphasis on research and development, and show an expenditure of 4–5% of sales in this activity. Drug (pharmaceutical) companies represent the portion of the chemical sector, which spends the largest fraction of sales, about 6%, on research and development programs [10]. This is probably a reflection of the greater costs involved in bringing new, human use drugs to market, as well as the generally higher rate of obsolescence of drugs compared to commodity chemicals.

1.2. TYPES AND SIGNIFICANCE OF INFORMATION

With the moderately high growth rate of the chemical industry and its high rate of obsolescence of both products and of the processes leading to them, the competition in this industrial area is vigorous. Technological and market success of a chemical company is a composite of the financial resources, raw material position, capabilities and motivation of staff, and the information resources that the company has at its disposal. The information resource is a particularly important one for the chemical processing industry. Information, or "know-how," may be derived from many kinds of prior experience. It may be generated from self-funded and practiced research or process development. It may also be purchased from appropriate other companies if this is available. Thus, sale of the results of research by a company, even if not used by that company to produce a product, may still produce an income for it in the form of licensing agreements, royalties per unit of product sold, and other considerations. In many ways this is a highly desirable component of a company's earnings since it does not require any capital investment, or raw material and product inventories, as are ordinarily required to generate an income from chemical processing.

Patents and the patenting system represent the orderly system of public documents used in most parts of the world to handle much of this kind of information. Patent protection is of substantial importance to chemical as well as other companies. Patents must be applied for in each country or region (e.g., the European Union) for which protection is desired. Otherwise, the subject of a patent may be practiced and the product sold without license in any country in which this precaution has not been taken. "Composition of matter" patents, which relate primarily to newly discovered chemical compounds, are issued on successful application by an inventor (individual or company). Utility (i.e., some type of useful function of the compound) must be demonstrated before a patent application of this type can be filed. In return this class of patent provides the best kind of protection for a new compound because the compound itself is protected from its sale by others for the 17- to 20-year life of a patent, regardless of the synthetic route developed to produce it.

"Process" patents are used to protect a new process or refinements to an established process, which is employed to produce an existing compound. This type of patent also provides useful protection against the commercial use by others of an improved, completely distinct process, which may be developed by a company. Process development may lead to lower product costs achieved from higher conversion rates or better selectivity, or more moderate operating conditions, and the like. In these ways, it provides the company with an economic advantage to practice this improvement.

Other patent areas are used by chemical and other companies, such as those covering machines and registered designs, trademarks and symbols, and copyright, but these are generally less fundamental to the operations of chemical companies than the composition of matter and process patent areas [11]. Trademarks and symbols are generally of more importance for sales, since company and product recognition comprise significant marketing factors. Trademarks and symbols have no expiry date, as long as the required annual maintenance fee is paid.

A patent comprises a brief description of the prior art (the narrow segment of technology) in areas related to the subject of the patent. Usually this is followed by a brief summary of what is being patented. A more detailed description of what is involved in the invention is then given, accompanied by descriptions of some detailed examples that illustrate the application of the invention. Usually at least one of the examples described is a description of an experiment, which was actually carried out, but they need not all have been actually tested. Differentiation between actually tested examples and hypothetical examples described in the body of the patent is made on the basis of the tense used in the description. If it is described in the past tense (i.e., "was" is used throughout), then it is a description of a tested example. If it is given in the present tense, it describes a hypothetical example. To be able to differentiate the two types of examples is of particular interest to synthetic chemists, for example, who are likely to be more successful if they follow a procedure of a tested rather than a hypothetical example. The last, and most important part of a patent is the claims section. Here, numbered paragraphs, each of which by custom is written all in one sentence, cover the one or more novel areas to be protected by the patent in order of importance. In the case of any contest of the patent by other parties, these claims must be disproved in reverse order, (i.e., the last and least important claim first followed by the others) if the last claim is successfully contested.

The granting of a patent confers on the holder a time-limited monopoly in the country of issue, for a period of about 17 years, to cover the novel composition of matter or advance in the art that is claimed by the patent. This country-by-country process has been recently simplified by the availability of European Union (EU) patents, which now cover all the member countries with one application [12]. During this time, the company or individual may construct a plant using the patented principles, which may take 6 or 7 years. Once production has begun, a product can be marketed from this plant at a sufficiently high price that the research and development costs involved prior to patenting, as well as reasonable plant write-off expenses, may be met. This stage of marketing can proceed without competition from others for the 10–11 years remaining from the original patent interval. Or a company may choose to license the technology to collect product royalties from another interested company. Or it may follow both options simultaneously, if it reasons that the market will be large enough to sustain both. For these reasons, the patent system encourages a company to carry out its own research since it provides a reasonable prospect of the company being able to recover its early development costs while it is using the new art, protected from competition.

Seventeen years (20 years in European countries) from the date of issue of a patent, however, the subject matter of the patent comes into the public domain. In other words, it becomes open to any other person or company who wishes to practice the art described in the patent and *sell* a product based on this technology. At this time, the price of the product will normally fall somewhat as the product starts to be produced competitively by others. But the originating company still has some production and marketing advantages from its longer experience in using the technology, from having one or more producing units, which may be largely paid for by this time, and from having already developed some customer confidence and loyalty.

The new regulatory requirements that must be met before marketing new drugs and pesticides are now taking up to 7–8 years to satisfy. This has increased the new product development costs, simultaneously decreasing the period of time available for monopoly marketing to allow recovery of development costs. Realization of this has led to moves in the U.K. and in the U.S.A. to extend the period of monopoly protection granted by the patent by the length of time required by a company to obtain regulatory clearance. These moves should at least encourage maintenance of the current level of research and development effort by companies even if it does not increase innovation.

Patent protection for an idea is for a limited time only, but even during the protected time the information in the patent becomes public knowledge. There may be some technological developments, which a company wishes to keep completely to itself, or which are so close to prior art (already practiced) that there is some doubt of the likelihood of success of a patent application. Information falling into these categories may be simply filed in company records for reference purposes and not be patented or otherwise publicized at all. This type of know-how is termed "proprietary information," useful to the company but kept within the company only. Agreements signed by all new employees working for a company ensure that this proprietary information does not become public knowledge. In return for risking possible eventual leakage and use of this information by others, the company gains the advantageous use of the information. In the meantime, it saves patenting costs (even if feasible) and avoids the certainty of public disclosure on issuance of a patent covering this information. But the ideas involved are not directly protected from use by others, whether or not the knowledge is lost via "leaks" or via independent discovery by a second company working on the same common knowledge premises as the first company, hence the value of the patent system in providing this assurance of protection.

A second approach to decrease the impact of public disclosure when a patent is filed is to apply for many patents on closely related technologies simultaneously. Some will relate to the core technology for which protection is desired. The others serve as distractors to those who would wish to discover and explore the new technology competitively.

1.3. THE VALUE OF INTEGRATION

Integration, as a means of consolidation by which a company may improve its competitive position, can take a number of forms. Vertical integration can be "forward" to carry an existing product of the company one or more stages closer to the final consumer. For instance, a company producing polyethylene resin may also decide to produce film from this resin for sale, or might decide to produce both film, and garbage bags from the film. By doing this, more "value-added" manufacturing stages are undertaken within the company. If these developments are compatible with the existing activities and markets of the company, they can significantly enhance the profitability of its operations.

Vertical integration may also be "backward" in the sense that the company endeavors to improve its raw material position by new resource discoveries and acquisition, or by purchase of resource-based companies strong in the particular raw materials of interest. Thus, it can explore for oil, or purchase an oil refinery to put itself into a secure position for ethane and ethylene. Or it can also purchase land overlying beds of sodium chloride or potassium chloride with mineral rights, or near sodium sulfate rich waters and develop these to use for the preparation of existing product lines. Either of these routes of backward integration can help to secure an assured source of supply and stable raw material pricing, both helpful in strengthening the reliability of longer term profit projections.

Horizontal integration is a further type, where the technological or information base of the company is applied to improve its competitive position in this and related areas. When a particular area of expertise has been discovered and developed, this can be more fully exploited if a number of different product or service lines are put on the market using this technology. For instance, Procter & Gamble and Unilever have both capitalized on surfactant technology in their development of a range of washing and cleaning products. Surface-active agents of different types have also been exploited by the Dow Chemical Company with its wide range of ion exchange resins, and cage structures by the former Union Carbide (now owned by Dow) with its molecular sieve-based technology. It can be seen from these examples that judicious application of one or more of these forms of integration can significantly strengthen the market position and profitability of a chemicals based company.

1.4. THE ECONOMY OF SCALE

The size or scale of operation of a chemical processing unit is an important competitive factor since, as a general rule, a large-scale plant operating at full capacity can make a product at a lower per unit cost. This is the so-called "economy of scale" factor. How does this lower cost product from a larger plant arise? First, the labor cost per unit of product is lower for a very large plant than for a small one. This is because proportionally fewer staff are required per unit of product to run a 1,000 tonne/day plant than, say, a 100 tonne/day plant. Secondly, the capital cost of the plant per unit of product is lower, if the plant is operating at full capacity.

Reduced labor costs result from the fact that if one person is required to control the raw material flows into a reactor in a 100 tonne/plant; in all likelihood, they can still control these flows in a 1,000 tonne/day plant. In fact an empirical expression has been derived by correlation of more than 50 types of chemical operations which, knowing the labor requirement for one size of plant, allows one to estimate with reasonable assurance the labor requirement for another capacity [13] (Eq. 1.1).

$$M = M_0(Q/Q_0)^n, \qquad\qquad 1.1$$

where M = labor requirement for plant capacity Q of interest,
$\quad\quad\quad\ M_0$ = known labor requirement for a plant capacity Q_0, and
$\quad\quad\quad\ n$ = exponent factor, normally about 0.25, for the estimation of labor requirements.

If 16 staff are required to operate a 200 tonne/day sulfuric acid plant, this expression allows us to determine that only about 24 staff ($16 \times (1,000/200)^{0.25}$) should be needed to operate a 1,000 tonne/day plant. Thus, when operating at full capacity, the larger plant would only have three-tenths the labor charge of the smaller plant, per unit of product.

The lowered plant capital cost per unit of product comes about because of the relationship of capital costs of construction to plant capacity, which is an exponential, not a linear relation (Eq. 1.2).

$$\text{capital cost} \propto (\text{plant capacity})^{2/3} \qquad\qquad 1.2$$

The approximate size of the fractional exponent of this expression results from the fact that the cost to build a plant varies directly as the area (or weight) of metal used, resulting in a square exponent term [14]. At the same time, the capacity of the various components of the processing units built increases in relation to the volume enclosed, or a cube root term. Hence, the logic of this approximate relationship.

In actual fact, a skilful design engineer is generally able to shave just a bit off this descriptively derived exponent, making capital cost relate to scale more closely in accord with Eq. 1.3 for whole chemical plants.

$$\text{capital cost} \propto (\text{plant capacity})^{0.60} \qquad 1.3$$

In order to use (Eq. 1.3) to estimate the capital cost of a larger or smaller plant, when one knows the capital cost of a particular size of plant, one has to insert a proportionality constant (Eq. 1.4).

$$C = C_0(Q/Q_0)^n, \qquad 1.4$$

where C = capital cost for the production capacity Q of the plant to be determined,

C_0 = is the known capital cost for production capacity Q_0, given in the same units as C, and

n = scale exponent, which is usually in the 0.60 to 0.70 range for whole chemical plants.

Thus, if it is known that the capital cost for a 200 tonne/day sulfuric acid plant is $1.2 million ($1.2 mm) then, using this relationship, it is possible to estimate that the capital cost of an 1800 tonne/day plant will be somewhere in the range of $4.49 mm to $5.59 mm (Eq. 1.5).

$$\begin{aligned} C &= \$1.2\,\text{mm} \ (1{,}800/200)^{0.60} & C &= \$1.2\,\text{mm} \ (1{,}800/200)^{0.70} \\ &= \$1.2\,\text{mm} \ (3.7372) & &= \$1.2\,\text{mm} \ (4.6555) \qquad 1.5 \\ &= \$4.485\,\text{mm} & &= \$5.587\,\text{mm} \end{aligned}$$

From construction cost figures the actual capital cost of construction of an 1,800 tonne/day sulfuric plant is about $5.4 mm, when taken at the time of these estimates. This figure agrees quite well with the two values estimated from the known cost of the smaller sized plant.

Of course, if one has recent capital cost information on two different sizes of plant for producing the same product, this can enable a closer capital cost estimate to be made by determination of the value of exponent n from the slope of the capital cost versus production volume line plotted on log–log axes for the two sizes of plant. For the particular example given, this experimentally determined exponent value would be 0.685. Note also that this capital cost estimation method is less reliable for plant sizes more than an order of magnitude larger or smaller than the plant size for which current costs are available [15].

From a comparison of the foregoing capital cost figures, it can be seen that nine times as much sulfuric acid can be made for a capital cost of only 3.7 to 4.7 times as much as that of a 200 tonne/day plant. Obviously if the large plant is operated at full capacity, the charge (or interest) on the capital which has to be carried by the product for sale by the larger plant is only about half (4.7/9.0) or even less than half (3.7/9.0) of the capital cost required to be borne by the 200 tonne/day plant, per unit of product.

To make the decision regarding the size of plant to build in any particular situation, careful consideration has to be given to product pricing, market size and elasticity, and market growth trends. Also it is a useful precaution to survey the immediate geographical area and public construction announcements for any other plans for a plant to produce the same product. The final decision should be based on a scale of operation which, within a period of 5–7 years, could reasonably be expected to be running at full capacity. That is, it should be possible to stimulate a sufficient market, within this period of time, to sell all of the product that the plant can produce. If the final size of plant built is too small, not only are sales restricted from inadequate production capacity but also the profit margin per unit of product is smaller than it potentially could have been if the product were being produced in a somewhat larger plant. If the final result is too large, and even after 10 years or so the plant is required to operate at only 30% of capacity to provide for the whole market, then the capital and frequently also labor costs per unit of product become higher than they would have been with say one-half or even one-quarter of the plant size. In this event, planning too optimistically can actually *decrease* the profitability of the operation. It is the significance of decisions such as these as to the financial health of a chemical company that justify the handsome salaries of its senior executives.

One remaining point to consider regarding scale is that the capital cost exponential factor of 0.60 to 0.70 relates to most whole plants. If considering individual processing units this factor can vary quite widely (Table 1.2). With a jaw crusher, for example, a unit with three times the capacity costs 3.7 times as much. Obviously, here, scaling up imposes greater capital costs per unit of product for a larger than for a smaller unit. But other associated costs may still be reduced. A steel vent stack of three times the height costs about three times as much, (i.e., there is no capital cost economy of scale here), and these capital cost increases with height may still have to be borne by the plant.

TABLE 1.2   Typical Values for the Exponent Scale Factor and How These Relate to the Cost Factor for Chemical Processing Equipment[a]

Type of equipment	Typical value of exponent n	Cost factor for three times scale
Jaw crusher	1.2	3.74
Fractionating column, bubble cap	1.2	3.74
Steel stack	1.0	3.00
Fractionating column, sieve tray	0.86	2.57
Forced circulation evaporator	0.70	2.16
Shell and tube heat exchanger	0.65	2.04
Jacketed vessel evaporator	0.60	1.93
Stainless steel pressure reactor, 300 psi	0.56	1.85
Industrial boiler	0.50	1.73
Drum dryer, atmospheric pressure	0.40	1.55
Storage tank	0.30	1.39

[a]Exponent values for use with Eq. 1.4, $C = C_o(Q/Q_o)^n$, and selected from those of Peters and Timmerhaus [15] and Allen [16].

However, for very simple components of processing units such as storage tanks, the value of this exponent is small, about 0.30, which allows a tank of three times the capacity to be built for only about 1.4 times the price. Thus, a composite of the scale-up exponent factors for individual units averages out to the 0.60 to 0.70 range for a whole chemical plant.

1.5. CHEMICAL PROCESSING

The chemistry side of the chemical process industries is concerned with the change of raw materials into products by means of *chemical conversions.* A single-reacted starting material rarely gives only pure product, so it is usually necessary to use physical separations such as crystallization, filtration, distillation, or phase separation to recover the product(s) from the unreacted starting materials and by-products. By-products are materials other than the product, which are obtained from reacted starting materials. These physical separation processes are often called *unit operations* to distinguish this from the chemical conversion. Similar features of unit operations may be compared from process to process, unlike chemical conversions [17]. The combination of the chemical conversion step, with all of the unit operations (physical separations) that are required to recover the product of the chemical conversion, is collectively referred to as a *unit process.*

Unit processes may be carried out in single-use (dedicated) equipment used solely to generate the particular product for which it was designed. Or they may be carried out in multiuse equipment used to produce first one product, followed in time by switches to produce one or more related products that have similar unit process requirements. Single-use equipment is invariably used for large-scale production, when 90% of full-time usage rates are required to obtain sufficient products to satisfy the market requirements. Multiuse equipment is more often chosen for small-scale production, and particularly for more complicated processes such as required for the manufacture of some drugs, dyes, and some speciality chemicals.

Proper materials of construction with regard to strength and toughness, corrosion resistance, and cost must all be kept in mind at the design stage for construction of a new chemical plant. Early experiments during the conception of the process will usually have been conducted in laboratory glassware. Even though glass is almost universally corrosion resistant (and transparent, and thus useful in the lab), it is too fragile for most full-scale process use. Mild steel is used wherever possible, because of its low cost and ease of fabrication. But steel is not resistant to attack by many process fluids or gases. In these cases titanium, nickel, stainless steel, brass, Teflon, polyvinylchloride (PVC), wood, cement, and sometimes even glass (usually as a lining) among other materials may be used to construct components of a chemical plant. The final choice of construction material is based on a combination of experience and accelerated laboratory tests. Small coupons of the short-listed candidate materials are suspended in synthetic mixtures prepared to mimic those to be found in the process. These are then heated to simulate anticipated plant conditions. Preliminary tests will be followed by further tests during small-scale process test runs in a pilot plant, wherever possible. Even when the final full-scale plant is completed, there may still be recurring corrosion failures of a particular component, which may require construction material changes even at this stage of process development.

1.5.1. Types of Reactors

Industrial reactor types can use the analogy between laboratory manipulations and a full-scale production plant. Very often in the laboratory a synthesis will be carried out by placing all the required reactants in the flask and then imposing the right conditions, heating, cooling, light, etc., on the contents to achieve the desired extent of reaction. At this stage, the contents of the flask are emptied into another vessel for the product recovery steps to be carried out. Operating an industrial process in this fashion is termed a *batch process* or *batch operation.* Essentially this situation is obtained when all starting materials are placed in the reactor at the beginning of the reaction, and remain in the vessel until the reaction is over, when the contents are removed. This mode of operation is the one generally favored for smaller scale processes, for multiple use equipment, and for new and untried, or some types of more hazardous reactions.

On the other hand, an industrial process may be operated in a *continuous mode*, rather than in a batch mode. To achieve this, either a single or a series of interconnected vessels may be used. The required raw materials are continuously fed into this vessel or the first vessel and the reaction products continuously removed from the last so that the volume of material in the reactor(s) stays constant as the reaction proceeds. The concentrations of starting materials and products in the reactor eventually reach a steady state. One or more tanks in series may be used to conduct the continuous process. Another option for a continuous process is to use a pipe or tube reactor, in which the starting material(s) is fed into the tube at one end, and the product(s) is removed at the other. In this case, the reaction time is determined by the rate of flow of materials into the tube divided by the length of the tube.

Since, in general, the labor costs of operating a large-scale continuous process are lower than for a batch process, most large-scale industrial processes are eventually worked in a continuous mode [18]. However, because of the more complicated control equipment required for continuous operation, the capital cost of the plant is usually higher than for the same scale batch process. Thus, the final choice of the mode of operation to be used for a process will often depend on the relative cost of capital versus labor in the operating area in which the plant is to be constructed. Most developed countries opt for a high degree of automation and higher capital costs in new plant construction decisions. For Third World nations, however, where capital is generally scarce and labor is low cost and readily available, more manual and simpler batch-type operations will often be the most appropriate. A smaller scale of operation could be sufficient to supply the smaller markets in these economies. Maintenance and repair operations for batch operations in less developed economies are also more easily accomplished than with the more complex control systems of continuous reactors.

There are several common combinations within this broad division into batch and continuous types of reactors, which use minor variations of the main theme. The simplest and least expensive of these subdivisions is represented by the straight batch reactor, which is frequently just a single stirred tank. All the raw materials are placed in the tank at the start of the process. There is no flow of materials into or out of the tank during the course of the reaction (i.e., the volume of the tank contents is fixed during the reaction) (Table 1.3). Usually there is also some provision for heating or cooling of the reacting mixture, either via a metal jacket around the reactor or via coils placed inside the reactor through which water, steam, or heat exchange fluid may be passed for temperature control. However, the temperature is not usually uniform in this situation since the initial concentrations and reaction rate of the two (or more) reactants are at a maximum, which taper to lower values as the reaction proceeds. Thus, heat evolution (or uptake) is going to be high initially and then gradually subside to coincide with a slowing of the reaction rate. At the end of the reaction, the whole of the reactor contents is pumped out for product recovery.

TABLE·1.3 A Qualitative Comparison of Some of the Main Configurations of Batch and Continuous Types of Liquid Phase Reactors

| Type of reactor | Illustration of concept | | Uniformity of | |
		Composition with time	Composition within reactor[a]	Temperature throughout process[b]
a. Batch		no	yes	no
b. Semi-batch	feeds	no	yes	yes
c. Continuous stirred tank (CSTR) sometimes "tank flow reactor"	feeds, product	yes	yes	yes
d. Multistage CSTR	feeds, product	yes	partly	partly
e. Tubular flow, sometimes "pipe reactor" or "plug flow reactor"	feeds, product	yes	no	no

[a]Meaning the composition within the reactor at any particular point in time.
[b]Referring to temperature constancy during the whole of the reaction phase.

A semibatch reactor is a type of batch configuration used particularly for processes, which employ very reactive starting material. Only one reactant, plus solvent if required, is present in the reactor at the start of the reaction. The other reactant(s) is then added gradually to the first, with continued stirring and control of the temperature. Through control of the rate of addition of one reactant, the temperature of the reacting mixture may be kept uniform as the reaction proceeds.

Continuous reactor configurations are generally favored for very large-scale industrial processes. If the process is required to produce only 2 million kg/year or less, the economics of construction will generally dictate that a batch process be used [19]. If, however, 9 million kg/year of product or more is required, there is a strong incentive to apply some type of continuous reactor configuration in the design of the production unit.

The stirred tank is the main element of the simplest type of continuous reactor, the continuous stirred tank reactor (CSTR). Continuity of the process is maintained by continuous metering in of the starting materials in the correct proportions, and continuous withdrawal of the stream containing the product from the same, well-stirred vessel. In this way the concentration and temperature gradients shown by simple batch reactors are entirely eliminated (Table 1.3). This type of continuous reactor is good for slow reactions, in particular, since it is a large, simple, and cheap unit to construct. However, reactors of the CSTR type are inefficient at large conversions [19]. For a process proceeding via first-order kinetics and requiring a 99% conversion a seven times larger reactor volume is needed than if only 50% conversion is desired.

A solution to the large volume requirement for high conversions in a single reactor is to use two or more CSTRs in series [19]. Use of two CSTRs in series allows the first reactor to operate at some intermediate degree of conversion, the product of which is then used as the feed to the second reactor to obtain the final extent of reaction desired (Fig. 1.2). From the diagrams it can be readily seen that the total reactor volume required to achieve the desired final degree of conversion is significantly reduced compared to the volume required to achieve the same degree of conversion in a single reactor. Carrying this idea further, it can also be seen that increasing the number of CSTRs operating in series to three or more units contributes further to the space-time yields and allows further reductions in reactor volume to be made to still obtain the same final degree of conversion. Therefore, multiple CSTRs operating in series allow either a reduction in the total reactor volume used to obtain the same degree of conversion as with a single CSTR, or a higher degree of conversion for a given total reactor volume. In either case, the engineering cost to achieve these changes is in the additional connecting piping and fluid and heat control systems required for multiple units, over a single reactor. This generally is the factor that limits the extent to which increased numbers of reactors are economic to use for improvement of process conversions.

Taking the multiple CSTR concept to its logical extreme, a very large number of small tank reactors connected in series can be likened to carrying out the same process in a very long, narrow-bore tube, referred to as a *tubular flow*, or *pipe reactor*. By placing sections or all of this tube in externally heated

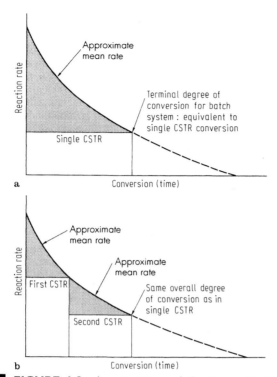

Reaction rate

Approximate
mean rate

Terminal degree of
conversion for batch
system : equivalent to
single CSTR conversion

Single CSTR

a   Conversion (time)

Reaction rate

Approximate
mean rate

Approximate
mean rate

First CSTR

Same overall degree
of conversion as in
single CSTR

Second CSTR

b   Conversion (time)

FIGURE 1.2 A comparison of the "space–time" yields, or saving in reactor volume achieved by carrying out a continuous process to the same degree of conversion in a single stirred tank reactor, versus two CSTRs operating in series. (Adapted from Wynne [19], with permission.)

(or cooled) sections, any desired processing temperature of the fluid mixture flowing inside the pipe can be obtained. If a high enough flow rate is used to cause turbulent flow, or if in-line mixers (streamline flow splitters employing an interacting series of baffles) are used, components of even immiscible mixtures may be made to mix thoroughly as they move down the tube. Turbulent flow conditions also help to obtain good heat transfer between the liquid mixture flowing inside the tube, and the tube wall. If the heat transfer fluid or gas flowing outside the tube is also moving vigorously, the temperature difference between the external heat transfer medium and the contents of the tube is also kept to a minimum.

The concentrations of starting materials flowing in a tube reactor decrease, and the concentration of product increases, as the mixture flows down the tube and the reaction proceeds. Thus reaction times for the raw materials flowing into a tube reactor can be calculated from the relation (distance from the inlet)/(reactant velocity). The induced turbulence in the tube occurs mostly in the cross-sectional dimension and very little along the length of the tube (i.e., there is little or no "backmixing," or mixing of newly entering raw materials) with raw materials that have already reacted for some time. This feature has led to the names *plug,* or *plug flow* reactor as other descriptive synonyms for tubular flow or pipe reactor.

Most industrial processes using the interaction of fluids to obtain chemical changes can be classified into one, or sometimes more of the preceding five liquid reactor types. Variations on these themes are used for gas–gas, gas–liquid, or gas–solid reactions, but these variations parallel many of the processing ideas used for liquid–liquid reactors [20]. A new continuous, spinning disk reactor concept has recently attracted interest for some intrinsically fast organic reactions and for possible application in crystallizations [21]. Modular microreactors have also become of interest to fine chemicals producers and pharmaceutical companies for their faster reactions, ease of scale-up, and low cost [22].

1.5.2. Fluid Flow Through Pipes

To understand the mechanism of the turbulent mixing process occurring in pipe reactors, we have to consider first some of the properties of fluid flow in pipes. Resistance to fluid flow in a pipe has two components, the viscous friction of the fluid itself within the pipe, which increases as the fluid viscosity increases, and the pressure differential caused by a liquid level difference or a pressure difference between the two vessels.

At relatively low fluid velocities, particularly for a viscous fluid (where turbulence is damped) in a small pipe one will normally obtain streamline flow (Fig. 1.3a). Under these conditions, the fluid is in a continuous state of shear with the fastest flow in the center of the pipe with low to zero flow right at the wall. The fluid velocity profile, along a longitudinal section of the pipe, is parabolic in shape.

At high fluid velocities, particularly for low viscosity fluids in large diameter pipes, small flow disturbances create eddies in the fluid stream, which fill the whole of the cross-sectional area of the pipe (Fig. 1.3b). Only a residual boundary layer against the inside wall of the pipe will maintain

Streamline flow throughout
a

Parabolic velocity profile

Main stream turbulent flow (eddies)
Streamline flow only in boundary layer
b

Uniform flow profile

FIGURE 1.3 Fluid flow characteristics and profiles of fluid flow in pipes: (a) At low Reynolds numbers, where streamline flow is obtained throughout the cross section. (b) At high Reynolds numbers, where turbulent flow is obtained for most of pipe volume. Streamline flow is only obtained in a thin boundary layer adjacent to the pipe wall where the influence of the wall and viscous forces control turbulence.

streamline flow under these conditions. This turbulent condition is more common for fluid movement in pipes since smaller pipes, which cost much less, may be used when high fluid velocities are used [23]. The cost saving obtained by using smaller pipe usually exceeds the small increase in pumping cost required to achieve the higher fluid velocities. There are also other reasons for this [22].

The development of turbulent flow depends on the ratio of viscous to inertial (density and velocity) forces, a ratio known as the Reynolds number, R_e (Eq. 1.6).

$$R_e = \frac{u\rho d}{\mu},\qquad\qquad 1.6$$

where u = fluid velocity,
  ρ = fluid density,
  d = pipe diameter, and
  μ = fluid viscosity.

Either metric or English units (i.e., the g/cm sec, or the lb/ft sec system) may be used for substitution, as long as the usage is kept consistent. Also, the same concept applies whether the "fluid" is a liquid or a gas [24]. In either case the result comes out to the same, dimensionless (unitless) Reynolds number, that is, a pure ratio. Whenever the Reynolds number for fluid flow in a pipe exceeds about 2100, one obtains turbulent flow. However, this is not a sharp dividing line. The division between streamline and turbulent flow situations is also dependent on factors other than those used in the Reynolds number calculation, such as the proximity of bends and flow-obstructing fittings, and the surface roughness of the interior of the pipe. So, normally, a Reynolds number range is given. If it is 2,000 or less, this is indicative of a streamline flow situation. If it is 3,000 or more, turbulence is expected [25].

Tubular or pipe reactors are designed to take advantage of this phenomenon to obtain good mixing. This means that relatively small bore tubes and relatively high flow rates are used for this type of continuous reactor. The dependence of good mixing on high flow rates may set a lower limit on the fraction of the design production rate at which the plant can operate. Turbulent flow of the raw materials in the pipe not only contributes good mixing, but also assists in maintaining good heat transfer conditions through the pipe wall separating the reactants flowing in the pipe from the jacketing fluid.

1.5.3. Controlling and Recording Instrumentation

Many kinds of sensors are needed to measure the process parameters important for effective operation of any type of chemical process. The principles of manual or automated process control require, first of all, an appropriate variable, which needs to be measured—temperature, pressure, pH, viscosity, water content, etc.—in order to study the progress of the reaction or separation process. For the measured variable to be significant in the control of the process it must represent a control parameter, such as steam flow, pump speed, or acid addition rate which, when altered, will cause a response in the measured variable. Finally, there must be some actuating mechanism between the variable, which is sensed and the process condition which requires adjustment. In simple processes and where labor costs are low the actuation may be carried out by a person who reads a dial or gauge, decides whether the parameter is high, low, or within normal range, and if necessary adjusts a steam valve, pump power or cooling water to correct the condition. For automated plants, the means of actuation may be a mechanical, pneumatic, electric, or hydraulic link between the sensor and the controlled parameter.

The amounts of materials fed to a chemical reactor are usually sensed by various types of flowmeters (Fig. 1.4). The proportions of raw materials reacting are known from the ratios of the metered flow rates of raw materials moving into the reactor, which may be adjusted if necessary. Today, flow control and many other variables are designed to be proportional, to control a process more readily. Valves may be set to different flow rates, pump speeds

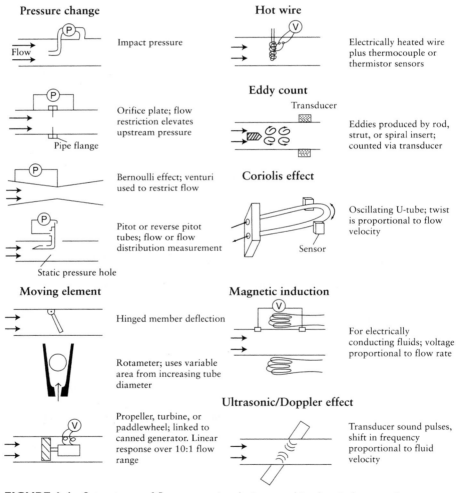

FIGURE 1.4 Some types of flow measuring devices used in chemical processing.

may be altered, conveyors moving solids may be adjusted, and heat input as electrical energy or steam or cooling water flows may be varied at will. The development of proportional control of process variables has greatly improved the refinement of process operation that is now possible. Occasionally, raw material quantity measurement is carried out using measuring tanks or bins, with a float, sight tube, or electrically activated tuning fork as level indicators. Sometimes, piezoelectric mass measurement of the bin plus contents is used, which is more like the usual laboratory method used for mass proportions.

Actuation of process controls in response to a measured process variable was initially pneumatic, using low pressure air, because of the reliability and inherent ignition safety of this system. However, with the growth of computer process control interfacing with electric or electronic actuation is easier, and the improved reliability and safety of these systems against ignition hazards have contributed to the growth in their use. Increased use of computer-based technology for plant automation has helped to refine process operations, improve yields and product quality and also provide savings in labor costs.

Automation of plant control using a computer to match ideal process parameters to the readings being taken from measured process variables allows close refinement of the operating process to the ideal conditions (Fig. 1.5). Manually, a process reading may be compared to the ideal condition every hour or half hour as a reasonable operating procedure. However, under computer control, it is possible to program the operating system so that 20 (or more) variables are monitored, compared to their ideal ranges, and process parameters adjusted, if necessary, every minute or even shorter intervals as required. The computer is given override management of the main process loop. Very short monitoring intervals provided by computer control make the process control easier because the process is never allowed to deviate far from ideal operating conditions.

FIGURE 1.5 Outline of scheme used for computer control of a chemical processing unit.

The frequent monitoring capability of the computer has also provided a stimulus to the design and implementation of rugged, more sophisticated online process analyzers such as infrared and mass spectrometers, turbidity sensors, and gas chromatographs, that can provide quicker and more frequent information on the progress of a reaction than is possible from conventional sampling and laboratory analysis. Conventional analysis could take an hour or two before reporting of a laboratory analysis. Thus under manual analysis conditions the process could deviate substantially from ideal causing a reduction in yields and the quality of product.

Whether equipped with online analyzers or not, all chemical processes also rely heavily on the information obtained from periodic manual laboratory analyses. These provide the resilience of a check on the performance of the online analyzers. More importantly, they also provide quality control checks on all raw materials that move into the plant site and all products and waste streams that move out. Sampling frequency retention time will be geared to the size of the shipment lot, the method of delivery, and the time to when all shipped product is likely to be consumed. Process and product quality monitoring will depend on the operating stability of the process and the quality control requirements (allowed impurity concentrations) of the product.

A part of the analytical planning for a chemical complex is the setting up and maintenance of a "sample library," where analyzed samples from each tank car load or reactor lot are stored for reference purposes. The retention time for these samples is set to exceed the probable delay between the time of product shipment and time of final consumption. Thus, if a customer is having difficulty with a particular batch of product, the retained samples enable the company to check the specifications and render rapid technical assistance. Reanalysis of the sample may also be used by the company to accept or reject claims regarding the quality of a product shipment. Thus, proportional process control, raw materials and product, analysis, sample retention, and careful record-keeping all comprise important parts of an operating chemical complex necessary for the maintenance of product quality and customer satisfaction.

1.5.4. Costs of Operation

Since the primary purpose for the existence of a chemical company is to make a profit for its owners and shareholders, it is vital to be able to determine accurate operating costs so that product pricing and marketing can achieve this objective. Today, of course, this profit picture is complicated by the fact that it has to be achieved while other company obligations relating to the welfare of its employees, to safety, to environmental quality, and to the community and country of operation as a corporate good citizen are also met.

Since the tragic events at Bhopal in 1984 and other accidents in the chemical industry, the first three of these objectives have received increased emphasis. The concept of "Product Stewardship," which originated in the late 1960s, was broadened and adopted by more practitioners with the development of the "Responsible Care" program launched in 1985 by the Canadian

Chemical Producers Association [26]. Other related policies such as a drive towards improved safety and sustainability of industry [27], and a reduction of the environmental impact of chemical processing contribute to this concept [28].

It is, of course, possible to produce an uneconomic chemical commodity, such as when required under the exigencies of war or with particular political objectives. But to achieve this objective requires artificial inducements such as tax concessions, subsidies, or the economics of an organized economy.

As a rough rule of thumb, the chemical process industries aim at an 8 to 10% after-tax profit (earnings), stated as percent of sales [29, 30]. If the financial decision regarding construction of a new plant relates to a process for a new product, the economic projections required for construction to proceed will require a slightly higher margin than this, for construction to proceed. If, however, the plant is to produce a well known commodity chemical that is already a large volume product (i.e., a less risky venture) then lower profit projections may be acceptable. This is particularly true if the company is intending to use a substantial fraction of the product in its own operations since its transfer costs are low. This "captive market" is a favorable economic "value-added" practice. However, these are just profit projections. The ever variable nature of business cycles places significant perturbations onto the realization of these projections so that actual, after-tax profit margins more usually range from 4 or 5% up to 12% in any particular year, and sometimes outside both extremes including a possible net loss.

What should be remembered, however, is that a profitable company earns not only an income in its own right (part of which goes to investors who put up the money to construct the plant), it also provides jobs and salaries to its employees who spend much of their earnings locally, stimulating further business activity. The company also pays local and federal taxes on a corporate basis and through employee income taxes, together providing large and direct and indirect sources of income to different levels of government. Sales taxes levied against many types of chemical commodities also provide government with direct income. These "multiplier effects" provide a large local and country-wide benefit from the operation of a profitable commodity-based company.

To accurately determine the costs of operation for any particular process, the effective stoichiometry, or quantity of product(s) to be expected from the raw materials consumed by a process must be accurately known. It can be helpful to know something about the mechanism or chemical pathway to the materials being produced since this knowledge help suggest process changes, which can increase reaction rates or raise process yields. However, it is not unusual to have a process which operates profitably producing salable product long before anything significant is known about the mechanism.

As examples of this, the early facilities to produce phenol by chlorobenzene hydrolysis and by cumene oxidation were both constructed when the stoichiometry demonstrated acceptable economics. It was long after the product had been on the market before anything was known about the respective mechanisms involved [31]. Secondary aspects of the process, such as capital costs, heat requirements, electric power, labor, and water needs must also be quantified to determine the product pricing structure necessary for profitable operation.

1.5.5. Conversion and Yield

Description of the quantitative aspects of a reaction often differ, using the same facts, when discussed in a research or academic context as compared to discussion in an applied, or industrial setting when detailed economic aspects are far more important. Probably the best way to understand these distinctions is to define the various terms used, employing a general example, Eq. 1.7.

$$
\begin{array}{lllll}
\text{A} & \rightarrow & \text{B} & + & \text{A} & + & \text{by-products} \\
1.00\ \text{mol} & & 0.60\ \text{mol} & & 0.30\ \text{mol} & & 0.10\ \text{mol} \\
\text{starting} & & \text{product} & & \text{unreacted} & & \text{from reacted A} \\
\text{material} & & & & \text{starting} \\
& & & & \text{material}
\end{array} \qquad 1.7
$$

The yield which would be reported in a research or academic setting would be based on the definition given in the word Eq. 1.8.

$$
\%\ \text{research yield} = \frac{(\text{moles of B formed}) \times 100}{(\text{theoretical moles of B which could be formed from moles of A charged})} \qquad 1.8
$$

The wording of the denominator of this equation takes into account the fact that not all reactions yield 1 mole of product for each mole of a starting material charged (placed in the reactor). Hydrolysis of a carboxylic anhydride, for example, may yield 2 moles of product per mole of raw material, and cracking reactions, two or more. Polymerizations usually yield very much less than a mole of product per mole of raw material (monomer).

Calculating a yield from the information of Eq. 1.7 using this definition gives a value of 60% (($0.60/1.00$) $\times$ 100) for the result, which would be reported using this system. This result takes no account of any starting A which did not react. This is in keeping with the primary quantitative objective in a research setting, the synthesized product of interest. The amount of any unreacted starting material in the residues from a reaction is seldom measured, and is usually discarded with any by-products, etc., once the product of interest has been isolated.

In contrast, the definition of yield in an applied or industrial setting differs somewhat from that described above. Here, the amount of unreacted starting material remaining after a reaction is carried out is measured, and is nearly as important in the economics of the process as the amount of product obtained. This parameter forms part of the definition used for industrial yield (Eq. 1.9).

$$
\begin{array}{l}
\%\ \text{industrial yield} \\
\text{(or selectivity)}
\end{array} = \frac{(\text{moles of B formed}) \times 100}{(\text{theoretical moles of B which could be formed from moles of A consumed})} \qquad 1.9
$$

Unreacted starting material is important in an industrial process because it is usually recovered from the product(s) and by-products. It is then recycled to the front end of the process with the fresh starting material just entering the process so that less new starting material is required. Its value is not lost (Fig. 1.6). Again, using the information from the example just given, the industrial yield works out to 85.7% $[(0.60/(1.00 - 0.30) \times 100]$ or, rounded, 86%. Thus, when taking into account the unreacted starting material, a more favorable yield picture is presented.

Having just worked through the industrial yield example allows one to visualize another important aspect. It tells us the fraction of the converted starting material that is product, a sort of efficiency term for avoidance of byproducts. For this reason, "selectivity" or "efficiency" are often used synonymously with industrial yield. From sustainability, economic, and emission control points of view, the selectivity of a process is a very important concept, and well worth the research effort toward maximization.

To an applied chemist or an engineer, one further piece of information related to the performance of a reaction is needed to estimate the quantitative results of a process, and that is the conversion. A good working definition of conversion is given by Eq. 1.10.

$$\% \text{ conversion} = \frac{[(\text{initial moles of A}) - (\text{final moles of A})] \times 100}{(\text{initial moles of A})} \qquad 1.10$$

Using this equation, a value of 70% $((1.00 - 0.30) \times 100/(1.00))$ is obtained for the conversion for the sample problem. In practical terms, what this means is that 70% of the initial raw material is no longer starting material and has been converted to something. It doesn't tell us what it has been converted to. The industrial yield (selectivity) tells us that.

It is worth noting that because percent conversion deals only with one reactant (usually the limiting one), exactly the same numerical value is obtained by rewriting Eq. 1.10 in mass terms, and using the resulting Eq. 1.11 for calculations.

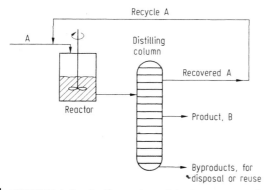

FIGURE 1.6 An illustration of the importance of recycle of recovered starting material in the industrial definition of yield.

$$\% \text{ conversion} = \frac{[(\text{initial mass of A}) - (\text{final mass of A})] \times 100}{(\text{initial mass of A})} \qquad 1.11$$

One needs to use the definitions of both the industrial conversion and the yield in order to determine how much of product B we can expect. This can be determined when having only the amount of starting A and the conversion and yield data for the process of interest. By multiplying the two fractions together, one obtains the fractional yield of product to be expected from a batch process, or the "yield per pass" (yield on one passage of the raw materials through the process) for a continuous process (Eq. 1.12).

$$\begin{array}{ccc} (\text{fractional} & \times\ (\text{fractional} & = (\text{fractional product} \\ \text{industrial yield}) & \text{conversion}) & \text{recovery}) \\ 0.86 & 0.70 & 0.60 \end{array} \qquad 1.12$$

Again, using the example, when the fractional product recovery is multiplied by the initial number of moles of A, one obtains a value of 0.60 ($(0.70 \times 0.86) \times 1.00$ moles of A) moles of B, as the amount of product to be expected. This is in agreement with the quantities specified in the original example, and, it will be noted, is the same as the academic yield specified on a fractional basis. Thus, we can write down the form of an additional relationship, specified in fractional terms, which is often useful in quantitative calculations which relate to industrial processes (Eq. 1.13).

$$(\text{industrial yield}) \times (\text{conversion}) = (\text{research yield}) \qquad 1.13$$

This can be further rearranged to another useful expression for calculations involving process efficiencies (Eq. 1.14).

$$\% \text{ selectivity} = \frac{\text{research yield (fractional)} \times 100}{(\text{fractional conversion})} \qquad 1.14$$

It is useful to consider the significance of the ability to carry out yield and conversion manipulations. It is, of course, desirable to have any industrial process operate with high yields and high conversions. If both of these conditions prevail, more product will be obtained from each passage of raw materials through a given size of reactor, and there will be less starting material to separate and recycle, than if this were not true. Many industrial processes do in fact operate with this favorable situation.

It is possible, however, to make a success of an industrial process which only achieves low conversions, as long as high yields are maintained. Very few industrial processes operate with industrial yields (selectivities) of less than 90%, and many operate with yields of 95% or better. Yet some of these, for example the vapor phase hydration of ethylene to ethanol and the ammonia synthesis reaction, both of which have low conversions in the 5 to 15% range. If one only had research yield information about these processes, 4 to 5% and 15 to 20%, respectively, neither would appear to be promising candidates for commercialization. However, both of these processes are operated on a very large scale because they achieve selectivities of better than 95% for the desired product. Thus, while it is desirable for an industrial process to obtain high conversions with high yields (selectivity), it is *vital* for a successful industrial process to have a high selectivity for the desired product.

1.5.6. Importance of Reaction Rate

Fast reactions, in general, are conducive to obtain a large output from a relatively small volume of chemical processing equipment. For example, the ammonia oxidation reaction, which is the first stage of production of nitric acid from ammonia, is essentially complete in 3×10^{-4} seconds at 750°C. This is sufficiently rapid so that the catalytic burner required to do this occupies only about the volume of a file cabinet drawer for the production of some 250 tonnes of nitric acid daily. Except for the cost of the catalyst inventory (which is platinum), the fabrication cost of the ammonia burner itself is relatively low. Follow-up reactions for the process are much slower than this so that the volume of equipment required to contain these parts of the process are much larger and more costly (Chap. 11).

Ammonia oxidation represents a process with which it was realized, early in the design stage, that carrying out this step at 600–700°C instead of at near ambient temperatures speeded up the process sufficiently to allow large conversion volumes to take place in a relatively small reactor. This same philosophy is followed whenever feasible with all chemical processes (i.e., a saving in reactor volume decreases capital costs). With some processes, such as the esterification of glycerin with nitric acid, technical complications put an upper limit on feasible reaction temperatures. This effectively prevents the use of higher temperatures to increase the rate of the reaction. Consequently, under the normal operating temperature of about 5°C this process, has a 60- to 90-min reaction time requirement. So to produce even 20 tonnes of nitroglycerin/day would require a batch reactor of 2 tonnes or so capacity, much larger than the ammonia oxidation unit required for a 250 tonne/day nitric acid plant.

These examples illustrate the principle that, wherever feasible, reaction conditions, catalysts, etc., are selected and developed in such a way that the rate of a commercial process is maximized. In doing so the size of the processing units required for a given volume of production is reduced, in this way decreasing the costs of construction. Reducing the capital costs also reduces the capital charge per unit of product, which decreases the price required from the product to still operate at a profit. In these ways, improvement of the rate of a chemical process becomes a further contributing factor in the market competitiveness of the chemical industry.

1.6. CHEMICAL VOLUME PERSPECTIVES

The chemicals listed in Table 1.4 are presently produced on the largest scale and are examples (Table 1.4, refs. [32–36]) of the so-called commodity, heavy, or bulk chemicals. Sodium chloride is not always classified as a *produced* chemical since most salt production is basically extractive in nature. Sulfur, too, is sometimes ranked with produced chemicals and sometimes with

TABLE 1.4   World and American Production of Large Scale Chemicals, in Millions of Metric Tonnes[a]

U.S. rank, 1995	Chemical	United States		World data	
		1980	1995	1980	1995
1	Sulfuric acid	40.1	43.3	132.4	87.7
2	Nitrogen	15.5	30.9	–	–
3	Oxygen	15.6	24.3	–	–
4	Ethylene	12.5	21.3	ca. 60[b]	77.4[b,c]
5	Lime	16.0	18.7	118.0	108.9
6	Ammonia	17.2	16.2	89.4	121.4
7	Phosphoric acid	9.9	11.9	139.6[d]	130.0[d]
8	Sodium hydroxide	10.3	11.9	31.5	192
9	Propylene	6.2	11.7	9.1	41.5
10	Chlorine	10.2	11.4	27.9[e]	36.4[e]
11	Sodium carbonate	7.5	10.1	28.1	30.3
12	Methyl-t-butyl ether	<0.1	8.0	1.5	15.0[f]
13	Ethylene dichloride	4.5	7.8	–	–
14	Nitric acid	7.8	7.8	29.7	50–60[g]
15	Ammonium nitrate	7.8	7.3	6.7	8.62[h]
16	Sugar	4.3[c]	7.0	84.4	116.8

[a]Data compiled from *Chemical and Engineering News* [32, 33], *U.N. Statistical Yearbook* [34], *Minerals Yearbooks* [35], and the Kirk-Othmer Encyclopedia [36].

[b]Estimated for 1982 from [36].

[c]For 1993.

[d]Stated as phosphate rock, for lack of phosphoric acid data.

[e]Estimated from sodium hydroxide production.

[f]Methyl-t-butyl ether *http://www.inchem.org/documents/ehc/ehc/ehc206.htm* by WHO.

[g]For 1999.

[h]From FAO Statistical Database. Available: *http://aps1.fao.org/servlet3?* and follow the links.

extracted minerals, depending on the origin of the sulfur. If one leaves these two chemicals aside, sulfuric acid emerges as the leading volume chemical product, both in the U.S. and worldwide. The availability of world production data for many chemicals is somewhat sporadic, but the world ranking for most of these lies near the U.S. ranking. American production of most of the chemicals on this list represents the largest single contribution to the world figure. The U.S. ranking data can often be used to estimate both world rankings and world production levels when these data are not available directly.

A number of other interesting observations can be made concerning these particular bulk chemicals. First, they are all not far removed, in terms of processing steps, from the natural raw materials from which they are derived. Virtually, all of the oxygen and nitrogen and a significant proportion of the salt, sulfur, and sodium carbonate are all obtained relatively directly from natural sources. Also, these commodities interrelate quite closely to one another, in chemical terms. Thus sulfuric acid, largely produced from sulfur is, in turn used, to produce phosphoric acid from phosphate rock. A large fraction of the nitrogen produced goes into the production of synthetic ammonia. Ammonia, in turn, is used for nitric acid production and also is combined with much of the nitric acid product in an acid-base reaction for the preparation of ammonium nitrate. Chlorine and sodium hydroxide are mostly obtained from the electrolysis of a solution of sodium chloride in water. It is interrelationships of this kind, coupled with very large world fertilizer markets for some of the secondary and tertiary products of these sequences, in particular ammonia, ammonium nitrate, and phosphoric acid (as salts), that keep many of these chemicals on this large volume production list.

If all American chemical companies are ranked by annual chemical sales, one obtains a list which includes a significant number of oil companies (Table 1.5). This exercise shows that in 1995, six of the 15 largest US chemical companies were oil companies which were also producing chemicals. By 2002, the number of oil companies in this group was reduced to two from mergers and reorganizations, which also substantially affected their rankings in this list. Listing the world's largest chemical companies in 2002 produces a similar picture (Table 1.6). The dominant positions of Germany, the U.S., and the U.K. in chemicals production is evident from this ranking, with three companies from each of these countries represented on this list. Four oil companies, each producing chemicals as a small fraction of their total sales, also show up on this list. Total, Exxon/Mobil, Shell, and BP had gross sales in 2002 of 97.14, 205.1, 178.9, and 179.0 billion US$, respectively [43]. Clearly, oil and gas production and processing are the dominant business areas of these companies.

TABLE 1.5   The Fifteen Largest Chemical Processing Companies in the United States Ranked on the Basis of Chemical Sales^a in 2002

Overall rank				Chemical Sales (10^9 US$)	
1980	1995	2002	Company	1995	2002
2	1	1	Dow Chemical^b	19.2	27.6
1	2	2	Du Pont	18.4	26.7
3	3	3	Exxon/Mobil^c (2002)	11.7	16.4
n/a	6	4	General Electric	6.6	7.6
n/a	14	5	Huntsman Chemical	4.3	7.2
n/a	40	6	PPG Industries	1.6	6.0
–	–	7	Equistar Chemicals	–	5.5
–	18, 26^d	8	Chevron Phillips	6.7^d	5.5
17	11	9	Eastman Chemical	5.0	5.3
–	24	10	Praxair	3.1	5.1
–	22	11	Air products	3.5	5.1
–	16	12	Rohm and Haas	3.9	5.0
–	28	13	Lyondell Chemical	2.5	3.3
–	–	14	Honeywell	–	3.2
5	5	15	Monsanto	7.2	3.1
21	7	–^{b}	Mobil Oil	6.2	–^{b}
4	8	–^{c}	Union Carbide	5.9	–^{c}

^{a}Compiled from Chemical and Engineering News [40], Peaff [41], and Tullo [38].
^{b}Merged with Exxon.
^{c}Union Carbide merged with Dow Chemical.
^{d}Rankings of the separate companies, and their total chemical sales.

■■■■ **TABLE 1.6   The World's 12 Largest Chemical Companies Based on 2002 Chemical Sales**[a]

Rank, 2002	Rank, 1995	Company	Chemical business area	2002 chemical sales (10^9 US$)
1	3	Dow Chemical[b]	Chemicals, plastics	27.6
2	5	Du Pont (U.S.A.)	Chemicals	26.7
3	1	BASF (Germany)	Chemicals, plastics	25.3
4	39	Total (France)	Chemicals	18.3
5	4	Bayer (Germany)	Chemicals, drugs	17.8
6	8	Exxon[c]/Exxon Mobil (U.S.A.)	Chemicals	16.4
7	6	Shell (U.K., Netherlands)	Chemicals	15.2
8	33	BP (U.K.)	Chemicals, plastics	13.1
9	49	Degussa (Germany)	Chemicals	11.1
10	20	Akzo-Nobel (Netherlands)	Chemicals, resins	9.4
11	7	ICI (U.K.)	Chemicals, plastics	9.2
12	43[d]	Mitsui Chemicals (Japan)	Chemicals	8.4

[a]Compiled from Layman [42] and Short [43].
[b]Merged with Union Carbide since 1995.
[c]Merged with Mobil since 1995.
[d]Mitsui Toatsu Chemicals.

REVIEW QUESTIONS

1. What minimum flow rate in cm/sec is required in a 10-m length of open straight pipe of 2 cm inside diameter to obtain a uniformly mixed paint from two components of combined density 1,100 kg/m^3 and 1.912 centipoise viscosity?

2. A 100 tonne/day electrolytic chlorine plant, complete with caustic soda facilities costs about 12 million dollars.
   (a) What would be the approximate capital cost of a 600 tonne/day facility?
   (b) What capacity chlorine plant could be built for $6 million?
   (c) If an operating staff of eight is required to run a 100 tonne/day plant, what staffing would be needed for plants with capacities of 1,000 tonne/day and 2,500 tonne/day?

3. The labor requirement for a chemical processing unit can also be related to size (capacity). If 32 people are required to operate a 100 tonne/day sulfuric acid plant, what would be the estimated labour requirement for a 1,000 tonne/day plant?

4. A tube reactor is to be used to contact an aqueous sugar solution with 5% by volume of a solvent of density 0.780 g/cm^3 for extraction. The aqueous solution has a density of 1.080 g/cm^3, and a viscosity of 1.201 centipoise.
   (a) If the tube to be used is 2 cm in diameter and the mixture of the aqueous solution and solvent is to flow at a combined velocity of 50 cm/sec, in the tube, would there be efficient contact (i.e., turbulent

flow) between the sugar solution and the other liquid phase? Very briefly describe the criterion for your answer.

(b) If the liquids to be contacted have a combined bulk density of 68 lb ft^{-3} and a viscosity of 7.923×10^{-4} lb ft^{-1} sec^{-1}, what flow velocity in feet per second would be required to ensure turbulent flow conditions in a 1-in. pipe (inside diameter)?

(c) What would be the daily rate (kg/24 hr) of sugar production in part (a) if the sugar concentration was 2% by weight and extraction efficiency was 99.9%?

5. The capital costs of 200 million kg/year and 500 million kg/year styrene plants are $18 million and $28 million.

(a) From a plot of log capital costs versus log capacities from this data, determine the specific value of the exponent n which could be used for more accurate estimation of costs for construction of other capacities of styrene plant.

(b) Using the exponent value determined in part (a), what would be the approximate capital cost of a 100 million kg/year plant?

(c) What would have been the estimated capital cost of a 100 million kg/year plant using the 0.60 mean value for the exponent n and each one of the two known capital costs given in the preamble?

6. Naphthalene ($C_{10}H_8$) can be converted microbiologically to salicylic acid ($C_7H_6O_3$) on a 94%, weight for weight basis, in one pass. (CO_2 and H_2O are also produced.)

(a) At 100% conversion, what is the percent selectivity (industrial yield) of this process?

(b) What is the "academic" or research yield for this process?

(c) Air is 21% oxygen (take as mole ratio 1 O_2:4 N_2). Assuming 25% conversion of the oxygen content of the air in this process, what mass of air would be required to produce 1,000 kg of salicylic acid? Ignore the oxygen consumed in the formation of byproducts.

7. To produce acetic acid (CH_3CO_2H) from the oxidation of acetaldehyde (CH_3CHO) with air, conversions of 26% and industrial yields (selectivities) of 95% on acetaldehyde are obtained.

$$CH_3CHO + 1/2O_2 \longrightarrow CH_3CO_2H \ (+ CH_3CHO + O_2 + \text{by-products})$$

$$\text{ca. } 70°C$$

Reactor charge

acetaldehyde	2.200 tonne
Mn(OAc)$_2$ (catalyst)	6.6 kg
Air	4.48×10^3 m^3(1 atm.,0°C)

(a) Calculate the acetic acid and acetaldehyde content expected in the reactor at the end of the reaction.

(b) Assuming that air is exactly 1 mole:4 moles, O_2:N_2, what would be the percent conversion of oxygen to acetic acid in accordance with the equation given?

FURTHER READING

American Chemical Society, "Chemical Abstracts," Columbus, OH, published continuously since January, 1907. Chemistry, technology, and patent indices for all countries.

"Applied Science and Technology Index." H.W. Wilson Company, Bronx, NY, since 1958; formerly The Industrial Arts Index, published since 1913.

"Business Periodicals Index." H.W. Wilson Company, New York, published since January, 1958. Current economic data.

"Chemical Market Reporter" (formerly Oil, Paint, and Drug Reporter, & Chemical Marketing Reporter) Schnell Publishing Co., New York, published 2 volumes per year since October 1871, gives weekly chemical prices.

Library Association, "British Technology Index," London, published since January 1962.

P.J.T. Morris, W.A. Campbell, and H.L Roberts, eds., "Milestones in 150 years of the Chemical Industry," The Royal Society of Chemistry, Cambridge, U.K., 1991.

Patent Office, "Patent Office Record," Patent Office, Ottawa, Ontario, published since 1872. Available: http://patents1.ic.gc.ca/intro-e.html

P. Tundo and P. Anatas, "Green Chemistry: Challenging Perspectives," Oxford University Press, Oxford, 2000.

U.S. Patent Office, "Official Gazette," Washington, DC, published weekly since 1872, (replaces "Patent Office Reports." Available: http://www.uspto.gov/patft/

L. Wilson and D. McCutcheon, "Industrial Safety and Risk Management," The Canadian Society for Chemical Engineering, Ottawa, 2003. p. 176.

REFERENCES

1. F.A. Lowenheim and M.K. Moran, "Faith, Keyes and Clark's Industrial Chemicals," 4th ed. Wiley-Interscience, New York, 1975.
2. M. Trowbridge, *Chem. Brit.* **11**(1), 15, Jan. (1975).
3. H.S. Sloan and A.J. Zurcher, "Dictionary of Economics," 5th ed., p. 459. Barnes and Noble, New York, 1971.
4. A.C.H. Cairns, *Roy. Inst. Chem. Rev.* **2**(1), 41, Feb. (1969).
5. Organization for Economic Cooperation and Development, "The Chemical Industry," OECD, Paris, 1970, and 1981.
6. Calculated for 1999 from the "OECD STAN Database for Industrial Analysis", STAN Indicators Database, vol. 2004, release 04, using currency conversions from the Web for 1999. Raw data available 6/28/2004 from: http://fiordiliji.sourceoecd.com/vl = 17758529/.
7. R. Stevenson, DuPont Steps into a New Era, *Chem. Brit.* **38**(8), 27–30, Aug. (2002).
8. A. Tullo, Growing it alone, *Chem. Eng. News*, **82**(17), 18, Apr. 28 (2004).
9. M.S. Reisch, C&EN talks with J.A. Miller Jr., *Chem. Eng. News*, **82**(19), 17, May 10 (2004).
10. Facts and Figures for Chemical R and D, *Chem. Eng. News*, **60**(30), 38, July 26 (1982).
11. Most Patents Continue to be in Chemical Field, *Chem. Eng. News*, **60**(14), 24, Apr. 5 (1982).
12. P.L.Short, European Patents Set for Revamp, *Chem. Eng. News*, **81**(18), 22–24, May 5 (2003).
13. F.P. O'Connell, *Chem. Eng.(N.Y.)*. **69**(4), 150, Feb. 19 (1962).
14. J.P. Stern and E.S. Stern, "Petrochemicals Today." Edward Arnold, London, 1971.
15. M.S. Peters and K.D. Timmerhaus, "Plant Design and Economics for Chemical Engineers," 3rd ed. McGraw-Hill, New York, 1980.
16. D.H. Allen, *Chem. Ind. (London)*. 98–100, Feb. 5 (1977).
17. G.T. Austin, "Shreve's Chemical Process Industries," 5th ed. McGraw-Hill, New York, 1984.
18. K. Denbigh, "Chemical Reactor Theory." University Press, Cambridge, U.K., 1965.
19. M.D. Wynne, "Chemical Processing in Industry." Royal Inst. of Chemistry, London, 1970.
20. H.S. Fogler, ed., "Chemical Reactors," ACS Symp. Ser. No. 167, American Chemical Society, Washington, DC, 1981.

21. C. Brechtelsbauer, N. Lewis, P. Oxley *et al.*, Evaluation of a Spinning Disk Reactor for Continuous Processing, *Org. Proc. Res. & Devel.* **5**, 65–68 (2001).

22. A.M. Rouhi, Microreactors Eyed for Industrial use, *Chem. Eng. News*, **82**(27), 18–19, July 5 (2004).

23. J.G. Raith, ed., "Chemical Engineering; a Special Study," Nuffield Foundation, Penguin, Harmondsworth, England, 1971.

24. R.V.Smith. Determining Velocity and Reynolds Number in Gas Pipelines and Wells, *Technology*, **81**, 108–112, Dec. 12 (1983).

25. R.H. Perry and D.W. Green, eds., "Perry's Chemical Engineer's Handbook," 6th ed. McGraw-Hill, New York, 1984.

26. J.M. Bélanger, Responsible Behavior: The Underpinning of True Partnership, *Can. Chem. News*, **53**(2), 34–35, Feb. (2001).

27. J. Johnson, Simply safer, *Chem. Eng. News*, **81**(5), 23–26, Feb. 3 (2003).

28. J. Clark, The Greening of Chemistry, *Chem. Brit.* **34**(10), 43–45, Oct. (1998).

29. Facts and Figures for the Chemical Industry, *Chem. Eng. News*, **60**(24), 31, June 14 (1982).

30. Facts and Figures for the Chemical Industry, *Chem. Eng. News*, **61**(24), 26, June 13 (1983).

31. P. Wiseman, "Industrial Organic Chemistry." Wiley-Interscience, New York, 1972.

32. Top 50 Chemical Products and Producers, *Chem. Eng. News*, **59**(18), 35–42, May 4 (1981).

33. Facts and Figures for the Chemical Industry, *Chem. Eng. News*, **74**(26), 38–79, June 24 (1996).

34. "United Nations Statistical Yearbook 1996," 43rd ed., United Nations, New York, 1997, and earlier editions.

35. "1996 Minerals Yearbook," Vol. I, U.S. Dept. of the Interior, Bureau of Mines, Washington, DC, 1998, and earlier editions.

36. "Kirk-Othmer Encyclopedia of Chemical Technology," 3rd ed., Vol. 9, p. 872. John Wiley, New York, 1984.

37. G. Peaff, Dow replaces DuPont to lead top 100 U.S. producers, *Chem. Eng. News*, **74**(19), 15–20, May 6 (1996).

38. A. Tullo, Top 50, *Chem. Eng. News*, **82**(20), 25–29, May 17, (2004).

39. P.L. Layman, Global top 50 Chemical Producers Shift Rankings…, *Chem. Eng. News*, **74**(30), 29–31, July 22 (1996).

40. P.L. Short, Global top 50, *Chem. Eng. News*, **81**(30), 17–19, July 28 (2003).

2
AIR QUALITY MEASUREMENT AND EFFECTS OF POLLUTION

We have first raised a dust and then complain we cannot see.

—*Bishop Berkeley (1685–1753)*

2.1. SIGNIFICANCE OF HUMAN ACTIVITY ON ATMOSPHERIC QUALITY

In the early days of habitation of this planet, when the human population was small and its per capita consumption of energy was primarily in the form of food (8,400–12,600 kJ/day; 2,000–3,000 kcal/day), the total human demands on the biosphere were relatively insignificant. Early requirements of goods were minimal and simple (requiring little fashioning) so that this early society's total demands and wastes were easily assimilated by the biosphere.

Advances in technology today provide an ever-increasing range of goods and services. This has been estimated to increase the individual consumption of energy some 100-fold from the requirements of primitive man. Of the 1 million kJ/person/day (230,000 kcal/person/day) that this consumption now represents, more than half, or 645,000 kJ (154,000 kcal) is estimated to be consumed by society's industrial, agricultural, and transportation needs. About a quarter is consumed per person through the medium of electricity generation and consumption, either for their own use, or by the prorated industrial power consumption on their behalf. High-technology agriculture, industrial production, and thermal electricity generation all use combustion processes to provide a major fraction of their energy requirements. When this gross increase in per capita energy consumption is coupled to the global population growth of more than a thousand times the population of primitive societies, it is easy to see how the total human demand placed on the biosphere by modern industrial societies has become so significant.

Most of the world's population growth has taken place since the Middle Ages. Industrial development has accelerated since about 1850, so that most of the increase in demand for energy and materials has occurred during the period when these two component increases coincided. Human activities now contribute similar volumes of some minor gaseous constituents of the atmosphere as do natural processes. From a world fossil fuel consumption of about 5 billion tonne/year, we now contribute about 1.5×10^{10} tonnes of carbon dioxide to the atmosphere annually. This is a significant fraction of the natural contribution of about 7×10^{10} tonne/year. Annual human contributions of sulfur-containing gases to the atmosphere are now about 10^{10} tonnes, oxidized and reduced nitrogen compounds about 50 million tonnes, and carbon monoxide about 200 million tonnes. All are now a similar order of magnitude to the natural contributions, which has invalidated the old jingle "The solution to pollution is dilution." There is just too large a total mass of atmospheric contaminants being discharged, and too small an atmosphere to accept these, to be able to obtain sufficient dilution. It becomes particularly noticeable when natural air movement is sluggish for any reason so that there is little pollutant mixing. Discharged pollutants build up in the local air mass, greatly exaggerating the immediate effects of the discharge.

2.2. NATURAL CONTAMINANTS

Natural sources of many common air contaminants make a contribution to the overall atmospheric pollutant loading. Oceans contribute large masses of saltwater spray droplets to the air as a result of wave action. As the water evaporates from these droplets very fine particles of salts are left suspended in the moving air, contributing to sea "smell" and atmospheric chemistry. Some 13 million tonnes of sulfate ion and similar masses of chloride are contributed to the atmosphere annually in this manner.

Volatile organic compounds are contributed to the atmosphere by many forms of plant life, by way of conifers such as cedar, pine, and eucalyptus, and aromatic plants such as lavender, mint and sage. The pine forests of New England and the eucalyptus forests of the Blue Mountains in Australia contribute large masses of terpenes to the air above them. Terpenes are plant products biosynthetically derived from isoprene and have a formula of the type $(C_5H_8)_n$, where n is based on the number of isoprene units in the compound (Eq. 2.1).

$$CH_2=C(CH_3)-CH=CH_2$$

isoprene α-pinene   limonene

All terpenes have one or more reactive double bonds. Reactions of terpene vapors in the presence of sunlight and air photochemically generate a blue haze over these forests when there is little air movement; hence the name Blue

Mountains. The same naturally occurring reactive compounds have been found to also contribute to urban photochemical smogs [1]. Estimates of the global atmospheric contribution by plants of terpenes and oxygenated terpenes range from 2×10^8 to 10^9 tonne/year.

The world's deserts contribute significant masses of dust and particulate matter to the atmosphere, some transported across considerable distances. More than 900 tonnes of meteoritic dust is estimated to be collected by the atmosphere annually. Active volcanoes contribute many orders of magnitude more than this. For example, some 90 million tonnes of dust and particles were estimated to have been discharged into the atmosphere by the eruption of Mount St. Helens in Washington State [2]. Not infrequently, several eruptions take place simultaneously, measurably increasing the dust loading in the global atmosphere. These contributions are long lived since some is forcefully injected as high as 40 km into the atmosphere. As a result, the global influence of these events has much more significance on atmospheric quality, even to the extent of lowering average world temperatures.

Sulfur dioxide and hydrogen sulfide from volcanic activity also contribute pollutants to the atmosphere on the scale of 1–2 million tonne/year. Other contaminating gases such as metal vapors are also discharged in significant quantities during these events [3]. Vapors of the more volatile metallic elements such as mercury are also lost continuously over ore bodies. The extent of these losses is sufficient to permit the use of sensitive vapor detection instruments as a prospecting method to locate the extent of the ore body by using this mercury vapor "halo" [4].

Biological contaminants in the atmosphere may not represent a large mass but are, nevertheless, an important component of atmospheric pollution because of their potent effect. For example, the pollens from many wild flowers and grasses such as golden rod and ragweed severely affect a large proportion of the population. Bacteria, viruses, and the living spores of some of the common molds are distributed through air, and can cause problems, particularly when a rapid, localized rise in numbers of the organism occurs. For example, marine waters near sewer outfalls can become contaminated by microorganisms. These can be dispersed by vigorous wave action on these waters, changing a water pollutant into an aerosol [5].

Little can be done about controlling most natural sources of atmospheric contaminants. However, it is still important to catalog these to quantitatively relate their importance to atmospheric quality. Only in this way is it possible to compare the relative significance of the two sources of contaminants. This is a necessary preliminary to provide appropriate guidance for the our own activities to decrease any negative impact on atmospheric quality when necessary.

2.3. CLASSIFICATION OF AIR POLLUTANTS

Air pollutants can be classified into one of three main categories based on their physical characteristics. This enables potential emissions to be grouped so as to assess appropriate measures for avoidance of production, or emission control. On this basis, one or more types of emission control devices may be selected for use, based on their mode of action.

The first of these classes of air pollutants is called coarse particulate matter (PM), and comprises solid particles or liquid droplets which have an average diameter greater than about $10\,\mu m(10^{-3}\,mm)$, (i.e., $> PM_{10}$). Particles or droplets of this class of contaminants are large enough to fall more or less rapidly out of the air of their own accord.

The second group of air pollutants is the aerosol class. This can also comprise solid particles or liquid droplets, but they are limited to a size range generally less than about $10\,\mu m$ average diameter (e.g., a median diameter of $2.5\,\mu m$, $PM_{2.5}$). This class has particles or droplets small enough in size that there is a strong tendency for them to stay in suspension in air [6]. Powders of the denser solids, such as magnetite, would have to have a particle size of $2.5\,\mu m$ or less to stay in suspension. This size range is also referred to as the Respirable Fraction, since these are not readily captured by ciliated mucous of the nasal passages and penetrate to the unciliated alveoli of the lungs [7]. A suspension of a finely divided solid in air is referred to as a "fume," and that of a finely divided liquid as a "fog."

The gases comprise the third major classification of air pollutants, which includes any contaminant in the gaseous or vapor state. This comprises the more ordinary "permanent" gases, such as sulfur dioxide, hydrogen sulfide, nitric oxide (NO), nitrogen dioxide (NO$_2$), ozone, carbon monoxide, carbon dioxide (pollutant?), etc., as well as the less common ones such as hydrogen chloride, chlorine, tritium ($^3_1$H) and the like. It also includes materials which

TABLE 2.1   Gravitational Settling Velocity for Spheres of Unit Density in Air at 20°C^a

Examples	Particle diameter (μm)	Particles per microgram	Micrograms per particle	Terminal velocity^b (mm/sec)
Carbon black	0.1	10^9	1×10^{-9}	8.5×10^{-4}
Clay	0.5			1.0×10^{-2}
Clay	1.0	10^6	1×10^{-6}	3.5×10^{-2}
Paint pigments	5.0			0.78
Silt, fog, flour	10	10^3	1×10^{-3}	3.0
Fine sand, flour	20			12
Pollens	50^c			72
Medium sand, pollens	100^d	1	1	250

^{a}From Barrett [8], Spedding [9], and calculated results.

^{b}The final speed of a particle through are from the force of gravity, slowed by the drag imposed by air.

^{c}Roughly equivalent to a powder which would pass through a No. 325 sieve (325-mesh, $54\,\mu m$ average opening size) and be retained on a No. 400 sieve (400-mesh, $45\,\mu m$ average opening size).

^{d}Roughly equivalent to a powder which would pass through a No. 170 sieve (170-mesh, $103\,\mu m$ average opening size) and be retained on a No. 200 sieve (200-mesh, $86\,\mu m$ average opening size).

are not ordinarily gases, such as hydrocarbon vapors, and volatile nonmetal or metal vapors (e.g., arsenic, mercury, zinc) when these are in the vapor state.

The dividing line between the particulate/aerosol classes, and the gaseous classes is clear enough because of the phase difference. However, the position of the dividing line between the particulate and the aerosol classes is less obvious, since it is based on whether or not a second phase stays in suspension in air. Consideration of the terminal velocities or speed of fallout of particles of differing diameters helps to clarify this dividing line. Table 2.1 illustrates that a significant terminal velocity in air begins to be observed at particle diameters of about 10 μm for a substance with a density of $1\,g/cm^3$ and larger. This is the physical basis of the approximate dividing line between these two classes. Figure 2.1 gives examples of typical particle size ranges for some common industrial and domestic substances that may become airborne.

2.4. PARTICULATE MEASUREMENT AND IDENTIFICATION

The particulate class represents particles or droplets which more or less rapidly settle out of air, and is also the easiest class to measure. For a source particulate determination, that is, if the particulates in the flue gases of a chimney or exhaust gases of a vent stack are to be sampled, then special holes are required in the ductwork. Probes with associated equipment and a means of reaching the sampling holes are necessary. More details of this procedure are given in connection with aerosol determination.

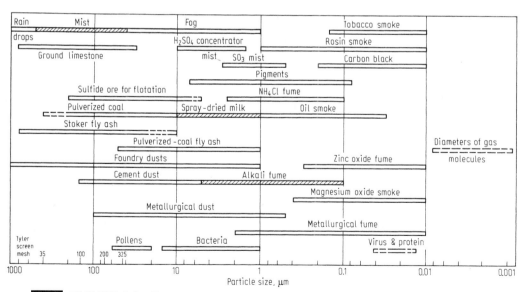

FIGURE 2.1 Typical particle size ranges of some types of precipitation, industrial and mineral processing streams and ambient air biologicals. (From Munger [10], with permission from McGraw-Hill.)

It is also possible, however, to gain a useful semi-quantitative record of source discharge information by observation of the opacity of a plume. The Ringelmann system was set up by Maximilian Ringelmann in about 1890 to accomplish this for black smokes [11]. Ringelmann numbers 1, 2, 3, 4, and 5 represent 20, 40, 60, 80, and 100% opaque plumes, respectively. A circular card with shaded opacities corresponding to these Ringelmann numbers which surround a central viewing hole is available to assist in these quantitative visual assessments. A trained observer is able to reproducibly estimate plume opacity to within half a Ringelmann number ($\pm$10% opacity) without this assistance. The obscuring properties of white smokes are referred to simply on a percent opacity basis.

A quantitative method for particulate determination uses fallout from ambient air as an indicator of the particle loading. This method can be as simple as a series of glass jars placed in flat collection areas without obstructions for sites for which dustfall measurements are desired. The jars may be used dry, or they may contain a liquid collecting agent to prevent any fallout from being swept out again by turbulence. After an interval of usually 30 days, the collected material is filtered (if wet collection was used), dried, and weighed. The weight obtained may be used to calculate a fallout value for each of the areas in which the collections were carried out. Older data was usually specified as short tons per square mile per month. Today, $mg/m^2/day$ is more common.

Essentially, the same procedure is used with a plastic dustfall canister on a special stanchion, with a few collection refinements to increase the reliability of the results. The stanchion places the top of the canister 4 ft above the base, and must be sited a reasonable distance away from trees, buildings, or roof fixtures to minimize the interference of fallout collection by anomalous air currents. A bird ring is used to discourage contribution to the collection by birds resting on the rim of the canister itself. A series of vertical spikes is welded to the bird ring for areas frequented by larger species such as gulls and crows, to prevent their roosting on the rim of the canister or the bird ring.

The amount of fallout obtained in relatively undisturbed open country is often a surprise to those who have not worked in the area of air pollution measurement. For example, collection of 100 mg of dustfall during a 30-day period by a glass container having a 10-cm-diameter opening, does not sound like much. But in conventional fallout units this amounts to $424\,mg/m^2$ day, or more impressively as $36.4\,ton/(mile)^2$ month (Eqs. 2.2 and 2.3).

$$(100\,mg \times 10{,}000\,cm^2/m^2) \div (3.1416 \times (5\,cm)^2 \times 30\,days) \qquad 2.2$$

$$(0.10\,g \times (2.54\,cm/in)^2 \div (12\,in/ft)^2 \times (5280\,ft/mile)^2)$$
$$\div (3.1416 \times (5\,cm)^2 \times 9.07 \times 10^5\,g/tonne) \qquad 2.3$$

This is actually quite a heavy fallout. Table 2.2 gives ranges of some recent fallout values for the U.K. and Canada, to give a basis for comparison. The Canadian figures are somewhat higher since they were obtained for an industrialized area with a relatively high population density.

An early Canadian particulate guideline stipulated that no discharge would be permitted in an area that would cause the fallout to exceed 20 tons per square mile per month.

TABLE 2.2 Average Values for the Particulate Fallout Experienced by Areas of Differing Urban Activity in the United Kingdom and Canada[a]

Type of area	United Kingdom, 1962–1963 Particulate fallout		Canada, 1968 Particulate fallout	
	mg/m² day	ton/mi² month[b]	mg/m² day[c]	ton/mi² month
Industrial	159	13.6	233–350	20–30
High-density housing	116	9.95	–	–
Town center	112	9.60	210–256	18–22
Town park	90	7.7	–	–
Low-density housing	82	7.0	117–163	10–14
Suburbs, town edge	70	6.0	–	–
Open country	38	3.3	82–140	7–12

[a]Compiled from Barrett [8] and International Joint Commission [12].
[b]Calculated using the factor $(ton/mi^2/month) \div (mg/m^2/day) = 0.08556$.
[c]Calculated from the values in $(tons/mi^2/month)$ by multiplying by the factor 11.68 $(1/0.08556)$.

2.5. AEROSOL MEASUREMENT AND IDENTIFICATION

Sampling and measurement of this class requires dynamic sampling equipment since this class of air pollutants represents particles and droplets too small to fall out of their own accord. Work has to be done on the gas to force it through the recovery or analytical equipment to capture the suspended matter. The analysis can be a source test for which a stack or waste vent is sampled directly, or it can be an ambient air survey, where the general condition of the outside air is determined.

Source testing requires the placing of a probe into the stack or waste vent through a sampling port in the side. A vacuum pump at the exhaust end of the sampling train draws the test gas sample into the train (Fig. 2.2). Good source sampling technique requires that the gas flow rate into the end of the probe is the same as the gas flow rate in the stack at the point being sampled, referred to as

FIGURE 2.2 Simplified layout of an impactor sampling train. (From Pilat et al. [13], with permission.)

iso-kinetic sampling [14]. Without this, some sorting of the particles sampled can occur which would produce unrepresentative results. For large ducts, different points along a cross section of the stack must be sampled to correct for any variation in particle concentrations across a section, a procedure referred to as "equal area sampling" [15]. This is necessary because even a long straight vertical duct will have lower gas flow rates closer to the wall than the center due to frictional effects. This flow rate difference will affect the relative particle loadings in the gases moving at the center, and at the edges of the duct.

As the sampled gases flow into the first series of large jets in the impactor used for collection, large particles with significant momentum strike the impactor plate and remain stuck there for later analysis (Fig. 2.3). Smaller particles with less momentum are diverted away from the plate by the diverted gas stream. As the gases proceed further into the impactor, smaller and smaller jet sizes force the gas to move at higher velocities. As the gas velocity increases, smaller particles receive sufficient kinetic energy to impinge on the collector plates, and stick there. The tendency to stick may be increased by application of a thin wipe of petroleum jelly. In this manner, collected particles are roughly classified as to size. On completion the impactor is carefully disassembled. The sorted collected material will give some information about the particle size distribution of the tested source, or the ambient air, whichever is being sampled. The collected material is also available for microscopic examination, or wet chemical and instrumental tests if desired. Good quantitative information for the whole range of particle sizes is not

FIGURE 2.3   Enlarged longitudinal section of a portion of a cascade impactor for source testing. (From Pilat *et al.* [13], with permission.)

possible with this system unless it is fitted with a fine pore filter (e.g., membrane) for final collection from the sampled gas. This type of filter has a large pressure drop across it so that it is rarely used for this purpose.

Filters are also useful on their own for the sampling of ambient air for suspended matter. The high-volume ("high-vol") sampler is one of the most common devices used for this purpose. An air turbine driven by a vacuum cleaner motor provides the suction to pull an air sample of $2000\,m^3$ or more through a thick 12×15-cm filter sheet during a predetermined test period, usually 24 hr. A recording flowmeter is used to keep track of the volume of air filtered, since the flow rate gradually declines during the test period as collected material gradually impedes air flow. The quantities (flow rate) × (time) give the volume of air filtered, and the gross weight minus the tare weight of the filter, dried to standard conditions, gives the mass collected. A quantitative mass per unit volume result for ambient air is thus obtained, only limited at the lower end of the particle size range by the pore size of the filter used. The result is usually specified in units of mass per unit volume (e.g., mg/m^3 or $\mu g/m^3$). A refinement of the high-vol sampling technique enables separation of the collected sample into coarse ($2.5-15\,\mu m$) and fine ($<2.5\,\mu m$) particle fractions with the help of Teflon filter elements with different pore sizes. Much slower ambient air sampling rates of about $1\,m^3/hr$ are claimed for this device.

None of these methods are efficient at retaining particles of $1\,\mu m$ diameter or less. For this particle size range, ultrafiltration on dense cellulose or molecular membrane filters of cellulose acetate (e.g., Millipore, Isopore) is necessary [16]. Cellulose is fine for qualitative filtration work but is difficult to use in quantitative studies because of a sensitivity to moisture adsorption which affects filter mass. Cellulose acetate is much less affected by moisture and is also close to being 100% efficient at retaining particles larger than about $0.1\,\mu m$. It has the further advantage that the filter element can be directly examined under a light transmission microscope for size distribution or particle identification studies. The fine pore size of this filter requires a pressure differential of nearly an atmosphere to obtain adequate air filtration rates, and a good filter support to enable the membrane to tolerate these pressures without rupture. Even using these conditions, only about $0.03\,m^3$ of air per minute can be filtered through the standard 5-cm diameter disk, one of the limitations of the method.

An analytical scale electrostatic precipitator developed by the Mine Safety Appliance Corporation, provides a technique for collection of fine aerosol material without requiring such a large pressure drop. This is particularly useful under low loading conditions. The removable glass tube of this device is tared prior to starting the test. Weighing again on completion of the test enables determination of particle loadings as low as $10\,\mu g/m^3$ after passage of $10\,m^3$ of air. It is easy to increase the sensitivity by using larger volumes of air.

Identification of many particles and aerosols can be made using direct microscopic examination coupled with a knowledge of the collection circumstances. Particle features such as transparent/opaque, colorless/colored, rounded/angular, isotropic/anisotropic, density, and refractive index may all be determined in this way. When these properties are combined with the general visual features such as those cataloged in one of the particle atlases [17], it can very often identify particles unequivocally. Particles below about 0.2 μm in diameter, which is the limit for the best light microscopes, may be studied using electron microscopy [18]. Various destructive wet chemical, or instrumental techniques such as ion chromatography, neutron activation analysis, or atomic absorption are also used to confirm identities established by the visual process. Details are available from a variety of sources [19].

2.6. ANALYSIS OF GASEOUS AIR POLLUTANTS

A wide variety of wet chemical, instrumental, and biological procedures exist for the analysis of gaseous air pollutants. The method of choice will depend on the objective (accuracy and number of results required), budget, equipment available, and level of training of available staff. As an example of the possibilities within these choices, an air (and later also water) pollution survey was conducted in the U.K. during 1972–1973, which was publicized by the *Sunday Times*. Hardware and methods literature were made available at cost through the Advisory Centre for Education, Cambridge. The unique aspect of this survey was that it was conducted predominantly by school children. Yet the number and geographical distribution of the results obtained has been classed as one of the most comprehensive overviews of low-level ozone pollution ever conducted in the British Isles.

The chief difference between the two "particle classes" of air pollutants and the gaseous air pollutants, as far as analysis is concerned, is that for the latter normal filtration methods cannot be applied. Physical adsorption, such as on activated charcoal, or silica, or absorption into matrices such as silica gel, rubber, or zeolites, are used to capture many gaseous air pollutants [20]. Recovery of the gases or vapors required for later analysis, or for recycle depending on the scale, is frequently achieved by warming the bed of adsorbent or absorbent while passing a slow air or an inert gas stream through the bed, sometimes assisted by reduced pressure. Wet absorption methods are also used, with a suitable reagent dissolved in liquid (usually water) to obtain a reaction with the pollutant of interest to trap it in the liquid, as the air containing it is sparged (bubbled) through it. Sulfur dioxide is not well retained when contaminated air is sparged through water alone, [21] but is efficiently trapped when a solution of hydrogen peroxide or sodium hydroxide in water is used (Eqs. 2.4–2.6).

$$SO_2 + H_2O \rightarrow H_2SO_3 \qquad\qquad 2.4$$

$$H_2SO_3 + H_2O_2 \rightarrow H_2SO_4 + H_2O \qquad\qquad 2.5$$

$$H_2SO_3 + 2NaOH \rightarrow Na_2SO_3 + 2H_2O \qquad\qquad 2.6$$

2.6.1. Concentration Units for Gases in Air

The concentration of a gas or vapor in air or any other gas phase may be specified on a volume-for-volume basis, a weight for volume basis, or on a partial pressure basis. Each of these concentration units has one or more features which makes its use convenient and informative for particular situations.

Reporting of concentrations of one gas in another on a volume for volume basis is normally corrected to 25°C and 1 atm (760 mm Hg) pressure [22, 23]. Under this system, relatively high concentrations are specified in percent so that 3% sulfur dioxide in air would correspond to 3 mL of sulfur dioxide mixed with 97 mL of air, both specified at 25°C and 1 atm. This also corresponds to 3 parts by volume of sulfur dioxide in 100 parts by volume of the sulfur dioxide/air mixture, both specified in the same volume units. Lower concentrations are specified in smaller units, ppm (or ppmv, v for volume) for parts per million (1 in 10^6), ppb for parts per billion (1 in 10^9), and even ppt for parts per trillion (1 in 10^{12}), now that adequate sensitivity has been developed for the analysis of some air pollutants at these low concentrations.

The volume for volume system of specifying concentrations of a gas in a gas is dimensionless (without units) (i.e., it represents a pure ratio of volumes in the same volume units) which, thus, allows the units to divide out. To a chemist, this system has the further advantage that comparisons made on a volume for volume basis are also on a molecular, or molar basis (i.e., a mole of any gas at normal temperature and pressure–NTP or STP; 0°C and 1 atm) occupies 22.41 L. This volume for volume comparison is true for most gases and vapors when existing in the form of mixtures at or near ambient conditions of pressure and temperature, and only deviates significantly from this molar equivalency when the conditions (pressure and/or temperature) become extreme. When using this system it must be remembered that volume for volume data are corrected to 25°C and 1 atm, when a molar volume corresponds to 24.5 (24.46) L (298 K/273 K × 22.41 L).

Specifying concentrations on a volume for volume basis is also convenient because a change in the temperature or pressure does not change the concentration, within reasonable limits. Any correction necessary due to a change in the measuring conditions is generally smaller than the experimental error of the analysis, since many air samples are analyzed at temperatures close to 25°C and 1 atm. The insensitivity of volume for volume units to small changes in temperature or pressure mean that these do not normally need to be specified when using this method.

Frequently, however, the concentration of a gas (or vapor) in air is specified on a mass (weight) per unit volume basis, common units for which are mg/L, mg/m^3, and μg/m^3, when 1 mg/L = 1000 mg/m^3 = 10^6 μg/m^3. Knowing the concentration of an air pollutant specified in these units allows easier determination of mass rates of emission, which are important for regulatory purposes and for exposure hazard calculations. However, these units do not provide any molecular basis for comparison of the concentrations of any two gases, which makes it more difficult to visualize chemical relationships with this system. Also, the concentration of a gas in a gas, specified in these units, changes with changes in temperature or pressure. This happens because the mass (or weight) of a minor component per unit volume in this system becomes smaller as the temperature of the gas mixture is raised, whereas the volume of the air (or other main component) in the mixture becomes larger. Both influences tend to make the specified concentration smaller with a rise in temperature. As already stated, standard temperature for the mass per unit volume measure is generally taken to be 25°C [22, 23]. However, 0°C has also been used [24]. Therefore, to avoid ambiguity in specifying results in these units, the temperature basis used *must* also be specified. Data presented in this format are also normally corrected to 1 atm pressure.

Since both volume for volume and weight for volume units are in common use for regulations and quoting air pollution results, it is necessary to be able to interconvert between these units. To obtain a value in mg/m^3 from a value in ppm one has to multiply the ppm value by the molecular weight of the component of interest in grams, and divide by 24.46, the molar volume at 25°C (Eq. 2.7).

$$mg/m^3 = \frac{1/10^6 \ (ppm) \times mol.wgt. \ (g/mol) \times 10^3 \ (mg/g)}{24.46 \ (L/mol) \times 10^{-3} \ (m^3/L)} \qquad 2.7$$
$$= (ppm \times mol.wgt.(g/mol))/24.46 \ (L/mol), \ at \ 25°C$$

The conversion process is similar for µg/m^3 and mg/L from ppm, again both at 25°C (Eq. 2.8, 2.9).

$$\mu g/m^3 = (ppm \times mol.wgt. \ (g/mol) \times 10^3)/24.46 \ (L/mol) \qquad 2.8$$

$$mg/L = (ppm \times mol.wgt. \ (g/mol) \times 10^{-3})/24.46 \ (L/mol) \qquad 2.9$$

For weight for unit volume values at temperatures other that 25°C, the molar volume for the temperature of interest must be used in place of 24.46 L/mol.

The third gas concentration method is the partial pressure system, which is occasionally used to specify the concentration of a gas in a gas (e.g., Eq. 2.10).

$$ppm \ of \ constituent = \frac{(partial \ pressure \ of \ constituent) \times 10^6}{(total \ barometric \ pressure)} \qquad 2.10$$

The same pressure units are used for the partial pressure of the component of interest and for the total barometric pressure readings, so that the units divide out. Thus, the partial pressure system is also dimensionless (a pure ratio) so that in all important respects it is equivalent to the volume per unit volume system for specifying concentrations of a gas in a gas. This system is particularly useful for making up synthetic mixtures of gases on a vacuum line when the partial pressure of each component is known. The carefully measured mixture(s) are useful for such purposes as calibration of instrumentation.

2.6.2. Wet Chemical Analysis of Gases

Sometimes the chemical capture reagent actually forms a part of the wet chemical analytical scheme to be used. For example, if hydrogen peroxide is used as the sulfur dioxide capture reagent, then titration using standard alkali gives the final sulfuric acid concentration obtained in the capture solution (Eq. 2.5). Relating the measured acid concentration back to the original volume of air passed through the absorbing solution then gives the sulfur dioxide concentration originally present in the air. The answer obtained by this method is subject to errors when other acidic or basic gases (e.g., NO, NO_2, NH_3) or aqueous aerosols such as sulfuric acid, nitric acid, or ammonium hydroxide are present in the air being sampled. The influence of these interferences can be decreased if the determination is carried out gravimetrically by the addition of barium chloride rather than by titration (Eq. 2.11).

$$H_2SO_4 + BaCl_2 \rightarrow 2\ HCl + BaSO_4 \downarrow \qquad 2.11$$

When sodium hydroxide in water is used as the medium to trap sulfur dioxide, it allows a separate result to be obtained for the sulfur dioxide content, independent of any sulfur trioxide or sulfuric acid which may be present in the sampled gas. This distinction is not possible when using hydrogen peroxide. To obtain this as a separate result requires addition of an involatile acid to the collection medium after collection. This is followed by heating, or sparging of the acidified solution with inert gas while heating, to release the sulfur dioxide from solution (Eq. 2.12).

$$Na_2SO_3 + H_2SO_4 \rightarrow Na_2SO_4 + H_2O + SO_2 \uparrow \qquad 2.12$$

Any sulfur trioxide forms sulfate which is involatile on acidification, unlike sulfite. Capture of the released sulfur dioxide in a fresh aqueous base, followed by titration with standard iodine solution, then gives the concentration of sulfur dioxide present in this second solution (Eq. 2.13, 2.14).

$$SO_2 + 2\ NaOH \rightarrow Na_2SO_3 + H_2O \qquad 2.13$$

$$\underset{\text{brown}}{SO_3^{2-} + I_2} \rightarrow \underset{\text{colorless}}{SO_3 + 2\ I^-} \qquad 2.14$$

Similar wet chemical techniques have been developed for many other polluting gases, but are beyond the scope of this outline. The investment in equipment and materials required for wet chemical analysis is generally low, but the skill level required of the analyst, particularly for some of the tests, is high. While it is possible to streamline some of the wet chemical methods to enable many results to be obtained in a working day, these methods are generally slow relative to other alternatives. Details of suitable methods are available from many sources [25].

One gas analysis technique using sealed adsorbent tubes borrows from the colorimetry methods of manual procedures. Draeger and now also the Gastec and the Matheson-Kitagawa systems all use glass tubes packed with a solid support which is coated with an appropriate colorimetric reagent for the gas of interest. Company literature assesses the coefficient of variation for the tubes with most readily discernable color changes as 10%, and for the less efficient tubes as 20–30% [26] (Eqs. 2.15–2.17).

$$\text{Standard deviation} = [(x_1^2 + x_2^2 + \ldots + x_n^2)/(n-1)]^{0.5}, \qquad 2.15$$

where x_1, x_2, ... x_n are the deviations of individual determinations from the mean of all determinations in the series being measured, and n represents the number of individual determinations in the series.

$$\text{Mean deviation} = (x_1 + x_2 + \ldots + x_n)/n, \qquad 2.16$$

where only the absolute values of the individual deviations are used, without regard to sign. It is also referred to as the average deviation.

$$\text{coefficient of variation} = \frac{\text{(standard deviation)}}{\text{(mean deviation)}} \qquad 2.17$$

This level of precision is sufficient for most purposes, even for the less common gases, and is much more convenient than most alternatives when the number of tests required is small (Table 2.3). In the early days of the development of these systems a tube for testing carbon tetrachloride vapor was questioned as to its capability to achieve a precision within 50% of stated values [29]. Even if true, this level of accuracy is still adequate for many regulatory and industrial hygiene requirements.

2.6.3. Instrumental Methods for Gas Analysis

Instrumental analysis of air samples can be frequently conducted directly on the air or gas sample, particularly when either the concentration of the contaminant is high or with a sensitive analytical method. Failing this, prior concentration of the component of interest is necessary. This may be accomplished by adsorption onto a substrate from low concentrations, to be released at higher concentrations later by heating, or by a gentle stream of inert gas. Or the concentration step may be accomplished by absorption from the air, as fine bubbles discharged under the surface of an entrapping liquid. The sampling

TABLE 2.3 Some Examples of the Gases Which may be Determined by Prepacked Colorimetric Tubes[a]

Gas or vapor	Measurement[b] range	
	Minimum	Maximum
Ammonia	5 ppm	10%
Benzene vapor	5 ppm	420 ppm
Carbon monoxide	5 ppm	7%
Chlorine	0.2 ppm	500 ppm
Ethylene oxide	5 ppm	3.5%
Formaldehyde	0.5 ppm	40 ppm
Hydrogen sulfide	1 ppm	7%
Mercury vapor	0.1 mg/m^3	2 mg/m^3
Methane thiol (methyl mercaptan)	2 ppm	00 ppm
Nitrogen dioxide	0.5 ppm	50 ppm
Oxygen	5 %	23%
Ozone	0.05 ppm	300 ppm
Perchloroethylene	5 ppm	1.4%
Phenol	5 ppm	–[c]
Sulfur dioxide	0.1 ppm	2,000 ppm
Tetrahydrofuran	100 ppm	2,500 ppm
Vinyl chloride	0.5 ppm	50 ppm
Xylene	5 ppm	1,000 ppm

[a]Compiled from Leichnitz [26], Matheson-Kitagawa [27], and Gastec Corp. [28].

[b]Concentrations are specified by volume (see text). Wider concentration ranges are covered with two to five detector tubes of differing sensitivities.

[c]Not specified.

method will be selected on the basis of the type of information required from the sample and the instrumentation that is available for analysis.

Ultraviolet and infrared spectrophotometric methods are useful for air analysis in a wide variety of formats. Either solution or gas sample cells may be used. For good ultraviolet sensitivity, the component of interest must have two or more double bonds present, preferably conjugated. Thus, sulfur dioxide, acrolein ($H_2C=CH-CHO$), and benzene all give very good ultraviolet sensitivity, when measured near their absorption maxima. In contrast, acetone, phosgene, and gasoline all have direct atmospheric measurement detection limits of only about 5 ppm, too high to be useful for many situations. Even if direct measurement sensitivities are poor, however, a colorimetric method coupled with ultraviolet spectrophotometry may be used for some pollutants to obtain improved sensitivity. For example, the purple complex formed by the reaction of formaldehyde with chromotropic acid is significantly more sensitive to ultraviolet measurements than is formaldehyde itself. Using this method solution detection limits of $3\,\mu M$ (0.1 mg/L) can be achieved (Eq. 2.18).

$$H_2CO \text{ (colorless)} + \text{chromotropic acid} \rightarrow \text{purple complex} \qquad 2.18$$

For a 10-L air sample this sensitivity corresponds to an ambient air detection limit of $0.1\,\text{mg/m}^3$ (0.08 ppm).

Ultraviolet absorption by sulfur dioxide may also be usefully applied in the field for direct plume observations. In a novel technique developed for the observation of the colorless discharge of sulfur dioxide and water vapor from a natural gas cleaning plant vent stack, the use of silica optics and a special UV sensitive film allowed the clear photographic observation of plume behavior from the ultraviolet absorption of the sulfur dioxide component [30].

The infrared spectra of many air pollutants are usually more complex than their ultraviolet spectra, since the infrared usually reveals several absorption maxima. However, for many single components the pattern of maxima obtained, its "fingerprint," can permit positive identification while the concentration is determined from the absorbance readings. All organic compounds absorb infrared energy, although the intensity and pattern of the absorption varies greatly from compound to compound. This affects the relative sensitivity of the method for different compounds. However, recently developed compact cells compensate for this by providing absorption path lengths of 20 m or more via multiple internal reflections (Fig. 2.4). Typical infrared detection limits for this system are 1.2 ppm for carbon monoxide, 1.5 ppm and 0.08 ppm for nitric (NO) and nitrous (N_2O) oxides, respectively, 0.1 ppm for sulfur dioxide, and 0.05 ppm for carbon tetrachloride. The wavelengths at which different compounds absorb vary, a feature which is used in dispersive infrared instruments to identify the absorbing species. Some of these infrared instruments are portable and fitted for 12-volt DC power, suitable for operating from an ordinary car battery. These field capabilities avoid the sample deterioration problems caused by poor field collection sample containment systems, or sample instability.

Rapid multiple scans, signal storage, plus Fourier transform capabilities have combined to push infrared detection limits to about 1/100 of the

FIGURE 2.4 Diagram of the operating details of a portable infrared gas analyzer, which uses a long path, multiple internal reflection gas cell. Adjusting the angle of the objective mirror alters the number of internal reflections obtained before the beam enters the detector. (From Wilks Scientific [31], courtesy of the Foxboro Company.)

concentrations possible with the more conventional infrared instruments just discussed, but at about 10 times the cost [32].

The most significant recent spectroscopic developments that relate to air pollutant analysis are closely linked to progress in laser technology. The coherence and high resolution possible with a laser source coupled with the high power available with some configurations makes inherently high sensitivities possible, particularly where the match between laser and absorbing frequencies is good. For example, these methods enable sensitivities of the order of 1 ppb for nitrous oxide, using a 100-m path length. Even lower sensitivities are possible for more strongly absorbing gases [33]. With high source powers, lasers may be used for remote sensing of air pollutants in real time in smog situations, discharge plumes, or even for ambient air pollutant concentrations in open countryside [34].

Various laser instrument arrangements are possible depending on the particular remote sensing application (Fig. 2.5). The direct absorption mode (Fig. 2.5a) is the simplest, but suffers from the inconvenience that source and detector units must be sited separately, and aligned for each particular test situation. Retroreflection, either from a surface-silvered mirror or from a geographical feature such as a tree or rock (Fig. 2.5b) at least allows the source and receiver to be operated from the same site. In the backscatter mode (Fig. 2.5c), the reflections from aerosol particles which have passed through the target air sample are picked up by the receiver-detector operating at the same location. In this situation, the laser is tuned to a high-absorption frequency of interest and then to a low-absorption frequency near this. The concentration of the component of interest may be computed from a comparison of the relative absorption signals, the basis of the so-called DIAL (differential absorption lidar) system [35].

Gas chromatography (GC) is another nonspectroscopic instrumental technique which is used for the analysis of gaseous air pollutants, particularly when it is coupled to a mass spectrometer. Gas chromatography provides efficient separation and detection of many common air pollutants, and can

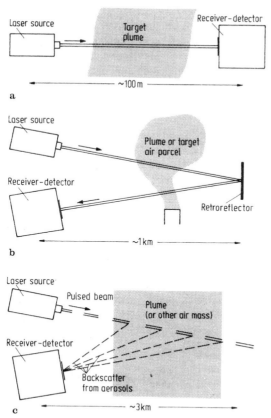

FIGURE 2.5 Some possible configurations of laser source and receiver detectors for use in remote sensing of air pollutants: (a) Simple direct absorption mode. (b) Retroreflective direct absorption mode. (c) Backscatter from aerosols mode.

give a measurable signal with as little as 10^{-12} g of separated component. However, unequivocal identification of the separated component is not possible with gas chromatography alone. Coupling a quadrupole mass spectrometer to the exit of a gas chromatograph allows both functions to be performed simultaneously; the GC separations, and the rapid mass spectrometer providing identification of each constituent from their fragmentation patterns or molecular ions. This combination of instruments is able to provide rapid answers for the real time tracking of gases, vapors, and fumes released from a train derailment, or other emergencies which could involve released chemicals [36].

Many other instrumental methods are also used for gas analysis, the details of which are available elsewhere [37]. Development, standardization, and calibration of the various methods used for air pollutants of interest are continuing and ongoing processes.

Real-time measurements of gas concentration changes in plume chemistry can involve some ingenuity in the delivery of instrumentation to the site of interest. Full sized, fixed wing aircraft or helicopters cause extensive mixing of the plume which would make the concentrations measured unrepresentative with considerable risk to the aircraft and stack. A small, unmanned radio-controlled aircraft carrying a compact gas chromatograph can take measurements without disturbing the plume and without risk. Freely rising helium-filled balloons can be used to lift instrumentation through the various layers of the atmosphere in order to obtain a snapshot of readings. A continuous profile of readings from one location and approximate altitude may be obtained from a Kevlar tethered balloon (to 2 km), or a box-kite (to 10 km) [38]. Freely floating tetroons designed to have neutral buoyancy with their instrument load can give longer term results at an approximately constant altitude over a horizontal distance which is determined by wind speed and direction.

2.6.4. Biological Methods for Air Pollution Assessment

High concentrations of air pollutants are known to kill many annual plants and trees. This amounts to a coarse biological indicator. Less severe exposures can cause premature senescence (early leaf drop) of sensitive species of trees and shrubs [39], which can be used as an indicator [40]. For example, the aspen poplar, *Populus tremuloides*, drops its leaves after exposure to as little as 0.34 ppm of sulfur dioxide for 1 hr. Early leaf drop slows tree growth, which can be qualitatively assessed by examining the growth ring widths from the tree cores. Several leaf drops in quick succession can kill trees or annual plants.

An odor warning of air pollutant exposure is provided to man and many animals by the olfactory membranes, another example of a biological indicator. In specially trained dogs, the sense of smell can rival the sensitivity of the most sophisticated instrumentation [41, 42]. Tamed bees have also been used to locate mines and explosives without hazard, and can be trained to seek out target chemicals in waste dump sites [43].

Since many toxic gases also possess an odor, an avoidance response can help to reduce exposure. Some examples of gases for which this warning effect is pronounced and some for which it is nonexistent are given in Table 2.4. One must keep in mind that the sensitivity of the sense of smell can vary widely with individuals and also that olfactory membranes are subject to narcosis (anesthesia) by some of these gases. Hydrogen sulfide and formaldehyde, for example, lose their odor for a gradual increase of concentration or for longer term exposures, which removes their safety warning.

Lichens can be a generic useful, semi-quantitative indicator of pollution levels over time and space in both city and rural study areas [44]. Lichens comprise a slow-growing symbiosis between an alga, which photosynthetically manufactures carbohydrate, and a fungus, which aids the alga in water storage in an otherwise inhospitable site and uses the manufactured carbohydrate for its own metabolism and growth. Their very slow growth, and their high reliance on the air for most of their nutrient requirements, makes for efficient uptake and retention of many air pollutants. Hence, the level of air pollution affects the growth and reproduction of lichens more significantly than that of most other, faster growing forms of plants which rely more on a

TABLE 2.4 Examples of Extent of Warning Given from Relationship of Human Olfactory Thresholds of Some Common Gases to the Time Weighted Average Industrial Hygiene Standards[a]

Gas or vapor	Olfactory threshold (OT)	Industrial hygiene standard[a] (HYG)	Warning effect, (HYG) ÷ (OT)
H_2S	25 ppb	10 ppm	400
O_3	<10 ppb	100 ppb	10
SO_2	0.3 ppm	2 ppm	7
NH_3	16 ppm	25 ppm	1.6
H_2CO	1 ppm	1 ppm	1
CO_2	1%	1%	1
CCl_4	540 ppm	2 ppm	0.004
CO	odorless	35 ppm	0

[a]No effect level for a healthy adult for time-weighted average exposures for an 8-hr working day over a 40-hr work week.

soil substrate. For these reasons, heavy metal fumes or other particulate air pollutant exposures also tend to accumulate to higher concentrations in lichens than in most other forms of plants.

In the hands of skilled botanists, a lichen diversity and distribution study of an area can give useful long-term exposure information regarding continuous or intermittent fumigation by pollutants such as sulfur dioxide which would be difficult to obtain in any other way. Even in the hands of a relatively inexperienced but careful observer, it is possible to determine upto six or seven zones (levels) of relative air pollutant exposures by recording the diversity and types of lichens found growing in an area. Thus, in a heavily polluted area, few or no healthy lichen varieties will be found. Those more resistant varieties that do occur in this exposed situation, such as species of *Pleurococcus*, will be squamulose (recumbent) forms which are closely bound to their support, a habit that minimizes exposure of the lichen to air pollutants. At the other extreme, in areas of little air pollutant exposure, both the diversity of species and the number of specimens of each species found will be larger. There will also be a better representation of the more pollution sensitive foliose (leafy) forms such as species of the genera *Parmelia* and *Letharia*, or the fruiticose (shrubby) forms of species of the genera *Usnea* or *Alectoria*. Biological methods based on lichen surveys such as this represent an important complement to the information obtained from wet chemical and instrumental analysis for determining the integrated levels of air pollution over time for an area. The latter methods usually provide more reliable instantaneous (snapshot) information.

2.7. EFFECTS OF AIR POLLUTANTS

Much has been said and written about the effects of air pollutants on plants and animals and so only a brief summary needs to be presented here. The health effects on man probably comprise the most direct, even if somewhat anthropocentric, concern with air pollutants. The primarily nuisance effect of the smell emanating from a fish-packing plant or brewery represent examples at the minor end of the effects range. However, since the average person takes in some 22,000 breaths per day, amounting to some 14–16 kg (30–35 lb) of air per day, and this contacts the large area of efficient moist exchange membranes of the alveoli in the lungs, even a relatively low concentration of many air pollutants is sufficient to have a noticeable effect. For this reason alone the air environment has to be the most important of the biospheric elements to be worth striving to maintain or improve in quality.

One logical approach which may be used to organize the consideration of air pollution problems and solutions is to start from the individual level of exposure with treatment of indoor air pollution, where the problems are usually more readily identified, and are of a scale more or less under the control of individuals or small groups of people. From these we can move to consideration of larger and larger scale macroscopic effects which become progressively more complex to understand and model. The scale and scope of these problems also become progressively more difficult for individuals, countries, or even the whole world to control. Topics covering this progression include smogs, acid rain, arctic haze, climate effects, and damage to the stratospheric ozone layer.

2.7.1. Indoor Air Pollution

Right from the time when people used natural shelters and lit poorly vented fires for warmth and light, to the building of small or large and complex facilities for living space, offices, factory operations, and warehousing, indoor air pollution has existed to varying degrees. For structures of all kinds in which involuntary air infiltration (drafts) and/or people and goods traffic, significant air exchange occurs and the quality of indoor air is seldom a problem. But with recent energy conservation concerns, tighter construction of new buildings and adjustments to the ventilation equipment of existing structures to provide lower levels of air exchange is in some instances sacrificing indoor air quality to save energy. Both in the area of the enclosed space of buildings and in the provision of fresh air in the transportation sector, (e.g., the passenger cabins of modern jet aircraft), there are instances where a small savings in energy obtained by one stakeholder is being far exceeded by the larger system costs (general malaise, loss of cognitive ability and attention span, etc.) that are being experienced [45].

We can use conditions common to the individual home or apartment to illustrate the concepts involved within a framework which most can relate to their own experiences. The degassing of composite woods such as chipboards, plywood, and hardboard as well as vapor loss from the more aromatic solid woods such as cedar and pine can contribute to the vapor contaminants in home air. The finishes on any of these woods or on wall surfaces also make contributions, particularly in the period shortly after application. Carpeting and the many easy-care flooring materials together with soft furnishings also release small quantities of vapors, particularly when new. Human presence and activities, as well as those of some pets in the residential space, also add to these static sources. Oxygen is consumed and carbon dioxide, moisture, and heat are released by respiration in proportions which vary somewhat with the composition of foodstuff being metabolized [40] (e.g., Eqs. 2.19 and 2.20).*

$$(C_6H_{10}O_5)_n + 6n\ O_2 \rightarrow 6n\ CO_2 + 5n\ H_2O \qquad \qquad 2.19$$
starch

$$(C_{51}H_{92}O_6)_n + 71\ O_2 \rightarrow 51\ CO_2 + 46\ H_2O \qquad \qquad 2.20$$
a typical fat

Heat generated by a conscious person at rest is approximately 90 W. This means that excessive heating is probably the most common discomfort factor that builds up in the air of densely occupied indoor space such as auditoria, but heat build-up from this source is seldom a problem in private accommodation.

Entering the space, movement, and cleaning activities within the space raises dust levels, and cooking activities put moisture, heat (and with gas stoves, carbon dioxide and additional moisture), aerosol droplets of fat, etc., into the air. Some of the pollutants (e.g., the smell of baking bread) are appreciated by most. Even so, the U.S. Environmental Protection Agency has required large commercial bakeries to install control equipment for the volatile organic compounds produced to help reduce their contributions to photochemical smog. Unvented combustion units such as gas stoves and space heaters contribute their combustion gases to the indoor air, for which control is usually recommended via ventilation. Less pleasant cooking smells may be masked by room deodorizers or room ozonizers, which make their own contributions. As a final component, plants exposed to light initiate photosynthesis, causing a net consumption of carbon dioxide (and some pollutants) and production of oxygen. But at night, metabolism continues in the plants consuming oxygen and producing carbon dioxide without the benefit of oxygen from photosynthetic processes which occurs during the day.

Various indoor air quality guidelines exist. To determine whether or not a guideline is being met, one can use the ventilation rate formula to evaluate the situation (Eq. 2.21).

$$\text{Allowable concentration of contaminant} = \frac{\text{rate of contaminant loss}}{\text{rate of dilution air supply}} \qquad 2.21$$

If the allowable concentration is specified in volume for volume units (e.g., %, ppm, ppb, etc.), then, that is used (times the appropriate factor e.g., 100, 10^6, 10^9, etc.) on the left-hand side. In this instance volume per unit time in the same units should be used for both of the substitutions of the right-hand side. The ventilation rate equation is also valid for use when contaminant concentration allowed is specified in mass per unit volume units such as mg/L, mg/m^3, μg/m^3, etc. In this case, mass per unit time units are used in the numerator and volume per unit time in the denominator of Eq. 2.21. To use Eq. 2.21 in either format does not require knowledge of the volume of enclosed space being ventilated. However, it does assume perfect uniformity of the gas mixture, and perfect mixing of the air with the contaminant.

Sometimes the rate of air exchange of an enclosed space is specified as air changes per hour (ACH, Eq. 2.22).

$$\text{air changes per hour} = \frac{\text{ventilation rate per hour}}{\text{volume of enclosed space}} \qquad 2.22$$

In this instance, one can use the ventilation formula structured to use the air changes per hour term (Eq. 2.23).

$$R = k_{ex}[C_i]v, \qquad 2.23$$

where R   = rate of production of gas or vapor,
       K_{ex} = air changes per hour,
       C_i = interior equilibrium concentration, and
       v   = volume of enclosed space.

Use of this equation requires the same assumptions as Eq. 2.21. It is only valid if there is essentially zero concentration of the contaminant of interest in the outside air being used for ventilation.

If there is a significant concentration of the gas or vapor of interest in the outside air, for example carbon dioxide, then Eq. 2.24 rather than 2.23 must be used to account for this.

$$R + k_{ex}[C_o]v = k_{ex}[C_i]v \qquad 2.24$$

where C_o is the concentration of gas or vapor in the outside air. Again, the same assumptions as for Eq. 2.21 are necessary.

How are these equations used? Let us consider a propane-powered forklift truck that is operating in a warehouse and producing 2.0 g/min carbon monoxide. The industrial hygiene standard for carbon monoxide is 35 ppm. What is the minimum ventilation rate required for the warehouse to ensure safe working conditions? To convert the mass of CO per minute to a volume, we can assume conditions of 1 atm pressure, 20°C.

$$v = \frac{2.0 \,\text{g/min CO}}{28.01 \,\text{g/mol}} \times \frac{0.0821 \,\text{L atm K}^{-1}\text{mol}^{-1} \times 293 \,\text{K}}{1 \,\text{atm}} = 1.72 \,\text{L/min CO}$$

The ventilation rate required for 35 ppm concentration is:

$$35 \,\text{ppm} = \frac{1.72 \,\text{L/min CO}}{x \,\text{L/min air}}$$

$$\frac{35}{10^6} = \frac{1.72 \,\text{L/min CO}}{x \,\text{L/min air}}$$

where $x = 4.91 \times 10^4$ L/min air, or 49.1 m^3/min air

The same ventilation rate would be required for a small or a large warehouse. A smaller warehouse would simply reach the equilibrium carbon monoxide concentration more quickly than the large one, assuming the same ventilation rates.

To convert the ventilation rate in m^3/min to units of air changes per hour requires:

$$49.1 \,\text{m}^3/\text{min} \times 60 \,\text{min/hr} = 2948.6 \,\text{m}^3/\text{hr}$$

For a nearly empty warehouse of $25 \times 40 \times 6$ m a ventilation rate of 2948.6 m^3/hr requires:

$$\frac{2948.6 \text{ m}^3/\text{hr}}{20 \text{ m} \times 40 \text{ m} \times 6 \text{ m}} = 0.614 \text{ air changes per hour}$$

As an example of the use of Eq. 2.24, use the 1000 ppm carbon dioxide generally accepted comfort guideline for interior air quality. The same propane-powered forklift produces 20 L/min carbon dioxide. What air change rate would be required to maintain the carbon dioxide guideline for the warehouse? Would it be the carbon monoxide or carbon dioxide requirement that would dictate the ventilation rate required? First we rearrange Eq. 2.24 to the form:

$$k_{ex} = \frac{R}{v([C_i] - [C_o])}$$

Taking the concentration of carbon dioxide in the outside air as 350 ppm (350×10^{-6}), and substituting:

$$k_{ex} = \frac{20 \text{ L/min} \times 10^{-3} \text{ m}^3/\text{L} \times 60 \text{ min/hr}}{4{,}800 \text{ m}^3 \times ((1000 \times 10^{-6}) - (350 \times 10^{-6}))}$$

$$= 0.3846, \text{ or } 0.385 \text{ air changes per hour}$$

Comparing the 0.385 ACH required for carbon dioxide control with the 0.614 ACH required for safe control of carbon monoxide levels, it is clear that control of the carbon monoxide concentration dictates the safe ventilation rate required.

2.7.2. Classical and Photochemical Smogs

Localized air pollution episodes tend to occur in areas subject to inversions. These stagnant air events will tend to occur in regions with low or no winds. A valley location or a plain bounded by mountains will tend to increase the occurrences and persistence of inversions. In the normal daytime situation, the sun warms the surface of the earth and the air mass immediately above it. Warming lowers the density of this surface air causing it to rise and in the process it mixes with the upper air levels. This dilutes any pollutants discharged into the surface air.

At night, or when there is fog or another meteorological event which brings an inversion to the area, the situation changes. During the inversion, the air close to the surface 100 to 300 m is relatively more dense than the layers immediately above it, which creates a stable situation. There is little or no mixing of any pollutants which may be discharged into the surface air. This situation causes a gradual buildup of any discharges to the surface layer to uncomfortable or dangerous concentrations in this layer.

The elevated air pollutant levels which occur during localized classical smog episodes tend to severely impact on the chronically ill, the young, and the old. They can cause effects ranging from watering of the eyes and restricted breathing, to an aggravation of respiratory illnesses, and even to a noticeable rise in the death rate recorded for the affected area during the most severe episodes [46]. London, England, the Donora valley, Pennsylvania, and the Meuse valley, Belgium, are just a few of the documented older examples of occurrences of this last level of severity. A classical smog can also serve to drop the pH of wetted surfaces sufficiently to seriously damage statuary and stone or masonry buildings from a simple solution reaction of limestone or marble stonework with the acidic water (e.g., Eq. 2.25).

$$CaCO_3 + 2\ HNO_3 \rightarrow \underset{\text{soluble}}{Ca(NO_3)_2} + H_2O + CO_2 \qquad\qquad 2.25$$

Today, two clearly differentiated types of smogs are recognized. In the classical variety of smog represented by the pollution episodes mentioned above, the problem was caused by the accumulation of primary air pollutants such as sulfur dioxide, particulates, and carbon monoxide contributed by smoke, usually complicated by the presence of fog, hence the term "smog" (Table 2.5). The fog component also tended to slow air pollutant dispersal by cutting off sunlight, preventing the warming of air close to the earth's surface by the sun. The characteristics and time of occurrence of a classical smog differ markedly from the more recent phenomenon of photochemical smog which was first described for Los Angeles, but which is now also experienced by Tokyo, Mexico City, and many other major urban centers that provide similar conditions for its formation [48].

In photochemical smog episodes, secondary air pollutants such as ozone, nitrogen dioxide, aldehydes, and peroxyacetyl nitrate are formed as a result of the chemical interaction of the primary air pollutants, principally nitric oxide and hydrocarbon vapors, with sunlight and air (Fig. 2.6) [49]. This interpretation of the processes involved has been verified by smog chamber experiments (Fig. 2.7), and has since been confirmed by field measurements as the sensitivity of ambient air instrumentation has improved [50, 51]. In photochemical smog episodes, it is the secondary pollutants that cause severe eye irritation and upper respiratory effects felt by people and at the same time causes serious damage to plants.

The authorities of the London area were quite successful in reducing emissions of sulfur dioxide and particulates sufficiently via passage of their

TABLE 2.5   Distinguishing Features of Classical and Photochemical Smogs[a]

	Classical smog	Photochemical smog
Location example	London, 1950s	Los Angeles
Peak time of occurrence	winter	summer
Conditions	early a.m., 0 to 5°C, high humidity plus fog	around noon, 22–35°C, low humidity, clear sky
Atmospheric chemistry	primarily reducing, SO$_2$, particulates, carbon monoxide, moisture	oxidizing, nitrogen dioxide, ozone, peroxyacetyl nitrate
Human effects	chest, bronchial irritation	eye irritation, (bronchial restriction) asthmatic reaction
Underlying causes	fog plus stable high, surface inversion, dispersal of primary pollutant emission is prevented, accumulate of hydrocarbons	sheltered basin, frequent stable highs, accumulation of secondary pollutants from photochemical oxidation

[a]Compiled from Williamson [47] and Kerr *et al.* [48].

Initiation (at sunrise, with high reactive hydrocarbon concentration):

$$RH \text{ (hydrocarbon)} + O_2 \xrightarrow[\substack{[O]}]{O_3, \text{ or}} R^{\cdot} + HOO^{\cdot}$$

$$HOO^{\cdot} + NO \rightarrow HO^{\cdot} + NO_2$$
$$NO + NO_2 + H_2O \rightarrow 2 \, HONO$$
$$HOO^{\cdot} + NO_2 \rightarrow HONO + O_2$$
$$HONO \xrightarrow[\substack{290\text{--}400 \text{ nm}}]{h\nu} HO^{\cdot} + NO$$

Propagation (converts reactive hydrocarbons to oxidized products):

$$HO^{\cdot} + RH \rightarrow H_2O_2 + R^{\cdot}$$
$$R^{\cdot} + O_2 \rightarrow ROO^{\cdot}$$
$$HOO^{\cdot} + RH \rightarrow H_2O_2 + R^{\cdot}$$

Termination (dominates at sundown):

$$R^{\cdot} + R^{\cdot} \rightarrow R\!-\!R$$
$$HO^{\cdot} + R^{\cdot} \rightarrow ROH$$

$$CH_3 - \overset{\displaystyle O}{\overset{\|}{C}} - OO^{\cdot} + NO_2 \rightarrow CH_3 - \overset{\displaystyle O}{\overset{\|}{C}} - OONO_2 \text{ (PAN)}$$

FIGURE 2.6 Generalized radical chain reactions for typical processes which occur during a photochemical smog episode.

Clean Air Act so that classical smogs are a thing of the past in this area. But the more frequent and intense sunshine experienced in the city as a result of this improvement coupled with a rise in hydrocarbon concentrations from an increase in automobile traffic means that London, too, now experiences the irritating effects of photochemical smog events [53].

Similar processes cause the attack of wrought iron and other metalwork, and accelerate the decay of exposed wood. Even aluminum, which is highly favored for use in exterior metalwork because of its resistance to corrosion under ordinary conditions, is now showing the classic pits and erosion scars of corrosion when used in areas which have severe local air pollution.

2.7.3. Acid Rain

The normal pH of rain or melted snow in equilibrium with 360 ppm of atmospheric carbon dioxide is 5.6 (Eqs. 2.26 and 2.27).

$$H_2O + CO_2 \rightleftharpoons H_2CO_3 \qquad\qquad 2.26$$

$$H_2CO_3 + H_2O \rightleftharpoons H_3O^+ + HCO_3^- \qquad\qquad 2.27$$

When pHs of less than this value occur in precipitation above the freezing point it is referred to as acid rain.

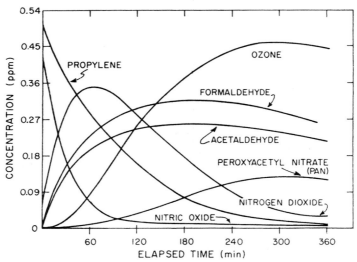

FIGURE 2.7 Results of a typical smog chamber experiment in which initial 0.5 ppm propylene, 0.45 ppm NO, and 0.05 ppm NO₂ in otherwise clean air is irradiated with artificial sunlight and analyzed at intervals. (From Lloyd et al. [52], with permission.)

This lowered pH is caused by rainout and washout of nitrate and sulfate from the atmosphere (Table 2.6) [54]. pHs of as low as 2.4, about the acidity of lemon juice or vinegar, have been measured for the rainwater of an individual storm in Pitlochry, Scotland. An annual pH value for precipitation of 3.78 has also been recorded for the Netherlands. It is evident that these acid pHs can cause structural damage to buildings of the type just described for classical smogs. Lowered pH precipitation is also seriously affecting the biota of those lakes which have a limited carbonate-bicarbonate natural buffering capacity [56]. When the ionic exchange capacity of soils becomes exhausted it too can drop in pH to result in reduced growth, or actual die-back of forests [57]. Liming to raise the pH of affected lakes into the normal range has been practiced successfully on an experimental scale [58]. But this measure can only be a local and temporary solution.

TABLE 2.6 Contributions of Sulfuric and Nitric Acids to the Acidity of Acid Rain[a]

Substance	Concentration (mg/L)	Contribution to total acidity[b]
H_2SO_4	5.10	57
HNO_3	4.40	39
NH_4^+	0.92	51
H_2CO_3	0.62	20
All others	ca. 0.4	ca. 12

[a]Data selected from Likens [55]. Results determined for a sample of rain of pH 4.01 collected at Ithaca, New York, in October 1975.

[b]Microequivalents per liter. Ammonium ion by titration to pH 9.0.

2.7.4. Arctic Haze

Related to smogs but only relatively recently noticed by the scientific community is the phenomenon of Arctic haze, a brownish turbidity occurring in the atmosphere of Arctic regions from late fall to March or April of each year. A suspended aerosol of primarily sulfates $(2 \, \mu g/m^3)$, organic carbon $(1 \, \mu g/m^3)$, and black carbon $(0.3 - 0.5 \, \mu g/m^3)$ reduces the visibility in the region to 3 to 8 km, over an area of several thousand square kilometers and up to an altitude of about 3,000 m during this period [59]. The novelty here is that this represents an accumulation of pollutants, remote from the original points of discharge, which by the idiosyncracies of atmospheric movements and conditions resides over and adversely influences sites not responsible for the emissions.

This tendency of aerosols to reduce visibility during Arctic haze episodes is also a very general effect of atmospheric particles. It is caused by the scattering of light by particles or droplets in the aerosol size range. Apart from aesthetic considerations, high loadings of solid and liquid (fogs) atmospheric aerosols also influence flight conditions at airports, and have been implicated in anomalous rainfall patterns of the St. Louis area.

2.7.5. Human Effects of Particulate Exposure

Another important consideration related to atmospheric loadings and particulate size distributions concerns the potential for human effects on inhalation. The body's defenses in the upper respiratory tract are adequate to trap more than 50% of particles larger than about 2 μm in diameter when present in the air breathed. However, particles smaller than this are not efficiently captured by the upper respiratory tract. These "respirable particulates" penetrate the lungs to the alveolar level. Here, the more vigorous Brownian motion of the small aerosols increases their collision frequency with the moist walls of the alveoli, thus trapping a large proportion of them at these sites. This occurs in an un-ciliated region of the respiratory system which is unable to degrade or flush out accumulated material, particularly if the inhaled aerosol is inert or insoluble. Hence, any physiological effect of the presence of these foreign substances is aggravated, which tends to increase the incidence of respiratory illness experienced in areas that have high concentrations of polluting aerosols in this size range. The efficiencies of many types of air pollution control equipment for the lower aerosol size range is poor, an important consideration when selecting from process emission control options.

2.7.6. Climatic Effects

Proceeding from the local and regional effects of air pollutants to the global scale, our escalating use of fossil fuels coupled with the widespread cutting of forests have contributed to a steady rise in the atmospheric carbon dioxide concentration, which now averages about 0.7 ppm per year (Fig. 2.8) [60]. Carbon dioxide is virtually transparent to the short-wavelength (ca. 1 μm) maximum in the incoming solar radiation, but has a substantial absorption

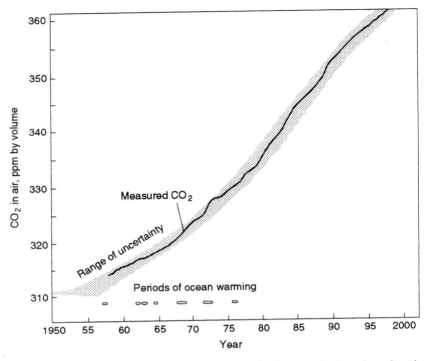

FIGURE 2.8 Past perspectives and present trends of atmospheric carbon dioxide concentrations at the Mauna Loa Observatory, Hawaii, compared with the recent periods of high solar flux and ocean warming intervals. (Assembled from the data of Lepkowski [61], Kerr [62], and the extrapolated carbon dioxide concentration data of Keeling *et al.* [63].)

band in the region of the longer wavelength infrared irradiation emanating from the much lower surface temperature of the earth. Thus, an increase in the concentration of atmospheric carbon dioxide does not affect the energy gained by the earth's surface from incoming solar radiation, due to the atmospheric absorption "window" in this region. However, carbon dioxide and some of the other atmospheric trace gases absorb a significant fraction of the radiant energy in the outgoing long wavelengths that would otherwise be lost (Table 2.7). This capture of infrared energy, dubbed the greenhouse effect, is what keeps the biosphere of the earth sufficiently warm to sustain life as we know it and is, therefore, vital for our continued existence. However, this is a delicate energy balance, and as the concentrations of global warming gases rise it is predicted from current climate models that this warming effect will become incrementally greater. It is the extent and importance of this influence in relation to a number of other conflicting factors that is still not well understood.

Ice cores from Arctic and Antarctic ice caps have provided atmospheric carbon dioxide concentration data for periods of up to 160,000 years before the present [71]. During most of this period, when humans contributed little to the atmospheric carbon dioxide, there were nevertheless wide fluctuations in concentration [72, 73]. Climate studies have correlated warm periods with

TABLE 2.7 Concentrations, Trends, and Warming Effect of the Greenhouse Gases[a, b]

Gas	Present atmospheric concentration (by vol.)	Rate of increase per year (%)	Atmospheric lifetime [70]	Contrib. to Warming Effect[b,c]	Relative to CO_2
CO_2	370 ppm	1.2 ppm (0.35)	ca. 500 year[d]	49%	1
CH_4	1.7 ppm	0.018 ppm (1.2)	87 ± 17 year	18%	20
$O_3 < 12$ km	0.02–0.1 ppm	variable	< few hr	see CCl_4 below	–
N_2O	310 ppb	0.58 ppb (0.019)	122 ± 24 year	6%	200
CFC 11 ($CFCl_3$)	230 ppt	8.9 ppt (0.056)	75 year	14%	1,000
CFC 12 (CF_2Cl_2)	380 ppt	5.0 ppt (0.018)	100 ± 32 year		
Methyl chloroform (CH_3CCl_3)	130 ppt	5.8 ppt (0.058)	34 ± 7 year	–	–
CCl_4	120 ppt	2.2 ppt (0.017)	34 ± 7 year	ca. 13% (incl. O_3)	–
Halons[e]	<100 ppt	–	25–110 year		–
Perfluoroalkanes:					
CF_4 (1997)	74 ppt	1.2 ppt (1.6 ± 0.6)	>2,300 year	<2%	3,600–8,500[f]
C_2F_6	2.9 ppt	–	–	–	3,600–8,500[f]
C_3F_8	0.2 ppt	–	–	–	3,600–8,500[f]

[a]Calculated and compiled from Ramanathan [64], Hileman [65], Johnston [66], Volk et al. [67], Khalil et al. [68], Zurer [69], and Fabian et al. [70].
[b]Water vapor and clouds are estimated to have twice the effect of all the gases listed here.
[c]Estimated in proportion to other greenhouse gases.
[d]Turnover rate is about 20% per year.
[e]Used in fire extinguishers and fire extinguishing systems. Bromine component provides powerful extinguishing action by quenching the propagating radicals produced by a fire. Halon 1211 is CBr_2F_2, Halon 1301 is $CBrF_3$, and Halon 2402 is $C_2Br_2F_4$.
[f]3600 GWP for 20 years after release, and 8500 GWP for 500 years after release, from Zurer [69].

the periods of high atmospheric carbon dioxide (180–300 ppmv), methane (350–650 ppbv), and nitrous oxide (N_2O, 190–280 ppbv) but the data are inadequately resolved to establish which caused which. Regardless of causation, it is clear that the earth's climate has had substantial warming and cooling trends since long before the substantial impacts of human activity. Since other major factors are also known to influence climate, it is possible that one or several of these exerted the primary influence and that carbon dioxide, methane, and nitrous oxide concentrations as well as climate changes simply tracked this.

Computer-based climate models are constantly being refined as computing power increases, to try to provide more reliable predictions. Cloud cover and atmospheric moisture content are currently estimated to have about twice the warming effect of all the gases of Table 2.7, or about two-thirds of the contribution from all atmospheric contributions. The effect of the moisture component on warming is one of the more difficult factors to model [74]. However, atmospheric moisture is not a realistically controllable variable.

Many other complex interactions are involved in climatic effects, among them lateral and vertical perturbations of ocean currents, changes in prevailing winds, periodicity in the earth's tilt (1 1/4°, 21,000-year cycle), position of elliptical orbit (97,000-year cycle), and changes in the solar flux (correlated with sunspots, ca. 13-year-cycle) [75]. As the periods of these factors differ, the warming effect of some will be augmented by in-phase peaks at times, and will be decreased or eliminated by out-of-phase peaks at others. So, the net effect is at best difficult to predict. Again, like atmospheric moisture, these variables are beyond our control.

An increase in average global temperatures from the greenhouse effect of only a Celsius degree or less could still cause a decrease in water storage at the earth's ice caps. This, in turn, would decrease reflectivity of the polar regions, and could contribute to a rise in the world's ocean levels causing changes to low lying coastlines [76]. An increase in average global temperatures could also increase the rate and extent of desertification of marginally arable land from changes in wind and rainfall patterns, reducing an already limited agricultural resource. Imperfect models predict the noticeable onset of some of these effects by the year 2050. Thus, the possible effects are sufficiently serious that we should take remedial action now that might decrease these effects, at least with the less drastic measures.

Many nations are involved in active discussion concerning a decrease in atmospheric carbon dioxide, which is estimated to contribute about 1/2 of the warming effect of all the green house gases, or about 1/6 of the total atmospheric contribution to global warming. The recent agreement struck by 38 industrialized countries in the Kyoto Accord gave a timetable for reduction of carbon dioxide emission reduction [77, 78]. Examination of the annual atmospheric fluxes of carbon dioxide into and out of the atmosphere establishes that human activity probably contributes less than 5% of the total (Fig. 2.9). Consequently, truly drastic reductions in fossil fuel and biomass combustion would be necessary to have a significant effect, if this was the only measure undertaken (see Section 3.6.6).

CO₂ INPUTS		ATMOSPHERIC	CO₂ OUTPUTS	

CO₂ INPUTS

Human:
Fossil fuels 22.0
Agriculture 7.3
Soils,
Land plants: 128.0
Oceanic
processes: 355.0
Total 512.3

ATMOSPHERIC CONTENT
2300

CO₂ OUTPUTS

Soils,
Land plants: 128
Oceanic
processes: 374
Total 502

HUMAN
SOIL
OCEAN
SOIL
OCEAN

FIGURE 2.9 Estimated annual atmospheric fluxes of carbon dioxide into and out of the atmosphere in billions of metric tonnes carbon dioxide. (Compiled from the data of Bolin [79] and McHale [80].)

2.7.7. Stratospheric Ozone

The other potential global problem from increased atmospheric pollutant levels, and of chlorofluorocarbons, stable chlorocarbons, and fixed nitrogen in particular, concerns the destructive impact of these gases on the stratospheric concentrations of ozone [81, 82]. The ozone layer, consisting of ozone (O_3) concentrations of about 10^{11} to 10^{13} molecules/cm^3, exists largely in the stratosphere, 11–50 km above the earth's surface. Ozone formation in this region occurs by the interaction of molecular oxygen with sunlight, followed by the reaction of the atomic oxygen with further oxygen (Fig. 2.10). The details have recently been surveyed [83]. Its value to life on earth is the powerful filtering function that it passively performs on the short-wavelength UV-B (ca. 280–320 nm region) of sunlight before it reaches the earth's surface (Fig. 2.10). The equilibrium concentrations of ozone present are a resultant of these primary ozone formation and photolytic removal processes.

Chlorofluorocarbons (CFCs) were first produced as the working fluid for air conditioning and refrigeration systems because of their safety and stability compared to NH_3 and SO_2 which were the common agents earlier. It is this very stability that led to their contribution to damage of the ozone layer which could not have been foreseen. Most waste gases and vapors discharged in the troposphere are destroyed or returned to the earth's surface in particulate form long before they reach the stratosphere. The high stability of the CFCs permits them to survive passage through the troposphere, to reach the stratosphere. Here, exposure of the CFCs to short-wavelength ultraviolet light from the sun causes bond homolysis, releasing the destructive chlorine atoms. These chlorine atoms, in their turn, contribute many cycles of ozone removal reactions before ultimately forming a stable product which leads to their own removal.

It took some years to prove that this interference would cause a decrease in the equilibrium ozone concentration present [81, 82]. A decrease in the

Normal ozone formation:

$$O_2 + h\nu \rightarrow 2O$$
$$O + O_2 + M \rightarrow O_3 + M \text{ (energy sink)}$$

Ozone filter function:

$$O_3 + h\nu \rightarrow O_2 + O(^1D) \text{ (electronically excited oxygen)}$$

NO_x involvement:

$$NO + O_3 \rightarrow NO_2 + O_2$$
$$NO_2 + O \rightarrow NO + O_2 \text{ (a cycle)}$$

_____

$$O_3 + O \rightarrow 2 O_2 \text{ (net reaction)}$$

Chlorofluorocarbon (CFC) involvement:
Primary (initiation)

$$CCl_3F + h\nu \rightarrow Cl + CCl_2F$$
(CFC 11)
$$CCl_2F_2 + h\nu \rightarrow Cl + CClF_2$$
(CFC 12)

Ozone destruction (propagation)

$$Cl + O_3 \rightarrow ClO + O_2$$
$$NO + O_3 \rightarrow NO_2 + O_2$$
$$ClO + NO_2 \rightarrow NO_3 + ClONO_2 \text{ (chemically stable)}$$
$$ClONO_2 + h\nu \rightarrow NO_3 + Cl$$
$$NO_3 + h\nu \rightarrow NO + O_2$$

_____

$$O_3 + O_3 \rightarrow 3 O_2 \text{ (net reaction)}$$

FIGURE 2.10 Outline of representative normal processes and some interfering reactions which affect the equilibrium concentration of stratospheric ozone.

stratospheric ozone concentration was expected to cause an increase in the global exposure to short ultraviolet. This, in turn, was anticipated could cause a general increase in mutations in exposed species of the biosphere and more particularly in the incidence of human skin cancer (melanomas) [84]. Fortunately, the downward trend in atmospheric chlorofluorocarbon concentrations in recent years resulting from international regulations which curtailed production, use, and the introduction of methods to capture any CFCs while servicing refrigeration units promise to minimize the influence of these effects [85]. Newer working fluids for cooling units, the hydrochlorofluorocarbons or hydrofluorocarbons, which either contain hydrogen or contain no chlorine have been developed. These have much lower or zero ozone-destroying potential as compared to the earlier CFCs [86].

REVIEW QUESTIONS

1. (a) Calculate the gas or vapor concentrations, in milligrams per cubic meter (mg/m^3) at 25°C and normal atmospheric pressure,

equivalent to 1.0 ppm (by volume) of ozone, nitrogen dioxide, sulfur dioxide, carbon monoxide, and mercury vapor.

(b) To what concentration, in mg/m^3 also at 400°C, would a flue gas sulfur dioxide concentration of 1000 ppm at 400°C correspond? Assume 1 atm pressure and ideal gas behavior.

(c) The same flue gas from part (b), when cooled to 25°C, still contains about 1000 ppm sulfur dioxide. What would be the sulfur dioxide concentration of this gas now, in mg/m^3 at 25°C?

2. A sea level air pollutant analysis is carried out by passing $100 \, m^3$ of air, measured at 0°C and 760 mm Hg, through an impinger containing 100 mL of 0.10 M aqueous sodium hydroxide. Subsequently, the entire contents of the impinger gave a titer to pH 8 of 35.6 mL of 0.100 M sulfuric acid.

(a) Assuming that sulfur dioxide is the only acid gas constituent in the air sampled that is titrated at the pH range given, what concentration of sulfur dioxide, in mg/m^3, would these results correspond to?

(b) When further acid was added to bring the pH to about 4, bubbles were observed to form in the glass titration vessel. What gas are these bubbles likely to be?

(c) Consider an initially dry impinger tube, cooled in ice/water to chill the incoming air to 0°C for the analysis and placed ahead of the tube containing absorbing solution. The tube contains several milliliters of "water" at the end of the passage of the air sample. Might this observation affect the accuracy of the result quoted from part (a), and if so, how? What very simple test could you use to check this qualitatively?

3. A certain old automobile uses 1 L of oil for every 500 km traveled. Assuming that oil contains 0.70 kg of carbon per liter, and that one-half the oil loss ended up as carbon monoxide, how many liters of carbon monoxide (at 0°C and 760 mm Hg pressure) would this car produce from the oil alone per 10,000 km of travel?

4. The average 1980 automobile discharges 1 g of nitric oxide (NO) per kilometer traveled. Assuming a gasoline (C_8H_{18}, density $0.9 \, g/cm^3$) consumption of 10 km/L under stoichiometric combustion conditions, what volume concentration of nitric oxide would be discharged in the exhaust? Take air to be exactly 1 mol:4 mol, oxygen:nitrogen.

5. The threshold limit value (TLV) for 1,2-dichloroethane (CH_2ClCH_2Cl) vapor in the air of work environments has been set at 100 ppm by volume.

(a) What ventilation rate would be required to maintain safe working conditions in a dry cleaning establishment losing $1.0 \times 10^{-3} \, m^3$ of solvent vapor per minute, assuming uniform composition and perfect mixing of the air in the establishment?

(b) Is the required ventilation rate affected by the size of the building? Explain.

(c) Assuming 20°C and 1 atm conditions, what does $1 \times 10^{-3} \, m^3 min^{-1}$ of dichloroethane vapor correspond to as a mass rate of emission?

(d) What is a TLV of 100 ppm by volume for 1,2-dichloroethane equivalent to, in mg m^{-3} (at 25°C, 1 atm)?

6. The TLV for ethyl alcohol (CH_3CH_2OH) vapor is set at 1900 mg m^{-3}.

    (a) What ventilation rate is required to maintain this standard in a distillery that is experiencing alcohol loss at the rate of 10 mg min^{-1}.

    (b) What is the TLV of ethyl alcohol vapor in volume for volume units, if it is 1900 mg m^{-3}?

7. (a) Briefly explain the significance of sunlight in the formation of photochemical smogs.

    (b) Calculate the elemental percent composition by weight and the percent by weight of nitrogen dioxide in peroxyacetyl nitrate ($CH_3CO_3NO_2$), a more stable and irritating end product of this process.

8. In the Biosphere Experiment concluded on September 26, 1993, eight people stayed in a materially closed environment for 2 years. Assume that each person consumed 500 g dry weight of starch ($C_6H_{10}O_5$)$_n$ daily during this experiment, and completely metabolized this component of their food to carbon dioxide and water.

    (a) What masses of carbon dioxide and water would the Biosphere closed system have to deal with annually from the human metabolism of starch?

    (b) Based on the assumptions above what annual mass of carbon dioxide would be produced annually by the current world population of 4×10^9 people with the same daily per capita starch metabolism as given above?

    (c) What fraction of the carbon dioxide produced by the annual global combustion of 11.4×10^9 tonnes of coal-equivalent fossil fuels (assume C content 85%) does the estimated human metabolic contribution represent?

9. An adult at rest, seated in a $3 \times 5 \times 6$ m room at 20°C and 1.0 atm, produces about 15.0 L of carbon dioxide from respiration in an 1.0 hr. The carbon dioxide content of ambient air is 370 ppm by volume (370 mL CO_2 in 10^6 mL (air + CO_2)). How long would it take for the air in the room to reach the 1000 ppm air quality standard for offices, if there was no air change? Assume ideal gas relationships.

10. A power station burns powdered anthracite coal which analyzes 88% C, 4% S, and 8% ash. If this facility used 20% more air than theoretically required to ensure complete combustion, and the sulfur was entirely converted to sulfur dioxide, what concentration of sulfur dioxide (ppm by volume) would be present in the flue gas if there were no emission controls? Assume air is 20% O_2, 80% N_2 by volume.

11. Coal is burned at the rate of about 21,000 tonne/day in the Sundance Steam Electric Plant in Alberta, to produce 1725 MW of power.

    (a) If the coal contains 0.6% sulfur and 11% ash, what masses of sulfur dioxide and ash would be produced daily?

(b) Assuming that the remainder of the coal is carbon and that combustion of the carbon and sulfur is accomplished with the stoichiometric amount of air, what volume concentration of sulfur dioxide and carbon dioxide would be present in the flue gases produced? Assume air composition to be $1\,O_2:4N_2$.

FURTHER READING

J.W. Berks, J.G. Calvert, and R.E. Sievers, eds., "The Chemistry of the Atmosphere: Its Impact on Global Change." American Chemical Society, Washington, DC, 1993.

Eula Bingham, Barbara Cohrssen, Charles H. Powell, eds. "Patty's Toxicology," 5th ed. Wiley, New York, 2001. In 9 vols.

P. Brimblecombe, "Air Composition and Chemistry," 2nd ed. Cambridge University Press, Cambridge, UK, 1996.

J.H. Seinfeld, and S.N. Pandis, "Atmospheric Chemistry and Physics: from Air Pollution to Climate Change." Wiley, New York, 1998.

H.B. Singh, ed., "Composition, Chemistry, and Climate of the Atmosphere." Van Nostrand-Reinhold, New York, 1995.

W.H. Smith, "Air Pollution and Forests: Interactions Between Air Contaminants and Forest Ecosystems." Springer-Verlag, New York, 1981.

State of the Environment Reporting, "The State of Canada's Environment." Government of Canada, Ottawa, 1991.

REFERENCES

1. Trees Contribute Significantly to Urban Smog, *Chem. Eng. News* **66**(38), 21, Sept. 19 (1988).
2. R. Findley, *Nat. Geogr.* **159**(1), 3, Jan. (1981).
3. M.B. Hocking and J.A. Jaworski, eds., "Effects of Mercury in the Canadian Environment." National Research Council, Ottawa, 1979, p. 62.
4. S.H. Williston, *J. Geophys. Res.* **73**(22), 7051 (1968).
5. S.A. Berry and B.G. Notion, *Water Res.* **10**, 323 (1976).
6. "Atmospheric Aerosol: Source Air/Quality Relationships," ACS Symp. Ser. No. 167. American Chemical Society, Washington, DC, 1981.
7. Urban Air Review Group, Chair R.M. Harrison, Airborne Particulate Matter in the United Kingdom, University of Birmingham, 1996. pp. 7–14 (of 175).
8. C.F. Barrett, *Roy. Inst. Chem. Rev.* **3**(2), 119, Oct. (1970).
9. D.J. Spedding, "Air Pollution." Clarendon Press, Oxford, 1974.
10. H.P. Munger, The spectrum of particle size and its relation to air pollution. In " Air Pollution," (L.C. McCabe, ed.), Chap. 16. McGraw-Hill, New York, 1952.
11. J.A. Dorsey and J.O. Burckle, *Chem. Eng. Prog.* **67**(8), 92, Aug. (1971).
12. "Joint Air Pollution Study of St. Clair-Detroit River Areas for International Joint Commission, Canada and the United States." International Joint Commission, Ottawa and Washington, DC, 1971.
13. M.J. Pilat, D.S. Ensor, and J.C. Bosch, *Atmos. Environ.* **4**, 671 (1970).
14. L. Svarovsky, *Chem. Ind. (London),* p. 626, Aug. 7 (1976).
15. A.W. Gnyp, S.J.W. Price, C.C. St. Pierre, *et al.,* *Water and Pollut. Control* **111**(7), 40, July (1973).
16. "Detection and Analysis of Particulate Contamination." Millipore Corp., Bedford, MA, 1966.
17. W.C. McCrone, J.G. Delly, and S.J. Palenik " The Particle Atlas; An Encyclopedia of Techniques for Small Particle Identification." 2nd ed., Vol. 5th. Ann Arbor Sci. Publ., Ann Arbor, MI, 1973.
18. R. Weaver, What is it? Enter the microscope. *Am. Lab.* **36**(8), 17–25, Apr. (2004).

19. W. Strauss, ed., " Air Pollution Control," Part III: Measuring and Monitoring Air Pollutants, Wiley-Interscience, New York, 1978.

20. Tenax for Trapping, *Gas-Chromatogr. Newsl.* **21**(2), 8, June (1980).

21. M.B. Hocking and G.W. Lee, Calculated Sulfur Dioxide Equilibrium at Low Concentrations Between Air and Water, *Water, Air, Soil Pollut.* **8**, 255 (1977).

22. G.D. Clayton and F.E. Clayton, eds., " Patty's Industrial Hygiene and Toxicology," 3rd ed., Vol. 2A. Wiley, New York, 1981; and 1st and 2nd (1963) editions of this work, F.A. Patty, ed.

23. S.J. Williamson, "Fundamentals of Air Pollution," pp. 411–412. Addison-Wesley, London, 1973.

24. Committee on Environmental Improvement, " Cleaning Our Environment, A Chemical Perspective," 2nd ed., p. 114. American Chemical Society, Washington, DC, 1978.

25. M.B. Jacobs, "The Chemical Analysis of Air Pollutants." Interscience, New York, 1960.

26. K. Leichnitz, " Detector Tube Handbook," 3rd ed. Draegerwerk AG, Lubeck, W. Germany, 1976.

27. "Matheson-Kitagawa Toxic Gas Detector System." Matheson, Lyndhurst, NJ, 1980.

28. "Gastec Precision Gas Detector System." Gastec Corp., Tokyo, 1974.

29. R.M. Ash and J.R. Lynch, *J. Am. Ind. Hyg. Assoc.* **32**(8), 552 (1971).

30. Photographing SO_2 Emissions, *Chem. in Can.* **26**(11), 16, Dec. (1974).

31. "Quantitative Analysis with the Miran Gas Analyzer," Seminar. Wilks Scientific, South Norwalk, CT, 1973.

32. J.N. Pitts, Jr., B.J. Finlayson-Pitts, and A. M. Winer, *Environ. Sci. Technol.* **11**(6), 568 (1977).

33. J.R. Alkins, *Anal. Chem.* **47**(8), 752A (1975).

34. W.F. Herget and W.D. Connor, *Envir. Sci. Technol.* **11**(10), 962 (1977).

35. E.D. Hinkley, *Environ. Sci. Technol.* **11**(6), 564 (1977).

36. D.A. Lane and B.A. Thomson, *J. Air Pollut. Control Assoc.* **31**, 122, Feb. (1981).

37. R. Perry and R.M. Harrison, *Chem. Brit.* **12**(6), 185, June (1976).

38. *Chem. Eng. News* **72**(42), 88, Oct.17 (1994).

39. Hocking, D. and Hocking. M.B. 1977. Equilibrium Solubility of Trace Atmospheric Sulfur Dioxide in Water and its Bearing on Air Pollution Injury to Plants. *Environmental Pollution*, (London), **13**, 57–64

40. T. Sawidis, A. Marnasidis, G. Zachariadis, *et al.*, A Study of Air Pollution with Heavy Metals in Thessaloniki City (Greece) Using Trees..., *Arch. Environ. Contam. Toxicol.* **28**(1), 118–124, Jan. 1995.

41. S. Feast, Odor Detection Technology for Characterizing the State of Chemical Systems, *Am. Labor.* **32**(6), 38–42, Mar. (2000).

42. K.M. Reese, Canine Olfaction Tried for Detecting Pollutants, *Chem. Eng. News* **63**(37), 84, Sept. 16 (1985).

43. R. Stevenson, Sniffing Out Danger, *Chem. Brit.* **36**(5), 36–40, May (2000).

44. M. E. Conti and G. Cecchetti, Biological Monitoring: Lichens as Bioindicators of Air Pollution Assessment: A review, *Environ. Pollut.* **114**(3), 471–92 (2001).

45. M.B. Hocking, Trends in Cabin Air Quality of Commercial Aircraft: Industry and Passenger Perspectives. *Rev. Environ. Health* **17**(1), 1–49, Apr. (**2002**).

46. K. Ito, G.D. Thurston, C. Hayes *et al.*, Associations of London, England, daily Mortality with Particulate Matter, Sulfur Dioxide and Acidic Aerosol Pollution, *Arch. Environ. Health* **48**(4), 213–220, July–Aug. (1993).

47. S.J. Williamson, "Fundamentals of Air Pollution." Addison-Wesley, Reading, MA, 1973.

48. J.A. Kerr, J.G. Calvert, and K.L. Demerjian, The mechanism of photochemical smog formation, *Chem. Brit.* **8**(6), 252, June (1972).

49. J.N. Pitts, Jr. and B.J. Finlayson, Mechanisms of Photochemical Air Pollution, *Angew. Chem., Int. Ed. in Engl.* **14**(1), 1–15 (1975).

50. R.M. Harrison and C.D. Holman, *Chem. Brit.* **18**(8), 563, Aug. (1982).

51. R.A. Harley, A.G. Russell, G.J. McRae *et al.*, Photochemical Modeling of the Southern California Air Quality Study, *Envir. Sci. Technol.* **27**(2), 378–388, Feb. (1993).

52. A.C. Lloyd, W.P. Carter, and J.L. Sprung, *Calif. Air Environ.* **4**(3), 1, Spring (1974).

53. R.M. Harrison and C.D. Holman, Ozone Pollution in Britain, *Atmos. Environ.* **13**, 1535 (1979). Also *Chem. Brit.* **18**(8), 563–570, Aug. (1982).

54. E.B. Cowling, *Envir. Sci. Technol.* **16**(2), 110A (1982).

55. G.E. Likens, *Chem. Eng. News* **54**(48), 29, Nov. 22 (1976).

56. B. Hileman, *Envir. Sci. Technol.* **15**(10), 1119 (1981).

57. D.H. DeHayes, P.G. Schaberg, and G.J. Hawley, Acid Rain Impacts on, Forest Health, *BioScience* **49**(10), 789–800, Oct. (1999).

58. M. Ghosh, Effect of Liming on a Fish Population in an Acidified Lake..., *Appl. Math. Comput.* **135**(2–3), 553–560, Mar. 10 (2003).

59. L.A. Barrie, Arctic Air Pollution: An Overview of Current Knowledge, *Atmos. Environ.* **20**(4), 643–663 (1986).

60. C.S. Wong, *Science* **200**, 197 (1978).

61. W. Lepkowski, *Chem. Eng. News* **55**(42), 26, Oct. 17 (1977).

62. R.A. Kerr, *Science* **269**, 633 (1995).

63. C.D. Keeling, R.B. Bacastow, A.F. Carter *et al.*, "Aspects of Climate Variability in the Pacific and the Western Americas," *Geophys. Monogr., Am. Geophys. Union,* **55**, 1989.

64. V. Ramanathan, The Greenhouse Theory of Climate Change: A test, *Science* **240**, 293–299, Apr. (1988).

65. B. Hileman, Global Warming, *Chem. Eng. News* **67**(11), 25–44, Mar. 13 (1989).

66. H.S. Johnston, Human Effects on the Global Atmosphere, *Ann. Rev. Phys. Chem.* **35**, 481–505 (1984).

67. C.M. Volk, J.W. Elkins, D.W. Fahey *et al.* Evaluation of Source Gas Lifetimes..., *J. Geophys. Res. Atmos.* **102**(D21), 25,543–25,564, Nov. 20 (1997).

68. M.A.K. Khalil, R.A. Rasmussen, J.A. Culbertson *et al.*, Atmospheric Perfluorocarbons, *Environ. Sci. Technol.* **37**(19), 4358–4361, Oct. 1 (2003).

69. P. Zurer, Perfluorocarbons Use, Emissions Face Restrictions, *Chem. Eng. News.* **71**(32), 16, Aug. 9 (1993).

70. P. Fabian, R. Borchers, B.C. Kruger *et al.*, CF4 and C2F6 in the Atmosphere. *J. Geophys. Res. Atmos.* **92**(D8), 9831–9835, Aug. 20 (1987).

71. J.M. Barnola *et al.*, Vostock Core Provides 160,000 year Record of Atmospheric CO_2, Nature **329**(6138), 1–7, Oct. (1987).

72. B.J. Palmer, *Environ. Letters* **5**(4), 249 (1973).

73. CO_2 buildup's Effect on Climate Explored, *Chem. Eng. News* **55**(31), 18, Aug. 1 (1977).

74. V. Ramanathan, B.R. Barkstrom, and E.F. Harrison, Climate and the Earth's Radiation Budget, *Physics Today* **42**(5), 22–33, May (1989).

75. R.A. Kerr, A Fickle Sun Could be Altering the Earth's Climate, *Science* **269**, 633, Aug. 4 (1995).

76. V. Gornitz, C. Rosenzweig, and D. Hillel, Is Sea Level Rising or falling? *Nature* **371**(6497), 481, Oct. 6 (1994).

77. B. Hileman, Kyoto Climate Conference, *Chem. Eng. News.* **75**(51), 20–22, Dec. 22 (1997).

78. B. Hileman, Greenhouse Gas Economics, *Chem. Eng. News* **76**(9), 28–31, Mar. 2 (1998).

79. B. Bolin, The Carbon cycle, *Sci. Am.* **223**, 124 (1970).

80. J. McHale, "The Ecological Context." G. Braziller, New York, 1970.

81. M.J. Molina and F.S. Rowland, *Nature (London)* **249**, 810 (1974).

82. R.M. Baum, *Chem. Eng. News* **60**(37), 21, Sept. 13 (1982).

83. T. Peter, The Stratospheric Ozone Layer: An overview, *Environ. Pollut.* **83**, 69–79 (1994).

84. T.M. Sugden and T.F. West, eds., " Chlorofluorocarbons in the Environment: The Aerosol Controversy." Society of Chemical Industry, Ellis Horwood, Chichester, U.K., 1980.

85. R.A. Rasmussen, M.A.K. Khalil, and R.W. Dalluge, *Science* **211**, 285 (1981).

86. L.E. Manzer, The CFC–Ozone Issue: Progress on the Development of Alternatives..., *Science* **249**, 31–35 (1990).

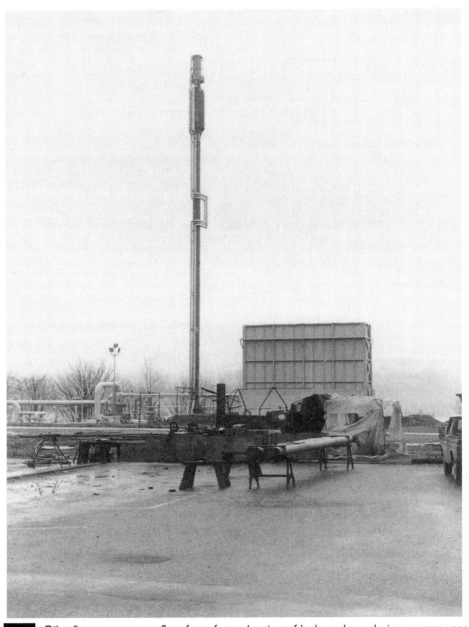
Oil refinery emergency flare for safe combustion of hydrocarbons during process upsets.

3
AIR POLLUTION CONTROL PRIORITIES AND METHODS

There is not always good cheer where the chimney smokes.

—Thomas Shelton, 1620

3.1. AIR POLLUTANT INVENTORIES

To establish the most effective pollution control strategies, it is important to start with an inventory of the sources, mass emission rates, and types of pollutants being discharged in the area of concern. This is accomplished by a detailed analysis of the emissions from each major point source of the inventory area. Data obtained by actual measurements are used whenever possible. But when these are not available, chemical and engineering theory is applied to come up with the best possible emission estimates within the time frame required for the inventory. Moving sources, such as automobiles, buses, and aircraft, are tallied by averages per unit, times the number of kilometers per unit, times the number of units to come up with good estimates for the contributions from these.

The pollutant inventory obtained gives an overview of the total current emission picture for the study area. Table 3.1 presents this data for Los Angeles, where the inventory was conducted to assist in determining the causes of the photochemical smog problem. For comparison, parallel data are presented for the whole of Canada which also has smog problems in some of the major centers. A more recent inventory conducted for Madrid displays a similar picture [4]. The best foundation for effective planning of abatement measures is obtained from inventories of this kind, aided by continual refinement of the information [5, 6].

From the tabulated data it is evident that automobiles rather than industrial activity are by far the largest source of pollutants overall (Table 3.1).

TABLE 3.1 Air Emission Inventory for Los Angeles in November 1973 Compared on a Percentage Basis to the Annual Inventory for Canada for 1974 and 1985[a]

Emission source	Emissions to the atmosphere (tonnes/day)						Percent of total Canada[b]	
	Hydrocarbons	Nitrogen oxides	Sulfur dioxide	Carbon monoxide	Totals, by source	Los Angeles	1974	1985
Automobiles	1,750	445	27	9,375	11,597	86.9	47.9	41.0
Organic solvent use	450				450	3.4		10.2
Oil refining	200	41	41	154	436	3.2	27.1	20.8
Chemicals production	50		60		110	0.8		
Combustion of fuels[c]	13	245	280	1	53	4.	9.1	16.9
Miscellaneous	37	27	5	14	21	1.	15	10.5
Totals, by pollutant	2,500	758	413	9,674	13,34	100.	100	100.0

[a]From Acres [1], The Clean Air Act [2], and Environment Canada [3].
[b]Canadian air pollutant inventory totals for the listed categories for 1974 and 1985 was 25.8 million tonnes, respectively. In 1985 road vehicles produced 30.4, 51.4, and 55% of the hydrocarbons, NO_x and CO, respectively.
[c]From stationary sources.

Road transport consists of multiple moving sources under the control of individuals, which makes emission abatement for this classification difficult to achieve. Nevertheless, by fundamental engine design changes coupled to accessory control units installed at the manufacturing stage, and by the application of emission control regulations which have to be met annually with vehicle license renewal, significant reductions have been achieved.

3.2. AUTOMOTIVE EMISSION CONTROL

As much as 70% of the hydrocarbons, 98% of the carbon monoxide, and 60% of the nitrogen oxides (NO_x), have originated from the operation of automobiles (Table 3.1). Hence, these are the principal automotive emissions of concern. The hydrocarbons and carbon monoxide arise from the fact that the most power for a given engine capacity (displacement) of early engine design was obtained with enough air for only 70–80% complete combustion. Nitrogen oxides are formed from the reactions of atomic oxygen and atomic nitrogen, which are formed against hot metal surfaces at high temperatures, with the corresponding elements (Eqs. 3.1 and 3.2).

$$N_2 + [O] \rightarrow NO + [N] \hspace{3cm} 3.1$$

$$O_2 + [N] \rightarrow NO + [O] \hspace{3cm} 3.2$$

These emission problems are not easy to solve. To run with a leaner tuned engine (a higher air to fuel ratio) obtains more complete fuel combustion and in so doing decreases the hydrocarbon and carbon monoxide concentrations in the exhaust, as would be expected. But at the same time the higher combustion temperatures obtained in the process raise the concentrations of nitrogen oxides obtained in the exhaust, and also decreased the power output per liter of engine displacement [1] (Fig. 3.1). It also affected drivability, or smooth engine response, at various throttle settings.

Early emission control methods were based on the use of a thermal reactor for hydrocarbon and carbon monoxide oxidation, combined with exhaust gas recirculation (EGR) for reduction of nitrogen oxide emissions (Fig. 3.2a). Hydrocarbons and carbon monoxide in the hot exhaust fed to the reactor, once heated, were rapidly oxidized to carbon dioxide and water by the additional pumped air which was fed to the reactor (e.g., Eqs. 3.3 and 3.4).

$$HC \, (= \text{hydrocarbon}) + O_2 \rightarrow CO_2 + H_2O \hspace{2cm} 3.3$$

$$2 \, CO + O_2 \rightarrow 2 \, CO_2 \hspace{3cm} 3.4$$

Some of the relatively inert exhaust gas, before it entered the reactor, was used to dilute the air fed to the carburetor by some 15%. Doing this decreased the peak combustion temperatures during normal engine operation and also decreased the concentrations of nitrogen oxides formed. These combined measures achieved significant emission reductions (Table 3.2), but at the same time caused a noticeable increase in fuel consumption and loss of performance.

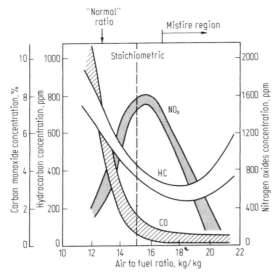

FIGURE 3.1   Plot of the approximate effect of various air to fuel ratios on the composition of automotive exhaust gas. Compiled from the data of Acres [1] and Stoker and Seager [7].

Two-stage catalytic exhaust purification approached these auto emission problems in a different way. Here the first stage of control, NO_x reduction, is achieved catalytically using the reserve chemical reducing capacity of the residual hydrocarbons and carbon monoxide already present in the exhaust (Fig. 3.2b; e.g., Eqs. 3.5–3.8).

TABLE 3.2   Typical Automotive Emissions Before and After Regulation, Related to U.S. Legislated Standards[a]

	Grams per mile				Gasoline evaporation (g/test)
	Carbon monoxide	Hydrocarbons	Nitrogen oxide	Particulates	
Before controls	80	11	4	0.3	60
1970 average	23	2.2	4–5	–	–
1975 standard	15	1.5	3.1	0.1	2
1980 standard	7.0	0.41	3.1	–	–
Thermal reactor + EGR[b]	4.5	0.2	1.3		
Catalytic, + cyclone	0.85	0.11	1.65	0.1	1–2
1994 model year[c]	3.4	0.25	0.4		
2001, clean fuel fleets (proposed)[c]	3.4	0.075	0.2		

[a]Compiled from Acres [1], Mikita and Cantwell [8], Chemical and Engineering News [9], and Acres and Cooper [10].

[b]EGR short for EGR, referring to the practice of recycling of 10–15% of exhaust gases to the intake of the engine to decrease peak combustion temperatures.

[c]From new sequences under more severe test conditions, which include measurements from a cold start.

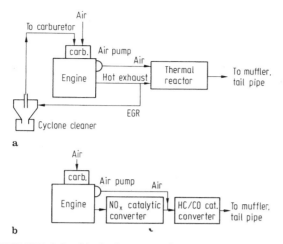

FIGURE 3.2 Block diagrams of two automotive emission control configurations. (a) Thermal reactor plus EGR. (b) Catalytic exhaust purification.

$$2 \, NO + 2 \, CO \rightarrow N_2 + 2 \, CO_2 \qquad 3.5$$

$$2 \, NO_2 + 4 \, CO \rightarrow N_2 + 4 \, CO_2 \qquad 3.6$$

$$2 \, NO + HC \, (= hydrocarbon) \rightarrow N_2 + CO + H_2O \qquad 3.7$$

$$2 \, NO_2 + HC \rightarrow N_2 + CO_2 + H_2O \qquad 3.8$$

All of these NO_x-reducing reactions are much more rapid and effective than the catalyzed direct redox reaction (Eq. 3.9), and hence convert most of the NO_x to innocuous gases. After NO_x reduction, secondary air injection,

$$2 \, NO \xrightarrow{\text{cat.}} N_2 + O_2 \qquad 3.9$$

and passage of the hot mixture into a second catalytic converter accomplish complete oxidation of the residual hydrocarbons and carbon monoxide not used for NO_x reduction in the first converter (Eqs. 3.7 and 3.8). One or two grams of platinum or palladium finely distributed on a ceramic matrix provide the catalytic activity of both converters. This combination exhaust treatment achieves better control of the hydrocarbons and carbon monoxide, and almost as good control of nitrogen oxides as achieved with the thermal reactor, and at the same time restores much of the lost fuel economy and performance (Table 3.2). If a limited EGR circuit is integrated with this dual converter system, the NO_x emission rate can be further decreased.

Since lead-containing antiknock additives of leaded gasolines rapidly destroy the activity of the catalytic emission control systems, unleaded gasolines have had to be used in automobiles fitted with this or related systems. However, since using this fuel modification also eliminates one of the main sources of lead emissions to the atmosphere, this trend also has a highly favorable additional impact [11, 12]. Before unleaded gasolines were on the market, it was common to find high lead levels in the core areas of busy cities and alongside freeways, a trend which is gradually reversing as the proportion of automobiles on the road equipped to burn unleaded gasolines increases.

While it is possible to reduce significantly particulate lead discharges from automobiles using leaded gasolines by employing cyclone-type particle collectors in the exhaust train [13], these methods are now outmoded. However, bioactivity of platinum group metals has prompted a concern regarding the risks of their dispersion as a result of their use in current automotive emission control systems. This has recently been reviewed [14]. A focus to this concern is the sensitization risk of palladium, since very low doses are sufficient to cause allergic reactions in susceptible individuals. This aspect should prompt a continued need for monitoring.

One of the early problems with catalytic control of automobile exhaust emissions was during the few minutes immediately after starting the engine when the cold catalytic systems did not function. This was solved by developing a porous zeolite which traps the unburned hydrocarbons while the catalysts are still cold [15]. Once the catalysts have warmed up, the zeolite canister also warms, releasing the trapped hydrocarbons to the catalytic systems to perform their important control reactions.

Today, to maintain or further improve the gaseous emission characteristics of internal combustion engines the trend is toward integral improvements in engine design, such as computer-controlled precision fuel injection and various stratified charge modifications which accomplish the chemistry of the add-on units but as an integral part of engine operation [16]. These developments are leading to more drivable, fuel efficient automobiles.

Combinations of these control measures have resulted in a decrease in ambient carbon monoxide concentrations, and in the severity and frequency of photochemical smog incidences, where they have been applied, even though there has been a concurrent increase in the number of automobiles on the road. Nevertheless, during severe and prolonged photochemical smog episodes it may be necessary to restrict automobile use temporarily as a public safety measure, as was used in Mexico City during such an event in October–November 1996.

Electric automobiles may be a viable option. In the late nineteenth century period of automobile development, they actually outnumbered gasoline-powered autos. Current technologies employing lead acid batteries charged by fossil-fueled power stations would not necessarily result in improvement. However, methods which involve the refinement of higher charge density, longer lived batteries, or new innovations in fuel cell technology could improve the environmental merits of this alternative. Prototype electric automobiles which use solar cells for battery charging have been demonstrated, but the high costs of solar cells and the variability of access to sunlight from latitude, season, and cloud factors are likely to keep these experimental for the time being.

3.3. AIR POLLUTANT WEIGHTING

Accumulation of emission data on a mass basis is a required first step to assess the overall impact of air pollutant discharges of a region prior to decisions on any necessary action. For development of the most cost-effective control strategy for an area, the contributing pollutants should be considered on the basis of their relative impacts in terms of health effects, or smog occurrences, and not solely on a mass basis. Two of the various weighting factors which have been proposed to do this are given in Table 3.3. While the weightings of these systems differ significantly from each other, especially for hydrocarbon vapors, they both assign the lowest weighting to carbon monoxide [17, 18].

If the weighting factors based on California standards are used to recalculate the relative significance of the Los Angeles emission sources, transportation still remains as the largest single sector, but only just larger than the combustion sources sector Table 3.4. This treatment redistributes the significance of emission control strategies among the transportation, oil refining and chemicals production, and combustion source sectors to obtain maximum effectiveness of ambient air improvement for this control area. It can be seen that a similar treatment of the Canadian data would give a higher emphasis to the need to control emissions from the industrial and power production sectors than found for Los Angeles.

3.4. METHODS AND LIMITATIONS OF AIR POLLUTANT DISPERSAL

Chimneys and vent stacks have been and still are popular for waste gas discharge from all types of stationary sources, large and small. Thermal power stations, smelters, refineries, and even domestic heating appliances all use discharge and dispersal methods to dispose of their waste gases. Serious problems can arise, however, if the source is a very large one (i.e., has a high mass rate of emission) discharging into a stable air mass. With the occurrence of a long-term inversion a temporary shutdown of the facility may be necessary to protect public health.

Simple dispersal of the flue gases from large emission sources is less acceptable today because of the limited capability of the atmosphere of highly industrialized regions of the globe to accept further loading. Discharged

TABLE 3.3 Examples of Weighting Factors Used for Primary Air Pollutants

Contaminant	Weighting factors based on	
	California ambient air standards[a]	U.S. federal secondary air quality standards[b]
Carbon monoxide	1.00	1.00
Hydrocarbons	2.07	125
Sulfur oxides, SO_x	28.0	21.5
Nitrogen oxides, NO_x	77.8	22.4
Particulates	106.7	37.3
Oxidants	186.9	-

[a]Used as the basis of the proposed Pindex air pollutant rating system [17].
[b]Based primarily on federal standards with health effect considerations [18].

TABLE 3.4 Comparison of Actual vs Weighted Daily Air Pollution Inventory for Los Angeles, November 1973[a]

Emission source	Total mass of gaseous primary air pollutants		Estimated mass of particulates and aerosols tonnes/day[b]		Weighted mass of primary pollutant emissions[c]	
	tonnes/ day	% of total	Actual	Weighted	tonnes/ day	% of total
Automobiles	11,597	86.9	74	7,896	56,271	41.5
Organic solvent use	450	3.4	0	0	930	0.7
Oil refining	436	3.2	123	13,124	19,714	14.6
Chemicals production	110	0.8				
Combustion of fuels	539	4.1	245	26,142	53,108	39.2
Miscellaneous	213	1.6	27	2,881	5,371	4.0
Totals	13,345	100.0	469	50,043	135,394	100.0

[a]Weightings from California standards, Table 3.3. See also Babcock [17].
[b]Includes hydrocarbon vapors, nitrogen oxides, sulfur oxides, and carbon monoxide.
[c]Includes the primary gaseous air pollutants, plus particulates and aerosols.

particulate matter only falls out again on the immediate area. Discharged fumes, fogs, and mists, although they are more widely dispersed, also eventually return to the earth's surface as they coagulate or coalesce into larger agglomerated particles, or are washed out of the atmosphere in precipitation. Discharged gases have the best prospect of efficient dispersal, but also eventually return adsorbed onto, or reacted with other gases, particles, or water and precipitation. So a higher chimney merely spreads the combined fallout from these processes over a wider area; it does not decrease the gross atmospheric loading or fallout rate.

Multiple high chimneys in a single area tend to produce plume overlap some distance downwind of the original discharge points. This effect serves to negate the reduction in ground level concentrations of air pollutants, the original objective of the high stack, because of the reduced possibility of dilution by horizontal diffusion (Fig. 3.3a).

If stack dispersal is to be used, it is important that it be sited in such a way that adjacent buildings or natural features, such as hills or gullies do not trap discharged waste gases close to the ground. Also to be avoided are backwash, a downdraft obtained on the leeward side of any large surface feature, or eddies which may occur in the same plane around the corners of obstructions. Any of these factors can cause plume impingement with the ground much closer to the point of discharge than anticipated, and hence can cause higher concentrations of the components of the discharged exhaust gases close to the site of release than anticipated.

A primary objective of stack discharge of waste gases is to obtain the minimum or zero elevation of the ground level concentrations of the discharged gases in the immediate area of the stack. The stack discharge point should be high enough that sufficient diffusion of the discharged gases occurs

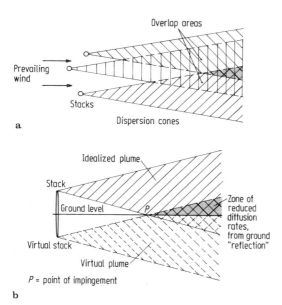

FIGURE 3.3 (a) Horizontal, and (b) vertical sections of stack discharges showing how overlapping plumes and virtual stack effects increase ambient air concentrations of pollutants at ground level above that to be expected from simple diffusion cone calculations.

to dilute these gases to make their concentrations acceptable at the point the diffusion cone intersects the ground. For any given mass emission rate, Eq. 3.10 may be used as a rough guide to the effect on the ground level

$$\text{ground level concentration} = K(m/uH^2) \qquad 3.10$$

where
m = mass rate of emission
u = wind velocity
H = *effective* stack height

concentration of a discharged gas which would be expected for different stack heights and wind velocities at the point of discharge. In this expression the "effective stack height" corresponds to the physical stack height plus the plume rise. The plume rise is obtained from a combination of thermal (buoyant) rise and momentum rise. The height component from this factor is defined as the vertical distance between the top of the stack and the point at which the center line of the plume levels off to within 10° of the horizontal. In the case of a warm or fan-forced discharge the plume rise can add considerably to the actual stack height, substantially raising the effective height of the discharge. Most real situations, however, are far more complex than can be reasonably approximated using this expression, so that more complex expressions are usually required to accommodate the greater number of variables [19].

The point of impingement for particulates, or the location on the ground where discharge products of the stack are at a maximum, corresponds to about three or four times the effective stack height. For gases, this corresponds to about 10 times the effective stack height. The influence of the "virtual stack effect" prevents further downward diffusional dilution of flue gases beyond the point of impingement because of the presence of the ground (Fig. 3.3b). Therefore, at points further from the stack than this the concentration of any discharged gases at ground level still continues to decrease, but is found to be roughly double what would be predicted from simple diffusion cone calculations.

The effective height of a chimney or the efficiency of effluent dispersion may be improved by any of a number of measures. Addition of a jockey stack, a smaller diameter supplementary stack installed on top of an existing installation, can help. Or improvement may be gained by raising the flue gas temperature or velocity to obtain greater plume rise. Valley inversions may be avoided by running ductwork up a hillside for discharge via a stack at the top, taking advantage of both the increased height above the inversion layer and the higher wind speeds for more efficient dispersal. Diffusion and dispersal rates under stable air conditions may also be improved by the use of vortex rings or super high stacks. But ultimately, particularly with very large installations discharging in areas where inversions are frequent, containment measures of one sort or another in conjunction with stack discharge have to be considered.

3.5. AIR POLLUTION ABATEMENT BY CONTAINMENT

With the increasing dilution and dispersal burdens being placed on a finite volume of atmosphere, it is becoming clearly evident that containment or neutralization of air pollutants is required *before* discharge in order to prevent further deterioration in air quality. Air emission controls which fall into this category may be precombustion or predispersal measures, which clean up a fuel or combustion process or modify a chemical process in such a way that the pollutant of concern is never produced or dispersed into the exhaust gases in the first place. Or they may comprise predispersal measures taken to remove or neutralize the particulates, aerosols, and gases to be discharged *after* combustion or chemical processing but before discharge. The choice of which action or actions to be taken depends on whether the process is already in operation or in the planning stage for construction, and the type of pre-, or postcombustion emission control strategies which are available or readily developed. These options are each discussed separately.

3.5.1. Precombustion Removal Methods

If the technology is accessible and the cost of applying it is not too great, precombustion removal of potential air pollutants has substantial merit. At this stage the offending substance (or substances) present in the raw material has not as yet been burned or reacted to form the process exhaust gases, which should make recovery easier.

Consider, for example, the incombustibles in coal. These can contribute significantly to the particulate content of the flue gases produced on combustion, as well as to the bottom ash volume. Incombustibles are now routinely removed before burning for more than half of the coal produced in North America. This is accomplished by using one or more of the following. The coal is first finely ground. It may then be washed with water on riffle boards, from which the less dense coal particles are carried by the water stream while the more dense rock particles tend to sink and are captured in the cavities of the riffles. In air jigs the powdered mixture is suspended on a bed of air in a fluidized form. The heavier rock particles tend to sink and may be drawn off the bottom, and the cleaned coal is drawn off the top. Or liquid or fluidized dense solid may be used to obtain a more direct sink-float or froth flotation separation of the coal, of density about $1.5\,\text{g/cm}^3$, from the much denser rock particles and other impurities.

Precombustion cleaning of powdered coal also removes a significant proportion of the sulfur. About half of the sulfur compounds present in uncleaned coal resides in inorganic pyrite and sulfates, which are removed by the particulate cleaning methods. The half that is present in organic sulfur compounds remains in the coal. This level of improvement can be further increased to some 80% or so of the total sulfur present by using various aqueous leaching chemicals, such as ferric sulfate or sodium hydroxide in the coal cleaning procedure (Eqs. 3.11 and 3.12).

Leaching:

$$FeS_2 + 8\,H_2O + 14\,Fe^{3+} \rightarrow 15\,Fe^{2+} + 2\,HSO_4^- + 14\,H^+ \qquad 3.11$$

Regeneration of leach solution:

$$2\,FeSO_4 + 1/2\,O_2 + H_2SO_4 \rightarrow Fe_2(SO_4)_3 + H_2O \qquad 3.12$$

Extraction with an organic solvent, such as molten anthracene or a sequential series of solvents, after conventional coal cleaning is also used [20, 21]. Microbial methods of coal desulfurization have also been tested. Precombustion cleaning measures such as these dramatically improve the emission characteristics of coal-burning installations. Coal conversion to a liquid or gaseous fuel permits much greater improvement in the residual precombustion impurities present, but also at greater cost.

The same kinds of considerations apply in the precombustion emission control measures used for the nondistilled fractions of petroleum, the residual fuel oils, which are used as utility fuels for ships and power generation. The residual fuel oils contain all of the involatiles of the original crude oil, concentrated by a factor of 6–10, since the volatile fractions have been removed. As a result, these fuels possess a significant potential for particulate and sulfur oxides emission on combustion. Normal refinery crude oil desalting by water washing and phase separation is capable of removing 95% or more of the particulate discharge problem before the crude is ever distilled. To lower the sulfur dioxide emissions requires either use of only low sulfur crude oils for residual fuel formulation, or more complicated and expensive catalytic desulfurization processing of the residual feedstocks, as discussed in Chap. 18 [22].

Even the methane present in natural gas is not without its sulfur gas polluting potential if burned directly in the form obtained from the well-head. This potential is removed by thorough desulfurization using an amine scrub before it is piped to industrial and domestic consumers (Section 7.5). This is another example of a precombustion cleaned fuel.

3.6. POSTCOMBUSTION EMISSION CONTROL

3.6.1. Particulate and Aerosol Collection Theory

Aerosol-sized contaminants are much more difficult to control than particulates because they do not readily settle out of a gas stream. The separation of the aerosol fraction of nongaseous air pollutants may be greatly improved if they are agglomerated (coagulated or coalesced) to the larger particles or droplets before removal is attempted. There are four primary means by which this may be achieved. Awareness of these contributing factors can suggest measures which could be taken to accelerate the process, or make it more efficient.

Brownian agglomeration, an important contributing process to the natural removal of aerosols from the atmosphere, is the first of these (background, see Chap. 2). The microscopic particles of fumes and fogs are so small that collision of gas molecules with them causes them to move in an erratic path, commonly called Brownian motion. When this occurs with high concentrations of aerosol particles (i.e., high particle *numbers* per unit volume), it results in frequent particle collisions. Collisions usually result in agglomeration by poorly understood mechanisms to form loose clusters or chains with solid particles, and larger drops from aerosol-sized liquid droplets.

The coagulation rate for any particular aerosol system may be predicted to a significant extent (Eq. 3.13),

$$\text{Coagulation rate} = K_b c^2, \qquad\qquad 3.13$$

where K_b is the rate constant for Brownian coagulation of the particular type of particle, and c is the number of particles per cm^3. The coagulation rate is proportional to the square of the particle concentration. Thus, the number of particles per unit volume is the single most important factor affecting the rate of Brownian coagulation. Table 3.5 illustrates this aspect, and gives the indication that concentrations of aerosols above about $10^6/cm^3$ are unstable. Concentrations higher than this will rapidly drop to near this concentration range by Brownian agglomeration processes. As particles larger than ca. 5- to 10-μm form (i.e., in the particulate rather than the aerosol size range), they are more susceptible to removal by sedimentation. It is a combination of processes such as these which contribute to keeping the aerosol concentration in the earth's atmosphere in the range below about 10^4 to 10^5 particles/cm^3.

Turbulent coagulation occurs when particles in multiple flow line intersections collide, a process stimulated by turbulent flow or eddy conditions.

TABLE 3.5 The Relationship of Coagulation Time to the Original Particle Concentration[a]

Initial number concentration of particles per cm^3	Time for number of particles to decrease to 1/10 initial concentration
10^{12}	0.03 sec
10^{11}	0.3 sec
10^{10}	3 sec
10^9	0.5 min
10^8	5.0 min
10^7	50 min
10^6	8.3 hr
10^5	17 hr
10^4	1.3 days

[a]Data calculated using the methods of the National Academy of Sciences [11], cited by Magill et al. [23].

The turbulent coagulation rate is again proportional to the square of the particle concentration (Eq. 3.14),

$$\text{Coagulation rate} = K_s c^2, \qquad 3.14$$

where K_s is the rate constant for turbulent coagulation. The magnitude of K_s, the proportionality constant for turbulent coagulation, is in the range of 10–4000 times K_b. Hence, turbulent coagulation can be used to substantially accelerate the natural agglomeration processes to aid in emission control.

When wood is burned out-of-doors, a smoke containing 10^6 particles/cm^3 may be generated. By a combination of Brownian and turbulent coagulation processes the average particle size in this smoke can increase from 0.2 to 0.4 μm before it has traveled 1000 m (Table 3.5).

Sonic agglomeration uses high-intensity sound waves to accelerate particle coagulation [24]. This procedure is based on the premise that particles of different sizes will tend to vibrate with different amplitudes, increasing the opportunities for collision and coagulation. Also, the particle concentration in the compression zone of a high-intensity sound wave will temporarily be raised, artificially increasing Brownian coagulation rates. Applying this technique involves the exposure of a particle-laden gas stream for a fraction of a second to a high-intensity siren or oscillation piston, operating at 300–500 Hz and 170 decibels in a sound insulated chamber. After exposure the particulate collection efficiency of downstream containment devices is significantly improved from the greater proportion of large particles produced by this process.

Electrostatic methods, the fourth class, may also be used to obtain efficient coagulation of particles of the aerosol size range and larger. By passing a particle-laden gas stream past the negative side of a high DC voltage corona (intense electric field), the particles become negatively charged. They are then efficiently attracted and coagulated against positively charged plates which are in close proximity in electrostatic precipitators. Periodic rapping of the plates or a trickle of water is used to dislodge the large particles into collection hoppers. People involved with stack sampling of gases downstream of an operating electrostatic precipitator should remember that the 20,000–30,000 volts normally used is high enough to leave a hazardous residual charge in the very small fraction of particles not retained by the precipitator.

3.6.2. Particulate and Aerosol Collection Devices

Gas cleaning devices vary in their removal efficiencies, but almost invariably they are more efficient at removing particulate, rather than aerosol size range material. Even so, the least efficient device for removing small particle sizes, the gravity settling chamber (Fig. 3.4), still plays an important role in emission control. When the gas stream to be cleaned has only large particles present which can be effectively removed by the device, then this device alone represents an inexpensive method of control. When the particle size range to be

FIGURE 3.4 Simplified diagrams of some of the common types of particulate and aerosol emission control devices. Some of these are also useful for gas or vapor emission control; see text for details.

collected is wider than this, and a dry cyclone is to be used for final particle collection, very large particles (100–200 μm) *should* be removed prior to the gas stream entering the cyclone. Otherwise the interior of the cyclone is subjected to very high abrasion rates.

Since the efficiency of gas cleaning devices is almost inevitably better for larger particles than for smaller and since the number of particles represented by any given mass goes up exponentially as the particle size goes down, mass removal, rather than particle removal efficiencies have become the standard method of specifying air cleaning capabilities. In one case a particulate control device with a 91% mass efficiency worked out to have only 0.09% particle count efficiency for the same hypothetical dust-laden gas sample. Thus, to be fully aware of the collection characteristics of an air cleaning device one must have both mass removal efficiency information, *plus* the particle size range over which the mass removal efficiency was measured. Typical characteristics of a number of particle collection devices have been collected in Table 3.6.

The operating characteristics of some common particulate emission control units are illustrated in Fig. 3.4. Cyclones, which are extensively used for this purpose, are most efficient in small diameters which force the entering gas stream through higher tangential velocities. It is high tangential velocities which most efficiently separate particles from the gas. To obtain the same high efficiencies with very large gas volumes a large number of small cyclones are connected in parallel, rather than using a single, very large cyclone.

Absolute filters, developed by the U.S. Atomic Energy Commission for containment of radioactive particles, are extremely efficient for retention of very fine particles but require considerable operating energy for the forcing fans because of a large pressure drop across the filter element.

TABLE 3.6 Examples of Typical Collection Efficiencies for Some Common Types of Particulate and Aerosol Gas Cleaning Devices[a]

Air cleaning device	Particle size range for 90% removal (μm)	Remarks
Gravity settling chamber	50–100	Effective only at low gas velocities, < 200 m/min
Baffle chamber	20–100	Gas velocity ca. 300 m/min
Dry cyclone 2–750 cm diameter,	10–50	Gas velocity 300–1500 m/min, smaller are more efficient
Spray washer	5–20	Gas velocity 70–100 m/min
Packed tower scrubber	1–5	Gas velocities to 4000 m/min, high resistance to gas flow (pressure drop)
Intensive rotary scrubber	0.1–1	High resistance to flow, energy intensive
Cloth filter	0.2–1	Used in baghouses or continuous roll filters
Absolute filter	0.01–0.1	High efficiency, but high resistance to gas flow
Electrostatic precipitator	0.01–10	Require low gas flow velocities, very low pressure drop, very large units (> 3000 m^3/min)
Activated charcoal	Molecular	For control of gases, vapors, and odors; also retains particles, about equals cloth filters

[a]Compiled from data of Williamson [19], Mednikov [24], Stern [25], and Patty [26].

Electrostatic precipitation is also very efficient for retention of very fine particles, as long as these are not highly electrically conductive. The structural features of these units dictate that they are only cost effective for particulate and aerosol emission control for low pressure drop clean-up of very large volumes of gas; hence their extensive use for the treatment of combustion gases of fossil-fueled power stations. Next generation developments have been reviewed [27].

3.6.3. Hydrocarbon Emission Control

Adoption of automotive control devices has had the most significant effect on the gross hydrocarbon emissions for regulated areas, because of the large contribution to hydrocarbons originating from automobiles (Tables 3.1 and 3.2). But this measure on its own will not necessarily be effective for photochemical smog abatement since it is the concentration of reactive hydrocarbons, rather than total hydrocarbons, which is of greatest significance.

Use of floating roofs tightly sealed to the sides of storage tanks can reduce the ventilation losses of hydrocarbons from tank farms and potential losses from other points of oil refinery operations, as discussed in Chap. 18.

In the dry cleaning business or in vapor degreasing plants, losses of hydrocarbon or chlorinated hydrocarbon vapors occur from the use and recovery of solvents. These losses can be prevented by the use of activated adsorbent systems to capture the organic vapors from the vents or hoods of these devices. Activated carbon is one of the best adsorbents for this purpose because of its selectivity for organic vapors, and can provide up to $1000\,m^2/g$ (ca. 8 acres/oz) of adsorption area. Solvent present in high concentrations in air may also be recovered by cooling the air to liquid nitrogen temperatures to condense the solvent.

Air contaminated with lower concentrations of organic vapors which can arise from a number of processes may be economically controlled by using this air stream as boiler feed air. Any combustibles present are burned to provide fuel value, in the process destroying any hydrocarbons present before discharge [28]. This method is particularly appropriate when the concentration of organic vapors in air is too small for economic containment using adsorption or condensation methods.

The paint industry has a high dependence on solvents and chemically active, film-forming components, some of which come into the regulatory "reactive hydrocarbon" class. It has also been hard pressed to meet air quality tests [29]. By the development of water-based coatings technologies, even for many metal-finishing applications [30], and by innovations into such areas as high solids or even dry powders, aided by electrostatic coating technologies the industry is decreasing its dependence on organic solvents for many finishing applications [31]. In turn these measures decrease the emission of hydrocarbons when the coating is applied, and simultaneously result in much less waste of coating solids.

3.6.4. Control of Sulfur Dioxide Emissions

Total sulfur dioxide discharges for the U.S.A. for 1981 amounted to 22.5 million tonnes. Combustion sources concerned with power generation produced about 60% of this, and a further 37% was evenly split between other stationary combustion sources (mainly space heating) and industrial processes. More than half of that originating from industrial processes, or about 11% of the total, arose from the smelting of sulfide minerals and metal refining activities. Only about 3% of the total originated with the transportation sector.

The extent of sulfur dioxide emission is of importance for the direct effect it has on the ambient air levels, for which guidelines have been laid down to protect public health (Table 3.7). The atmospheric half-life of discharged sulfur dioxide is estimated to be short, of the order of 3 days. This rapid return of sulfur dioxide to the earth's surface, both as sulfur dioxide itself and as its oxidized and hydrated products, is the reason for its significance as a contributor to the acidity of rainfall about which comment has already been made. "Sulfurous acid," from hydrated sulfur dioxide, and sulfuric acid, formed from the oxidation of sulfurous acid or the hydration of sulfur trioxide, both contribute to low pH rain. Sulfur trioxide forms through the gas phase and heterogeneous oxidation of sulfur dioxide on particulate matter. This occurs to a significant extent both in the plumes downwind of large chimneys, where the concentrations are relatively high, and in the ambient air.

The origins of two of the major sulfur dioxide discharges in Canada are combustion sources and sour gas plants, which produce waste gas streams which contain 0.15–0.50% sulfur dioxide. These are too low to be economically attractive for containment. Smelter sources produce roaster exhaust gas streams containing 2–5% sulfur dioxide, or with modern equipment up to about 15%, concentrations which are economically favorable for capture of the sulfur

TABLE 3.7   Ambient Air Quality Standards for Canada and the U.S.A. in μg/m³, and Parenthetically in ppm[a]

Pollutant	Averaging time	Primary standards (welfare related, or maximum acceptable)		Secondary standards (health related, or maximum desirable)	
		Canada	U.S.A.	Canada	U.S.A.
Total suspended	Annual[b]	70	75	60	60
particulate	24 hr	120	260	–	150
Sulfur dioxide	Annual[c]	60(0.02)	80(0.03)	30(0.01)	–
	24 hr	300(0.11)	365(0.14)	150(0.06)	–
	3 hr	–	–	–	1,300(0.50)
	1 hr	–	–	450(0.17)	–
Carbon monoxide	8 hr	15(13)	10(9)	6(5)	10(9)
	1 hr	35(30)	35(40)	15(13)	35(40)
Nitrogen dioxide	Annual[c]	100(0.05)	100(0.053)	60(0.03)	100(0.053)
	24 hr	200(0.11)	–	–	–
Ozone	24 hr	50(0.025)	–	30(0.015)	–
	1 hr	160(0.08)	235(0.12)	100(0.05)	235(0.12)
Lead	3 month	–	1.5	–	1.5

[a]Compiled from U.S. EPA [32] and *Canadian Chemical Processing* [33].
[b]Annual geometric mean.
[c]Annual arithmetic mean.

dioxide. This may either be sold as liquid sulfur dioxide (boiling point −10°C), or may be converted to sulfuric acid for sale. Smelter grade acid is less pure than sulfuric acid from sulfur burning sources, and so fetches a much lower market price. However, it is still a useful product for such less demanding applications as metal pickling and the manufacture of fertilizers.

A number of solutions exist for decreasing the sulfur dioxide emissions from sources where the concentrations present are too low for economic containment. The choice in any particular situation depends on the existing ambient air quality (particularly with regard to public health) in the operating area, and the time frame within which abatement action has to be taken.

Temporary shutdown or a reduced level of operation are energy curtailment measures which may be required of a fossil-fueled power station or a smelter to avoid producing local, dangerously high ambient concentrations of sulfur dioxide during a severe inversion. With local experience this type of temporary curtailment may be anticipated, to avoid "brown-outs". A power station, for instance, can arrange to have separate stockpiles of low-cost, relatively high sulfur content coal for periods of power generation during normal atmospheric conditions. Stockpiles of alternative more expensive low sulfur coal, or even coke or natural gas (if suitably equipped), may be burned during severe inversion episodes. With these choices available, costly and inconvenient emission curtailment by reduction of power production or temporary shutdown will be required less often.

Some emission reduction measures, such as process modifications to avoid generating the sulfur dioxide-containing waste gases, have longer lead times. For smelters this might mean adoption of hydrometallurgical technology, such as is now available for the processing of copper, nickel, and zinc, in order to bypass the older roasting methods of sulfur removal from ores [34]. For power stations this option might mean the installation of more rigorous precombustion sulfur removal techniques for their coal or residual fuel oil, or increased emphasis on natural gas firing. Or it might mean contributions from hydroelectric, nuclear or the solar-related or wind power-based technologies in the utility's long-term plans for power generation.

A third emission reduction choice is to remain with the existing front end process, which continues to produce a sulfur dioxide-containing waste gas stream, and move to some system which can effectively remove the sulfur dioxide from this waste gas before it is discharged. Many methods are available, each with features which may make one more attractive than the others for the specific sulfur dioxide removal requirements (Table 3.8). Some of the selection factors to be considered are the waste gas volumes and sulfur dioxide concentrations which have to be treated and the degree of sulfur dioxide removal required. It should be remembered that the trend is toward a continued decrease in allowable discharges. The type of sulfur dioxide capture product which is produced by the process and the overall cost are also factors. Any by-product credit which may be available to offset process costs could also influence the decision. Finally, the type of treated gas discharge required for the operation (i.e., warm or ambient temperature, moist or dry, etc.), also has to be taken into account. Chemical details of the processes of Table 3.8 are outlined below.

Only one control procedure, ammonia injection, relies solely on gas phase reactions. The moist flue gas containing hydrated sulfur dioxide undergoes a gas phase, acid–base reaction to produce particulate solids (Eqs. 3.15 and 3.16).

$$SO_2 + H_2O + NH_3 \rightarrow NH_4HSO_3 \qquad\qquad 3.15$$
$$\text{particulate}$$

TABLE 3.8   The Chemical Details of Examples of Stack Gas Desulfurization Processes[a]

Reaction phases	Process examples	Regeneration method	Product(s)
Gas–gas	Homogeneous:		
	gaseous ammonia injection plus particle collection	none	$(NH_4)_2SO_3$, $(NH_4)_2SO_4$
	Heterogeneous:		
	SO_2 reduction with H_2S + catalyst (Claus)	none	S
	SO_2 reduction with CH_4 + catalyst to H_2S, then amine scrub (e.g., Girbotol) or aqueous AQ (Stretford)[b] recovery	none air	conc. H_2S S
	SO_2 catalytic oxidation with air (chamber or contact process), plus hydration	none	H_2SO_4
Gas–liquid	Absorption, plus chemical reaction:		
	by aqueous ammonia solution	none	$(NH_4)_2SO_3$, $(NH_4)_2SO_4$
	dimethylaniline solution (ASARCO)	heat	conc. SO_2
	aqueous sodium sulfite (Wellman–Lord)	heat	conc. SO_2, Na_2SO_4
	citrate process, U.S. Bureau of Mines	H_2S	S
	citrate process, Flakt–Boliden version	heat	conc. SO_2
	eutectic melt (Na_2CO_3/K_2CO_3), gives dry plume	water gas (H_2, CO)	conc. H_2S
	limestone (or lime) slurry scrubbing, inexpensive, throw-away slurry product	none	$CaSO_3 \cdot nH_2O$ $CaSO_4 \cdot 2H_2O$
Gas–solid	Physical adsorption:		
	SO_2 onto activated charcoal or alkalized alumina	heat	conc. SO_2
	Adsorption, plus chemical reaction:		
	powdered limestone injection, 20–60% efficient capture with particlecollection (precipitator), waste product	none	dry $CaSO_3$ and $CaSO_4$
	CuO (or other metal oxides)	methane	conc. SO_2
	Chemical reaction:		
	SO_2 reduction in heated coal bed, Resox process	none	S

[a]Compiled from sources cited in the text where further details are available, plus the reviews of Kaplan and Maxwell [35], Siddiqi and Tenini [36], and Engdahl and Rosenberg [37].
[b]AQ, short for anthraquinone disulfonic acids, see text for details.

$$SO_2 + H_2O + 2\,NH_3 + 1/2\,O_2 \quad \rightarrow \quad (NH_4)_2SO_4 \qquad 3.16$$
$$\text{catalysis}$$

The solid product, now a separate phase, can be readily captured by electrostatic precipitation or any other particulate collection device which is already normally in place for fly-ash control.

Heterogeneous reduction processes still involve the reaction of gases, but in these cases the reaction occurs in the presence of a suitable solid phase catalyst. Sulfur dioxide may be reduced to sulfur with hydrogen sulfide, if this is available, and the sulfur vapor condensed out of the gas stream by cooling, as in the second half of the Claus process (Eq. 3.17).

$$SO_2 + 2\,H_2S \xrightarrow{\ Fe_2O_3\ } 3\,S + 2\,H_2O \qquad 3.17$$

Or the sulfur dioxide may be reduced catalytically with methane or other hydrocarbons to hydrogen sulfide. The hydrogen sulfide produced by this method is captured by amine scrubbing of the reduced gas stream (Eqs. 3.18–3.20).

$$3\,SO_2 + 2\,CH_4 \xrightarrow{\ Al_2O_3\ } 2\,H_2S + 2\,CO_2 + 2\,H_2O + S \qquad 3.18$$

Absorption:

$$H_2S + H_2NCH_2CH_2OH \rightarrow HS^- + H_3\overset{+}{N}CH_2CH_2OH \qquad 3.19$$

Regeneration:

$$HS^- + H_3N^+CH_2CH_2OH \xrightarrow[\text{heat}]{} H_2S + H_2NCH_2CH_2OH \qquad 3.20$$

The high concentration of hydrogen sulfide obtained by heating the amine salt may then be easily and economically converted to elemental sulfur via the Claus process. The Stretford process, which may also be used for hydrogen sulfide capture, accomplishes both capture and conversion using an aqueous mixture of 1,5-, and 1,8-disulfonic acids of 9,10-anthraquinone (AQ) to sulfur in a single step [38] (Eq. 3.21).

Absorption–chemical conversion:

$$AQ{=}C{=}O + H_2S \rightarrow AQ{=}C(SH)(OH) \rightarrow AQ{=}C(H)(OH) + S \qquad 3.21$$

Regeneration of the quinone from the quinol is accomplished with air (Eq. 3.22).

Regeneration:

$$2\,AQ{=}C(H)(OH) + O_2 \rightarrow AQ{=}C{=}O + 2\,H_2O \qquad 3.22$$

Oxidation of sulfur dioxide with air, via the first stage of the contact or the chamber process to sulfuric acid, also serves to improve the collection efficiency of the sulfur oxides. Sulfur trioxide has a very strong affinity for water, unlike sulfur dioxide, so that its collection by direct absorption into water is extremely efficient, and the product sulfuric acid is a salable commodity.

Sulfur dioxide containment by gas–liquid interactions can be as simple as flue, or process gas scrubbing with dilute ammonium hydroxide (Eqs. 3.23 and 3.24).

$$2\,NH_4OH + SO_2 \rightarrow (NH_4)_2SO_3 + H_2O \qquad 3.23$$

$$2\,NH_4OH + SO_2 + 1/2\,O_2 \rightarrow (NH_4)_2SO_4 + H_2O \qquad 3.24$$

The ammonium salt products crystallized from the concentrated spent scrubber liquors may be used as valuable constituents of fertilizers. Or they may be first reduced to ammonia and hydrogen sulfide with natural gas, followed by conversion of the hydrogen sulfide to sulfur by a Claus-type sequence [39]. In this French-designed modification, the ammonia is recycled to the scrubbing circuit.

Similar acid–base chemistry is involved in the American Smelting and Refining Company's (ASARCO's) sulfur dioxide capture process using aqueous dimethylaniline (Eq. 3.25).

$$PhN(CH_3)_2 + SO_2 + H_2O \rightarrow Ph\,\overset{+}{N}(H)(CH_3)_2 + HSO_3^- \qquad 3.25$$

In this case the sulfurous acid-amine salt is heated to regenerate the much more expensive dimethylaniline solution for reuse, and to obtain a concentrated sulfur dioxide gas stream. The more concentrated sulfur dioxide becomes economic for acid production.

The Wellman–Lord process uses the effective sodium sulfite–sodium bisulfite equilibrium to capture sulfur dioxide from flue gases [40] (Eqs. 3.26 and 3.27).
Absorption:

$$SO_2 + H_2O + Na_2SO_3 \rightarrow 2\,NaHSO_3 \qquad 3.26$$

Regeneration:

$$2\,NaHSO_3 + heat \rightarrow SO_2 + H_2O + Na_2SO_3 \qquad 3.27$$

Most of the sodium bisulfite produced is converted back to sodium sulfite on regeneration, but some sodium sulfate inevitably forms from irreversible oxidation. This is crystallized out, dried, and sold as a wood pulping chemical ("salt cake").

The citrate process, in which much development work has been invested by the U.S. Bureau of Mines and by Pfizer, Inc., uses an aqueous solution of citric acid to capture sulfur dioxide (Eqs. 3.28 and 3.29).
Ionization:

$$HOC(CH_2CO_2H)_2CO_2H \rightarrow H_2Cit^- + H^+ \qquad 3.28$$
$$\text{citric acid (first ionization} = Cit^{-1})$$

Absorption:

$$H_2{\cdot}Cit^{-1} + HSO_3^- \rightarrow (HSO_3{\cdot}H_2Cit)^{2-} \qquad 3.29$$

Regeneration of the uncomplexed citric acid at the same time as formation of an elemental sulfur product from the bisulfite anion is obtained by treating the absorption solution with hydrogen sulfide [41] (Eq. 3.30).
Regeneration:

$$(HSO_3 \cdot H_2Cit)^{2-} + H^+ + 2\,H_2S \rightarrow 3\,S + H_2Cit^- + 3\,H_2O \qquad 3.30$$

If hydrogen sulfide is not conveniently available from local process sources it may be produced on site by reducing a part of the sulfur product of the process with methane (Eq. 3.31).

$$4\,S + CH_4 + 2\,H_2O \xrightarrow{\text{Al}_2\text{O}_3} 4\,H_2S + CO_2 \qquad\qquad 3.31$$

The Flakt–Boliden version of the citrate process uses thermal regeneration of the citrate absorbing solution to obtain stripped citrate solution and a stream of up to 90% sulfur dioxide. Recovery of a sulfur dioxide product gives flexibility to the final stage of processing as to whether liquid sulfur dioxide, sulfuric acid, or sulfur are obtained as the final product. This process has recently been reexamined [42].

A further gas–liquid interaction process to a useful product is the sodium carbonate-potassium carbonate eutectic melt process. This operates at a temperature of about 425°C, well above the melting point of the eutectic [43]. Sulfur dioxide absorption takes place with loss of carbon dioxide (Eq. 3.32).

$$Na_2CO_3 + SO_2 \rightarrow Na_2SO_3 + CO_2 \qquad\qquad 3.32$$

The original eutectic is regenerated plus a hydrogen sulfide product obtained in a separate unit by treating the sodium sulfite with "water gas" (Eq. 3.33).

Regeneration:

$$Na_2SO_3 + CO + 2\,H_2 \rightarrow Na_2CO_3 + H_2S + H_2O \qquad\qquad 3.33$$

The hydrogen sulfide initial product can be subsequently converted to more useful sulfur via the Claus process. In return for the high operating temperatures required for this process, it gives a substantially dry plume, unlike the other gas–liquid interaction processes mentioned. This may be an important consideration if the process is to be operated in a fog susceptible area.

A finely powdered limestone or lime slurry in water used in a suitably designed scrubber is an effective and relatively low-cost sulfur dioxide containment method (Eqs. 3.34–3.37).

$$CaCO_3 + SO_2 \rightarrow CaSO_3 + CO_2 \qquad\qquad 3.34$$

$$CaCO_3 + SO_2 + 1/2\,O_2 \rightarrow CaSO_4 + CO_2 \qquad\qquad 3.35$$

$$Ca(OH)_2 + H_2O + SO_2 \rightarrow CaSO_3 + 2\,H_2O \qquad\qquad 3.36$$

$$Ca(OH)_2 + H_2O + SO_2 + 1/2\,O_2 \rightarrow CaSO_4 + 2\,H_2O \qquad\qquad 3.37$$

Variables affecting this process have recently been reviewed [44]. With either reagent, however, a throwaway product is obtained. Land has to be allocated for lagoon disposal of the spent scrubber slurries, or other systems have to be set up to handle the waste solid. Recent variations of this approach are to employ a zeolite prepared from fly ash [45] or the alkalinity of fly ash itself in water slurry as means to capture sulfur dioxide.

The simplest gas–solid containment systems conceptually are the direct adsorption ones. These accomplish adsorption on solids such as activated carbon, or alkalized alumina at relatively low temperatures and ordinary pressures [46]. In a separate unit a more concentrated sulfur dioxide stream is produced when the saturated absorbent is regenerated by heating. This is a more economically attractive feed to an acid plant or for liquefaction or sulfur generation.

Among the simplest systems, at least for power stations that already burn powdered coal is the use of powdered limestone injection. Essentially the same burner assembly is required, and the combustion temperature of the coal is sufficient to form calcium oxide (lime) from the injected limestone (Eq. 3.38).

$$CaCO_3 \rightarrow CaO + CO_2 \qquad 3.38$$

The alkaline lime, in a gas–solid phase reaction, reacts with sulfur dioxide in the combustion gases to form solid particles of calcium sulfite and calcium sulfate which are captured in electrostatic precipitators (Eqs. 3.39 and 3.40).

$$CaO + SO_2 \rightarrow CaSO_3 \qquad 3.39$$

$$CaO + SO_2 + 1/2\ O_2 \rightarrow CaSO_4 \qquad 3.40$$

The process is simple in concept, and to operate, but the stoichiometry of the gas–solid reaction is only about 20 to 60% of theory so that an excess of limestone is required for moderately efficient collection. This procedure also imposes a heavier solids handling load on the precipitators, and it yields a throwaway product. It does, however, produce a dry plume, which may be an advantage in some situations.

Other gas–solid containment systems, in which a significant amount of the original development work was invested by Shell, are based on copper oxide on an alumina support [47]. The absorption step of this system both oxidizes and traps the sulfur dioxide on the solid support as copper sulfate (Eq. 3.41).

Uptake:

$$CuO + SO_2 + 1/2\ O_2 \rightarrow CuSO_4 \qquad 3.41$$

Regeneration by methane reduction is conducted in a separate unit returning sulfur dioxide, but now in more useful high concentrations (Eq. 3.42).

Regeneration:

$$2\ CuSO_4 + CH_4 \rightarrow 2\ Cu + CO + 2\ H_2O + 2\ SO_2 \uparrow \qquad 3.42$$

The finely divided copper on the solid support is rapidly reoxidized to cupric oxide when it comes into contact with the hot flue gases on its return to the absorption unit (Eq. 3.43).

$$Cu + 1/2\ O_2(\text{hot flue gas}) \rightarrow CuO \qquad 3.43$$

Ferric oxide on alumina has also been tested in a very similar process.

A final gas–solid interaction process involves chemical reaction for sulfur dioxide containment. A bed of crushed coal kept at 650–815°C is used to reduce a stream of concentrated sulfur dioxide passed through it, to elemental sulfur and carbon dioxide (Eq. 3.44).

$$SO_2 + C \rightarrow CO_2 + S \qquad 3.44$$

3.6.5. Control of Nitrogen Oxide Emissions

Discharges of nitrogen oxides in the U.S.A. in 1981 totaled 19.5 million metric tonnes, only slightly less than the total discharges of sulfur dioxide for the same year. Stationary combustion sources, however, were still the most significant contributor to nitrogen oxides emission with a total of 10.2 million tonnes, followed by the transportation sector, with 8.4 million tonnes, and industrial sources with about 950,000 tonnes.

The primary and secondary ambient air quality standards for nitrogen oxides are somewhat higher than for sulfur dioxide because the needs for control measures are driven by very similar requirements, namely, health effects and their influence on the pH of precipitation (Table 3.7). The significant role of nitrogen oxides (NO_x) in the chemistry of photochemical smog episodes is a further factor used to determine ambient air quality standards.

The mechanism of formation of NO_x in combustion processes is a composite of the relative concentrations of nitrogen and oxygen present in the process and the combustion intensity (peak temperatures). Thus, small space heaters operating at lower temperatures typically produce combustion gases containing about 50 ppm NO_x, while large power plant boilers without abatement measures can produce flue gas concentrations as high as 1500 ppm [48]. The formation reactions involve a combination of atoms of oxygen with molecular nitrogen and atoms of nitrogen with molecular oxygen. Both elements have finite concentrations present in atomic form at high combustion temperatures (Eqs. 3.1 and 3.2). The higher concentration of atomic species present at high temperatures rather than low, and the very rapid reaction rates at high temperatures cause the equilibrium of these reactions to be established rapidly. A small further contribution to the nitrogen oxides in exhaust gases can come from the oxidation of nitrogen-containing organics (combined nitrogen) in the fuel.

To decrease nitrogen oxide emissions a single or a combination of measures may be used. In utility applications one of the simplest measures is to match combustion rate to load requirements which also serves to reduce average fuel combustion and hence the costs of power generation. Fuel combustion at any rate greater than the load requirement produces excessive boiler and flue gas temperatures and raises nitrogen dioxide concentrations.

More fundamental equipment modifications are necessary to adopt 2-stage combustion or larger flame volume as means of decreasing NO_x formation. Tests of systems which use a fuel-rich combustion core to produce reducing gases which partially reduce the nitrogen oxides which simultaneously form near the fuel-lean walls to nitrogen are also yielding promising results [49] (e.g., Eqs. 3.5–3.8 and 3.45).

$$2\ NO_2 + CH_4 \rightarrow N_2 + 2\ H_2O + CO_2 \qquad\qquad 3.45$$

Reducing the excess air normally used in combustion processes can also assist in decreasing nitrogen oxide formation. This requires continuous monitoring of the oxygen content of flue gases using oxygen analyzers coupled to automatic controls or manual reset alarms to ensure that the excess air stays within predetermined limits of the theoretical requirement. To change from a 15% excess combustion air to a 2–3% excess can alone decrease nitrogen oxide emissions by some 60%, and at the same time increase energy recovery efficiency.

Other process changes can also decrease NO_x emissions by decreasing its formation. Among these is recirculation of a fraction of the flue gas to decrease peak combustion temperatures, a measure which has also had some success in decreasing automotive NO_x emissions. Use of pure or enriched oxygen instead of air for combustion can also help when used with flue gas recirculation, by decreasing the concentration of nitrogen available for oxidation. Use of fuel cells for power generation could eliminate utility-originated NO_x emission, but the pure gas requirement for present fuel cell technology makes this option prohibitively expensive except in special circumstances [50].

Postformation nitrogen oxide emission control measures include selective catalytic and noncatalytic reduction with ammonia, which between them are used by some 900 power station installations worldwide [51]. The catalytic removal methods are 70–90% efficient at NO_x removal, but are more expensive to operate than the noncatalytic methods which are 30–80% efficient. Ammonia or methane noncatalytic reduction of NO_x to elemental nitrogen is also an effective method which is cost-effective for high concentration sources such as nitric acid plants (Chap. 11). NO_x capture in packed beds is less expensive, but this method is not particularly effective [23]. It is also not a very practical method either for utilities or for transportation sources. Two-stage scrubbing has also been proposed as an effective end-of-pipe NO_x control measure. The first stage uses water alone and the second uses aqueous urea.

3.6.6. Carbon Dioxide Emission Abatement

Concern about current levels of carbon dioxide discharges stems from its substantial contribution to global warming as one of the greenhouse gases, and not from toxicity or acid rain considerations. Let us put this contribution into perspective. Current climate change models indicate that about two-thirds of the greenhouse effect is from the presence of moisture in the atmosphere, both in vapor form and as clouds. About one-third of the effect is from all the other greenhouse gases and about one-half of this, or one-sixth of the total, is from carbon dioxide. So it is possible that factors easier to control than carbon dioxide could have a greater effect [52].

It has been estimated that the total greenhouse warming effect from water vapor and the suite of gases of Table 2.7 raises the global average temperature by about 30°C. Without most of this warming effect the earth would be inhospitable. A drop of average temperatures by as little as 5–6°C has been estimated was sufficient to bring on the ice ages. Therefore, the earth's biosphere is reliant on a rather delicate energy balance for mean temperatures as we know them. Venus, for example, has a very different atmosphere of mostly carbon dioxide and nitrogen at 92 bars (ca. 90 atm), and devoid of water vapor. It is also shrouded in sulfuric acid clouds. These conditions give Venus about 100 times the greenhouse warming effect of the earth producing a mean surface temperature of 460°C (735 K) [53].

The earth's atmospheric reservoir of gases contains about 2.6 trillion (2.6×10^{12}) tonnes of carbon dioxide. The total annual flux of carbon dioxide into this reservoir is about 0.61 trillion tonnes, of which human activities contribute about 29 billion tonnes or 4.7% of the total. The rate of increase in concentration of atmospheric carbon dioxide has been 1.2 ppm/yr for a number of years now. For this rate of increase, the annual amount of carbon dioxide required would be 7.7×10^9 tonnes/year (1.2 ppm/year $\times 2.6 \times 10^{12}$ tonnes $\div 360$ ppm current atmospheric concentration). What these approximations tell us is that 27% (7.7×10^9 t/yr $\div$ 29×10^9 t/yr $\times 100$) of our present annual discharges of carbon dioxide would have to be curtailed, other things being equal, just to keep the atmospheric concentration of carbon dioxide constant at its current level of 370 ppm. Since natural sources annually contribute 20 times as much carbon dioxide as do human activities, a relatively minor perturbation of the natural sources may augment or defeat any changes in atmospheric concentration of carbon dioxide expected as a result of human curtailment activity.

Most of the human contribution of carbon dioxide to the atmosphere is from fossil fuel and biomass combustion. Since this combustion is primarily for energy production, all forms of energy conservation would help. Better insulated buildings and housing, efficient mass transit, living close to one's workplace, increased efficiency of chemical processes, etc. could all contribute to this. Switching from carbon dioxide intensive fuels to less carbon dioxide intensive fuels can also help. Replacing coal by natural gas roughly halves the carbon dioxide produced per unit of energy (Table 3.9). The hydrogen listing here may be a bit misleading since in this role it is only acting as a medium of energy transfer, like electricity. While hydrogen combustion itself does not produce carbon dioxide, most current methods of producing it are carbon dioxide intensive (Chap. 11). Increased use of all forms of geothermal, tidal, hydroelectric, and solar (wind, photovoltaic, etc.) energy can obviously help, although each of these alternatives requires fossil fuels for construction. Nuclear power could help since it produces carbon dioxide-free energy (neglecting the energy consumption of uranium mining, processing, reactor construction, etc.) but has accompanying problems.

Another approach which could mediate the effects of continued energy generation via the combustion of fossil fuels would be to artificially store, or

TABLE 3.9 Approximate Energy Equivalencies and Carbon Dioxide Production from Combustion of Different Fuels[a]

Fuel	BTU/lb	kWh/kg	kgCO$_2$/kg fuel	kgCO$_2$/kWh
Coal	13,000	8.39	3.22[b]	0.38[b]
Furnace oil				
No. 5 equiv.	18,800	12.14	3.12	0.26
No. 2 equiv.	19,200	12.40	3.11	0.25
Gasoline	19,700	12.72	3.08	0.24
Propane	21,500	13.88	3.00	0.22
Natural gas	23,900	15.43	2.74	0.18
Methanol	9,500	6.13	1.37	0.22
Hydrogen	61,100	39.44	0.0[c]	0.0[c]

[a]Calculated from the known energy content and the stoichiometry of the respective combustion reactions.

[b]Also with a mean of 0.04 kg SO$_2$/kg fuel, or 0.005 kg SO$_2$/kWh energy.

[c]See text for explanation.

sequester at least a part of the carbon dioxide produced to offset the anthropogenic contribution to the atmosphere [54]. More contrived methods for sequestration of carbon dioxide, such as its injection into the ocean as "dry ice torpedoes" [55] or piped release and absorption deep into the sea [56], and storage in deep saline aquifers [57], have also been suggested. Thermodynamic considerations would tend to suggest that processes to do this would decrease ultimate energy recovery. This was confirmed by a detailed engineering study of different types of power plants which found that carbon dioxide sequestration reduced the efficiency of the power plants by 6–13%, and increased the production cost of electricity by 25–72% [58].

Another approach is to increase carbon dioxide uptake by forests to reverse the effects of severe deforestation of the last 150 years. It has been estimated that a rapidly growing rainforest can remove 4–7 kg/m^2 year of carbon dioxide from the atmosphere, as compared to a typical crop uptake of 0.8–1.6 kg/m^2 year. Thus, vigorous reforestation could assist in increasing the photosynthetic removal of carbon dioxide from the atmosphere [59]. Annual crops also perform photosynthetic uptake of carbon dioxide, but consumption and metabolism of the product(s) and prompt decomposition of the plant wastes promptly return the fixed carbon dioxide to the atmosphere [60].

Perhaps measures which promise to produce a relatively small favorable perturbation of the approximately 50% oceanic component of carbon dioxide uptake would have a chance of achieving a more substantial reduction in atmospheric carbon dioxide. Preliminary experiments have shown that provision of soluble iron can dramatically improve this process. Addition of iron sulfate at the 2 nM level to iron deficient but otherwise fertile regions of the oceans doubled the phytoplankton growth [61]. Great care should be taken to try to determine any possibly serious side effects before measures of this kind are undertaken on a larger scale. Promotion of an appropriate mix of these measures will help to limit the continued increase in atmospheric carbon dioxide.

3.6.7. Abatement of Methane Losses

Attention could also be directed to decreasing the discharges of the greenhouse gases more potent than CO_2, which have the potential to decrease greenhouse warming by a factor of from 21 (CH_4) to over 1000 (CFCs, etc.) times as much per molecule as the reduction of carbon dioxide discharges could achieve (Table 2.7). Methane losses to the atmosphere during the production of oil and natural gas in Alberta (Canada) alone amount to about 1.5 Mt per year, equivalent to 30 Mt per year of carbon dioxide equivalents [62, 63].

Globally over 100 billion cubic meters (67.6 Mt, at 0.676 kg/m^3 [64]) of gas are flared or vented each year, about 1% of world gas production [65, 66]. This may not seem like much, but would be enough to meet the total annual natural gas requirements of Germany and France. If we assume that half of this loss is by venting, then the annual global methane loss is the global warming equivalent of 710 Mt per year (67.6 Mt per year CH_4 × 0.5 × 21CH_4/CO_2 GW Potential) of carbon dioxide. This would add the equivalent of about 2.4%

(0.71 Gt CO_2 equiv. $\div$ 29 Gt anthropogenic CO_2 discharges/year $\times$ 100) to the global warming potential (GWP) of actual carbon dioxide discharges from human activity.

Oil and gas field methane loss control is estimated to cost less than Can\$5/t CO_2 equivalent, compared to estimates of Can\$75/tCO_2 to sequester carbon dioxide [63]. Methane "control" by flaring (burning) is preferable to venting, from the perspective of reducing emissions of greenhouse gases, because the carbon dioxide produced has a lower greenhouse warming potential than methane. But this produces a saving of a factor of only about 7.6 carbon dioxide equivalents because it produces 2.75 tCO_2/tCH_4 (Eq. 2.28).

Warming Potential from Venting vs Flaring of CH_4:

$$CH_4 \; + \; 2\,O_2 \; \rightarrow \; CO_2 \; + \; 2\,H_2O \qquad\qquad 2.28$$

Molec. wts	16	32	44	18	
Mass ratios	16	64	44	36	
Tonnes	1	4	2.75	2.25	
Relative CO_2 warming[a]		58	0	2.75	?[b]

Notes: a. Relative warming potential, $58 = 2.75 \times 21$
       b. Warming potential of questionable value, see Section 3.6.6.

Just venting the amount of methane included in the above calculation is equivalent to emitting 710 Mt CO_2 (see above). Flaring the same amount of methane emits 93 Mt CO_2 (67.6 Mt $CH_4 \times 0.5 \times 2.75$ t CO_2/t CH_4 burned). Dividing 710 Mt CO_2 by 93 Mt CO_2 gives us the ratio 7.6:1 benefit ratio. However, if the methane itself is captured and the energy of methane combustion *used*, the benefit is improved over flaring, but the improvement factor obtained in this case is less clear.

Methane losses from landfills and enteric fermentation (livestock) are of the same order of magnitude as natural gas venting, but are sources which would be more difficult to capture or control.

3.6.8. Halocarbon Loss Abatement

Loss reduction of the chlorofluorocarbons (CFCs) has been achieved at the front end by stopping production of these materials, and at the end of life by capture and reuse. Whilst these measures were introduced primarily to help preserve the ozone layer, they are also helping to reduce global warming. The atmospheric concentrations of CFCs 11 and 12 are 6 orders of magnitude less than carbon dioxide. However, they are estimated to have about 1000 times the global warming effect per molecule of carbon dioxide (Table 2.7). So each molecule of CFCs 11 and 12 that is prevented from release to the atmosphere reduces the potential global warming effect by 1000 times that of a molecule of carbon dioxide. In the early 1980s 1.5 million metric tonnes of CFCs were being produced annually. On release this would have contributed a GWP of the equivalent of about 1.5 billion metric tonnes of carbon dioxide. Fortunately production of these refrigerants has been discontinued for ozone layer protection reasons, so this is no longer adding to the GWP of the CFCs already present in the atmosphere.

Consider just one more representative example of the effect of inadvertent discharges, the perfluorocarbons tetrafluoromethane and hexafluoroethane. These byproducts of aluminum production are nonflammable, essentially nontoxic, and make no contribution to ground level smog or stratospheric ozone depletion. But they do have a GWP per molecule of 3500–8500 times that of carbon dioxide [67]. The estimated 30,000 tonnes of perfluorocarbons produced each year thus annually contribute the global warming equivalent of about 105–255 million tonnes of carbon dioxide. Control of these discharges could decrease the GWP by the equivalent of about 100 million tonnes of carbon dioxide annually. The value of this level of GWP reduction would have to be weighed against the cost of control of perfluorocarbon loss vs the equivalent carbon dioxide emission reduction or sequestration.

REVIEW QUESTIONS

1. A coal-fired power station stack of an effective height of 20 m is discharging flue gases at a constant rate containing 1500 ppm sulfur dioxide and 8% carbon dioxide, both expressed on a volume for volume basis.

    (a) For a ground level concentration of sulfur dioxide found to be 20 ppm, 2000 m away when the wind speed was 25 km/hr, what would be the value of the constant in the equation:

    $$\text{Ground level conc.} = K \cdot \frac{m}{uH^2}$$

    where m = mass rate of emission (or concentration, if invariant), u = wind velocity, and H = effective stack height?

    (b) What would be the expected concentration of carbon dioxide at the same point on the ground under the same conditions as in Part (a)?

    (c) What would be the expected ground level concentrations of both gases, 2000 m away; if the effective stack height was raised to 100 m which simultaneously raised the wind speed at the top of the stack to 35 km/hr?

2. To decrease carbon dioxide emissions to the atmosphere, it has been proposed that power generation be switched from No. 2 fuel oil (take as $C_{15}H_{32}$; 12.40 kWh/kg) to methane (CH_4; 15.43 kWh/kg) or methanol (CH_3OH; 6.13 kWh/kg).

    (a) How many kilograms of carbon dioxide would be produced on complete combustion of a kilogram of each of these fuels?

    (b) What mass of carbon dioxide would be produced per kWh of energy produced by each of these fuels?

    (c) What fuel choice(s) would produce even less carbon dioxide?

3. The national air quality objective for "suspended particulate matter" is $60\ \mu g/m^3$.

    (a) What would this correspond to in ng/L?

    (b) Can this $60\ \mu g/m^3$ be converted to ppm by volume? Explain your answer.

4. (a) Natural gas (take as 100% methane) is to be used to fuel a boiler. Taking air as exactly 1:4, mol O_2:mol N_2, what volume ratio of air to natural gas would theoretically be just sufficient to burn the methane completely?

    (b) What volume ratio of air to natural gas would be required to allow 10% excess air?

5. (a) What aspect of the presence of sulfur dioxide in flue gases makes it attractive to capture the SO_2, and if this requirement is met name any 4 useful capture products that may be produced?

    (b) The cost(s) of what raw material has to be absorbed by the SO_2 capture processes that produce a throwaway product?

6. (a) Outline any 3 methods by which the CO_2 produced by fossil fuel production may be captured (sequestered) on a large scale to decrease greenhouse warming from energy production.

    (b) What aspects of CO_2 sequestering can substantially raise the cost of energy production?

    (c) Explain why use of hydrogen as a fuel may still not help to decrease carbon dioxide emissions.

    (d) What changes *could* capitalize on the clean burning characteristics of hydrogen?

7. (a) What operating adjustments of thermal plants can help to decrease NO_x production?

    (b) Outline the details of a catalytic method of NO_x control that may be used for large scale combustion sources.

    (c) Explain the method of operation of a chemical reduction method which may be used to decrease NO_x emissions.

8. A resting adult flying in a commercial aircraft consumes 20.0 L/hr oxygen and produces 17.7 L/hr carbon dioxide, both at 20°C and 1 atm pressure.

    (a) What ventilation rate would be required to maintain a carbon dioxide concentration of 1000 ppm in this passenger's cabin space, assuming 20°C and 1 atm conditions and a carbon dioxide concentration of 350 ppm in the outside air?

    (b) For an aircraft cabin air space of 1000 L per passenger, how many air changes per hour would be required to maintain 1000 ppm by volume carbon dioxide?

    (c) What ventilation rate would be required at 30,000 ft altitude, where the captain is able to maintain conditions of 20°C and a pressure of 0.690 atm?

    (d) What happens to the oxygen consumed that does not produce carbon dioxide? (*Hint:* Consider metabolism of carbohydrates.)

FURTHER READING

American Chemical Society, "Cleaning Our Environment: A Chemical Perspective," 2nd ed. Washington, DC, 1978.

H. Brauer and Y.B.G. Varma, "Air Pollution Control Equipment." Springer-Verlag, New York, 1981.

T.E. Graedel and B.R. Allenby, "Industrial Ecology." Prentice Hall, Upper Saddle River, NJ, 1995.

Organization for Economic Cooperation and Development, "Motor Vehicle Pollution: Reduction Strategies Beyond 2010." Paris, 1995.

C.P. Straub, "Practical Handbook of Environmental Control." CRC Press, Boca Raton, Florida, 1989.

REFERENCES

1. G.J.K. Acres, *Chem. Indus. (London)*. p. 905, Nov. 16 (1974).

2. The Canadian Environment, "The Clean Air Act, Annual Report 1977–1978, Ottawa" (cited Tby M. Webb). W.B. Saunders Co., Can. Ltd., Toronto, 1980.

3. Environment Canada website data. Available from: http://www.ec.gc.ca/pdb/cacbk_e.html

4. M.T.F. Jimenez, A. Climent-Font, and J.L.S. Anton, Long Term Atmospheric Pollution Study at Madrid city (Spain). *Water Air Soil Pollut.* **142**(1–4), 243–260 (2003).

5. M. Amann, Emission Inventories, Emission Control Options...: An Overview of Recent Developments. *Water Air Soil Pollut.* **130**(1–4), 43–50 (2001).

6. R.M. Cushman, Additivity of State Inventories of Greenhouse-Gas Emissions. *Environ. Manag.* **31**(2), 292–300, Feb. (2003).

7. H.S. Stoker and S.L. Seager, "Environmental Chemistry: Air and Water Pollution," Scott Foresman, Glenview, Illinois, 1972, p. 20.

8. J.J. Mikita and E.N. Cantwell (Dupont), Exhaust Manifold Thermal Reactors—A Solution to the Automotive Emissions Problem. *68th Ann. Meet., Nat. Pet. Refiners Assoc.* San Antonio, TX, Apr. 5–8, 1970.

9. Auto Emissions Control Faces New Challenges, *Chem. Eng. News*, **58**(11), 36, Mar. 17 (1980).

10. G.J.K. Acres and B.J. Cooper, *Platinum Met. Rev.* **16**(3), 74, July (1972).

11. "Airborne Lead in Perspective." National Academy of Sciences, Washington, DC, 1972.

12. M. Chiaradia and F. Cupelin, Behaviour of Airborne Lead and Temporal Variations of Its Source Effects...*Atmos. Environ.* **34**(6), 859–971 (2000).

13. K. Habibi, *Environ. Sci. Technol.* **7**(3), 223 (1973).

14. J. Kielhorn, C. Melber, D. Keller *et al.*, Palladium—A Review of Exposure and Effects to Human Health. *Int. J. Hyg. Environ, Health*, **205**(6), 417–432 (2002).

15. M. Jacoby, Getting Auto Exhausts to Pristine. *Chem. Eng. News*, **77**(4), 36–44, Jan 25 (1999).

16. J. Daniels, *Recherche*, **11**(114), 938 (1980).

17. L.R. Babcock, Jr., *J. Air Pollut. Control Assoc.* **20**(10), 653 (1970).

18. E.G. Walther, *J. Air Pollut. Control Assoc.* **22**(5), 353 (1972).

19. S.J. Williamson, "Fundamentals of Air Pollution," Addison-Wesley, Reading, MA, 1973.

20. New Way to Cope with Sulfur Oxides, *Hydrocarbon Process.* **51**(9), 19, Sept. (1972).

21. A. Das and D.K. Sharma,...Desulfurization of Assam Coal and Its Sulfur-Rich Lithotypes by Sequential Solvent Extraction.... *Energy Sources*, **23**(8), 687–697, Oct. (2001).

22. Desulfurization Refinery Capacities, *Envir. Sci. Technol.* **7**(6), 494 (1973).

23. P.L. Magill, F.R. Holden, and C. Ackley, eds., "Air Pollution Handbook," McGraw-Hill, New York, 1956.

24. E.P. Mednikov, "Acoustic Coagulation and Precipitation of Aerosols" (transl. by C.V. Larrick), Consultants Bureau, New York, 1965.

25. A.C. Stern, ed., "Air Pollution," 2nd ed., Vol. 3. "Sources of Air Pollution and Their Control," Academic Press, New York, 1968.

26. F.A. Patty, ed., "Industrial Hygiene and Toxicology," 2nd ed., Vol. 1. Interscience, New York, 1958.

27. J.S. Chang, Next Generation Integrated Electrostatic Gas Cleaning Systems. *J. Electrostatics*, **57**(3–4), 273–291, Mar. (2003).

28. G.D. Arnold, *Chem. Indus. (London).* p. 902, Nov. 16 (1974).
29. L.A. O'Neill, *Chem. Indus. (London).* p. 464, June 5 (1976).
30. C.M. Hansen, *Ind. Eng. Chem., Prod. Res. Dev.* **16**(3), 266 (1977).
31. J. Cross, *Chem. Brit.* **17**(1), 24, Jan. (1981).
32. "National Air Quality and Emissions Trends Report, 1981," EPA 450/4-83-011. Research Triangle Park, NC, 1983.
33. Ottawa Sets Timetable for Cleanup, *Can. Chem. Process.* **58**(3), 19, Mar. (1974).
34. R. Derry, *Chem. Indus. (London).* p. 222, Mar. 19 (1977).
35. N. Kaplan and M.A. Maxwell, *Chem. Eng.* (N.Y.) **84**, 127, Oct. 17 (1977).
36. A.A. Siddiqi and J.W. Tenini, *Hydrocarbon Process.* **56**(10), 104, Oct. (1977).
37. R.B. Engdahl and H.S Rosenberg, *CHEMTECH.* **8**(2), 118, Feb. (1978).
38. Hydrogen Sulphide Removal by the Stretford Liquid Purification Process. *Chem. Indus. (London).* 883, May 19 (1962).
39. P. Bonnifay, R. Dutriau, S. Frankowiak *et al.*, *Chem. Eng. Prog.* **68**(8), 51, Aug. (1972).
40. B.H. Potter and T.L. Craig, *Chem. Eng. Prog.* **68**(8), 53, Aug. (1972).
41. F.S. Chalmers, *Hydrocarbon Process.* **53**(4), 75, Apr. (1974).
42. B. Skrbic, J. Cvejanov, and R. Paunovic, A...Gas–Liquid Equilibrium-Model for Sulfur-Dioxide Absorption in Aqueous Sodium-Citrate Solution. *Chem. Eng. Sci.* **46** (12), 3314–3317 (1991).
43. SO_2 Removal, *Chem. Eng. News*, **49**(6), 52, Feb. 8 (1971).
44. L. Chang, W. Bao, and K. Xie,...Factors Influencing Limestone Capturing Sulfur During Coal Combustion. *Energy Sources*, **23**(3), 287–293, Apr. (2001).
45. A. Srinivasan and M.W. Grutzeck, The Adsorption of SO_2 by Zeolites Synthesized from Fly Ash. *Environ. Sci. Technol.* **33**(9), 1464–1469, May 1 (1999).
46. J.H. Russell, J.I. Paige, and D. L. Paulson, Evaluation of Some Solid Oxides as Sorbents of Sulfur Dioxide. *Rep. Invest.—U.S., Bur. Mines*, **RI-7582** (1971).
47. F.M. Dautzenberg, J.E. Nader, and A.J.J. Ginneken, *Chem. Eng. Prog.* **67**(8), 86, Aug. (1971).
48. D.R. Bartz, *Calif. Air Environ.* **5**(1), 10, Fall (1974).
49. Low NO_x Burner to be Demonstrated in Utah, *Chem. Eng. News*, **59**(17), 16, Apr. 27 (1981).
50. Zeroing in on Flue Gas NO_x, *Environ. Sci. Technol.* **3**(9), 808 (1969).
51. M. Radojevic, Opportunity NO_x, *Chem. Brit.* **34**(3), 30–33, Mar. (1998).
52. Anon, Reducing Air Pollution to Limit Global Warming. *Chem. Eng. News*, **78**(34), 39, Aug. 21 (2000).
53. M.A. Bullock and D.H. Grinspoon,...Evolution of Climate on Venus. *Icarus*, **150**(1), 19–37, Mar. (2001).
54. P. Freund and W.G. Ormerod, Progress Toward Storage of Carbon Dioxide. *Energy Conv. Manag.* **38**, 199–204 (1997).
55. W. Seifritz, Methanol as the Energy Vector of a New Climate Neutral Energy System. *Int. J. Hydrogen Energy*, **14**(10), 717–726 (1989).
56. R.L. Rawls, Sequestering CO_2. *Chem. Eng. News*, **78**(38) 66–69, Sept. 18 (2000).
57. R.G. Bruant, A.J. Guswa, and M.A. Celia, Safe Storage of CO_2 in Deep Saline Aquifers. *Environ. Sci. Technol.* **36**(11), 240A–245A, June 1 (2002).
58. M. Gambini and M. Vellini, CO_2 Emission Abatement from Fossil Fuel Power Plants by Exhaust Gas Treatment. *J. Eng. Gas. Turbines Power-Trans. ASME* **125**(1), 365–373, Jan. (2003).
59. A.J. Plantinga, The Cost of Carbon Sequestration in Forests....*Crit. Rev. Environ. Sci. Technol.* **27**, 5185–5192 (1997).
60. T.O. West, O. Tristram, and W.M. Post, Soil Organic Carbon Sequestration Rates by Tillage and Crop Rotation....*Soil Sci. Soc. Amer. J.* **66**(6), 1930–1946, Nov./Dec. (2002).
61. R. Rawls, Ironing the Ocean, Iron, a Micronutrient for Marine Plants. *Chem. Eng. News*, **74**(45), 40–43, Nov. 4 (1996).
62. B.R. Peachy, Methane Emissions Reduction in Oil and Gas Operations, Paper 32, Abstr. of the 53rd Can. Chem. Engin. Conf., Hamilton, Ont., Oct. 26–29, 2003.
63. Personal communication, B.R. Peachy, New Paradigm Engineering Ltd. Edmonton, AB, Sept. 2004.

64. From US EPA, "Reference Tables and Conversions", Available at http://www.epa.gov/globalwarming/publications/emissions, April 2002.
65. U.S. Gov. Accountability Office Report GAO-04-809, "Natural Gas Flaring and Venting," July 2004, 32 p. Available at www.gao.gov/cgi-bin/getrpt? GAO-04-809.
66. Much Natural Gas Wasted..., *Chem. Eng. News*, **84**(34), 27, Aug. 23. (2004).
67. P. Zurer, Perfluorocarbons Use and Emissions May Face Restrictions. *Chem. Eng. News*, **72**(32), 16, Aug. 9 (1993).

4 WATER QUALITY MEASUREMENT

Water is H₂O, hydrogen two parts, oxygen one, but there is also a third thing, that makes it water and nobody knows what that is.
                    —D.H. Lawrence (1885–1930)

4.1. WATER QUALITY, AND SUPPLY OVERVIEW

Water is a vital commodity to industry as a process feedstock (reacting raw material), as a solvent, and for cooling purposes, just as it is to individuals for all our personal water requirements. The global supply appears to be so extensive that many people take it for granted. However, when one considers that only 3% of this total resource exists as freshwater, the concept of this as a globally limited and exhaustible resource becomes more real (Table 4.1). If one also considers that roughly three-quarters of this 3% is frozen from immediate use by the ice caps and glaciers of the world, then an appreciation of the limited extent of the available global freshwater becomes more apparent.

If the total available surface freshwater supply of some 126,000 km³, which excludes the ice caps and glaciers, was evenly distributed over the total nonfrozen land area of the Earth, it would amount to only some 1.1 m in depth. Addition of the net annual land-based precipitation, after evaporation, would add only a further 0.8 m in the depth. The combined total certainly does not represent a limitless resource to serve the agricultural, industrial, and personal needs of a world population of almost six billion.

The world's freshwater resources are also not uniformly distributed over the land surface, even in the form of lakes and water courses. This occurs partly from the uneven distribution of glaciers and their meltwaters, and partly from the wide disparities in rainfall over the Earth's surface. Total rainfalls of above 11 m are experienced in some years in the Mt. Waialeale area of Hawaii and at Cherrapunji, India, while less than 0.2 cm falls in the same period in Arica, Chile, or at Wadi Halfa in the Sudan [1]. This uneven

TABLE 4.1 Estimated Distribution of the Global Water Resource[a]

Location	Volume, 10^3 km^3	Fraction of total	Percent of fresh
Overall:			
Oceans[b]	1,319,000	97.2	–
Ice caps, glaciers	29,200	2.15	76.8
Freshwater on land or air	8,500	0.65	23.2
Total	1,358,000	100.00	100.0
Freshwater:[c]			
Antarctic ice cap	26,900	1.99	71.33
Greenland ice cap	21,200	0.15	5.57
Glaciers	210	0.02	0.56
Groundwater, to 4 km depth	8,360	0.62	22.17
Freshwater lakes	125	0.009	0.33
Rivers, average	1.25	0.0001	0.003
Atmosphere, average	13.0	0.001	0.03
Total	37,709	2.8	100.0
Precipitation, annual:			
Over oceans	320		
Over land area	100		
(less evaporation)	70		
Net, to land	30		

[a]Compiled from van der Leiden [1].
[b]Includes saline lakes, principally the Caspian Sea, with a water volume of 10^4 km^3.
[c]Instantaneous values. Does not consider precipitation, which balances out evaporation, etc., on a long-term basis.

distribution of source waters has already made it necessary to reuse much of the available natural supply. It has also stimulated the development of economical desalination techniques and the construction of large-scale desalination plants to purify brackish or the much larger marine sources water [2].

Multiple usage and reuse of available supplies, and sometimes poor liquid and general waste disposal practices have combined to severely degrade the quality of the surfacewaters of many parts of the world. As an example, the Cuyahoga River, which flows into Lake Erie at Cleveland, Ohio, was declared a fire hazard in 1969 from the accumulation of combustible organics floating on its surface [3]. Shortly thereafter, it actually caught fire and the resulting heat seriously damaged two steel railway bridges. Also, news reports from Mexico City in 1976 announced that the city had found it necessary to post "No Smoking" signs beside a promenade, which ran alongside the Tlalnepantla River, motivated by similar risks. The Rhine River in Europe, which flows through several different federal jurisdictions with confused responsibilities for water quality regulation, has also had its share of degradation. Such surfacewater quality trends plus a contribution from aesthetics have led to a strong growth from piped to bottled water for drinking purposes, even in the developed world.

In Europe, about 2% of the surface freshwater is in lakes and 98% is in rivers. In North America, Asia, and Africa, about 75% is in lakes and 25% in rivers. Stationary bodies of water such as the Great Lakes, the Caspian and Black Seas, Lake Baikal and Lake Victoria are also undergoing quality problems as a result of their much slower water exchange rates than rivers, even though they represent much larger volumes of water. The continuous exchange provided by rivers maximizes their ability to recover from waste discharges.

Australia, as the driest continent and partly because it is also the flattest, has different water supply problems. Long periods of drought prevail for 20 years or more over much of the interior, punctuated by brief periods of flooding. These factors contribute to the coastal location of all major Australian cities. Lake Eyre, a great salt lake of 9,300 square km, which occupies a shallow basin in the interior, a part of which lies 15 m below sea level, is usually dry. It completely fills only about twice a century, and completely dries up again within about 2 years.

The Mediterranean and Baltic Seas also have little external water exchange, causing high sensitivity to pollution. The high level of industrial activity on their shores, coupled with the waste disposal practices, have combined to severely affect their water quality [4]. The very size of the main oceans provides a large buffer capacity for waste disposal before the water quality is noticeably affected. But even these extended ultimate waste sinks are showing signs of degradation, particularly around busy ports and coastal industrial cities as well as in the regions of major shipping lanes in the open ocean [5].

With cooperation and a determined effort, it is possible to reverse the trends toward deterioration in the water quality in any of these areas. As an example, the Thames River through London, England, as recently as 1958, showed no fish at all in the 60 km stretch of estuary between Fulham and Gravesend. Only pollution-tolerant tubifex worms and eels were found to be living in this region [6]. This low-species diversity was mainly the result of the high temperatures, a high biochemical oxygen demand (BOD), and a low dissolved oxygen content. By 1982, however, the sweeping powers granted to the Thames Water Authority enabled control of most of the causes of this poor water quality, and returned dissolved oxygen concentrations to an average of 50% of saturation. Many species of fish had by this time returned to this habitat, and plans were being made to restock the river with salmon. These aspects, together with the return of abundant waterfowl, were all signs of a dramatic improvement.

Contamination of the Great Lakes in North America was also clearly recognized in the late 1960s, as was evidenced by increased nutrient and pesticide levels, and decreased catches of commercial species of fish coinciding with increased populations of coarser (more pollutant-tolerant) species [7]. But fortunately, this was recognized early enough by both the U.S.A. and Canada to enable a joint effort to bring about a noticeable improvement [8]. Full restoration of the former water quality in these lakes is likely to take longer than it took with the Thames River because of the much slower water exchange rates. Some 70 years would be required for the St. Lawrence River to drain the water of the Great Lakes system once.

4.2. WATER QUALITY CRITERIA AND THEIR MEASUREMENT

A surface water resource of very poor quality is easily recognized by sight by anyone, and frequently also by smell. The need for improvement is also self-evident. Even though these assessments are valid, they are subjective. They do not place the assessment into categories, nor quantify them to enable determination of the steps necessary to obtain improvement. Also, a body of water may appear to be "all right" to the senses but still be of poor quality for some kinds of uses. As examples, groundwater supplies from springs and wells are known, which would superficially appear pristine but which are high in toxic arsenic [9], fluorides, or sulfides [10]. Recreational lakes in southern Norway and the Adirondacks (U.S.A.) are known which have pHs of 4 and less [11]. For these reasons, it is useful to have a set of quantifiable criteria, which may be used to measure the water quality. The values obtained for each criterion allows a valid quantitative assessment of the water quality of different sources to be made. They also permit identification of the appropriate action, which may be required to improve the quality of the surface water. If the surfacewater source is to be used as a municipal or industrial supply, the values obtained for these criteria also establish the complexity of the treatment measures necessary.

4.3. SPECIFYING CONCENTRATIONS IN WATER

There are six methods in use to specify the concentration of a substance in water, three of them in common use and the other three used less frequently. The simplest system, and probably the most widely used, is the weight fraction (Eq. 4.1). Using the weight fraction, or, in dilute solutions,

$$\text{Weight fraction} = \frac{w_A}{(w_A + w_B)} \qquad 4.1$$

approximately $= w_A/w_B$ (i.e., when w_A is small), or mass fraction gives a dimensionless concentration term if the same mass units are used for the numerator and denominator. Typical values expressed in these terms are given in Table 4.2.

Another common system used to specify aqueous concentrations is the weight of constituent(s) per volume of solution, for which common volume units are liters (L), or cubic meters (m^3). Again, for dilute solutions or suspensions in water, which have a density under ambient conditions of approximately 1 kg/L, these units would result in the same approximations as given in Table 4.2.

Molarity is a third common unit used for aqueous solutions (Eq. 4.2).

$$\text{Molarity} = M = \frac{(\text{moles of substance})}{(\text{L of solution})} \qquad 4.2$$

Molarity it is only useful when the compositions of the constituents in water are known, since to evaluate molarity requires dividing the mass of each constituent in grams (per liter of solution) by the respective molecular weights. This system is, therefore, more complicated to evaluate but the results can be

■ **TABLE 4.2** **Common Weight Fraction Units for Specifying Concentrations in Water**

Unit	Abbreviation	Factor times wt. fraction	Approximations[a]
Percent	%	100	
Parts per thousand	‰	1000	g/L
Parts per million	ppm	10^6	mg/L
Parts per billion[b]	ppb	10^9	μg/L
Parts per trillion[b]	ppt	10^{12}	ng/L
Parts per quadrillion[b]	ppq	10^{15}	pg/L

[a]Approximation is closer than experimental error for very dilute (<1 ppm) solutions.
[b]These terms may have different exponents for European data (e.g., ppb factor, a million million) is 10^{12}, ppt (a million million million) is 10^{18}, and ppq (million[4]) is 10^{24}.

more useful. Increased value is obtained, for example, because reacting ratios of solutions for water or wastewater treatment can be more readily calculated from molarity information than from weight fraction information.

Less commonly used systems are the volume fraction, mole fraction, and molality (Eqs. 4.3–4.5).

$$\text{Volume fraction} = \frac{V_A}{V_{(solution)}} \qquad 4.3$$

A common difficulty is that $V_{(solution)} \neq V_A + V_B$.

$$\text{Mole fraction of B} = \frac{\text{moles B}}{\text{moles A} + \text{moles B}} \qquad 4.4$$

$$\text{Molarity} = \frac{\text{moles A}}{1000\,\text{g of solvent}} \qquad 4.5$$

Each of these systems has special applications which require their use (e.g., mole fraction in heats of mixing calculations, and molality for osmotic pressure calculations). But none of these is in common use for the evaluation or remediation of water pollution problems.

4.4. SUSPENDED SOLIDS

Insoluble matter in surfacewaters may be partly settleable, consisting of fairly large particles such as fine sand, or they may be nonsettleable, falling into the colloidal size range of silts and clays. The dividing line by diameter between these two classifications differs with the density of the particle. For spherical particles of density $2\,\text{g/cm}^3$, diameters larger than about 100 μm (0.10 mm) settle out more or less quickly; diameters smaller than this are slower to settle (Table 4.3). The suspended solids content of a water sample may be determined by filtering an appropriate measured volume through a tared, fine glass fiber filter mat, and then rinsing any dissolved salts from the mat with a small portion of distilled water. Very fine sintered glass filters may also be used, but paper filters perform poorly in this application [13]. The solids collected are

TABLE 4.3   Terminal Settling Velocities Versus Diameter of Sphere of Specific Gravity 2, in Water at 25°C[a]

Particle diameter (μm)	Terminal velocity (mm/sec)	Characteristic description[b]
1000	100	coarse sand
500	60	coarse sand
200	30	coarse/fine sand
100	6	fine sand
50	1.5	fine sand
20	0.2	fine sand/silt
10	0.06	silt
5	1.5×10^{-3}	silt
2	2.2×10^{-4}	silt/clay
1	6.0×10^{-5}	clay

[a]Settling velocities estimated from Weast [12].
[b]Approximate dividing lines indicated by joint descriptions.

dried at a standard temperature, usually 100–105°C. The dry weight thus obtained is then related to the original volume filtered for give the result. Depending on the pore size of the filter medium used, this method gives a result, which includes the settleable and much of the nonsettleable suspended solids present in the sample. The size distribution of the collected particles may be determined by passage of a resuspension of the particles in a liquid through a Colter counter [14].

Centrifugation may also be used for suspended solids determination using the conical end centrifuge tubes, which can be graduated for convenient volume determination. After centrifuging the sample, the supernatant water is decanted from the precipitate, and the precipitate is rinsed with a small portion of distilled water to remove any dissolved salts. Centrifuging again, decanting, and drying the residue as before gives the suspended solids result. A smaller proportion of the nonsettleable suspended solid is retained by this method rather than by filtration, a factor which has to be considered when comparing results obtained by the two systems.

An Imhoff cone is used to determine the settleable solids content of treated waste waters to check efficiency of treatment. This cone-shaped measuring device, of 1L capacity and made of transparent glass or plastic, is graduated down the side to units of tenths of a milliliter at its lower apex. The sample is placed in the cone and allowed to stand for a period of 1 hr. Then the volume of the solids layer settled is read directly from the graduations on the side.

In situ turbidity determinations using a Secchi disk are related to, but do not strictly correspond to suspended solids determinations [15]. The disk is a circular plate of about 20 cm in diameter made of sheet steel or other metal, and is painted with alternating black and white quadrants on its upper face. A rope or chain calibrated in meters is attached to the disk via an eyebolt at the center of the disk, so that the disk hangs horizontally. For a reading, the disk is lowered into the water until the black and white painted quadrants appear to be uniformly gray due to the turbidity of the water, and the depth at which this occurs is recorded. It is then lowered a few meters more and then gradually raised again, noting the depth at which the black and white quadrants just become discernible again. The average of these two readings gives the Secchi depth in meters. This can range from 0.2 m or less for a silt-laden river in spring flood, to 25 m or more for an oligotrophic (geologically young) pristine mountain lake. An estimated extinction coefficient may be obtained by dividing 1.7 by the Secchi depth in meters (k = 1.7/Secchi depth), or by using a submersible light meter to make actual measurements at two depths.

4.5. DISSOLVED SOLIDS

Water quality and monitoring programs also have an interest in the dissolved solids or salts content of water systems. Among the monovalent cations, there is a particular interest in the concentrations of sodium, potassium, and ammonium ions (and neutral ammonia, which is in equilibrium). Among the polyvalent cations, calcium and magnesium are the main ones of concern because of their strong tendency to precipitate and form useless curds with natural soaps, and also form adhering deposits or scale on the walls of water-heating appliances from domestic kettles to industrial boilers. Occasionally, this category might also include iron, aluminum, or other polyvalent cations, since, if these are present, they can contribute to this tendency. This property common to polyvalent cations is referred to as hardness. Thus, hard waters, with a high polyvalent cation content, do not launder well with ordinary soaps, whereas soft waters do (Table 4.4). Soft waters also heat more cleanly in water-heating appliances without depositing scale.

Chloride, sulfate, carbonate, and fluoride are among the principal anions of interest in water treatment programs. High chloride concentrations are of concern because of their tendency to accelerate the corrosion rates of pipelines and local water distribution systems. Sulfate at concentrations above 150 ppm can cause severe digestive upset, essentially the symptoms of diarrhea, especially in nonacclimatized people. The concentration of carbonate present has important consequences in relation to hardness, about which more is said later. Fluoride concentrations of ca. 1 ppm in water supplies are beneficial in

TABLE 4.4 A Rough Guide to the Scale of Hardness of Natural Waters According to Content of Calcium Carbonate (or the Equivalent)[a]

Description	Hardness, as ppm (mg/L) of $CaCO_3$
Soft	0–50
Moderately soft	50–100
Slightly hard	100–150
Hard	200–300
Very hard	>300

[a]Data from Klein [13].

reducing dental caries. Natural concentrations much above this, which occur occasionally can give rise to toxic symptoms [16].

The concentrations of the nutrient anions of interest in water monitoring programs include nitrogen, primarily as nitrate, NO_3^-, and nitrite, NO_2^-, and phosphorus, primarily as phosphate, PO_4^{3-}. Nitrite is also of independent interest because of its high toxicity, particularly to infants [17]. Since nitrate is also reduced to nitrite in the digestive system, the presence of either ion at concentrations above 10 mg/L in a water supply is cause for concern. During 1945–1969 in the U. S., 328 cases of methemoglobinemia (blue baby syndrome) and 39 fatalities were reported for infants from this cause [18]. Both groups of ions are nutrients for the growth of algae and other aquatic plants in water systems. If the concentration of phosphate is allowed to rise above 0.015 mg/L (calculated as P) and nitrate above 0.3 mg/L (as N) in surface waters, then algal blooms (prolific algal growth) are likely to occur [19, 20].

The specific conductance of a water sample provides a simple method to determine the total dissolved ionic solids present in the sample. It is also an inexpensive technique, which lends itself to continuous monitoring of a river or waste stream for the total ion content (Fig. 4.1), and can be easily used to check the accuracy of analyses conducted for specific ions. Specific conductance is measured via a pair of carefully spaced platinum electrodes, which are placed either directly in the stream to be measured or in a sample withdrawn from it [22]. The water temperature should be 25°C, or the result corrected to this temperature. Voltages in the 12 to 14 range, and frequencies of 60 to 1000 Hz AC are used, plus a Wheatstone bridge circuit to obtain a conductivity reading in μmho/cm or μS/cm (microsiemen/cm). The response obtained is linear with the total ion content over a wide range of concentrations (Fig. 4.1). Examples of the conductance ranges and seasonal variation of some typical Canadian rivers are

FIGURE 4.1 Total dissolved solids concentration (by analysis) versus specific conductance, shown by Gila River (Bylas, Arizona) water samples taken over a 1-year period. (Reprinted from Hem [21].)

FIGURE 4.2 Distribution of the specific conductance readings for some Canadian rivers during 2-year data collection periods. (Data from Thomas [23].)

presented in Fig. 4.2. Rivers of both low- and high-dissolved solids, and with narrow or wide seasonal variations are clearly evident from these plots.

More specific and detailed information about the particular ions of interest may be obtained through the application of conventional analytical techniques. For example, complexation titration using a stable complexer such as EDTA (ethylenediaminetetraacetic acid) is useful for the determination of dissolved calcium, as well as certain other ions [24]. Since EDTA is a hexacoordinate complexer it forms very stable 1:1 complexes with many metal ions, including calcium. An 0.01 M standardized EDTA solution, usually as the disodium salt Na_2EDTA, is first made up. Exactly 1.00 mL of this solution is equivalent to 400.8 µgCa, or 1000 µgCaCO$_3$. Then titration of a sample containing calcium ion in the presence of a weakly complexing dye indicator gives displacement of the calcium from the indicator plus a corresponding change in color of the solution, which is used as the end point of the determination (Eqs. 4.6 and 4.7).

$$M^+ + dye \rightarrow M^+ \cdot dye \text{ (weak complex)} \qquad 4.6$$

$$M^+dye + EDTA \rightarrow dye + M^+ \cdot EDTA \text{ (strong complex)} \qquad 4.7$$
$$\text{colorless}$$

Murexide, or Eriochrome blue-black R are suitable indicators for this analysis. A number of other metal ions may also be determined by related complexometric titrations.

Ion selective electrodes provide a simple and accurate method for the determination of many ions in solution. These have been developed using the same electrochemical principles as the pH electrode, which is basically an ion selective electrode specific for hydrogen ion. Thus, concentrations of Na^+, K^+, Ca^{2+}, and Pb^{2+} as examples of cations, and F^-, Cl^-, Br^-, I^-, S^{2-} and CN^- as anions may all be measured using this method. With the appropriate measuring precautions and attention to possible interferences, most cations can be determined at concentrations as low as 10^{-5} to 10^{-6} M, and lead for example down to 10^{-7} M.

The groups of nutrient ions of current concern are classed as "nitrogen," which refers to "combined nitrogen," (i.e., nitrite, nitrate, ammonia or ammonium salts, and occasionally amines), and "phosphorus," which generally refers to phosphate, PO_4^{3-}. Dissolved nitrogen gas is not of concern from a nutrient standpoint, since this is not an accessible source to most aquatic and land plants. However, blue-green algae are able to "fix" elemental nitrogen dissolved in water to obtain nutrient nitrogen. Elemental white phosphorus is only very slightly soluble in water (<0.002 mg/L) and for this reason it is also not a problem from a nutrient standpoint, although it may be a problem for other reasons (see Phossy water).

Nitrite concentrations are determined by, first, diazotization of sulfanilamide. The acidified waste sample to be analyzed is the source of the nitrous acid (Eq. 4.8).

$$H_2N\!-\!\!\bigcirc\!\!-\!SO_2NH_2 + HNO_2 + H^+ \longrightarrow 2\,H_2O + N_2^+\!-\!\!\bigcirc\!\!-\!SO_2NH_2$$

$$\underset{\text{sulfanilamide}}{} \qquad \underset{\substack{\text{nitrous}\\\text{acid}}}{} \qquad \underset{\text{diazonium salt (= I)}}{} \qquad 4.8$$

When the diazotization is conducted in the presence of N-(1-naphthyl) ethylenediamine, the reactive diazonium salt couples with the amine to form an intenselycolored, reddish purple azo compound (Eq. 4.9).

$$H_2NCH_2CH_2N(H)\!-\!\!\bigcirc\!\!-\!N\!=\!N\!-\!\!\bigcirc\!\!-\!SO_2NH_2 \qquad 4.9$$

$$\underset{\text{reddish purple}}{}$$

The concentration of azo compound obtained can be determined either colorimetrically, using a comparator, or spectrophotometrically. The nitrite-detection limit of this method is 0.02 mg/L of nitrite, specified as N, which is equivalent to 0.02 ppm or 20 ppb N in fresh waters [24].

Nitrate is determined on a separate water sample by first reducing the nitrate to nitrite with either hydrazine sulfate or with an easily oxidized metal such as cadmium (Eq. 4.10).

$$NO_3^- + Cd \rightarrow CdO + NO_2^- \qquad\qquad 4.10$$

The sensitive nitrite method is then used to determine the total nitrite, which will now be equivalent to nitrite plus the original nitrate. Subtraction of the original nitrite from this total nitrite result gives the value for the nitrate concentration.

Reduced nitrogen, such as ammonia or ammonium salts, may be analyzed by spectrophotometric determination of the absorbance of the yellow-brown colloidal suspension formed on the addition of Nessler's reagent, which is potassium mercuri–iodide (Eq. 4.11).

$$2\,K_2HgI_4 + 2\,NH_3 \rightarrow NH_2Hg_2I_3 + KI + NH_4I \qquad 4.11$$

$$\underset{\text{yellow-brown}}{}$$

If amines or amides are present, then the determination requires prior Kjeldahl decomposition of the amine to ammonium hydrogen sulfate first using boiling sulfuric acid (Eqs. 4.12 and 4.13).

$$RCH_2NH_2 + H_2O + H_2SO_4 \rightarrow RCH_2OH + NH_4HSO_4 \qquad 4.12$$

$$RCONH_2 + H_2O + H_2SO_4 \rightarrow RCOOH + NH_4HSO_4 \qquad 4.13$$

Following this, the solution is made alkaline with sodium hydroxide and the ammonia content measured using Nessler's reagent. Again, subtraction of the initial Nessler reagent reading yields an amine concentration independent of the original ammonia content.

Phosphate concentrations may be measured gravimetrically, by weighing the yellow precipitate formed on the addition of a solution of ammonium molybdate to the water sample acidified with nitric acid (Eq. 4.14) [25]. The product stoichiometry for this process is certain, although the structure is not.

$$PO_4^{3-} + 12\,(NH_4)_2MoO_4 + 24\,HNO_3 \rightarrow \qquad 4.14$$
$$(NH_4)_3PO_4 \cdot 12MoO_3 \cdot 3H_2O + 9\,H_2O + 21\,NH_4NO_3 + 3\,NO_3^-$$

If higher sensitivities are desired then the ammonium phosphomolybdate product may be reduced with stannous chloride or ascorbic acid in order to give soluble molybdenum blue [24]. Spectrophotometric determination of the blue absorbance at 885 nm gives a detection limit of 0.003 mg/L phosphate, stated as phosphorus (3 ppb in freshwater) or 0.0092 mg/L as expressed phosphate [24].

4.6. TOTAL SOLIDS OR RESIDUE ANALYSIS

Analysis of a water sample for total solids content (suspended and dissolved) requires evaporation of the water from a measured volume of sample, usually 1 L. For water samples, the residue is usually dried at 180°C, to put calcium sulfate in the anhydrous state, and magnesium sulfate as the monohydrate, $MgSO_4 \cdot H_2O$, in the residue [13]. The level of hydration of many salts varies with the final drying temperature, so this should be stated with the results.

To determine the total solids of river muds or sewage sludges, which may have a high organic content, lower drying temperatures of 100–105°C are normally used to reduce the risk of volatilization or thermal decomposition of the organics. For this reason, azeotropic drying is often convenient for water removal from a weighed amount of the mud or sludge sample. When perchloroethylene or toluene is used as the azeotroping solvent, the temperature required for water removal decreases to 88.5°C or 85°C, the boiling points of the respective water azeotropes. The water removed from the sample is read from a graduated solvent/water separator, from which the solvent layer from the condensate of the process is returned to the sample flask.

4.7. DISSOLVED OXYGEN CONTENT

Oxygen, nitrogen (as N_2), carbon dioxide, and the other trace gases of the atmosphere are in a dynamic equilibrium between the upper layers of the surface waters of the earth and the air. Gases in the surface layer of water move through (e.g., during mixing induced by river flow) or are largely restricted from moving through (e.g., in a stable, thermally-stratified lake) the water column below it. The solubility of oxygen in water is low, but its presence is necessary for the existence of most kinds of water organisms, and is essential to water quality.

The solubility of any gas in a liquid with which it does not react is proportional to the pressure of that gas (i.e., the partial pressure) above the liquid (e.g., Eq. 4.15 for oxygen).

$$P_{o_2} = k_{o_2} \cdot X_{o_2} \qquad\qquad 4.15$$

where P is the partial pressure of oxygen above the water in atmospheres, k_{o_2} is the Henry's law constant, and X_{o_2} is the mole fraction of oxygen in the water phase. This solubility relation, also known as Henry's law, is quite closely followed for pressures less than, say, an order of magnitude different from normal atmospheric pressure, and for any given temperature. Oxygen comprises about 21% (actual value 20.95%) of the atmosphere by volume; therefore it contributes 0.21 atmosphere partial pressure above surface waters at sea level. Its proportion remains at about 21% of the total to at least 50 km altitude, and the partial pressure drops in proportion to the lower pressures that exist at higher altitudes.

Gas solubility in a liquid is also inversely proportional to temperature (Eq. 4.16).

$$T = \frac{k_t}{X_{o_2}} \qquad\qquad 4.16$$

where k_t is the proportionality constant at absolute temperatures. The effect of this factor is a decrease in the solubility of oxygen in surface-waters if these are warmed, either by the sun or by the discharge of warmed wastewaters into a river or lake.

The combined influence of temperature and pressure on the solubility of some common gases important in water quality studies is given in Table 4.5. Some interesting features to note are that oxygen, with a Henry's law constant of 1.91×10^7 mol/(mol mm Hg), is more than twice as soluble in water as nitrogen. Oxygen has a very similar molar solubility to argon, a reflection of the diatomic nature of oxygen gas, which has a molecular weight of 32 g/mol, similar to the 39.95 g/mol of monatomic argon. Also evident here is the rationale for the use of compressed oxygen-helium mixtures for deep-sea diving [27]. The much lower solubility of helium in water and in blood reduces the risk of gas bubble formation in the bloodstream on the return of a deep-sea diver to ordinary atmospheric pressure. There is a risk of the "bends" if ordinary compressed air is used for diving under these circumstances, from the much higher solubility of nitrogen.

The presence of other solutes in the water also affects the solubility of oxygen or other gases in water. A high concentration of dissolved solids in the

▮ **TABLE 4.5   Henry's Law Constants for the Solubility of Some Dry Gases in Water^{a,b}**

Temperature (°C)	Henry's law constants/10^7				
	Argon	Carbon dioxide	Helium	Nitrogen	Oxygen
0	1.65	0.0555	10.0	4.09	1.91
10	2.18	0.0788	10.5	4.87	2.48
20	2.58	0.108	10.9	5.75	2.95
30	3.02	0.139	11.1	6.68	3.52
40	3.49	0.173	11.0	7.60	4.14
50	3.76	0.217	10.9	8.20	4.50

^{a}For the equation p = kX, where p = partial pressure of the gas in mm of Hg, X = mole fraction of the gas in water, and k = the Henry's law constant (e.g., 1.91×10^7 for oxygen at 0°C).
^{b}Values selected from Mason [26].

water phase decreases the solubility of oxygen. Other things being equal, therefore, marine waters tend to have a lower dissolved oxygen content than freshwaters.

The combined influence of all of these solubility factors on the actual oxygen content of ordinary surface waters is presented in Table 4.6. From this data it can be seen that the air-saturated oxygen content of seawater is normally about 20% less than the oxygen content of freshwater, at the same temperature. Also evident is the effect of elevated temperatures on oxygen solubility. At 30°C one obtains only about half the oxygen content in the water phase at saturation as at 0°C for either freshwater or saltwater. Lakes at higher altitudes, exposed to lower atmospheric pressures and hence lower oxygen partial pressures, will tend to have lower concentrations of dissolved oxygen, other things being equal. Similarly the salt lakes, which lie below sea level, such as the Dead Sea and Salton Sea, will tend to have higher concentrations of dissolved oxygen than would be predicted for similarly saline waters at sea level.

The Winkler test, or minor variations of it, is the standard wet chemical procedure for measuring the concentration of dissolved oxygen in water samples [24, 28, 29]. It uses a standard 300 mL BOD sample bottle, which is closed with a glass stopper with a polished cone-shaped end. This closure makes it possible to completely fill the bottle with the water sample and avoids any interference from bubbles of air which might otherwise be trapped in the bottle. Initially a white precipitate of manganous hydroxide is prepared in the BOD bottle (Eq. 4.17).

$$MnSO_4 + 2\ KOH \rightarrow Mn(OH)_2 \downarrow + K_2SO_4 \qquad 4.17$$
$$\text{white}$$

If there is any dissolved oxygen in the water sample being tested, it oxidizes the manganous hydroxide to a brown suspension of manganic hydroxide in the test solution (Eq. 4.18).

$$2\ Mn(OH)_2 + O_2 \rightarrow 2\ MnO(OH)_2 \downarrow \qquad 4.18$$
$$\text{brown ppt}$$

TABLE 4.6   Oxygen Content of Water in Equilibrium with Air Saturated with a Water Vapor at 760 torr (mm Hg)[a]

Temperature (°C)	Oxygen content (mg/L)		Water vapor pressure (mm Hg)
	Fresh water	Sea water, 3.5% salinity	
0	14.6	11.22	4.58
10	11.3	8.75	9.21
20	9.2	7.17	17.54
30	7.6	6.10	31.82
40	6.6	5.13	55.32
50	5.6	–	92.51

[a]Compiled from Klein [13] and American Public Health Association [24].

If no brown coloration is observed at this stage of the test, then there is no dissolved oxygen in the water sample being tested. The presence of any oxygen causes a brown coloration from quantitative formation of manganic hydroxide. The amount of manganic hydroxide is accurately determined by first putting it into solution with sulfuric acid (Eq. 4.19).

$$MnO(OH)_2 + 2\ H_2SO_4 \rightarrow Mn(SO_4)_2 + 3\ H_2O \qquad\qquad 4.19$$

To determine the original levels of oxygen present, excess aqueous potassium iodide is added, releasing iodine equivalent to the manganic sulfate present in the most rapid reaction of this sequence (Eq. 4.20).

$$Mn(SO_4)_2 + 2\ KI \rightarrow MnSO_4 + K_2SO_4 + \underset{\substack{\text{brown}\\\text{solution}}}{I_2} \qquad\qquad 4.20$$

The iodine generated is prevented from being lost as vapor by ensuring that an excess of potassium iodide is present, which forms the stable ionic complex, KI_3. The liberated iodine is measured by the familiar sodium thiosulfate titration, using starch as an indicator near the end point of the titration to sharpen its observation (Eq. 4.21).

$$I_2 + 2\ S_2O_3^{2-} \rightarrow 2\ I^- + \underset{\substack{\text{colorless}\\\text{solution}}}{S_4O_6^{2-}} \qquad\qquad 4.21$$

Each mole of liberated iodine is equivalent to a half mole of dissolved oxygen in the original water sample, so that each milliter of 0.0250 M thiosulfate is equivalent to 200 µg of original dissolved oxygen content.

The Winkler test is reliable and sensitive to oxygen concentrations as low as 20 µg/L (20 ppb) for a 200 mL sample of ordinary clean surface waters. It is somewhat less accurate for sewage and industrial effluents when interfering substances may be present. It is also a primary analytical determination method, which means that it may be used to calibrate or check other instrumentation such as dissolved oxygen meters based on the oxygen electrode. The performance of these instruments has been recently compared [30]. They are generally reliable as long as they are regularly recalibrated, and operated at or near their calibration temperatures. Periodic performance checks should also be carried out [31]. Water samples of a range of known dissolved oxygen levels

FIGURE 4.3 Variation of dissolved oxygen content of pond water with aquatic plants as affected by photosynthetic activity accompanying sunshine, and metabolic oxygen consumption during dark hours. Data plotted from Klein [13].

may be readily prepared for this purpose. A 0 mg/L (ppm) test may be obtained by sparging a quantity of freshly boiled distilled water with nitrogen for 3 or 4 hr in a thermostat bath of the required test temperature. Sparging air through a water sample at 20°C for the same period would give 9.2 mg/L, and sparging pure oxygen through should give 44.0 mg/L at this temperature.

For meaningful results over time with waters having prolific weed or algae growth, care must be taken to sample or measure the dissolved oxygen *in situ* at the same time each day, and note the ambient temperature. This is because photosynthetic production of oxygen toward the end of a sunny day can almost double the dissolved oxygen content of the water to well above saturation levels, as compared to the values in the very early morning after a cloudy day (Fig. 4.3) [32]. Metabolic consumption of oxygen under these circumstances can reduce the dissolved oxygen concentration to near zero. A lack of appreciation of the diurnal variation in dissolved oxygen content from these causes has caused serious misinterpretation of the effect of waste stream discharges [33].

Dissolved oxygen concentrations in surface waters below about 5 mg/L or 50% of saturation are generally unsatisfactory for a diverse biota. In fact, game (or sport) fish require more like 60 to 70% of saturation to do well. Lower dissolved oxygen concentrations than these tend to limit the habitat to the growth and reproduction of coarser species of fish. Zero dissolved oxygen eliminates the survival of gill breathers and discourages many other species. Only organisms capable of air breathing, such as eels and certain *Tubifex* worms, can survive under these conditions.

In water, the dissolved air, or more particularly the dissolved nitrogen, can sometimes be too high for fish well-being. Even 10% supersaturated or 110%

of saturation is already sufficient to put them at some risk. When gill-breathers are exposed to waters supersaturated in air, their blood rapidly becomes supersaturated too, through gill action. The oxygen component of this supersaturation is not a problem since it is metabolically consumed to below saturation in the blood. But at 120% of saturation, the excess nitrogen in the bloodstream, which is not metabolically consumed, comes out of solution behind the membranes of the eye sockets causing "pop-eye," and also behind the membranes of the fins, tail, and mouth causing gas blisters ([34], and references cited therein). These symptoms are collectively referred to as nitrogen narcosis, and can cause acute stress or death of the fish. Exposure to 120 to 140% of saturation for any length of time is sufficient to cause symptoms to appear, together with some fish mortality. Exposure to concentrations above 140% of saturation causes high mortalities.

Air supersaturation of natural waters can occur from the warming of a cold, air-saturated stream as it flows into a shallow reservoir, or either from the sun or by its use as a source of cooling water, which amounts to thermal causes [35]. It is also a common occurrence while releasing reservoir water into a plunge basin, at the bottom of which the pressure of air bubbles entrained in the water may be briefly raised to 2 or more atmospheres from the pressure of the head of water over the bubbles. This increased pressure is sufficient to supersaturate the water, causing supersaturation problems here and some distance downstream of the plunge basin. Strangely, the air in excess of saturation does not quickly dissipate in the manner of the excess carbon dioxide in a soft drink bottle. Plunge basin supersaturation can be avoided by the use of "flip-lip" or "flip-bucket" spillway designs, which spread spilled water widely and as a fine spray so as to avoid carrying bubbles to depth.

Discharge of a relatively large volume of water containing oxidizable wastes into surface waters will result in decreased dissolved oxygen content, as oxygen is consumed by the oxidizable material. Discharge of any oxygen-consuming waste into surfacewaters always causes a decrease in the dissolved oxygen content. However, if this oxidizable waste discharge coincides with a low dissolved oxygen in the receiving waters, or the volume of the discharge is high-relative to the volume of the receiving water, then the resultant drop in oxygen concentration will be greater and the recovery of the stream to normal oxygen concentrations will be less rapid. Also, it is known that fish and other aquatic animals that are stressed by low-dissolved oxygen concentrations are more susceptible on exposure to any additional stressors such as high temperatures, the presence of toxic substances, and significant deviations from a pH of 7.

4.8. RELATIVE ACIDITY AND ALKALINITY

The pH and dissolved solids content of a water supply are important factors, which relate to the corrosivity of the supply. Corrosivity is of concern in the lifetime of municipal water distribution systems, as well as for the feed water of power boilers and process cooling water. pHs in the neutral to slightly alkaline range are generally preferred for these applications. Most aquatic creatures require a pH in the 5 to 8.5 range for optimal growth and reproduction, although they may survive for a time at pHs somewhat outside this range.

Natural waters in the acid range can arise by absorption of acidic atmospheric gases (Chap. 2) or by the accumulation of humic acids on percolation through peaty soils. They have the potential to mobilize elements of the rocks and soils, through which they flow. In limestone, dolomites, and similar rock formations, calcium, magnesium, and other elements may be dissolved, in the process increasing the pH. At the same time, these processes raise the hardness of water (e.g., Eqs. 4.22 and 4.23).

$$CaCO_3 \; + \; H_2CO_3 \rightarrow Ca(HCO_3)_2 \qquad\qquad 4.22$$
$$\text{insoluble} \qquad\qquad\qquad \text{soluble}$$

$$MgCO_3 + 2\,HNO_3 \rightarrow Mg(NO_3)_2 + H_2O + CO_2 \qquad 4.23$$
$$\text{insoluble} \qquad\qquad\quad \text{soluble}$$

Industrial liquid waste streams, in general, have the greatest potential of any of the common aqueous waste streams to influence the pH of receiving waters. Their pH sometimes differs substantially from neutral conditions and they can involve very large volumes. For example, pulp mills producing fully bleached kraft pulp generate some very acidic and some strongly alkaline waste streams by their processes. Drainage waters from coal mines and from metal mines working sulfide ores can also be quite acidic from the acid generated by the bacterial oxidation of sulfides (e.g., Eqs. 4.24–4.26).

$$2\,FeS_2 + 7\,O_2 + 2\,H_2O \rightarrow 2\,FeSO_4 + 2\,H_2SO_4 \qquad 4.24$$

$$2\,FeS_2 + 8\,O_2 + H_2O \rightarrow Fe_2(SO_4)_3 + 3\,H_2SO_4 \qquad 4.25$$

$$Fe_2(SO_4)_3 + 6\,H_2O \rightarrow 2\,Fe(OH)_3 + 3\,H_2SO_4 \qquad 4.26$$

Further industrial examples are the chemical plants producing phosphatic fertilizers or phosphoric acid, which may also have quite acidic discharges, and ammonia plants or coal-coking plants can generate quite alkaline streams.

The pH of waste streams or surfacewaters is most frequently monitored using a pH meter. Accurate calibration of the meter is required, using buffers at slightly higher and lower pHs than those of the samples to be measured. Also, the temperature calibration should be properly adjusted and the electrode given sufficient time in the sample being measured to come to thermal as well as pH equilibrium before taking a reading [36]. Using a meter allows determination of pH free from any interferences due to color, colloidal matter, coarse turbidity, free chlorine, or the presence of other oxidants or reductants. However, pH measurement of distilled water or high-purity natural waters may give inaccurate readings from too low conductivities, (i.e., when the resistance of the water sample is of the same order of magnitude as the impedance of the meter). This can be corrected by addition of a small amount of a neutral salt, such as 1 g/L potassium chloride, to raise the total ionic strength to about 0.1 M [37]. If the pH measurement is not being conducted *in situ*, it should be measured promptly after sampling to avoid errors due to gas exchange or biochemical processes that may alter the pH after collection. The pH of eutrophic waters can be highly time dependent from the influence of variable rates of metabolic activity on carbonate-bicarbonate buffer. Marine waters, and any other streams with a relatively high sodium ion content will require a sodium ion correction, especially if the measurement is being conducted at high pH.

For low-cost occasional pH measurement and for measurements in situations unfavorable to the placing of a fragile glass pH electrode, narrow-range single indicator and wide-range multiple indicator pH papers are convenient and sufficiently accurate for most purposes. Some of these include eight or more indicators and are capable of being read to 0.5 pH unit and some others to 0.3 pH unit.

4.9. TOXIC SUBSTANCES

Here we consider a number of potential water contaminants which, for want of space for details, are grouped together. Included are candidates such as toxic heavy metals, pesticides, water- and weed-treatment chemicals, radioactive particles, and the like.

The relative toxicities to fish of some appropriate examples of toxic substances are given in Table 4.7. It should be remembered that aquatic toxicity is difficult to pin down to an exact value since it depends not only on the particular species and the ages and state of health of the exposed individuals, but also on the time exposed, water temperature, simultaneous presence of other contaminants, oxygen content, prior acclimatization, hardness, and other associated factors [39]. On the whole, though, the high sensitivities of fish to the substances tabulated are the consequence of the good blood/water

TABLE 4.7 Approximate Acute Toxicities of Some Potential Water Contaminants to Freshwater Fish[a]

Substance	Approximate lethal concentration [mg/L (ppm)]
Chlorine	0.05–0.2
Chloride	ca. 6000
Copper	0.05
Cyanide	0.04–0.10
DDT	0.02–0.10
Fluorides	2.6–6.0
Malathion	13.0
Mercury, Hg^{2+}	ca. 0.01
Phenols	1–10
Pentachlorophenol[b]	5
Natural soap	10–2[b]
(in hard water)[c]	(900–1000)
Synthetic detergent	6–7
Toxaphene	0.01
Zinc, Zn^{2+}	0.15–0.60

[a]Exposure times and toxicity criteria vary. Compiled from Klein [17], Mason [26], and Ryckman *et al.* [38].

[b]Sodium oleate ($NaO_2C(CH_2)_{16}CH_3$), which is much less toxic in hard water, parenthesized value.

[c]Sodium dodecylbenzenesulfonate ($NaO_3S(CH_2)_{12}C_6H_5$).

exchange processes necessary in gill-breathing animals for their respiration, a parallel in human exposures would be our high sensitivity to many types of atmospheric contaminants.

More than 20 metals or metalloids, Al, Sb, As, Ba, Be, Bi, Cd, Co, Cu, Ce, In, Pb, Hg, Mo, Ag, Te, Tl, Sn, Ti, W, U, and Zn are significant in industrial hygiene programs [40, 41]. Therefore significant water contamination by any of these could be potentially hazardous.

Some wet chemical sample preparation such as pH or oxidation state adjustment is normally required for most metal ion determinations. Then complexometric titration using EDTA, as already mentioned, or diphenylthiocarbazone ("dithazone"; Eq. 4.27) may be used for cadmium, copper, lead, mercury, or zinc determinations.

$$Ph-N=N-C(S)-NH-NH-Ph \qquad\qquad 4.27$$

With the latter reagent, the concentration of the colored complex obtained may be measured colorimetrically, and compared to solutions made up from complexes of known concentrations of the metals of interest. This method is easily applied in the field. Or it may be determined somewhat more accurately, but at greater expense, using a spectrophotometer in the laboratory.

Other versatile and sensitive techniques for determination of metal ions in water are atomic absorption, which uses the attenuation of a beam of light of the appropriate wavelength by the atomized metal as the measure of concentration, and anodic stripping voltammetry, which is an electrochemical technique. Further details of all these procedures are available in standard texts ([20], and see also this chapter's Further Reading).

Determination of pesticide and herbicide content, or the presence of other organics in surfacewaters, usually involves a preliminary concentration step by extraction with an immiscible organic solvent such as hexane or heptane. The concentrations of these substances present are often so low that special highly purified "pesticide" grades of solvent are required to avoid analytical interference from contamination of the sample by traces of pesticide already present in the solvent. After extraction, the small volume of organic phase obtained is then dried and carefully concentrated. Analysis is often by injection of a small sample into a highly sensitive gas chromatograph (e.g., [42]). A series of peaks of differing retention times is obtained on a strip chart. From this, probable identities of the compounds represented by each peak may be frequently established by comparison with the retention times for a solution of standard reference materials separately injected into the chromatograph. More unequivocal and rapid peak identification may be obtained from a mass spectrometer coupled to the outlet of the gas chromatograph. The mass spectrometer can be used to establish accurate masses and fragmentation patterns of the constituents of each of the peaks obtained on the chromatograph trace. This is often sufficient to identify each material.

Radioactive particles are generally present in surfacewaters at such low levels that a preliminary concentration of a large sample is necessary for determination. Then, gross counting of the concentrated sample is carried out to give an overall measure of the radionuclide content. The result obtained can then be related back to the original sample volume. Related to this, it should be mentioned that deliberate radiolabeling at trace levels can provide a safe and very sensitive tracer for tracking of water flow and for determining the dispersion patterns of waste discharges [43].

4.10. MICROORGANISMS

Knowledge of the types and populations of microorganisms present in water is an important part of any water quality considerations. For this purpose, microorganisms can be conveniently grouped into two main classes. The indigenous, or naturally-occurring organisms, such as ambient types of bacteria, phytoplankton and zooplankton, and algae and diatoms, comprise one of these. This class is relatively harmless in water supplies. The second class are the bacteria, viruses, and parasites, which can arise in surface water from the excreta of a number of warm-blooded animals or from human sewage contamination. This group includes examples, which are harmless (i.e., they form a component of the normal human intestinal flora), and some examples which are pathogenic (disease-causing), such as typhoid fever. Detection of the presence of any of this second group of organisms in water supplies should be taken as a warning of the risk of contamination by pathogens.

A survey of the microorganism status of a water sample can be taken by means of a standard plate count. To do this, a portion of the sample itself plus several dilutions are inoculated onto a nutrient agar medium in separate petri dishes, which are shallow plates with loosely-fitted lids. The spotted dishes are then incubated for a period of time, after which the spots of growth in each are counted. Very high microorganism numbers in the water sampled tend to cause a merging of spots, making accurate counting impossible, and agglomerations or clumping of organisms tends to give lower counts. Both problems are minimized by a suitable number of sample dilutions before inoculation of aliquots of each dilution onto the plates.

Since it is the degree of contamination by the bacteria of warm-blooded animals that is most significant from a water quality standpoint, this is measured by means of the differential ratio test [44] (Eq. 4.28).

$$\text{Differential ratio test} = \frac{(20°C\ count)}{(37°C\ count)} \qquad 4.28$$

If > 10, reasonably good supply
If < 10, indicative of contamination

The count at 20°C is taken for a 48-hr incubation of the plates at this temperature. The 37°C count is taken for a 24-hr incubation at this temperature, to approximate the propagation conditions provided by warm-blooded animals. If the ratio obtained by the differential ratio test is above 10, it is taken as an indication of a reasonably good water supply.

Use of different nutrients in the petri dish, as well as different incubation conditions and various staining and slide-making techniques, permits positive identification of the particular micro-organisms collected [45]. Tests giving results comparable to the differential ratio test may also be carried out by using different plating out media for inoculation, or by membrane filtration techniques [46]. There is also a simple dipstick method, which may be used for more qualitative bacterial monitoring, the Coli-Count water tester. This uses a filter paper matrix already factory-impregnated with dry nutrient with a grid marked on it to facilitate counting after incubation. The dipstick is immersed into the water to be sampled, which both inoculates and activates the medium. Incubation for the required time in the sterile container provided, followed by close examination, gives a convenient indication of microbiological water quality.

Analysis of waters for viruses is more difficult because of their small size, about 10 to 300 nm average diameter. Also, viruses need susceptible living cells such as chick embryos or tissue cultures for cultivation and identification in the laboratory, which make them more difficult to work with. Nevertheless, viruses represent an important microbiological class for water and wastewater-monitoring programs since serious waterborne diseases such as polio and hepatitis are transmitted in this way.

4.11. TEMPERATURE

Higher than normal surface water temperatures are most often caused by the warmed discharges from industrial process or thermal power station cooling requirements. This may seem to be a relatively superficial parameter. However, water temperatures are related to several other important water quality parameters. Increasing the temperature of the surface water of a lake, for example, increases the rate of evaporation of water from the lake, helping to cool it (Fig. 4.4). The net water consumption caused by the increased

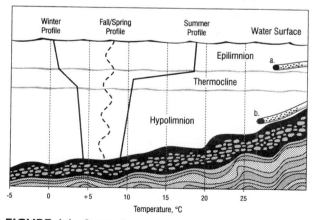

FIGURE 4.4 Seasonal, and warmed discharge effects on the change in temperature with depth of a lake. (a) Primarily surface effects resulting from a warm discharge into the epilimnion. (b) A larger mixing volume and normal temperature profile disruption obtained from a warmed discharge into the hypolimnion.

evaporation from the cooling load of a 100-MW thermal power station has been estimated to equate to the consumptive load (water demand) of a city of 100,000 people [47]. This increased evaporation also increases the concentrations of dissolved solids and nutrients in the water system [48]. These effects, in turn, can increase turbidities due to algal blooms, and nuisance weed growth as a result of the increased nutrient concentrations. Higher average year-round temperatures can also decrease the average concentrations of dissolved oxygen, from the decreased solubility of oxygen at higher water temperatures.

High thermal loadings also affect the distribution of fish populations, since coarse fish such as suckers, carp, and catfish can tolerate higher temperatures and lower dissolved oxygen concentrations better than game fish such as salmon, trout, or other commercially valuable species. Even for coarse fish, exposure to temperatures above 40°C for a brief period is usually fatal. Warm water temperatures can also favor year-round spawning of some coarser species of fish, which tends to further improve their competitive position in this habitat. These effects can cause coarse fish to gradually displace any preexisting populations of sport fish.

Temperature measurement of surface water using ordinary thermometers is relatively straightforward. Measurement at depth to determine thermal gradients is more difficult. Flip thermometers may be used, which are lowered to the depth of interest in the normal, vertical temperature-sensing position. At the depth of interest after waiting for a time for the temperature to register, the thermometer is flipped upside down. This breaks a slender portion of the column of mercury in the expansion capillary, preserving the temperature reading for later examination at the surface. Thermocouples are more robust and may be used for continuous temperature measurement at the surface or at one or more depths simultaneously. This system also lends itself to continuous automatic recording. A thermocouple uses the electrical potential generated by a pair of dissimilar metal junctions, commonly iron/constantan or chromel/alumel, to obtain a millivolt reading which can be translated to temperature using a table. Alternatively, the same sensors can give a direct temperature readout via a meter or strip chart using solid-state circuitry. The fast response of this system facilitates the taking of a series of readings while gradually lowering the sensor through a single-depth profile. This process may be used for extensive data gathering required to monitor the spread of warmed waters with distance.

Thermistors (thermal resistors) are semiconductor devices, which give a very large drop in resistance with temperature, of the order of 20,000 ohms per degree over a 500°C temperature range. These may also be used for temperature studies requiring quick response times. Aerial infrared photography (an aerial thermogram) or ERTS (Earth Resources Technology Satellites, Landsat and Seasat) studies can reveal more details more quickly for field overviews of a river or a lake receiving warmed discharges than the majority of surface methods [49]. Satellite studies can also give an instantaneous thermal picture of very large areas of water, combined with the distributions of algae and the like, in a manner not possible by any other means.

4.12. OXYGEN DEMAND

Any dissolved oxidizable material present in water places a demand on the dissolved oxygen content of the water as biochemical processes consume oxygen in order to utilize it. Therefore, the oxygen demand of surface waters or waste streams discharged into them can have a substantial effect on the concentration of dissolved oxygen. Biochemical oxygen demand (BOD), the chemical oxygen demand (COD), and the total organic carbon content (TOC) tests are used to measure this parameter, named after the measurement methods used. Each of these tests is complementary to the others in defining the oxygen demand profile of a water sample. Each also has special situations in which it is a more useful or practical measure of oxygen demand than the other tests.

4.12.1. Biochemical Oxygen Demand

The biochemical oxygen demand is a measure of the biochemically oxidizable material present in the water sample expressed in terms of the oxygen required to consume it. More precisely, BOD is defined as the number of milligrams of oxygen taken up by a 1-L sample of water on incubation in the dark for 5 days at $20°C$ [13]. For freshwater samples, 1 mg/L equates to 1 ppm (weight for weight), so that both sets of units have been in common use. To avoid potential ambiguity, however, since about 1970, the use of mg/L (or multiples of this) has been growing. In marine waters, because of density differences, it is more accurate to stick to mg/L units. The 5 subscript sometimes used with the BOD label refers to the 5-day test period, which is normally assumed. It is said that the 5-day test period came about during the development phase of this test in England, when it was assumed that a waste discharge to any English river would be deposited to the sea within 5 days. If the test period differs from this (e.g., 10 days or 20 days), then it should be stated by the subscript used, BOD_{10} or BOD_{25}. The BOD test is the most lengthy of the oxygen demand tests to perform, and the answers obtained may be difficult to interpret. However, it is also a useful test in the sense that the result comes closest to reproducing the natural oxidation and recovery conditions in a river or lake [50]. This rationale is the reason for the standard test temperature of $20°C$. The test can be performed with relatively simple, inexpensive equipment requiring little operator time, even though there is a lengthy delay before the result is obtained.

There are three possible approaches to biochemical oxygen demand measurement. The oxygen required to oxidize the organic matter in the water sample can come solely from the dissolved oxygen content of the sample itself or from added aerated dilution water, procedures which are referred to as direct and dilution techniques, respectively. Or the oxygen can come from a closed air space above the water sample to be analyzed, in which case the procedure is referred to as a manometric, or respirometer method. It is difficult to measure the small volume of gas phase oxygen consumption obtained from a low BOD water sample. For this reason, the manometric method is normally reserved for use with sewages and high oxygen demand industrial waste streams where the oxygen consumption will be large enough to obtain a valid reading.

The direct technique is appropriate for ordinary clean river waters. It requires complete filling of two 300-mL BOD bottles with the sampled water, and measurement of the oxygen content of one within 15 to 20 min of collection. The other bottle is incubated in the dark, to avoid photosynthetic contribution to the oxygen content of the water, for 5 days at 20°C. The extent of biochemical oxidation of any organics in the sample is then determined by measuring the oxygen content of the second bottle. The difference between the measured oxygen content of the first and second bottles is a direct measure of the BOD of the river sampled (Eq. 4.29).

Direct biochemical oxygen demand:

$$BOD_5, mg/L = DO_1 - DO_5, \qquad\qquad 4.29$$

where DO_1 is the dissolved oxygen content (mg/L) after 15 minutes, and DO_5 is the dissolved oxygen content after 5 days.

Since the oxygen content of freshwaters is ordinarily limited to a maximum of 9.2 mg/L (Table 4.5) and will usually be somewhat lower than this, BOD measurement by *filling* both BOD bottles with the sampled waters is limited to the measurement of BODs, which are less than the dissolved oxygen content of the water. For sewage and some types of industrial effluents, where BODs of from 100 up to 20,000 are not uncommon, the wastewater stream must be diluted with distilled aerated dilution water containing added nutrient salts before the BOD test is carried out. For a sample with an unknown BOD, two BOD bottles will be filled for each of several dilution ratios, so as to ensure that a test with at least one of the dilutions gives a consumption of about 50% of the initial dissolved oxygen content. Usually, the right types of bacteria required will be present naturally. But if not, a small amount of treated municipal sewage (stored frozen) will also be seeded into each bottle to inoculate it. Again the dissolved oxygen of one bottle of each dilution will be measured immediately and of the other after 5 days in the dark at 20°C. The BOD of the sample from the test will be calculated using Eq. 4.30, or 4.31 if a sewage seed was used.

Biochemical oxygen demand with dilution:

$$BOD_5 = \frac{(DO_1 - DO_5) \times 300\,mL}{(\text{volume of sample per bottle, mL})} \qquad\qquad 4.30$$

$$BOD_5 = \frac{(DO_1 - DO_5 - \text{seed correction})\,300\,mL}{(\text{volume of sample per bottle, mL})} \qquad\qquad 4.31$$

where the seed correction $= \dfrac{(DO_1 - DO_5)}{(\text{mL of seed per bottle})},$

which is determined in a separate test using only the seed plus dilution water in a BOD bottle.

Various refinements to these direct measurement techniques have been proposed. Substances such as glucose or potassium hydrogen phthalate, which are completely oxidized during the normal 5-day period of the BOD test, may be used as test substances to check the experimental technique of the BOD method used [24] (Eqs. 4.32 and 4.33).

Dissolved oxygen (% of saturation)	River pollution status[b]	BOD loading (mg/L)[c]
90 or more	very clean	1 or less
ca. 90	Clean	2
75–90	fairly clean	3
50–75	moderately polluted	5
25–50	heavily polluted	10 to 20
<25	severely polluted	20 or more

TABLE 4.8 Classification of River Quality Based on Dissolved Oxygen Content, and Independently Based on Biochemical Oxygen Demand[a]

[a]Compiled from Klein [13, 17], and Woodiwiss [51].

[b]Oxygen content of the river water is a result independent from the BOD result, even though it may be the first part of the two steps required for BOD measurement. While the two readings are not necessarily connected, frequently there will be a low oxygen saturation level coinciding with high BOD occurrences.

[c]Whole-river BOD loadings of at least 70 mg/L have been recorded for the Trent River [51], and exceeding 450 mg/L for the River Irwell (Radcliffe, U.K.) [17].

$$C_6H_{12}O_6 + 6\ O_2 \rightarrow 6\ CO_2 + 6\ H_2O \qquad 4.32$$

$$C_8H_5O_4K\ (= Ph(CO_2H)(CO_2K)) + 15/2\ O_2 \rightarrow$$
$$8\ CO_2 + 2\ H_2O + KOH \qquad 4.33$$

Parallel classifications of river water quality based on their BOD loadings and independently based on their dissolved oxygen content have been proposed (Table 4.8). As a rough rule of thumb intended to maintain river quality a guideline has been suggested that no discharge to a river should bring the BOD of the river to more than 4 mg/L.

The manometric method of BOD determination measures the volume of oxygen uptake by a measured volume of the undiluted sample, placed in a brown glass bottle to prevent any photosynthetic influence (Fig. 4.5). After filling, the system is closed and the mercury manometer is set to a zero reading. Stirring is continued during the test to facilitate gas transfer across the air-water interface while bacterial attack of the organic materials present consumes oxygen and produces carbon dioxide (Eq. 4.34).

$$C(organics) + O_2 \rightarrow CO_2 \qquad 4.34$$

However, there would be little or no gas volume change without some means of taking up carbon dioxide as it is formed. This is accomplished by a wick, moistened with aqueous potassium hydroxide and placed in a stainless steel cup above the surface of the liquid (Eq. 4.35). The net oxygen uptake is read directly from a mercury manometer.

$$CO_2 + 2\ KOH \rightarrow K_2CO_3 + H_2O \qquad 4.35$$

This procedure makes it easy to take intermediate readings during the test to obtain a profile of the rate of oxygen consumption during the 5 days, or for a longer period if desired. The manometer reading obtained after the elapse of 5 days' stirring is the BOD for the sample tested. Samples of a higher BOD

FIGURE 4.5 Diagram of a manometric apparatus for determination of biochemical oxygen demand [52]. (Courtesy of the Hach Company, Ames, IA.)

than the range of the manometer would require dilution before loading the apparatus, to obtain a valid result.

4.12.2. Chemical Oxygen Demand

In contrast to the BOD test, which is designed to measure the oxygen demand of only the biochemically oxidizable carbon compounds present in the sample, the COD test gives a close measure of the total oxygen demand of the sample. It is also a practical method to obtain a valid oxygen demand result when the sample contains toxic substances, which cannot be easily neutralized. A standard BOD test conducted under these conditions would give a low or zero result from the toxicant action on the microorganisms, even with biochemically oxidizable material present.

If strong chemical oxidants such as potassium permanganate [53] or potassium dichromate [24] are used under strongly acidic conditions, both the easily oxidized and the more resistant organic materials are converted to carbon dioxide and water (Eq. 4.36).

$$C_nH_aO_b + c\ Cr_2O_7^{2-} + 8c\ H^+ \rightarrow n\ CO_2 + [(a + 8c)/2]\ H_2O + 2c\ Cr^{3+}$$

$$4.36$$

where $c = 2/3\ n + a/6 - b/3$

During the process, any inorganic salts present are also converted to the oxidized forms.

If we use formaldehyde as an example of an oxidizable substrate, the chemistry of what happens by dichromate oxidation is given by Eq. 4.37.

However, the final result is specified as the mg/L of oxygen, which would be consumed equivalent to the amount of chemical oxidant required, (i.e., each mole of formaldehyde would consume one mole of oxygen) (Eq. 4.38).

$$3\,CH_2O + 2\,Cr_2O_7^{2-} + 16\,H^+ \rightarrow 3\,CO_2 + 11\,H_2O + 4\,Cr^{3+} \qquad 4.37$$

$$3\,CH_2O + 3\,O_2 \rightarrow 3\,CO_2 + 3\,H_2O \qquad 4.38$$

Under the strong oxidizing conditions of the test, many organic compounds give 95 to 100% of the theoretical oxygen consumption, although some of the more stable aromatics such as benzene and toluene, give lower oxidant consumptions than this. A micro semiautomated method of COD determination of surface and wastewaters has been developed to process large numbers of samples through this test.

4.12.3. Total Organic Carbon And Oxygen Demand Estimation

Total organic carbon determination of water samples is an instrumental method originally developed by the Dow Chemical Company [54] and sold commercially by Beckman under license from Dow. Either normal surface-water samples or waste streams high in organics may be analyzed, and samples of only $10-30\,\mu L$ are required. The result may be used to estimate oxygen demand from the TOC reading [55].

The procedure involves, first, acidification of the bulk sample to pH 2 with hydrochloric acid, and then sparging of the sample with nitrogen or helium to remove any carbon dioxide or any other inorganic carbon present (Eqs. 4.39 and 4.40).

$$Na_2CO_3 + 2\,HCl \rightarrow 2\,NaCl + H_2CO_3 \qquad 4.39$$

$$H_2CO_3 \rightarrow H_2O + CO_2 \downarrow \qquad 4.40$$

A $10-30\,\mu L$ sample is then injected into the instrument. Initially, the water and any volatile organic carbon compounds present are vaporized at 200°C. The volatilized carbon compounds are first completely oxidized to carbon dioxide, and then all the carbon dioxide is catalytically converted to methane, in a hydrogen atmosphere (Eq. 4.41).

$$CO_2 + 4\,H_2 \xrightarrow[\text{Ni}]{350°C} CH_4 + 2\,H_2O \qquad 4.41$$

The final step to obtain a volatile organic carbon reading is a quantitative analysis of the methane produced in a gas chromatograph using a sensitive flame ionization detector, which is unaffected by the water present. This instrumental sequence gives a very good linear response to the quantity of methane produced, which is the reason for the initial conversion of a mixture of organic compounds, which would have a different molar detector responses, to methane which gives only a single quantitative response.

■ **TABLE 4.9 Comparison of the Experimental and Theoretical Oxygen Demand with Estimates from Total Organic Carbon Reading for Some Representative Compounds in Water**[a]

| Compound | BOD (g O_2/g compound) | | COD (g O_2/g compound) | | TOC | |
	5 day	20 day	$K_2Cr_2O_7$[b]	Theoretical	g C/g comp'd	g O_2/g comp'd[c]
Acetone	0.8	1.6	1.6–2.2	2.2	0.62	1.65
Adipic acid	0.6	1.2	1.3	1.42	0.49	1.31
Benzene	0.3–1.0	1.5–2.0	1.9	3.08	0.97	2.59
Chloroform	0	0	0.02	0.34	0.10	0.268
Ethanol	0.8–1.5	1.5–1.6	1.8–2.0	2.08	0.52	1.39
Heptane	1.9	-	0.06–0.2	3.52	0.84	2.24
Methanol	0.8–1.1	1.3	1.4–1.5	1.50	0.38	0.99
Phenol	1.4–1.9	1.9–2.1	2.4	2.38	0.77	2.04
Styrene	0.5–1.5	1.7–2.4	2.9	3.08	0.92	2.46

[a]Specified as g/g to assist understanding (i.e., a pure ratio, or dimensionless). Data compiled from Verschueren [56], and calculated from first principles.

[b]Or $KMnO_4$. Either may be used as the chemical oxidant.

[c]That is, an estimate of the oxygen demand based on the TOC reading. Compare with the theoretical COD readings.

The TOC instrument also has provision for the analysis of refractory organic carbon, the less volatile and more oxidation resistant forms of carbon, which may be present. To do this, a vaporization and oxidation mode at 850°C is used, which gives a separate reading. The total organic carbon reading for the sample is the sum of the two-component readings (Eq. 4.42).

$$TOC = VOC + ROC, \qquad 4.42$$

where VOC is the volatile organic carbon and ROC is the refractory organic carbon. The result is specified in mg/L carbon. This may be readily translated into an estimate of oxygen demand for the carbon content *only*, by calculating the oxygen consumption required to convert this to carbon dioxide (i.e., by multiplying the TOC reading by $32.00 \div 12.01$) (Eq. 4.43).

$$\begin{array}{ccc} C & + \ O_2 & \rightarrow \ CO_2 \\ 12.01 & 32.00 & 44.01 \end{array} \qquad 4.43$$

Each of these measures of oxygen demand will give somewhat different results because of the differences in the analytical methods used (Table 4.9). But each can be generally related to the others by a factor, depending on the particular kind of waste analyzed. Thus, the COD for ordinary domestic sewages, for example, is frequently about twice the BOD. For industrial wastewater streams, the COD : BOD ratio could be substantially different, depending on the difficulty of oxidation of any dissolved organics. The COD reading minus the BOD reading is sometimes referred to as the refractory organic component (i.e., referring to the organics, which are more resistant to oxidation).

4.13. BIOLOGICAL INDICATORS

From the number of different species of fauna (diversity) and an appreciation of their relative sensitivity to pollutants, much can be learned about the condition of fresh or marine waters and sediments [57, 58]. Field experiments to determine the presence or absence of pollution-sensitive species of plants relative to the abundance of tolerant species can also be used [59]. They can also help to determine the bioavailability and acute toxicity of pollutants in bottom sediments.

Sessile (more-or-less stationary) organisms are sometimes preferred for biological assessments since they are unable to avoid discharged wastes. These can be sampled by scooping up portions of the bottom material for examination and counting of sensitive vs. tolerant species. Alternately, artificial substrates may be placed in the streambed for colonization and later lifted for examination for diversity in a similar fashion. Monitoring of the behaviour of tubifex worms has been suggested as a method of providing a sensitive early warning system for a wide variety of toxic substances [60]. Mussels, and macroinvertebrate communities have also been examined for their ability to serve as biological indicators [61, 62]. It is also possible to conduct these assessments at the microscopic level using diatoms, which have been employed to assess the biological condition of some large American rivers [63].

Using solely biological methods to assess water quality, if properly quantified against appropriate control stations or against surveys conducted prior to discharges, is said to give a superior indication than either chemical or physical data alone.

REVIEW QUESTIONS

1. What are the significant factors which influence the dissolved oxygen content of surface waters?
   (a) What are the wt/wt (ppm) and wt/vol (mg/L) oxygen concentrations for surface waters, which contain 1.50×10^{-4} M dissolved oxygen?
   (b) Is this an acceptable oxygen concentration in surface waters for the respiration requirements of aquatic life?
2. Surface water temperature changes also significantly affect a number of other water quality parameters. What are they, and in which direction is each of these affected? Explain.
3. (a) Calculate the COD (chemical oxygen demand) for a refinery aqueous waste stream containing 900 mg/L phenol (C_6H_5OH).
   (b) Assuming that phenol is fully biodegradable, what would be the BOD_5 for this waste stream?
   (c) What instrumental TOC (total organic carbon) reading (mg/L) should be obtained?

(d) What would be a good estimate of the oxygen demand based only on the TOC reading?

4. A BOD_5 test run on a sample of river water without dilution using a standard 300-mL BOD bottle gave an initial DO (dissolved oxygen) reading of 9.1 mg/L O_2 (15 min after sampling) and a final DO reading of 2.7 mg/L O_2.
(a) What BOD would these results indicate for the river water?
(b) Based on established guidelines, what appropriate comment could be made concerning the water quality of this river based on the BOD reading only?

5. A 5-mL sample of municipal sewage was placed in a standard BOD bottle (300 mL) and topped up with aerated dilution water. Initial dissolved oxygen was 4.7 mg/L, and final (after 5-day test period) was 1.3 mg/L.
(a) What was the BOD of the sewage sampled?
(b) Additional precautionary BOD runs of 1-, 2-, 10-, and 20-mL aliquots of sewage were also carried out in separate BOD bottles, each using the same aerated dilution water as given above. What final dissolved oxygen (after the test period) should be observed for each of these runs?

6. Potassium hydrogen phthalate ($C_8H_5O_4K$) has been suggested for use as a primary standard to check BOD determination procedures. A standard solution of this reagent is made up containing a weighed 50 mg/L. Assuming complete biochemical oxidation during the 5-day test period, what BOD reading (mg/L) should one obtain for this solution?

7. A wash water stream from a frozen french fries processing plant gives a TOC (total organic carbon) reading of 1.1% (by weight) as it enters the extended aeration facilities.
(a) Assuming full biodegradability of this primarily starch plus hydrolyzed starch suspension, what would be the BOD of this stream?
(b) Treated effluent from these facilities, analyzing 245 mg/L organic carbon, is to be discharged into a municipal sewerage system. What BOD [same assumptions as Part (a)] would this correspond to?
(c) The volume from this operation is to comprise 5% of the total effluent being treated by the municipal sewage works under maximum flow conditions. Is this likely to upset the sewage treatment operation? If so, why? If not, why not?

8. (a) Calculate the TOC for each of the following contaminants in aqueous solutions:
    20 mg/L benzene, 20 mg/L ethanol, 20 mg/L glycerol ($C_3H_8O_3$), and 20 mg/L glucose ($C_6H_{12}O_6$).
(b) What would be the estimated oxygen demand of each of the above solutions based only on the TOC reading?
(c) Generalize the effect of degree of oxidation of a contaminant on the oxygen demand estimate based on TOC.

FURTHER READING

M.L. Hitchman, "Measurement of Dissolved Oxygen," Wiley-Interscience, New York, 1978.

K.J.M. Kramer, ed., "Biomonitoring of Coastal Waters and Estuaries," CRC Press, Boca Raton, FL, 1994.

W. Leithe, "Analysis of Organic Pollutants in Water and Wastewater," (translated by STS, Inc.). Ann Arbor Sci. Publ., Ann Arbor, MI, 1973.

L.J. Thibodeaux, "Chemodynamics, Environmental Movement of Chemicals in Air, Water, and Soil," Wiley, New York, 1979.

REFERENCES

1. F. van der Leiden, compiler-editor, "Water Resources of the World," Water Information Center, Port Washington, NY, 1975.
2. D.G. Downing, R. Kunin, and F.X. Pollio, *Chem. Eng. Prog., Symp. Series.* **64**(90), 126 (1968).
3. G. Young and J.P. Blair, *Nat. Geogr.* **138**(6), 743 (1970).
4. Meditteranean Conference, *Chem. Eng. News,* **54**(8), 19, Feb. 23 (1976).
5. W. Marx, "The Frail Ocean," Coward McCann, New York, 1967.
6. A. Wheeler, *Science Journal*, **6**(11), 28–33, Nov. (1970).
7. Internat. Lake Erie Water Poll'n Board *et al.* "Pollution of Lake Erie, Lake Ontario, and the International Section of the St. Lawrence River," International Joint Commission, Windsor, Ontario, 1969, in 3 vols.
8. Great Lakes Water Quality Board, "1987 Report of Great Lakes Water Quality," International Joint Commission, Windsor, Mar. 1989, in 3 vols.
9. M. Sharpe, Deadly Waters Run Deep: The Global Arsenic Crisis, *J. Envir. Monitoring*, **5**(5), 81–85N (2003).
10. S.K. Gupta and K.Y. Chen, *J. Water Pollut. Control Fed.* **50**(3), 493 (1978).
11. D. O'Sullivan, *Chem. Eng. News*, **54**(25), 15, June 14 (1976).
12. R.C. Weast, ed., "Handbook of Chemistry and Physics," 56th ed. CRC Press, Cleveland, OH, 1975.
13. L. Klein, "River Pollution", Vol. 1. "Chemical Analysis," Butterworth, London, 1959.
14. G.P. Treweek and J.J. Morgan, *Envir. Sci. Technol.* **11**(7), 707 (1977).
15. K.M. Reese, Volunteers Measure Water Transparency in 38 States, *Chem. Eng. News*, **73**(31), 48, July 31 (1995).
16. D. Rose and J.R. Marier, "Environmental Fluoride," Report No. 16081, National Research Council, Ottawa, 1977.
17. L. Klein, "River Pollution, Causes and Effects," Vol. II. Butterworths, London, 1962.
18. "Cleaning Our Environment: A Chemical Perspective," 2nd ed. American Chemical Society, Washington, DC, 1978.
19. C.N. Sawyer, *J. N. Engl. Water Works Assoc.* **61**, 109 (1947).
20. C.N. Sawyer, *Sewage Ind. Wastes*, **26**, 317 (1954).
21. J.D. Hem, Study and Interpretation of the Chemical Characteristics of Natural Water. *Geol. Surv. Water-Supply Pap. (U.S.).* **1473**, Washington, DC, 1959.
22. Metrohm, "Measurement in Chemistry, 660 Conductometer," Metrohm Ltd., Herisan, Switzerland, 1990.
23. "Surface Water Quality in Canada—An Overview," Inland Waters Directorate, Water Quality Branch, Ottawa, 1977.
24. "Standard Methods for the Examination of Water and Wastewater," 19th ed. American Public Health Assoc., American Water Works Assoc., & Water Environment Fed'n., Washington, DC, 1995, and earlier editions.
25. I.D. McKelvie, D.M.W. Peat, and P.J.Worsfold, Techniques for the Quantification and Speciation of Phosphorus in Surface Waters, *Anal. Proc. Incl. Anal. Commun.* **32**, 437–445 (1995).

26. C.F. Mason, "Biology of Freshwater Pollution," Longmans, London, 1981.

27. E.A. Hemmingsen, *Science*, **167**, 1493 (1970).

28. L.W. Winkler, *Berichte Dtsch. Chem. Ges.* **21**, 2843 (1888).

29. R. Carignan, A.-M. Blais, and C. Vis, Measurement of.... Community Respiration in Oligotrophic Lakes Using the Winkler Method, *Can. J. of Fisheries & Aquatic Sci.* **55**(5) 1078–1084, May (1998).

30. R.T. Wilkin, M.S. McNeill, C.J. Adair *et al.* Field Measurement of Dissolved Oxygen: A Comparison of Methods, *Ground Water Monit. Remed.* **21**(4), 124–132, Fall (2001).

31. M.L. Hitchman, "Measurement of Dissolved Oxygen," Wiley-Interscience, New York, 1978.

32. G.J. Shroepfer, *Sewage Works J.* **14**, 1030 (1942); cited by Klein [10].

33. R. Hall, Pollution Control Problems Solved with High-tech Telemetry, *Process Industries Canada*, **70**(5), 20–22, Oct. (1986).

34. M.J.R. Clark, "Annotated Extracts of Some Papers Dealing with Various Aspects of Dissolved Atmospheric Gases," Water Resources Service, Victoria, B.C., 1977, plus Appendix 2, reference list for the above.

35. B.G. D'Aoust and M.J.R. Clark, *Trans. Am. Fish. Soc.* **109**, 708 (1980).

36. M.S. Fraut, The Effect f-Temperature on pH Measurements, *Amer. Labor.* **27**(11), 18–23, July (1995).

37. H.L. Youmans, *J. Chem. Educ.* **49**(6), 429 (1972).

38. D.W. Ryckman, A.V.S. Prabhakara Rao, and J.C. Buzzell, Jr., "Behavior of Organic Chemicals in the Aquatic Environment," Manufacturing Chemists Assoc. Washington, DC, 1966.

39. G. Koller, K. Hungerbühler, and K. Fent, Data Ranges in Aquatic Toxicity of Chemicals, *Envir. Sci. Pollut. Res.* **7**(3), 135–143 (2000).

40. G.S. Fell, *Chem. Brit.* **16**(6), 323, June 1980.

41. F.A. Patty, ed., "Industrial Hygiene and Toxicology," 2nd ed. Vol. 2. Interscience, New York, 1963.

42. F.W. Kawahara, *Envir. Sci. Technol.* **5**(3), 235 (1971).

43. K.E. White, *Chem. Brit.* **12**(12), 375, Dec. (1975).

44. P.V. Scarpino, In: "Water and Water Pollution Handbook," (L.L. Ciaccio, ed.), Vol. 2, p. 639. Dekker, New York, 1971.

45. R. Hare, "An Outline of Bacteriology and Immunity," 3rd ed. Longmans, London, 1967.

46. "Biological Analysis of Water and Wastewater," Millipore Corp., Bedford, MA, 1973.

47. T.L. Brown, "Energy and the Environment," Merrill, Columbus, OH, 1971.

48. R.J. Allan and D.J. Richards, "Effect of a Thermal Generating Station on Dissolved Solids and Heavy Metals in a Prairie Reservoir," Inland Waters Directorate, Regina, Sask., 1978.

49. T.M. Lillesand and R.W. Kiefer, "Remote Sensing and Image Interpretation," pp. 393, 583, Wiley, New York, 1979.

50. T.J. McGhee, R.L. Torrens, and R.J. Smaus, *Water Sewage Works*, **119**, 58, June (1972).

51. F.S. Woodiwiss, *Chem. Ind. (London)* 11, 443 (1964).

52. "Operating Manual, 6 Bottle Manometric Apparatus, Model 2173B," 3rd ed. Hach Company, Ames, IA, 1982.

53. D.W. Ryckman, A.V.S. Prabhakara Rao, and J.C. Buzzell, Jr., "Behavior of Organic Chemicals in the Aquatic Environment," Manufacturing Chemists Assoc., Washington, DC, 1966.

54. C.E. Van Hall and V.A. Stenger, *Anal. Chem.* **39**(4), 503 (1967).

55. R.H. Jones, *Water Sewage Works*, **119**(3), 72, Mar. (1972).

56. K. Verschueren, "Handbook of Environmental Data on Organic Chemicals," Van Nostrand Reinhold, Toronto, 1977.

57. F.S. Woodiwiss, The Biological System of Stream Classification..., *Chem. & Industry (London)*, 443–447, Mar. 14 (1964).

58. Anon., Biological method tests river pollution, *Chem. Eng. News,* **49**(23), 31–32, June 7 (1971).

59. J.L. Doust, M. Schmidt, and L.L. Doust, Biological Assessment of Aquatic Pollution: A Review, with Emphasis on Plants..., *Biol. Rev. of the Cambridge Philos. Soc.* **69**, 147–186, May (1994).

60. M. Leynen, T. Van den Berckt, and J.M. Aerts, The Use of Tubificidae in a Biological Early Warning System, *Envir. Poll'n.* **105**(1), 151–154 (1999).

61. A. Viarengo and L. Canesi, Mussels as Biological Indicators of Pollution, *Aquaculture*, **94**, 225–243, May 1 (1991).

62. B. Pinel-Alloul, G. Methot, and L. Lapierre, Macroinvertebrate Community As a Biological Indicator of Ecological and Toxicological Factors..., *Envir. Poll'n.* **91**(1), 65–87 (1996).

63. L.S. Fore and C. Grafe, Using Diatoms to Assess the Biological Condition of Large Rivers in Idaho (U.S.A.), *Freshwater Biol.* **47**(10), 2015–2037, Oct. (2002).

5

RAW WATER PROCESSING AND WASTEWATER TREATMENT

It's no good throwing away dirty water until you get clean.

—V. Bridges, 1922

5.1. WATER QUALITY RELATED TO END USES

Water quality requirements differ according to the application for which the water is required. While the highest quality surface waters can generally be used for any freshwater need, many applications do not have as stringent requirements [1]. A detailed consideration of water quality needs by use is an extensive exercise beyond the scope of this treatment [2]. Table 5.1 illustrates how the required levels of water quality parameters change with the anticipated end use. Livestock require a low level of bacteria to avoid illness, while certain types of irrigation can tolerate substantially higher levels because of the sterilizing effect of exposure to sunlight. Recreational waters permit low levels of bacteria even though drinking water must have none. This is because contracting an illness takes a threshold number of bacteria, usually greater than one, which varies with the particular pathogen and the state of health and immunity of the individual. For example, in a carefully monitored study, various doses of *Salmonella typhosa* (Quailes strain) were used to induce infection in 210 subjects. At a dose of 10^3 per subject, no one contracted the disease. At dose rates of 10^5 and 10^9 per subject, 28 and 95%, respectively, became ill [4].

There may be appropriate end uses for almost any type of surface water supply. The exact requirements for any particular end use will vary with local conditions and with government objectives. Setting of standards for this purpose is a complicated process, which requires consideration of factors such as toxicity, effects on biota, agricultural needs, risk of illness, and aesthetics (e.g., for recreational use). It also may require consideration of

███ **TABLE 5.1   The Water Quality Requirements for Various Primary Classes of End Use**[a]

Agricultural Uses
Livestock: Low bacteria, <40/100 ml, and low concentrations of toxic substances (e.g., F^-)
Irrigation: Low dissolved solids, <500 mg/L, (to avoid increased soil salinity)
　　　　Total bacteria: allowable 100,000/100 mL; desirable <10,000/100 mL
　　　　Low heavy metals

Fish, Aquatic Life, Wildlife Requirements
　　　　Concentrations of toxic substances low
　　　　pH near neutral, 6.5–8.5
　　　　Low BOD, 1–2 mg/L or less
　　　　High dissolved oxygen: cold, 6–7 mg/L at 15°C or less;
　　　　　　warm, 4–5 mg/L at ca. 20°C
　　　　Low temperature, turbidity, etc.

Industrial Uses
Cooling: Low hardness, <50 ppm of $(Mg^{2+} + Ca^{2+})$, (usually as SO_4^{2-} and CO_3^{2-})
　　　　Low corrosivity

Food processing, brewing, and soft drinks:
　　　　As public drinking water, but F^- <1 ppm

Thermal power:
　　　　Total dissolved solids <0.1 ppm (and lower for boiler pressures above 135 atmospheres)

Public Recreational Requirements
　　　　Free of color, odor, taste, and turbidity
　　　　Total bacteria <1,000/100 mL, coliforms <100/100 mL[b]
　　　　Low nutrients, to avoid nuisance algal growths

Public Drinking Water (treated)
　　　　No bacteria
　　　　Low nitrates, nitrites <10 ppm
　　　　Very low pesticides, none >0.05 ppm
　　　　Fluoride, allowable to 2.4 ppm
　　　　Toxic substances (metals etc.) below criteria levels
　　　　Total dissolved solids <500 ppm

[a]Condensed from McNeeley *et al.* [1] and Canadian Public Health Association [3].
[b]Guidelines are based on the premise that these concentrations would be unlikely to cause illness. Above these levels the area is closed to swimming.

the effects of short- or long-term consumption and use, the availability and efficiency of processing methods to remove impurities, any seasonal variations in supply characteristics together with possible alternatives, and wastewater treatment methods. To look after this multiplicity of requirements involves the skills of many disciplines. Guidelines are available to assist in interpreting regulatory standards for this purpose.

5.2. TREATMENT OF MUNICIPAL WATER SUPPLIES

The primary physiological water requirement of an adult human not subjected to heat stress is only 2.5–3 L/day. Yet in households equipped with facilities for running water the per capita consumption for such uses as cooking, bathing, washing/cleaning, human waste transport, and garden irrigation is 95–380 L/day (25–100 U.S. gal/day) [5]. When the very large water demand of many industries is added to domestic requirements, it can easily be seen why the facilities for the production of a safe and convenient water supply rank among the largest scale materials processing operations.

5.3. SIMPLE MUNICIPAL WATER TREATMENT

Depending on the raw water quality and the seasonal stability of its characteristics many municipalities are able to use a simple two-stage supply treatment involving preliminary filtration, followed by a disinfection step. Fast filtration methods use a pressure differential to force the raw water through a bed of finely granulated clean sand, crushed anthracite coal, or sometimes a mixture of media, to remove suspended solids present in the supply (e.g., see Fig. 5.1). As the filtration proceeds, accumulated solids in the filter bed cause a gradual increase of the resistance of water flow through the filter. For periodic cleaning, the filter is back-flushed using filtered water at a sufficiently high flow rate to lift the filter medium and free the accumulated algae, diatoms, silt, etc. The flushing water is discarded to the sewer. This need for periodic filter cleaning requires a water treatment plant to have either a large enough holding basin for filtered water to provide for both back-flushing and normal filtered water requirements while a single filter is being cleaned, or two filters, one of which can continue operation while the other is in the back-flush mode.

There is also a "slow filtration" variant of filtration used by some water treatment plants, which combines both a physical separation of suspended solids plus some biological consumption or adsorption of undesirable dissolved substances in the water supply in a single-, two-, or three-stage unit [6].

FIGURE 5.1 Flow diagram for an intermediate complexity municipal water treatment plant.

Bacteria, algae, and diatoms that accumulate on the coarse sand layer of the filter bed metabolize nutrients, etc., from the supply, thereby removing them. Among the benefits of slow filtration is removal of up to 50% of any chlorinated pesticide content of the influent water [6]. Slow filtration may require a prior aeration step to ensure that the biochemical processes remain aerobic, since anaerobic operation can form sulfides and amines from sulfate and nitrate, which would contribute bad odors or tastes to the supply.

After suspended matter has been removed a disinfection step, usually with chlorine, is necessary to ensure that the supply is free of any viable pathogenic organisms. The active bactericide, undissociated hypochlorous acid, is formed immediately on the contact of chlorine gas with water (Eqs. 5.1 and 5.2).

$$Cl_2 + H_2O \rightarrow H^+ + Cl^- + HOCl \qquad\qquad 5.1$$

$$\underset{\text{active}}{HOCl} \rightarrow \underset{\text{relatively inactive}}{H^+ + ClO^-} \qquad\qquad 5.2$$

Sufficient gaseous chlorine is added to the water to leave a "residual chlorine" concentration of 0.1–0.2 mg/L, after the normal consumption of a part of the added chlorine in reactions with any dissolved or residual suspended matter in the supply has taken place [6]. A preferred contact time of 1–2 hr is recommended, but at least 20–30 min of contact time should be ensured before use. The residual chlorine content is necessary to maintain safe transmission of the treated water supply through a local piping system.

Since the dissociation constant for hypochlorous acid is very small, in neutral solutions undissociated hypochlorous acid is the dominant species of this equilibrium (Eq. 5.3).

$$K = \frac{[H^+][OCl^-]}{[HOCl]} = 2.95 \times 10^{-8} \text{ at } 18^\circ C \qquad\qquad 5.3$$

Under acid conditions the higher concentration of hydrogen ion depresses hypochlorous acid dissociation (Eq. 5.2), which consequently tends to raise the fraction of undissociated hypochlorous acid above that in near neutral solutions (100% at pH 5 or lower). Thus, the effectiveness of chlorine disinfection is maintained or enhanced at low pH.

At a pH of 7.5, near neutral conditions, the hypochlorous acid will be roughly 50% dissociated and disinfection will still be quite effective. However, in disinfections of somewhat alkaline water supplies, the decreased hydrogen ion concentration will tend to depress the concentration of undissociated hypochlorous acid. At a pH of 10 or higher, hypochlorous acid would be present almost entirely (ca. 99.7%) as hypochlorite anion (e.g., see Section 15.5.2). This would substantially decrease the effectiveness of chlorine disinfection under these conditions [6], which could be remedied by pH adjustment of the supply before disinfection, by a higher chlorine dose rate, or by a longer contact time before water use. However, in practice, any source of supply above a pH of 8.5 would be adjusted to 8.5 before disinfection for other reasons.

When the time between disinfection and delivery at the householder's tap is long, such as when treatment is carried out at a reservoir some distance away, ammonia may be added at the same time as the chlorine. The more stable chloramines formed give a longer term residual chlorine to the supply than is possible with chlorine alone (Eqs. 5.4–5.6).

$$HOCl + NH_3 \rightarrow NH_2Cl + H_2O \qquad\qquad 5.4$$

$$HOCl + NH_2Cl \rightarrow NHCl_2 + H_2O \qquad\qquad 5.5$$

$$HOCl + NHCl_2 \rightarrow NCl_3 + H_2O \qquad\qquad 5.6$$

However, chloramines also require a longer disinfection period than chlorine to obtain the same effectiveness, and so are only used in situations like this.

Calcium or sodium hypochlorite (e.g., household bleach) is also effective for disinfection of water supplies, and is a convenient choice for smaller volumes when required, such as for a single household, a hamlet, or for small campsites [7, 8]. Hypochlorite solutions are somewhat more expensive to buy than compressed chlorine gas, but the equipment required for metering it into the supply is simpler. With hypochlorite salts, similar equilibria operate to rapidly establish a concentration of hypochlorous acid, which corresponds to the amount of hypochlorite salt added (Eq. 5.7, Table 5.2).

$$Ca(OCl)_2 + 2\,H_2O \rightarrow Ca(OH)_2 + 2\,HOCl \qquad\qquad 5.7$$

mol. wts.	143	18	74	52.5
moles	12	1	2	
mass ratios	143	36	74	105

The "% available chlorine" from calcium hypochlorite by definition includes Cl_2, $HClO$, and OCl^- (but not Cl^-), is 99.2%, or nearly the same effectiveness on a weight for weight basis as treating with elemental chlorine [9]. As long as the residual chlorine level used with hypochlorite salt disinfections is the same as with chlorine gas, there is no difference in relative effectiveness.

Chlorine is the most common disinfectant for public water supplies. Its effectiveness, simplicity of application, and low cost have been important

TABLE 5.2 "Available Chlorine" Content of Various Chlorine-containing Water Disinfectants[a]

Disinfectant	Molecular weight	Moles of equivalent chlorine[b]	Percent by weight	
			Actual chlorine	Available chlorine[c]
Chlorine	70.9	1	100.0	100.0
Hypochlorous acid	52.5	1	67.7	135.0
Sodium hypochlorite	74.5	1	47.7	95.2
Calcium hypochlorite	143.0	2	49.6	99.2
Chloramine (NCl_3)	120.4	3	88.4	176.7[d]

[a]Data recalculated from White [9].

[b]The number of moles of hypochlorite (oxidizing chlorine), which could form on dissolving 1 mole of the disinfectant in water.

[c]This equates to the calculated weight of elemental chlorine required to produce the same amount of hypochlorous acid in water as the given disinfectant; e.g., for calcium hypochlorite

$$\frac{(70.9\,\text{g/mol} \times 2\,\text{mol/mol} \times 100)}{143.0\,\text{g/mol}} = 99.2$$

[d]Action is very slow.

factors in the widespread adoption of chlorination as a public health measure. Questions have been raised about mutagenic risks from the presence of chloroform and other halomethanes formed from chlorination of substances in public drinking water supplies [10]. However, mutagenic risks from this source are small relative to the health risks from not disinfecting the water supply [11–13].

Other methods that are used for water disinfection, including hypochlorites, have a lower tendency to produce halomethanes than chlorine. Chlorine dioxide (ClO_2) has a negligible chlorination tendency, and is also an effective disinfectant at high pHs. It is favored in Germany [14]. Other disinfectants such as ozone, iodine, ultraviolet irradiation, and gamma irradiation avoid the use of chlorine entirely. Ozone, which is favored in France, is an effective disinfectant and can remove odor, taste, and color because of its more general oxidizing properties. This could be an important consideration with a poor-quality raw water supply [15]. However, it is not possible to maintain a disinfection residual in the water supply with ozone because of its instability, which makes it necessary to back up the ozone treatment with a low chlorine dosage to provide this transmission safeguard [6]. Iodine, as crystals, or more conveniently as a tincture, as an equilibrium solution in water, or as iodine-releasing tablets, is an effective field disinfection method but is too expensive for municipal water treatment. The remaining methods mentioned are also effective, but have cost, maintenance, or lack of experience problems attached to their use.

5.4. MORE ELABORATE MUNICIPAL WATER TREATMENT METHODS

When there is less control over raw water quality, such as when the supply is downstream of several towns already using a river, or when the natural supply from a groundwater source is poor, more complicated treatment methods are required (e.g., Fig. 5.1). Following the preliminary treatment steps outlined in the following paragraphs, filtration plus chlorination, or chlorination alone can be used to disinfect the finished supply. When filtration is not used to supplement disinfection, the U.S. EPA recommends that two different disinfection methods be used to ensure supply safety, (e.g., uv irradiation and chlorination).

It is generally desirable, if not always economic, to remove excessive hardness (>150 ppm) from water supplies before distribution. Synthetic detergents for laundry use have circumvented much of the problem from the use of carboxylate soaps in hard waters. For personal bathing, however, ordinary soaps are usually preferred. Hardness removal is also not necessary for drinking water since a degree of hardness in the water appears to reduce the risk of heart attack [16]. However, excessive hardness still causes a rapid deposition of scale on water-heating devices of all kinds, from kettles and domestic water heaters to commercial power boilers.

Carbon dioxide in natural waters solubilizes the calcium carbonate of limestone or marble by forming soluble calcium hydrogen carbonate (bicarbonate) (Eqs. 5.8 and 5.9). This is the process, which puts bicarbonate hardness into solution in surface water or groundwater supplies.

$$CO_2 + H_2O \rightarrow H_2CO_3 \qquad 5.8$$

$$H_2CO_3 + 2\ CaCO_3 \rightarrow 2\ (CaHCO_3)_2 \qquad 5.9$$
$$\text{insoluble} \qquad \text{soluble}$$

Bicarbonate hardness is also referred to as temporary hardness, because it may be removed by heating the water. This reverses the reactions, which put the calcium and magnesium into solution (Eqs. 5.10 and 5.11).

$$Ca(HCO_3)_2 \xrightarrow[\text{heat}]{} CaCO_3 \downarrow + H_2O + CO_2 \uparrow \qquad 5.10$$

$$Mg(HCO_3)_2 \rightarrow MgCO_3 \downarrow + H_2O + CO_2 \uparrow \qquad 5.11$$

They also deposit insoluble calcium or magnesium carbonate onto heated surfaces to cause scale buildup, so that heating for removal is not of practical use. Interestingly, addition of calcium in the form of slaked lime ($Ca(OH)_2$) can efficiently remove much of the temporary hardness of natural water supplies (Eqs. 5.12–5.14).

$$Ca(HCO_3)_2 + Ca(OH)_2 \rightarrow 2\ CaCO_3 \downarrow + 2\ H_2O \qquad 5.12$$

$$Mg(HCO_3)_2 + Ca(OH)_2 \rightarrow MgCO_3 + CaCO_3 + CaCO_3 \downarrow + H_2O \qquad 5.13$$

$$MgCO_3 + Ca(OH)_2 \rightarrow Mg(OH)_2 \downarrow + CaCO_3 \downarrow + H_2O \qquad 5.14$$

The second equivalent of calcium hydroxide is necessary to remove magnesium temporary hardness because the magnesium carbonate product obtained using one equivalent of calcium hydroxide is still somewhat soluble.

Permanent hardness is caused by the presence of the sulfates (or other anions), rather than the bicarbonates of calcium and magnesium [17]. Neither calcium hydroxide nor heat is effective for removal of permanent hardness. However, treatment with sodium carbonate (soda ash) is effective (Eq. 5.15). Removal of both temporary and permanent hardness requires addition

$$Na_2CO_3 + CaSO_4 \rightarrow CaCO_3 \downarrow + Na_2SO_4 \qquad 5.15$$

of sodium hydroxide (Eq. 5.16).

$$Ca(HCO_3)_2 + 2\ NaOH \rightarrow CaCO_3 \downarrow + Na_2CO_3 + 2\ H_2O \qquad 5.16$$

The sodium carbonate formed from the removal of temporary hardness is available to precipitate any calcium or magnesium salts present as permanent hardness. Thus, the usual water treatment for removal of hardness involves adding a mixture of quicklime (CaO) and sodium carbonate in accordance with the amount of permanent hardness present. Sometimes a coagulant such as alum ($Al_2(SO_4)_3 \cdot 14H_2O$) or ferrous sulfate will be added to help coagulate and settle any finely divided hardness in suspension [18].

The same steps may be conducted at the household level to decrease the undesirable effects of a hard water supply on laundry by adding of washing soda ($Na_2CO_3 \cdot 10H_2O$) or household ammonia to the laundry supply water. In some cases, a water softener which contains beads of a sulfonated polystyrene resin in the sodium form may be installed (Fig. 5.2). As hard water passes through, sodium ions in the resin are exchanged on a charge equivalent basis with any calcium, magnesium, or other polyvalent cations in the supply.

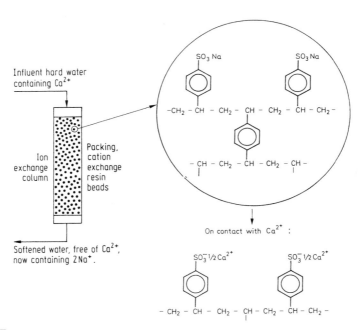

Influent hard water
containing Ca^{2+}

Ion
exchange
column

Packing,
cation
exchange
resin
beads

Softened water, free of Ca^{2+},
now containing $2\,Na^+$.

FIGURE 5.2   Details of operation of a household water softener.

In the process, the household water is softened, and the water softener resin gradually accumulates polyvalent cations until its capacity is exhausted. The resin is regenerated by passing a saturated sodium chloride brine solution through the softener, which displaces the polyvalent ions by sodium ion. During the regeneration step, the outlet of the softener is connected to the sewer. After a short period of flushing the resin with freshwater to remove the residual brine used for regeneration, the hardness removal operating cycle begins again.

Special techniques are required for the removal of other metal ions from raw water supplies, although lime treatment followed by flocculation does remove some heavy metal content by adsorption onto the precipitated material [6]. Excessive fluoride (i.e., >2.4 ppm in the supply), may also be decreased by lime treatment. Fluoride precipitation is greatly improved by the presence of 50 mg/L or more of magnesium ion in the water at the time of treatment (Eq. 5.17).

$$2\,F^- + Ca(OH)_2 \;\rightarrow\; CaF_2 \downarrow \qquad\qquad + 2\,OH^- \qquad 5.17$$
$$\text{adsorbed on } Mg(OH)_2$$

Bone ash, and synthetic apatite, which are essentially calcium phosphate $(Ca_3(PO_4)_2)$, or a synthetic apatite, calcium hydroxide mixture, are also effective methods for fluoride removal because of their affinity for these phosphate salts.

A taste or smell in the water supply may be removed by single or double aeration, if the causative agent is volatile. Otherwise adsorption on activated carbon, or chemical oxidation with ozone or chlorine, or both methods used in series may be required. Many organic toxic substances, such as pesticides, are also efficiently adsorbed by activated carbon and may be removed in this manner. Lime treatment, if not already practiced for other reasons, is also efficient at removing dissolved colored substances, such as humic acids, when these are present in the supply. Sometimes lime addition may have to be accompanied by activated carbon adsorption or coagulation plus filtration for effective color removal.

5.5. MUNICIPAL WATER BY DESALINATION

Coastal communities in arid areas may need to produce a potable supply from seawater or brackish supplies. These sources require more extensive and expensive methods. The dissolved solids content of seawater is roughly 3.5%, 35% (i.e., 35 "parts per thousand") or 35,000 mg/L. This needs to be reduced to about 1% of this figure in order to meet the dissolved solids requirement of a potable water supply. Brackish waters, which have a lower salt concentration than seawater are a preferred raw water source when available, since the energy cost for salt removal is less [19]. It is possible under the right conditions to use solar-power for this purpose [20].

Distillation can accomplish salt removal with good thermal efficiency, if it is carried out under reduced pressure using several stages in series [21] (Fig. 5.3). Up to six stages have been found to be economic for maximum energy utilization at current energy costs. More than 50% of the world's existing desalination capacity in 1997, representing a total capacity of 9.6 million m³/day (2.3 billion U.S. gal/day), utilize multistage flash distillation for saline water conversion [23]. The largest capacity plants are installed in Kuwait (22,500 m³/day) and Oman, the coastal arid states of the U.S.A., Russia, and the Netherlands.

Membrane processes of the reverse osmosis (hyperfiltration) or electrodialysis types are used, but usually for smaller scale facilities (Fig. 5.4). Reverse osmosis units use high pressures of brackish water or seawater charging on one side of a semipermeable membrane, sufficient to exceed the osmotic

FIGURE 5.3 Schematic of the mode of operation of a multistage flash distillation unit. (Modified from Pryde [22], and reprinted courtesy of Cummings Publishing Co.)

Spiral wound
membrane, see detail
below

Fresh water
permeate

Porous
backing material

Permeate
tube

Positive terminal

Brackish water

Negative
terminal

Mesh spacer

Membrane

Glue line

Brackish water

Brine   Brine

Fresh water

Reverse Osmosis   Electrodialysis

FIGURE 5.4 Principles of operation of two types of membrane water desalination units: reverse osmosis and electrodialysis. (From Pryde [22], and reprinted courtesy of Cummings Publishing Co.)

pressure of the deionized water on the product side of the membrane [24]. Thus, water free of ions permeates through the membrane at a rate proportional to the applied pressure (Table 5.3). Desalination by electrodialysis again uses membranes, but this time, of a type permeable to ions. The driving force for ion removal in this instance is provided by the opposing electrochemical potentials on either side of the membrane. Electrochemical removal of ions is continued in several stages of electrodialysis cells operating in series until the salt concentration in the product water is decreased to the required level.

Differential freezing methods may prove to be particularly useful in water-short polar regions, where the energy costs for freezing would be low. Reducing the brine temperature to below freezing generates substantially salt-free ice crystals, which are separated from the brine mother liquor (Fig. 5.5). For efficient energy use, the heat removal capability of the thawing ice may be used to prechill incoming brine, prior to freezing.

TABLE 5.3 A Comparison of the Performance Characteristics of Three Commercial Reverse Osmosis Membrane Configurations[a]

Characteristic	Hollow fiber cellulose acetate	Hollow polyamide	Spiral wound sheets of cellulose acetate
Mandatory pretreatment	pH control	softener or pH	pH control
pH range (ideal)	4–7.5 (5.5)	4–11 (4–11)	3–6 (5.5)
Prefiltration, μm	5	5	25
Maximum Cl_2, ppm	1.0	0.1	1.0
Maximum temperature,°C	35	35	35
Normal operating pressure	2,700 kPa (400 psig)	2,700 kPa (400 psig)	2,700–4,100 kPa (400–600 psig)
Max. product back pressure	550 kPa (80 psig)	275–340 kPa (40–50 psig)	275–340 kPa (40–50 psig)
Rated TDS level[b]	2,000 ppm	2,000 ppm	2,000 ppm
TDS rejection rates[b]	90–95%	90–95%	90–98%
Water recovery	75%	50–75%	22–75%
Max. suspended solids,	1.0 JTU[c]	1.0 JTU[c]	1.0 JTU[c]
Ease of cleaning	Good	poor	very good
Surface water performance	Good	poor	excellent
Well water performance	excellent	good	excellent
Packing density, m^2 membrane/m^3 module	29,500	–	660
Flux, L/m^2/day (pressure)	7.3 (4,100 kPa)	–	730 (5,500 kPa)
Flux density, L/m^3 module/day	220,000	–	480,000

[a]Properties compiled from McBain [25].

[b]TDS = total dissolved solids. Newer membranes can handle up to 35,000 ppm, i.e., ordinary seawater.

[c]Jackson turbidity units, on an arbitrary scale [26].

5.6. WATER QUALITY REQUIREMENTS OF INDUSTRY

Most industrial cooling water requirements may be met by the usual characteristics of the surface waters available, as long as the hardness and the concentrations of chloride or other substances, which may cause corrosion are low. Bacteriological and color, dissolved oxygen, etc., requirements for this use are minimal, and zero-dissolved oxygen is desirable to reduce the corrosion risk. The food-processing industry is one exception to this generalization; however, to cool processed foods in bulk or in final packages of bottles or cans, potable water quality is required. Otherwise, a pinhole leak in the sterile package could cause cooling water and possible organism leakage into the contents.

A potable quality water supply or better is also required for the soft drinks and alcoholic beverage industries, for medical and pharmaceutical applications, and for the water needs of many types of research institutions. The beverage industry has a fluoride requirement of <1 mg/L, lower than that recommended for ordinary potable water supplies, to accommodate a possible very high individual consumption rate. Deionized or distilled grades of water quality are sufficient for the majority of the other applications.

FIGURE 5.5 Schematic of differential freezing as a method for salt removal from sea-water. (From Pryde [22], and reprinted courtesy of Cummings Publishing Co.)

The feedwater requirements of modern high pressure steam boilers are among the most stringent of all water quality standards [27]. By 1970, high pressure boiler operation already required direct conversion of feedwater to superheated (dry) steam at 160 atm (2,400 psi) in the boiler tubes. Any dissolved solids present gradually accumulated in the tubes, contributing to eventual failure. Reliable operation of this type of boiler requires water with a total impurity level of 0.03 mg/L (0.03 ppm (0.000003%)), or less. Water of this standard amounts to the highest purity of any commercial chemical, more pure than the best grades of most analytical reagents. An objective has just recently been set to further raise the boiler water standard up to a requirement of 0.01 mg/L (10 ppb) or less total impurities.

The boiler feedwater quality requirements equate to or exceed the purity requirements of the electronics industry, which currently has among the most stringent requirements. Basic minimum specifications are: neutral salts, 0.02 mg/L; total dissolved solids, 0.02 mg/L; free base or mineral or organic acid, nil; and a conductivity of 0.05 μS/cm or better. The processing steps necessary to meet these standards require a composite of all the best existing water purification technology in a single processing train (Fig. 5.6).

FIGURE 5.6 Schematic flowsheet of the water treatment system used by the electronics industry to obtain the required level of finished water quality (see text). (From Arden [27], reprinted courtesy of the Royal Society of Chemistry.)

5.7. TREATMENT OF MUNICIPAL WASTEWATERS

The principal chemical and physical characteristics of the domestic or sanitary sewage effluent must first be known in order to be able to decide the most efficient treatment and disposal options (Table 5.4). While the tabulated list of impurities appears long it should be remembered that the total impurities present still constitute less than 1% of the sewage volume to be treated. From a chemical standpoint, municipal sewage, thus, represents a dilute solution and suspension of a heterogeneous mixture of constituents not, an easy stream to treat. Nevertheless, since untreated domestic sewage represents about one-quarter of the total biochemical oxygen demand (BOD) waste loading discharged to surface waters, at least for the United States, the rigor of the treatment and/or disposal options selected can have a significant influence on surface water quality [30].

5.8. DISCHARGE REQUIREMENTS AND POSTDISCHARGE REMEDIES

Discharge of untreated sewage of the properties outlined, if permitted, could only be satisfactory if two composite requirements are met: (1) There is

TABLE 5.4 Approximate Values for the Principal Physical and Chemical Characteristics of Municipal Wastewaters[a]

Characteristic	Domestic sewage			Urban storm waters
Volume	70–200% of supply (from infiltration)			0–5000 + % of supply
Temperature:	1–2° above supply			ambient
BOD	ca. 100–300 mg/L			1– >700 mg/L
COD	ca. 2 × BOD			5–3,100 mg/L
Microorganisms, (as bacteria, viruses, etc., number per 100 mL)	ca.100–3,000 × 10⁶ (mostly nonpathogenic)			200, to 146 × 10⁶
Total solids: mg/L	655			450–14,600
Solids distribution,	Inorganic	Organic	BOD[b]	
Suspended*	65	170	110	2–11,300
Dissolved**	210	210	30	450–3,300[c]
Total	275	380	140	
*Suspended solids: (by volume)	1/3 nonsettleable (<50 μm)			
	2/3 settleable (>100 μm diameter)			
**Dissolved solids:	Approx. twice municipal supply (dry weather) N 50 mg/L (mostly NH₃ and urea), P 30 mg/L (mostly as PO₄³⁻), plus NaCl, etc.			

[a]Compiled from Klein [28], American Chemical Society [29, 30], and Field and Fan [31].
[b]Distribution of biochemical oxygen demand (BOD) between the suspended and dissolved matter of sewage with a total BOD of 140 mg/L.
[c]At high end of range found in areas using highway deicing chemicals ($CaCl_2$, NaCl, etc.).

sufficient dilution by the receiving stream; and (2) there is good mixing with this stream on discharge. The dilution requirement means that, post discharge, the BOD of the river should not exceed 2 mg/L, and the dissolved oxygen should not be depressed below about 4 mg/L (5 mg/L, Alberta; 3 mg/L, New York), to meet the criteria of a clean river. To discharge raw sewage of an average BOD of about 200 mg/L, thus, demands a river flow of more than 200 times the sewage volume at low river flow periods, and a predischarge river BOD of 1 mg/L or less, to meet the dilution requirement alone. Also, there must be no other wastewater discharges in close proximity. The good mixing requirement for this scenario is also difficult to achieve and maintain, even for a winding river course. So there are very few situations where the requirements for raw sewage discharge to rivers could be compatible with these criteria, even if it were aesthetically acceptable.

Similar considerations apply to the discharge of untreated domestic effluent to the sea [32]. Raw sewage discharge into shallow, poorly mixed saltwaters can give the same problems with anaerobic decomposition producing odors and microbial contamination on the shoreline and surrounding air, as would be experienced with a small river or a lake [33].

It is possible to remedy temporarily depressed oxygen levels or high BODs in a river or lake. This can happen for short periods during times when stream flow is very low, or when the process upsets temporarily increase the BOD of discharged effluent. Electrically driven aerators when placed on floats across, the stream have been used for the Lippe River in Germany. Using air of oxygen partial pressure 0.21 atm, this system was found capable of boosting the dissolved oxygen content by about 1 kg/900 watt hours (2 lb/horsepower hour) in the Upper Passaic River, New Jersey [34].

Alternatively, to take advantage of the increased oxygen exchange rates provided by increased air pressure, the whole of the low flow river water may be passed through a U-tube bored 14 m or more beneath the river bed. Passage of air bubbles into the downflow side boosts the oxygen partial pressure in the bubbles 2.5 times to about 0.50 atm, greatly accelerating oxygen dissolution rates (Fig. 5.7). For the Red Deer River in Alberta, this procedure more than doubles the dissolved oxygen content when it is used at times of low flow. This is particularly useful in winter when the ice cover prevents reaeration. When the river flow is high and aeration, therefore, unnecessary, the bulk of the water flows over the weir bypassing the U-tube.

Use of pure oxygen, (i.e., their UNOX system), was once proposed by Union Carbide for the rejuvenation of highly eutrophic lakes. Certainly, this system provided excellent oxygen transfer conditions, and was made feasible with the availability of low-cost processes for oxygen enrichment of air. More recently, the UNOX system has been used for rapid treatment of high BOD waste streams from a fully bleached kraft pulp mill before discharge [36].

5.9. STREAM ASSIMILATORY CAPACITIES

Adequate sewage treatment is necessary to minimize the frequency of the less efficient, post discharge solutions to improve river water quality. The degree

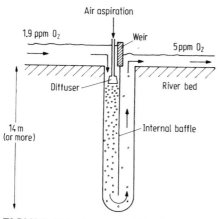

FIGURE 5.7 Automated U-tube river aeration, at times of low flow. A 1.3-m head drives the whole flow through the U-tube, which in practice can consist of a single or several parallel boreholes with an internal baffle [35].

of treatment required is dependent on the capacity of the receiving body of water to accommodate oxidizable wastes without degradation, its assimilatory capacity [37]. In general, the oxidation capacity will be higher for a river, which has continuous water turnover than for a lake. For the same water volume a shallow, fast-flowing river will have a higher assimilatory capacity than a deep, slow-flowing river. This is because the initial content of dissolved oxygen is likely to be higher, and the oxygen exchange rate with the air will also be more rapid for the shallow river. Shallow fast-flowing stretches of a river have two or three times the reaeration capacity of deeper, slow-moving pools. These characteristics, which can be measured using tracer techniques [38], have led to the tabulation of stream assimilatory capacities as related to their volume of flow, depth, and oxygen exchange rates (Table 5.5).

Knowing the approximate assimilatory capacity of a stream, which equates to its recovery or reaeration rate, allows an estimate of the acceptable

TABLE 5.5 Waste Assimilatory Capacity Related to River Size[a]

Stream class	Common examples[b]	Assimilatory capacity[c]	Average depth, m	Average flow, m^3/sec
I	Numerous local	0.023	0.17	0.4
II	Allegheny, Kansas, N. & S. Saskatchewan, Rhone, Po	0.56	1.5	300–1,500
III	Mississippi, St. Lawrence, Orinoco, Ganges, Yenisei	2.90	13.7	20,000

[a]Compiled mostly from the data of Todd [5] and Canadian National Committee [39].
[b]Based on flow rates of first example of each class.
[c]Round number estimates in tonnes BOD per kilometer per unit oxygen deficiency for the physical data as given. Originally stated in units of tons BOD/mile/unit oxygen demand: Class I, 0.04; Class II, 1.0; Class III, 5.2.

BOD loadings, which could be discharged to the river without severely degrading river quality. Assuming a discharge of 2.4×10^6 L/day of untreated sewage of 200 mg/L average BOD, which could reasonably arise from a small town of 24,000 inhabitants, this would equate to a daily BOD loading of about 480 kg/day (Eq. 5.18).

$$200 \, \text{mg/L} \times 2.4 \times 10^6 \, \text{L/day} = 4.8 \times 10^8 \, \text{mg/day} \qquad 5.18$$

Discharge of this volume of oxygen demand into a Class I river, if evenly distributed through the day, would result in a river BOD loading of 14 mg/L (Eq. 5.19).

$$\frac{4.8 \, \text{m}^3/\,\text{sec} \times 10^8 \, \text{mg/day}}{0.4 \, \text{m}^3/\text{sec} \times 86,400 \, \text{sec/day} \times 1,000 \, \text{L/m}^3} = 13.9 \, \text{mg/L BOD} \qquad 5.19$$

This is a very heavy loading sufficient to severely degrade river quality, even if good mixing is achieved. However, if the BOD of this sewage was decreased to 20 mg/L by suitable treatment before discharge, i.e., (to one-tenth of the raw BOD values), then discharge could occur without severe river degradation.

Another approach to this problem is to use a figure for the oxygen demand loading per capita, estimates of which range from 54 to 115 g/day [30]. For the community example of 24,000, this method gives a range for the gross BOD discharge of 1,296–2,760 kg/day for untreated sewage, much higher than the estimate based on the previous method. If discharges on this scale were actually practiced to a Class I river, this would correspond to very severe river degradation from gross BOD loadings of 37.5 to 80 mg/L.

From Table 5.5, the background calculations allow a quick estimate for a Class I river that 23 kg of BOD discharge would cause a drop in river BOD of 1 mg/L, one kilometer below the point of discharge. It can also be seen that, although the flow volume of a Class III river is of the order of 50,000 times greater than a Class I river, the assimilatory capacity is only 130 times more. This fact reflects the much slower reaeration rate of a deep, slow-flowing river with vertical stratification as compared to shallow, faster flow rates. It also is a reminder that when one river has 10 times the average flow volume of another does *not* necessarily mean that it can accept 10 times the oxidizable waste volume without degradation: hence, the need for sewage treatment.

5.10. PRIMARY AND SECONDARY SEWAGE TREATMENT

Methods for municipal sewage treatment may be conveniently grouped into three categories according to the stage of wastewater processing and the types of treatment used. Primary treatment applies mechanical processes, which amount to physical separation procedures to initial sewage cleanup. Secondary sewage treatment uses biochemical and biological methods to remove and consume wastes, and provides a substantial improvement of effluent quality over that obtained by primary treatment alone. Tertiary, or advanced treatment, involves one or more of biological, chemically based, or physical separation processes for a further improvement of effluent quality. The level

FIGURE 5.8 Plots of plausible degrees of degradation and recovery rates for oxygen-consuming discharges to a stream, which heavily degrades, temporarily degrades, or stays within reasonable guidelines to maintain stream quality.

of treatment required in any particular situation is decided on the basis of the population loadings to be served by the system and the assimilatory capacity of the receiving body of water. If this is a river, the level of treatment required has to reflect the presence of any towns above or below the discharge point, the times and volumes of lowest flows, periods of ice cover, and whether it is also to be used for recreational or other purposes. A plot of the interrelationships between the dissolved oxygen and biochemical oxygen demand helps to make the influence of these considerations clearer (Fig. 5.8).

Primary treatment involves the use of a combination of one or more of the following types of processing units (Fig. 5.8). First the effluent is screened by self-cleaning bar screens of 2–5 cm spacing to remove any coarse material, followed by a set of finer screens, which are usually self-cleaning. These preliminary steps protect the downstream pumps and sludge handling equipment from fine gravel, sand, and any other large suspended matter, which could enter the plant. This risk is present because it is common to collect both sanitary and storm runoff waters in the same piping, so-called combined sewerage systems. After screening, the waste stream passes through an electrically driven comminutor, which chops any residual coarse material ready for further processing. Following this, the wastewaters are passed into quiescent channels with underflow discharge which remove floatables, grease, hair, plastics, scum, etc. These underflow channels feed into a large sedimentation tank (clarifier) with a retention time of a few hours. The supernatant overflow from this tank is the primary treated sewage effluent. These processing steps achieve about a 60% reduction in suspended solids and a 30% reduction in oxygen demand compared to the influent sewage, plus smaller improvements in other parameters (Table 5.6). Sludge collected from the bottom of the primary settling tank contains 2.5–5% solids, and has to be dewatered and further treated for disposal. Details of disposal options are discussed later.

Biochemical consumption, immobilization, and further coagulation plus settling of residual wastes in the sewage stream are the objectives of secondary treatment. Very high concentrations of microorganisms are employed to bring

TABLE 5.6   Approximate Cumulative Waste Removal Efficiencies of Various Sewage Treatment Procedures, in Percent[a]

	Primary	Primary + secondary	Shallow lagoons[b]	Chemical coagulation[b]	Reverse osmosis or electrodialysis[b]
BOD	35%	90%	95%	95%	95%
COD	30	80	90+	85	95
Refractory organics	20	60	85	80	90
Suspended solids	60	90	95	95	95+
Total N	20	50	85	60	90–95
Total P	10	30	85	85	90–95
Dissolved minerals	1–2	5	10	10	50–90
Incremental cost, US$ per $100\,m^{3c}$	0.80–1.05	1.85–4.00	1.60–3.20	3.20–6.60	33–100
Bacteria	35%	90%	99%		
Postdisinfection[d]	90%	99%	99.9+		

[a]Data from American Chemical Society [30], Babbitt and Baumann [40], and Weinberger *et al.* [41]. Costs are approximate, and can vary significantly with scale.

[b]A tertiary treatment option. Percent improvements given are from primary + secondary + indicated tertiary treatment measure.

[c]$100\,m^3$ equates to roughly 22,000 Imperial gallons or 26,400 U.S. gallons.

[d]Approximate values postdisinfection, usually by chlorination [40].

about accelerated biochemical and biological consumption of residual wastes in the primary effluent, aided by a vigorous supply of air (for oxygen). The BOD and suspended solids are both decreased to about 1/10th of influent values and other qualities of the effluent are also improved somewhat over that provided by only primary treatment (Table 5.6). With augmented microorganism populations and artificial aeration the treatment time required is reduced to a matter of hours, instead of the several days that the same processes would take in a natural setting. Thus, the effluent improvement is conducted in much more compact equipment than would otherwise be possible. The detailed methods used to achieve this rapid waste utilization by microorganisms vary widely with country and any local conditions, which may have to be met [42].

In North America, and to a lesser extent in Europe, activated sludge plants or variations of this process are common for secondary sewage treatment, as shown schematically in Fig. 5.9. High bacterial populations are maintained by the return of settled active sludge and artificial aeration is maintained by compressed air or "brush" aerators to stimulate a high level of aerobic microbial activity in this tankage. Both dissolved and suspended matter is rapidly converted to settleable bacterial biomass. Biomass plus absorbed and adsorbed impurities are removed in a secondary sedimentation tank to produce a supernatant sewage effluent treated to the secondary level, of BOD 20–30 mg/L, plus a settled secondary sludge (Table 5.6).

Other variants to obtain efficient microbiological action include the trickling filter, common in Europe, in which primary treated effluent is passed over a bed of coarse gravel with air spaces between pebbles. A microbially-active

FIGURE 5.9 Schematic flow diagram for the principal steps involved in municipal sewage treatment to the secondary level. (A composite of components given by the American Chemical Society [30] and Field and Fan [31].)

layer on the gravel oxidizes BOD and consumes nitrogen and phosphorus in the wastewater. The biodisk or rotating biological contactor (RBC), uses similar principles and is suitable for smaller installations [44]. Both systems use efficient air exchange on a microbial support medium, rather than using compressed air, to obtain rapid aerobic conversion of wastes and consumption of some nitrogen and phosphorus from the effluent. Where land is available, and the sewage volume is not large, the Dutch oxidation ditch can provide effluent roughly equivalent to secondary effluent quality. Deep shaft treatment method, originally developed by ICI (U.K.) can be used when land is scarce. This accomplishes the same type of improvement in a bore hole of 0.3 to 10 m in diameter and 150–300 m deep [45]. The pressure of the head of water boosts aeration rates to 10 times that possible with a regular activated sludge plant, and produces half the volume of sludge. Apparently, the very high surface to volume ratio of the working bacteria and their high intracellular–extracellular gas exchange rates, protects them from cell rupture (the "bends") under these conditions.

Bacterial counts in the supernatant liquor from a secondary treatment plant are decreased to about 10% of those in the influent raw sewage, but are still high. Therefore, if adequate improvement in effluent quality is obtained at this level of treatment, it is normally disinfected by a fairly heavy dose of chlorine (ca. 10 ppm) before discharge. Postchlorination reduces the bacteria count to about 1% of that in raw sewage [41]. Each 1 ppm of post chlorination also decreases the BOD of the effluent by about 2 ppm from the oxidizing action of the chlorine, which consumes about 50 to 80% of the chlorine added. The higher chlorine dose rates used for sewage disinfection, together with the higher concentration of organics also present, are a more significant source of chlorinated organics to surface waters than that contributed from the disinfection of municipal water supplies. Gamma irradiation, ultraviolet light irradiation, and bromine chloride are other effective measures, which have been used for disinfection of treated sewage effluent [46]. The first two of these disinfection methods do not contribute any halogenated organics or additional toxicity to the effluent.

If the final chlorine-disinfected effluent is to be discharged to a sensitive watercourse, dechlorination may be necessary to avoid problems from the toxicity of the residual chlorine. After a 20- to 30-min holding time for completion of the disinfection to take place, the treated effluent is dechlorinated by adding a reducing agent such as sulfur dioxide to convert chlorine to chloride, followed by pH neutralization with lime (Eqs. 5.20 and 5.21).

$$HOCl + SO_2 + H_2O \rightarrow HCl + H_2SO_4 \qquad 5.20$$

$$2\ Ca(OH)_2 + 2\ HCl + H_2SO_4 \rightarrow CaCl_2 + CaSO_4 + 4\ H_2O \qquad 5.21$$

Dechlorination does decrease chlorine-induced toxicity but does not remove any chlorinated organics, which may have formed during the disinfection step.

One of the remaining problems of this sequence of sewage treatment is the wide range of influent volumes, which result from the use of combined sewerage systems. Combined systems are those in which domestic sewage and storm runoff are collected in the same piping system for economy of installation. Combined systems may have outflows of 50 or more times normal domestic flows at times of heavy rain, far exceeding the design capacity of the treatment plant. To deal with this situation, the total *initial* effluent from a storm is captured in holding tanks for later gradual treatment. Overflow points are provided at critical junctions in the collection system to release later combined sewage flows directly to surface waters after the initial captured, highly polluted flow of a storm. At this time, the outflows become relatively dilute, low-toxicity wastes for which direct discharge constitutes a less serious problem. Initial storm flow from a combined system can have characteristics significantly worse in terms of its effect on discharge compared to untreated domestic sewage (Table 5.4). Street accumulation of soil, vehicle residues, animal droppings, pesticides, and fertilizers during the preceding dry spell, as well as material sedimented out of domestic sewage during periods of slow flow, will all be flushed out of the system in the initial period of a vigorous rainstorm.

Combined sewerage systems problems can be solved by a gradual separation of the two systems and provision of separate storm-water treatment facilities as the construction costs for these facilities become feasible. Or the capacity of holding basins at the treatment facility can be increased to retain a larger proportion of the highly contaminated initial storm flows for later gradual treatment, combined with the capability to discharge later, less contaminated flows from a prolonged storm.

5.11. TERTIARY, OR ADVANCED SEWAGE TREATMENT

If the effluent quality produced by secondary treatment is still not adequate, for example for discharge to a protected watershed or to a small stream, then one or more of the tertiary treatment options may have to follow the primary and secondary steps. Tertiary treatment is sometimes referred to as advanced treatment and covers a variety of methods for the upgrading of secondary effluent quality [47].

One or more shallow lagoons in series having a total secondary effluent holding time of the order of days can alone accomplish significant further improvement. They have to be shallow to ensure good sunlight penetration and efficient oxygen exchange to the bed of the lagoon, assuring aerobic bacterial action as well as photosynthetic nutrient utilization. Rooted and free-floating plants and bacterial action remove nutrient ions plus C, H, O, and S compounds in the secondary effluent (Table 5.6). Lagoons also decrease the suspended solids and bacterial counts by a factor of about 10 by a combination of soil adsorption and coagulation/settling [48]. However, these steps are all slower, which is why a longer residence time is required for this step than for primary or secondary treatment. For these reasons, tertiary lagoons do have a significant land requirement for all except the smallest settlements.

Chemical methods of effluent improvement are generally quicker, have a smaller treatment volume and land requirement, but have chemical costs attached to their use. Coagulation of suspended solids may be accomplished using an inorganic coagulant, such as alum. Strictly speaking, alum is $K_2SO_4 \cdot Al_2(SO_4)_3 \cdot 24H_2O$. But since it is the aluminum ion of this double salt, which is the active coagulant, "alum" for sewage treatment normally refers to aluminum sulfate itself, $Al_2(SO_4)_3$, usually used as a solution in water. Alum treatment promotes agglomeration of the residual colloidal matter to larger particles, which then settle. It also can remove 90–95% of the dissolved phosphate, dropping this from 35–55 mg/L to 0.5 mg/L or less [49] (Eqs. 5.22 and 5.23).

$$Al^{3+} + \text{negatively charged colloids} \rightarrow \text{settleable floc} \qquad 5.22$$

$$Al_2(SO_4)_3 + 2\,Na_3PO_4 \rightarrow 2\,AlPO_4 \downarrow + 3\,Na_2SO_4 \qquad 5.23$$
$$\text{soluble} \quad\;\; \text{insoluble}$$

To achieve this requires alum dosages of 200–300 mg/L, which contributes a significant, quite acidic inorganic content to the sludge, which will add to ash volume, and could contribute to corrosion in some sludge disposal systems.

Hydrated lime (calcium hydroxide) may be employed at similar dosages as alum, and has about the same phosphate removal efficiency (Eq. 5.24).

$$3\,Ca(OH)_2 + 2\,PO_4^{3-} \rightarrow Ca_3(PO_4)_2 \downarrow + 6\,OH^- \qquad 5.24$$

Lime treatment is generally lower in cost than alum, but it still has the disadvantage of adding a substantial inorganic (hence incombustible) fraction to the sludge disposal system.

Some commercial organic polyelectrolytes (charged water soluble polymers) are very efficient for colloid flocculation [50]. These, being combustible, may be readily burned along with the sludge. They are more expensive than inorganic flocculants on a weight for weight basis. However, they are also active at dosages of only 1–2 ppm, which makes them competitive on a treated sewage volume basis. If used in conjunction with ferrous chloride or sulfate, which are available as by-products of the steel pickling industry, they too are capable of decreasing the phosphate concentration by some 80% [51] (e.g., Eq. 5.25).

$$3\ FeCl_2 + 2\ Na_3PO_4 \rightarrow \quad Fe_3(PO_4)_2 \cdot 8\ H_2O \downarrow \ + 6\ NaCl \qquad 5.25$$
$$\text{vivianite (finely divided)}$$

The flocculated insoluble vivianite, which forms is removed by subsequent settling.

It is also possible to remove phosphate from secondary effluent by passage over aluminum electrodes carrying an AC current [52]. With high effluent flow rates and low treating currents, about 1.4 g of aluminum is consumed per gram of phosphate removed.

If nitrogen and phosphate removal are required, combinations of chemical and biological action are required. Secondary effluent is treated with lime to decrease the pH and then subjected to anaerobic digestion to convert up to 90% of the combined nitrogen compounds present to ammonia (Eq. 5.26). The ammonia formed in solution is air-stripped from the effluent

$$NO_3^-, NO_2^-\ (\text{anaerobic digestion}) \rightarrow NH_3\ \text{in solution} \qquad 5.26$$

before discharge. This procedure was adopted for the treatment plant in the sensitive Lake Tahoe, California, watershed.

Another alternative for removal of combined nitrogen is by denitrification of nitrate and nitrite using appropriate bacteria in the presence of a carbon source such as methanol (Eq. 5.27).

$$6\ NO_3^- + 5\ CH_3OH + 6\ H^+ \rightarrow 3\ N_2 + 5\ CO_2 + 13\ H_2O \qquad 5.27$$

Converting combined nitrogen into a nitrogen gas end product totally removes the combined nitrogen from the ecosystem. Air stripping of an ammonia product has the potential for a return of the ammonia in subsequent precipitation.

Finally, other dissolved, nonnutrient salts may have to be removed for water reuse in some water conservation programs. This may best be accomplished using reverse osmosis, or electrodialysis [53], or ion exchange [54]. The operating details of these processes are discussed with municipal water treatment.

For some water reuse applications, activated carbon may be employed to adsorb 90 to 98% of any residual degradation-resistant organics when necessary. The resultant effluent quality is adequate for many industrial coolant or irrigation applications. After minimal further treatment, such as by reaeration and chlorination, it could even be reused for potable purposes in an emergency [55]. However, not enough is known about the potential for accumulation of trace toxins to recommend this procedure for long-term potable water use.

5.12. SLUDGE HANDLING AND DISPOSAL

The large volume of sludge produced during sewage treatment and its high water, low solids content pose a substantial secondary environmental problem, its ultimate disposal. Frequently, the costs of sludge treatment and ultimate disposal or destruction equal all the other costs of sewage treatment [56].

Primary sludges containing 2.5–5% solids, and the sludges from trickling filters and activated sludge plants containing 0.5–5% and 0.5–1% solids, respectively, are particularly difficult to dewater because much of the water is tied up intracellularly by microorganisms and in sedimented flocs. In terms of disposal mass, roughly 1 tonne dry weight of sludge is produced daily for each 10,000 population connected to the sewerage system.

Anaerobic sludge digestion at 30–35°C reduces sludge volume by about 90% while producing a medium-grade fuel gas of methane and carbon dioxide, in the mole ratio of roughly $2CH_4 : CO_2$ (e.g., Eqs. 5.28–5.31, Fig. 5.9).

$$CO_2 + 4\,H_2 \rightarrow CH_4 + 2\,H_2O \qquad\qquad 5.28$$

$$4\,HCOOH \rightarrow CH_4 + 3CO_2 + 2\,H_2O \qquad\qquad 5.29$$

$$CH_3COOH \rightarrow CH_4 + CO_2 \qquad\qquad 5.30$$

$$2\,CH_3CH_2OH \rightarrow 3\,CH_4 + CO_2 \qquad\qquad 5.31$$

The gas produced may provide energy for the sewage treatment plant itself, or it may be relatively easily upgraded by carbon dioxide removal (water washing) and added to a town gas distribution system [57]. Small volumes of digested sludge may be disposed of by plowing into land, though safety questions remain, which relate to residual pathogens and possible high concentrations of heavy metals, particularly in the sludges from heavy industrial areas.

Sludge may be disinfected by chlorination, accomplished by addition of 2–4 mg/L chlorine, or by sludge electrolysis, which produces chlorine *in situ* from sodium chloride already present in wet sludge [58]. Gamma irradiation has also been used for sludge disinfection [59], and various heat treatments are also effective for this purpose [60]. For example, dewatered sludge may be heated to 70°C, and then held at this temperature for 30 min from the exotherm obtained by mixing in the correct proportion of burned lime (CaO). The water uptake by the hydration reaction heats and sterilizes the sludge, and dries it further (Section 7.3).

Any of these disinfection alternatives improves the safety of land disposal of sludges, providing that this is to a level area without surface water courses, that the heavy metal content of the sludge is low, and that the land application is not too frequent or heavy. For the year following application to agricultural land, the site should be clearly marked, and the land used only for pasture, fallow, or forage crops, not for produce or dairy cattle. This sludge disposal option has been used for reclamation of strip-mined areas [61], or it may be composted with the organic content of municipal waste or wood chips to produce a useful soil conditioner. Or the sludge may be dried on its own to enable economic shipping and marketing as a low analysis fertilizer, as is promoted by the city of Milwaukee, WI, with their Milorganite.

If none of the disposal options above can be practiced, then wet or dry incineration are the only alternatives. Wet oxidation with air under pressure at 300–350°C, the Zimmerman process, yields liquid effluent containing a much reduced volume of easily settled ash [62]. Dry incineration in multiple hearth units uses much of the heat generated by combustion to complete the drying of the entering sludge [63]. The final product, a sterile ash, produced is only 3–5% of the original dewatered volume for easier ultimate disposal. Appropriate precautions should be taken for the possibility of metal elution from the ash in the landfill. Alternatively, the sludge can be more thoroughly dewatered by filtration or centrifugation and combined with milled coal to form briquets in a process developed by BASF, and then burned in a coal-fired power station as a low sulfur energy source [64].

5.13. INDUSTRIAL LIQUID WASTE DISPOSAL

Industrial liquid wastes can only be considered for discharge into municipal sewerage systems if they meet certain strict criteria, because of possibly severe incompatibility problems. They may have very high BODs, in the tens of thousands instead of 100–300 mg/L, corresponding to carbon loadings of 1% or more. Oxygen demands that are this high can seriously overload the BOD removal capabilities of the usual municipal sewage treatment plant. Industrial waste streams may contain toxic constituents such as cyanide ion, heavy metals, or toxic organics, which could kill the active microorganisms employed for secondary sewage treatment. They may also contain dissolved or immiscible solvents, which are hazardous to both the sewage treatment plant and to the sewage collection system. An accident with released hexane, for instance, left the main street of Louisville, Kentucky, pock-marked with a regular sequence of 7-m craters as a result of sewer explosions [65]. Therefore, many kinds of industrial liquid wastes require preliminary treatment before acceptance by a municipal system and many are not acceptable to a municipal system under any circumstances.

With any by-product or waste stream, utilization if possible, is a far more attractive proposition than paying for ultimate treatment and disposal [66]. For instance, diphenyl ether is obtained as a by-product of the chlorobenzene to phenol process roughly in the proportions of (Eq. 5.32). It is not the desired product. Diphenyl ether production may be decreased by recycling

$$20 \, C_6H_5Cl + 20 \, NaOH \rightarrow 18 \, C_6H_5OH + (C_6H_5)_2O + 20 \, NaCl + H_2O \quad 5.32$$

a part of it back into the process. However, markets have been developed to use this as a component of high-temperature heat transfer fluids and in brake fluid formulations. In these applications, diphenyl ether performs a useful function and earns a profit rather than imposing a disposal cost burden on the process.

As other examples of this philosophy, the anhydrous hydrogen chloride in excess of market requirements produced during chlorinations of hydrocarbons to halocarbon solvents, or during the manufacture of toluene diisocyanate, may be usefully employed in the production of phosphoric acid from phosphate rock (Chap. 10). Fluorides associated with the production of phosphoric acid and phosphatic fertilizers may be used to manufacture synthetic cryolite (Na_3AlF_6), useful in aluminum smelting (Chap. 12), or fluorspar, valuable in iron and steel metallurgy (Chap. 14). The iron-containing red muds from alumina purification prior to smelting, and the spent acids produced in the pickling of steel plate with sulfuric or hydrochloric acid can be used to prepare chemicals useful in water and sewage treatment [67]. Aluminum itself can also be economically recovered from fly ash [68].

A large number of examples of these types of uses can be found for process by-products. However, each new process developed poses a new by-product problems to test the ingenuity of the developers to come up with production avoidance strategies, or uses for them, rather than generate a waste disposal problem. Thoughts of this kind were behind the "Product Stewardship" ideas advanced within the Dow Chemical Company in 1971, to promote this concept at an early stage of new process development [69], and have recently been adopted by the Chemical Specialties Manufacturers Association.

Ingenuity in the handling and utilization of these kinds of wastes can be in-house [70], or may be facilitated by good communication between companies with potentially useful waste streams available. The setting up of waste trading exchanges [71], or the establishment of an independent waste disposal company [72] can help to achieve optimum codisposal options for a wide variety of wastes.

5.13.1. Aqueous Wastes with High-suspended Solids

Liquid waste streams with a high-suspended solids content can be cleaned up by solids removal in clarifiers, thickeners, and liquid cyclones and by accelerated settling by inclined "Chevron" settlers or the like [73]. For waste streams with very finely divided solids in suspension (i.e., less than about 100 μm) dissolved air flotation techniques have been shown to be more efficient than methods employing sedimentation. Final dewatering of the sludges obtained may be carried out on a continuous filter or a centrifuge. The clarified water product can be accepted for more potential options of reuse or final disposal options than untreated water, and the separated solids may be burned or discarded to landfill, as appropriate [74].

5.13.2. Aqueous Wastes Containing an Immiscible Liquid

Liquid waste streams containing an insoluble liquid can arise from extraction processes, from steam ejectors operating on solvent distillation systems, or from the loss of heat exchange fluid from a heat exchanger. These should be phase-separated before final disposal measures are undertaken. A simple settler, or a unit such as an American Petroleum Institute (API) separator can be used to accomplish this step. Coupling the initial separator to an entrained or dissolved air flotation unit can reduce the concentration of residual organics further [75]. The recovered organics can be recycled via a further cleanup if required, and the water phase more safely discarded.

5.13.3. Heated Effluent Discharges

Various options exist to deal with high thermal loads to water [76]. If the cooling water supply is plentiful the impact of the heated discharge can be decreased by reducing the temperature of the effluent. Increasing the cooling water flow rate to existing heat exchangers, or installation of larger heat exchangers can accomplish this. If the cooling water supply is limited, it may be reused by passage through a cooling pond or through convective or forced air circulation evaporative cooling towers (Fig. 5.10). When heated effluent is available near an urban center it can be used for community heating, at least in winter. Or it can be employed for the warming of greenhouses or soil heating, or for accelerated fish rearing. If water supplies for this purpose are very limited, indirect cooling towers, which do not use evaporation, or air cooling options may have to be adopted. These options may also be necessary if evaporative cooling causes severe fogging in the vicinity.

5.13.4. Aqueous Waste Streams with a High Oxygen Demand

Aqueous waste streams with BODs of 5,000 to 20,000, which are essentially nontoxic may be produced by a food processing industry or by a distillery, may be efficiently treated by extended aeration in lagoons, or in deep shaft

FIGURE 5.10 Diagrams of direct (evaporative) and indirect cooling towers designed to shed heat loads to air rather than surface waters.

aeration units [77]. With added nutrient ammonium phosphate to promote bacterial waste utilization, BODs may be decreased by 90–95%. Spray irrigation of this type of waste stream with or without prior BOD reduction has also been advocated, in which case infiltration to the soil and the activity of soil bacteria are relied on to deal with the oxygen demand. With this system, it is important not to overload the land disposal area to avoid producing anaerobic conditions, which could interfere with soil recovery.

5.13.5. Highly Colored Wastewaters

Activated carbon has been tested for the cleanup of highly colored waste streams such as result from dyeing operations [78]. Coal carbons are 99% effective for dye removal and accept dye loadings as high as 0.40 kg dye/kg carbon. Lignite carbon is also useful, but less effective.

A combination of lime (10 g/L) and ozone (at 0.9 g/L) were found to be optimum to remove color from fermentation effluents, when neither alum nor biochemical oxidation were effective [79].

5.13.6. Fluid and Solid Combustible Wastes

Combustible wastes, which are fairly fluid at ordinary temperatures may be simply atomized and burned for disposal. The combustion units to handle organics containing a heteroatom such as nitrogen or chlorine are designed to handle more corrosive combustion conditions. They would also require a water scrubber or other gas emission controls to clean combustion gases before discharge. Very viscous or solid combustible wastes are normally destroyed by burning in fluidized bed or multiple hearth combustors [74] (Fig. 5.11). Energy recovery can be practised to improve the economics of this method of waste destruction [80].

Aqueous solutions of volatile combustible organics such as phenols, alcohols, and ketones can also be burned for disposal, but if dilute may have to be assisted by an external fuel. Fairly fluid combustible wastes burned at the same time as these aqueous combustibles may be discarded simultaneously without requiring purchased fuel to assist the process.

5.13.7. Neutralization and Volume Reduction of Intractible Waste Streams

Noncombustible or toxic liquid wastes that do not lend themselves to disposal by any of the means just outlined are the most difficult to dispose of safely. These kinds of wastes provide the strongest incentive for a producer to develop a use of the waste stream to minimize ultimate disposal costs. Thus, acidic pickling plant wastes, which contain iron plus unused sulfuric or hydrochloric acids may be used to precipitate phosphate from secondary sewage effluent (e.g., Eq. 5.25). Or they may be blended with aqueous alkaline wastes containing phenolate and unexpended sodium hydroxide to neutralize the pH extremes of both streams.

Wastes containing cyanide ion and various heavy metals may be partially detoxified (cyanide neutralized) by combining these in proper proportions with the toxic waste streams from chlorohydrin-based ethylene oxide or

FIGURE 5.11 Multiple hearth combustion unit with emission controls, for the disposal of sludges and combustible solid wastes. Rakes around air preheat tube at center of multiple hearth unit rotate slowly around tube to move initial sludges or solids across each tray and down, eventually to exit as ash from the base of the unit. (Adapted from Sebastian and Cardinal [74]. Excerpted by special permission from Chemical Engineering [Oct. 1998]. Copyright © 1968, by McGraw-Hill, Inc., New York, NY 10020.)

propylene oxide manufacture, which will usually contain residual hypochlorite (Eq. 5.33).

$$2\,NaCN + Ca(OCl)_2 \rightarrow 2\,NaCNO + CaCl_2 \qquad 5.33$$

At higher ratios of oxidant to cyanide, complete conversion to carbon dioxide and nitrogen can be achieved [81] (Eq. 5.34).

$$4\,NaCNO + 3\,Ca(OCl)_2 + 2\,H_2O \rightarrow N_2 + 4\,NaCl + CaCl_2 + 2\,Ca(HCO_3)_2 \quad 5.34$$

Since a slight excess of chlorine is usually employed to obtain cyanide destruction, the waste stream may still be quite toxic from the residual chlorine present. Dechlorination using sulfur dioxide can be used to reduce any residual chlorine to chloride, which removes the toxicity before discharge (Eq. 5.20).

The metals content of an aqueous waste stream can be substantially decreased by complexation or adsorption methods and the recovered metal used to offset a part of the treatment costs [82]. Unexpended hypochlorite can also be neutralized by contacting this stream, in the correct proportions, with spent scrubber liquor from sulfur dioxide emission control systems (Eq. 5.35).

$$Ca(OCl)_2 + 2\,CaSO_3 \rightarrow CaCl_2 + 2\,CaSO_4 \qquad 5.35$$

Decreasing the volume of waste streams containing a high concentration of dissolved salts by such means as ion exchange or reverse osmosis can also help reduce the ultimate disposal costs of the solutions of salts produced by these neutralization reactions by decreasing disposal volumes.

5.13.8. Ultimate Destruction or Disposal of Hazardous Wastes

Incineration at the high continuous operating temperatures of a cement kiln [83] or by large ocean-going incinerators such as the Chemical Waste Management, Inc., and Bayer vessels, *Vulcanus* and *Vesta*, operating at sea, are two ultimate disposal options for more than usually hazardous wastes. Monitoring of combustion gases for complete destruction is important. Chemical fixation into absorbent solids by proprietary processes such as those trade-named Sealosafe, Chemfix, or Poz-o-tec, and then use of the fixed product as landfill or as foundation material is a further alternative [84]. Direct land-filling with proper site management, recording, and site posting has also been used for hazardous waste disposal [72]. A clay lining to the disposal site decreases the transmission of larger molecules through the soil, but is still quite permeable to smaller organics [85]. More rigorous backup containment by a twin membrane, such as the Robertson Barrier [86], in which the membranes are separated by a tough polymer mesh, or by a cement lining supplemented by a drainage system under the lining for additional resilience, may be more useful in some situations.

Deep well disposal, generally to a brine aquifer at depth, has also been used, particularly for waste brines (primarily $CaCl_2$, NaCl, and similar). To use this method without damage to groundwater resources requires a great deal of care. The wells must be located in a seismically stable region, and double pipe lining must be used to deliver the discharge to a porous stratum that is capped by one or more impermeable strata to ensure retention of the waste in this stratum. Drilling the well itself is expensive, US$30–100/meter. Also, suspended solids must be removed before discharge to avoid plugging and injection pressures kept low to ensure safe operation [87] (Fig. 5.12). The

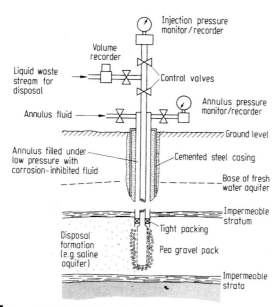

FIGURE 5.12 Details of an injection installation for deep-well disposal of liquid wastes.

potential long-term costs of a poorly operated deep well disposal installation if it should cause groundwater contamination are so great that licensing and monitoring of these operations should always be required. There is a potential for contamination of groundwater from deep well disposal operations. Parallel risks from bacteria and viruses also exist in groundwater recharge operations using treated sewage [88], or from toxins leached from poorly placed or poorly operated municipal or industrial waste dumps [89, 90].

REVIEW QUESTIONS

1. (a) Where does the dividing "line" come between particle diameters, which settle relatively rapidly and particle diameters, which settle slowly in water? How does this dividing line compare with the parallel dividing line for particles in air or other gases?
   (b) Briefly, what effect do you expect the above comparison to have on the relative ease of removal of particles from wastewater streams as compared to gas streams?

2. Stearic acid, $C_{17}H_{35}COOH$, may be used to provide a monolayer to reduce the rate of water loss by evaporation from the surface of a reservoir. The molecule floats at the surface with the hydrophobic hydrocarbon chain upward and the hydrophilic carboxylic acid downward, and in so doing covers roughly a circular area of radius 2.82 Å (0.282 nm) per molecule. How many kilograms of stearic acid would be required to form a solid monolayer over the surface of the Elk Lake reservoir, which covers an area of 289 acres? (1 acre $= 4,047\,m^2$)

3. (a) When chlorine is added to water of initial pH 5 for disinfection, all of the active hypochlorous acid formed remains undissociated. What is the percent weight ratio, hypochlorous acid formed to initially added chlorine, under these circumstances?
   (b) If calcium hypochlorite $(Ca(OCl)_2)$, for convenience, is used for water disinfection instead of chlorine, what percent ratio of hypochlorous acid to the initially added weight of calcium hypochlorite could optimally be formed?
   (c) Considering your answer to part (b), what rate of anhydrous calcium hypochlorite would have to be added to water to equate to the effectiveness of treating with 2 ppm of elemental chlorine?
   (d) What dose rate, in mg/L, would have to be used for a calcium hypochlorite bleach solution of 5 1/4% available chlorine (by weight) to equate to treating with 2 ppm of elemental chlorine?
   (e) Addition of potassium hypochlorite to water of pH 5 for disinfection produces what percent weight ratio, hypochlorous acid to initially added potassium hypochlorite, using this disinfecting agent?
   (f) Explain at least two factors, which would favor use of sodium or calcium hypochlorite over potassium hypochlorite for water disinfection.

4. Give two reasons why sparging a refinery wastewater stream with small air bubbles (froth flotation) should provide such an effective way

of removing further hydrocarbon impurities in a refinery wastewater stream, even after phase separation in an API separator.

5. Chlorine and calcium hydroxide (slaked lime), used to ensure high pH conditions to avoid generation of extremely toxic hydrogen cyanide gas, are used in combination to chemically neutralize waste aqueous solutions of sodium cyanide. In this way the much less toxic cyanate ion (CNO^-) is produced for discharge by oxidation of the cyanide ion.
   (a) Write the equations, which would govern the chemical reactions required to carry out this neutralization.
   (b) What would be the required stoichiometric amounts of chlorine and calcium hydroxide to neutralize 1,000 L of effluent containing 100 mg/L sodium cyanide?

6. Sulfide ion (S^{2-}), which is toxic, may be present in the wastewaters from froth flotation of sulfide ores or from refinery operations. It may be neutralized by the addition of the stoichiometric (not excess) amount of chlorine to oxidize sulfide to sulfate.
   (a) Give the equations, which describe the chemistry of this neutralization.
   (b) What rate of chlorine addition (g/hr) would be required to neutralize an alkaline waste stream of 35 L/hr containing 1,200 mg/L sulfide?
   (c) What problems, apart from excessive reagent cost, might the operator of such a neutralization plant face on the addition of a large excess of chlorine to the sulfide-containing waste stream?

7. With a 2.5:1 mole ratio of sodium hypochlorite to potassium cyanide it is possible to convert the toxic cyanide ion to harmless nitrogen gas:

$$2\ KCN + 5\ NaOCl + H_2O \rightarrow N_2 + 5\ NaCl + 2\ KHCO_3$$

What amounts of sodium hydroxide and chlorine would be required to neutralize 100 L of effluent containing 30 mg/L potassium cyanide in this way?

8. Seawater at 25°C, assumed to contain only NaCl at 3.00% by weight (density 1.0197 g/mL), is to be desalted to 10% by volume of a freshwater permeate still containing ca. 200 ppm by weight salt.
   (a) What would be the approximate osmotic pressure (Π) of the raw seawater used (i.e., the pressure, which would have to be exceeded to produce desalinated water)?

$$\Pi = i[S]RT(\text{approx.}),$$

where i = no. of ions (species) produced on dissolving solute in water, [S] = molarity of solute, R = 0.08206 L atm/deg mol, and T = absolute temperature, K.
   (b) What would be the molar concentration of salt in the rejected brine (i.e., in the 90% that is rejected), and the approximate osmotic pressure of the reject?
   (c) If 50% of the raw seawater was to be retained as freshwater permeate, what would be the salt concentration, and the approximate osmotic pressure of the rejected brine?

(d) Give two reasons why 10% freshwater retention may be more practical for routine reverse osmosis operations than 50% retention.

9. A good municipal raw water supply with a total organic carbon (TOC) content of 2.4 mg/L and bromide ion concentration of 20 μg/L is disinfected with 1.5 mg/L chlorine.

(a) Demonstrate whether the measured chloroform concentration in the range of 2.3–14 μg/L in the treated water is consistent with the disinfection method used.

Possible approach: Write an equation assuming a 1.00:1.50 mole ratio of C (organic): Cl_2, and Cl_2: $CHCl_3$ (the optimum possible chlorine utilization ratios). Using the mass ratio from this equation, estimate the maximum chloroform concentration that would be expected from the initial TOC and chlorine contents.

(b) Does the 2.5 mg/L chloride ion that is also present in the raw water have any role in trihalomethane formation? Explain.

(c) Bromodichloromethane at 0.3–0.8 μg/L is also found in the treated drinking water. Describe a feasible pathway, starting with bromide ion that could describe how this might happen. Using the mass ratios from an equation employing C (organic): Br_2, and Br_2: $BrCl_2CH$ mole ratios of 1.0:0.5, demonstrate whether or not the measured bromodichloromethane concentration is feasible from the raw water analytical data, and the given chlorine disinfection rate.

(d) Other trihalomethanes are found in chlorine-disinfected municipal water when there is a ten times higher bromide ion concentration in the raw water supply. What are they?

10. (a) Using the methods outlined in 9.(a), estimate the maximum possible chloroform concentration from treated domestic wastewater containing 20 mg/L organic carbon (TOC), when this is disinfected with 50 mg/L chlorine. Base your answer on the concentrations and the elemental proportions present, to produce $CHCl_3$ as the sole halomethane product.

(b) What role might the timing of postdisinfection SO_2 neutralization of excess chlorine play in the final chloroform concentration found in the final treated wastewater? Use equations to outline the processes involved.

11. Chlorine, at 1–3 ppm, and bromine, at 3–5 ppm, are recommended disinfectants for use with hot tubs and spas. Their mechanism of action is "much the same."

(a) Write the appropriate equations to describe the formation of the active bactericide from a solution of bromine in water.

(b) Assuming that the mode of action is much the same, why should one need 3–5 ppm of bromine when 1–3 ppm is sufficient with chlorine?

(c) Calculate the weight/volume, and the molar concentrations as halogen (unreacted), of 3 ppm Cl_2 in water and 5 ppm Br_2 in water, assuming a final solution density of 1.000 g/cm^3 for each.

(d) Suggest a plausible reason for the higher disinfectant concentrations called for (whatever disinfectant is used) in this application than the 1.0 ppm chlorine recommended for drinking water.

11. (a) What absolute air pressure would just be sufficient to allow aeration by compressed air fed to a perforated pipe laid at the 10-m-deep dredged floor of a stagnant freshwater harbor. (Assume water density 1.00 g/mL.)

(b) For a given volume of air injected, would small or large bubbles give better oxygen absorption. Why or why not?

(c) For an optimum bubble size in the same aeration system with both gases, what volume of pure oxygen would be theoretically required to obtain equivalent dissolved oxygen in the harbor waters as with 1,000 L of air under the same conditions?

(d) Why might oxygen bubbles be completely absorbed, under microbubble conditions, when air microbubbles continue to break the surface?

FURTHER READING

G.F. Connell, "The Chlorination/Chloramination Handbook." American Water Works Association, Denver, CO, 1996.

EPA, "Guiding Principles for Constructed Treatment Wetlands Providing for Water Quality and Wildlife Habitat." Environmental Protection Agency, Office of Water, Washington, DC, 2000.

C. Gottschalk, J.A. Libra, and A. Saupe, "Ozonation of Water and Wastewater: A Practical Guide to Understanding Ozone and its Application." Wiley-VCH, Weinheim, 2000.

M.J. Hammer, "Water and Wastewater Technology." Prentice-Hall, Upper Saddle River, NJ, 2001.

R.D. Letterman, "Water Quality and Treatment, A Handbook of Community Water Supplies." McGraw-Hill, New York, 1999.

D. Mara and N.J. Horan, "The Handbook of Water and Wastewater Microbiology." Academic Press, San Diego, CA, 2003.

"Waterborne Pathogens," American Water Works Association, Denver, CO, 1999.

N.C. Thanh and A.K. Biswas, eds., "Environmentally-sound Water Management." Oxford University Press, New York, 1990.

REFERENCES

1. R.N. McNeeley, V.P. Neimanis, and L. Dwyer, "Water Quality Source Book: A Guide to Water Quality Parameters." Inland Waters Directorate, Water Quality Branch, Ottawa, 1979.

2. J.E. McKee and H.W. Wolf, "Water Quality Criteria," 2nd ed., Publ. No. 3-A. California State Water Quality Control Board, Sacramento, CA, 1971.

3. "Canadian Drinking Water Standards and Objectives 1968." Joint Committee on Drinking Water Standards, Advisory Committee on Public Health Engineering, and Canadian Public Health Assoc., Ottawa, 1969.

4. R.B. Hornick, S.E. Griesman, T.E. Woodward et al., Typhoid Fever: Pathogenesis and Immunologic Control, N. Engl. J. Med. 283(13), 686–691, Sept. 24 (1970).

5. D.K. Todd, "The Water Encyclopedia." Water Information Center, Port Washington, NY, 1970.

6. "Water Treatment Handbook" (translated from the French by D.F. Long and Co.). Degremont Co., London, 1973.

7. A Refresher on Sodium Hypochlorite, *Water Pollut. Control*, **115**(1), Jan. 11 (1977).

8. M.B. Hocking, Treatment of Drinking Water, *Chem. Eng. News*, **61**(1), 50, Jan. 3 (1983).

9. G.C. White, "Handbook of Chlorination," p. 189. Van Nostrand-Reinhold, New York, 1972.

10. B. Hileman, Cancer Risk Found from Water Chlorination, *Chem. Eng. News*, **70**(28), 7–8, July 13 (1992).

11. L. Raber, Chloroform is Safe at Levels Found in Tap Water, *Chem. Eng. News*, **76**(23), 28, June 8 (1998).

12. C. Anderson, Cholera Epidemic Trace to Risk Miscalculation, *Nature*, **354**, 255, Nov 28 (1991).

13. S.E. Hrudey, Risk ranking: Beware the Dangers of Misplaced Confidence, *Envir. Risk Manag.* **2**(1), Mar. 3 (1995).

14. E.J. Calabrese, G.S. Moore, and R.W. Tuthill, *J. Environ. Health*, **41**(1), 26, July/Aug. (1978).

15. M. Peleg, *Water Res.* **10**, 361 (1976).

16. Calcium May be Key to Health, *Chem. Eng. News*, **55**(16), 17, Apr. 18 (1977).

17. R.A. Horne, "The Chemistry of Our Environment." Wiley-Interscience, New York, 1978.

18. J.-Q. Jiang and N. Graham, Clearly Cleaner, *Chem. Brit.* **34**(3), 38–41, Mar. (1998).

19. J. Passanisi, Project Compares Brackish Water Desalination Technologies. Part 1, *Water/Engin. & Manag.* **149**(2), 14–17, Feb. (2002).

20. S. Kalogirou, Survey of solar desalination systems . . . , *Energy* (Oxford, England). **22**, 69–81, Jan. (1997).

21. H. Ettouney, H. El-Dessouky, and F. Al-Juwayhel, Performance of the Once-through Multistage Flash Desalination Process, Proc. of the Inst. of Mech. Eng. Part A, *J. Power Energy*, **216**(A3), 229–241 (2002).

22. L.T. Pryde, "Environmental Chemistry: An Introduction." Cummings, Menlo Park, CA, 1973.

23. N. Wade and K. Callister, Desalination: The State of the Art, *Water Environ. Manag. J.* **11**, 87–97, Apr. (1997).

24. K.B. Franca, H.M. Laborde, and H. Neff, ... Solar Powered Water Desalination Systems, Utilizing Reverse Osmosis, *J. Solar Energy Engin.* **122**(4), 170–175, Nov. (2000).

25. D. McBain, *Chem. Brit.* **12**(9), 281, Sept. (1976).

26. A.D. Eaton, L.S. Clesceri, A.E. Greenberg *et al.*," Standard Methods for the Examination of Water and Wastewater," 20th ed., Amer. Public Health Assoc., Amer. Water Works Assoc., & Water Envir. Federation, Washington, DC, 1998, (and earlier editions).

27. T.V. Arden, *Chem. Brit.* **12**(9), 285, Sept. (1976).

28. L. Klein, "River Pollution," Vol. 2. Butterworth, London, 1962.

29. "Cleaning Our Environment: A Chemical Perspective," 2nd ed. American Chemical Society, Washington, DC, 1978.

30. "Cleaning Our Environment: The Chemical Basis for Action." American Chemical Society, Washington, DC, 1969.

31. R. Field and C.-Y. Fan, *J. Environ. Eng. Div. (Am. Soc. Civ. Eng.)* **107**, 171, Feb. (1981).

32. D.V. Ellis, "Sewage Disposal to the Sea," West. Geogr. Ser., Vol. 12, p. 289. University of Victoria, Victoria, BC, 1977.

33. S.A. Berry and B.G. Notion, *Water Res.* **10**, 323 (1976).

34. W. Whipple, Jr., No. 97, *Chem. Eng. Prog. Symp. Series*, **65**, 75 (1969).

35. A First for Alberta, Aeration by U-tube, *Water Pollut. Control*, **110**(4), 42, Apr. (1972).

36. B. O'Connor, T. Kovacs, S. Gibbons *et al.*, . . . Paper Mill Effluents from Oxygen-activated Sludge Treatment . . . A Source of Toxicity to Fish, *Water Qual. Res. J. Canada*, **35**(2) 189–200 (2000).

37. R.J. Benoit, Self purification in natural waters, In "Water and Water Pollution Handbook," (L.L. Ciaccio, ed.), Vol. 1, pp. 141–215. Marcel Dekker, New York, 1971.

38. C.C.K. Lui and Y-S Fok, . . . Assimilatory Capacity Analysis Using . . . Tracer Techniques, *Water Res. Bull'n.* **19**(3), 439–445, June (1983).

39. "Discharge of Selected Rivers of Canada." The Secretariat, Canadian National Committee for the Internat. Hydrological Decade, Ottawa, 1972.

40. H.E. Babbit and E.R. Baumann, "Sewerage and Sewage Treatment," 8th ed. Wiley, New York, 1958.
41. L.W. Weinberger, D.G. Stephan, and F.M. Middleton, *Ann. N. Y. Acad. Sci.* **136**, 131, July 8 (1966)
42. L. Smith and R.V. Daigh, *J. Water Pollut. Control Fed.* **53**(8), 1272 (1981).
43. "Introduction to Popular Treatment Methods for Municipal Wastes and Water Supplies," Ontario Water Resources Commission, Toronto, ca. 1972.
44. J.A. Chittenden and W.J. Wells, Jr., *J. Water Pollut. Control, Fed.* **43**(5), 746 (1971).
45. O.C. Collins and M.D. Elder, *J. Inst. Water Pollut. Control*, **79**(2), 272 (1980).
46. K.L. Murphy, *Water Pollut. Control*, **112**(4), 24, Apr. (1974).
47. H.E. Maynard, S.K. Ouki, and S.C. Williams, Tertiary Lagoons: A Review of Removal Mechanisms and Performance, *Water Res.* **33**(1) 1–13, Jan. (1999).
48. D.W. Hendriks, F.J. Post, and D.R. Khairnar, *Water, Air, Soil Pollut.* **12**, 219 (1979).
49. R.E. Finger, *J. Water Pollut. Control Fed.* **45**(8), 1654 (1973).
50. M.B. Hocking, K.A. Klimchuk, and S. Lowen, Polymeric Flocculants and Flocculation, *J. Macromol. Sci.: Rev. Macromol. Chem. Phys.* C39 (1), 175–201 (1999).
51. Fe Wastes Help Remove Phosphates, *Can. Chem. Process.* **58**(6), 44, May (1974).
52. L.A. Campbell and A.J. Horton, *Water Pollut. Control*, **110**(3), 28, Mar. (1972).
53. Reverse Osmosis Studied for Sewage Treatment, *Water Pollut. Control*, **112**(4), 15, April (1974).
54. R. Kunin and D.G. Downing, *Chem. Eng. (N.Y.).* **78**, 67, June 28 (1971).
55. E.J. Middlebrooks, ed., "Water Reuse." Ann Arbor Sci. Publ., Ann Arbor, MI, 1982.
56. J.L. Jones, D.C. Baumberger, Jr., F.M. Lewis *et al.*, *Environ. Sci. Technol.* **11**(8), 968 (1977).
57. D.W. Osborne, *J. Proc. Inst. Sewage Purif.* **3**, 195 (1962); cited by L. Klein, "River Pollution." Vol. 3. Butterworth, London, 1966.
58. Spinoff Process Aids Sewage Treatment, *Environ. Sci. Technol.* **5**(9), 756 (1971).
59. S.I. Borrely, A.C. Cruz, N.L. Del Mastro *et al.*, Radiation Processing of Sewage and Sludge. A Review, *Prog. Nucl. Energy*, **33**(1–2), 3–21 (1998).
60. Porteous Unit Ready for Start-up, *Environ. Sci. Technol.* **2**(11), 1068 (1968).
61. W.E. Sopper, Reclamation of Mine Land Using Municipal Sludge, *Adv. Soil Sci.* **17**, 351–431 (1992).
62. Wet Air Oxidation Processes, *Chem. Can.* **29**(5), 15 (1977).
63. J. Werther and T. Ogada, Sewage Sludge Combustion, *Progr. Energy Combust. Sci.* **25**(1), 55–116 (1999).
64. Energy from Sludge, *Chem. Brit.* **19**(2), 98, Feb. (1983).
65. Ralston Purina Suspect in Louisville Blast, *Chem. Eng. News*, **59**(8), 11, Feb. 23 (1981).
66. M. Campbell and W. Glenn, "Profit from Pollution Prevention." Pollution Probe, Toronto, 1982.
67. V. Orescanin, K. Nad, V. Valkovic *et al.*, Red Mud and Waste Base: Raw Materials for Coagulant Production, *J. Trace Microprobe Tech.* **19**(3), 419–428 (2001).
68. Process Extracts Aluminum from Flyash, *Chem. Eng. News*, **60**(16), 28 (1983).
69. Annual Report, Dow Chemical Company, Midland, MI, 1975.
70. M. Reisch, Du Pont Reaps Profit from Waste Streams, *Chem. Eng. News*, **90**, 13, Oct. 22 (1990).
71. Waste Exchanges turn Problems into Solutions, *Water Pollut. Control*, **128**(1), Feb. 6. (1990).
72. D.C. Wilson, *Chem. Brit.* **18**(10), 720 (1982).
73. M.B. Hocking and G.W. Lee, *Fuel*, **56**(7), 325 (1977).
74. F.P. Sebastian and P.J. Cardinal, Jr., *Chem. Eng. (N.Y.).* **75**, 112, Oct. 14 (1968).
75. A.E. Franzen, V.G. Skogan, and J.F. Grutsch, *Chem. Eng. Prog.* **68**(8), 65, Aug. (1972).
76. F.L. Parker and P.A. Krenkel, "Engineering Aspects of Thermal Pollution." Vanderbilt University Press, Nashville, TN, 1969.
77. CIL to Test New Sewage Unit, *Can. Chem. Process.* **59**(10), 32, Oct. (1975).
78. P.B. Dejohn and R.A. Hutchins, *Text. Chem. Color.* **8**(4), 34, Apr. (1976).
79. B. Inanc, F. Ciner, and I. Ozturk, Colour Removal from Fermentation Industry Effluents, *Water Sci. Tech.* **40**(1), 331–338 (1999).
80. L. Bebar, P. Martinak, J. Hajek *et al.*, Waste to Energy in.... Thermal Processing of Waste, *Appl. Therm. Eng.* **22**(8), 897–906, June (2002).

81. G.R. Mapstone and B.R. Thorne, *J. Appl. Chem. Biotechnol.* **28**(2), 135, Feb. (1978).

82. A.B. Wheatland, C. Gledhill, and J.V. O'Gorman, *Chem. Indus. (London).* p. 632. Aug. 2 (1975).

83. Cement Kilns Can Destroy Toxic Chlorinated Compounds, *Can. Chem. Process.* **61**(5), 6, May (1977).

84. Waste solidification, *Chem. Brit.* **10**(3), 105, Mar. (1974).

85. W.J. Green, G.F. Lee, and R.A. Jones, *J. Water Pollut Control Fed.* **53**(8), 1348 (1981).

86. To save and protect, *Can. Chem. News,* **43**(6), 9, June (1991).

87. K.P. Saripalli, M.M. Sharma, and S.L. Bryant, Modeling Injection Well Performance During Deep-well Injection of Liquid Wastes, *J. Hydrology,* **227**(1–4), 41–55, Jan. 31 (2000).

88. B.H. Keswick and C.P. Gerba, *Environ Sci. Technol.* **14**(11), 1290 (1980).

89. J.R. McBride, E.M. Donaldson, and G. Derkson, *Bull. Environ. Contam. Toxiol.* **23**, 806 (1979).

90. Waste Sites Pose Risk to Water Supply, *Chem. Eng. News,* **58**(40), 6, Oct. 6 (1980).

6

NATURAL AND DERIVED SODIUM AND POTASSIUM SALTS

Ye are the salt of the earth: but if the salt have lost its savour, wherewith shall it be salted?
                    —Matthew, ca. 65 A.D.

6.1. SODIUM CHLORIDE

Sodium chloride, or common salt, is one of the earliest chemical commodities produced. Its production from seawater was prompted by essential dietary needs, and later for its value as a food preservative caused by the scattered accessibility of land-based sources. The word salary itself is derived from the Roman "salarium," which was a monetary payment given to soldiers for salt purchase to replace the original salt issue. While the initial production and harvesting of sodium chloride was from dietary interests, food needs today represent less than 3% of consumption, and uses as a chemical intermediate far exceed this (Table 6.1). The wide availability of sodium chloride has contributed to the derivation of nearly all compounds containing sodium or chlorine from this salt, and to the establishment of many large industrial chemical operations adjacent to major salt deposits. Three general methods are in common use for the recovery of sodium chloride, which in combination were employed for the worldwide production of 225 million tonnes of this commodity in 2000 and 183 million tonnes in 1990 (Table 6.2).

6.1.1. Solar Salt

Outdoor recovery of sodium chloride by evaporation of sea water or natural brines is carried out in areas, which have a high evaporation rate and low rainfall. Slow-evaporation rate in temperate countries restricts the use of natural evaporation to only a small fraction of the production in these

TABLE 6.1   Change in Sodium Chloride Use Profile With Time for the U.S.A.[a]

Application	1976		2000	
	Thousand tonnes	**% of total**	**Thousand tonnes**	**% of total**
Highway snow and ice control	8,101	20.1	19,700	36.5
Chlorine and sodium hydroxide production	19,003	47.2	21,300	39.4
Sodium carbonate (soda ash)	3,684	9.1	0	0
All other chemicals	1,742	4.3	1,080	0.2
Food processing	1,080	2.7	1,760	3.3
Livestock feeds	1,746	4.3	1,870	3.5
Meatpacking, tanning	524	1.3	484	0.9
Pulp and paper	193	0.5	106	0.2
Table salt	905	2.2	823	1.5
Textile and dyeing	185	0.5	209	3.9
Miscellaneous uses	3,146	7.8	6,668	12.3
Total	40,311	100	54,000	100

[a]Calculated from data in *Minerals Yearbooks* [1, 2], and the *U.N. Statistical Yearbook* [3].

areas. However, the low capital, energy, and labor costs of this process still make this a dominant and attractive method for world salt recovery. About 45% of the world's salt is produced by solar evaporation. Some examples of contributors to this total, with percentages of their production, are: Namibia (southwest Africa), 100%; Ethiopia, 88%; Portugal, 42%; Spain, 41%; France, 22%; Italy, 15%; Germany, 11%; U.S.A., 4%; and Canada, 0% [5].

TABLE 6.2   Production of the Major World Salt (NaCl) Producers[a]

	Thousands of metric tonnes			
	1976	**1980**	**1990**	**2000**
U.S.A.	40,114	36,607	37,000	45,510
China	30,000	17,280	20,000	31,280
U.S.S.R.	14,000	14,500	14,700	3,200[b]
U.K.	7,900	6,586	6,430	5,800
Germany	7,495[c]	12,970[c]	15,700	15,700
France	6,416	7,100	6,610	7,000
Canada	6,398	7,029	11,300	12,164
Poland	5,470	3,357	4,060	4,200
Australia	5,350	5,315	7,230	8,778
Mexico	4,591	5,990	7,140	8,884
India	4,480	7,262	9,500	14,453
Italy	4,012	5,270	4,430	3,600
Other Countries	31,199	35,488	38,900	68,631
Total	167,425	164,754	183,000	225,000

[a]From 1982, 1994 and 2000 *Canadian and U.S. Minerals Yearbooks* [1, 2, 4].
[b]Estimated, as Russia.
[c]For the former West Germany.

Feed brines for solar salt recovery may be derived from seawater with salinity of about 3.5%, or from one of the enclosed seas or natural salt lakes of higher salinities (Table 6.3). The Caspian Sea is not used for solar salt production because of its much lower salinity (1.3%) and its temperate location. Also, its salts content is about 25% lower in chloride and about three times higher in sulfate than ordinary sea salt.

Processing involves moving the brine, with tidal assistance if feasible, into large shallow (15–60 cm) initial evaporation ponds where solar evaporation is allowed to proceed until it reaches a density of $1.03 \, g/cm^3$, when calcium carbonate crystallizes out. Dyes may be added assist absorption of solar energy. Further evaporation to a density of about $1.10 \, g/cm^3$ causes calcium sulfate to crystallize out. Sodium chloride crystallization starts to occur when the density reaches about $1.21 \, g/cm^3$ (Fig. 6.1) [11]. Phase diagrams are available [8]. The brine density is allowed to flow or is pumped to the harvesting (crystallizing) ponds where sodium chloride crystallization is carried out until the brine density reaches 1.25 to $1.29 \, g/cm^3$. At this point, some 75% of the sodium chloride will have crystallized (Table 6.4). The mother liquor, or bittern, is then run off, and the residual crystallized salt scraped into long piles or windrows to drain. This product is washed with brine and then centrifuged to remove most of the brine. It is frequently marketed in moist form (8–10% water) [12]. Simple drying, either by natural or artificial means, produces an industrial-grade solar salt of about 95% sodium chloride. The brine-washed and dried salt may be as high as 99% NaCl [13]. Yields range from about 50 tonne/hectare/year (about 20 tonne/acre) in the San Francisco Bay area [14], 150–175 tonne/hectare/year for the May–September season from the Great Salt Lake, and 12–185, or more from seawater for areas with higher evaporation rates [15].

To obtain a food or dairy grade of product, the solar salt may be recrystallized. Pure salt is hygroscopic and tends to cake with changes in relative humidity. To avoid this, salt crystals are coated with 0.5–2% of finely powdered hydrated calcium silicate, magnesium carbonate, or tricalcium phosphate to give the crystals free-flowing characteristics. Iodized salt for table use has 0.01% potassium iodide plus stabilizers added to provide a human trace nutrient. Cattle blocks comprise highly densified and dyed aggregations of salt, to which various micronutrients are added according to the local soil deficiencies of the marketing area.

The bittern (spent brine) from solar salt production contains 300–400 g/L dissolved solids relatively enriched in the less concentrated salt impurities. This may be either discarded or further worked to recover other elements of value. Brine from the Great Salt Lake, for instance, is processed for magnesium chloride hexahydrate recovery [10], which occurs at a density of $1.26 \, g/cm^3$. This is later converted to metallic magnesium [12]. The Dead Sea brines are processed primarily for potassium chloride (potash), but are also worked for sodium and magnesium chlorides and derived products such as bromine and hydrochloric acid [16] (Sections 6.2.2 and 8.8).

Properly run, solar salt works can substantially boost the ecological diversity of the saline wetlands required to operate them [17]. Microscopic algae and seaweeds predominate in the lower salinity basins, replaced by brine

TABLE 6.3   Salinities and Compositions, in Percent by Weight of Dissolved Salts, for Some Major Brine Sources of Solar Salt

	Seawater[a]	Red Sea[b]		Searles Lake, California[c]		Dead Sea[a,d]		Great Salt Lake,[e] Utah
		Surface	Below 2000 m	Upper	Below 40 m	Upper	Below 100 m	
Salinity, %:	1–5	3.6–4.1	25.6	34.83	34.60	19.5–25	26.7	13.8–27.7
Average	3.0	3.8						25.0
Varies with	place, depth	depth		depth		time, depth		
Density, g/mL	1.025	–	–	1.30	1.31	1.17	1.24	1.10–1.22
NaCl	77.8	77.8	89.1	46.66	46.97	31.3	40.2	84.4
MgCl$_2$	10.9	10.9	–	–	–	48.4	51.5	3.8
MgSO$_4$	4.7	4.7	trace	–	–	–	–	8.4
CaSO$_4$	3.6	3.6	trace	–	–	0.56	0.02	–
KCl	2.5	2.5	1.6	13.92	8.67	4.35	4.69	3.1
CaCO$_3$	0.34	0.34	–	–	–	–	–	0.3
MgBr$_2$	0.22	0.22	2.4	–	–	1.94	2.29	–
CaCl$_2$	–	–	5.3	–	–	13.50	14.33	–
Na$_2$SO$_4$>	–	–	–	20.67	19.51	–	–	–
Na$_2$CO$_3$	–	–	–	13.35	18.35	–	–	–
Other	–	–	trace[f]	1.48[g]	1.52[g]	–	–	traces[b]
Total salts	8.44 × 10^{16i}	–	–	–	–	1.15 × 10^{10j}	–	5.44 × 10^9

[a] Calculated from data in *Encyclopaedia Brittannica* [5].
[b] Calculated from data of Degens and Ross [6].
[c] Data from Shreve and Brink [7]. Composition of the nearly dry Searles Lake salts is very similar to that of the nearby Owens Lake, from which it was separated by evaporation [8].
[d] Calculated data [3] agree quite closely with the analyses reported by Epstein.
[e] Calculated from data of Rankama and Sahama [8], Hutchinson [9], and Kirk-Othmer [10]. Traces of magnesium carbonate and other more minor constituents are also present.
[f] Predominantly of other sulfates.
[g] Includes predominantly Na$_2$B$_4$O$_7$, Na$_3$PO$_4$, and NaBr.
[h] Includes ionic strontium and borate.
[i] Corresponds to a volume of 1.87 × 10^7 km^3.
[j] It has been estimated that a total of some 910,000 tonnes of salts per year is entering via the Jordan River.

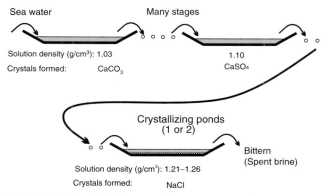

Evaporating ponds in series
(90% of pond area)

Sea water   Many stages

Solution density (g/cm³): 1.03   1.10
Crystals formed: CaCO₃   CaSO₄

Crystallizing ponds
(1 or 2)

Bittern
(Spent brine)

Solution density (g/cm³): 1.21–1.26
Crystals formed: NaCl

FIGURE 6.1 Conditions used for evaporation for impurity removal, and the crystallization stage(s) for solar salt (sodium chloride) production from seawater. Operating details from Korovessis and Lekkas [11].

shrimp and brine fly larvae at higher salt concentrations, providing food for shore birds. As they rise to about 22% salt, red halophilic bacteria proliferate, providing further food for birds, and aiding solar energy absorption.

6.1.2. Sodium Chloride by Conventional Mining

By the evaporation of large inland seas or land-locked lakes of large drainage basins in geologic history, extensive subterranean deposits of sodium chloride and other salts in layers of cumulative thickness as great as 400 m have been laid down in many parts of the world. Deposits in well-consolidated strata and less than 500–600 m below the surface are usually economic to work by conventional mining techniques. In Canada, these accessible deposits amount to about 74% of the total and in the U.S.A., about 35% of the total [1, 2].

The extent and quality of the deposit is determined from a pattern of core drillings (samples of the subterranean rock brought to the surface), which are

TABLE 6.4 Properties of Sodium Chloride and Aqueous Solutions

Density:	$2.165\,\text{g/cm}^3$
Melting point:	800.8°C
Boiling point:	1465°C
Solubility:	35.7 g per 100 g water at 0°C
	39.8 g per 100 g water at 100°C

Saturated sodium chlorine brine has a boiling point of 108.7°, and contains 28.41% NaCl. At 25°C it has a specific gravity of $1.1978\,\text{g/cm}^3$, and contains 26.48% NaCl.

obtained for the specific area of interest. Access to the horizontal or near horizontally bedded salt is obtained by sinking a vertical shaft of about 5 m in diameter. The salt is then mined by undercutting, drilling, blasting and underground haulage to move the rock salt to the shaft and then lift it to the surface. Deposits of this type are quite dense so that underground cavities can range upto 15 m in each dimension. However, stout supporting pillars of 20–60 m^2 are required at regular intervals because of the plastic (flows under pressure) nature of sodium chloride. In this way salt recoveries from the conventional mining operation usually range from 25 to 40% and sometimes as high as 60% of that present in the deposit.

Crushing and screening of the mined salt to size may be carried out underground or at the surface. Mechanical and electronic sorting then gives an unrefined but beneficiated (enriched) product of better than 95% NaCl (occasionally, better than 99%). Most Canadian and American mined rock salt is employed as mined in highway deicing. A substantial fraction of the American rock salt is also consumed for the production of chlorine, sodium hydroxide, and other chemicals (Table 6.1). For the small quantities of higher purity sodium chloride, the beneficiated material is dissolved in water, concentrated, and crystallized. The purified sodium chloride crystals are filtered and dried. Further finishing steps for marketing are as described in Section 6.1.1.

6.1.3. Solution Mining of Sodium Chloride

To recover salt from underground deposits located more than about 600 m below the surface, or with poorly consolidated strata above the deposit, solution mining is usually the method of choice. This procedure provides most of the brine for Canadian chloralkali operations. Low-production rates may be accomplished by sinking a single drilled well, protecting it with a casing and then fitting a smaller bore concentric pipe inside this casing. Piping is usually made of fiberglass-reinforced plastic to avoid corrosion problems. Pumping water down one tube displaces brine up the other. For larger rates of production, two (or more) boreholes will be sunk into the salt formation. To initiate production, water will be pumped at high rate and pressure into one hole to introduce an expansion fracture to connect the two wells, a process called "fracturing." These fractures allow water passage between the wells and starts brine recovery from one of them.

Wells in Canada recover brine from deposits as deep as 2000 m, and may have as many as 20 wells for saturated brine production from a single site. At intervals, new wells will be bored some distance removed from both the initial solution mining site and surface-producing facilities. This leaves intervening support columns of salt underground to reduce the risk of subsidence of the overlying strata [18]. Saturated brine emerging from the underground cavity will contain 26% by weight sodium chloride, small amounts (0.01–0.1%) of calcium and magnesium salts and possibly traces of ferrous and sulfide ions. Insoluble matter will have largely remained underground. This brine may be suitable as obtained, or it may require some specific chemical treatment to purify it before use (see Section 8.1).

High-purity crystalline salt product is obtained from this solution by aeration to remove most of the sulfide as hydrogen sulfide and then chlorinated to oxidize any residual sulfide and ferrous ions (Eqs. 6.1–6.3).

$$Cl_2 + H_2O \longrightarrow HCl + HClO \qquad \qquad 6.1$$

$$HCl + HClO + 2\,FeCl_2 \longrightarrow H_2O + 2\,FeCl_3 \qquad \qquad 6.2$$

$$4\,HClO + S^{2-} \longrightarrow SO_4^{2-} + 4\,H^+ + 4\,Cl^- \qquad \qquad 6.3$$

Calcium, magnesium, and ferric ions are then removed as their hydroxides by treating the brine with small amounts of sodium carbonate (soda ash) and sodium hydroxide (caustic soda) and filtering or allowing the mixture to settle (Eqs. 6.4–6.6) [14].

$$CaCl_2 + Na_2CO_3 \longrightarrow CaCO_3 \downarrow + 2\,NaCl \qquad \qquad 6.4$$

$$MgCO_3 + 2\,NaOH \longrightarrow Mg(OH)_2 \downarrow + Na_2CO_3 \qquad \qquad 6.5$$

$$FeCl_3 + 3\,NaOH \longrightarrow Fe(OH)_3 \downarrow + 3\,NaCl \qquad \qquad 6.6$$

Water is then removed from the treated brine by multiple-effect vacuum evaporation with low-pressure steam providing the heat for the first effect (Fig. 6.2).

To evaporate water from brine by heating with steam, one can only obtain somewhat less water evaporation than the mass of steam consumed (a heat transfer efficiency limitation). However, application of a partial vacuum to the heated brine in the first stage and the use of hot water vapor from this stage to heat the next stage, and so on allows far more efficient use of steam. Progressively, more water is evaporated for each tonne of primary steam used as the number of stages is increased (Fig. 6.3). Further thermal efficiencies are introduced by countercurrent movement of brine and steam (Chap. 1), and by use of the still warm exit brine to prewarm incoming brine. In combination, these measures can give evaporation efficiencies of the order of 1.75, 2.5, 3.2, and 4.0 tonnes of water per tonne of steam from two, three, four, and five stages of evaporation, respectively [19]. Capital costs increase with the number of stages, and the percent improvement obtained with each additional effect decreases so that the maximum practical number of effects is 12. Partial vacuum for the last effect (stage of evaporation) is economically provided without moving mechanical devices by the use of cooling water with a barometric condenser, and a steam or water ejector for removal of the small amount of noncondensible gases (Fig. 6.4).

A stream of a saturated brine suspension of crystals is continuously withdrawn from each evaporator–crystallizer, and the salt crystals separated on a continuous rotary filter with return of the brine to the evaporator. Much of the salt may be marketed in moist condition, or it may be passed through a drier moving countercurrently to heated air to give "vacuum salt" of typically 99.8–99.9% purity [21].

There must be some net throughput of brine through the evaporator circuit to waste to avoid gradual buildup of impurities. In some operations, in fact, when the feed brine is of consistently high purity, the bleed rate of brine mother liquor can be controlled at an appropriate level by continuously keeping track of calcium and magnesium concentrations, to give a finished sodium chloride purity of 99.5% or better, without requiring brine pretreatment [14].

FIGURE 6.2 Forced circulation evaporator such as used for simultaneous brine evaporation–crystallization. Mechanical movement of brine past the heat-exchange surface avoids decreased heat transfer efficiency from crystallization on this surface. Constructed of Monel or Monel-clad steel for parts contacting brine. (Reprinted from Kirk-Othmer [10], with permission.)

The crystal size of vacuum salt is quite small because of the high evaporation rates and turbulence in the evaporator crystallizers, which makes this product less suitable for some uses. For this reason, about 1% of the brine evaporation business is still conducted at atmospheric pressure in open pans, the "Grainer" process. Here, after special treatment to remove calcium sulfate, the brine is evaporated slowly at 95°C to give large crystals of high-surface area and purity suitable for the dairy industry. This process uses more heat per unit product than the vacuum process. Alternatively, for the large dense crystals required for the regeneration of ion-exchange water softeners, quantities of vacuum salt may simply be heated to about 815°C and the melt cooled and broken up to large pieces for this purpose.

6.1.4. New Developments in Sodium Chloride Recovery

As pressure on existing freshwater supplies tightens, seawater desalination plants, using multieffect vacuum distillation or reverse osmosis, are required in increasing numbers for provision of freshwater. The residual evaporated brines from these plants contain much higher salt concentrations than ordinary seawater and this is also obtained near potential salt markets.

FIGURE 6.3 Quadruple effect evaporator–crystallizer. Pressure and boiling temperature decrease from left to right, temperature decreases about 20–30°C across the four effects. Feed brines to each effect may be directly from brine wells, or may be fed cascade fashion from the filtrate to the salt brine slurry of the previous effect, or a combination of these methods. (Adapted from Kirk-Othmer [10], with permission.)

These features are likely to encourage further development of vacuum or solar evaporation salt recovery operations to work these brines in proximity to the desalination plants. In this way, various salts may be more profitably recovered from these artificially enriched seawaters for reasons similar to the present incentives to use the rich natural brine sources for sodium chloride production (Table 6.3). Similar energy savings should be obtained.

FIGURE 6.4 (a) Detail of a water ejector using a supply pressure of $1.4–1.7 \times 10^5$ kilopascals (gauge pressure; 20–25 psig), or higher. (b) Diagram of a barometric condenser plus steam-driven ejector to remove and condense vapor and provide the reduced pressures required for the last stage of evaporation. With steam pressures of $2.8–3.5 \times 10^5$ kilopascals (gauge pressure; 40–50 psig) or higher and six stages of ejection absolute pressures in the range of 0.005–0.50 mm Hg are available [20].

6.2. POTASSIUM SALTS

6.2.1. Potassium Carbonate

Potash, as potassium carbonate is referred to in the industry, is the name derived from the methods originally used for recovery of this potassium salt from wood ashes. About 1 tonne of crude potassium carbonate was recovered from the water leachate of the ash generated from the burning of some 400 tonnes (more than 200 cords) of hardwood [22]. Perhaps because of the availability of a large wood supply, Canada was the world's largest potash exporter during much of the first half of the 19th century.

At this time, purified potassium carbonate was primarily of value as an ingredient in glass making. The discovery of mineral potassium chloride at Stassfurt, Germany, in 1852, rapidly provided competition to the wood ash leachate industry since this mineral could be readily converted to potassium carbonate by the Leblanc process, which was already in commercial scale oper-

TABLE 6.5 Major World Producers of Potash Given as the Mass of K_2O Equivalent (see text)[a,b]

	1976	1990	2000	
	Thousand tonnes	Thousand tonnes	Thousand tonnes	% of total
Belarus	–	–	3,786	14.3
Canada	4,524	6,990	9,107	34.4
East Germany	2,873	2,650	3,407[d]	12.9[d]
France	1,453	1,290	360	1.4
Israel	615	1,310	1,748	6.6
China	440	29[e]	380[e]	1.4
Spain	480	686	653	2.5
U.S.A.	3,630	1,710	1,300	4.9
U.S.S.R.	7,480	9,000	3,700[c,e]	14.0
West Germany	1,840	2,310	3,407[d]	12.9[d]
Other	987	1,525	2,059	7.8
World Total	24,322	27,500	26,500	100

[a]Data from 1976, 1991, and 2001 *Minerals Yearbooks* [1, 23].
[b]Dominant product is potassium chloride (e.g., for Canada 99% of the potash produced). Small amounts of potassium sulfate and potassium nitrate also produced are included here.
[c]Russian Federation.
[d]Figures for reunited Germany.
[e]Estimated by *Minerals Yearbooks*.

ation (see Chap. 7). The mineral potash is also more specifically referred to as "muriate of potash" (muriatic acid = hydrochloric acid) for potassium chloride, and "sulfate of potash" for potassium sulfate in reference to these particular salts.

6.2.2. Potassium Chloride Production and Use Pattern

Prior to 1960, world production of potassium chloride was dominated by the U.S.S.R., U.S.A., East and West Germany, and France (Table 6.5). But since the incidental discovery of potash mineralization in Saskatchewan during oil prospecting in 1943, and the first commercial production there in 1962, Canada became the world's leading producer in 1968 and 1969. Canada is at present the leading producer and exporter, followed by Belarus as the second largest producer of potash.

World production volume doubled in the 10 years from 1961 to 1971 from 9.82 to 19.1 million tonnes of K_2O equivalent, and rose to 25.8 million tonnes in 1976. Since then, however, production growth appears to have levelled, with 27.5 million tonnes produced in 1990. At the 1976 levels of production, the present Canadian reserves of 107×10^9 tonnes of potassium chloride (67×10^9 tonnes, K_2O equivalent), primarily in sylvinite and carnallite minerals, would be sufficient to supply the world demand for some 2600 years. West German and Russian reserves appear to be of a similar order of magnitude [24]. The process technologies available for potassium chloride recovery, including illustrations of equipment used, have recently been reviewed [25].

Some 95% of the potassium chloride is consumed directly in fertilizer applications. Most of the rest is used to make potassium hydroxide (see Chap.

8), which is ultimately employed to manufacture special types of glass. It is also used for the saponification of fats to produce liquid soaps, among a number of other small scale uses.

6.2.3. Potassium Chloride Recovery from Natural Brines

While it would seem feasible to use seawater bitterns from solar salt facilities (Section 6.1.1) as the raw material for potassium chloride recovery, there are as yet no commercial processes operating on this basis [26]. Virtually, all present production facilities use natural brines from sources already relatively richer in potassium than raw seawater (Table 6.3), and all have to cope somehow with the solubility differentials with temperature (Table 6.6).

In general, it is expected that mixtures of salts, such as in seawater, will tend to come out of solution in the reverse order of their solubilities, though this will also be strongly affected by their initial relative concentrations and somewhat by common ion effects. Thus, a precipitation sequence something like calcium carbonate > gypsum ($CaSO_4 \cdot 2H_2O$) > sodium chloride > magnesium sulfate ($MgSO_4 \cdot 6H_2O$) > magnesium chloride ($MgCl_2 \cdot 6H_2O$) > potassium chloride > sodium bromide would normally be expected. For this reason, recovery of the less-concentrated minerals, particularly if they are more soluble than sodium chloride, requires progressively more complex processing than is required for sodium chloride recovery.

Searles Lake, California, is a remnant of a former salt lake, which is now worked for minerals. It consists primarily of beds of salts containing brine in the interstices, the brine lying from 15 cm below to 15 cm above the beds of salts depending on season and extent of flooding [28]. By comparison with seawater, these brines contain far higher concentrations of sodium and potassium ions, much of it present as the sulfate, not chloride, and negligible concentrations of calcium and magnesium (Table 6.3). Initial triple-effect evaporation allows separation of mainly crystals of sodium chloride and Burkeite ($Na_2CO_3 \cdot 2Na_2SO_4$) with a small amount of dilithium sodium phosphate. With a complex series of subsequent steps, sodium sulfate (Glauber's salt) and sodium carbonate (soda ash) are recovered, and the lithium salt is converted to lithium carbonate and phosphoric acid for recovery of both

TABLE 6.6   Change of Aqueous Solubility of Potassium Sulfate with Temperature[a]

Temperature (°C)	Solubility (g/100 mL)
0	7.35
20	11.11
40	14.76
60	18.17
80	21.40
100	24.10

[a]Data selected from *Chemical Engineers' Handbook* [27].

components. Brines from the initial evaporation are then rapidly chilled (by cooling water and reduced pressure) to bring down potassium chloride crystals on a scale of some 675 tonne/day, and leaving a supercooled solution of sodium borate. Seed crystals and more prolonged crystallization time produce crystalline sodium borate as a further product. Full details have been described [15].

Solar evaporation, from primary and secondary ponds of 100 and 30 km² in extent, the initial stage for potassium chloride recovery from the Dead Sea brines [26]. The smaller number of constituent ions present in these waters significantly simplifies salts recovery, and the fact that they contain nearly twice the relative potassium chloride concentration of seawater also improves profitability. Developed from a process, which was first operated in 1931, evaporation in the first pond reduces the volume of the brine to about one-half of the initial volume and brings down much of the sodium chloride together with a small amount of calcium sulfate (Fig. 6.5). The concentrated brines are then transferred to the secondary pond where evaporation of a further 20% of the water causes carnallite ($KCl \cdot MgCl_2 \cdot 6H_2O$) and some further sodium chloride to crystallize out. With care, a 95% potassium chloride product on a scale of some 910,000 tonne/year is obtained either by countercurrent extraction of the carnallite with brines, or by hot extraction of potassium chloride from the sylvinite matrix followed by fractional crystallization for its eventual recovery [16].

The "end brines" from the secondary evaporation pond, consisting largely of concentrated calcium and magnesium chlorides, are mostly returned to the sea. However, some 73,000 tonnes of sodium chloride, 900 tonnes of magnesium chloride and with some chemical conversion 23,000 tonnes of magnesia

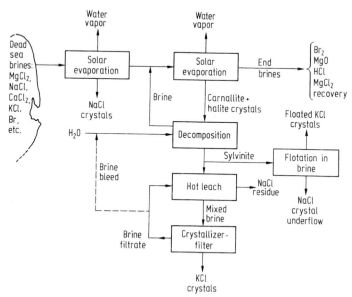

FIGURE 6.5 Flow sheet outlining details of potassium chloride recovery from the Dead Sea brines.

(MgO), 91,000 tonnes of hydrochloric acid, and 18,000 tonnes of bromine are also obtained annually from these brines. Magnesium chloride is obtained from further processing of the sodium chloride residual bittern.

Potassium is also recovered from the Great Salt Lake, Utah, but mainly as the sulfate [12].

6.2.4. Potassium Chloride Recovery by Conventional Mining and Froth Flotation

Dominant minerals, which occur in potash-bearing deposits are halite (NaCl), sylvite (KCl), and carnallite (KCl $\cdot$ MgCl$_2$ $\cdot$ 6H$_2$O), in which the two salts appear as separate crystals ranging in size from a few microns (μm) to 25 mm. Halite and sylvite minerals also occur together as physical mixtures known as sylvinite (n NaCl $\cdot$ KCl; n normally about 2). Sylvite and carnallite occur both separately and in closely associated mixtures with one another. All of these potash-bearing combinations also contain varying amounts of insolubles such as clays, anhydrite (CaSO$_4$), dolomite (CaCO$_3$ $\cdot$ MgCO$_3$) and calcite (CaCO$_3$), which variously affect potash beneficiation steps.

Recovery of sylvinite from the shallower deposits of about 1,000 m depth at the northern margin of the sloping 30-m-thick deposits in Saskatchewan, Canada, is entirely by the shaft and room method. One natural obstacle to overcome was an underground 60-m-thick bed of poorly consolidated, water-saturated sand (Blairmore, and other formations) lying below the 380 m level, which introduced considerable access difficulty. This was eventually solved by prolonged chilling of these formations with lithium chloride brine at -46°C after which the frozen formation was then safely penetrated by rapid excavation and placement of watertight caissons through the unstable formation before it thawed. Apart from the access difficulties for this particular deposit, conventional mining of sylvinite is conducted in a parallel manner to the mining of salt (Section 6.1.2).

The high grade of sylvinite obtained from deposits in Saskatchewan (about 40% KCl; n about 2) favored the application of the differential surface activity between the finely ground potassium and sodium chlorides present to effect a separation of the two minerals, rather than the considerably more energy-intensive solution of using fractional crystallization technology [14, 29]. This process marks an unusual application of froth flotation technology, commonly applied to the concentration of metal ores (Chap. 13), and here employed in the beneficiation of a water-soluble mineral. Mined sylvinite lumps are first dry crushed in gyratory or roll crushers and then ground to a fine pulp in saturated brine in a ball mill (Fig. 6.6). After classification in a cyclone, the coarse material is returned for regrinding and the fine pulp is deslimed (fine clays removed) by rigorous agitation with brine and removal of the clay-laden brine fraction. Brine is recycled after clay removal in thickeners (Fig. 6.7). The cleaned pulp is then treated with guar gum, a suppressant, which coats any remaining clay to prevent adsorption of the amine collector by clays. For each tonne of ore processed, 100 g of a tallow amine (a long alkyl chain, primary amine) is added to induce attachment of the potassium chloride particles to air bubbles, which does not affect the surface properties of sodium chloride. Also, 110 g/tonne of a polyethylene glycol is added to the

FIGURE 6.6 Flow diagram of the separation of fertilizer grade potassium chloride (sylvite) from sylvinite by froth flotation in saturated brine.

FIGURE 6.7 Diagram of a single deck mechanical thickener for producing a clear supernatant and a solids enriched slurry from a pulp suspension in water. Slow-moving rakes transport sedimented solids toward thickened pulp outlet, and an exit valve for this stream allows control of the solid/liquid ratio.

mixture to encourage formation of a stable froth. An anionic (fatty acid salt) collector may be used if it is desired to reverse the process and float sodium chloride to give a potassium chloride underflow [22].

Froth flotation of the surface-sensitized pulp by vigorous aeration and agitation in saturated brine (density ca. 1.18 g/mL), first in a series of "rougher" cells with further refinement in "cleaner" units, produces a stable froth consisting almost entirely of beneficiated potassium chloride and an underflow of sodium chloride (particle densities of 1.984 and 2.165 g/mL), respectively. Potassium chloride recoveries from sylvinite by this procedure are 90 to 95% complete [22].

FIGURE 6.8   Solid bowl continuous centrifuge for separation of a solid suspension from a liquid. With sodium chloride or potassium chloride slurries in brine it is capable of producing a dry cake of 92–99% solids.

Flotation streams are centrifuged (Fig. 6.8) for separate recovery of the potassium chloride and sodium chloride and the brine is returned, usually via a thickener, to the flotation circuit for reuse. At present, the sodium chloride is simply dumped. Potassium chloride is dried in gas-fired rotary kilns, and then sized by sieving into one of the four commercial grades: Granular, Coarse, Standard, and Soluble, all of which are 60.2–60.7% K_2O equivalent purities. Potash dust from the drying operation is captured, dissolved in hot brine, and chilled under vacuum to form purified potassium chloride crystals. After filtration and drying, this produces a further Soluble grade of 62.5% K_2O purity. Resolution of this material and crystallization from water again gives Refined, 99.9% pure potassium chloride, which is used primarily in chemical (not fertilizer) applications.

The bulk of the fertilizer market demand is for the two Coarse grades, necessitating some compaction of the finer grades to produce these. Passage of the finer Standard and Soluble grades of material through closely spaced steel rollers under pressure forms a fused sheet of potassium chloride, because of the plastic nature of this salt. This fused sheet is then broken up and rescreened to raise the proportion of Granular and Coarse products produced.

Potassium chloride as a component in fertilizers is specified in terms of its "K_2O equivalent." On this basis, 100% pure potassium chloride is equated to 63.18% K_2O equivalent. The approximately 96% pure potassium chloride product of the flotation separation is thus equivalent to 60.7% K_2O (see Chap. 9). Chemical grades correspond to the once-crystallized, soluble product of about 99.5% KCl, and 99.95% KCl, which is the Refined, twice crystallized material. The price differential, 60–62.4% K_2O equivalent at 108–112 US$/tonne, and 99.95% KCl at 115–137 US$/tonne (1995 prices [30]) is sufficient to cover the cost of the additional processing for Chemical grade potash.

6.2.5. Solution Mining of Potassium Chloride

Innovative technology has been successfully applied to the preferential solution mining of potassium chloride from the sylvinite deposits 215 m thick and lying some 1600 m below the surface situated under much of southern Saskatchewan. These beds would at best be only marginally profitable to mine by

**TABLE 6.7 Aqueous Hot and Cold Solubility Differentials
Between Potassium Chloride and Sodium Chloride[a]**

	Solubility (g/100 mLH$_2$O)	
Temperature (°C)	Potassium chloride	Sodium chloride
0	27.6	35.7
20	34.7	36.0
40	40.0	36.6
60	45.5	37.3
80	51.1	38.4
100	56.7	39.8

[a]Data selected from *Chemical Engineers' Handbook* [27].

conventional means. This unique operation takes advantage of the differential, which exists between the hot and cold water solubilities of sodium chloride and potassium chloride (Table 6.7), to recover a larger proportion of the potassium chloride than the sodium chloride from the deposit [31]. The present capacity is 1.35 million tonne/year [32].

This process is distinguished from conventional solution mining by its application of a heated, mixed sodium chloride/potassium chloride brine, rather than freshwater at ambient temperature for salt dissolution. The warm extracting solution is pumped into the ore body via a concentric pair of pipes placed in a drill hole such that the reservoir brine temperature is maintained at 45°C or higher (Fig. 6.9). Natural formation temperatures aid in maintaining these temperatures in a large reservoir. Some sodium chloride but significantly more potassium chloride is dissolved from the formation and brought to the surface via the "up-brine" piping for potassium chloride recovery. Any suspended sediment is removed from the mixed brine in thickeners plus a filtration circuit. Following this it is concentrated to 99% saturation at 100°C in two sets of quadruple-effect vacuum evaporators operated

FIGURE 6.9 Potassium chloride from subterranean sylvinite by solution mining.

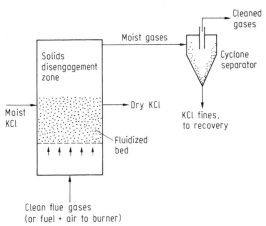

FIGURE 6.10 Fluidized bed drier. Indirectly heated air rather than the combustion gases from burning natural gas may also be used for bed fluidization and for cleaner drying action.

in a series–parallel arrangement (Fig. 6.3). Sodium chloride and any precipitated calcium sulfate are removed, hot, in a Bird solid bowl centrifuge (Fig. 6.9). The saturated, mixed brine from the centrifuges is chilled by a combination of vacuum and cooling water, which produces a suspension of potassium chloride crystals in a predominantly sodium chloride brine. The crystals are separated on horizontal table filters and dried by countercurrent application of the clean flue gases produced from gas-fired steam generation, in a fluid bed drier (Fig. 6.10). This product, 99+% potassium chloride (62.5% K_2O equivalent), is somewhat higher purity than the crude froth flotation product.

Part of the centrifuged sodium chloride is processed for table and cattle block salt production. Unused sodium chloride is slurried to a 162-hectare (1.62-km^2) impounding basin, where wet storage minimizes windage losses. The bulk of the mother liquor, together with a small amount of make-up water, is reheated and returned to the formation for resaturation with potassium chloride. Any wastewater has been disposed of underground to a water-bearing formation at the 1,200–1,500 m level. Dust from drying operations and the Tyler–Hummer screens, which are used for grade classification, is captured in water scrubbers. The brines produced by dust capture are returned to process streams for salts recovery.

6.2.6. Environmental Aspects of Sodium and Potassium Chloride Recovery

Processes, which recover sodium or potassium chlorides from natural brines originating from the ocean or salt lakes probably, have the least salination impact on surrounding lands. Since many of these operations use solar evaporation, they also have a low external energy requirement. The ratio of reserves to annual production rates, even for the salt lakes, is so large that there is not likely to be any noticeable salinity decrease for many years.

Estimates for the Dead Sea indicate that the present production rate exceeds the rate of influx by about 23%. However, these types of operations are not close to many major markets. This provides the incentive to also recover sodium and potassium salts from inland deposits.

Purification of mined inland sodium chloride has created problems with the disposal of the waste mother liquors obtained from the purification process. Deep-well disposal into a deep brine aquifer has been practised to minimize subterranean impact [33]. River discharge used to be employed, but with growing freshwater demands and the significant negative impact of a salt discharge on freshwater uses, this method has been discontinued.

Concerns exist regarding the sodium chloride stockpiles and waste disposal basins from the potash recovery operations in Saskatchewan, which are presently accumulating sodium chloride at nearly 4 million tonne/year [34]. Precautions are being taken to minimize the contamination of the limited surface and aquifer freshwater courses of the area by surrounding the waste disposal basins with impermeable clay dikes [35]. Some aeolian salt contamination of the surrounding soils is still occurring, despite windage control measures such as stockpile wetting and polymeric spray coatings. Salt-sensitive plant life in the vicinity may be affected [36].

"No particular health hazards have been demonstrated..." to workers, from potassium chloride dust in mines and mills, although a comprehensive baseline study has not been conducted [37].

6.2.7. New Developments in Potassium Chloride Recovery

Recovery of potassium salts from natural brines using solar evaporation ponds has been made less dependent on the weather at Great Lakes Minerals and Chemicals Corporation by a chloride to sulfate conversion process followed by recovery of the sulfate [1, 10]. Highly absorbing dyes added to the brine at later stages of production may enable increased production from the same evaporation area by improved solar energy absorption. Artificial sylvinite from evaporation of marine waters is less subject to potassium chloride and sodium chloride crystal agglomeration if a small amount of a crystal habit modifier, such as sodium ferrocyanide, is added during carnallite decomposition stages [23]. This improves flotation recoveries of potassium chloride.

In the processing of mined potash ores, clay in the pulped ore causes higher consumption of collector reagents by adsorption of these, and reduced sylvite recoveries. The clay may be separated from the pulped ore prior to the addition of the amine collector by a desliming decantation, which is a gentle, preflotation collection of the clays in a clay-laden froth. The brine from the clay/brine froth is recycled to the flotation circuit after flocculation of the clay followed by removal of the floc in a thickener [22].

Collector adsorption on residual clays has been decreased by employing various triethanolamine derivatives (molecular weight < 5000) as additional blocking agents for the clay active sites, in addition to the usual starches or guar gums generally used. When combined with the use of straight-chain fatty amines (and not amine salts) as the potassium chloride collector, recoveries are improved. To provide a closer particle size match between the crude product and market demands, more emphasis could be placed on pneumatic flotation assisted by more viscous collector oils, ideally of the butylpolyoxypropylene type, $C_4H_9-(O-CH_2-CH(CH_3))_n-OH$, n = 2–5. Potash of larger average particle size would be floated. This might permit abandonment of roller fusion of potash fines.

Modified fracturing techniques and multiple well sitings could raise potassium chloride recoveries from solution mining to 60–70% of in-place potash. (About 40% is usual for shaft mining.) Improvements may also be achieved by employing an oil blanket, with or without a gas head, and solvent density stratification to allow greater control of zones of dissolution [38]. Cavity interconnections may be detected by dye injection into down-brine wells followed by monitoring, to assist in avoiding subsidence. Further solution of sylvite may be prevented by a coating of sodium chloride crystallized out on the walls of the cavern.

Improved selectivity of potash recovery may be obtained by high ammonia concentrations in water, but this has not been commercially adopted. Solution mining of potash has matured from the pioneering work of Kalium Chemicals at Belle Plaine, Saskatchewan [39], to encourage adoption by at least two operations formerly based on conventional mining [40]. These operations are sited to use solar evaporation for potash recovery. One uses 162 hectares ($1.62 \, km^2$) of PVC-lined evaporation ponds to produce 236,000 tonnes of potassium chloride per year [29]. Experiments with solar absorbing dyes may lead to further increases in production.

6.3. SODIUM SULFATE

Synthetic or manufactured sodium sulfate was originally produced in Canada in the late 1800s [41], as a by-product of the Mannheim furnace method for the production of hydrochloric acid [42]. This process consists of an acidulation step, where sodium chloride is treated with sulfuric acid at high temperatures, followed by a purification step where the crude sodium sulfate is recrystallized from water (Eqs. 6.7 and 6.8).

$$2 \, NaCl + H_2SO_4 \xrightarrow{\quad 600–800°C \quad} 2 \, HCl + Na_2SO_4 \qquad \text{6.7}$$
$$\text{crude salt cake}$$
$$\text{(ca. 98\% on NaCl)}$$

$$Na_2SO_4 + 10 \, H_2O \longrightarrow Na_2SO_4 \cdot 10H_2O \qquad \text{6.8}$$
$$\text{Glaubers salt (ca. 98\% recovery)}$$

This chemistry is the basis of one operating plant in the U.S.A. It is still a significant source of synthetic sodium sulfate in Europe, and is the origin of

TABLE 6.8 Major World Producers of Natural and Synthetic Sodium Sulfate, Including Saltcake and Detergent Grades[a]

	Thousands of tonnes			
	1972	1980	1990	1997[c]
Belgium	–	249	250	250
Canada	460	454	347	305
Germany	300[b]	202[b]	167	120
Japan	410	290	253	195
Mexico	128	400	650	160
Spain	210	309	590	680
U.S.A.	1,205	1,142	665	580
U.S.S.R.	420	603	490	–
Other countries	627	854	1,458	3,230
World total	3,760	4,502	4,870	5,520

[a]Calculated from *Minerals Yearbooks* [1, 23].
[b]For the former West Germany.
[c]Estimates.

the term "salt cake," which is used synonymously with sodium sulfate to specify industrial grade material.

6.3.1. Production and Use Pattern for Sodium Sulfate

World production of sodium sulfate has averaged about 4 million tonnes/year for at least 25 years, which is approximately one-sixth the scale of potassium chloride production (Table 6.8). Spain and the U.S.A. are among the larger producers, but do not dominate world production. Almost nearly half of the total is from sodium sulfate recovered from natural sources, generally from captive lake basins in areas with high evaporation rates or from aquifers with a high dissolved sodium sulfate content. The remainder is recovered as by-product material from other industrial processes. In the U.S.A., in 1990, more than 50% of the sodium sulfate was consumed as a filler in powdered laundry detergents. This partially replaced sodium tripolyphosphate to decrease phosphate releases to surface waters. Its use in 1990 as make-up for the chemical losses incurred in the kraft process for the production of pulp and paper is still some 78% of the total supply in Canada and the U.S.A., as it has been for the last 20 years. Quantities of about 4 and 10%, respectively, are currently used in Canada and the U.S. as ingredients in glass making.

6.3.2. Recovery from Natural Brines

In recent years, only about one-half of the sodium sulfate in the U.S. has been produced from natural sources (e.g., see Table 6.3), while almost all the Canadian product is obtained in this way. Some plants use a floating dredge to mine lake bottom crystal beds of the minerals mirabilite ($Na_2SO_4 \cdot 10H_2O$), laid down by successive natural seasons of evaporation and chilling, or thenardite (Na_2SO_4), which tends to crystallize out in the presence of significant concentrations of sodium chloride [37]. Processing of the recovered natural salts involves dehydration by melting the decahydrate (melting point 32.4°C) and evaporation of the water of hydration in a submerged combustion unit. More than one-half of the weight of the crude product is water. Kiln drying of the wet crystal mass at higher temperatures produces a "salt cake" grade of product, with a minimum 97% Na_2SO_4 content, suitable for kraft process use and selling at about $90/tonne in 1995 [27].

For a "detergent grade" product, brines may be ponded for natural evaporation or may be put through multiple-effect evaporators for concentration, and then chilled to 0°C or below to induce crystallization of the decahydrate, Glauber's salt. More frequently, however, by-product sodium sulfate, which is frequently recovered from process liquors by a crystallization step, is produced on a sufficient scale and purity (up to 99.77% Na_2SO_4) to supply the small detergent grade market at $100–139/tonne in 1995 [27].

6.3.3. By-product Sodium Sulfate

By-product sources include material available from the Mannheim furnace process for the production of hydrogen chloride, which directly yields by-product sodium sulfate. This may be sold in the form obtained as a "salt cake" product, or it may be recrystallized after neutralization and removal of insoluble matter to give a detergent grade product. Careful recovery of the sodium sulfate solution generated from rayon spinning (1.1 kg of sodium sulfate is obtained for each kilogram of rayon spun) gives Glauber's salt on a scale second only to natural brines, and a mother liquor still containing some sodium sulfate, which may be returned to the spinning process. Dehydration of the Glauber's salt from this source yields a detergent grade sodium sulfate.

At present, the Hargreaves–Robinson process to produce hydrogen chloride provides a significant proportion of the by-product sodium sulfate in Europe, and at least one plant in the U.S.A. still uses this process. Air, steam, and sulfur dioxide are passed over a heated bed of porous salt granules causing a heterogeneous reaction, which yields hydrogen chloride and sodium sulfate (Eq. 6.9). The crude salt cake by-product may be purified for detergent markets in a

$$4\ NaCl + 2\ SO_2 + O_2 + 2\ H_2O \longrightarrow 2\ Na_2SO_4 + 4\ HCl \qquad 6.9$$
$$93\ to\ 98\%\ yields$$

manner similar to that used for the crude Mannheim furnace product.

REVIEW QUESTIONS

1. (a) What are the virtues and the restrictions that relate to sodium chloride recovery from natural sources?
   (b) How do these considerations affect the selection of the best sodium chloride recovery method for a recovery plant located at any particular latitude?
2. From a Handbook (Lange's, or the Chemical Rubber Co.), determine the measured solubilities of calcium sulfate, calcium carbonate, sodium chloride, and magnesium chloride.
   (a) How does the ranking of these solubilities compare with the various recovery stages of solar salt production?
3. (a) How is it possible to practice froth flotation in water as a means of separation of water-soluble minerals?
   (b) What measures can be used to improve the ability of froth flotation to differentiate between the minerals one wishes to separate?
4. What factor is used to give preferential recovery of potassium chloride from a solution mined underground mineral mixture of potassium chloride and sodium chloride, and how does this operate?
5. (a) What masses of anhydrous hydrogen chloride and crude salt cake are theoretically possible on the acidulation of 1.000 tonne of sodium chloride in the Mannheim process?
   (b) With a 100% conversion, and a selectivity of 96% to HCl, and 98% to crude sodium sulfate (both based on NaCl), what actual masses of the two products would be obtained?
   (c) Is this a hydrogen chloride process or a sodium sulfate process?

FURTHER READING

M. Hancer and J.D. Miller, The Flotation Chemistry of Potassium Double Salts: Schoenite, Kainite, and Carnallite, *Min. Eng.* **13**(14–15), 1483–1493, Dec. (2000).

H.W.J. Lee, "Optimal Control of the Solar Salt Production Process." CSIRO, Canberra, ACT, 1993.

"Potassium Chloride." Environment Canada, Environmental Protection Service, Ottawa, 1985.

"Sodium Chloride." Technical Services Branch, Environmental Protection Programs Directorate, Environmental Protection Service, Ottawa, 1984 (environmental aspects, toxicology, etc.).

REFERENCES

1. "1976 Minerals Yearbook," Vol. I, pp. 1136, 1190. U.S. Dept. of the Interior, Bureau of Mines, Washington, DC, 1978, and 2000 edition published 2002.
2. "Canadian Minerals Yearbook 1977," pp. 382–388. Energy, Mines and Resources Canada, Minister of Supply and Services, Hull, 1979, and 2001 edition published 2003.

3. "United Nations Statistical Yearbook," 29th ed. United Nations, New York, 1977.

4. D.S. Kostick, Salt, in USGS Minerals Yearbook, Vol. 1, 2001. Available: http://minerals, usgs.gov/minerals/pubs/commodity/saltmyb01. pdf; and earlier years.

5. "Encyclopaedia Brittannica," 15th ed., Vol. 16, p. 192. Macropaedia, London, 1974.

6. E.T. Degens and D.A. Ross, *Sci. Amer.* **222**(4), 32 (1970).

7. R.N. Shreve and J.A. Brink, Jr., "Chemical Process Industries," 4th ed., p. 266. McGraw-Hill, Toronto, 1977.

8. K. Rankama and T.G. Sahama, "Geochemistry," pp. 283–287. University of Chicago Press, Chicago, 1950.

9. G.E. Hutchinson, "A Treatise on Limnology," Vol. 1, p. 569. Wiley, New York, 1957.

10. "Kirk-Othmer Encyclopedia of Chemical Technology," 2nd ed., suppl. pp. 438, 444. Interscience, Toronto, 1968.

11. N.A. Korovessis and T.D. Lekkas, Solar Saltworks Production Process Evolution—Wetland Function in Saltworks: Preserving Saline Coastal Ecosystems, Proc. of the 6th Conf. on Environ. Sci. & Technology, Pythagorion, Samos, pp. 11–30, Sept. 1, 1999.

12. Chemicals output rises, *Chem. Eng. News*, **55**(11), 10, Mar. 14 (1977).

13. S.A. Hellenic Saltworks, Conserving Nature We Produce Salt Throughout Greece, Athens, 1994.

14. F.A. Lowenheim and M.K. Moran, "Faith, Keyes and Clark's Industrial Chemicals," 4th ed., p. 723. Wiley-Interscience, Toronto, 1975.

15. R.M. Stephenson, "Introduction to the Chemical Process Industries," p. 38. Reinhold, New York, 1966.

16. J.A. Epstein, *Hydrometallurgy*, **2**, 1 (1976).

17. B.J. Javor, Industrial Microbiology of Solar Salt Production, *Ind. Microbiol. Biotechnol.* **28**(1), 42–47, Jan. (2002).

18. F.G. Bell, T.R. Stacey, and D.D. Genske, Mining Subsidence and Its Effect on the Environment . . . , *Environ. Geol.* **40**(1–2), 135–152, Dec. (2000).

19. B.G. Reuben and M.L. Burstall, "The Chemical Economy," p. 149. Longmans, London, 1973.

20. W.J. Meade, ed., "The Encyclopedia of Chemical Process Equipment," p. 329. Reinhold, New York, 1964.

21. R.W. Thomas and P.J. Farago, "Industrial Chemistry," p. 79. Heinemann, Toronto, 1973.

22. C.J. Warrington and B.T. Newbold, "Chemical Canada," p. 5. Chemical Institute of Canada, Ottawa, 1970.

23. J.P. Searls, Potash, in USGS Minerals Yearbook, Vol. 1, 2001, and earlier years.

24. V.A. Arsentiev and J. Leja, *Can. Min. Metall. Bull.* **70A**, 154, Mar. (1977).

25. C.F. Perucca, Potash Processing in Saskatchewan—A Review of Process Technologies, *CIM Bulletin*, **96**(1070), 61–65, Apr. (2003).

26. J.A. Epstein, *Chem. Indus. (London)*. **55B**, 572 (1977).

27. R.H. Perry, ed., "Chemical Engineers Handbook," 4th ed., pp. 3–92, 3–93. McGraw-Hill, Toronto, 1969.

28. R.W. Mumford, *Indus. Eng. Chem.* **30**, 872 (1938).

29. C. Poole and S.A. Beeson, Separation of KCl values from Canadian sylvinite and carnallite potash ores by selective flocculation, *CIM Bulletin*, **93**(1040) 81–85, May (2000).

30. Prices Calculated in U.S. Dollars per Tonne of KCl, Based on "Unit K2O Equivalent," as quoted in "Chemical Marketing Reporter," pp. 31, 32. Schnell Pub. Co., New York, 1995.

31. J.B. Dahms and B.P. Edmonds, *Can. Pat.* **672**, 308 (to Pittsburgh Plate Glass Co.) (1963).

32. Potash Product Profile, *Can. Chem. Process.* **60**(3), 58, Mar. (1976).

33. H.F. Lund, ed., "Industrial Pollution Control Handbook," pp. 7–34. McGraw-Hill, Toronto, 1971.

34. J.A. Vonhof, "A Hydrogeological and Hydrochemical Investigation of the Waste Disposal Basin at IMCC (Canada) Ltd." K2 Potash Plant, Esterhazy, Saskatchewan, National Hydrology Research Institute, Ottawa, 1980.

35. A. Vandenberg, "An Unusual Pump Test Near Esterhazy, Saskatchewan," Tech. Bull. No. 102. Inland Waters Directorate, Water Resources Branch, Ottawa, 1978.

36. R.S. Edwards, The Effects of Air-borne Sodium Chloride and Other Salts of Marine Origin on Plants in Wales. In "Proceedings of the 1st European Congress on Air Pollution Effects on Plants and Animals," p. 99. Centre for Agric. Publ. and Documentation, Wageningen, 1969.

37. John Markham, cited in *Chem. Can.* **27**(9), 14, Oct. (1975).

38. H.H. Werner, *Can. Pat.* **838**, 477 (1970).

39. First Solution-mined Potash, *Chem. Eng.* (*N.Y.*). **71**(22), 84, Oct. 26 (1964).

40. R.L. Curfman, *Min. Congr. J.* **60**, 32, Mar. (1974).

41. C.J.S.Warrington and R.V.V. Nichols, "A History of Chemistry in Canada," p. 213. Pitman, Toronto, 1949.

42. F.A. Lowenheim and M.K. Moran, "Faith, Keyes and Clark's Industrial Chemicals," 4th ed., p. 762. Wiley-Interscience, Toronto, 1975.

7
INDUSTRIAL BASES
BY CHEMICAL ROUTES

I sometimes dig for buttered rolls
Or set limed twigs for crabs;...
　　　　　—*Lewis Carroll (1832–1898)*

From the lime-kiln into the coals.
　　　　—*Quintus Septimius Tertullian, ca. 150* A.D.

7.1. CALCIUM CARBONATE

The dominant source of calcium carbonate is limestone, the most widely used of all rocks. This often occurs in nature with traces of clay, silica, and other minerals, which may interfere with some applications. However, high-calcium limestone can consist of 95% or more calcium carbonate. White marble is a metamorphosed form of pure limestone with a dense crystal structure and is chemically suitable, but is usually of higher value for other uses. Dolomitic limestones consist of calcium and magnesium carbonates present in a near one-to-one molar basis, though the ratio can vary widely. For many applications, the magnesium carbonate does not interfere with the use of the calcium component of the dolomite. Other natural sources of calcium carbonate are chalk, which comprises the shells of microscopic marine organisms, bivalve shells, which are employed as an industrial feedstock on the shores of the Gulf of Mexico, and corals, which are the massive, sub-marine fused skeletons of multiple stationary organisms.

To obtain higher purity and higher density grades of calcium carbonate, synthetic calcium carbonate is also produced on a large scale [1]. As early as 1850, J.&E. Sturge Ltd. of Birmingham, England was treating a calcium chloride Solvay waste stream with sodium or ammonium carbonates to produce a high grade of calcium carbonate (Eqs. 7.1 and 7.2).

$$CaCl_2 + Na_2CO_3 \rightarrow 2\ NaCl + CaCO_3 \downarrow \qquad\qquad 7.1$$

$$CaCl_2 + (NH_4)_2CO_3 \rightarrow 2\ NH_4Cl + CaCO_3 \downarrow \qquad\qquad 7.2$$

In the U.S. by 1900, Solvay by-product calcium chloride, causticizer sludges, and direct carbonation of slaked lime were all used to produce fine grades of calcium carbonate (Eqs. 7.3 and 7.4).

$$Ca(OH)_2 + Na_2CO_3 \rightarrow 2\ NaOH + CaCO_3 \downarrow \qquad\qquad 7.3$$

$$Ca(OH)_2 + CO_2 \rightarrow H_2O + CaCO_3 \qquad\qquad 7.4$$

Industrial uses of natural calcium carbonate include very large-scale consumption in the cement, iron and steel, and other metal-refining industries as well as the Solvay ammonia soda process. It is also used directly or indirectly in the manufacture of at least 150 other chemicals. In recent years, use in emission control measures for combustion processes has taken a growing fraction of limestone. The higher purities and tailored physical properties of the synthetic grades of calcium carbonate command higher prices, which restricts their uses. More than 75% of the U.S. synthetic calcium carbonate is applied as a smooth, white surface coating to paper. Smaller amounts are used in paints to give flatness and low gloss, in rubber as a reinforcing agent, in plastics as a filler, and in many other consumer items such as foods, cosmetics, toiletries, pharmaceuticals, and the like.

7.2. CALCIUM OXIDE

Calcium oxide, burned lime, quicklime, or simply lime as it is known in industry, is the largest single use for limestone and dolomite [2]. It is obtained by heating (calcining) calcium carbonate to a temperature of 900–1,000°C to cause loss of carbon dioxide (Eq. 7.5, Table 7.1).

$$CaCO_{3(s)} \xrightarrow{\ 900-1,000°\ } CaO_{(s)} + CO_{2(g)}$$
$$100.09 \qquad\qquad 56.08 \quad 44.01\ \text{mol. wt.} \qquad\qquad 7.5$$

$$\Delta H = +177.8\,\text{kJ}\ (+42.5\,\text{kcal})$$

TABLE 7.1   Calcium Carbonate Dissociation Pressures[a]

Temperature (°C)	Pressure (mm Hg)	Temperature (°C)	Pressure (mm Hg)
500	0.1	852	381
550	0.41	898	760
605	2.3	950	1,490
701	23.0	1,082.5	6,758
749	72	1,157.7	14,202
800	183	1,241	29,711

[a]Compiled from the data of Kirk-Othmer [1] and Weast [3].

Lower temperatures slow the rate of carbon dioxide loss and can leave unburned limestone in the product (Table 7.1). Also, if a moderate particle size rather than finely ground limestone is used to reduce dust carryover in the kiln, additional carbon dioxide partial pressure must be developed in the larger granules to ensure carbon dioxide diffusion to the outside.

In Canada, and probably elsewhere in the early 19th century, burned lime manufacture was carried out as a batch process using pot kilns heated with wood or coal [4]. These produced 6–27 tonne/cycle, as did the draw kilns of the late 19th century, which were designed to use fuel more efficiently. Today, a variety of vertical flow kilns are in common use, mainly designed to operate with lower energy requirements than the early models [5, 6]. Newer designs feature the separation of the function of gas flow required for combustion from the limestone heat requirement, to obtain a gas stream as high in carbon dioxide concentration as possible. This is desirable when the process is operated as a component of the Solvay ammonia–soda process (Section 7.4.1).

Probably the rotary horizontal kiln is the most versatile, since it allows a feed of lumps or fines of limestone or marble, or wet or dry calcium carbonate sludges (Fig. 7.1). The main component of this calcination system is a 2.5- to 3.5-m diameter by 45- to 130-m long firebrick-lined inclined steel tube. Heat is applied to the lower end of this via oil, gas, or coal burners [7]. The feed to be calcined is fed in at the top end. Slow rotation of the tube on its axis gradually moves the feed down the tube, as it tumbles countercurrent to the hot combustion gases. In this way, wet feed is dried in the first few meters of travel. Further down the tube, carbon dioxide loss begins as the temperature of the feed rises. By the time the solid charge reaches the lower, fired end of the kiln it reaches temperatures of 900–1,000°C and carbon dioxide evolution is virtually complete. Normally the temperature of the lower end of the kiln is not allowed to go much above this as it reduces the life of the kiln lining. It also adversely affects the crystal structure of the lime product since it produces a "dead-burned" or "overburned" lime. Over-burned lime is difficult to slake to convert it to calcium hydroxide and raises

FIGURE 7.1 Details of operation of a rotary horizontal lime kiln. Temperature of calcium carbonate charge rises as it moves down the kiln. The hot calcium oxide, as it cools, is used to simultaneously preheat the firing air. Fuel may be natural gas or oil.

fuel costs unnecessarily. The calcined product is normally packed in airtight sheet steel barrels or bulk containers for shipment to keep it in anhydrous form.

With a specific heat of about 0.90 J/g°C at 100°C (0.214 cal/g°C; varies with temperature) calcium carbonate requires some 7.92×10^5 kJ/tonne of sensible heat simply to bring it up to the final kiln temperature.

$$0.90 \, \text{J/g°C} \times 10^6 \, \text{g/tonne} \times (900 - 20)°C = 7.92 \times 10^8 \, \text{J/tonne}$$

The endotherm of the reaction consumes a further 1.78×10^6 kJ/tonne, which, when combined with the sensible heat requirement and an allowance of a further 52×10^5 kJ/tonne for imperfect heat transfer efficiencies gives a total heat requirement for this process of about 2.8×10^6 kJ/tonne.

$$177.8 \, \text{kJ/mol} \times \frac{106 \, \text{g/tonne}}{100.09 \, \text{g/mol}} = 1.776 \times 106 \, \text{kJ/tonne}$$

Production estimates for different fuels range from 2.5 to 3.4 kg of lime/kg of coke [7] or coal [8], approximately equivalent to 4.5 kg of lime/kg of Bunker C fuel oil [9]. Alternative fuels have been examined [10, 11].

Improved energy efficiencies are obtained by using heavy "recuperator chains" loosely slung inside the kiln to aid in heat transfer from hot gases to the drying and reacting solids, and by using the heated air obtained from lime cooling as the firing air for the kiln. Some other kiln designs provide better heat transfer and recovery than the rotary kiln but have feed restrictions, which do not affect the rotary horizontal kiln [5]. A recent survey examines kiln operating problems and suggests solutions [12]. A potential energy saving is possible from the use of oxygen-enriched air, which is now available at low cost. Using oxygen enrichment decreases the volume of associated nitrogen, which has to be heated. There is little point in utilizing pure oxygen since one of the functions of gas throughput is to displace the dissociated carbon dioxide.

7.2.1. Lime Kiln Emission Control

Environmental concerns of lime kiln operation chiefly relate to exhaust gas dust control and are usually solved by water scrubbing. The slaked lime (calcium hydroxide) produced by the scrubber can be employed as a base in the operations of a chemical complex, or sold. Electrostatic precipitation of precooled gases is also used, sometimes in series with water scrubbing. The dissociated carbon dioxide discharged is not ordinarily regarded as a pollutant. The amount discharged from this source is far less than the carbon dioxide produced by fossil fuel combustion. Lime kilns associated with Solvay ammonia–soda plants may be able to recycle a part of the dissociated carbon dioxide, particularly if the concentrations are raised by the use of oxygen-enriched combustion air. Recently, interest has been shown in the collection of carbon dioxide from large industrial sources such as lime kilns for maintaining pressure in the producing zone in enhanced oil recovery operations. Energy and mass balances of lime kiln operations have recently been reviewed [13].

7.2.2. Uses of Calcium Oxide

The low cost and desirable chemical properties of calcium oxide (burned lime) are reflected by its consumption on a very large scale by both the cement and the iron and steel industry. In both cases *in situ* lime preparation from limestone by heat is an integral part of the process. The metal smelting industry also employs lime as a basic refractory lining material. Its basicity is also useful in processes for the heterogeneous absorption of acid gases produced from combustion or metallurgical roasting processes. The strong affinity of lime for water is applied to the industrial drying of alcohols as, for example, to remove water from the glycerin obtained from the saponification of fats, or from the 5% water azeotrope of ethanol. Most of the remaining uses are indirect since they employ slaked lime (calcium hydroxide), the product of the reaction of lime with water.

7.3. CALCIUM HYDROXIDE

Calcium oxide (burned lime) can be slaked with a theoretical amount of water to form calcium hydroxide (slaked lime) as a white powder, but this must be conducted with care. The highly exothermic reaction can produce sufficiently high temperatures that it could slow the rate of hydration [7] (Eq. 7.6).

$$CaO + H_2O \rightarrow Ca(OH)_2 \quad \Delta H = -64.9\,kJ(-15.5\,kcal) \qquad 7.6$$

This process is also sufficiently exothermic, so it is dangerous to store large quantities of calcium oxide in combustible bags or even in nonfireproof buildings if there is any risk of contact with water. The high heat of hydration coupled with the relatively low specific heats ($CaO = 0.226\,cal/g$; $Ca(OH)_2 = 0.286\,cal/g$ [9]) could produce a temperature rise of $730°C$ above ambient. This is more than enough to ignite combustible materials.

Other variables in the preparation of slaked lime powder by the addition of water to lime are the original calcining temperature and the magnesia (MgO) content. Very high calcining temperatures produce a very dense "dead-burned" lime, which can retard the hydration reaction. Any significant concentration of magnesia, such as obtained from calcining a dolomitic limestone ($CaCO_3 \cdot MgCO_3$; a "dolime") also shows this effect. However, the hydration retardation shown by burned dolimes is regarded as a working advantage in its application as a plaster.

Continuous rotary and batch hydrators are used to produce calcium hydroxide (slaked lime) powder. The continuous version uses a slightly inclined, slowly rotating steel cylinder of about 1 m diameter by 6–7 m in length. The calcium oxide is fed into the upper end and the correct proportion of water is sprayed in, followed by tumbling in the cylinder to produce a uniform product. Open trough-type batch hydrators, in which the reacting components are mechanically combined, provide greater control of hydration rate and temperature and some improvements in the quality of the product. An "explosion process" in which hydration is conducted in a pressure vessel produces a slaked lime with better flow characteristics, and a smaller mean particle size more suitable for uses such as filters [14].

Coping with the large exotherm is less of a problem when lime is slaked with a liberal amount of water because of the high heat capacity of the excess water present. In the preparation of cement mixes, mortars, and plasters the low solubility of slaked lime in water (1.84 g/L at 0°C; 0.77 g/L at 100°C) does not pose any problems. However, this factor does have to be considered in applications of calcium hydroxide as an industrial base.

7.3.1. Uses of Calcium Hydroxide

Some of the uses of slaked lime are direct and some are indirect via calcium oxide addition to water or to an aqueous medium. These range from its employment as a component of insecticides, to its value as a component of water treatment programs. The old but still viable use as an active pesticide component in lime–sulfur sprays and its substrate utility in many insecticidal dusts are also important. Its chemical basicity is used in medicinal applications as an antacid, and also prompts its use for pH adjustment in animal food formulations, and for the neutralization of overly acid soils. Large quantities are employed as a slurry in water to absorb acid contaminants from flue gases and from roaster off-gases of small smelters. It is also used in papermaking. In the sulfite process it provides the active cation of the pulping liquor (Chap. 15). In the kraft pulping process it is used indirectly for the regeneration of the sodium hydroxide component of the pulping liquor from sodium carbonate recovered from the combustion of spent pulping liquor.

Calcium hydroxide frequently performs a dual role in water treatment. First, it aids the removal of colloidal suspended solids by encouraging coagulation and settling. Second, it reacts with any dissolved calcium, magnesium, or other cations, which are present as their bicarbonates to precipitate these ions and in this way softens the water supply for domestic and industrial uses (Eqs. 7.7–7.9).

$$Ca(OH)_2 + Ca(HCO_3)_2 \rightarrow 2\ CaCO_3 \downarrow + 2\ H_2O \qquad 7.7$$

$$Ca(OH)_2 + Mg(HCO_3)_2 \rightarrow MgCO_3 + CaCO_3 \downarrow + 2\ H_2O \qquad 7.8$$

$$Ca(OH)_2 + MgCO_3 \rightarrow Mg(OH)_2 \downarrow + CaCO_3 \qquad 7.9$$

7.4. SODIUM CARBONATE

Up until the end of the 18th century, the only commercially available sodium carbonate was a low grade product containing only 5–20% of this salt. It was obtained from the burning of kelp or barilla (a stout, berry-bearing Spanish shrub), followed by extraction of the ashes with water [15].

7.4.1. Synthetic Sodium Carbonate

In early 19th century, the availability of sodium carbonate was enhanced by the introduction of the Le Blanc process, which increased the potential supply and produced a more pure product (Eqs. 7.10 and 7.11). The early versions of this process were significant contributors to local air pollution

$$2\,NaCl + H_2SO_4 \rightarrow Na_2SO_4 + 2\,HCl \qquad\qquad 7.10$$

$$Na_2SO_4 + 2\,C + CaCO_3 \xrightarrow{\text{heat}} CaS + Na_2CO_3 + 2\,CO_2 \qquad 7.11$$

since about three-fourth of a tonne of hydrogen chloride was discharged for each tonne of sodium carbonate produced. Fortunately, the scale of operations at that time was small. In 1836, it was discovered that the acid discharge could be controlled by absorption in water, but it was 1863, before this control measure was made compulsory [15]. The other negative feature of Le Blanc process operations of that time was the large piles of waste calcium sulfide and calcium oxysulfide, which were generated and discarded nearby, reclamation of which was very difficult.

The ammonia–soda process to sodium carbonate, which is still practiced on a large scale today, was known in 1822 [15] (Eqs. 7.12-7.15). However, this early version had no ammonia

$$NH_3 + H_2O \rightarrow NH_4OH \qquad\qquad 7.12$$

$$NH_4OH + CO_2 \rightarrow NH_4HCO_3 \qquad\qquad 7.13$$

$$NH_4HCO_3 + NaCl \rightleftharpoons NaHCO_3 \downarrow + NH_4Cl \qquad\qquad 7.14$$

$$2\,NaHCO_3 \xrightarrow{175°C} Na_2CO_3 + CO_2 + H_2O \qquad\qquad 7.15$$

recovery phase so that the cost of the ammonia required to work this process was prohibitive. For this reason, the process was not commercialized until 1863 when E. Solvay devised a method, which employed a suspension of slaked lime in water to release, and allowed recovery of the ammonia (Eqs. 7.16 and 7.12).

$$2\,NH_4Cl + Ca(OH)_2 \rightarrow 2\,NH_3 \uparrow + CaCl_2 + 2\,H_2O \qquad 7.16$$

Modern practice of the Solvay ammonia–soda process involves initial successive saturations of a purified saturated solution of sodium chloride in water, first with ammonia gas and then with carbon dioxide. The carbonation helps to drive reaction (Eq. 7.14) to approximately 73% to the right. Sodium

TABLE 7.2 **Quality of Typical Sodium Carbonate from the Ammonia–Soda (Solvay) Process**[a]

Component	Analysis (%)
Na_2CO_3	99.50 (58% Na_2O)
NaCl	0.20–0.25
Na_2SO_4	0.02
Insolubles	0.02
Water	Balance

[a]Data from Kent [16].

bicarbonate, being the least soluble salt component of this mixed solution, is precipitated and filtered off. It is then dried and calcined under mild conditions in a rotary kiln to yield the sodium carbonate product (Table 7.2). Conducting the calcination in a sealed rotary kiln enables recovery of the liberated carbon dioxide (Eq. 7.15), which provides some of that required for the primary carbonation (Fig. 7.2). The remaining carbon dioxide is obtained from a separate lime kiln, which produces calcium oxide. The calcium oxide (burned lime) is required to produce calcium hydroxide for ammonia recovery, and also may be used for carbon dioxide recovery from boiler flue gases if necessary.

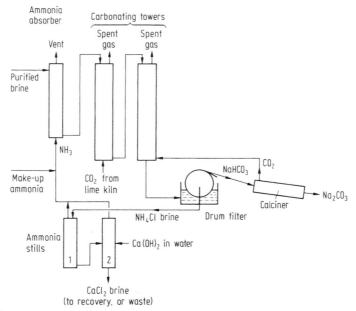

FIGURE 7.2 Flowsheet for the ammonia–soda (Solvay) process to provide sodium carbonate. Units numbered 1 and 2 are ammonia stills, the first being used to recover free ammonia and the second to recover that produced by the addition of slaked lime (Eq. 7.16).

The ammonia recovery system of modern configurations of Solvay plants is quite efficient. Only about 3 kg of ammonia make-up is required to maintain the 300 kg or so of absorbed ammonia, which is needed for each tonne of sodium carbonate produced [17]. This high ammonia-recovery efficiency was the feature that gradually spelled the end of the commercial Le Blanc sodium carbonate plants, the last of which closed down in the period 1920–1930 [15].

The crude sodium carbonate from the Solvay process is quite pure, and is a low density (about $0.5\,g/cm^3$) fluffy material and referred to as "light soda ash." The density may be increased to the $1.0\,g/cm^3$ required for some uses by conversion to the monohydrate, $Na_2CO_3 \cdot H_2O$. The anhydrous light ash is tumbled with the appropriate quantity of water and then dried at $<100°C$. Sal soda (washing soda) is the decahydrate, $Na_2CO_3 \cdot 10H_2O$, which is obtained by crystallization of the light ash from water and drying at ambient temperatures ($<35.5°C$).

The sodium bicarbonate ($NaHCO_3$) initial product of the Solvay process is not sold, principally because of impurities such as ammonia. The market for sodium bicarbonate is satisfied by the carbonation of a saturated solution of sodium carbonate, the final Solvay product, and filtration to recover the precipitated sodium bicarbonate [5] (Eq. 7.17). This product is the household chemical "baking soda." This is also used to supply the sodium bicarbonate used for

$$Na_2CO_3 + CO_2 + H_2O \rightarrow 2\,NaHCO_3 \qquad 7.17$$

the manufacture of baking powder, for the preparation of carbonated mineral waters, and also as the filling for one type of dry powder fire extinguisher.

7.4.2. Natural Sodium Carbonate

Sodium carbonate production from natural sources of solid beds of trona ($Na_2CO_3 \cdot 2Na_2SO_4$), trona and sodium carbonate-rich brines, and nahcolite mineral became important much more recently since the discovery of major deposits of these minerals in the U.S. For those nations with well defined, suitable deposits of these minerals the incentives for sodium carbonate production from these sources are substantial: a halving of the capital costs of a similarly sized Solvay plant, and much less potential for environmental problems. For these reasons, the U.S., with proven mineral reserves containing the equivalent of over 23 billion tonnes of sodium carbonate [18], has been the world's largest producer of sodium carbonate in recent years, with 10.5 million tonnes in 2002.

Trona may be recovered from the deposit by conventional dry, or solution mining depending on the nature and accessibility of the source [19]. Fractional crystallization of solutions of trona [20] or the sodium carbonate-rich brines separates the sodium carbonate for recovery. Sodium carbonate has also been produced from trona in a spray drier [21].

Nahcolite mineral is naturally occurring sodium bicarbonate (sodium hydrogen carbonate), which is named from its constituent elements, Na, H, C, O, and "lite" for stone. It is dry mined and recrystallized for a sodium bicarbonate product, or heated to form sodium carbonate (Eq. 7.15).

Fallout from the lower cost and decreased polluting potential of sodium carbonate recovery from natural deposits was seen in Fall 2000, with the closure of the last North American Solvay plant (450,000 tonne/year), which had been operational in Amherstburg, Ontario since 1919 [22]. The last American Solvay plant closed in 1986.

7.4.3. Environmental Aspects of Sodium Carbonate Production

A difficulty of present-day Solvay operations is related not so much to the technology of the process but to the basic chemistry, which dictates that more than a tonne of calcium chloride is produced for each tonne of sodium carbonate obtained. For producers operating in areas with dissipative markets for calcium chloride in applications such as high early strength cement, winter ice and snow removal for highways, or dust control on unimproved roads, the coproduct can be sold at reasonable prices. Non-dissipative uses of calcium chloride such as employment as the dissolved salt of an aqueous heat transfer fluid in artificial ice rinks or as a component of solar energy storage systems [23], are not large markets. Producers who cannot sell this coproduct are faced with high brine-disposal costs since the waste, acidic calcium chloride brines cannot be discharged to surface fresh-water courses. Some processors have adopted calcium chloride brine discharge into underground brine aquifers as a disposal method (deep-well disposal, Section 5.13.8).

Brine-disposal problems may be avoided by proposed modifications to the Solvay process, which employ magnesium hydroxide for ammonia recovery [17]. The magnesium chloride formed could be recycled and hydrogen chloride obtained as a coproduct by contacting the magnesium chloride with superheated steam (Eqs. 7.18 and 7.19). However, this process remains unproven.

$$MgCl_2 + H_2O \xrightarrow[\text{heat}]{} MgO + 2\,HCl \qquad\qquad 7.18$$

$$MgO + H_2O \rightarrow Mg(OH)_2 \qquad\qquad 7.19$$

A shift to the processing of substantial alternate natural mineral sources of sodium carbonate in the U.S. has eliminated the calcium chloride disposal problems of the Solvay process for sodium carbonate production. Coupling this advantage to the much lower capital cost of a natural sodium carbonate plant has contributed to the shift away from synthetic sodium carbonate in the U.S. [13] (Table 7.3). Kenya is the only other country reported to be recovering natural sodium carbonate and was operating at 260,000 metric tonnes in 2001 [19]. China, the second largest producer, and all other world producers still rely heavily on the ammonia–soda process [17, 24] (Table 7.4).

TABLE 7.3 Growth of Natural Sodium Carbonate Production at the Expense of Synthetic Material in the U.S.[a]

Year	Sodium carbonate produced (thousands of tonnes)		Natural, as percentage of total
	Manufactured	Natural	
1940	2,744[b]	118	4.1
1950	3,621	318	8.1
1960	4,134	734	15
1970	4,004	2,439	38
1978	1,360	6,160	82
1990	–	9,160	100
2000	–	10,200	100

[a]Calculated from the data of *Minerals Yearbooks* [24].
[b]A further 22,680 tonnes was made in this year by the carbonation of electrolytic sodium hydroxide.

An additional source of sodium carbonate occasionally used is via the carbonation of electrolytic sodium hydroxide as it is formed in the cell, or separately, later (e.g., [26]). The precipitated sodium hydrogen carbonate is calcined to obtain sodium carbonate in a manner similar to the last step of the Solvay process (Eq. 7.15).

TABLE 7.4 Major World Producers of Sodium Carbonate (Soda Ash, Thousands of Metric Tonnes)[a]

	1960	1970	1980	1990	2000
Bulgaria	128	300	1,630	1,050	800
China, Mainland	–	–	1,778	3,750[b]	8,343
France	848	1,419	1,800	1,180	1,000
East Germany	1,117	1,334	955	2,290[c]	1,400[c]
India	145	446	660	1,400[b]	1,500
Italy	–	662	105	610[b]	1,000
Japan	516	1,230	1,494	1,130	1,000
Poland	522	644	129	968	1,081
Romania	180	582	1,033	632	550
U.S.S.R.	1,793	3,485	5,269	4,360	2,199[d]
U.K.	–	–	1,500	1,000[b]	1,000
U.S.A.	4,135	4,004	1,360	9,160	10,200
West Germany		1,540	1,555	2,290[c]	1,400[c]
Other	2,096	2,649	5,269	4,470	4,127
World Total	13,961	16,610	26,173	32,000	34,200

[a]Does not include natural sodium carbonate, except for the U.S. for 1990 and 2000, when the total production was from natural sources (Table 7.3). Data calculated from *Minerals Yearbooks* [24] and from *U.N. Statistical Yearbooks* [25].
[b]Estimate.
[c]For the reunited Germany.
[d]Russia.

7.4.4. Uses of Sodium Carbonate

Nearly half of the sodium carbonate produced in the U.S.A., whether from synthetic or natural sources, is consumed by the glass industry. A further quarter of the product goes into general chemicals production. The detergent industry, pulp and paper, and water treatment processes each consume somewhat less than 10% of the total. Some industries, such as kraft paper production, produce and consume large quantities of sodium carbonate internally [24]. For this reason, the figures for this captive production by the pulp industry do not appear in the reported consumption data.

7.5. SODIUM HYDROXIDE BY CAUSTICIZATION

Prior to 1850, sodium hydroxide was not available commercially. Processes such as soap making, which required sodium hydroxide (caustic soda) had to obtain this by causticization of a solution of purchased sodium carbonate with lime (Eq. 7.20). Sodium hydroxide itself was also made

$$Na_2CO_{3(aq)} + Ca(OH)_{2(aq)} \rightleftharpoons 2\ NaOH_{(aq)} + CaCO_{3(s)}$$
$$\Delta H = -8.4\,kJ\ (-2.0\,kcal)$$

7.20

in this way until about 1890 when the first electrolytic product started to be produced [15]. Despite the advantages of electrolytic production of caustic soda, the causticization or "lime-soda" process dominated the production of trade sodium hydroxide in the U.S. until at least 1940, and is still of commercial significance in the regeneration of spent pulping liquors in the manufacture of pulp and paper by the kraft process [27]. It is also used for the production of sodium hydroxide without chlorine when the market demand for sodium hydroxide and chlorine exceeds the stoichiometric ratio available from electrolysis of sodium chloride solutions.

The success of this process depends on the low solubility of calcium carbonate, which is also what allows this process to be used for small-scale production of sodium hydroxide by a batch process. For batch operation, the functions such as slaking, mixing, and settling may be carried out in the same wooden (or steel) vessel. Separate units are required for these steps in the continuous process. A further chemical feature important to the recycle of the spent lime of this process is the relatively easier thermal loss of carbon dioxide from calcium carbonate than from sodium carbonate. Since sodium bicarbonate ($NaHCO_3$) may be calcined at 175°C to obtain carbon dioxide loss, it might be expected that sodium hydroxide could be made by heating sodium carbonate at a higher temperature followed by hydration of the resulting oxide (Eqs. 7.21 and 7.22).

$$Na_2CO_3 \xrightarrow[\text{heat}]{} Na_2O + CO_2$$

7.21

$$Na_2O + H_2O \rightarrow 2\ NaOH$$

7.22

While this series of reactions is possible, much higher temperatures are required for sodium carbonate dissociation to sodium oxide and carbon

TABLE 7.5   Carbon Dioxide Dissociation Pressures of Sodium Hydrogen Carbonate and Calcium and Sodium Carbonates[a]

NaHCO$_3$			CaCO$_3$	Na$_2$CO$_3$
Temperature (°C)	Pressure (mm Hg)	Temperature (°C)	Pressure (mm Hg)	Pressure (mm Hg)
30	6.2	700	22.6	0.99
50	30.0	820	237	3.0
70	120.4	880	595	9.9
90	414.3	990	3,324	12.0
100	731.1	1,100	8,490	21
110	1,252.6	1,150	13,440	28
115.5	1,654.6	1,180	18,068	38
		1,200	21,534	41

[a]Dissociation pressures for sodium salts calculated from International Critical Tables [28], and for calcium carbonate obtained by interpolation of data from Weast [29].

dioxide, than for calcium carbonate dissociation. Hot, concentrated sodium hydroxide is also very corrosive. Therefore, it is easier to go through the same sequence with calcium carbonate, which has much higher decomposition pressures at all temperatures (Table 7.5). This dissociation can then be followed by the slaking and the causticization reactions (Eqs. 7.6 and 7.20).

A hot (80–90°C) 20% solution of sodium carbonate is stirred with a slight stoichiometric excess of a slurry of calcium hydroxide in water ("milk-of-lime"). Since this reaction is only slightly exothermic (Eq. 7.20), the equilibrium is not significantly affected by operation of the reaction at high temperatures. The reaction is also faster and produces larger, more readily precipitated particles of calcium carbonate at high rather than at ambient temperatures. An additional advantage is gained by operation at high temperatures, since higher concentrations of sodium carbonate may be employed (Table 7.6). This step alone raises the concentration of sodium hydroxide obtained directly from the reaction to about 12%, sufficient for use in pulping liquor preparation [31]. It also decreases evaporation costs for any applications, which demand higher sodium hydroxide concentrations.

However, operation of a causticizer at 80°C or higher is not without an attendant drawback. While the solubility product for calcium carbonate (Eq. 7.23) is low in pure water, it is higher in water containing sodium hydroxide, which, in effect, displaces the equilibrium of the formation reaction somewhat to the left.

$$K_{sp} (CaCO_3): [Ca^{2+}][CO_3^{2-}] = 0.87 \times 10^{-8} \text{ at } 25°C \qquad 7.23$$

$$K_{sp} (Ca(OH)_2: [Ca^{2+}][OH^-]^2 = 1.3 \times 10^{-6} \text{ at } 25°C \qquad 7.24$$

In practical terms this means that at initial 2, 5, 10, 15, and 20% sodium carbonate concentrations, one obtains approximately 99.4, 98.9, 96.5, 91.5,

TABLE 7.6   Solubilities of Sodium Carbonate and Sodium Hydrogen Carbonate in Water[a]

| Temperature (°C) | Solubility (wt %) | |
	Sodium carbonate	Sodium hydrogen carbonate
0	6.5	6.5
10	11.5	7.5
20	17.7	8.8
25	27.8	9.4
30	31.4	10.0
40	–	11.3
42	32.7	–
50	–	12.6
60	–	13.8
70	31.7	–
80	31.4	–
90	–	–
100	31.1	19.1

[a]Data from Kirk-Othmer[1] and Stephen and Stephen[30].

and 83.5% conversions of the sodium carbonate to sodium hydroxide, respectively [32]. Industrial yields on sodium carbonate for a lime recycle process on the basis outlined amounts to about 90% [8], giving a yield per batch, or yield per pass for a continuous process of about 75% (0.835 × 0.90) when using a 20% initial sodium carbonate concentration.

The situation described above, showing gradually diminishing sodium carbonate conversions for increasing initial concentrations, is the concern that limits the practical sodium carbonate concentrations used in this process to well below what would be technically feasible. An empirical expression, which may permit estimation of sodium carbonate conversions from the known solubility products and the initial sodium carbonate concentrations has been derived [33], but has been found to be unreliable [34]. This lack of rigor is partly because the solubility of calcium hydroxide is too high to obtain a solubility product constant, but also so low that only partial ionization precludes the application of ionic equilibria to determine ion concentrations. This expression predicts higher sodium carbonate conversions for higher excess ratios of slaked lime to sodium carbonate, whereas in practice this has been found to make very little difference [22].

The insoluble calcium carbonate is separated from the sodium hydroxide solution by simple settling in a batch process, or by settling in a thickener (Fig. 6.6) followed by filtration of the more concentrated slurry from the thickener in a continuous process (Fig. 7.3).

Continuous filtration involves aqueous sodium hydroxide removal from the sludge by drawing this solution through a filter element fastened to the periphery of a slowly rotating drum, which is perforated on its curved surface. The driving force for fluid flow through the filter is the reduced pressure maintained in the interior of the drum (Fig. 7.4). Normally in the larger-scale

FIGURE 7.3 Flowsheet giving the details of a continuous causticization process for the production of sodium hydroxide. Slaker water is normally derived from lime mud washing. The process may also be operated in a batch mode requiring a smaller number of process vessels, and with or without lime recycle.

causticization processes, the calcium carbonate filter cake or "lime muds" will be calcined to obtain calcium oxide. After slaking, this regains the calcium hydroxide required for causticization (see Section 7.3).

The crude sodium hydroxide solution is about 12% by weight, commercially acceptable if used nearby. This could be for the chemical pulping of wood, or for soap-making applications. The crude product will contain 2–3% impurities, mainly sodium carbonate with traces of sodium chloride.

Shipping costs of the sodium hydroxide are reduced for distant markets if most (or all) of the water is removed first. Evaporation is carried out either batchwise, in corrosion-resistant close-grained cast iron pots over an open

FIGURE 7.4 Cross section through the drum and slurry trough of a drum filter, used for the continuous separation of a solid from a liquid suspension. This version includes a water wash for scavenging the last traces of a solution of value. Sodium hydroxide washings are retained in the process by collecting them separately in the filter and using this to dissolve fresh sodium carbonate, and to slake burned lime. (Modified from Codd *et al.* with permission [35]).

flame, or in continuous caustic evaporators heated with an industrial heat exchange fluid (Section 8.2.1). When 50% sodium hydroxide is cooled, sodium carbonate and sodium chloride residues are less soluble and crystallize out. These salts are removed from the 50% product by settling or filtration. For distant shipping or for small-scale customers, all the water may be removed from the solution. The strong affinity of sodium hydroxide for water requires temperatures of 500–600°C, or reduced pressures to accomplish this, leaving molten sodium hydroxide containing less than 1% water [36]. If this sodium hydroxide is manufactured from Solvay sodium carbonate, it would give roughly the following analysis [30]: NaOH, 98.62%; Na_2CO_3, 0.90%; NaCl, 0.30%; and Na_2SO_4, 0.18%. Detailed use profiles for sodium hydroxide are given in Chap. 8.

7.5.1. Emission Control, Causticization Process

Emission problems in a causticization operation arise chiefly in the area of dust control of the exit gases from the lime kiln since the water circuit is virtually self-contained. Effective containment is obtained by the use of scrubbers, which achieve some 99% mass removal efficiency from the exhaust gases. Spent scrubber liquor may be returned to the causticization circuit for recycle. Using spent scrubber liquor, either for the slaking of lime or to prepare fresh sodium carbonate solutions for causticization, avoids creating a water emission problem from this aqueous waste stream. It also improves the raw material balance of the process.

Some facilities use electrostatic precipitation for lime kiln dust control, which avoids generation of a wastewater stream. The captured dust can be recycled to the slaker to avoid dust disposal costs and assess the material balance of the operation, reducing raw material costs.

REVIEW QUESTIONS

1. The Solvay process to produce sodium carbonate ordinarily operates with about 90% conversion and 75% yield (industrial definition) on sodium chloride.
   (a) Give the molar quantities of sodium carbonate, sodium chloride, and miscellaneous sodium compounds (by-products, $NaHCO_3$, etc.) to be expected from one passage (i.e., batch basis) of 2 moles of sodium chloride through this continuous process.
   (b) What molar quantities of sodium carbonate, sodium chloride, and sodium-containing by-products (expressed as moles of Na^+) would be expected if the unreacted sodium chloride was recycled to extinction (i.e., on an annual basis)?
2. The sole Canadian Solvay plant produced 500,000 tonnes (1 tonne = 1,000 kg) of anhydrous sodium carbonate in 1999.
   (a) What quantity of sodium chloride, in tonnes, would be theoretically required to produce this amount of sodium carbonate assuming quantitative reactions?

(b) Current prices for limestone for producing the calcium hydroxide, salt, and sodium carbonate monohydrate ($Na_2CO_3 \cdot H_2O$) are 22, 31, and 62 dollars per tonne, respectively. What gross (*before* deduction of operating costs, labor, etc.) annual profit should this operation be currently making, assuming quantitative reactions?

3. Batch causticization of 8,000 L of a hot, 2.0 M solution of sodium carbonate is accomplished by the addition of 10% excess preslaked calcium hydroxide as a slurry in 2,000 L of water. Under these conditions, conversion and yield (selectivity) on sodium carbonate are 95 and 90%, respectively.
   (a) What would be the conversion on calcium hydroxide if its yield (selectivity) was 100%?
   (b) Assuming that no residual water is lost with the precipitated calcium carbonate as it is removed, what resultant molar concentrations of sodium hydroxide and sodium carbonate would be obtained?
   (c) Evaporation in stages, with intermediate removal of impurities (with no loss of the associated sodium hydroxide), should yield what mass of 100% sodium hydroxide?

4. Sodium hydroxide solution, 50% by weight, is available commercially at $225/tonne of solution.
   (a) What theoretical profit or loss, on a raw material cost basis only, would be incurred to produce each tonne of anhydrous sodium carbonate from this sodium hydroxide and sell the product at $210/ tonne? Carbon dioxide for carbonation is to be obtained from flue gas; assume essentially zero cost.
   (b) Sodium carbonate monohydrate and sodium carbonate decahydrate are also commercial grades, selling at $290 and $430 per tonnes, respectively. Give the relative dollar profit margins per tonne of contained Na_2CO_3 for each of these grades, neglecting any costs other than those outlined in part (a). Which of the three grades would be the most profitable?

5. (a) What masses of ammonia and carbon dioxide would be theoretically required by a batch ammonia–soda process to produce 1.00 tonne (1,000 kg) of sodium carbonate?
   (b) What mass of calcium hydroxide would be required to free the theoretical amount of ammonia from 1.00 tonne of ammonium chloride, allowing a 5% excess to complete the ammonia formation reaction?

FURTHER READING

V.S. Ramachandran, "Calcium Chloride in Concrete," Applied Science Publishers, London, 1976.
Environmental Protection Service, "Sodium Hydroxide." Technical Services Branch, Ottawa, 1984.
J.A.H. Oates, "Lime and Limestone: Chemistry and Technology, Production and Uses," VCH, Weinheim, New York, 1998.
R.D.A. Woode, Sodium Carbonate. In: "Industrial Inorganic Chemicals: Production and Uses" (R. Thompson, ed.), pp. 123–148. The Royal Society of Chemistry, Cambridge, UK, 1995.

REFERENCES

1. "Kirk-Othmer Encyclopedia of Chemical Technology," 2nd ed. Vol. 4, pp. 2, 7. Wiley, New York, 1964.
2. M. Miller, Lime, *Mining Eng.* **53**(6), 43–44, June (2001).
3. R.C. Weast, ed., "Handbook of Chemistry and Physics," 51st ed. pp. F–64, B–78. Chem. Rubber Pub. Co., Cleveland, OH, 1970.
4. C.J.S. Warrington and R.V.V. Nichols, "A History of Chemistry in Canada," p. 96. Sir Isaac Pitman and Sons, Toronto, 1949.
5. R.N. Shreve and J.A. Brink, Jr., "Chemical Process Industries," 4th ed. pp. 166, 211. McGraw-Hill, Toronto, 1977.
6. Anon., New Vertical Lime Kiln, *ZKG International*, **55**(7), 10 (2002).
7. R.W. Thomas and P. Farago, "Industrial Chemistry," p. 63. Heinemann, Toronto, 1973.
8. W.L. Faith, D.B. Keyes, and R.L. Clark, "Industrial Chemicals," 3rd ed. pp. 483, 694. Wiley, New York, 1965.
9. R.H. Perry, C.H. Chilton, and S.D. Kirkpatrick, eds., "Chemical Engineers' Handbook," pp. 9–6, 3–116. McGraw-Hill, Toronto, 1963.
10. N. McCubbin, Alternatives to Fossil-fuel for the Lime Kiln, *Pulp Pap.-Can.* **86**(9), 85–88 (1985).
11. L. Ernstbrunner and S. Lechner, The Use of Secondary Fuels in the European Lime Industry, *ZKG Internernational,* **55**(5), 54–62 (2002).
12. E. Notidis and H. Tran, Survey of Lime Kiln Operation..., *TAPPI J.* **76**(5), 125–131, May (1993).
13. V.R. Scholz, R. Jeschar, R. Jennes *et al.*, Environmental Aspects of the Manuf. and Use of Lime Products, *Zement-Kalk-Gips.* **47**(10), 571–581, Oct. (1994).
14. Corson (of Lime Hydration Process) Honored, *Chem. Trade J.* **156**, 414 (1965).
15. F.S. Taylor, "A History of Industrial Chemistry," Reprint ed., p. 182. Arno Press, New York, 1972.
16. J.A. Kent, ed., "Riegel's Industrial Chemistry," p. 129. Reinhold, New York, 1962.
17. F.A. Lowenheim and M.K. Moran, "Faith, Keyes and Clark's Industrial Chemicals," 4th ed. pp. 706, 714. Wiley-Interscience, Toronto, 1975.
18. D.S. Kostick, Soda Ash, In: U.S. Geolog. Survey, Mineral Commodity Summaries, Jan 2004.
19. D.S. Kostick, Soda Ash, 2001. Available: http://minerals.usgs.gov/minerals/pubs/commodity/soda_ash/sodamyb01.pdf
20. H.W. Haynes, Solution Mining of Trona, *In Situ*, **21**(4), 357–394 (1997).
21. M. Dogan, C. Goldur, G. Dogu *et al.*, Soda Ash Production from Trona in a Spray Drier, *J. Chem. Technol. Biotechnol.* **68**(2), 157–162, Feb. (1997).
22. M. McCoy, General Chemical...Solvay Plant Closure, *Chem. Eng. News*, **78**(50), Dec. 11 (2000).
23. G.A. Lane, J.S. Best, E.C. Clarke *et al.*, "Solar Energy Subsystems Employing Isothermal Heat Sink Materials," ERDA Contract No. NSF-C906. The Dow Chemical Company, Midland, MI, 1976.
24. "US Geological Survey Minerals Yearbook, Metals and Minerals," Vol. I, 2002. U.S. Dept. of the Interior, Bureau of Mines, Washington, DC, 2004, and earlier editions.
25. "United Nations Statistical Yearbook 2001," 48th ed. United Nations, New York, 2003, and earlier editions.
26. Hooker, Brandon, Manitoba, *Can. Chem. Proc.* **59**(5), 8, May (1975).
27. R. N. Shreve, "The Chemical Process Industries," p. 274. McGraw-Hill, New York, 1945.
28. E.W. Washburn, ed., "International Critical Tables," Vol. VII, p. 305. McGraw-Hill, New York, 1930.
29. R.C. Weast, ed., "Handbook of Chemistry and Physics," 56th ed. p. F-86. Chem. Rubber Publ. Co., Cleveland, OH, 1975.
30. H. Stephen and T. Stephen, eds., "Solubilities of Inorganic and Organic Compounds," Vol. I, p. 115. Pergamon, New York, 1963.
31. J. A. Kent, ed., "Riegel's Industrial Chemistry," p.158. Reinhold, New York, 1962.
32. J.C. Olsen and O.G. Direnga, *Indus. Eng. Chem.* **33**, 204 (1941).

33. L.F. Goodwin, *J. Soc. Chem. Indus.* **45**, 360T-361T (1926).
34. R.M. Stephenson, "Introduction to the Chemical Process Industries," p. 56. Reinhold, New York, 1966.
35. L.W. Codd *et al.,* eds., "Chemical Technology: An Encyclopaedic Treatment," Vol. 1, p. 23. Barnes and Noble, New York, 1968.
36. F.A. Lowenheim and M.K. Moran, "Faith, Keyes and Clark's Industrial Chemicals," 4th ed. p. 741. Wiley-Interscience, Toronto, 1975.

Shop construction of closely stacked titanium screens used for the dimensionally stable anodes of one modern diaphragm cell.

A row of diaphragm cells in the cell building of a chloralkali plant. Brine inlet is visible at the top foreground of first cell, and the sight glasses for level control, plus funnels to catch the broken streams of cell effluent (product) are visible along the right hand side of the row.

8

ELECTROLYTIC SODIUM HYDROXIDE, CHLORINE, AND RELATED COMMODITIES

...apparatus for the production of Chloride gas and of metallic Sodium through the agency of dynamic electricity.

—A.L. Nolf, 1882

8.1. ELECTROCHEMICAL BACKGROUND AND BRINE PRETREATMENT

All electrolytic routes to sodium hydroxide and chlorine from sodium chloride brine have to contend with keeping the highly reactive products separated. Sodium and chlorine products would react vigorously together to return starting salt (Eqs. 8.1 and 8.2).

$$Na^+ + 1e^- \rightarrow Na \text{ (or, with water} \rightarrow NaOH + 1/2\ H_2) \qquad 8.1$$

$$Cl^- - 1e^- \rightarrow 1/2\ Cl_2 \qquad 8.2$$

A sodium hydroxide product, without separation, would react vigorously with chlorine to give sodium chloride and sodium hypochlorite (Eqs 8.3, 8.4), so that similar precautions are required with either option.

$$2\ Na + Cl_2 \rightarrow 2\ NaCl \qquad 8.3$$

$$2\ NaOH + Cl_2 \rightarrow NaCl + NaClO + H_2O \qquad 8.4$$

Three main methods are used to keep these electrolytic products apart. One involves the separation of the electrochemical cell into two compartments by a porous vertical diaphragm, which permits the passage of brine and ions, but keeps the products separated. Another employs a flowing mercury cathode to continuously carry sodium, in the form of an amalgam, away from the brine and chlorine streams of the cell. The third, and most recent development, is an ion-selective membrane (e.g., Nafion) separator between the compartments, which allows only sodium ions to pass through, not chloride ions nor brine. Diaphragm cells run efficiently with natural brines (solution-mined brines) and account for some three-quarters of North American chloralkali production. Mercury cells require a higher degree of brine purity for efficient operation, and so are used by European chloralkali producers who mostly use vacuum pan or recrystallized salt to make up the electrolyte. Membrane cells are used for some 80% of chloralkali production in Japan, where a mercury-based toxicity episode (Minimata disease) in the early 1950s drove legislation to eliminate the operation of mercury cells at a time when membrane technology was emerging.

The initial stages of brine pretreatment are quite similar for all three cell types except that the mercury and membrane cells have more stringent requirements. If calcium sulfate is present, its removal requires the addition of a barium carbonate suspension (Eq. 8.5). If calcium is present as the chloride, it is removed by the addition of sodium carbonate (Eq. 8.6). Magnesium ion

$$CaSO_4 + BaCO_3 \rightarrow CaCO_3 \downarrow + BaSO_4 \downarrow \qquad 8.5$$

$$CaCl_2 + Na_2CO_3 \rightarrow CaCO_3 \downarrow + 2\ NaCl \qquad 8.6$$

requires a higher pH for effective removal, which is accomplished by the addition of small amounts of sodium hydroxide (Eq. 8.7). The bulk of the precipitated salts is removed by settling,

$$MgCl_2 + 2\ NaOH \rightarrow Mg(OH)_2 \downarrow + 2\ NaCl \qquad 8.7$$

either in the treating unit itself or in a separate clarifier. This is followed by a polishing filtration to give a brine containing 5 ppm or less of Ca^{2+} and Mg^{2+}. Acidification with hydrochloric acid to about pH 2 produces a brine, which is suitable as a feed for a diaphragm cell.

Brine purity requirements for the mercury cell are more strict, since traces of transition metal ions can cause severe process upset by lowering the hydrogen overvoltage on mercury. Simple adsorption of interfering ions onto the precipitated calcium and magnesium flocs will normally achieve sufficient control. Iron is simultaneously precipitated as the hydroxide (Eq. 8.8). If this is inadequate,

$$FeCl_3 + 3\ NaOH \rightarrow Fe(OH)_3 \downarrow + 3\ NaCl \qquad 8.8$$

then the volume of precipitate will be increased by deliberate addition of calcium and/or iron salts until control of vanadium, chromium, manganese, and iron to 0.5 ppm or less is achieved. To further increase the hydrogen overvoltage on mercury, the feed is acidified with hydrochloric acid to pH 2.5–5 prior to use. From this point onward, details of the three processes differ and will be treated separately.

8.2. BRINE ELECTROLYSIS IN DIAPHRAGM CELLS

The first electrolytic production of chlorine was by the electrolysis of a potassium chloride brine with coproduction of potassium hydroxide by the

Griesheim Company, in Germany, in 1888 [1]. Sodium hydroxide and chlorine were first produced electrolytically in 1891 in Frankfurt, Germany, using technology advanced in the patents of Mathes and Weber. The same technology was first used in the U.S.A. at Rumford Falls, Maine, by the Electrochemical Company in 1892. Both of the early cells used an operating procedure, which amounted to a scaled-up version of a laboratory electrolyzer (Fig. 8.1a) and were operated in batch modes.

The cell was loaded with saturated potassium chloride brine, indirectly heated to 80–90°C with steam pipes, and then a current of about 3500 A was passed through the cell for a period of 3 days. During this time, the chlorine and hydrogen produced were collected, and the potassium hydroxide concentration in solution rose to about 7% (Eqs. 8.9 and 8.11).

$$\text{Anode:} \qquad 2\,Cl^- - 2\,e^- \rightarrow Cl_2 \qquad\qquad 8.9$$

$$\text{Cathode:} \qquad 2\,H^+ + 2\,e^- \rightarrow H_2 \qquad\qquad 8.10$$

$$\text{Overall:}\ 2\,K^+ + 2\,Cl^- + 2\,H^+ + 2\,HO^- \rightarrow Cl_2 \uparrow + H_2 \uparrow + 2\,KOH \qquad 8.11$$

At this time, the electrolyte was pumped out and processed to recover the potassium hydroxide by crystallization of unreacted potassium chloride, and the cell was then recharged with fresh potassium chloride and water to repeat the whole process again, on a batch basis.

The Griesheim method was successful and represented a significant advance in simplicity over the earlier chemically based causticization process used to produce potassium hydroxide and chlorine. At the same time, it was slow, and it also made for inefficient material handling with the frequent necessity for emptying and filling of cells. It also had a low electrical efficiency, partly because solution concentrations were not optimum and partly because of side reactions caused by diffusion of the products through the diaphragm. These side reactions consumed both of the electrolytic products, in the process producing sodium chlorate (Eqs. 8.12–8.16).

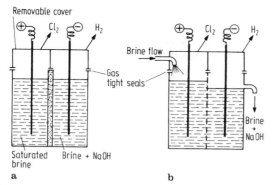

FIGURE 8.1 (a) Schematic diagram of a Griesheim-type cell which operated on a batch basis. The product was pumped out after electrolysis for about 3 days. (b) Schematic diagram showing the LeSueur percolating diaphragm, which changed the mode of cell operation to a continuous basis and decreased the concentration of impurities in the sodium hydroxide product.

$$2\,OH^- + Cl_2 \rightarrow 2\,HClO \qquad\qquad 8.12$$

$$HClO + H_2O \rightleftharpoons H_3O^+ + ClO^- \qquad\qquad 8.13$$

$$2\,ClO^- \rightarrow ClO_2^- + Cl^- \qquad\qquad 8.14$$

$$ClO_2^- + ClO^- \rightarrow ClO_3^- + Cl^- \qquad\qquad 8.15$$

$$Na^+ + ClO_3^- \rightleftharpoons NaClO_3 \qquad\qquad 8.16$$

Many of these difficulties were overcome by the Rumford Falls, Maine, choralkali operation in 1892 by incorporating improvements developed by E. A. LeSueur of Ottawa, Canada, just prior to this [2]. These innovations centered on a percolating diaphragm, which allowed the continuous passage of both ions and brine solution from the anode to the cathode compartment [3] (Fig. 8.1b). To obtain brine flow in the desired direction, a tube continuously fed saturated brine into the anode compartment. An outlet was placed at the desired cathode compartment solution level, lower than the anode compartment level, which allowed for the continuous removal of nonelectrolyzed brine and the caustic product. These measures provided a positive pressure in the anode compartment and continuous brine flow through the diaphragm to the cathode compartment. They also prevented back-diffusion of hydroxide ions through the diaphragm, which avoided both consumption of products and formation of impurities, and achieved continuous, rather than batch operation.

The features of the percolating diaphragm together with continuous brine entry and brine and caustic removal, are incorporated into virtually all modern diaphragm cell designs. This includes the North American Hooker, Diamond Alkali, and Dow filter press designs, which all use vertical diaphragms. They also feature in the European horizontal diaphragm designs such as the Billiter cell, and the I.G. Farben-industrie modification, which introduced large flutes to the diaphragm and cathode for increased capacity [4].

The percolating diaphragm of most modern cells consists of a thick asbestos fiber pad, supported on a crimped heavy iron wire mesh or a punched steel plate located on the cathode side of the diaphragm. The wire diaphragm support also serves as the cathode, which places it as close as possible to the anode for high electrical efficiency while remaining on the cathode side of the diaphragm (Fig. 8.2). The iron does not corrode under these severe conditions (contact with the hot saturated brine) because it is the cathode (which provides "cathodic protection") for its function in the cell. It continuously supplies electrons to the solution such that the metal itself is not electrochemically scavenged to do so (Eq. 8.17).

$$Fe_{(s)} - 3e^- \rightarrow Fe^{3+}_{(aq)} \qquad\qquad 8.17$$

In this severe service, however, the asbestos fiber pad of the diaphragm has to be replaced at intervals ranging from 1 to 3 yr. During this time, graphite particles from the anode and insoluble impurities in the brine gradually accumulate on the anode side of the diaphragm, reducing its permeability. Also, the passage of electrical power through the diaphragm causes the asbestos fibers of the pad to swell. To maintain the chlorine and caustic production rate under these conditions, the brine level in the anode compart-

FIGURE 8.2 Diagram of the components of a commercial asbestos diaphragm cell, and flowsheet for the electrolytic production of chlorine and sodium hydroxide. All commercial cells employ a multiple series of anodes and diaphragms in a single brine container, only one of which is shown here. The asbestos layer is located on the anode side of the steel wire grid cathode, and located relatively much closer than shown here to minimize operating voltage.

ment is raised, which raises the differential pressure across the diaphragm to maintain the brine flow rate through it. During this time, the operating voltage is also raised to help offset the decreased ion mobility through the diaphragm. Diaphragm permeability can be improved several times, while in service, by a weak acid wash which serves to remove "hardness." But eventually, there is no further adjustment possible and the cell is shut down and dismantled for overhauling. The spent asbestos is removed from the iron grid with high-pressure water jets, and the water slurry is collected in a disposal pit. New asbestos is then slurried in brine, placed in the anode compartment, and by means of a vacuum on the hydrogen exit of the cell it is applied as an even coating to the cathode grids. After fitting new anodes, if necessary, the cell is then reassembled and placed back in service.

The saturated (about 300 g/L NaCl) pretreated brine is preheated to about 90°C and fed to the anode compartment as a *broken* stream, broken in order to avoid electrical losses. Preheating the brine both increases cell conductivity by increasing ion mobility with the lower solution viscosity, and decreases chlorine solubility in the brine. In this way less dissolved chlorine is carried through the diaphragm in the percolating brine, which contributes to higher cell current efficiencies and lower concentrations of impurities in the cathode exit brine (Eqs. 8.12–8.16). It also tends, however, to result in increased graphite wear.

Anode electrode reactions produce chlorine gas, which is removed from the vapor space above the anode compartment by ducting made of polyvinyl chloride (PVC) or other chemically resistant material under slight negative pressure (Eqs. 8.18–8.20). High brine temperatures assist

$$Cl^-_{(aq)} \rightarrow Cl^\circ_{(aq)} + 1\ e^- \qquad\qquad 8.18$$

$$2\ Cl_{(aq)} \rightarrow Cl_{2(aq)} \qquad\qquad 8.19$$

$$Cl_{2(aq)} \rightarrow Cl_{2(g)} \qquad\qquad 8.20$$

removal of chlorine from the brine and the slight negative pressure both aids in chlorine removal and provides a safeguard in the event of any minor leaks in the chlorine collection system.

At the cathode, which is the crimped iron wire grid supporting the diaphragm, electrons are added to water to form hydrogen gas and hydroxide ions (Eqs. 8.21–8.23). The hydrogen gas

$$H_2O_{(aq)} + 1\ e^- \rightarrow H^\circ + OH^- \qquad\qquad 8.21$$

$$2\ H^\circ_{(aq)} \rightarrow H_{2(aq)} \qquad\qquad 8.22$$

$$H_{2(aq)} \rightarrow H_{2(g)} \qquad\qquad 8.23$$

is removed under slight positive pressure. It might be thought reasonable to expect that it would be sodium ions, which are electrolyzed, not water, since these are also present at high concentrations [5]. Sodium, if formed, would react with water under these conditions to yield sodium hydroxide and hydrogen gas, the observed products. But examination of the relevant electrochemical data shows that this is not so. The deposition potential for hydrogen is $-0.828\,V$, which together with the $-0.30\,V$ hydrogen overpotential on iron gives a total of $-1.128\,V$ required for hydrogen generation. The single electrochemical potential for sodium deposition is $-2.71\,V$, more than twice that required for hydrogen deposition [6]. Therefore, hydrogen and hydroxide ion are the direct, not indirect, cathode electrochemical products. So the overall products of brine electrolysis in a diaphragm cell are sodium hydroxide, chlorine, and hydrogen (Eq. 8.24).

$$NaCl + H_2O \xrightarrow{\text{electrol.}} NaOH + 1/2\ Cl_2 + 1/2\ H_2 \qquad\qquad 8.24$$

In practice, brine and current flows to a diaphragm cell are synchronized to only electrolyze about 30–40% of the sodium chloride present in the feed brine solution. To electrolyze a larger fraction of the sodium chloride present than this would appear to be advantageous in that less salt would have to be removed from the cell effluent to produce sodium hydroxide suitable for sale. However, to do so would require an increased current flow for the same rate of brine flow through the diaphragm, which would, in turn, necessitate a higher operating voltage. The higher voltage requirement would result in higher electrical power cost for each tonne of product made. Alternatively, it would require that the current flow be maintained at its original value while the brine flow rate is decreased. Decreasing the brine flow would increase hydroxide ion back-diffusion through the diaphragm with two undesirable consequences. Loss of both hydroxide ion and chlorine cell products would occur, resulting in an impurity in the sodium hydroxide (Eqs. 8.12–8.16). Also, the graphite consumption would increase from the increased rate of reaction of the anode material with oxygen from the electrolyzed hydroxide ion (Eqs. 8.25–8.26a, b).

$$2\,OH^- \rightarrow H_2O + 1/2\,O_2 + 2\,e^- \qquad\qquad 8.25$$

$$2\,C + O_2 \rightarrow CO\uparrow \qquad C + O \rightarrow CO\uparrow \qquad\qquad 8.26a$$

$$CO + O \rightarrow CO_2\uparrow \qquad 2\,CO + O_2 \rightarrow 2\,CO_2\uparrow \qquad 8.26b$$

Even with only about 30–40% electrolysis of the brine, anode life is limited to about 2–3 year from graphite consumption at the rate of 2–3 kg/tonne of chlorine.

8.2.1. Purification of Diaphragm Cell Products

Chlorine and hydrogen as obtained directly from the diaphragm cell are both hot and wet, and the 8–12% sodium hydroxide liquid cell effluent still contains 12–19% dissolved ("undecomposed") sodium chloride. Some applications of these products can use them in the form obtained directly from the cell. For example, propylene chlorohydrin may be ring-closed to propylene oxide using the crude sodium hydroxide solution still containing sodium chloride (Eq. 8.27). However, majority of uses require at least a simple cleanup of the crude products prior to use.

$$
\underset{\text{CH}_3 - \text{CH} - \text{CH}_2}{\overset{\overset{\displaystyle Cl}{|} \quad \overset{\displaystyle OH}{|}}{}} + NaOH/NaCl \rightarrow \underset{\text{CH}_3 - \text{CH} - \text{CH}_3}{\overset{\displaystyle O}{\diagup \diagdown}} + 2\,NaCl \qquad 8.27
$$

Elemental chlorine under ambient conditions is a heavy (dense) greenish-yellow gas, which may be stored as a liquid at temperatures below $-35°C$, its atmospheric pressure boiling point. Important considerations for its collection and purification are that it is highly toxic (hence the slight negative pressure for collection) and it is highly reactive with many ordinary metals, especially when wet, and with organic substances generally. It is also slightly soluble in water, with which it partially reacts. The first step in the cleanup is indirect cooling in process water-cooled heat exchangers constructed of polyvinyl chloride, titanium, or fiberglass-reinforced plastic, which removes much of the water vapor and entrained brine droplets from the chlorine. It is then contacted countercurrently to concentrated sulfuric acid, which has a strong dehydrating action. This takes place by passage through one or more fiberglass-reinforced plastic towers packed with chemical stoneware. The packing provides a large surface area for efficient contact of the chlorine with the sulfuric acid. Acid consumption is 2–5 kg/tonne of chlorine. Once dried, steel piping may safely be used to move chlorine. However, its reactivity must still be kept in mind during compression and liquefaction of the gas since chlorine can spontaneously react with iron at temperatures above 150°C. Care must also be taken to maintain anhydrous conditions during these steps.

Much of the chlorine produced will be simply compressed and pipe-lined to the customer if it is to be consumed locally. It may be consumed by the same company producing it ("captive consumption") or by another. But if chlorine has to be shipped across any distance, it will normally be liquefied by compression and cooling to minimize storage and shipping volumes. Early compressors were reciprocating, and operated in a similar fashion to a piston engine but with graphite piston rings and using no cylinder lubricant to avoid the reaction of chlorine with hydrocarbons. Larger facilities today use several stages of centrifugal turbine compressors [7]. The compressed hot gases are first air or water cooled and then further chilled either by Joule-Thomson effect cooling obtained by expansion of a portion of the cooled compressed gas, or occasionally by using liquid ammonia or other refrigerant in a separate circuit. Liquefaction may be achieved at 700–1,200 kPa (about 7–12 atm) pressure at 18°C, or at 100 kPa at −40°C, just below the atmospheric pressure boiling point. Most producers settle for intermediate liquefaction conditions of about 200 kPa and 4°C [8].

Small amounts of noncondensible gases, called "blow gas" or "sniff gas," consist of traces of hydrogen, oxygen, carbon dioxide, and air, which accumulate during chlorine liquefaction. This gas contains some residual chlorine, so it is vented through an aqueous sodium hydroxide or sodium hypochlorite stream to capture the traces of chlorine before discharge. In mercury cell plants, the aqueous sodium hypochlorite stream produced in this way may be used for mercury recovery. Hydrogen gas, which tends to get more concentrated in the chlorine gas stream as liquefaction proceeds, must not be allowed to rise to more than 5% in the liquefaction or venting systems or there is risk of an explosion from the sudden reaction of chlorine with it.

The main hydrogen stream from the cell is also hot and wet. Since hydrogen has a low solubility in water, it is usually cooled and cleaned of chlorine, salt, etc., by a direct, countercurrent contact with a cool spray of dilute sodium hydroxide in water. Clean hydrogen exits from the top of the cooler through baffles or plastic mesh placed so as to remove entrained water droplets.

Crude 10% sodium hydroxide containing sodium chloride is purified in a similar manner to the product of the causticization process. The water is evaporated in nickel or nickel-clad steel (to reduce corrosion) multiple-effect evaporators to about 50% sodium hydroxide concentration. At this concentration, sodium chloride is only about 1% soluble (2%, on a dry basis) in the more concentrated caustic so that the bulk of it crystallizes out and is filtered off. This quite pure sodium chloride is recycled to the cells. For many purposes, such as for pulp and paper production, this purity of 50% sodium hydroxide is quite acceptable. If higher purities are required, sodium hydroxide may be separated from residual water and salt by chilling to the double hydrate crystals $NaOH \cdot 2H_2O$, m.p. about 6°C, or as $NaOH \cdot 3.5H_2O$, with a m.p. of about 3°C, or by counter-current extraction [9]. The sodium hydroxide obtained after these steps contains 2–3 ppm sodium chloride, equivalent to the purity of the mercury cell product ("rayon grade") [10]. Concentrations of 73% and 100% sodium hydroxide (see details, Section 7.5) are also marketed.

8.3. BRINE ELECTROLYSIS IN CHLORATE CELLS

A cell similar in principle to the diaphragm cell described above and operated without a diaphragm gives the same initial products but allows these to react with each other (Eqs. 8.24, 8.12–8.16). This is the basis of the chlorate cell. The initial electrochemical products ultimately form sodium chlorate as the final cell product from the electrolysis of aqueous sodium chloride [11]. Part of the product formation involves the "chemical chlorate" formation just outlined (Eqs. 8.12–8.16), and part of it forms from "electrolytic chlorate" formation (Eq. 8.28).

$$6 \; ClO^- + 3 \; H_2O - 6 \; e^- \rightarrow 2 \; ClO_{3^-} + 6 \; H^+ + 4 \; Cl^- + 3/2 \; O_2 \qquad 8.28$$

So the overall process may be summarized as given in Eq. 8.29 [12].

$$NaCl_{(aq)} + 3 \; H_2O \xrightarrow[\text{6 faradays}]{\text{electrolysis}} NaClO_{3_{(aq)}} + 3 \; H_2 \qquad 8.29$$

Hydrogen bubbles as they form, move the brine solution to be electrolyzed through the chlorate cell and up to the chemical chlorate reactor without any moving parts. Chlorate cells are operated with sodium dichromate inhibitor present in the brine to prevent corrosion from hypochlorite ion.

Theoretical and actual operating voltages are 1.71 V and about 3.5 V, respectively, and current efficiency is about 94%. These electrochemical parameters give a power consumption of 4,950–6,050 kWh/tonne of sodium chlorate for these cells [11].

The major market for sodium chlorate is for the preparation of the chlorine dioxide used for bleaching of wood pulp (Chap. 15). This market has recently risen by an order of magnitude every 10 years from the demand to replace of part, or all of the chlorine previously used for bleaching [13]. Other smaller uses of sodium chlorate are as a weed killer, for the making of matches and fireworks, and for the tanning of hides. A small market exists for sodium (and potassium) perchlorate, made by electrolysis of sodium chlorate using platinum electrodes (Eq. 8.30).

$$4 \; NaClO_3 \xrightarrow{\text{electrol.}} 3 \; NaClO_4 + NaCl \qquad 8.30$$

8.4. ELECTROCHEMICAL ASPECTS OF BRINE ELECTROLYSIS

The overall chemical changes involved in the electrolysis of aqueous sodium choride can be expressed as given in Eq. 8.31.

$$NaCl_{(aq)} + H_2O_{(\ell)} \xrightarrow{\text{electrol.}} NaOH_{(aq)} + 1/2 \; H_{2(g)} + 1/2 \; Cl_{2(g)} \qquad 8.31$$

To enable thermodynamic insight into the overall energetics of this process, it may be broken down to its component enthalpy changes as follows (Eqs. 8.32–8.34).

	$\Delta°H$	
$Na_{(s)} + 1/2\ Cl_{2g} \rightarrow NaCl_{(aq)}$	-407 kJ $(-97.0$ kcal$)$	8.32
$H_{2(g)} + 1/2\ O_{2(g)} \rightarrow H_2O_{(\ell)}$	-286 kJ $(-68.3$ kcal$)$	8.33
$Na_{(s)} + 1/2\ O_{2(g)} + 1/2\ H_{2(g)} \rightarrow NaOH_{(aq)}$	-286 kJ $(-68.3$ kcal$)$	8.34

To obtain a net enthalpy change for the process of Eq. 8.31, the signs of the first two processes have to be changed to correspond to these in the opposite direction to that given. Then summing the three components $(+407 + 286 - 469)$ kJ gives a net enthalpy change of $+224$ kJ (i.e., it is an endothermic process).

The net enthalpy change of a process may be used to obtain the theoretical voltage required for electrochemical equilibrium by using the Gibbs-Helmholtz equation [11]. It also can be summed for the two component electrochemical changes taking place (Eqs. 8.35 and 8.36).

$$2\ Cl^- \rightarrow Cl_2 + 2\ e^- \qquad\qquad -1.360\ V \qquad\qquad 8.35$$
$$2\ H_2O + 2\ e^- \rightarrow H_2 + 2\ HO^- \qquad -0.828\ V \qquad\qquad 8.36$$

In both cases, a value of close to 2.20 V is obtained, which is defined as the "theoretical cell voltage" or "theoretical decomposition voltage." Actual operating voltage is 3.7–4.0 V. Factors such as the finite spacing required between electrodes, the cathode over-potential requirement for deposition of hydrogen, the resistance of the hot brine, and the resistance to ion diffusion imposed by the presence of the diaphragm (and other minor components) all contribute to this higher than theoretical operating voltage. The ratio of the theoretical cell voltage to the actual operating voltage is defined as the "voltage efficiency" of the cell. Expressed as a percentage ($\times 100$) the voltage efficiency normally lies in the 55 to 60% range. Voltage efficiencies usually drop somewhat with increased production rates, as the anodes wear (increasing electrode separation) and diaphragm permeability decreases.

The current consumption, or electrical flow rate in amperes required to operate the cell, is the factor that directly relates to the amount of electrochemical product collected. One faraday (1 mol of electrons, after the original discoverer) of electricity is required to deposit 1 g equivalent weight of a substance in an electrochemical reaction (Eq. 8.37).

$$1\ \text{faraday} = 96{,}494\ \text{coulombs/(g equiv. wt.) or,}$$
$$1\ \text{faraday} = 96{,}494\ \text{ampere sec/(g equiv. wt.)} \qquad 8.37$$
$$\text{since 1 coulomb (C)} = 1\ \text{ampere second (A} \cdot \text{sec)}$$

So chloride, being a monovalent ion, would theoretically require 3.15×10^4 amperes to be converted to chlorine at the rate of 1 tonne/day (Eq. 8.37). Side reactions from hydroxide ion

$$\frac{96{,}494\ A \cdot sec/(g\ equiv.wt.)}{(24 \times 60 \times 60)sec/day} \times \frac{10^6\ g/tonne}{35.453\ g/(g\ equiv.\ wt.)} \qquad 8.37$$
$$= 3.1502 \times 10^4\ \text{ampere days/tonne chlorine}$$

back-diffusion across the diaphragm and from the presence of some dissolved chlorine in the brine percolating from the anode to the cathode compartment, as well as the production of small amounts of oxygen at the anode, consume electrical power without contributing to the primary products of the cell. These factors reduce the cell efficiency to somewhat less than theory, so that about 32.5 kA is actually required for a daily tonne of chlorine. The ratio of the theoretical current requirement to the actual current consumed is called the "current efficiency," and ($\times 100$) is thus about 92–96%. By convention, the current efficiency is normally understood to refer to the cathode current efficiency, which for a diaphragm chloralkali cell is nearly the same, though marginally higher than the anode current efficiency. Ordinarily, cell operators expect current efficiencies to fall slightly with increasing decomposition efficiencies of the cell (decreasing salt to sodium hydroxide ratios in effluent brine) for the reasons outlined earlier.

Current density refers to the total current flow in kiloamperes divided by the anode electrode area in square meters, expressed as kA/m^2. High-current densities are desirable, particularly for electrochemical processes, which yield unstable products. With current densities of $2–3 \, kA/m^2$, electrolytic products of the diaphragm cell are rapidly moved from the sites of formation, which decreases side reactions and maximizes current efficiencies [14]. High-current densities, however, increase heat generation, anode wear, and the operating voltage so that lower current densities (and more cells) are better if the cells can be made cheaply.

The overall "energy efficiency" of the cell is the voltage efficiency times the current efficiency. This definition relates to the gross *power* consumption for any given quantity of product (e.g., a tonne of chlorine). By lowering the energy efficiency, an increased operating voltage directly increases the power cost per tonne of product. For example, the mean power requirement for chlorine production in a diaphragm cell amounts to some 2,700–2,900 kWh/ tonne. But if the cell is operated below capacity, at 3.2 V, it would only require some 2496 kWh/tonne ($3.2 \times 32.5 \times 24 \, hr$). Operating the cell at near the end of its diaphragm life and/or pushing the cell to produce somewhat beyond its rated capacity by running at 4.2 V would require 3276 kWh/ tonne, an additional 31% more power (and power cost) for the same mass of chlorine as at 3.2 V. Since the power costs some 60 to 70% of the total chlorine production costs, these voltage increments add significantly to operating cost [15]. Nevertheless, for temporarily slack or tight markets operating cells at 5–10% below or above rated capacities (or actually taking cells out of service) may be used to help adjust the production rate to match demand.

Finally, the electrochemical "decomposition efficiency" is defined as the equivalents (of product) produced divided by the equivalents charged (or entering) the electrochemical cell (Eq. 8.39).

$$\begin{matrix} \text{Decomposition} \\ \text{efficiency} \end{matrix} = \frac{\text{(equivalents produced)}}{\text{(equivalents charged)}} = \frac{\text{(moles electrolyzed)}}{\text{(moles fed to cell)}} \qquad 8.39$$

This quantity equates to the proportion of the salt present in the incoming solution, which is electrolyzed on its passage through the cell. For the reasons previously discussed, in a diaphragm cell this electrolyzed fraction is about 0.30–0.40, or ($\times 100$) 30–40%. Since this represents a mole (not weight)

TABLE 8.1   Typical Operating Characteristics of Relatively Small and Large Diaphragm Chloralkali Cells

	Rated current flow (A)	
	30,000[a]	150,000[b]
Current efficiency	96.5	96
Current density[c], kA/m^2	1.29	2.7
Operating voltage[c] (nominal)	3.82	3.48
Operating temperature, °C	90	85–90
Cell effluent:		
Temperature, °C	95	–
% NaOH	10.5	10.5
$NaClO_3$, g/L	0.07	0.05–0.25
Cell output, tonnes per day:		
Chlorine	0.92	4.60
NaOH	1.03	5.19
Power,[c] kWh/tonne Cl_2	3000	2740

[a]Data calculated from Sconce [1].

[b]Data selected from Puschaver [15] and Thompson [16].

[c]The improved performance of the larger cell is at least partly the consequence of the dimensionally stable anodes used (see Section 8.5).

fraction electrolyzed, the crude 10–12% sodium hydroxide in water produced also contains 12–15% by weight dissolved sodium chloride.

The interrelationships between these various operating and electrolytic characteristics for diaphragm cells are given in Table 8.1.

8.5. BRINE ELECTROLYSIS IN MERCURY CELLS

Use of the knowledge that an electrolytic cell employing mercury as a cathode would cause sodium deposition in the mercury, rather than hydrogen generation, was made simultaneously in 1892 by H.Y. Castner in the U.S.A., and independently by Karl Kellner in Austria [1]. Their designs employed a rocking cell to move mercury from the sodium amalgam (sodium/mercury alloy or solution of sodium in mercury) forming side of the cell to the sodium amalgam hydrolysis side (Fig. 8.3). This design became a commercial success after further development by Solvay et Cie. This method of brine electrolysis gave no direct contact between the brine and the sodium hydroxide streams, so it was possible to produce a high-purity 50% sodium hydroxide solution directly. No concentration step, or sodium chloride crystallization, etc., was required. This provided such clear advantages for the high-purity sodium hydroxide market that Castner–Kellner rocking cells were used at an Olin–Mathieson chloralkali operation at Niagara Falls until as recently as 1960, when they converted to a new mercury cell design. These newer designs now use variations on the theme of a long shallow sloping flat-bottomed trough in which the electrolysis is conducted. Pumps are used to circulate mercury continuously, as a thin film flowing on the bottom of the trough.

FIGURE 8.3 Vertical section of a Castner–Kellner rocking mercury cell, which demonstrates the principle of operation. The cell had provision for brine entry and removal of depleted brine from the left compartment, and for purified water entry and removal of sodium hydroxide from the right compartment. The eccentric revolved slowly to move the mercury on a regular basis between the two compartments. A cell divider extended close enough to the bottom of the cell to maintain a mercury seal, which prevented any mixture of the water phases of the two compartments.

Pretreatment of brine to be used as feed for a mercury cell is the same as for diaphragm cell feed plus precautions to ensure that molybdenum, chromium, and vanadium ions are maintained at levels below $1\,\mu g/L$ (1 ppb) and that (magnesium + iron) salts are kept below $1\,mg/L$ (1 ppm) to avoid reducing the hydrogen overvoltage on mercury relative to the sodium deposition voltage. If one or more of these ions exceeds these limits, then hydrogen gas evolution in the electrolyzer (rather than sodium deposition in mercury) can become significant. Hydrogen contamination of the chlorine, the principal product of the electrolyzer, introduces an explosion hazard at the time the chlorine is compressed (Eq. 8.40). This is especially true if compression is accompanied by chlorine liquefaction, which would raise the concentration of hydrogen in the gas phase.

$$H_{2(g)} + Cl_{2(g)} \rightarrow 2\,HCl_{(g)} \quad \Delta H = -92.3\,kJ/mol\,(-22.1\,kcal/mol) \qquad 8.40$$

Generally, the concentrations of the interfering transition metal ions in the brine are decreased sufficiently by simple adsorption onto the colloidal precipitates of calcium carbonate, magnesium carbonate, and barium sulfate produced by conventional brine pretreatment. The brine is then acidified with hydrochloric acid to about pH 4. This helps to maintain a high hydrogen overvoltage on mercury in order to minimize the formation of hydrogen in the

TABLE 8.2 A Comparison of Mercury Cell Gas Composition for Graphite versus Dimensionally Stable Anodes[a]

	Graphite anode (%)	Dimensionally stable anode[b] (%)
Chlorine	97.5–98.5	97.5–98.5
Hydrogen	0.4–0.6	0.4–0.6
Oxygen	0.1–0.3	0.2–0.5
Nitrogen	0.3–0.7	0.3–0.7
Carbon dioxide	0.4–0.9	0.3–0.7

[a]Data from Puschaver [15].
[b]For details, see Section 8.7.

FIGURE 8.4 Mercury cell process for the electrolytic production of chlorine and sodium hydroxide. Packing in the vertical decomposer is graphite lumps. For treatment of cell products, see Fig. 8.2.

electrolyzer (Table 8.2). The brine is then preheated to about 65°C before passage to the electrolyzer, for the same efficiency reasons as heating of the diaphragm cell feed.

Modern mercury cells comprise two key parts, the electrolyzer and the decomposer (Fig. 8.4). The electrolyzer consists of a slightly inclined (about 1 in 100) ebonite (a hard rubber) lined steel trough with an unlined, smoothly machined steel floor. This steel floor is connected to the negative side of a DC potential, and hence is the mercury which covers it makes it the cathode. The upper anode elements of the electrolyzer consist of graphite blocks with spaces between them to allow the escape of chlorine. They are suspended with a 3- to 4-mm gap between the mercury film and the graphite. Both saturated brine and mercury enter at the top of the electrolyzer and move cocurrently through the electrolyzer while a high current flow is maintained. In its passage through the electrolyzer, the sodium chloride concentration in the brine is decreased by electrolysis from an initial 300 g/L or so, to 260–280 g/L. In the process, chlorine gas is generated and collected (Eqs. 8.18–8.20), and a solution of 0.3–3% sodium metal in the mercury stream is produced (Eq. 8.41).

$$Na^+ + Hg + 1 \ e^- \rightarrow Na_{(Hg)} \qquad 8.41$$

Since the electrolyzed fraction of the sodium chloride feed concentration is much lower than for the diaphragm cell, the decomposition efficiencies are correspondingly less, about 7–14%. The low decomposition efficiencies help to keep the electrolyzer operating voltage low.

The equilibrium electrochemical potential, or theoretical voltage requirement for the electrolyzer reaction (Eq. 8.42) is 3.13 V, significantly higher than the 2.20 V required by the diaphragm cell. However, better internal cell conductivities keep the operating potential to

$$NaCl_{(aq)} + Hg_{(\ell)} \xrightarrow{\ electrol.\ } Na_{(Hg)(\ell)} + 1/2 \ Cl_{2(g)} \qquad 8.42$$

only about 4.3 V, which gives a voltage efficiency of about 70%.

The sodium amalgam cathode product of the electrolyzer of the mercury cell is the chief difference between this and the diaphragm cell. The production of sodium hydroxide from this stream uses a separate set of electrochemical reactions conducted in a decomposer or denuder. This unit is usually located below the electrolyzer so as to allow gravity feed of the sodium amalgam from the electrolyzer to the top end of the decomposer (Fig. 8.4). Deionized water is fed into the bottom of the decomposer to provide countercurrent flows of sodium amalgam and water. Mercury, stripped (or denuded) of sodium, is continuously drawn off the bottom of the decomposer, and a hot solution of 50–70% sodium hydroxide in water, plus hydrogen gas, from the top.

The decomposer is packed with graphite lumps or plates, which are in direct contact with each other and with the metal shell of the decomposer. These provide an electrochemical short circuit for the potential generated by the sodium solution reaction. Thus, in this unit, the sodium amalgam becomes the anode, electrons being given up by the sodium to the mercury matrix (Eq. 8.43) in the process of going into solution in the water phase.

$$Na_{(Hg)} \rightarrow Na^+ + Hg + 1 \ e^- \qquad 8.43$$

At the same time, hydrogen evolution occurs at the graphite surfaces, which provide the electrons for this process by the direct contact with the mercury phase, which simultaneously generates hydroxide ions in solution (Eq. 8.44).

$$H_2O + 1 \ e^- \rightarrow OH^- + 1/2 \ H_2 \qquad 8.44$$

There have been many attempts to utilize the approximately 1.8 V generated by the electrochemical reactions of the decomposer. However, it has not been found possible to do this *and* to maintain high concentrations of sodium hydroxide and low residual sodium in the stripped mercury. The sodium hydroxide product obtained from the decomposer of a mercury cell is very pure, containing 0.001% or less sodium chloride. This product is referred to as "rayon grade caustic" because the high purity and low sodium chloride content makes it particularly suitable for rayon manufacture. This is achieved without the special purification steps required for the diaphragm cell product.

The largest mercury cells are capable of much higher production rates per cell than the largest commercially available diaphragm cells. Typical operating characteristics are given in Table 8.3.

Current efficiencies of the two types of cells are approximately the same, so that the approximately 20% additional power requirement per tonne of chlorine of the mercury cell results almost entirely from the higher operating voltage. This is not regarded as a significant penalty, since there is considerable energy saving in the areas of caustic (sodium hydroxide) evaporation and purification with the mercury cell. However, if the impure caustic product of the diaphragm cell can be utilized right at the plant site, there is a significant power saving for these products. This is why some chloralkali producers operate both types of cells, so that the lower cost product may be economically used locally. Mercury loss in the products and waste streams of the mercury cell has also become a significant negative factor now in the choice of the chloralkali process used. This has increased the volatility of mercury prices.

TABLE 8.3 Typical Operating Characteristics of Small and Large Mercury Chloralkali Cells

	Rated current flow (A)	
	30,000[a]	**400,000**[b]
Current efficiency	95	96
Current density,[c] kA/m^2	4.8	12
Operating voltage[c] (nominal)	4.5	4.32
Cell output, tonnes per day:		
Chlorine	0.95	
NaOH		12.2
	1.08	13.8
Power, kWh/tonne Cl$_2$	3,580	3,400
Mercury: Inventory, kg	1,410	–
Loss rate, g/tonne Cl$_2$	250[d]	–
Graphite consumption, kg/tonne Cl$_2$	2.6	
Electrode gap, mm	–	3–4

[a]Data calculated from Sconce [1].

[b]Data selected from Warrington and Newbold [3].

[c]Improved operating characteristics of the larger cell achieved by the use of dimensionally stable anodes (see Section 8.7).

[d]Current cells show mercury loss rates of 1 g or less per tonne of chlorine.

8.6. BRINE ELECTROLYSIS IN MEMBRANE CELLS

Conventional diaphragm cell technology has a significant disadvantage over mercury cell technology in the sense that the former produces a relatively dilute sodium hydroxide product mixed with sodium chloride, as opposed to the pure sodium hydroxide solution obtained in high concentrations directly from the mercury cell. The recently developed ion-selective membrane cells now used in the industry substantially remove this handicap [17]. These cells use a polytetrafluoroethylene, ion permeable membrane (e.g., Nafion) through which no percolation occurs, to replace the usual nondiscriminating asbestos (or synthetic polymer) fiber brine percolating diaphragm (Fig. 8.5). By using carboxylate or sulfonate groups on the surface of the membrane, it becomes selectively permeable to sodium ions and rejects chloride ions [18]. With this modification to a diaphragm cell design, it becomes possible to produce a sodium hydroxide product stream that is low in sodium chloride.

As the saturated brine passes through the anode compartment of a membrane cell it becomes depleted in sodium chloride through the formation of chlorine gas (Eqs. 8.18–8.20) and through diffusion of sodium ions through the microporous membrane. The negatively charged groups on the surface of the membrane sheet prevent both the forward diffusion of chloride ion and the backward diffusion of hydroxyl ion. Purified water, added to the cathode compartment, is partially electrolyzed to hydrogen gas and hydroxide ions which, when combined with the diffused sodium ions, gives the sodium

FIGURE 8.5 (a) Schematic diagram showing a vertical section of a membrane cell for the electrolysis of brine. (b) Details of a section of membrane illustrating the ion selection mechanism.

hydroxide solution cathode product. Since there is neither chloride diffusion nor brine percolation through the diaphragm, the product is a nearly pure sodium hydroxide solution in water. Depending on electrical current and water flow rates, concentrations of sodium hydroxide of around 15% are common, and concentrations of up to 35% is usually possible from continuously operating membrane cells. This caustic product contains only about 50 μg/g sodium chloride, as compared to about 30 μg/g in 50% sodium hydroxide from mercury cells. Higher concentrations than this of hydroxide ion from membrane cells lead to decreased current efficiencies from hydroxide ion back-diffusion, the reason for the present sodium hydroxide concentration limitation of these cells.

Other remaining technical concerns with membrane cells relate to somewhat lower current efficiencies and to relatively short membrane lifetimes. At present, this is limited to 2–3 year of operation when coupled to much more careful brine pretreatment than is required for conventional asbestos diaphragm cells. A combination of mercury cell and membrane cell technologies has been recently tested for commercial feasibility [19]. The economics of the three primary chloralkali technologies have also been reviewed [20].

8.7. EMISSION CONTROL

8.7.1. Emission Control Aspects of Diaphragm Cells

Conventional diaphragm cells have potential environmental losses from operating materials. Fiber dispersion into the cell products occurs from the asbestos diaphragm, and steps have been taken to monitor the degree of hazard [21]. Cell streams are all generally either in closed circuits, or are fed as components to the following processes so that the risk of outside asbestos dispersal from these sources during normal operation is low. During diaphragm changes, however, when the fiber is dispersed in water or brine for removal, precautions are taken to avoid both personnel exposure and dispersal of waste fiber on final disposal. Little is known about the hazards of oral ingestion of asbestos fiber.

Another emission control aspect of diaphragm cell operation concerns the use of the crude cell product, still containing sodium chloride, to carry out base-catalyzed reactions such as ring closure of propylene chlorohydrin (Eq. 8.27) or hydrolysis of chlorobenzene (Eq. 8.45).

$$C_6H_5Cl + NaOH + H_2O \rightarrow C_6H_5OH + NaCl + H_2O \qquad 8.45$$

It is desirable that the water phase from these reactions, now high in sodium chloride, be reused in chloralkali cells, but the traces of organics present after use interfere with efficient cell operation. By providing the right metabolic conditions in ponds and seeding this waste stream with a heterogeneous bacterial population using sewage sludge, it has been possible to remove the organic constituents in what are referred to as "Bio-ox" units [22]. After clarification, the brine may be resaturated and fed to electrolysis cells without causing problems. The discharge alternative, rather than bio-oxidation and recycle, would yield a waste stream with a high oxygen demand (BOD) and high salt loading.

8.7.2. Emission Control Aspects of Mercury Cells

Any of the products of brine electrolysis, chlorine, sodium hydroxide, and hydrogen can be hazardous if released. When releases do occur, it is usually from process upsets or breakdowns, which may be minimized by the construction of fail-safe plants, proper maintenance, and by safe transport and storage practices. Probably of greater long-term concern is the mercury loss experienced through the process streams of a mercury cell chloralkali operation. These losses can also carry over to the products of the diaphragm cell, even though this does not use mercury, if a common brine well or common salt dissolver is used for both sets of cells.

Precautions have always been practiced to contain the mercury (Table 8.3) of the operating mercury cells, primarily for the value of the metal itself. The steel flasks of the metal are also covered with a layer of water to reduce mercury vapor loss during shipping and storage. However, in 1970, it was demonstrated that mercury ions and the free metal could be converted by natural processes to the far more toxic forms of mercury, methylmercury salts and dimethyl mercury even under water [23, 24]. Industrialists, toxicologists, and legislators alike were alarmed by this discovery, which led to the rapid installation of control measures to drastically reduce mercury loss rates in Europe, Japan, and North America [25] (Table 8.4).

Electrolyzer mercury losses occur in the spent brine in the form of finely divided mercury droplets, as dissolved mercuric chloride, and as the stable tetrachloromercury complex ($HgCl_4^{2-}$). Droplet entrainment in the brine stream occurs both directly, from the use of a high-speed centrifugal pump to return mercury from the decomposer to the upper end of the electrolyzer,

TABLE 8.4 Grams of Mercury Lost Per Tonne of Chlorine Produced by Mercury Cell Electrolysis Plants[a]

| Source of loss | Canadian practice prior to 1970 | Losses after abatement measures were instituted | | | | | |
| | | Canada 1971 | Sweden | | Britain 1976 | Canada 1974 | Japan 1974 |
			1969	Stenungsund			
In products:							
Chlorine	0.05	0.05	<0.5	Slight	1.4	0.6	0.024
Caustic soda	1–11.5	1.75	1.0	0.4–1.0	–	–	–
To sewer:							
Hydrogen coolers	0–100	–	–	–	–	–	–
Basement drains	15–150	0.15–1.25	<5.0	0.5	18	0.43	0.00
Brine sludges	1–25	–	<2.0	0.11	–	–	–
To atmosphere:							
End box vents	1–10	1.25	not given	–	10–25	–	–
Hydrogen	10–200	1.25	1.0	<0.5	3.6	5.58	0.76
Ventilation	1–10	2.5–5.0	variable	ca. 1.0	1.5–2.5	–	–
Losses not accounted for:	15	–	–	–	0.2	6.50	8.68
Solid wastes:							
From brine system	–	1.25–5.0	–	–	0.2–0.6	0.82	3.00
Caustic filter, cell room muds, effluent treatment.	–	1.25	–	–	–	–	–
Totals, g/tonne Cl₂:							
Range	32–510	9.45–16.8	ca. 9.5	2.51–5.62	34.6–51.3	ca. 14.0	12.5–116
Average experience	196	–	–	–	–	–	50.1 (5 y)
Prior experience	–	–	150–200	new facility	250	–	new facility
Other comparisons:	300, U.S. 50–250, World	0.77–3.58[b] (1972)	–	–	–	–	–

[a] From Hocking and Jaworski [26] with permission.
[b] In liquid effluents only.

and indirectly from the interaction of the two immiscible flowing liquids. Elemental chlorine present in the electrolyzing brine produces oxidizing conditions in the brine. Any contact of this hot, low-pH layer with finely divided mercury droplets oxidizes the mercury to mercuric chloride and tetrachloromercury dianion, which dissolve in the brine (Eqs 8.46, 8.47). Total mercury in the spent brine may be 3–5 μg/g (ppm)

$$Hg° + NaClO + H_2O + NaCl \rightarrow HgCl_2 + 2\ NaOH \qquad 8.46$$

$$Hg° + NaClO + H_2O + 3\ NaCl \rightarrow HgCl_4^{2-} + 2\ Na^+ + 2\ NaOH \qquad 8.47$$

amounting to a potential mercury loss rate of 2–15 g/tonne chlorine produced. However, since the depleted brine normally flows in a closed circuit either to underground wells or to a dissolver, this does not result in a mercury loss to the biosphere. Brine losses, which occur during process upsets and brine pretreatment sludges, however, must both be treated before discharge to avoid mercury losses.

If diaphragm and mercury cells were run on the same dissolver or brine wells, operators still had to cope with possible mercury losses in the diaphragm cell products since the resaturated brine from either source still contained from 0.2–1 μg/g (0.2–1 ppm) mercury. Some settling occurs during resaturation, reducing the mercury concentration only slightly from that leaving the cell. This problem could be solved by an additional sulfide treatment step for the brine pretreatment for the diaphragm cells (Eqs 8.48, 8.49) [27], [28].

$$S_n^{2-} + Hg° \rightarrow HgS \downarrow + S_{n-1}^{2-} \qquad 8.48$$

$$Na_2S + Hg^{2+} \rightarrow HgS \downarrow + 2\ Na^+ \qquad 8.49$$

Care was required, however, to avoid treating the brine with excess sulfide, a natural tendency, or resolution of mercuric sulfide would occur (Eq. 8.50).

$$HgS_{(s)} + S^{2-} \rightarrow HgS_{(aq)}^{2-} \qquad 8.50$$

Dechlorination of the brine by sparging it with air was also necessary, sometimes also with a reductant such as hydrazine present, prior to sulfide treatment. Otherwise, precipitate reoxidation and resolution would occur (Eq. 8.51).

$$HgS + 4\ ClO^- \rightarrow HgCl_4^{2-} + SO_4^{2-} \qquad 8.51$$

A more permanent and less troublesome alternative was simply to separate the diaphragm and mercury cell resaturation systems by installing an additional dissolver or by drilling additional brine wells.

Mercury droplets, mercuric chloride, and mercury vapor are present in the hot moist chlorine gas produced by the electrolyzer. Carryover of entrained mercury in the chlorine is decreased by passage through a "demister," a labyrinth filter of titanium ribbon. Any mercury vapor present is condensed along with water vapor in an indirect, process water-cooled heat exchanger. The condensate, which contains 0.1–0.2 μg/g mercury, may be recycled to the brine circuit or charged to a waste brine treatment system prior to discharge [25].

Mercury loss in the decomposer section occurs as finely divided metal droplets in the sodium hydroxide stream as a result of the vigorous interaction between the deionized water and sodium amalgam streams to ensure as complete sodium removal from the mercury as feasible. Mercury concentrations in the range of $0.5–10\,\mu g/g$ ($0.5–10$ ppm) in the sodium hydroxide as it leaves the decomposer represent a mercury loss rate of $1–20$ g/tonne of chlorine produced. This ultimate loss occurs indirectly through processes, which use the sodium hydroxide, such as pulp and paper production. Efficient centrifuging and filtration on porous carbon or other caustic resistant media (occasionally both) are procedures capable of reducing the mercury concentrations in 50% sodium hydroxde to about $0.1\,\mu g/g$.

A far more significant mercury loss used to occur from the hydrogen stream of the decomposer. Hydrogen has a relatively low degree of water solubility, so it used to be efficiently cooled and washed by direct counter-current contact with a water shower. It left the decomposer hot and nearly saturated in mercury vapor, at $60°C$ it contained about $300\,mg/m^3$ Hg in hydrogen. This loss rate represented about 92 g of mercury per tonne of chlorine. When this hydrogen comes into contact with the cooling water, most of the mercury is condensed into the water phase, which produces a water stream high in suspended mercury. The hydrogen stream, too, still contains sufficient mercury to represent a potential loss of $2–11$ g/tonne chlorine produced. Current practice dictates that hydrogen be cooled indirectly in two stages; sometimes including prior compression to $2–3$ atm to improve mercury recovery. The first stage uses process water to cool to about $20°C$ followed by a stage using a refrigerant such as ammonia, liquid propane, or a Freon to chill the hydrogen to -10 to $-20°C$. With good mercury mist control, these measures bring the mercury content down to $< 1\,mg/m^3$ and the loss rate in hydrogen down to <1 g/tonne chlorine.

Sludges result from the pretreatment of resaturated brine for removal of impurities, and from brine to be discharged, which was occasionally necessary because of water buildup in the brine circuit. These sludges contain $8–15$ mg/g (dry basis) mercury as a complex mixture of compounds. To recover the mercury, most of the water is removed, and then the sludge is resuspended in aqueous sodium hypochlorite. The hypochlorite oxidizes the sulfide and any elemental mercury present (Eqs. 8.46 and 8.51) in order to produce a concentrated aqueous stream of dissolved mercury salts. Insoluble components are then removed by filtration, and the solution is then returned to the brine circuit. When this reaches the electrolyzer, electrolytic reduction recovers the dissolved mercury present (Eq. 8.52).

$$Hg^{2+} + 2\,e^- \rightarrow Hg° \qquad\qquad 8.52$$

This is a neat example of the use of ingenuity to recycle mercury salts present in aqueous streams.

The residual solid filtered from the hypochlorite dissolving solution now contains less than $50\,\mu g/g$ mercury. This is discarded by blending it with some inert diluent material such as sand and then burial. Disposal areas are sealed using plastic film or layers of impermeable clay for containment, and have a ring of drainage tile placed around the perimeter with provision for treatment of any collected eluate. Addition of a chemical reductant such as zinc followed by vacuum distillation of the sludges for recovery of the mercury has also been used [29].

Mercury vaporization losses to the cell room air can amount to 1–5 g/tonne chlorine [30]. Therefore, ventilation is important to ensure safe working conditions, which require that the mercury vapor concentration be kept below $0.05 \, mg/m^3$. This is achieved by tight cell construction, localized hoods and venting of critical cell areas, and cell room ventilation rates of six to eight air changes per hour (e.g., [31]). Mercury cell chloralkali plants in moderate climates are able to operate their cells outside, which avoids these ventilating problems, but does not control potential emissions.

The provisions outlined above, together with good housekeeping practices, provide a safe working environment. However, the ventilation requirements have caused a noticeable elevation of the mercury concentration in the air and in the precipitation in the immediate area of an operating mercury cell chloralkali plant [32, 33]. It would be possible to employ the chemistry of Eqs. 8.46 and 8.47 in massively sized scrubbers to treat the ventilation air and reduce discharges. But the very high installation costs have meant that this measure has never been adopted. So, the best overall compromise has been to maintain vapor control at the cells as tightly as possible.

Complications of mercury containment, and awareness of its natural conversion in aqueous bioactive media into the much more toxic methylated forms [34], have convinced the North American chlorine and caustic producers to build new facilities, or at the time of a major overhaul to convert existing facilities to diaphragm cell technology [35, 36]. Nearly 60% of U.S.A. chloralkali production was by mercury cells in 1970, whereas by 1991 this had dropped to 18% [37]. In Japan, mercury cell chloralkali production has been phased out entirely. In Europe, these complications have not had such a dramatic effect on the cell choice since a large proportion of the total chlorine is still produced by mercury cells.

8.7.3. Emission Control Aspects of Membrane Cells

Since the synthetic membrane of membrane cells replaces the older asbestos fiber diaphragm cells, no control precautions are necessary. Graphite (or later, titanium) cathodes avoid the use of mercury eliminating the need for mercury control. However, if membrane cells are operated on the same site as mercury cells, they do require a separate brine circuit from the mercury cells to maintain their mercury-free status.

8.8. NEW DEVELOPMENTS IN BRINE ELECTROLYSIS

Dimensionally stable anodes (DSAs) made of a corrosion-resistant titanium screening have been adopted by both mercury cell and diaphragm cell operators. A baked-on rutile titanium dioxide paste used with these electrodes decreases the electrochemical overvoltage and raises practical operating current densities. While significantly more expensive than graphite, this anode type has 30 times the conductivity of graphite and allows the maintenance of closer tolerances on the anode, mercury film spacing in the operation of mercury cells, resulting in savings of both power and operator time [15]. For diaphragm cells, titanium electrodes mean much longer anode life than the 3- to 4-yr typical life of a graphite anode. Since dimensionally stable anodes do not continually shed particles, as graphite anodes do, the lifetime of the asbestos diaphragm is also extended somewhat. However, occasional problems with titanium corrosion do occur [38]. Further advantages with diaphragm cells are a decrease in the carbon dioxide content of the chlorine (Tables 8.2, 8.5) and a measurable (2–5%) increase in chlorine yields.

A further experimental improvement suggests the use of *tert*-butyl hydroperoxide as a diaphragm cell cathode depolarizer [39]. This cell would operate at only about 3 V, and would produce *tert*-butanol instead of hydrogen in a much more compact cell design.

Even highly resistant asbestos fiber is degraded and the fiber length shortened somewhat with time during normal diaphragm cell operation. Replacing the asbestos with a more inert synthetic polymer fiber allows longer periods of operation without cell overhaul. Teflon (polytetrafluoroethylene), in fiber form, has been successfully used to maintain diaphragm porosity and percolation capacity longer than is possible with asbestos.

A more drastic realignment of diaphragm cell technology, which accomplishes production of sodium hydroxide free of chloride from an electrolyte of a sodium chloride/zinc chloride melt has been successfully tested in experimental prototypes [40], and recently reviewed [41]. The salt mixture used gives electrolytic ion mobility via an eutectic depression of the melting point of salt to about 330°C, rather than the 801°C required to melt pure sodium chloride [42]. This anolyte melt is separated from the cathode compartment by a sodium ion permeable, β-alumina ceramic sheet (Fig. 8.6). Passage of an electric current through the cell requires a potential of about 5 V and produces dry chlorine. This cell uses a graphite anode, a sodium hydroxide melt (containing a small amount of water) cathode electrolyte, and a pure nickel cathode. The much higher temperatures required for operation of this cell are maintained directly from internal cell resistance. When the smaller surface to volume ratio is considered, together with the larger electrical current flow of

TABLE 8.5 A Comparison of Diaphragm Cell Gas Composition for Graphite Versus Dimensionally Stable Anodes[a]

	Graphite anode (%)	Dimensionally stable anode (%)
Chlorine	96.5–98.0	96.5–98.0
Hydrogen	0.1–0.4	0.1–0.4
Oxygen	0.5–1.5[a]	1.0–3.0[b]
Nitrogen	0.1–0.5[a]	0.1–0.5[b]
Carbon dioxide	1–2	0.1–0.3[c]

[a]Data from Puschaver [15].
[b]From air.
[c]From brine.

FIGURE 8.6   Operating details of a ß-alumina ceramic cell for the production of chlorine, hydrogen, and nearly anhydrous sodium hydroxide. The ß-alumina ceramic divider serves an ion-selective membrane function to the sodium chloride/zinc chloride analyte melt.

full-size commercial cells, this seems to be a feasible prospect. Commercial success of this cell rests on developments with the β-alumina ceramic to improve lifetime and ion flow capabilities, and on the completion of larger scale tests where voltage and current efficiencies may be determined more accurately at industrial rates of production.

8.9. BALANCING CHLORINE AND SODIUM HYDROXIDE PRODUCTION

Production figures for the major producers of sodium hydroxide together with the world totals are given in Table 8.6. From the growth of sodium hydroxide production on a country-by-country and a worldwide basis, it can be seen that the volume of this commodity chemical slightly more than doubled every 10 years from 1950 to 1970. Decades since 1970 show only a 30–50% increase, with the exception of China, which has shown 74 and 100% increases for the two most recent decades. The slowing of growth seen for most countries is probably the result of the commodity reaching "mature" status in those countries.

TABLE 8.6   Major World Producers of Sodium Hydroxide[a]

	Thousands of metric tonnes					
	1950	1960	1970	1980	1990	2000
Canada	134	339	860	1,459	1,651	1,094
China	–	–	–	1,923	3,354	6,679
France	242	263	1,094	1,325	1,426	3,710
Italy	159	426	999	–	1,111	1,838
Japan	195	843	2,606	3,159	3,917	4,471
U.K.	–	–	–	967	1,002	–
U.S.A.	2278	4510	9200	10,025	10,917	11,518
Germany	336[b]	776[b]	1,682[b]	3,187[b]	3,268	6,678

[a] Data selected from *U.N. Statistical Yearbooks* [43], and Facts and Figures Reports [44].
[b] Production of the former West Germany.

Electrolytic production of sodium hydroxide and chlorine from sodium chloride solutions so heavily dominates the supply of these chemicals now that the production of chlorine can be quite closely approximated by multiplying the sodium hydroxide figures by the theoretical ratio of chlorine to sodium hydroxide (70.906:79.996) of 0.886:1.000. For the U.S.A., for example, the actual production ratio of chlorine to sodium hydroxide is 0.953 to 1.000, quite close to the theoretical ratio.

The use pattern for both sodium hydroxide and chlorine is wide, reflecting the broad importance of these commodities to the chemical and other industries (Table 8.7). The largest use for both commodities is for the manufacture of organic chemicals, which consume about one-third of the sodium hydroxide and almost two-thirds of the chlorine. The pulping of wood for paper and the bleaching steps for pulp brightening together consume about one-seventh of the sodium hydroxide and chlorine. The production of inorganic chemicals accounts for only about one-tenth of each product. Water treatment and sewage disinfection, the traditional uses of chlorine, account for about 5% of the total.

Chlorine and sodium hydroxide production by the electrolysis of brine solutions necessarily locks the ratio of the two products to the theoretical ratio of the process [45, 46]. When the market for sodium hydroxide exceeds the market for chlorine, the causticization of sodium carbonate to sodium hydroxide (Section 7.3) may be used by some suppliers and consumers to supplement the available sodium hydroxide without producing large amounts of excess chlorine. Another expedient for large-scale chloralkali producers faced with this situation is to stimulate the chlorinated solvent or hydrochloric acid markets in an attempt to increase the consumption of chlorine to restore the balance. These measures are not usually rapid enough to respond over the short term unless the solvent plant is also operated by the chloralkali producer.

TABLE 8.7 Use Profile for Sodium Hydroxide and Chlorine in the U.S.A.[a]

Sodium Hydroxide		Chlorine	
Application	% of total	Application	% of total
Organic chemicals	34	Vinyl chloride	17
Inorganic chemicals	9	Propylene oxide	10
Petroleum refining	5	Methylene chloride	3
Food processing	1	$Cl_xC_2H_{6-x}$	14
Pulp and paper	15	Fluorocarbons	8
Soaps and detergents	6	Other organic chemicals	10
Alumina from bauxite	5	Inorganic chemicals	11
Rayon and cellulose	3	Pulp and paper	13
Textiles	3	Water and wastewater treatment	5
Other, miscellaneous	7	Other miscellaneous	9
	100		100

[a]Based on data from *Chemical and Engineering News* [44] and Dickert [7].

Occasionally, the market will swing the other way leaving chlorine in short supply. Fused sodium chloride is commercially electrolyzed in Down's cells to give chlorine and metallic sodium [11]. Sodium production in the U.S.A. has averaged 135,000–150,000 metric tonnes annually since 1968, which represents about a 2–3% contribution to the chlorine supply from this source. In the U.K., it is estimated that as much as 10% of the available chlorine arises from Down's cell technology. Potassium chloride solutions are also electrolyzed for commercial potassium hydroxide, but the contribution to the chlorine supply from this source is even less than from fused sodium chloride electrolysis.

Chlorine has also been made commercially until very recent years by the chemical oxidation of sodium chloride with nitric acid, the Salt process [1]. Nitrate is used to oxidize chloride to chlorine and in the process is converted to nitric oxide (Eq. 8.53).

$$6 \; NaCl + 12 \; HNO_3 + 2 \; Na_2CO_3 \rightarrow 3 \; Cl_2 + 10 \; NaNO_3 + \qquad \qquad 8.53$$
$$2 \; CO_2 + 2 \; NO + 6 \; H_2O$$

The coproduced sodium nitrate could be utilized and sold as a fertilizer constituent to offset the cost of chlorine production by this route.

Routes to chlorine from hydrogen chloride oxidation have been refined from the original commercial ventures of Weldon, which employed manganese dioxide (Eq. 8.54). The Weldon process permitted a maximum recovery of 50% of the chlorine in the hydrogen chloride consumed. Attempts to recycle the manganese (II) chloride met with mixed success [47].

$$4 \; HCl + MnO_2 \rightarrow MnCl_2 + Cl_2 + H_2O \qquad \qquad 8.54$$

The Deacon process solved these problems by replacing manganese dioxide by air as the oxidizer (Eq. 8.55). Further tuning of the Deacon process resulted in the development of the Kel Chlor Cu/heat

$$4 \; HCl + O_2 \underset{cooling}{\overset{Cu/heat}{\rightleftharpoons}} 2 \; Cl_2 + 2 \; H_2O \qquad \qquad 8.55$$

variant, which employs concentrated sulfuric acid as a dehydrating agent. Dehydration greatly assists the process by driving the equilibrium of Eq. 8.55 to the right [48]. The chlorine product of Eq. 8.55 may also be efficiently removed by reaction with an olefin, which has the same net effect on the equilibrium as water removal by sulfuric acid (Eq. 8.56). This version is known

$$2 \; HCl + CH_2{=}CH_2 + 1/2 \; O_2 \xrightarrow{\overset{Cu}{catalyst}} CH_2Cl{-}CH_2Cl + H_2O \qquad 8.56$$

as oxychlorination, and 1,2-dichloroethane, the chlorination product, is denser than water, and insoluble in water so is readily separated [49].

The steps outlined above may also be carried out separately by electrolysis of a hydrochloric acid solution, or hydrogen chloride gas in a fused eutectic of lithium chloride and potassium chloride to obtain hydrogen and chlorine [50]. The chlorine can then be used for chlorination and the hydrogen chloride produced recycled (e.g., Eqs. 8.57–8.59).

$$CH_4 + Cl_2 \rightarrow CH_2Cl_2 + 2\ HCl \qquad \qquad 8.57$$

$$CH_2{=}CH_2 + Cl_2 \rightarrow CH_2Cl{-}CH_2Cl \qquad \qquad 8.58$$

$$CH_2Cl{-}CH_2Cl \rightarrow \quad CH_2{-}CHCl \ + HCl \qquad \qquad 8.59$$
$$\text{vinyl chloride}$$

Another expedient to improve the balance in the supply of chlorine and caustic is to increase the consumption of sodium hydroxide by diverting part or all of this product to production of sodium carbonate (soda ash) [51]. This process reacts sodium hydroxide with the carbon dioxide of clean flue gas (Eq. 8.60). It also simplifies shipment of the product in a dry

$$2\ NaOH + CO_2 + H_2O \rightarrow Na_2CO_3 + 2\ H_2O \qquad \qquad 8.60$$

form. However, recovery of sodium carbonate from naturally occurring trona is generally less expensive. A combination of these methods allows a balance in the supply and demand of these commodities to be achieved.

The hydrogen coproduct from the cells of large chloralkali facilities may be used to contribute to that required for ammonia or methanol production (Eqs. 8.61 and 8.62).

$$N_2 + 3\ H_2 \rightarrow 2\ NH_3 \qquad \qquad 8.61$$

$$CO + 2\ H_2 \rightarrow CH_3OH \qquad \qquad 8.62$$

If the chloralkali plant is operating in the vicinity of an oil refinery, it can provide a part of the hydrogen gas requirements of hydrocracking or various hydrotreating processes used in refining. Hydrogen may also be burned in chlorine to profitably produce hydrochloric acid, either for captive use or for sale [52] (Eq. 8.63).

$$1/2\ H_{2(g)} + 1/2\ Cl_{2(g)} \rightarrow HCl_{(g)} \quad \Delta H = -92.3\ kJ\ (-22.1\ kcal) \qquad 8.63$$

When the chlorine market is slack, this option can also usefully consume chlorine, provided the hydrogen chloride can be used or sold. However, since the masses of hydrogen produced in all except the largest facilities are relatively small, many plants simply burn the hydrogen in air to produce energy to assist in steam raising. For a diaphragm cell plant producing 50% sodium hydroxide, this option can reduce the external fuel requirement by as much as one-third.

REVIEW QUESTIONS

1. Briefly, how do the brine pretreatment steps required for diaphragm cell electrolysis and mercury cell electrolysis compare, and what are the reasons for these differences?
2. One thousand liters per hour of a 10% sodium hydroxide solution (wt. of solute/wt. of solution; density 1.116) is to be concentrated to 50% (density 1.540) in a single-effect evaporator operating at a reduced pressure of 40 mm Hg. Atmospheric pressure at the site of operation is 760 mm Hg.

(a) How many kilojoules per hour would be required to evaporate at this rate under the conditions outlined, with caustic solution entry and exit at 20°C, and condensate discharge also at 20°C? Assume perfect heat transfer efficiencies.

(b) If the overall heat transfer efficiency, fuel oil combustion → steam raising → evaporation from caustic solution is only 25%, how many liters of fuel oil would need to be burned each hour to provide this energy? (No. 2 fuel oil ≡ 140,000 Btu per Imperial gallon; 1 Btu ≡ 1054 joules.)

3. (a) What current flow would theoretically be required to electrolytically produce chlorine from aqueous brine at the rate of 1 tonne (1 Mg) per day?

(b) Assuming 95% current efficiency for a chloralkali cell operating at 4.1 V, what would be the number of kilowatt-hours required per tonne of chlorine produced?

(c) What masses of sodium hydroxide (100% basis) and hydrogen would be coproduced along with the 1 tonne of chlorine?

(d) Commercial prices for the cell products are currently chlorine, $159/tonne; and 50% (by weight) sodium hydroxide, $225/tonne of solution. For a power cost of 5.5¢/kWh, what proportion of the sales value of these two products is paid for the electric power to produce them?

4. The rate of production of chlorine gas by electrochemical cells is sometimes determined by addition of oxygen to the chlorine stream at a known rate, and accurate determination of the oxygen concentration at a well mixed point downstream of the point of addition. To the chlorine produced by diaphragm cells, which already contains 0.8% by volume oxygen, additional oxygen is added at the rate of 1 kg min^{-1}. Analysis prior to compression (20°C, 1 atm) indicates a resultant oxygen concentration in the chlorine of 3.1%.

(a) What is the rate of chlorine production, in kg hour^{-1}?

(b) What mass of sodium hydroxide (kg, 100% basis) will be produced during the same interval?

(c) Give two origins for the source of the small percentage of oxygen already present in the chlorine cell product and explain how this arises. Give equations if reactions are proposed.

5. Sodium hydroxide solution, 50% by weight, is available commercially at $225/tonne of solution.

(a) What theoretical profit or loss, on a raw material cost basis only, would be incurred to produce each tonne of anhydrous sodium carbonate from this sodium hydroxide and sell the product at $210/tonne? Carbon dioxide for carbonation is to be obtained from flue gas. Assume essentially zero cost.

(b) Sodium carbonate monohydrate and sodium carbonate decahydrate are also commercial grades, selling at $290 and $430 per tonne, respectively. Give the relative dollar profit margins per tonne of contained Na_2CO_3 for each of these grades, neglecting any costs other than outlined in part (a). Which of the three grades would be the most profitable?

6. Four rail tank cars filled with 340 tons of liquid chlorine (boiling point is $-34.6°C$) fell from a barge carrying them in Malaspina Strait, British Columbia, in February 1975. The vapor pressure of chlorine is given by the expression:

$$\log_{10} P = \left(\frac{-0.2185A}{K}\right) + B \text{ where P} = \text{pressure in torr (mm Hg)},$$

$A = 5180.4$, $B = 7.5499$, and $K =$ absolute temperature. Calculate the critical depth in meters below which any chlorine escaping from the gradually corroding cars would do so more safely as a liquid. Temperature and density of the seawater in the area are $10°C$ and 1.025 g/ml, respectively.

7. (a) In a small-scale solvent plant, 40 tonnes of carbon tetrachloride and 10 tonnes of chloroform are produced daily, based on methane chlorination. Methyl chloride and methylene chloride are recycled. What daily mass of hydrogen chloride (in tonnes) would be coproduced by this process?
(b) Give the equations and calculate the mass ratios, hydrogen chloride to chlorine, theoretically possible from *each* of the three chemical processes for regenerating elemental chlorine from hydrogen chloride.

8. (a) What weights of sodium and chlorine (in kilograms) would theoretically be expected from the consumption of 10,000 ampere hours of electricity in a Down's cell (nonaqueous variant) electrolyzing molten sodium chloride with 90% current efficiency? Neglect the associated calcium.
(b) Give two reasons why the current efficiency for this electrochemical process is noticeably lower than the usual experience of chloralkali cells.

9. (a) What is the voltage efficiency of a Down's cell operating at 6 V, when the theoretical voltage required is 3.1 V?
(b) For current efficiency of 80% when operating at 6 V, what is the energy efficiency of this Down's cell?
(c) To push production higher, the Down's cell operation is to be raised to 7 V, which at the same time decreases the current efficiency to 78%. What is the energy efficiency of the cell under these conditions?
(d) For an electricity cost of 5¢/kWh what would be the respective power cost to produce 1 tonne of sodium (with associated chlorine) from each set of cell operating parameters?

10. The Salt process to chlorine using sodium chloride normally proceeds with about 90% conversion and 90% selectivity ("Industrial" yield) to chemically produce chlorine and sodium nitrate.
(a) What molar quantities of chlorine and sodium nitrate would be expected from the passage of 6 mol of sodium chloride once through this process?
(b) If the unreacted sodium chloride was recycled through the process until it was all converted (reacted) what total number of moles of chlorine and sodium nitrate would be obtained from the initial 6 mol of salt charged?

(c) What masses of sodium chloride and nitric acid would be theoretically required to produce 1 tonne of chlorine via the Salt process, and what mass of sodium nitrate would be coproduced?

FURTHER READING

L. Barregård, "Occupational Exposure to Inorganic Mercury in Chloralkali Workers: Studies on Metabolism and Health Effects," University of Göteborg, Göteborg, 1991.

R.W. Curry, ed., "Modern Chlor-Alkali Technology," Vol. 6, Special Publication No. 164. Royal Society of Chemistry, Cambridge, UK, 1995.

R. Thompson, ed., "Industrial Inorganic Chemicals; Production and Uses." The Royal Society of Chemistry, Cambridge, 1995.

M. Drabkin and E. Rissmann, "Waste Minimization Audit Report: Case Studies of Minimization of Mercury-bearing Wastes at a Mercury Cell Chloralkali Plant," Haz. Waste Eng. Res. Lab., Office of Res. and Devel., U.S. Envir. Protection Agency, Cincinnati, Ohio, 1988.

J. Greenblatt, "The Chloralkali Chemicals." Office of Industries, U.S. International Trade Commission, Washington, D.C., 1999.

L.C. Mitchell and M.M. Modan, Catalytic Purification of Diaphragm Cell Caustic, *Chem. Eng. (N.Y.)*, **86**, 88 (1979).

T. Sasano and W. Schlegel, W, "Technology and Economics of Chloralkali Membrane Cells." SRI International, Zürich, 1985.

REFERENCES

1. J.S. Sconce, ed., "Chlorine, Its Manufacture, Properties, and Uses," pp. 85–89. Reinhold, New York, 1962.
2. E.A. LeSueur, Brit. Pat. 5983 (1891); details cited by Warrington and Newbold [4].
3. C.J. Warrington and B.T. Newbold, "Chemical Canada," p. 38. Chemical Institute of Canada, Ottawa, 1970.
4. H.A. Sommers, *Chem. Eng. Prog.* **61**(3), 94 (1965).
5. C.J.S. Warrington and R.V.V. Nichols, "A History of Chemistry in Canada," p. 207. Chemical Institute of Canada/Pitman and Sons, Toronto, 1949.
6. R.C. Weast, ed., "Handbook of Chemistry and Physics," 56th ed., p. D-146. Chemical Rubber Publ. Co., Cleveland, OH, 1975.
7. E.J. Dickert, "Multistage Centrifugal Compressors," Reprint No. 116, Vol. 4, No. 1. Compressed Air and Gas Institute, Cleveland, OH, 1977.
8. "Kirk Othmer Encyclopedia of Chemical Technology," 3rd ed., Vol. 1, p. 709. Wiley, Toronto, 1978.
9. H.C. Twiehaus and N.J. Ehlers, *Chem. Indus. (N.Y.).* **63**, 230 (1948); *Chem. Abstr.* **45**, 4415 (1951).
10. R.N. Shreve and J.A. Brink, Jr., "Chemical Process Industries," 4th ed., p. 214. McGraw-Hill, Toronto.
11. K. Viswanathan and B.V. Tilak, Chemical, Electrochemical, and Technological Aspects of Sodium Chlorate Manufacture, *J. Electrochem. Soc.* **131**, 1551–1559, July (1984).
12. F.A. Lowenheim and M.K. Moran, "Faith, Keyes, and Clark's Industrial Chemicals," 4th ed., p. 716. Wiley-Interscience, Toronto, 1975.
13. S. Ainsworth, OxyChem ...Role as Sodium Chlorate Marketer, *Chem. Eng. News*, **67**(34), 10–11, Aug. 21 (1989).
14. R.L. Dotson, *Chem. Eng. (N.Y.).* **85**(16, Pt.1), 106, July 17 (1978).
15. S. Puschaver, *Chem. Indus. (London).* p. 236, Mar. 15 (1981).
16. R. Thompson, ed., "The Modern Inorganic Chemicals Industry," p. 106. Chemical Society, London, 1977.

17. W.N. Brooks, The Chloralkali Cell: From Mercury to Membrane, *Chem. Brit.* **22**(12), 1095–1098, Dec. (1986).
18. Chloralkali Membrane Data, *Chem. Eng. News*, **58**(30), 23, July 28 (1980).
19. F. Hine and A.J. Acioli Maciel, . . . the Amalgam Cell and the Membrane Cell Processes for Chloralkali Production, *J. Appl. Electrochem.* **22**, 699–704 (1992).
20. R.E. Means and T.R. Beck, A Techno/Economic Review of Chloralkali Technology, *Chem. Eng. (N.Y.)* **91**, 46–51, Oct. 29 (1984).
21. B. Strokova, S. Evstatieva, S. Dimitrova *et al.*, Study of Asbestos Exposure . . . in the Chemical Industry, *Int. Arch. Occup. Envir. Health*, **71**, 19–21, Sept. (1998).
22. Dow Bioclarification Test Looking Good, *Chem. Eng. News*, **54**(32), 24, Aug. 2 (1976).
23. S. Jensen and A. Jernelov, *Nature (London).* **223**, 753 (1969).
24. J.M. Wood, F.S. Kennedy, and C.G. Rosen, *Nature (London).* **220**, 173 (1968).
25. E.J. Laubusch, "Mercury Emissions Control in Scandinavia," Pamphlet No. R-105 (unpublished). Chlorine Institute, New York, 1970.
26. M.B. Hocking and J.F. Jaworski, eds., "Effects of Mercury in the Canadian Environment." National Research Council of Canada, Ottawa, 1979.
27. G.L. Bergeron and C.K. Bon, U. S. Pat. 2,860,952 (to the Dow Chemical Company) (1958).
28. D.M. Findlay and R.A.N. McLean, Can. Pat. 1,083,272 (to Domtar Inc., Montreal) (1979).
29. R.A. Perry, *Chem. Eng. Prog.* **70**(3) Mar. (1974).
30. F.L. Flewelling, *Chem. Can.* **23**(5), 14 (1971).
31. S.K. Dangwal, . . . Control of Mercury Vapor Exposure in the Cell House of Chloralkali Plants, *Environ. Research*, **60**(2), 254–258, Feb. (1993).
32. M.J. Bumbaco, J.H. Shelton, and D.A. Williams, "Ambient Air Levels of Mercury in the Vicinity of Selected Chloralkali Plants," Rep. EPS 5-AP-73-12. Environment Canada, Ottawa, 1973.
33. A. Jernelov and T. Wallin, *Atmos. Environ.* **7**, 209–214 (1973).
34. D. Lean, Mercury Pollution, a Mind Numbing Problem, *Can. Chem. News*, **55**(1),23–26, Jan. (2003).
35. M.B. Hocking and M. Gellender, Avert Mercury Pollution, *Chem. Intern.* p. 7–14, Apr. (1982).
36. J. Johnson, Where Goes the Missing Mercury, *Chem. Eng. News*, **82**(11), 31–32, Mar. 15 (2004).
37. U.S./World Chloralkali Industry, *J. Electrochem. Soc.* **138**(10), 3141–3143, Oct. (1991).
38. P. Kohl and K. Lohrberg, Material Problems in the Three Versions of Chloralkali Electrolysis. *J. Appl. Electrochem.* **19**, 589–595 (1989).
39. H.B. Johnson, cited in Brine Electrolysis: A More Efficient Cathode, *Chem. Eng. News*, **58**(14), 38, Apr. 7 (1980).
40. Y. Ito, S. Yoshizawa, and S. Nakamatsu, *J. Appl. Electrochem.* **6**, 361 (1976).
41. Y. Ito, . . . Novel Molten Salt Electrochemical Processes, *Electrochemistry*, **68**(2), 88–94 (2000).
42. I. Fukuura, Development of ß-Alumina Ceramics for a Separator for Molten NaCl Electrolysis. In: "Applications of Solid Electrolytes" (T. Takahashi and A. Kozawa, eds.),. pp. 71–74. JEC Press, Cleveland, OH, 1980.
43. "United Nations Statistical Yearbook 2001," 48th ed. United Nations, New York, 2003, plus earlier editions.
44. Facts and Figures for the Chemical Industry, *Chem. Eng. News*, **80**(25), pp. 42–82, June 24 (2002), and same feature articles from June of earlier years.
45. N. Botha, The Outlook for the World Chloralkali Industry, *Chem. Indus. (London).* (20), 832–835, Oct. 16 (1995).
46. R. Shamel and A. Udis-Kessler, Critical Chloralkali Cycles Continue, *Chem. Indus. (London).* 179–182, Mar. 19 (2001).
47. F.S. Taylor, "A History of Industrial Chemistry," p. 192. Arno Press, New York, 1972.
48. Chlorine Recovery (Kel-Chlor Process) Pullman Kellogg, *Hydrocarbon Process.* **56**, 139, Nov. (1979).
49. P.J. Thomas, *Chem. Indus. (London).* p. 249, Mar. 15 (1975).
50. Y. Ding and J. Winnick, Electrolytic Recovery of Chlorine from Hydrogen Chloride Gas with Fused Molten Salt Electrolyte. *J. Appl. Electrochem.* **26**, 143–146 (1996).
51. Chlorine Production is Higher, *Can. Chem. Process.* **59**(5), 8, May (1975).
52. More HCl Producers Turn to Chlorine Burning, *Chem. Eng. News*, **57**(27), 9, July 2 (1979).

Plate 9.1 Skyline view of the towers used for hydrogen sulfide stripping, Sulfinol solvent recovery, etc., in the cleaning of natural gas by the Shell Canada Jumping Pound Complex, Cochrane, Alberta.

9 SULFUR AND SULFURIC ACID

Vice and virtue are products like sulphuric acid and sugar.

—H.A. Taine (1828–1893)

9.1. COMMERCIAL PRODUCTION OF SULFUR

Sulfur is widely distributed in the earth's crust, but only to the extent of about 0.1% by weight. It is found here chiefly as the element, in the form of sulfides, and as sulfates. The annual mass of global sulfur produced is about twice that of sodium hydroxide, another important chemical commodity, so sulfur is significant in the chemical marketplace. However, consistent with the varied geological forms in which sulfur occurs, its nonuniform distribution in the crust and the differences in the degree of industrialization of a country, there is considerable variation in the level of production by country (Table 9.1). The influence of these diverse factors is reflected in the per capita level of production. Poland and Canada have natural factors that make sulfur production advantageous, and produce around 150 and 300 kg per capita per annum. These figures are much higher than the 25 to 50 kg per capita per year produced by Japan, the U.S.A., and the U.S.S.R., other large-scale producers. Thus, the level of sulfur production of a country is markedly influenced by natural and market factors, which makes this a poor indicator of the level of industrial activity of a country.

Diverse methods have been used to recover this element commercially to cope with the varied forms in which sulfur occurs (Table 9.2). For example, about 85% of Canada's sulfur production is from sulfides removed from natural gas to "sweeten" it. It results from, or is an involuntary by-product of natural gas production and not a product sought for its own sake. Poland, on the other hand, obtains about 80% of its annual sulfur by employing Frasch recovery of natural sulfur which is more directly responsive to markets in its volume of production.

■ **TABLE 9.1   Major World Producers of Sulfur**[a]

	Thousands of metric tonnes				
	1960	1970	1980	1990	2000
Canada	249	4,440	7,405	6,790	9,946
China	244	250	2,300	5,370	5,560
France	791	1,736	2,213	1,050	1,110
Japan	256	343	2,784	2,630	3,486
Mexico	1,349	1,381	2,252	2,410	1,325
Poland	25	2,711	5,535	4,900	1,813
Spain	42	1,278	1,236	1,150	708
U.S.A.	5,898	8,678	11,866	11,600	10,300
U.S.S.R.	863	1,600	11,000	9,030	5,900[c]
West Germany	84	176	1,775	1,550[b]	1,240[b]
Other	8,769	9,744	7,665	11,320	16,712
World	18,690	32,480	56,635	57,800	58,100

[a]Selected from data in *Minerals Yearbooks* [1].
[b]For the united Germany.
[c]Russia.

For the U.S., the world's largest producer of sulfur, the picture is much more diverse, with sulfur recovery from petroleum processing dominating the output. Natural gas processing makes the largest contribution to the Russian sulfur industry, similar to the picture for Canada.

The large-scale production of sulfur as a process is secondary to natural gas production developed in Canada since 1951 when the first plant was built [2]. Since this sulfur is a by-product of gas production its price is only loosely

■ **TABLE 9.2   Sources of Sulfur Production for Major Producers in 2000, as Percentages of the Total**[a]

	Canada	Poland[b]	U.S.A.	Russia	World
Primary elemental:					
Frasch	–	81.7	8.7	–	4.2
Native	–	–	–	0.9	1.2
Pyrite	–	–	–	5.9	7.7
By-product:					
Coal	–	–	–	–	–
Metallurgy	11.7	14.4	10.0	–	19.5
Natural gas		–	19.6	83.1	13.3
Petroleum	88.3	3.9	61.8	–	19.3
Tar Sands		–	–	–	27.4[b]
Miscellaneous	–	–	–	10.2	7.6
Totals, percent	100	100	100	100	100
(Thousand tonnes)	(9,946)	(1,813)	(10,300)	(5,900)	(58,100)

[a]Entries calculated from selected data from *Minerals Yearbook*, Sulfur, 2001 [1]. Blank entries low or zero.
[b]Includes undifferentiated contributions.

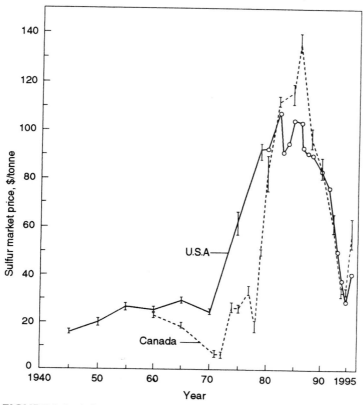

FIGURE 9.1 Influence of market and nonmarket factors on the U.S. dollar price of sulfur. Spot prices are given for the U.S. post-1978 to simplify the comparison.

connected to the cost of production. Figure. 9.1 shows that, within a transportation cost differential, Canadian involuntary sulfur production tended to hold American sulfur prices down. Rising energy costs have added significantly to the cost of Frasch sulfur, and the development of a more orderly marketing structure in Canada has served to bring North American sulfur prices to a more realistic level.

Traditionally, almost 90% of all sulfur produced has been converted to sulfuric acid [3]. While this is a broadly based market, it is not a volume-flexible one. The only other significant uses of sulfur are by the pulp and paper industry, about 5%, and in the manufacture of carbon disulfide, 2.5%, neither significant enough to have a major impact on sulfur markets. Hence, new viable uses for sulfur were sought, stimulated by the low prices and the widespread surplus of sulfur during the late 1960s and early 1970s. The sulfur (or sulfur dioxide) production associated with other industries such as natural gas processing or the smelting of sulfide ores has recently become an important contributor to the sulfur industry. This has occurred as sulfur gas containment, rather than discharge and dispersal has been increasingly imposed on these sectors with the growing recognition of environmental damage from this source (Chap. 3).

Closed cell-foam sulfur slabs poured *in situ* from a melt have been successfully tested for highway subgrade insulation in arctic regions for prevention of frost-induced damage [4]. As little as 10 cm of foam could replace 120 cm of gravel in this application. Incorporation of upto 40% sulfur by weight into asphalt formulations produces a finished paving with better cold flexibility and sufficient hardness to be useful under high summer temperatures [5]. A paving material is also possible in which the sulfur plus an "additive" can entirely replace petroleum-based asphalt as an aggregate binder [6]. Sulfur may also be used as a durable traffic paint, applied as a melt. This highway marker can accept traffic within a few seconds after application.

Construction applications for sulfur have also been tested. Interlocking bricks made predominantly of sulfur have been found to speed up construction, but have inherent fire hazards, which restrict their use. Concrete blocks impregnated with molten sulfur have shown a more than tenfold increase in compressive strength, from 5.9 to 83.9 MPa (megapascals; $1\,MPa \equiv 10^6\,N/m^2$), at the same time as an improvement in tensile strength from 1.3 MPa to 8.5 MPa, for a 13–15% sulfur content [7]. It also improves its chemical resistance. Sulfur-coated bamboo has been found to be an economical yet effective concrete reinforcing agent as a replacement for steel, in areas where bamboo occurs naturally.

Adding sulfur to a sulfur-deficient soil can boost crop yields by 1,000% or more, particularly for seed crops having a high-sulfur content [8]. A deficiency of sulfur in the soil has traditionally been corrected by the application of sulfate-containing fertilizers such as ammonium sulfate and superphosphate (Section 10.5). Today, however, methods of sulfur application have been extended to include finely divided sulfur itself in a weakly granulated form, or as a 10–12% solution of sulfur in anhydrous ammonia as ways of raising the sulfur analysis of the fertilizer formulation. Other variations along this theme include the coating of highly soluble nitrogenous fertilizers, such as urea, with sulfur to produce a slow release, and hence greater long-term effectiveness of the sulfur and other nutrients [9].

Some of these uses are evidently developing into viable markets for sulfur. However, a leveling off of the increases in the sulfur production from recovery processes and increased sulfur demand for traditional uses has offset this pressing need for new markets.

9.2. PROPERTIES OF ELEMENTAL SULFUR

Sulfur is an intriguing element in the multitude of forms in which it can exist. Under ordinary conditions, it occurs as a solid which exists in the well-known bright yellow form as either rhombic or monoclinic crystals, or as a dark, amorphous, moldable mass referred to as plastic sulfur (Fig. 9.2). The rhombic form is most stable at room temperature. The transitional equilibrium between the rhombic and the monoclinic forms occurs at 95.5°C. Just above this temperature, at 114.6°C, it melts to a transparent, pale yellow mobile liquid. In all the above forms, sulfur occurs as molecules of eight-membered, crown-shaped rings.

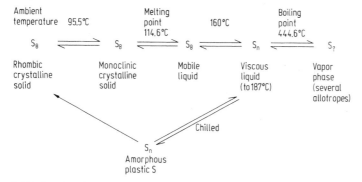

Soluble in carbon disulfide or anhydrous ammonia
Slightly soluble in strong bases
Insoluble in water
Density 2.07 g cm^{-3} at 20°C.

FIGURE 9.2 Allotropic forms and physical properties of sulfur.

At a little below 160°C, the rings start to break up into chains. At about 160°C, the ring-bonded form of sulfur is largely lost to a much darker, reddish-brown, viscous linear polymeric form, where n, the number of bonded sulfur atoms, is significantly more than eight. Above 187°C, depolymerization of this linear form occurs, decreasing the viscosity again. Rapid chilling of the viscous liquid gives the dark plastic sulfur at room temperature, thought to be a linear molecular sulfur occurring in helices. Slower cooling of liquid sulfur allows the transitions through the eight-membered ring, and the monoclinic and rhombic crystal forms, and yields the more familiar yellow powder modification of sulfur.

9.3. SULFUR RECOVERY BY MINING AND RETORTING

Conventional mining is both of historical interest and is still used to produce nearly a quarter of the world's sulfur (Table 9.2). Mining brings to the surface lumps of either a volcanic or one of the many pyritic forms of sulfur. Some of the pyritic forms are pyrite (FeS_2) itself, chalcocite (Cu_2S), and chalcopyrite ($CuFeS_2$). The sulfur content of the raw mineral is usually 25–35%, but may run as high as 50%. To obtain the sulfur in a separated form, the original procedure was to pile the lumps of ore outside and seal these with clay or earth. Burning a part of the contained sulfur sealed into these mounds, with careful control of the air generated sufficient heat to melt any elemental sulfur present, and thermally decompose the pyrite (Eqs. 9.1–9.3) [10].

$$3\ FeS_2 + 8\ O_2 \rightarrow Fe_3O_4 + 6\ SO_2 + heat \qquad 9.1$$

$$S + O_2 \rightarrow SO_2 + heat \qquad 9.2$$

$$8\ FeS_2 \rightarrow S_8 + 8\ FeS \qquad 9.3$$

This primitive method gave sulfur recoveries of about 50% of that contained in the ore, the rest being burned to provide the heat requirement of the process. More advanced sulfur recovery techniques using dual chamber furnaces are capable of recovering as much as 80% of the contained sulfur.

Sulfur dioxide coproduced by this method was generally converted to sulfuric acid in these early operations by using the chamber process. Using this source for the sulfur dioxide for acid making put all of the vaporized arsenic from the pyrites directly into the sulfuric acid product. At the time this aspect was not considered to be serious since most of the sulfuric acid was consumed in the Le Blanc process to produce sodium carbonate (Eqs. 9.4 and 9.5), causing arsenic vaporization and loss again, with the hydrogen chloride produced by the first step [11].

$$2 \; NaCl + H_2SO_4 \rightarrow 2 \; HCl + Na_2SO_4 \qquad\qquad 9.4$$

$$Na_2SO_4 + C + CaCO_3 \rightarrow CaS + Na_2CO_3 + CO_2 \qquad 9.5$$

If sulfuric acid production was the primary goal of the operation, the air supply to the piles of ore would not be restricted and the whole of the contained sulfur was converted to sulfur dioxide (Eqs. 9.1 and 9.2). On the other hand, if the sulfur dioxide could not be used on site it was also possible to practice sulfur dioxide reduction back to elemental sulfur by using incandescent coke [11, 12] (Eq. 9.6).

$$SO_{2(g)} + C_{(s)} \rightarrow CO_{2(g)} + S_{(g)} \qquad\qquad 9.6$$

9.4. FRASCH SULFUR

Native sulfur deposits were discovered in 1865 while prospecting for oil in a dome-like underground formation in Louisiana. These beds lay under a thick bed of unstable quicksand which eliminated conventional mining as an economic recovery method. However, from initial successful experiments in 1891, Herman Frasch developed a method using superheated water and air to recover sulfur from this deposit on a commercial scale by 1902 [13]. During the trying development period, a skeptical detractor promised to "eat every barrel of sulfur produced." Since this first large-scale production by the Frasch method the procedure has been the chief source of the success of the U.S. as a dominant sulfur producer. The state of Texas alone supplied 70% of the world's sulfur by this means in 1935, but its decline in the U.S. was almost as rapid. By 2000, U.S. Frasch sulfur production ceased [14]. Essentially, the same method has also been used for most of the sulfur produced by Poland in recent years, although it is not referred to as the Frasch process there [15].

A primary geological requirement for the success of the Frasch process lies in the unusual salt dome structures which occur in Louisiana, Texas, and parts of Mexico in the coast area of the Gulf of Mexico (Fig. 9.3). Only about 28 of the some 400 or so known dome formations in the area have produced commercial quantities of sulfur, because only a relatively few possess all the correct geological features to enable exploitation [16]. Nevertheless, enormous reserves are accessible from a single dome with the right features. One dome in Texas, for example, has produced 12 million tonnes of sulfur during

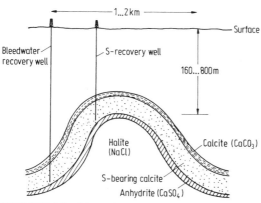

FIGURE 9.3 Vertical cross section of a typical salt dome formation occurring in the area around the Gulf of Mexico. Approximate placements of sulfur recovery and bleed water wells are shown.

its useful life [17]. Offshore Frasch capability is now tapping into reserves of 67 million tonnes of sulfur along with oil and gas recovery off the Louisiana coast [18].

Frasch sulfur recovery today is still practiced much as originally developed. It initially involves conventional oil well-drilling equipment to reach the sulfur-bearing calcite zone. After placing a string of three concentric pipes into the well (Fig. 9.4), superheated water at 150 to 165°C is pumped down to the formation in the outermost, 15–20 cm diameter pipe to melt the native sulfur in place [19]. After passage of hot water for some time, heated compressed air is forced down as well, inducing vigorous agitation of the superheated water and molten sulfur mixture in the formation. As a result, slugs of water, sulfur, and air are forced to enter the intervening sulfur return pipe. The outermost pipe carrying superheated water serves to keep the sulfur (melting point

FIGURE 9.4 Details of a conventional Frasch sulfur string. Superheated water (140–160°C, 17 bar) enters outermost pipe, and air at 150–200°C and 34 bar enters the innermost pipe.

Water

Air jet

FIGURE 9.5   The air lift principle shown applied to a U-tube which illustrates that the tube leg with a mixture of bubbles and water will rise to a higher level to counterbalance the leg having water alone.

114.6°C) molten in the return pipe, which is jacketed by it. At the same time the heated compressed air continues to agitate the mixture in the formation, and by means of the air lift principle (Fig. 9.5) and some air incorporation into the sulfur, assists in bringing slugs of molten sulfur (density $2.0 \, g/cm^3$) with water to the surface.

At the surface, hot tanks receive this mixture, still under pressure, and allow a lower liquid sulfur phase to separate from the water. Much of this product is simply filtered to remove any excess carbonaceous material while still molten, before it is shipped in this state via insulated tank trucks, rail tank cars, or barge. Traces of heavy petroleum occasionally occur with the sulfur from dome sources, and are removed by sublimation of the sulfur from the oil, if necessary.

For a solid sulfur product, the sulfur layer is pumped to large outside vats where it is allowed to cool. The simplest containment for this method consists of a vertical sheet metal rim which is used to hold the molten sulfur until it solidifies. Once the product is solid, the metal rim is moved upward ready to receive the next layer of molten sulfur. The large, compact dense masses of product thus formed represent one of the least expensive and best environmentally controlled forms of storage of excess sulfur inventory. For delivery, the large mass is broken up using small explosive charges or mechanical means.

In other sulfur recovery operations, better control of product particle size is obtained. The molten sulfur is chilled on a steel belt to produce a roughly 0.5-in.-thick flake sulfur, or sprayed into water or even tumbled in drums to obtain granulated sulfur. Pelletized sulfur, similar in particle size range to the granulated variety, is also obtained when a melt is sprayed into the top of a tower to form droplets of molten sulfur which harden into shot-sized beads as they fall through a current of air. In this form, it is referred to as prills [20]. The narrow size range of particles of prilled sulfur, as well as the negligible dust content in this form, makes it more convenient to use. It normally commands a price premium.

9.4.1. Environmental Aspects of Frasch Operations

A major cost item in Frasch processing is fuel for water heating, usually heavy oil or natural gas, since some 4 to 50 tonnes of superheated water are required for each tonne of sulfur recovered at the surface [21]. Seawater may be used for this purpose in areas where freshwater is scarce [18]. Some hot water is lost from the producing zone via undergound fractures and by slumping into the cavities from which sulfur (and some soluble salts) has been withdrawn [22]. However, much of this "leaked" water is recovered via "bleed wells" located downslope on the dome from the producing zone (Fig. 9.3). The residual heat content of bleed water and production waters which return sulfur to the surface may be salvaged by reheating the once-used water (with some withdrawal to allow a partial purge of impurities) to the required 140–165°C production temperatures. This is usually carried out indirectly using steam rather than by a direct firing boiler, to decrease the corrosion and fouling problems which can occur from the high total solids content of the recycle water. The scale of water heating capacities of these plants ranges from 4 to 40 million liters per day (1–10 million U.S. gallon/day).

Direct risk to boiler equipment and sulfur storage vats from overburden slumping is avoided by locating these surface facilities a kilometer or so away from the producing wells. Risk to other parties not involved in the Frasch operations is minimized by purchase of the surface rights overlying the areal extent of the mined dome. Production upsets from slumping or subsidence are minimized by providing double sets of connecting pipe and much of the associated equipment to minimize downtime (lost production) in the event of a rupture. The extended period of formation heating required before the sulfur production rate becomes steady and the problems associated with pipe blockage by solidified sulfur make it important that the injection of super-heated water be maintained as much as possible without interruption, once production from a particular well has been started.

Bleed water or process water in excess of injection requirements, which contains dissolved salts at similar concentrations to seawater, may be safely discarded to the sea after heat exchange energy recovery and treatment to remove toxic sulfides. Initial heat exchange from the discarded water to fresh process water not only conserves fuel but decreases calefaction risks (thermal pollution) of the receiving body of water. Prolonged aeration of these streams for both stripping and oxidation of sulfide has been used to decrease the sulfide toxicity of this stream before discharge. When stripping in the presence of nickel salts or mineral acid, the sulfide content is reduced to acceptable levels in 5–10 hr. Alternatively, hypochlorite (Eq. 9.7) or permanganate can be used to convert sulfide to much less toxic sulfate [23].

$$Na_2S + 4\ NaOCl \rightarrow Na_2SO_4 + 4\ NaCl \qquad\qquad 9.7$$

However, chemical oxidants impose higher reagent costs. Flue gas scrubbing of the waste water for sulfide removal may only be used when it is acceptable to vent hydrogen sulfide with the spent flue gases at the site of operations.

9.5. SULFUR FROM SOUR NATURAL GAS

In the same way as Frasch sulfur recovery rapidly became the dominant source of the world's sulfur in the early 1900s, sulfur recovered from the hydrogen sulfide contained in sour natural gas and other petroleum streams began to exceed that produced by the Frasch process in 1970 [24]. This dominance of sulfur from petroleum sources has been given added impetus by the substantially increased energy costs of the 1970s, which has had a negative impact worldwide on Frasch production. It has also been stimulated by the increased demand for natural gas, which has given a boost to the secondary, sulfur-recovery aspects of natural gas processing. Canada started processing of sour natural gas in 1951, and is now a major contributor to sulfur from petroleum sources (Table 9.2). Canadian production from this source probably peaked in 1976. However, since tar sands development began to supply a larger fraction of Canadian petroleum, sulfur recovery from these bituminous sources contributes a more dominant proportion of the Canadian sulfur supply. At Lacq, France, the large amounts of sulfur recovered from natural gas are leading to the development of a sulfur-based chemicals industry at this site.

Why is sulfur recovery so important to the natural gas industry? First, hydrogen sulfide in the presence of moisture is extremely corrosive to steel pipe lines. This makes it desirable to remove it as near as possible to the producing well head. Second, while some natural gas streams are virtually free of hydrogen sulfide as they emerge from the deposit, some gas wells yield a product containing 5–25% hydrogen sulfide by volume, occasionally as much as 85% H_2S [4]. The latter wells were usually capped because the high cost of sulfur recovery made the production of natural gas unprofitable. Natural gas has a reputation for being a clean easily controlled fuel. However, if burned with even 5–25% H_2S present, it would generate about the same concentrations of sulfur dioxide content in the flue gases. Also, the hydrogen sulfide itself and its sulfur dioxide combustion product are both very toxic so that there is a safety aspect to its removal as well. Hence, it is only after sour natural gas is treated to remove sulfur compounds prior to combustion to obtain a "sweetened" natural gas, a precombustion emission control measure, that it can truly be regarded as clean.

9.5.1. Amine Absorption Process for Hydrogen Sulfide Removal

Sulfur recovery from sour natural gas is conducted in two stages. The first stage involves removal of the reduced sulfide gases from the natural gas stream which is usually achieved by scrubbing the gases with an amine solution. The raw gases are first contacted with a counter current solution of monoethanolamine (MEA) in diethylene glycol (DEG) (about 1 MEA:2 DEG by volume). Monoethanolamine is a condensation product of ethylene oxide with ammonia (Eq. 9.8), and is a weakly basic organic liquid.

$$NH_3 + H_2C\overset{\displaystyle O}{\overset{\displaystyle /\ \backslash}{-}}CH_2 \rightarrow H_2N-CH_2CH_2-OH \qquad 9.8$$
$$\text{boiling point } 170°C$$

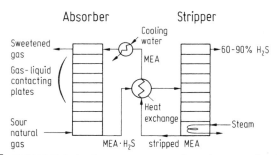

FIGURE 9.6 Flowsheet giving the schematic details of Girbotol and related sour gas sweetening processes. Twenty to twenty-four plates are used in the absorbers and strippers [25] for efficient contacting and desorption with monoethanolamine (MEA).

When contacted with hydrogen sulfide in natural gas under high pressures at near ambient temperatures in a tray-type absorber (Fig. 9.6), this solvent mixture forms a monoethanolammonium salt. This remains dissolved in the absorbing fluid to collect as a solution at the bottom of the absorbing unit (Eq. 9.9).

$$H_2N-CH_2CH_2-OH + H_2S \rightleftharpoons \overset{+}{H_3N}-CH_2CH_2-OH \qquad 9.9$$
$$HS^-$$

This scrubbing mixture also dries the incoming natural gas. The light hydrocarbon component of the gas stream is only very slightly soluble in the absorbing fluid and passes through the absorber unchanged.

If used on a one-pass basis, the volumes of monoethanolamine required would be prohibitively expensive. However, since the hydrogen sulfide is only weakly associated to the monoethanolamine, it is readily driven off and the MEA recovered by heating this solution indirectly with steam in a regenerator or stripper (Fig. 9.6; Eq. 9.10).

$$\overset{+}{H_3N}-CH_2CH_2-OH \xrightarrow{100-140°C} H_2N-CH_2CH_2-OH + H_2S \qquad 9.10$$
$$HS^-$$

The volatile hydrogen sulfide (b.p. −60.7°C) is then separated as an overhead gas stream. The monoethanolamine (b.p. 170°C) and diethylene glycol (b.p. 245°C) emerge as a regenerated solution, hot, from the bottom of the stripping column. To conserve heat, the hot, H_2S-lean monoethanolamine stream from the stripper is heat exchanged with the cooler H_2S-rich stream from the base of the absorber, sometimes with additional cooling with process water before it enters the absorber.

Most natural gas at the well head contains almost no carbon dioxide although some wells yield gas with up to 10% carbon dioxide and a few much more [26]. Carbon dioxide in the presence of moisture or other proton source is also acidic in nature. If present, it is also collected by the monoethanolamine with about the same efficiency as hydrogen sulfide, although it reacts more slowly (Eqs. 9.11–9.14; Table 9.3).

TABLE 9.3   Solubilities of Carbon Dioxide and Hydrogen Sulfide in 15.3% by Weight Monoethanolamine in Water[a]

Gas partial pressure (mm Hg)	Moles carbon dioxide per mole amine		Moles hydrogen sulfide per mole amine	
	40°C	100°C	40°C	100°C
1	0.383	0.096	0.128	0.029
10	0.471	0.194	0.374	0.091
50	0.582	0.299	0.683	0.203
100	0.576	0.347	0.802	0.279
500	0.672	0.458	0.959	0.536
1000	0.727	0.509	–	–

[a]Data selected from *Chemical Engineers' Handbook* [27].

Moist conditions:

$$CO_2 + H_2O \rightarrow H_2CO_3 \qquad\qquad 9.11$$

$$H_2CO_3 + H_2O \rightarrow H_3O^+ + HCO_3^- \qquad\qquad 9.12$$

$$H_2CO_3 + H_2N-CH_2CH_2-OH \rightarrow \; \overset{+}{H_3N}-CH_2CH_2OH \qquad 9.13$$
$$HCO_3^-$$

Dry conditions:

$$CO_2 + 2\ H_2N-CH_2CH_2-OH \rightarrow$$
$$HO-CH_2CH_2-NH-COO^- + HOCH_2CH_2-\overset{+}{N}H_3 \qquad 9.14$$

Carbon dioxide removal raises the fuel value of the natural gas. However, it also has the undesirable effect of raising the acid gas loading of the absorbing stream in the natural gas sweetening plant, decreasing the possible gas throughputs.

Tertiary amines, when dry, do not react with carbon dioxide and yet still rapidly form a salt with hydrogen sulfide. This selectivity allows preferential absorption of hydrogen sulfide over carbon dioxide from a natural gas stream. Realization of this has led to the use of methyldiethanolamine ($(CH_3N$ $(CH_2CH_2OH)_2$) as an absorbing fluid in situations where both acid gases occur in the natural gas stream. For wet gas streams, the formation of carbonic acid (Eq. 9.11) could cause absorption on the tertiary amine. However, it was found that the salt-forming reaction with carbon dioxide was relatively much slower than with hydrogen sulfide. Hence, a short natural gas contact time for the absorbing fluid with moist natural gas streams minimizes carbon dioxide absorption and yet still efficiently removes hydrogen sulfide.

A further potential concern in gas processing is the presence of carbon disulfide (CS_2) or carbonyl sulfide (COS) in the natural gas. These may be from natural sources or are occasionally contributed from the use of carbon disulfide to remove sulfur blockages from sour gas wells or components of the Girbotol plant. These sulfur compounds react with monoethanolamine to yield complex thiazolidine and oxazolidine heterocycles and polymerization products of these which are not dissociated in the regeneration step of the gas cleaning plant [28] (Eq. 9.15).

$$
\begin{array}{ccc}
\text{H}_2\text{C}-\text{CH}_2 & \text{H}_2\text{C}-\text{CH}_2 & \text{H}_2\text{C}-\text{CH}_2 \\
\text{S} \quad\text{N--H} & \text{O}\quad\text{N--H} & \text{O}\quad\text{N--H} \\
\quad\text{C} & \quad\text{C} & \quad\text{C} \\
\quad\text{S} & \quad\text{S} & \quad\text{O}
\end{array}
\qquad (9.15)
$$

thiazolidine-2-thione   oxazolidine-2-thione   2-oxazolidine

Diethanolamine ($HN(CH_2CH_2OH)_2$) and diisopropanolamine are far less susceptible to irreversible reaction than monoethanolamine and hence have largely replaced MEA at locations where carbon disulfide and carbonyl sulfide occur in the natural gas.

A minor variant to the amine scrubbing process described above is the Sulfinol process, which still uses an alkanolamine base, diisopropanolamine (35%), but in a solvent consisting of a mixture of sulfolane (40%), tetramethylene sulfone ($(CH_2)_4SO_2$, a good hydrogen sulfide solvent) and water [29]. Other processes are based on hydrogen sulfide absorption in aqueous alkaline carbonate solutions, such as the Catacarb and Benfield systems (Eqs. 9.16 and 9.17).

Absorption:

$$
Na_2CO_3 + H_2S \rightarrow NaHCO_3 + NaSH \qquad 9.16
$$

Regeneration:

$$
NaHCO_3 + NaSH \xrightarrow{\text{heat}} Na_2CO_3 + H_2S \uparrow \qquad 9.17
$$

Still further process variations use physical absorption of hydrogen sulfide by the solvent, rather than chemical reaction with it, to effect the sweetening of natural gas.

9.5.2. Claus Process Conversion of Hydrogen Sulfide to Sulfur

Hydrogen sulfide separated from natural gas by amine scrubbing is a highly odorous, toxic, low boiling gas and for these reasons is difficult to store or ship in large quantities. Hence, a Claus unit is closely associated with each amine scrubber to convert the hydrogen sulfide gas product of the scrubber to elemental sulfur. This is much easier and safer to store and ship in large quantities.

The technology of Claus conversion of hydrogen sulfide to sulfur was developed in Germany in about 1880. However, it was not until 1940 that this process was commercially adopted in the U.S. By 1967, annual sulfur production in the U.S.A., from this process had already reached 4.8 million tonnes, and by 2000, it was double the U.S. Frasch production (Table 9.2). Two reactions are employed in Claus units. The first is a simple combustion of one-third of the hydrogen sulfide stream in air, carried out in a waste heat boiler to capture the heat evolved as steam (Eq. 9.18).

$$
H_2S_{(g)} + 3/2\ O_{2(g)} \rightarrow SO_{2(g)} \quad \Delta H = -519\,kJ\ (-124\,kcal) \qquad 9.18
$$

Enough heat is retained in the boiler combustion gas stream for the second catalytic stage of the process to proceed using this (Fig. 9.7). The sulfur dioxide/steam output is blended with the remaining two-thirds of the

FIGURE 9.7 Schematic diagram of a three-stage Claus reactor unit showing steam generation both by direct combustion of one-third of the Girbotol output in a hydrogen sulfide fueled boiler, plus further steam recovery from the sulfur condensers. These are designed to remain hot enough to keep sulfur liquid.

hydrogen sulfide. This gas mixture is then fed to heated iron oxide catalyst beds where sulfur dioxide is reduced and the hydrogen sulfide is oxidized, both to a sulfur product, with further evolution of heat (Eq. 9.19).

$$SO_{2(g)} + 2\ H_2S_{(g)} \underset{300-320°C}{\rightleftharpoons} 3\ S_{(\ell)} + 2\ H_2O_{(\ell)} \quad \Delta H = -143\ kJ\ (-34\ kcal)$$

$$9.19$$

A portion of the heat obtained from this reaction is also recovered as steam which, together with that produced by the sulfur-burning boiler, is frequently sufficient to heat the stripper of the associated Girbotol (amine scrubber) plant. The net overall Claus chemistry, thus, involves oxidation of hydrogen sulfide with air to yield sulfur and water (Eq. 9.20).

$$H_2S + 1/2\ O_2 \rightarrow S + H_2O \qquad 9.20$$

In fact, in most Claus reactors the same overall reaction occurs in the first hydrogen sulfide combustion unit, so that a part of the final sulfur product is already present in the combustion gases of the first unit.

For the hydrogen sulfide to sulfur conversion part of this process (Eq. 9.19) the equilibrium lies very close to 100% on the sulfur side at temperatures below 125°C, but drops to barely 50% at 560°C [30] (Fig. 9.8). Above this temperature, the equilibrium climbs again, to reach almost 75% at 130°C. With suitable catalysts, the rate can be made acceptably fast at temperatures of 300°C or so, sufficient to keep the sulfur in the vapor state. This gives about 90% conversion of hydrogen sulfide to sulfur. (The theoretical equilibrium conversion at 300°C would be about 93%.) This would seem to give excellent sulfur recoveries. When it is considered that the Claus plants operating at Lacq in France and at Waterton and Aquitaine in Canada are producing sulfur on the scale of 2,000–3,000 tonne/day, a 10% sulfur discharge would still represent hundreds of tons of sulfur dioxide loss per day. For the

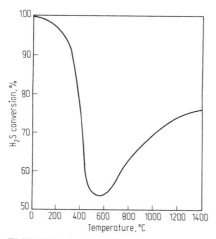

FIGURE 9.8 Equilibrium conversion of hydrogen sulfide to sulfur vapor based on thermodynamic calculations, one atmosphere total system pressure and with no sulfur removal [30, 31].

province of Alberta alone, losses from this source could amount to up to 5,000 tonne/day. However, if the sulfur vapor from one Claus stage is condensed out, and the residual gases are blended with further hydrogen sulfide and passed over catalyst, the equilibrium of Eq. 9.19 is displaced further to the right to obtain upto 94–95% conversion and containment of sulfur (Fig. 9.7). A third Claus reactor stage can similarly boost recoveries to the 96–98% range [32]. But even this is still not deemed to be adequate control in Japan and the U.S.A., or for large gas processing plants in Alberta, where recoveries of about 99.5% are required. This could theoretically be achieved with a fourth Claus stage. However, the small sulfur return for the large increased investment required for the additional stage makes other types of control devices more attractive.

9.5.3. New Developments and Emission Controls, Claus Technology

Primary Claus plant effluent gas from a single stage may still contain 2–3% sulfur dioxide, which would represent both a significant loss of feedstock and an emission problem if not further processed. Modeling experiments could help [33]. Both problems may be alleviated by the addition of extra stages of Claus reactors but only to approach the limit imposed by the equilibrium relation of Eq. 9.19 (Fig. 9.8) at Claus operating temperatures, or about 1% sulfur dioxide. The Sulfreen process takes advantage of Claus chemistry at temperatures of 125–150°C to use the higher equilibrium sulfur conversion accessible at these temperatures. Low reactor operating temperatures are combined with sulfur removal as a liquid from the gas phase, as it forms, which further displaces the equilibrium to the right. This gives the process a theoretical 100% conversion efficiency of sulfur dioxide to sulfur [34]. As a result of the condensed sulfur, which gradually accumulates on the surface of the catalyst, the reactor has to be purged with a periodic high-temperature sulfur vaporization cycle. Exit gas sulfur dioxide concentrations from the process are decreased to the 2,000–2,500 ppm range, as compared to the 1% (10,000 ppm) or so experienced by the normal Claus technology. Actual sulfur containments of 99.9% have been claimed for the Sulfreen process [35].

A related procedure is used in the Westvaco process, except that sulfur dioxide is catalytically oxidized to sulfur trioxide using activated carbon at 75–150°C. The sulfur trioxide is then hydrated to sulfuric acid which is absorbed onto the active carbon [36]. Sulfur recovery from the sulfuric acid is as sulfur dioxide, which is formed in a regenerator by raising the temperature of the carbon and adding hydrogen sulfide.

Operation of the initial hydrogen sulfide burner in a mode to favor the high-temperature Claus reaction in this unit rather than to optimize on percent sulfur dioxide produced has also been found to raise sulfur recoveries [37]. So has the installation of a continuous ultraviolet spectrophotometer to monitor the ratios of sulfur dioxide and hydrogen sulfide in the exit gases from a Claus first-stage reactor. This provides a better control basis for oxygen or reducing gas addition prior to the subsequent stages.

Other integral measures to improve sulfur recoveries include the use of oxygen-enriched air for primary hydrogen sulfide combustion [38], assisted by modelling experiments [33]. This would raise the partial pressure of the sulfur gases throughout the process, for the same total system pressure. The integral inclusion of water removal as a part of the sulfur-forming reaction should also serve to raise conversion efficiencies by displacing the equilibrium of Eq. 9.19 further to the right [31]. Thus far, neither of these alternatives appears to have been tested under process conditions.

A further option to consider is the suggestion that hydrogen sulfide be simply thermally decomposed to sulfur and hydrogen (Eq. 9.21) [39]. Thermodynamically, it appears that

$$2\ H_2S \rightarrow 2\ H_2 + 2/x\ S_x \quad \Delta H = +159\ kJ\ (+38\ kcal) \qquad 9.21$$

burning only a part of the hydrogen product should be sufficient to provide the energy required for thermal decomposition. Another suggestion proposes the use of focused sunlight as the source of the heat [40].

Add-on tail-end emission control options include the use of polyethylene glycol scrubbing (the Clauspol process [34]), and the use of methane or hydrogen as a sulfur dioxide reducing agent in a follow-up stage, rather than hydrogen sulfide (the SCOT process; Shell Claus off-gas treating [34, 35]). Hydrogen sulfide formed in the SCOT process is scrubbed and recycled to the front end of the Claus sequence. Another control option is to react the dilute sulfur dioxide stream with oxygen and water vapor to form sulfuric acid (Eq. 9.22), and absorb this onto activated carbon at 75–150°C [41].

$$SO_2 + H_2O + 1/2\ O_2 \rightarrow H_2SO_4 \qquad 9.22$$

A 90% sulfur dioxide stream, which is a high enough concentration for recycling, is then regained by a combination of heating in the absence of air and partial reduction with hydrogen sulfide, as already described in detail.

9.6. SULFURIC ACID

Sulfuric acid, though not often evident in the final product, plays some chemical or refining function in the preparation of a very large number of chemicals (Table 9.4). This has occurred because it is a strong, inorganic acid that is also low in cost. Prices have ranged from $50 gradually rising to $95 per tonne in the U.S.A [45]. This relatively low cost has meant that its chemical properties of high acidity, dehydrating action, ability to sulfonate (as oleum), and react with unsaturated hydrocarbons have all been employed commercially.

The large volume and wide application of sulfuric acid in the chemical and petroleum refining industries has meant that the per capita production of sulfuric acid is one of the better indicators of the industrial development of a country. Less circumstantial anomalies occur in a listing of annual sulfuric acid production (Table 9.5), a chemical product, than occurs with per capita sulfur production, an extractive product. Thus, average sulfuric acid production levels of the developed countries are from 50 up to 200 kg per capita per year compared to less than 5 kg per capita for Third World countries.

World sulfuric acid production has grown by a factor of 1.4–1.6 every 10 years since 1930, except for nearly doubling during the 1950–1960 period. A growth in a country's sulfuric acid production that exceeds this rate is

TABLE 9.4 Use Pattern for Sulfuric Acid in Industry

	Percent of total consumption			
	U.S.A.[a]	U.K.[a]	W. Germany[a]	U.S.A.[b]
	1975/78	1976	1976	2000
Phosphate fertilizers	38	32	15[d]	52.3
Ammonium sulfate	7	–	6	1.6
Other agric. chemicals	–	–	–	9.5
Detergents	–	11	[d]	<0.01
Fibers, cellulose film	–	9	46	0.04
Petroleum refining	5–8	–	–	14.4
Alcohols	7	–	[e]	–
Titanium dioxide	5	15[c]	22	3.2
Iron and steel pickling	3	–	11	0.12
Explosives	2.5	–	–	0.06
Other chemicals	10	16	[e]	1.7
Battery acid	0.5	–	–	0.08
Miscellaneous	19–22	17	[e]	17.2
	100	100	100	100

[a]Prepared from data of *U.N. Statistical Yearbooks* [42] and *Chemical and Engineering News* [43]. Due to differences in the breakdown of the listings used, a dash does not necessarily indicate zero consumption of a category.
[b]Calculated from data in the *U.S. Geological Survey Minerals Yearbook* [44].
[c]Includes other paint and pigment uses as well as titanium dioxide.
[d]Detergent consumption included in phosphate fertilizer category.
[e]Plastics, petrochemicals, and miscellaneous uses included with fibers and cellulose film.

TABLE 9.5 Major World Producers of Sulfuric Acid[a]

	Thousands of metric tonnes					
	1950	1960	1970	1980	1990	2000
Australia	617	1,776	1,762	2,175	1,464	n.a.
Belgium	880	1,403	1,794	–	1,906	709[b]
Canada	686	1,517	2,475	4,295	3,560[c]	3,804
China	–	–	–	7,640	11,969	23,650
France	1,215	1,983	3,682	4,943	3,771	4,781[d]
Germany	1,446[d]	3,170[d]	4,435[d]	4,108[d]	3,652[d]	8,742
India	104	354	1,189	–	3,272	5,540
Italy	1,276	2,299	3,327	2,822	2,038	3,696
Japan	2,030	4,452	6,925	6,777	6,887	7,059
Poland	285	685	1,901	3,019	1,721	1,505[b]
Spain	456	1,132	2,021	–	2,848	3,334
U.K.	1,832	2,745	3,352	3,376	2,000	1,854
U.S.A.	11,820	16,223	26,784	40,071	40,222	44,032
U.S.S.R.	n.a.	5,398	12,059	23,033	27,267	8,258[e]
Other	2,310	4,834	13,239	–	47,482	n.a.
World	25,000	48,220	86,180	–	135,918	n.a.

[a]Derived from data in *U.N. Statistical Yearbooks* [42] and June issues of *Chemical and Engineering News* [43]. Those countries that have had an annual production in excess of 2 million metric tonnes are listed.
[b]Mass for 1999.
[c]Mass for 1989.
[d]Production of the former West Germany.
[e]Russian Federation.

evidence of its development of a technological base. Those countries with close to the world's growth rate in acid production possessed well-developed technologies prior to 1950 (Table 9.5).

Two processes are used commercially to produce sulfuric acid, the contact process, where the oxidation of sulfur dioxide to sulfur trioxide is accomplished with air heterogeneously using a solid catalyst, and the chamber process, where the oxidation step is accomplished homogeneously with a gaseous catalyst. Of the two, the contact process is used to produce more than 95% of the supply of sulfuric acid at present, both in Europe and in North America.

9.6.1. Contact Process Sulfuric Acid

The original conception of the contact process is credited to a patent issued to P. Phillips in 1831 [46], but the practice of the principal components taught by this patent took nearly 50 years to bring to commercial success. The key step, the reaction of sulfur dioxide and air over a yellow-hot platinum surface to obtain sulfur trioxide, took extensive development work to obtain reasonable conversions. Coupling this initial catalytic oxidation to the hydration of the sulfur trioxide product was eventually achieved on the scale of 17,000 tonne/year by 1880, rising to 105,000 tonne/year by 1890, by BASF (Badische

Anilin-und-Soda-Fabrik), in Germany. Today single train plants of about 4,000 tonne/day (1.5 million tonne/year) are operating.

The chemical reactions involved in this process were outlined in the patent, which laid the basis for this industry. However, many improvements in practice have been achieved from a more detailed knowledge of the gas phase kinetics of the sulfur dioxide oxidation step, as well as from a better understanding of the gas-liquid equilibria associated with the hydration step.

Initially sulfur is burned in air to produce sulfur dioxide and heat (Eq. 9.23). Sulfur dioxide

$$S_{(g)} + O_{2(g)} \rightarrow SO_{2(g)} \quad \Delta H = -297 \text{ kJ } (-70.9 \text{ kcal}) \qquad 9.23$$

for this purpose may also be obtained via the oxidation of pyrites or from the roasting of other sulfidic ores. The additional capital cost involved in dust removal equipment to clean sulfur dioxide from these secondary sources makes them less important. Roaster sources of sulfur dioxide are more often driven by emission control incentives of metallurgical operations than by an interest in sulfuric acid production.

The second reaction, oxidation of sulfur dioxide to sulfur trioxide with air, is a somewhat less exothermic, equilibrium reaction (Eq. 9.24).

$$SO_{2(g)} + 1/2 \ O_{2(g)} \underset{}{\overset{V_2O_5}{\rightleftharpoons}} SO_{3(g)} \quad \Delta H = -98.2 \text{ kJ } (-23.5 \text{ kcal}) \qquad 9.24$$

It is usually catalyzed by a 6–10% vanadium pentoxide coating on a powdered pumice or kieselguhr support. Supported platinum used in many early contact acid plants is a more active catalyst than vanadium pentoxide. It gives the same sulfur trioxide for the same contact time at lower initial gas temperatures. However, its significantly higher cost and its susceptibility to poisoning have led to vanadium pentoxide virtually taking over this function [47]. A potassium hydroxide-promoted vanadium pentoxide catalyst is capable of virtually 100% sulfur dioxide conversion at 380–400°C with the correct initial gas proportions and sufficient contact time. It also has the ability to maintain near this level of activity over a 20-year life. A fraction of the catalyst is usually replaced annually to maintain high activity.

A catalyst is a substance, which speeds up a chemical reaction but is not itself consumed in the reaction. However, it cannot alter the equilibrium position of a chemical reaction (i.e., the relative proportions of sulfur dioxide and sulfur trioxide present after the reaction). Thus, the reaction rate for an equilibrium reaction, such as that represented by Eq. 9.24, is the speed with which equilibrium is reached (*not* the speed to complete conversion).

The equilibrium for a gas phase reaction may be written in a parallel manner to the operation of the law of mass action in solution equilibria (Eq. 9.25). The only difference is that the component concentrations are expressed in terms of partial pressures.

$$K_p = \frac{SO_3}{P_{SO_2} \times (P_{O_2})^{1/2}} \qquad 9.25$$

The value of the equilibrium constant for the correct ratio and concentrations of sulfur dioxide and oxygen (Eq. 9.25) has been determined experimentally

███ **TABLE 9.6 Partial Pressure Equilibrium Constants for the Oxidation of Sulfur Dioxide[a]**

Temperature (°C)	K_p
400	397
500	48.1
600	9.53
700	2.63
800	0.915
900	0.384

[a]Conditions, one atmosphere total system pressure, partial pressures of components given in atmospheres. Data selected from Bodenstein and Pohl [48].

at a number of temperatures (Table 9.6). The Le Chatelier principle states in essence that if a system in equilibrium is disturbed by a change in conditions, the position of the equilibrium will shift in the direction that will minimize the effect of the change on the system. Since this reaction is exothermic in the direction toward the sulfur trioxide product, from the Le Chatelier principle it would be expected that the equilibrium would lie more toward sulfur trioxide at low temperatures than at high temperatures.

Considering the equilibrium data, at 400°C, with a value of K_p of 397, the equilibrium of Eq. 9.24 is about 96% on the side of sulfur trioxide. However, at this temperature the time required for sulfur dioxide to react with oxygen is relatively long, which would require a large reactor and large catalyst volume (and, thus, higher capital costs to use these conditions) to obtain significant sulfuric acid production rates.

At 500°C the rate of reaction is about 100 times as fast as at 400°C, requiring a much smaller reactor volume for the same sulfuric acid throughput, but the equilibrium constant, K_p, drops to about 50. Hence, at this temperature only about 85% of the sulfur is present as sulfur trioxide.

At 600°C, the rate of reaction is some 30 to 50 times faster again, requiring an even smaller reactor for the same throughput, but the rate of dissociation of sulfur trioxide to sulfur dioxide becomes appreciable. The value of K_p drops to about 10, giving only about 60–65% of the sulfur as sulfur trioxide at this temperature, and the remainder as sulfur dioxide. For process purposes there is no point in considering the sulfur oxide equilibrium situation for any higher temperatures than this. With a promoted vanadium pentoxide catalyst bed at 600°C a 2–4 sec contact time is already sufficient to obtain essentially equilibrium concentrations at this temperature.

For process optimization, therefore, advantage is taken of the very fast reaction rates at 550 to 600°C to operate at these temperatures to about 60 to 65% conversion. The gas mixture is then cooled to 400–450°C, generating further steam, to take advantage of the more favorable equilibrium at this temperature, before being passed over three (or more) additional catalyst beds to reach about 97–98% sulfur dioxide conversion to sulfur trioxide. Occasionally additional air is added at this stage to assist in displacing the equilibrium further to the right.

Since 1.5 moles (volumes) of reacting gases are converted to 1 mole (volume) of sulfur dioxide, it would be expected that carrying out this process under pressure, by the principle of Le Chatelier, would tend to drive the equilibrium of the reaction more to the right. This has been confirmed in practice, but the improvement obtained has not been worth the additional capital costs required to operate the whole process under pressure [49]. The same sort of effect may also be achieved by raising the sulfur dioxide and oxygen concentrations, still in a total system pressure of 1.013×10^5 kP (equivalent to 1 atmosphere; $1\,\text{kP} = 1\,\text{N/m}^2$), as could be obtained by carrying out the sulfur combustion in oxygen-enriched air. But again, so far, economics have discouraged this option. New oxygen enrichment developments may change this.

These chemical principles are put into practice in a contact acid plant by first burning an atomized jet of filtered molten sulfur in a stream of dry air (Fig. 9.9). The air required is dried, prior to combustion, by upward countercurrent passage through a tower containing a stream of nearly concentrated sulfuric acid, which is trickled down an acid-resistant chemical stoneware packing. Preliminary drying of the air is necessary to avoid corrosion problems from the moist gases, which would otherwise result after combustion, and to decrease the problems from sulfuric acid mist formation on eventual disposal of the spent waste gases. With a stoichiometric excess of air for this reaction, a gas stream containing 9–12% sulfur dioxide in air at a temperature of nearly 1,000°C is obtained.

A fire tube boiler is used to reduce gas temperatures to the 400 to the 240°C range, simultaneously generating steam, before passage of the hot gases

FIGURE 9.9 Contact process for making sulfuric acid and oleum from sulfur. Additional heat control measures are usually present for the last three stages of conversion.

over the first catalyst bed of the four-pass converter. Sulfur dioxide is 60–65% converted to sulfur trioxide at this stage, simultaneously raising gas temperatures to about 600°C from the exotherm of the reaction and in so doing taking advantage of the very rapid reaction rates at these temperatures. Then, to take advantage of the higher equilibrium proportion of sulfur trioxide accessible at lower temperatures, the gases are again cooled to about 400°C, generating further steam, before passage over the second catalyst bed. Conversion to 80–85% sulfur trioxide occurs in the second catalyst bed accompanied by a smaller temperature rise. Cooling by gas–gas exchange and/or by addition of small amounts of dry air at ambient temperatures is used to bring the reacting gases to the 400°C range for each of the third and fourth catalyst stages. Under ideal conditions, these last two stages give an overall 98% conversion of sulfur dioxide to sulfur trioxide.

The sulfur trioxide concentration at this stage is about 10% by volume. After cooling to near ambient temperatures, this product is absorbed in concentrated or nearly concentrated sulfuric acid, where both absorption and hydration occur via countercurrent contact in a chemical stoneware packed tower (Eq. 9.26).

$$SO_3 + H_2O \rightarrow H_2SO_4 \quad \Delta H = -130\,kJ\ (-31.1\,kcal) \qquad 9.26$$

Using 98% sulfuric acid for hydration reduces the vigor (the exotherm) of this reaction by decreasing the water concentration from 55.5 molar (for pure water) to 1.1 molar. The concentration of sulfuric acid having the lowest vapor pressure (boiling point 338°C) is also 98%. So using this concentration for the hydration reaction also decreases the tendency to form sulfuric acid mists on absorption. The stream leaving the bottom of the absorber is then cooled, product acid is drawn off, and water is added to the remaining acid. Water addition to the process occurs from the diluted acid produced by the air drier and by the metered addition of pure water, to produce the concentration required for absorption. Commercial grades of sulfuric acid give an analysis of about 96% H_2SO_4 (freezing point, about −7°C) to reduce risks of freezing during storage and transport, since pure 100% sulfuric acid freezes at 10.3°C. The contact acid plant product is also quite pure, normally containing less than 20 µg/g (ppm) sulfur dioxide and less than 10 µg/g iron, or nitrogen oxides.

Oleum is a solution of sulfur trioxide in sulfuric acid. If this is the desired product from the contact plant, it is normally obtained from the first sulfur trioxide absorber. Absorption here, where the highest sulfur trioxide concentration is available in the entering gases, yields up to 20% oleum directly. This is equivalent to 20 kg of sulfur trioxide dissolved in 80 kg of 100% H_2SO_4. At this concentration, the oleum vapor pressure becomes so high that sulfur trioxide loss occurs almost as readily as absorption so that this is also the limit of oleum concentrations available by direct absorption. Higher concentration oleums may be obtained by distillation of the sulfur trioxide (boiling point 45°C) from 20% oleum, and condensing this now 100% sulfur trioxide vapor stream into 20% oleum until the desired concentration of oleum is obtained.

Several commercial grades of oleum are marketed, including 60% and 100%. At 60% and higher concentrations it is thought that a significant fraction of the mixture actually exists as pyrosulfuric acid, $H_2S_2O_7$, a different compound [50]. Pure sulfur trioxide, 100% oleum, is marketed with about 0.25% of stablizers such as sulfur, tellurium, carbon tetrachloride, or phosphorus oxychloride added to prevent crystallization or polymerization. Sulfur trioxide is easier to handle in the liquid state. Also, if either crystallization or polymerization were to occur during storage or shipping, these processes are sufficiently exothermic that they could raise the temperature of stored sulfur trioxide to above its boiling point. Should this happen, it risks sulfur trioxide release or vessel rupture from overpressure.

Operating a contact sulfuric acid plant essentially as described above, thus, allows straightforward production of concentrated sulfuric acid, oleums, and pure sulfur trioxide, which may be used in sulfonations or to chemically rejuvenate process-diluted sulfuric acid by simple addition. Typical raw material and utility requirements are given in Table 9.7, from which it can be seen that the contact process is actually a net *producer* of high-pressure steam, sometimes a useful feature in a chemical complex.

Provision of the sulfur dioxide feed gas for a contact plant is usually obtained by sulfur combustion, but there are many other potential sources. Pyrite, FeS_2, is burned (Eq. 9.27) and other sulfidic minerals are roasted (Eqs. 9.28 and 9.29), the latter primarily for their metal values rather than for the sulfur dioxide.

$$FeS_2 + 5/2\ O_2 \rightarrow FeO + 2\ SO_2 \qquad 9.27$$

$$CuS + 3/2\ O_2 \rightarrow CuO + SO_2 \qquad 9.28$$

$$ZnS + 3/2\ O_2 \rightarrow ZnO + SO_2 \qquad 9.29$$

Extensive dust removal facilities are required to clean up the sulfur dioxide stream from these sources, which adds substantially to the capital cost of the plant and offsets the raw material cost advantage obtained from these sulfur sources. Sometimes, hydrogen sulfide is simply burned to produce sulfur dioxide when the source of the hydrogen sulfide is near a producing sulfuric acid plant. If this method is used then the need for elemental sulfur production by the Claus process is bypassed.

If, however, none of these more direct sulfur sources is available and anhydrite ($CaSO_4$) is available on-site, or close by, then it is possible to

---

TABLE 9.7 Typical Raw Material Requirements and Utilities Consumed in the Production of 1 Tonne of 100% Sulfuric Acid[a]

	Contact plant, 180 tonne/day	Chamber process 50 tonne/day
Sulfur, kg	337–344	337
Nitrogen oxides, kg (or ammonia burned)	–	2–2.5
Water, L (process, plus cooling)	16,700	10,000
Electricity, kWh	5–10	15–16.5
Steam, kg	1,000 (credit)	–
Air, m³	7,800	8,600

[a]Compiled from Lowenheim and Moran [17], Bodenstein and Pohl [48], and Shreve [51].

practice thermal reduction of anhydrite with coke to obtain sulfur dioxide [52] (Eqs. 9.30 and 9.31).

$$CaSO_4 + 2\ C \xrightarrow[\text{heat}]{} CaS + 2\ CO_2 \qquad\qquad 9.30$$

$$CaS + 3\ CaSO_4 \rightarrow 4\ CaO + 4\ SO_2 \qquad\qquad 9.31$$

Again, the sulfur dioxide gas stream requires efficient dust removal prior to conversion. By using the lime clinker produced from this process as raw material for cement production (about equal amounts of sulfuric acid and clinker are produced), the economics have been sufficiently favorable to have 20–25% of U.K. (ICI) sulfuric acid produced via this means at one time.

9.7. CHAMBER PROCESS SULFURIC ACID

The chamber process for the production of sulfuric acid is by far the older of the two commercial processes. Records exist at least back to 1746, when Dr. Roebuck in Birmingham was burning sulfur and nitre (KNO_3) in the presence of steam in lead-lined rooms (chambers, hence the name) to produce sulfuric acid [11]. Less-detailed references exist to a similar process by B. Valentine as far back as the late 1400s. Less well-documented reports exist dating back to A.D. 1000 of the use of processes of this type.

The early chamber process operations were relatively small scale, and relied on experience that the desired product was obtained by carrying out certain steps, but with little understanding of the chemical details. The science of chemistry itself had not developed sufficiently to be able to determine the chemical details when this process was first practiced. Even if it had, the number and complexity of the reactions involved were sufficient to forestall complete understanding until about 1910. This was, however, the only existing process for sulfuric acid until the late nineteenth century.

Until as recently as 1945, it was used for close to one-half of the sulfuric acid produced in the U.K. and the U.S.A. Since then, however, the proportion of sulfuric acid produced via the chamber process has rapidly declined, in the U.S.A. to about 10% of the total in 1958, and further to only about 0.5% of the total in 1983, and has led to it being called obsolete. This happened mainly because this process is only able to produce about equal proportions of 62% and 78% sulfuric acid, and not the 98% product and oleums possible via the contact process (Tables 9.7, 9.8).

A key feature of the chamber process is that nitrous acid in the vapor state is the oxidant used to convert sulfurous acid to sulfuric acid (Eq. 9.32). Unlike platinum, or V_2O_5 catalysts of the

$$H_2SO_3 + 2\ HNO_2 \rightarrow H_2SO_4 + \underset{\text{nitric oxide}}{2\ NO} + H_2O \qquad\qquad 9.32$$

contact process, nitrous acid is immune to the catalyst poisons often present in roaster gases. This feature may yet prove of value for acid production from dilute sulfur dioxide obtained from roaster gases, as has been reported for molybdenum ore roasting [55]. Traces of oxides of arsenic, nitrogen, and selenium and of sulfates of iron, copper, mercury, zinc, and lead may be

▮ **TABLE 9.8 Relationship Between Degrees Baumé and Density for Some Common Sulfuric Acid Concentrations**[a]

Concentration % H_2SO_4 (wt.)	Degrees Baumé	Density at 15.6°C	Remarks
0.0	0.0	1.00	pure water
10.9	10.0	1.075	"dilute" acid
26.0	23.0	1.190	
29.8	26.0	1.219	
33.5	29.0	1.250	conc. of battery acid
37.5	31.80	1.281	
62.5	50.0	1.526	"chamber" or "fertilizer" acid
73.1	57.0	1.647	$H_2SO_4 \cdot 2H_2O$
77.7	60.0	1.706	"Glover tower" acid
84.5	63.2	1.773	$H_2SO_4 \cdot H_2O$
90.0	65.1	1.8144	
93.2	66.0	1.8354	"Oil of vitriol"
95.0	[b]	1.8337	ordinary concentrated acid
100.0	[b]	1.8305	"monohydrate" sulfuric acid

[a]Data compiled with some calculations from Faith et al. [53] and Lange [54].
[b]Baumé readings become unreliable indicators of sulfuric acid concentration because of a decrease in density at these high concentrations.

present in the product acid depending on the ore(s) roasted. For applications such as the production of fertilizers, these impurities can be tolerated. Full details of the chamber process are available in the 1st and 2nd editions of this book, and references cited therein.

9.8. EMISSION CONTAINMENT FOR SULFURIC ACID PLANTS

From the stoichiometry of the balanced combined equation representing the raw material requirements (Eq. 9.33), the theoretical sulfur, oxygen, and water requirement for any given quantity of sulfuric acid production may be calculated.

$$\text{S} \quad + \quad 3/2\ \text{O}_2 + \text{H}_2\text{O} \ \rightarrow \ \text{H}_2\text{SO}_4 \qquad 9.33$$

molecular weight	32.06	32.00	18.02	98.08
mole	1	×1.5	1	1
mass ratios	$\dfrac{1}{32.06}$	$\dfrac{\times 1.5}{48.00}$	$\dfrac{1}{18.02}$	$\dfrac{1}{98.08}$

The theoretical sulfur requirement works out to be 326.9 kg/tonne (Eq. 9.34), noticeably less than the 337–344 kg of sulfur per tonne experienced by many sulfuric acid producers (Table 9.7).

$$\frac{32.06\,\text{g/mol}}{98.08\,\text{g/mol}} \times 1\ \text{tonne} \times 1000\,\text{kg/tonne} \qquad 9.34$$

Material balance and conservation of matter considerations allow us to say that this 10–14 kg/tonne difference between the theoretical and actual sulfur consumption, less a proportion to account for any impurities, which may be present in the feed sulfur, represents the sulfur loss from the process.

The experience reported above represents 95–97% sulfur to sulfuric acid conversion efficiencies. Sulfuric acid plants of 200 tonne/day capacity are common and some large singe-train plants are now operating on the scale of 4,000 tonne/day [17], so even these high conversion efficiencies are not enough to avoid local emission problems. Thus, the U.K. has a requirement of 99.5% containment of the sulfur burned as a feedstock, while the U.S.A.'s 99.7% sulfur dioxide to sulfur trioxide conversion efficiency [56]. With modern process modifications and emission control devices in place, these requirements are being met.

9.8.1. Contact Process Sulfuric Acid Emission Control

The main pollutant losses from a contact acid plant occur through the absorber exit gases, but these losses are caused by a composite of several process variables. The absorber exhaust stream is composed mainly of nitrogen and oxygen. Prior to adoption of abatement measures, it also contained 0.13–0.54% with an average of 0.26% by volume sulfur dioxide, and $39\,\mathrm{mg/m^3}$ to $1730\,\mathrm{mg/m^3}$ (1.1–$48.8\,\mathrm{mg/ft^3}$) with an average of $457\,\mathrm{mg/m^3}$ ($12.9\,\mathrm{mg/ft^3}$) of sulfuric acid mists. Sulfuric acid mists are harmful to plant life and human populations. Effects of the sulfur dioxide were considered in Chap. 2. Since there is contact of nitrogen and oxygen with heated metal surfaces in a contact acid plant, there is also the potential for the formation and loss of nitrogen oxides (NO_x) in these exit gases. However, the concentration of NO_x would be expected to be less than that present in fossil-fueled power station stacks, and certainly less than obtained from an uncontrolled chamber acid plant.

To start with, the discharge of unconverted sulfur dioxide may be decreased by optimizing the sulfur dioxide to oxygen ratio entering the converter [57]. If, simultaneously, the thickness of the catalyst beds in the four-pass converter is increased, or the gas velocity through the converter is reduced, the increased gas to catalyst contact time will decrease the percent of unconverted sulfur dioxide. A regular maintenance schedule for cleanup and partial replacement of the catalyst of the first catalyst bed helps to maintain conversion efficiencies. A water or ammonia scrubber for tail gas cleanup from the last absorber (Fig. 9.10) coupled with the above steps also significantly decreases the emission level. Sale of the recovered ammonium sulfate from an ammonia scrubber spent liquor can help offset installation and operating costs.

But more than any other single measure, the interpass absorption (IPA) system has introduced a measure integral to the process, which reduces sulfur dioxide emissions to an order of magnitude below that obtained from a normal four-pass converter (i.e., to the range of 0.01 to 0.03% of the sulfur feed) [58]. This system employs intermediate cooling and sulfur trioxide absorption after the gas mixture has gone through only three stages of conversion. By removing as much as possible of the sulfur trioxide product of the conversion *before* the last catalyst stage, the equilibrium of the conver-

FIGURE 9.10 Floating bed scrubber, such as might be used to control sulfur dioxide and sulfuric acid mists. (Courtesy of European Plastic Machinery Mfg., A/S, Copenhagen.)

sion reaction of Eq. 9.24 is strongly shifted to the right. Any residual sulfur dioxide still present after absorption is then much more completely converted to sulfur trioxide as it passes through the last stage of the converter at a moderate 400–420°C (Fig. 9.11). The net effect of these modifications raises the sulfur conversion to the 99.7% range. Combining IPA with operation under a moderate pressure of about 5×10^5 kP (about 5 atm) for both conversion and absorption yields a further improvement in conversion efficiencies to about 99.8%, as practiced by the Ugine Kuhlmann process [59]. Sulfur

FIGURE 9.11 Schematic diagram of the converter section of a contact sulfuric acid plant employing an interpass absorption system for both better sulfur conversion and emission abatement. Product acid (or oleum) is cooled indirectly with process water prior to storage for sale.

consumption for this system is 330 kg/tonne, very close to the theoretical 326.9 kg.

The acid mist emission problem can arise from any of several factors. Water vapor in the air feed to the sulfur burner may cause mists because, as the water vapor plus sulfur trioxide stream drops below the dewpoint (condensation) temperature, sulfuric acid aerosol formation occurs. If this cooling happens in the absorbers, it may result in mist losses. Mist formation is minimized or eliminated by adequate drying of the air fed to the sulfur burner with concentrated sulfuric acid.

Water may get into converter gases from the combustion of hydrocarbons present in the sulfur (e.g., the dark grades of Frasch sulfur may contain up to 0.3% bitumen). When burned, this produces water vapor at a point in the process beyond where the air drying tower can help. The best that can be accomplished in this situation is to keep the gas mixture above the dewpoint throughout, until it enters the absorbers, to minimize corrosion problems in the intervening process equipment, and to rely on ancillary devices to minimize mist discharge after absorption.

Occasionally mist problems occur during startup when the concentration of acid used for air drying is too low or while oleum is being produced in the first absorber, both of which promote aerosol formation [60]. These causes are usually intermittent and discharge prevention relies on mist control equipment.

The most common control device is the Brink mist eliminator, which is an impaction filter made of fiberglass packed between stainless steel screens. These units are useful for the exit gas stream from the last absorber, and for the intermediate dried air stream leaving the sulfuric acid drier. They may also be used for the sulfur dioxide air stream after an IPA unit, if present. Mist eliminator mass efficiencies are 97–98%, which reduces sulfuric acid aerosol concentrations from 850 to 1275 mg/m^3 (24 to 36 mg per standard cubic foot (scf)) range to 18–21 mg/m^3 (0.5–0.6 mg/scf). Water or an alternative closed-cycle solvent may be used with scrubbers, sequenced either following a mist eliminator or to replace a mist eliminator in its function at the last absorber. Mass efficiencies of 90–99% have been reported for water-wash scrubbers with acid mist loadings of 3.5–7 g/m^3, even for particle size distributions containing 60% less than 3 μm diameter. Electrostatic precipitators have also been used for mist elimination [57], and methods of improving electrical gas cleaning have been examined [61].

9.8.2. Emission Control for Chamber Process Acid

Emission control for chamber process acid has been discussed in detail in the 1st and 2nd editions of this book (see Section 9.7).

9.9. RECYCLING OF SULFURIC ACID

Motivated partly by decreasing supplies and increasing costs of elemental sulfur, but more from the high costs of environmentally acceptable disposal methods for spent sulfuric acid, an increasing number of acid consumers are recycling used acid. The cost of disposal options such as barge transport of waste acid 180 km out to sea has now become comparable to the cost of recycling options [62], which has stimulated the adoption of recycling by major acid users. For example, in West Germany by 1976, some 13% of the total sulfuric acid used by each of the two largest use categories, organic chemicals and inorganic pigments (Table 9.4), was already being recycled. The percentage of acid recycle by these sectors is likely to increase further and adoption of acid recycle by smaller users is expected to grow. Cooperative recycle arrangements may help to reduce the individual cost of small-scale recycle options.

The properties of the spent sulfuric acid vary widely. It might be acid, which is diluted but otherwise uncontaminated, to acid, which is less concentrated and contaminated with metals or other impurities, to acid which is both diluted and contaminated. Measures for recovery and recycle therefore vary in complexity from simple reconcentration, to water removal accompanied by chemical purification steps. Occasionally, thermal destruction of impurities accompanied by dissociation of the acid and recycle of the sulfur content for acid remanufacture may be required. For these reasons the appropriate acid recycle method depends on the use of the original acid, and the condition of the spent acid obtained. Specific examples follow.

Regeneration of high acid concentrations from sulfuric acid, which has been only diluted by water, such as obtained from air or other gas drying functions, can be accomplished by boiling water in either a batch (pot) or continuous (heat exchanger) mode [63]. Temperatures of about 300°C are required for product acid concentrations of 95% or better at normal atmospheric pressure. If the pressure is reduced to 20 mm Hg, water removal may be accomplished at about 200°C. In both cases, lead or lead-lined equipment is necessary for the dilute acid stages to avoid corrosion problems. Steel may be used for containment of concentrations above 95% (65°Bé).

Regeneration of high concentrations of sulfuric acid may also be achieved by addition of oleum or sulfur trioxide to diluted acid. The inventory of acid in circulation is increased by a corresponding amount (Eq. 9.26). A third method of reconcentration, useful when the acid consumption occurs as a part of, or adjacent to a contact sulfuric acid plant, is to pass the diluted acid itself through the acid plant absorption tower. This amounts to on-site addition of sulfur trioxide, and the increased acid inventory obtained can be sold to markets through the normal producer channels.

Large-scale nitrations, which use a mixture of high concentrations of nitric plus sulfuric acids yield a spent acid, which is mainly diluted sulfuric acid with traces of nitric acid and organics. First the organics are removed by extraction with an immiscible solvent such as toluene. The aqueous acid raffinate is then sparged with steam to remove most of the residual nitric acid. Any traces of nitric acid left is destroyed by treatment with hydrogen peroxide and hydroxylamine. Reconcentration then regains a useful purity of high-strength sulfuric acid [64]. A less rigorous procedure for the recovery of sulfuric acid used in nitrations involves stirring the spent acid with activated carbon (1–3%) and diatomaceous earth (2–5%). This directly yields an acid ready for concentration by distillation. Denitration of spent nitration acid by addition of ammonium ion or urea, or by electrochemical means have also been tested [65].

Methods have also been developed for destruction of the organic content of spent acid using hypochlorite or hydrogen peroxide [66]. If the concentration of organics in the waste acid is not high, even the residual nitric acid content may be adequate to destroy these, in the process consuming much of the residual nitric acid [62] (Eqs. 9.35 and 9.36).

$$5 \text{ C (organic contaminants)} + 4 \text{ HNO}_3 \rightarrow 5 \text{ CO}_2 + 2 \text{ N}_2 + 2 \text{ H}_2\text{O} \qquad 9.35$$

$$\text{C (organic contam.)} + 2 \text{ ONOSO}_3\text{H} \longrightarrow \text{ CO}_2 + 2 \text{ H}_2\text{SO}_4 + \text{N}_2 \quad 9.36$$
$$\text{(unbalanced)}$$

Metal ion impurities in spent acid impose different recovery problems. For instance, iron residues in the spent acid from titanium dioxide manufacture may be removed by crystallization of first $\text{FeSO}_4 \cdot 7\text{H}_2\text{O}$ followed by crops of various iron(II) salts, or by electrodialysis [67]. Incidentally electrodialysis has also been used for sulfuric acid recovery from wastewater, either directly or after preconcentration on cation exchange resin [68].

Removal of copper ion from plating bath liquors has been accomplished electrochemically, by plating it out, and also by extraction with tributyl phosphate [69]. The extract brings nickel, antimony, arsenic, and copper into the organic phase, leaving a significantly cleaner acid raffinate. Tributyl phosphate has also been used to extract iron from pickling waste liquors.

Sometimes, however, the spent acid is so severely contaminated with inorganics or intractable tars that high-temperature oxidation of the impurities plus thermal dissociation of the sulfuric acid for sulfur dioxide recovery becomes the only feasible option to recycle the acid [70] (Eq. 9.37).

$$2 \text{ H}_2\text{SO}_4 \rightarrow 2 \text{ SO}_2 + 2 \text{ H}_2\text{O} + \text{O}_2 \qquad 9.37$$

Temperatures of about 1100°C are necessary in the presence of added air in order to promote complete oxidation of any organic impurities and thermolysis of the acid. Oxygen-enriched air may be used with advantage [71]. Subsequently, the sulfur dioxide is recovered and passed through a conventional contact plant for regeneration to sulfur trioxide and thence via rehydration to concentrated sulfuric acid (Eqs. 9.25 and 9.26).

If disposal appears to be the only practical alternative despite this range of recycle options, then pH neutralization prior to discharge is preferred, and is required by many jurisdictions. One inexpensive method employs the waste product from dry lime-based sulfur dioxide emission control systems [72] (Eq.9.38).

$$\text{CaO} + 1/2 \text{ O}_2 + \text{SO}_2 \rightarrow \text{CaSO}_4 \qquad 9.38$$

Only a fraction of the lime employed in such processes is chemically consumed. The residual calcium oxide, comprising up to two-thirds of weight of the precipitator catch, provides the inexpensive and convenient source of base for neutralization of waste but unrecoverable sulfuric acid. By using the residual chemical activity of both waste streams, the disposal problems and costs of dealing with each of them are decreased.

It is also possible to use ammonia to neutralize waste acid, which consumes a more costly base. However, the ammonium sulfate obtained can be sold as a valuable fertilizer constituent to help offset the cost of the base.

REVIEW QUESTIONS

1. It has been proposed that hydrogen sulfide be thermally decomposed to recover both hydrogen and sulfur products. Why might this be a commercially attractive alternative to the Claus process?
   Note:

   $$2\ H_2S_{(g)} \rightarrow 2\ H_2(g) + 2/x\ S_x(g) \quad \Delta H = +159\,kJ\ (+38\,kcal)$$

   $$H_{2(g)} + 1/2\ O_{2(g)} \rightarrow H_2O_{(g)} \quad \Delta H = -238\,kJ\ (-56.9\,kcal)\ at\ 100°C$$

2. (a) Iron pyrites (FeS_2) may be burned to form Fe_2O_3, and sulfur dioxide feed for a sulfuric acid plant. How many kilograms of pyrites would theoretically be required to produce one tonne of 100% H_2SO_4 in this way?
   (b) If pyrites and raw sulfur cost $5.00 and $25.00 per tonne, respectively, other factors being equal, which would be the more economical feed, and how much less would the raw material cost be per tonne of sulfuric acid produced?

3. (a) How many kilojoules of energy is theoretically available on the combustion of 1000 kg of sulfur initially at 25°C producing sulfur dioxide also finally at 25°C?
   (b) From a raw water supply at 20°C, what mass of 100°C steam could theoretically be produced from the energy available as calculated from part (a)?

4. (a) For a contact process sulfuric acid plant, using a stoichiometric ratio of dry air to sulfur in its sulfur burner, what percent by volume sulfur dioxide would be produced? (Dry air is 21% oxygen, 78% nitrogen, 1% argon by volume.) Assume ideal gas behavior.
   (b) If twice the stoichiometric requirement of air for the oxidation of sulfur dioxide to sulfur trioxide is now added, and a 97% conversion is obtained, what percent by volume concentrations of sulfur trioxide and sulfur dioxide would leave the four-pass converter?

5. A 200 tonne/day contact sulfuric acid plant burning elemental sulfur costs $1.5 million to build. Smelter by-product sulfur dioxide is available as an alternate feed at no raw material cost, but for the same capacity requires a total investment of $3.75 million to utilize this feed. Assuming continuous operation, what would the "breakeven" price for sulfur have to be to make the utilization of smelter sulfur dioxide attractive if the company has to pay 10% interest on capital borrowed to build either plant? Use estimates based on the first year of operation only. Assume the same labor costs.

6. (a) What three countries are currently the largest sulfur producers?

(b) What is the approximate current per capita production of the largest per capita sulfur producer?

7. (a) Outline any chemical conversions that might be involved in Frasch sulfur production.

(b) Why is Frasch sulfur production referred to as discretionary, whereas the majority of Claus sulfur production is not?

8. (a) Give the normal capture reaction for hydrogen sulfide removal from natural gas, and the reaction for regeneration of the capture reagent. Where does the energy for regeneration come from?

(b) How does a high carbon dioxide content in the natural gas affect hydrogen sulfide removal, and what chemical and processing remedies are there for this?

9. A natural gas stream at 50 atm absolute pressure and 40°C contains 600 mm Hg partial pressure of hydrogen sulfide, and enters a Girbotol scrubber at $100 \, m^3 min^{-1}$. Not more than 1 mm Hg partial pressure of hydrogen sulfide can remain after cleaning. Apply the data given below to a sketch of the Girbotol process for the removal of hydrogen sulfide from sour natural gas and, assuming ideal gas behavior, calculate the following:

(a) How many moles of H_2S per minute must be removed from this sour natural gas stream to meet these requirements? Assume ideal gas behavior.

(b) How many kilograms per minute of scrubber liquid comprising 15.3 wt. % (2.5 M) monoethanolamine in ethylene glycol which would be required to remove the hydrogen sulfide from this gas stream? The H_2S partial pressure at the top of the stripper is 300 mm Hg, and at the amine exit, 10 mm Hg.

	Moles hydrogen sulfide per mole amine	
H₂S partial pressure (mm Hg)	40°C	140°C
1	0.128	0.016
5	0.271	0.025
10	0.374	0.040
50	0.683	0.091
100	0.802	0.124
300	0.931	0.200
600	0.970	–

FURTHER READING

Two classics:

W.W. Duecker and J.R. West, "The Manufacture of Sulfuric Acid." Reinhold Publ. Corp., New York, 1959.

M. Sittig, "Sulfuric Acid Manufacture and Effluent Control." Noyes Data Corp., Park Ridge, NJ, 1971.

J.S. Kanwar, "Fertilizer Sulfur and Food Production." Kluwer Academic Publishers, U.S., and Canada, 1986.

W.E. O'Brien, W.L. Anders, "Potential Production and Marketing of FGD (Flue Gas Desulfurization) By-product Sulfur and Sulfuric Acid in the U.S." (1983 projection), Publication: U.S. EPA, Research Triangle Park, NC, 1979.

M.E.D. Raymont, ed., "Sulphur: New Sources and Uses." ACS Symp. Ser. No. 183. American Chemical Society, Washington, DC, 1983.

J.R. Shafer, R.W. Grendel, and D.R. Pogue, Pollution Control Practices; Sulfuric Acid Plants for Handling H_2S Gases, *Chem. Eng. Prog.* **74**, 62 (1978).

REFERENCES

1. "Minerals Yearbook, Metals and Minerals," Vol. 1, 2001. U.S. Dept. of the Interior, Bureau of Mines, Washington, DC, 2003, and earlier editions.

2. A.H. Vroom, *Hydrocarbon Process.* **51**(7), 79, July (1972).

3. "Canadian Minerals Yearbook 1978," p. 437. Department of Energy, Mines and Resources, Ottawa, 1980.

4. Sulfur-based Foam Protects, *Chem. Eng. News,* **55**(14), 24, Apr. 4 (1977).

5. Sulfur-asphalt Blend, *Chem. Eng. News,* **57**(42), 8, Oct. 15 (1979).

6. Test Road Paved, *Chem. Eng. News,* **58**(35), 25, Sept. 1 (1980).

7. M. Gellender, *Can. Chem. Process.* **62**(1), 14, Jan. (1978).

8. Sulfur Response, *Chem. Can.* **24**(7), 5, Summer (1972).

9. Sulphur-coating Urea in a Spouted Bed, *Sulphur.* **134**, 35, 40 (1978); cited by "*Minerals Yearbook 1978–79,*" p. 897. U.S. Bureau of Mines, Washington, DC, 1980.

10. "Kirk-Othmer Encyclopedia of Chemical Technology," 2nd ed., Vol. 19, p. 337. Wiley, Toronto, 1969.

11. F.S. Taylor, "A History of Industrial Chemistry," Reprint ed., pp. 184, 189. Arno Press, New York, 1972.

12. Coal Converts Sulfur Dioxide to Sulfur, *Des. News,* **34**, 18, Nov. 20 (1978).

13. M.C. Nagel, Herman Frasch, Sulfur King, *J. Chem. Educ.* **58**(1), 60–61 (1981).

14. On Purpose Sulfur Production to Cease in the U.S., *Chem. Eng. News,* **78**(31), 15, July 31 (2000).

15. Polish Sulphur Industry Revisited, *Sulphur.* **144**, 17 (1979).

16. "Thorpe's Dictionary of Applied Chemistry," Vol. IX, p. 218. Longmans, Green, Toronto, 1954.

17. F.A. Lowenheim and M.K. Moran, "Faith, Keyes, and Clark's Industrial Chemicals," 4th ed., p. 786. Wiley-Interscience, New York, 1975.

18. J.M. Ackerman, Main Pass – Frasch Sulphur Mine Development. *Mining. Eng.* **44**, 222–226, Mar. (1992).

19. D.D. Williams, M.K. Lee, J.E. Crawford *et al.,* ...Heat Transfer in...Aquifers Associated with Frasch Thermal Mining, *Ground Water,* **37**(4), 517–522, July–Aug. (1999).

20. Sulfur Prilling Tower Starts Operating, *Chem. Eng. News,* **58**(13), 37, Mar. 31 (1981).

21. R.N. Shreve, "Chemical Process Industries," 3rd ed., p. 322. McGraw-Hill, Toronto, 1967.

22. R.H. Aldabbagh and M. Aldabbagh, Environmental Effects of Wastewater Emanation from Sulfur Mining..., *Water Sci. Technol.* **24**(11), 181–187 (1991).

23. H.F. Lund, ed., "Industrial Pollution Control Handbook," pp. 14–19. McGraw-Hill, New York, 1971.

24. L.F. Hatch, *Hydrocarbon Process.* **51**(7), 75, July (1972).

25. R.F. Smith and A.H. Younger, *Hydrocarbon Process.* **51**(7), 98, July (1972).

26. "Acid Gas Content of Alberta Natural Gas," listing to April 30, 1974. Energy Resources Conservation Board, Calgary, Alberta, 1974.

27. J.H. Perry, ed., "Chemical Engineer's Handbook," 4th ed., pp. 14–10. McGraw-Hill, New York, 1969.

28. J.B. Osenton and A.R. Knight, "Reaction of Carbon Disulphide with Alkanolamines Used in the Sweetening of Natural Gas", Can. Nat. Gas Proc. Assoc. Meeting, Calgary, Alta., Nov. 2, 1970; and L.D. Polderman, C.P. Dillon, and A.B. Steele, *Oil and Gas J.* **54**(2), 180 (1955).

29. D.M. Considine, ed., "Chemical and Process Technology Encyclopedia," pp. 12–15. McGraw-Hill, New York, 1974.

30. Pollution Control in Claus Sulphur Recovery Plants, *Sulphur.* **109**, 36, Nov./Dec. (1973).

31. M.J. Pearson, *Hydrocarbon Process.* **52**(2), 81, Feb. (1973).

32. H. Krill and K. Storp, *Chem. Eng.(N.Y.).* **80**(17), 84, July 23 (1973).

33. W.D. Monnery, W.Y. Svrcek, and L.A. Behie, Modelling the Modified Claus Process Reaction Furnace..., *Can. J. Chem. Eng.* **71**, 711–724 (1993).

34. P. Grancher, *Hydrocarbon Process.* **57**, 257, Sept. (1978).

35. Sulfur Recovery Routinely Hits 99+%, *Can. Chem. Process.* **62**(5), 21, May (1978).

36. F.G. Ball, G.N. Brown, J.E. Davis *et al.*, *Hydrocarbon Process.* **51**(10), 125, Oct. (1972).

37. P. Grancher, *Hydrocarbon Process.* **57**, 155, July (1978).

38. B.M. Khudenko, G.M. Gitman, and T.E.P. Wechsler, Oxygen Based Claus Process for Recovery of Sulfur from H_2S Gases. *J. Environ. Engin.* **119**(6), 1233–1251 (1993).

39. M.E.D. Raymont, *Hydrocarbon Process.* **54**(7), 139, July (1975).

40. O.A. Salman, A. Bishara, and A. Marafi, An Alternative to the Claus Process for Treating Hydrogen Sulfide. *Energy (Oxford).* **12**(12), 1227–1232 (1987).

41. F.J. Ball, G.N. Brown, J.E. Davis *et al.*, *Hydrocarbon Process.* **51**(10), 125, Oct. (1972).

42. "United Nations Statistical Yearbook 2001," 48th ed. United Nations, New York, 2003, and earlier editions.

43. Facts and Figures for the Chemical Industry, *Chem. Eng. News*, **80**(25), 42–82, June 24 (2002); and June issues of earlier years.

44. Sulfur, by Joyce A. Ober, in: "U.S. Geological Survey Minerals Yearbook, 2001." USGS, Washington, DC, 2003.

45. *Chemical Market Reporter.* **260**(26), 50, Dec. (2001) (and 1954 issue).

46. "Kirk-Othmer Encyclopedia of Chemical Technology," 2nd ed., Vol. 19, p. 441. Wiley, New York, 1969.

47. A. Phillips, In "The Modern Inorganic Chemicals Industry" (R. Thompson, ed.), p. 183. Chemical Society, London, 1977.

48. M. Bodenstein and W. Pohl, *Z. Elektrochem.* **11**, 373 (1905); cited by Shreve 3rd ed. [21].

49. R.A. Bauer and B.P. Vidon, *Chem. Eng. Prog.* **74**, 68, Sept. (1978).

50. F.A. Cotton and G. Wilkinson, "Advanced Inorganic Chemistry," 3rd ed., p. 180. Interscience, New York, 1972.

51. R.N. Shreve, "Chemical Process Industries," p. 353. McGraw-Hill, New York, 1945.

52. D.M. Samuel, "Industrial Chemistry-Inorganic," 2nd ed., p. 80. Royal Institute of Chemistry, London, 1970.

53. W.L. Faith, D.B. Keyes, and R.I. Clark, "Industrial Chemicals," 3rd ed., pp. 744, 752. Wiley-Interscience, New York, 1965.

54. N.A. Lange, ed., "Lange's Handbook of Chemistry," 10th ed., p. 1156. McGraw-Hill, New York, 1969.

55. *Sulfur.* (157), 34, Nov.–Dec. (1981).

56. M.F. Tunnicliffe, *Chem. Brit.* **14**, 2 (1978).

57. J.R. Donovan, J.S. Palermo, and R.M. Smith, *Chem. Eng. Prog.* **74**, 51, Sept. (1978).

58. E.F. Spencer, Jr., In "Industrial Pollution Control Handbook" (H.F. Lund, ed.), pp. 14–2 to 14–4. McGraw-Hill, New York, 1971.

59. R.A. Bauer and B.P. Vidon, *Chem. Eng. Prog.* **74**, 68, Sept. (1978).

60. D.R. Duros and E.D. Kennedy, *Chem. Eng. Prog.* **74**, 70, Sept. (1978).

61. A.A. Moshkin and A.I. Izokh,...Improving Performance in Electrical Gas Cleaning in the Contact Production of Sulfuric Acid, *Chem. Petrol. Engin.* **35**(5–6), 306–309, May–June (1999).

62. R.F. Vaccaro, G.D. Grice, G.T. Rowe, and P.H. Wiebe, *Water Res.* **6**, 231 (1972).

63. H.R. Kueng and P. Reimann, *Chem. Eng. (N.Y.).* **89**, 72, Apr. 19 (1982).

64. C.M. Evans, Practical considerations in concentration and recovery of nitration-spent acids, In "ACS Symposium Series," vol. 623, pp. 250–268, 1996.

65. R.N. Gedye, Y.N. Sadana, and A.C.E. Edmonds, Electrochemical...Recycling of Nitration Waste Concentrated Sulphuric Acid, *J. Appl. Electrochem.* **17**, 731–736, July (1987).

66. E.I. Elbert, N.E. Tsveklinskaya, R.A. Bovkun *et al.*, U.S.S.R. Pat. 601, 222, (1978); *Chem. Abstr.* 89, P8414w (1978).

67. I. Smith, G.M. Cameron, and H.C. Peterson, Acid Recovery Cuts Waste Output. *Chem. Eng. (N.Y.)*. **93**(3), 44–45, Feb. 3 (1986).
68. B.R. Nott, *Ind. Eng. Chem. Process Res. Dev.* **20**, 170 (1981).
69. A. De Schepper and A. Van Peteghem, Can. Pat. 1,044,825, (1978); *Chem. Abstr.* **90**, P139717r (1979).
70. U. Sander and G. Daradimos, *Chem. Eng. Prog.* **74**, 57, Sept. (1978).
71. Anonymous, ICI's Sulphuric Acid Recovery Project Cuts out Dumping. *Process Eng.* **73**, 21, Feb. 1992.
72. P.-W. Lin, *Environ. Sci. Techol.* **12**(9), 1081 (1981).

10 PHOSPHORUS AND PHOSPHORIC ACID

...Marine Acid combined with Phlogiston [which] results in a kind of Sulphur...that...takes fire of itself upon being exposed to the open air.
—*Andrew Reid, 1758*

Without phosphorus there would be no thought.
—*Ludwig Buchner (1824–1899)*

10.1. PHOSPHATE ROCK DEPOSITS AND BENEFICIATION

Phosphorus occurrence in the lithosphere is predominantly as phosphates, PO_4^{3-}, although a rare iron–nickel phosphide, schreibersite ($(Fe, Ni)_3P)_8$ is also known in nature [1]. For this reason, phosphates are the primary source of elemental phosphorus for chemical process requirements. Only 0.20–0.27% phosphate (0.15–0.20% as P_2O_5; 0.07–0.09% as P) is present in ordinary crustal rocks.

Fortunately, the major phosphate deposits have a much higher P_2O_5 equivalent concentration than average crustal concentrations. About 80% of the current world production is derived from the sedimentary phosphorites of marine origin. These contain phosphate concentrations of 29–30% (as P_2O_5 equivalent) in unweathered deposits, and 32–35% in leached deposits [2]. These phosphates frequently occur as a variety of apatite in beds of loosely consolidated granules. Second in commercial importance are the igneous apatites (general formula: $Ca_5(PO_4)_3(Cl, F, OH)$), which average about 27% P_2O_5, and comprise some 15% of current phosphate production mainly from deposits in the former U.S.S.R., South Africa, Uganda, and Brazil. Guano sources, mostly resulting from deposition from sea birds, can range from 10 to 32% P_2O_5, depending on whether the deposits are relatively

■ TABLE 10.1 Major World Producers of Phosphate Rock[a]

Producer	Thousands of metric tonnes				
	1960	1970	1980	1990	2000
China	300	1,700	10,726	21,552	19,400
Israel	224	880	2,307	2,428	4,110
Jordan	362	913	3,911	5,925	5,526
Morocco	7,492	11,424	19,341	21,396	21,459
Nauru	1,248	2,200	2,087	926	500
Senegal	198	998	1,408	2,147	2,000
South Africa	268	1,685	3,185	3,165	2,796
Togo	—	1,508	2,933	2,314	1,400
Tunisia	2,101	2,969	4,502	6,259	8,339
U.S.S.R.	7,000	17,800	30,300	33,500	11,100[c]
U.S.A.	17,797	35,143	54,415	46,343	38,600
Others	1,875	3,411	22,158	11,271	16,073
World[b]	39,445	81,620	158,986	157,226	132,000

[a]Average phosphate content ranged from 29 to 36% (as P_2O_5). Data assembled from *U.N. Statistical Yearbooks* [3], Jasinsky [4], and Leyshon and Schneider [5].
[b]World totals include unlisted producers.
[c]Russian Federation.

modern or leached. The source of most of the material produced from Nauru, in the Western Pacific, and from Christmas Island in the Indian Ocean (Table 10.1), is from large reserves of this origin.

The United States, Morocco, and China are dominant producers (Table 10.1) of phosphate, and the state of Florida is by far the largest single producing area. From phosphorite deposits about 10-m thick covering an area of about $5000\,km^2$, Florida alone supplied one-third of the world total in 1979. Phosphorite deposits may require beneficiation (enrichment) by froth flotation for removal of slimes (clays and other finely divided material) to raise the phosphate content of the ore for market. Igneous apatite deposits require only ore selection and crushing to market size requirements. Guano deposits merely require loading to ocean freighters for transport. Thus, phosphate rock marketing is not very dependent on mineral technology; instead it is more dependent on the efficiency of solids material handling and shipping economies for competitive export. Prices in 1995 ranged from about US$ 25/tonne (run of mine) in Tampa, Florida to about US$ 50 per tonne in Casablanca (Morocco) at 77% TPL (triphosphate of lime, $Ca_3(PO_4)_2$).

10.1.1. End Use Areas for Phosphate Rock

About 87%, or about 32 million tonnes of the phosphate rock consumed in the U.S.A. in 1978 went to agricultural uses, the remainder going as feedstock for elemental phosphorus production [6]. Phosphate rock of fluorapatite stoichiometry ($CaF_2 \cdot 3Ca_3(PO_4)_2$), which is destined as a mineral supplement for animal feeds or as a bulking agent in fertilizers is normally defluorinated.

This is accomplished by heating in a rotary kiln with silica and steam (Eq. 10.1).

$$3\ CaF_2 \cdot 3Ca_3(PO_4)_2 + SiO_2 + H_2O \xrightarrow{ca.1500°C}$$
fluorapatite

$$SiF_4 \uparrow + 2\ HF + 3\ CaO + 9\ Ca_3(PO_4)_2$$

10.1

The volatile fluorides formed by this process are captured by water scrubbers (Section 10.4.3). However, about 58% of agricultural rock, or 50% of the total, ultimately goes into wet process phosphoric acid production (Section 10.4). Nonagricultural uses consume 13% of the total for elemental phosphorus production, from which both high-purity phosphoric acid and other phosphorus derivatives are made (Table 10.2). Nearly half of this, about 6% of the total, goes into the manufacture of phosphate builders for detergents, about a tenth into food and beverage additives, and the remainder into a multitude of small-scale applications.

10.1.2. Environmental Impacts of Phosphate Rock Processing

The principal environmental concerns of phosphorite surface mining relate to water consumption, storage of large volumes of waste slimes from beneficiation, and reclamation of the mined-out area. Some concern has also been expressed about exposure to slightly elevated radiation levels on reclaimed land from exposed uranium concentrations. Natural dewatering of the waste slimes is a slow process; it can take 2 years before the solids content reaches 25–30% [7]. This decreases the water recycle capability and increases the volume of the stored slimes, which at times can exceed the volume of the mined-out areas. Mining of igneous apatites, however, does not pose any significant environmental problems relevant to ore recovery. Guano sources typically involve surface mining and no beneficiation, so have only surface reclamation to look after as necessary.

TABLE 10.2 Relationship of P_2O_5 and P_4 Content to Phosphate Rock Containing Different Concentrations of Calcium Phosphate, $Ca_3(PO_4)_2$[a]

Calcium phosphate content (%) (Equivalent to % BPL, or % TPL[b])	Percent P_2O_5 equivalent	Percent P_4 equivalent
–	100.00	43.64
100	45.76	19.97
80	36.61	15.98
60	27.47	11.98
40	18.30	7.99

[a]Pure fluorapatite $CaF_2 \cdot 3Ca_3(PO_4)_2$ analyzes 92.26% $Ca_3(PO_4)_2$, 42.26% P_2O_5, and 42.26% P_2O_5, and 18.44% P_4 equivalent.

[b]BPL and TPL are still the commonly used trade designations for bone phosphate of lime and triphosphate of lime, respectively, which are used synonymously with calcium phosphate ($Ca_3(PO_4)_2$).

10.2. ELEMENTAL PHOSPHORUS

Phosphorus was first isolated in the seventeenth century by a procedure, which gave a recovery of about one ounce of phosphorus from a hogshead (large barrel, ca. 238 L) of urine [8]. In 1769, a more convenient method that made larger quantities accessible was developed by Scheele. This involved treating crushed bones with sulfuric acid and extraction of the phosphoric acid from the gypsum with small amounts of water. The water was then evaporated and the residual phosphoric acid mixed with charcoal, and then vigorously heated (Eqs. 10.2 and 10.3).

$$\underset{\substack{\text{from}\\\text{crushed bone}}}{Ca_3(PO_4)_2} + 3\,H_2SO_4 + 6\,H_2O \rightarrow \underset{\substack{\text{gypsum}\\\text{(insoluble)}}}{3CaSO_4 \cdot 2H_2O} \downarrow + 2\,H_3PO_4 \quad 10.2$$

$$4\,H_3PO_4 + 10\,C \xrightarrow[\text{heat}]{\text{white}} \underset{\substack{\text{white}\\\text{phosphorus}}}{P_4} \uparrow + 6\,H_2O + 10\,C \qquad 10.3$$

The phosphorus vapor formed was captured by condensation with water, which also provided a protective layer to cover the product, since white (sometimes called yellow) phosphorus spontaneously catches fire in air. Produced in this manner, phosphorus was a small-scale article of commerce in France and Britain in the eighteenth century, mainly for match production.

The allotrope (type) of elemental phosphorus obtained by this procedure is a soft pale yellow solid at ordinary temperatures. It forms tetrahedral molecules of four phosphorus atoms each, with three bonds to each atom. The six bonds of a "white" phosphorus molecule are at 60° to each other placing them under great strain. This is thought to be the origin of the high reactivity (e.g., the spontaneous flammability in air). In the dark, it luminesces, the property which originally gave this element its name. The other property important to both its original isolation and its commercial recovery is its boiling point of 280°C, low enough for easy vaporization as it is formed in the heated reduction mixture and yet high enough to permit condensation of the vapor by a shower of cold water. By being insoluble in and denser than water, direct contact (a shower "spray," rather than heat exchange) condensation is the most efficient and least expensive option for vapor capture. This also provides an immediate protective covering of water to avoid its contact with air. The moderate melting point permits easy transfer of phosphorus in the liquid state merely by keeping the protective water layer warm.

10.2.1. Electric Furnace Phosphorus

The forerunners of the modern electric furnace for phosphorus production were first operated in England by J.B. Readman (1888) and Albright and Wilson (1893), the latter unit with a production capacity of 180 tonne/year [1]. Electric arc heating of the charge allowed independent control of the atmosphere above the heated charge for the first time, unlike earlier methods, which used combustion to produce the white heat necessary. This development revolutionized commercial production of phosphorus.

Today, some 13–15% of American phosphate rock is consumed for elemental phosphorus production. Prior to charging to the phosphorus furnace, the phosphate rock of about 65–70% BPL (Bone Phosphate of Lime, $Ca_3(PO_4)_2$, from early origins) is sintered at 1,200–1,250°C in a rotary nodulizing kiln from a sand particle size to marble to golf ball-sized porous lumps. This preliminary step increases charge porosity in the furnace, which facilitates the release of phosphorus vapor and decreases entrained dust carried out of the furnace with the product vapor. The nodulized rock is fed in the proper ratio with silica, sand, and coke into the phosphorus furnace. This is made of steel lined with monolithic carbon and fitted with three adjustable carbon electrodes. Three-phase AC power at 200–300 V is fed to the electrodes, which heat the charge to 1,260–1,480°C, sufficient to fuse it and initiate the fluxing and the reducing reaction, which leads to phosphorus formation. This may be a composite of phosphate partial reduction by silica followed by reduction of the phosphorus pentoxide (P_4O_{10}) on contact with the incandescent coke (Eqs. 10.4 and 10.5).

$$2\ Ca_3(PO_4)_2 + 6\ SiO_2 \rightarrow \underset{\substack{\text{sublimes}\\\text{at } 300°C}}{P_4O_{10}} + CaSiO_3 \qquad\qquad 10.4$$

$$P_4O_{10} + 10\ C \rightarrow \underset{\text{vapor}}{P_4\uparrow} + 10\ CO \qquad\qquad 10.5$$

Or it may be that silica merely serves as a fluxing agent or solvent for the reduction, in which case the overall process is best represented by Eq. 10.6.

$$\underset{\text{BPL}}{2\ Ca_3(PO_4)_2} + \underset{\text{flux}}{6\ SiO_2} + 10\ C \xrightarrow{1,300-1,400°C} \underset{\text{vapor}}{P_4\uparrow} + \underset{\text{slag}}{6\ CaSiO_3} + 10\ CO$$

$$\Delta H = +2,840\ kJ\ (+680\ kcal) \qquad\qquad 10.6$$

Electrical energy provides the heat requirement of the endotherm of this reaction. This is an indirect electrical contribution to the chemical change, which is therefore called an electrothermal rather than an electrochemical process.

The carbon monoxide, phosphorus vapor mixture from the furnace is first fed to an electrostatic precipitator where any entrained dust is removed while the gases are still hot (Fig. 10.1). In this way, nonvolatile contaminants are excluded from the condenser to keep the product purity high. The water spray condenser captures most of the liquid white phosphorus and separates it from the noncondensible carbon monoxide. The direct contact condenser may also be followed by an indirect, tubular-type condenser to help improve phosphorus containment efficiency [9]. The product is stored in the liquid state in sumps under a layer of hot water, which maintains the phosphorus in liquid form for easy pumping and transfer via hot water-jacketed pipes. Typical product specifications of this phosphorus are P_4, >99.85%; Pb, <2 ppm; As_2O_3, <250 ppm; and F, <3 ppm. The cleaned carbon monoxide gas stream comprises about 93% of the gases leaving the furnace, and is used to fuel the phosphate rock nodulizing kiln (Fig. 10.1).

The white phosphorus allotrope product of the phosphorus furnace is only occasionally used on site. This is because shipping of the most important product, phosphoric acid, involves more than three times the mass of the

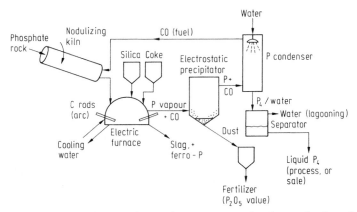

FIGURE 10.1   Flowsheet for the electric furnace process for the production of elemental phosphorus. Phossy water from the phosphorus phase separator (sump) is normally maintained in a closed-circuit holding pond system which also provides the feedwater for the phosphorus direct contact condenser. Some facilities have a further indirect, heat exchange condenser for the carbon monoxide stream, after the direct contact condenser.

element. Thus, molten white phosphorus is shipped by road or rail, in insulated tank cars of some 50 tonnes capacity, and by sea in mild steel, beehive-shaped tanks of 1,100 tonnes capacity, under both water and inert gas, for further processing.

Two more stable phosphorus allotropes are red and black phosphorus. Small amounts of these are are also produced for special purposes from the white phosphorus product of the electric arc furnace. Red phosphorus is obtained by heating white phosphorus at 400°C for several hours, which yields a complex polymeric material, more dense ($2.20\,g/cm^3$) and considerably more stable than the white variety. Red phosphorus is not only stable in air, but far less toxic than white phosphorus. Black phosphorus is more dense again (2.25–$2.69\,g/cm^3$), and has a different more complex structure. It is obtained by heating the white variety at 220 to 370°C for 8 days plus requires either a pressure exceeding $10^4\,kg/cm^2$ or a seed crystal of black phosphorus. This product has a structure resembling graphite, is a good electrical conductor, and can be lit with a match only with difficulty [10] (Table 10.3).

A single, present-day phosphorus furnace produces from 60 to 160 tonnes of phosphorus per day, almost the same as the *annual* production figures of the early arc furnaces, and requires a power supply of about 90,000 kW for a single, 160 tonne/day furnace. Power consumption per tonne of phosphorus produced varies with the % calcium phosphate in the rock (% BPL level) and furnace size among other factors but ranges around 12,000–14,000 kWh/ tonne (Table 10.4). Hence, power is a major cost component of electric furnace phosphorus production. This realization has prompted a reexamination of fossil-fueled (petroleum coke-based) sources of heat for rotary kiln combustion to provide the energy of the endotherm of the reaction [15].

Raw material addition and phosphorus vapor removal from the furnace occur on a more-or-less continuous basis. Removal of the molten calcium silicate slag, more than 7 tonnes of which is formed for each tonne of phos-

TABLE 10.3 Properties of the Allotropic Forms of Phosphorus

	Phosphorus allotrope		
	White	**Red**	**Black**
Density, g/cm^3	1.82	2.34	2.70
Melting point, °C	44	ca. 400	—
Flammability in air	spontaneous	when heated to 200°C	when heated to >550°C
Solubility	soluble in CS$_2$ and many organic solvents; about 3 mg/L^a in water at 15°C	insoluble	insoluble
Toxicity,^{b} acute (human, approx.)	<0.2 g	relatively harmless	relatively harmless
chronic	1 mg/day	—	—

^{a}Value from the *Handbook of Chemistry and Physics* [11]. More recent data gives 0.56 mg/L at 20°C.
^{b}Details given in Idler *et al.* [12].

phorus, and ferrophosphorus, which is formed from iron present in the charge, takes place periodically through appropriately placed tap holes on the lower furnace wall. The slag may be made into a rock wool insulating material or may be crushed for use in glass making or for the liming of soils. More often it is used as an aggregate for highway or railway roadbed construction. Ferrophosphorus is normally sold for steel making since it provides a safe and convenient source for phosphorus addition to alloy steels.

TABLE 10.4 Electric Furnace Operating Experience for Phosphorus Production, Per Tonne of Phosphorus Produced^a

Furnace size (kW)	15,000	25,000	Large	Large
Raw materials:				
Phosphate rock, kg	10,000	9,340	7,250	6–10,000
Silica, kg	1,500	1,560	2,660	0.5–3,500
Coke, kg	1,500	1,550	1,430	1,200–1,800
Power, kWh	14,320	11,830	11,850	11,600–16,000
Electrode carbon, kg	—	—	15	20–60
Water:				
Spray condenser, L	13,500	400	13,000	—
Products:				
Phosphorus, kg	1,000^b	1,000	1,000	1,000^b
Ferrophosphorus, kg	300	360	90	90–380
Slag, kg	900	8,100	7,100	7,100–8,900
Gases, as CO, kg	2,800	2,820	3,300	—

^{a}Data compiled from Kent [7], Ellwood [9], Kirk-Othmer [10], Thompson [13], and Canadian Chemical Processing [14].
^{b}For 87–90% elemental phosphorus recovery from phosphate rock.

10.2.2. Uses of Elemental Phosphorus

Early small-scale uses of white (or yellow) phosphorus were for the manufacture of matches and fireworks. However, the debilitating occupational illness of phossy jaw among workers, and the occasional hot pockets and singed purses from spontaneous ignition led to its discontinuance for matches by the early 1900s [12]. Today, either red phosphorus or phosphorus sesquisulfide (P_4S_3; Eq. 10.7), both much safer, are used for matches, flares, and other incendiary devices.

$$P_4 + 3\ S \xrightarrow[\text{heat}]{CO_2} P_4S_3 \qquad\qquad 10.7$$

These ignition components possess a lower vapor pressure, are much less toxic, and much less likely to ignite spontaneously than white phosphorus.

The phosphorus halides represent an important class of chemical intermediates and prepared by combining the elements (Eqs. 10.8–10.10).

$$P_4 + 6\ Cl_2 \rightarrow 4\ PCl_3 \qquad\qquad 10.8$$

$$2\ PCl_3 + O_2 \rightarrow 2\ POCl_3 \qquad\qquad 10.9$$

$$P_4 + 10\ Cl_2 \rightarrow 4\ PCl_5 \qquad\qquad 10.10$$

The products phosphorus trichloride, phosphorus oxychloride, and phosphorus pentachloride are useful intermediates for the preparation of phosphite and phosphate esters used as extractants and plasticizers, for the manufacture of gasoline and oil additives, and for phosphate-based pesticides. However, 90% of the phosphorus produced is used for the preparation of high purity grades of phosphoric acid (Section 10.3) and various phosphate salts for detergent formulation from this [16]. The decline in the phosphate content of detergents from concerns about phosphate pollution in water courses was the main cause of the steady decline in the U.S. production of phosphorus in the early 1970s, from about 540,000 tonnes in 1970 to about 390,000 tonnes in 1976 (Table 10.5). Since that time detergent consumption of phosphorus has stabilized at about 45% of the total. Prices range around $1.20/kg, depending on purity.

10.2.3. Environmental Aspects of Phosphorus Production

Electric furnace phosphorus production is relatively tolerant of the impurities present in the poorer grades (lower $Ca_3(PO_4)_2$ content) of phosphate rock. A high-silica content in the phosphate rock can be accommodated by a calculated decrease in the proportion of silica fed to the furnace, and the presence of carbonaceous matter may be allowed for by decreasing the coke. The presence of more or less iron oxide in the ore affects the yield of ferrophosphorus by-product but without seriously affecting operation. When the market for ferrophosphorus is strong, lumps of iron will be deliberately added to the furnace charge to boost its yield, since this is a valuable product.

In the working environment, a principal concern is to ensure that the exposure to elemental phosphorus in air is kept below $0.1\,\text{mg/m}^3$, which is the time weighted average permissible exposure limit for workers (TWA, PEL)

TABLE 10.5 Annual Production of Yellow Phosphorus by Selected Countries

	Thousands of metric tonnes[a]			
	1965	**1976**	**1992**	**1994/95**
Canada	(27)	(85)	(89)[b]	0
China	(ca. 18)	(35)	(150)	296
France	(14)	17	(15)	15
Italy	(13)		(15)	(15)
U.S.A.[c]	(555)	397	323[b]	264
U.S.S.R.	(ca. 100)	(250)	(425)	36[d]
W. Germany	(73)	(80)		(50)[e]

[a]Countries having a 1965 annual production capacity of more than 10,000 tonnes are listed. Sources include Kirk-Othmer [10, 17], Thompson [13], and *Chemical and Engineering News* [18]. When only capacity data was available, this is given in parentheses.
[b]Data for 1990.
[c]U.S. production in 1970 and 1994 was 542,000 and 231,000 tonnes, respectively.
[d]For Russian Federation. Kazakstan, of the former U.S.S.R., produced 468 thousand metric tonnes in 1995.
[e]For the reunited Germany.

[19]. This recommendation by the ACGIH (American Conference of Governmental Industrial Hygienists) and OSHA (Occupational Health and Safety Act) comes with an additional stipulation that exposures should not exceed $0.3\,mg/m^3$, even for short periods. A primary environmental concern in the operation of a phosphorus furnace is with potential losses of dissolved and suspended elemental phosphorus in water, since even low concentrations are toxic to freshwater and marine animal life. Care is also taken to avoid losses of fluoride to air or freshwater, which are more likely to occur during process upsets, since fluorides may cause damage to adjacent plant life or freshwater organisms [20].

The nodulizing of phosphate rock before it is charged to the furnace improves gas release from the heated furnace charge and decreases dust entrainment in the gaseous furnace products. If fluorapatite (theoretical 3.7%F) forms most of the phosphate rock feed used, this will contain about 2.8% fluoride. In the rotary nodulizing kiln, dusts are entrained in the hot gases and about 8% of the fluoride present is vaporized as hydrogen fluoride and silicon tetrafluoride (Eq. 10.1). The dust is collected from the kiln exhaust gas stream and the fluorides are then removed by scrubbing with water [21]. Hydrogen fluoride is extremely soluble in water and is effectively trapped in this form. Silicon tetrafluoride reacts with water to form soluble fluosilicic acid and colloidal silica (Eq. 10.11).

$$3\ SiF_4 + 2\ H_2O \rightarrow 2\ H_2SiF_6 + SiO_2 \qquad 10.11$$

Dust entry to the phosphorus condenser from the furnace gas stream is prevented by two-stage electrostatic precipitation of any involatiles from the carbon monoxide and phosphorus vapor mixture leaving the furnace. This measure raises the purity of the condensed phosphorus recovered from sumps and decreases the amount of phosphorus condensed in water in colloidal form rather than as more easily settled droplets, at the next stage. An indication of the effectiveness of these measures is given by source monitoring which indicated that fine fraction phosphate dust was the dominant species [22].

Condensation of phosphorus vapor by direct contact with a water shower economically provides the efficient heat transfer necessary for condensation and provides the protective water covering to prevent the phosphorus from burning in air. It also places some dissolved (solubility ca. 3 mg/L [23]) and some colloidal phosphorus into the water layer. To prevent loss of this toxic form of phosphorus to surface waters, condensers are normally operated on a total water recycle basis from a lagoon or a holding pond reservoir.

During electrothermal phosphate reduction, 80 to 90% of the fluoride contained in the rock remains with the slag as its calcium salt [21] (Eq. 10.12).

$$CaF_2 \cdot 3 \ Ca_3(PO_4)_2 + 9 \ SiO_2 + 15 \ C \rightarrow 3/2 \ P_4 \uparrow \ + \qquad\qquad 10.12$$
$$CaF_2 \ + 15 \ CO \uparrow +9 \ CaSiO_3$$

Since calcium fluoride is insoluble in water, loss of fluoride in this form is relatively benign. The fluoride present in the slag, however, may be mobilized into the water phase if the hot slag is contacted with water to fracture the material for use as fill. The fluoride not captured in the slag is converted to silicon tetrafluoride and hydrogen fluoride which leave the furnace with the phosphorus vapor and carbon monoxide. However, they are absorbed by the water stream of the phosphorus condensers (Eq. 10.11). Thus, the water from the phosphorus condenser contains dissolved and colloidal phosphorus and dissolved fluoride as the ion and as the complex fluosilicate anion.

For these reasons, the water outflow from the sumps fed by the phosphorus condensers normally contains about 1,700 ppm P_4. This is combined with phossy water displaced by incoming phosphorus in the water-blanketed storage tanks. This stream also contains the scrubbed fluorides [23]. These contaminants accumulate in the recycle pond water so that this may contain phosphates and phosphorus at 1.7 g/L (specified as P_2O_5), and fluoride and ammonium ion each at 10 g/L (when ammonia is used for pH control). The build-up of excess dissolved solids can affect the process and cause corrosion problems unless a portion of the pond water is continuously removed from the closed system. Percolation of discharged water through a pile of granulated slag decreases the fluoride concentration from about 1% (10,000 ppm) to 30 ppm by a reaction forming insoluble calcium fluoride [21]. This treated water may be retreated with lime in a second lagoon, if necessary, to decrease fluoride concentrations still further (Eq. 10.13).

$$2 \ NH_4F + Ca(OH)_2 \rightarrow CaF_2 \downarrow + 2 \ NH_4OH \qquad\qquad 10.13$$

Alternatively, the wastewaters may be treated with aluminum slags and/or sodium hydroxide to produce cryolite or ammonium cryolite of value for aluminum production (Eqs. 10.14 and 10.15).

$$6 \ NH_4F + NaAlO_2 + 2 \ NaOH \rightarrow Na_3AlF_6 + 6 \ NH_3 + 4 \ H_2O \qquad 10.14$$
$$12 \ NH_4F + Al_2(SO_4)_3 \rightarrow 2 \ (NH_4)_3AlF_6 + 3 \ (NH_4)_2SO_4 \qquad 10.15$$
$$\text{ammonium}$$
$$\text{cryolite}$$

Yet another option, magnesium fluosilicate, used in Portland cements, or sodium fluosilicate used for the fluoridation of water supplies may be prepared for sale (Eqs. 10.16 and 10.17).

$$MgO + H_2SiF_6 \rightarrow MgSiF_6 + H_2O \qquad\qquad 10.16$$

$$2\ NaOH + H_2SiF_6 \rightarrow Na_2SiF_6 + 2\ H_2O \qquad\qquad 10.17$$

The phosphorus content of the recycled pond water can vary widely depending on recycle rates, but seldom goes below about 23 ppm P_4 at the inlet and averages 0.3–0.5 ppm occasionally rising to 1.0–2.0 ppm at the exit. These concentrations are much lower than those directly from the phosphorus condensers, but are still not safe for discharge since phosphorus concentrations of a few parts per billion are toxic to fish [23].

Since phosphate (PO_4^{3-}) is nontoxic at these concentrations, it might be expected that aeration of phossy water should rapidly render it safe. However, several tests have shown that oxidation by aeration is relatively slow and ineffective [12, 23]. Ozone or potassium permanganate are effective oxidants, but are too expensive for commercial use. Chlorine is also effective but controlling the ratio relative to phosphorus is difficult; an excess of chlorine was not much better than the pollutant it was intended to neutralize. The most effective combination for phossy water treatment appears to be addition of lime and one or more chemical coagulants, coupled with settling and occasionally centrifugation which gives a waste water containing 50 μg/L (ppb) or less phosphorus (as P_4). If these measures are still inadequate, then holding of the treated wastewater in a large settling lagoon for 24 hr or longer provides a further safeguard by decreasing phosphorus concentrations to 2–3 μg/L [24]. More recently, it has been proposed that phosphorus-containing water can be safely incorporated into fertilizer formulations [25].

More general environmental development interests than the details above relate to the capability of recycling several types of phosphates in a phosphorus plant, as long as these sources are low in iron, copper and zinc [26]. An automated system for determining the energy-saving potential in phosphorus production has been developed [27].

10.3. PHOSPHORIC ACID VIA PHOSPHORUS COMBUSTION

Three main routes are employed for the commercial production of phosphoric acid. The "dry process," also called the "combustion process" yields a "furnace acid." It is obtained via the combustion of yellow phosphorus in air followed by hydration of phosphorus pentoxide product with water. The other two processes operate by acidulation of phosphate rock with strong acids, and are referred to as "wet processes." The Dorr process uses sulfuric acid, and the Haifa process uses hydrochloric acid. The wet processes are discussed separately later.

For combustion, liquid phosphorus is displaced from a closed, steam-heated storage tank by metering in hot water (Fig. 10.2), to a burner which uses compressed air to finely atomize the phosphorus. Additional combustion air then completes the oxidation to produce phosphorus pentoxide and a great

FIGURE 10.2 Diagram of process paths for the production of phosphoric acid by the combustion of elemental phosphorus. Graphite tubes are frequently employed to provide the initial cooling for the extremely corrosive combustion products.

deal of heat (Eq. 10.18). The initial product is named from P_2O_5, which was originally thought to be the molecular formula of the P^{5+} oxide.

$$4\,P + 5\,O_2 \rightarrow P_4O_{10} \qquad\qquad \therefore\ \Delta H = -3010\,kJ\ (-720\,kcal) \qquad 10.18$$
$$\text{melts } 580\text{–}585°C$$
$$\text{sublimes } 300°C$$

The combustion chamber may be operated as a separate unit and the phosphorus pentoxide vapor ducted via an intermediate cooler to the base of the hydration tower. Or the combustion chamber itself may form the base of the hydration tower. In either case, indirect cooling with water jackets absorbs the heat produced by the hydration reaction.

Hydration is accomplished in a steel tower filled with acid-proof packing, by passing in phosphorus pentoxide vapor at the base, and trickling in phosphoric acid and water at the top, thereby accomplishing both cooling and hydration (Eq. 10.19).

$$P_4O_{10} + 6\,H_2O \rightarrow 4\,H_3PO_4 \quad \Delta H = -188\,kJ\ (-45\,kcal) \qquad 10.19$$

Hydration, too, releases some additional heat. The properties of phosphorus pentoxide and the absorption process inevitably leave as much as 25% of the oxide plus a phosphoric acid mist in the exit gases from the absorption tower. These are captured on passage through an electrostatic precipitator. By variations of the process details and equipment, grades (concentrations) of phosphoric acid from 75–105% (ortho, or superphosphoric acid) H_3PO_4 may be made in this manner [28].

Phosphoric acid production by phosphorus combustion is usually accomplished in a stepwise manner as outlined, with intermediate isolation of the phosphorus. However, it may also be made by direct contact of phosphorus vapor from the furnace of a phosphorus plant with an air stream, and then passing the phosphorus pentoxide produced directly into a hydrator, without collection of the intermediate phosphorus as a liquid. Direct conversion to

TABLE 10.6   Partial Analysis of a Typical Food-grade Phosphoric Acid Produced by Phosphorus Combustion (75.0% H_3PO_4; 54.3% P_2O_5 equiv.)[a]

Component	Concentration ppm	Component	Concentration ppm
Na_2O	200	H_2S	0.1
O_2 consumed[b]	20	As_2O_3	0.05
Fe	2	Al,SO_4 each	0.0
Cl	2	CaO,K_2O each	0.0
F	0.4	SiO_2	0.0
Cr	0	Color	10[c]
Pb	0.2	Turbidity	1[c]
Cu	0.1	Odor	none

[a]Data from *Kirk-Othemr Encyclopedia of Chemical Technology* [10] and Beveridge and Hill [29].
[b]A measure of the phosphorus present in oxidation states lower than phosphate. Determined in a manner similar to that used for the measurement of BOD.
[c]In American Public Health Association (APHA) units.

phosphoric acid in this way is attractive because the bulk of white phosphorus produced is converted to phosphoric acid.

Furnace phosphoric acid is pure enough for most uses as obtained directly from the process. Food-grade applications require removal of traces of arsenious oxide (Table 10.6). Arsenic is present to the extent of 50–180 ppm (as As_2O_3 equivalent) in the feed phosphorus because of the similarity of its chemical properties to those of phosphorus (same group in the periodic table). This ends up in the product acid on oxidation and hydration [1]. It may be removed by the addition of the requisite amount of sodium sulfide or hydrogen sulfide to the furnace acid, which precipitates it as the arsenic III and V sulfides (e.g., Eq. 10.20).

$$As_2O_3 + 3\ H_2S \rightarrow As_2S_3 \downarrow + 3\ H_2O \qquad\qquad 10.20$$

Any excess hydrogen sulfide is subsequently removed from the acid by air stripping, and then clarified by passage through a sand filter. By this means, the arsenic content (as As_2O_3) is decreased to below $0.1\ \mu/g/g(<0.1\ ppm)$. The only other impurities present at levels above $1\ \mu g/g$ will be Na_2O, 0.02% ($200\ \mu g/g$), and iron and chloride at about $2\ \mu g/g$ each [10].

10.3.1. Environmental Factors of Furnace Phosphoric Acid

Losses of phosphoric acid mists or phosphorus pentoxide fumes through a corrosion-induced rupture of ductwork or a vessel, or from a process-related parameter, represent the most commonly experienced emission control problems with furnace acid plants. When this occurs, a white plume, or "ghost," of aerosol droplets or particles remains visible downwind long after the steam component has dissipated. Corrosion problems are now better understood so that more exposed components are constructed of stainless steel. If high temperatures are also involved, the equipment is lined with graphite [28].

Heat exchanger tubes for the cooler situated between the phosphorus burner and the hydrator towers are also made of graphite, and the gas path in the electrostatic precipitator consists of carbon tubes with cathode elements of stainless steel wire to ensure reliable operation.

Process parameters of importance for mist and fume control include measures to minimize the moisture content of the combustion air fed to the phosphorus burners. A high moisture content in this gas stream tends to increase mist formation. Solid phosphorus pentoxide has been found to be very difficult to dissolve in either water or phosphoric acid. This prompts temperatures in the hydrator to be maintained high enough so that absorption takes place from the vapor phase. Further studies have shown that uptake of phosphorus pentoxide vapor in 70% phosphorus acid is only about 60% and climbs with increasing acid concentrations up to about 88% phosphoric acid. Operating with two fiber beds operated in series, the first bed as an agglomerator, and the second as a collector, can also efficiently control mists and fumes from furnace acid plants [30]. Mass containment efficiencies of better than 99.9% were reported for a median aerosol particle diameter of 1.1 to 1.6 μm.

10.4. PHOSPHORIC ACID USING SULFURIC ACID ACIDULATION

The oldest, and still the lowest cost route to phosphoric acid (see Section 10.2) is via the addition of high concentrations of sulfuric acid to finely ground phosphate rock. This acidulation step releases phosphoric acid from the calcium phosphate salts present and produces insoluble gypsum (calcium sulfate dihydrate) which can be removed by filtration (Eq. 10.21).

$$CaF_2 \cdot 3Ca_3(PO_4)_2 + 10\ H_2SO_4 + 20\ H_2O \rightarrow \hspace{2cm} 10.21$$
$$6\ H_3PO_4 + 10\ CaSO_4 \cdot H_2O + 2\ HF$$

The fluoride ion usually present in the phosphate rock is of no commercial value, and adds substantial difficulties to phosphoric acid production by this method since strong acid acidulation releases virtually all of the fluoride originally present in the rock. Considering just the phosphoric acid-forming part of the overall acidulation reaction, this is a moderately exothermic process [1] (Eq. 10.22).

$$Ca_3(PO_4)_{2(s)} + 3\ H_2SO_{4(\ell)} + 6\ H_2O_{(\ell)} \rightarrow 2\ H_3PO_{4(\ell)} + 3\ CaSO_4 \cdot 2H_2O_{(s)}$$

BPL of 70–75% $\Delta H = -321$ kJ (-76.4 kcal)   10.22

Overall plant design and operational features emphasize the need to obtain as high a concentration of phosphoric acid as possible. Also important is the arrangement of acidulation and vapor control equipment to ensure safe containment of the incidental fluorides. These objectives are accomplished with adequate cooling measures to operate the whole process at low to moderate acidulation temperatures.

Keeping the temperature low is necessary to avoid the precipitation of anhydrite (anhydrous calcium sulfate) or calcium sulfate hemihydrate ($CaSO_4 \cdot 0.5H_2O$) which tend to form at high operating temperatures. Both tend to precipitate in a finely divided form which is difficult to filter [1]. If forms of calcium sulfate less hydrated than gypsum are formed, they may revert to gypsum at cooler downstream portions of the process, causing deposition and plugging of equipment. Gypsum is the coarser dihydrate of calcium sulfate which is easily filtered. While it may form at higher temperatures, it is only stable below about 40°C and then too for phosphoric acid concentrations below 35% (by weight). Hence, the desirability of keeping acidulation temperatures low.

Severe corrosion conditions which exist throughout this process have led to the use of wood, lead-lined wood, polyester-fiberglass, and rubber-lined steel for construction of much of the equipment, with dense carbon or a nickel alloy being favored for heat exchangers.

10.4.1. Operation of the Acidulation Process

Phosphate rock of 70 to 75% BPL, or as high as is reasonably obtainable, is finely ground in a ball mill and then mixed with cooled recycled phosphoric acid-gypsum slurry in a digestion tank (Figs. 10.3 and 10.4). At this stage the only reaction which occurs is between acid and any carbonates present in the rock, and between phosphates and low concentrations of sulfuric acid which may be present, in this way, minimizing foaming (Eq. 10.23) and heat evolution on initial acid contact.

$$3 \ CaCO_3 + 2 \ H_3PO_4 \rightarrow Ca_3(PO_4)_2 + 3 \ CO_2 + 3 \ H_2O \qquad 10.23$$

The newly formed mixed slurry is then moved into the next digestion tank where fresh sulfuric acid mixed with relatively dilute recycle phosphoric acid (about 50 to 55% sulfuric acid concentration) is added and vigorous mixing continued. Acidulation continues through two or more additional digestion tanks combined with several passes of the partially reacted slurry through a slurry cooler. In the cooler, direct cooling occurs by evaporation of a part of the water in the slurry under reduced pressure as one of the measures to avoid anhydrite formation. This also helps to keep the final phosphoric acid concentration high. Digestion requires a period of 4 to 8 hr while keeping temperatures to below 75–80°C, after which time the slurry proceeds to the product acid recovery system [16]. Under these acidulation conditions gypsum, the desired hydrate, is initially precipitated. Provided the slurry is cooled or that filtration is accomplished after a reasonably short time at these temperatures, the gypsum does not revert to troublesome anhydrite or the hemihydrate.

Recovery of phosphoric acid from the fully digested slurry is by continuous countercurrent filtration and washing in moving belt (e.g., Fig. 10.5), moving pan, or horizontal rotary filters [1]. Monofilament polypropylene cloth is a durable filter element in this service. All filtration methods initially filter the cooled, fully digested gypsum phosphoric acid slurry to yield a filtrate of 40–45% (29–33% P_2O_5 equivalent) phosphoric acid. This is the maximum concentration directly obtainable by this process. The gypsum filter cake, still damp with 40 to 45% phosphoric acid, is washed with dilute phosphoric acid collected from a later wash stage and this filtrate is used to dilute the fresh incoming sulfuric acid before it contacts the powdered rock.

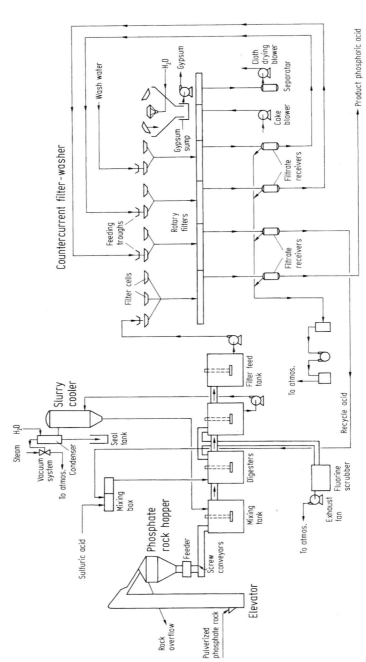

FIGURE 10.3 Process outline for phosphoric acid production by sulfuric acid acidulation of phosphate rock. Fume control and slurry evaporative cooling systems are shown, together with a countercurrent filtration and washing system which serves to maximize product acid concentrations accessible without evaporation. (From Shreve and Brink [30 a] by permission, McGraw-Hill.)

FIGURE 10.4 Details of operation of a slurry vacuum cooler, in which a small water jet exhauster maintains the vacuum, largely produced by direct steam condensation and the barometric leg of the barometic condenser. (From Slack [55], page 275 by permission, Marcel Dekker Inc.)

This wash sequence is repeated from first once to three or four times, each time using the filtrate from the subsequent wash stage as the wash liquor, until the last stage where pure water is used. In this way, the gypsum is virtually freed of adhering residual acid and in the process the countercurrent reuse of the wash water also minimizes the amount of dilute phosphoric acid to be incorporated into the product. Thus, the product acid concentration is kept as high as is feasible.

10.4.2. Purification of the Product Acid

The crude phosphoric acid obtained from this process is not very concentrated, nor very pure. Concentration by boiling the water also reduces the solubility of many of the dissolved impurities present, such as calcium, fluoride ions, aluminum, and sulfate ions causing precipitation and scaling of these materials on heat transfer surfaces. As a result, rather specialized heat transfer equipment is used to achieve this without needing frequent equipment cleaning. Submerged combustion units having no intervening heat transfer surfaces, or the direct bubbling of hot air or flue gases through carbon nozzles

FIGURE 10.5 Diagram illustrating the mode of operation of a moving belt filter to obtain phosphoric acid recovery and permitting countercurrent washing of the gypsum filter cake.

discharging into a shallow stream of acid in an open trough both successfully achieve this [10, 31]. Multiple effect evaporation has more recently also been shown to be feasible, as long as it is accompanied by forced circulation [32].

Concentration alone serves to remove most of the impurities through precipitation, and about 40% of the dissolved fluoride by volatilization, and produces about 70% phosphoric acid $(50 + \%P_2O_5)$ suitable for many uses. For solution fertilizers, where residual impurities may cause precipitation on long-term storage, or for the manufacture of a food-grade acid, further purification steps are necessary [33]. Fluoride is stripped to less than $50\,\mu g/g$ as volatile hydrogen fluoride, by sparging the crude concentrate with superheated steam. Or it may be removed by the addition of powdered silica, which forms volatile silicon tetrafluoride. This is removed by sparging with air or steam [10] (Eq. 10.24).

$$SiO_2 + 4\,HF \quad \rightarrow \quad SiF_4 + 2\,H_2O \qquad\qquad 10.24$$

$$\text{b.p.} \quad 19°C \quad -86°C$$

Arsenic is also extracted from the phosphate rock by acid, and is removed by precipitation as the sulfide using sodium sulfide [34] (Eq. 10.20). Addition of a slight excess of sulfuric acid precipitates most of the calcium (Eq. 10.25), after which, filtration of the treated acid through a sand bed produces a clarified, substantially pure product.

$$Ca(H_2PO_4)_2 + H_2SO_4 + 2\,H_2O \rightarrow CaSO_4 \cdot 2H_2O \downarrow + 2\,H_3PO_4 \qquad 10.25$$

To produce a food-grade phosphoric acid, a few additional purification steps may be required (Table 10.5). Overall phosphate recoveries from the rock are normally about 90%.

10.4.3. Recent Developments in Sulfuric Acid Acidulation

The kinetics and the variables affecting initial separation of calcium sulfate from phosphoric acid as the hemihydrate have been carefully examined [35]. The value of separating the phosphate dissolution stage (rapid stirring) [36], from the calcium sulfate crystallization step via the hemihydrate (slow agitation) has been tested on a laboratory scale [37].

In another variant of the hemihydrate process, the foam layer formed by reaction with rock carbonates (Eq. 10.26) is used to distribute a more concentrated sulfuric acid acidulant [38].

$$2\,CaCO_3 + 2\,H_2SO_4 + H_2O \rightarrow 2\,CaSO_4 \cdot 0.5H_2O + CO_2 \uparrow \qquad 10.26$$

Temperatures of 105–120°C are employed, which favor rapid reaction and hemihydrate formation, and yield a more concentrated phosphoric acid, saving on the energy cost of phosphoric acid concentration.

A single, isothermal reactor design more reliably yields gypsum crystals than multiple-reactor designs with poorer temperature control [39]. This benefit has also been claimed for a two-vessel loop system. The effect of surfactants on crystallization of phosphogypsum has been examined [40]. Higher reactor circulation rates and increased agitation coupled with other refinements also improve acid recovery [41].

Incorporation of methanol at the rate of 2 kg/kg phosphorus pentoxide in the dried acidulate significantly improves the purity of the phosphoric acid product in a two-stage process. This modification is especially advantageous with a low-grade phosphate rock [42]. Acetone has also been found of value in an experimental variant which enables much less expensive sulfurous acid, rather than sulfuric acid, to be used as the acidulant [43]. In this case, the solvent forms an α-hydroxysulfonic acid with the sulfurous acid (Eq. 10.27) which is sufficiently strong to attack phosphate rock at a reasonable rate.

$$\begin{array}{c} CH_3 \\ \diagdown \\ C=O + SO_2 + 2H_2O \rightarrow \\ \diagup \\ CH_3 \end{array} \quad \begin{array}{c} H_3C \quad OH \\ \diagdown \diagup \\ C \\ \diagup \diagdown \\ H_3C \quad SO_3^- \end{array} + H_3O^+ \qquad 10.27$$

Post acidulation applications of a solvent either as an extractant [44] or as a precipitant [32] are valuable for acid purification, most recently using methyl isobutyl ketone [45].

Methods of decreasing the fluoride concentrations of phosphoric acid have been surveyed [46]. Apart from endorsement of the addition of a reactive form of silica prior (Eq. 10.22) to stripping, acid dilution prior to reevaporation has been recommended. The additional water vapor removal assists purification by entraining hydrogen fluoride and silicon tetrafluoride as they vaporize.

Uranium recovery from the wet process phosphoric acid filtrate has been of interest for many years [47]. U.S. operations have recovered a total of 200,000 kg/yr of U_3O_8 (yellow cake) from this source [48]. Prior cleanup of dissolved organics in the raw phosphoric acid with activated carbon improves the ease of uranium extraction by providing clean phase separation. Fractionation of the natural radioactivity present in wet process acid has been examined [49].

10.4.4. Emission Control Measures for Wet Process Acid

Contact of the fluoride present in fluorapatite phosphate rocks with a strong mineral acid mobilizes much of the fluoride to hydrofluoric acid (HF: boiling point, 19°C; Eq. 10.21) and silicon tetrafluoride (SiF_4; boiling point, −86°C; Eq. 10.24). Less-reactive sodium fluoride has an acute lowest oral toxic dose in humans and in dogs of about 75 mg/kg body weight, and can also show severe skeletal effects on from chronic exposure [50]. For these reasons, volatile fluorides are easily the most important components requiring control in a wet process phosphoric acid plant. Industrial hygiene requirements for workers have been set at 2.5 mg F/m^3 for dusts and 3 ppm for hydrogen fluoride.

From an initial approximately 3.8% fluoride which may be present in the phosphate rock (e.g., for fluorapatite), the fluoride is distributed to virtually all of the acid product streams. However, the concentrations in each stream vary widely depending on the origin of the rock and the process details. One study found that, of the original fluoride present in the rock, about 29%

remained in the gypsum filter cake (much less than captured in the slag of a phosphorus plant), 15% was lost to air, and about 55% ended up in the crude phosphoric acid. The gypsum-bound fluoride is volatile since rates of fluoride emissions to air of 0.43 kg/hectare have been demonstrated from gypsum settling ponds when the pH in this study was 1.6 to 1.8, and the fluoride concentration was 2,800–5,100 mg/L. This information would suggest that separation of the fluoride disposal from the gypsum disposal functions, or liming of the ponds, could significantly decrease discharges to air. More recent studies suggest that it may be possible to bind up to 85% of total fluoride in the gypsum [51].

Fluoride losses to air occur in dusts raised in phosphate rock grinding operations, which can be controlled by a baghouse. Small amounts of further dust loss occur from the first digester, along with fluoride loss mainly as silicon tetrafluoride vapor from the whole digestion train (1,200–3,500 mg F/m^3) as well as from the filtration and vacuum areas. A system of covers for the digestion vessels and hoods over the filtration area, all exhausted to a venturi scrubber system using low water flow rates on a recycle basis, best meet these control requirements [52]. Pond water equilibrium fluoride concentrations of 0.2% can result from the scrubber fluoride loading. Excess pond water volumes may be discarded safely by fluoride removal via single or double liming the effluent in one or two smaller separate basins before discharge (Eq. 10.13). The calcium fluoride may be readily recovered.

Crude 28 to 41% phosphoric acid (20 to 30% P_2O_5 equivalent) may contain 1 to 2% fluoride, or about 55% of the fluoride originally present in the rock, mostly as fluosilicic acid (H_2SiF_6) or convertible to fluosilicic acid [53]. Normal evaporative concentration serves to vaporize about two-fifths of this, which can be increased by variations of this technique [1] (see Section 10.4.1). The vapor control procedures just outlined effectively capture this vaporized fluoride (Fig. 10.5). If the crude acid is not required to be concentrated, however, it may be purified by the addition of sodium carbonate. Sodium fluosilicate precipitates may be filtered, leaving phosphoric acid containing only about 0.1% fluoride (Eq. 10.28).

$$H_2SiF_6 + Na_2CO_3 \rightarrow Na_2SiF_6 \downarrow + H_2O + CO_2 \qquad 10.28$$

Sodium fluoride is soluble. The sodium fluosilicate may be purified and marketed directly for purposes such as fluoridation of water supplies, or it may be used to prepare other salts, for example, synthetic cryolite (Na_3AlF_6), valuable for aluminum smelting.

Gypsum is a relatively inert by-product of wet process acid production. However, it may pose an environmental problem to wet process acid operations simply because some 5 to 7 tonnes are produced for each tonne of phosphoric acid (100% basis). Years ago, disposal by some facilities was simply into the nearest watercourse via a slurry pipeline. However, with the realization that even inactive suspended solids in freshwater streams can have severe smothering and siltation effects on both water quality and the diversity of stream life, this practice has been largely discontinued. Use of the gypsum for wallboard manufacture has been studied [54], and is used where supplies of natural gypsum are scarce, such as in the U.S. [55, 56]. Application of phosphogypsum as a source of sulfur dioxide for sulfuric acid production (Section 9.6.1) has not attracted much interest. Thus, inland facilities have largely settled on gypsum dispersal into extensive holding ponds, or a sequence of lagoons laid out with landfill objectives. Some coastal facilities in Europe dump waste gypsum into the sea [57]. One operation found a 90-m outfall satisfactory for the daily dispersal of 400 to 500 tonnes of gypsum, assisted by the co-discharge of some 550 m^3/hr of seawater used for cooling.

10.5. PHOSPHORIC ACID USING HYDROCHLORIC ACID ACIDULATION

Nitric acid and hydrochloric acid have also been used to obtain phosphoric acid from phosphate rock. Process difficulties in obtaining a pure phosphoric acid using nitric acid have resulted in this chemistry only being utilized to prepare granular fertilizers [33] (Eqs. 10.29 and 10.30), rather than acid preparation.

Acidulation:

$$CaF_2 \cdot 3Ca_3(PO_4)_2 + 20\ HNO_3 \rightarrow 6\ H_3PO_4 + 10\ Ca(NO_3)_2 + 2\ HF \quad 10.29$$

Ammoniation (full):

$$6\ H_3PO_4 + 10\ Ca(NO_3)_2 + 18\ NH_3 \rightarrow 6\ (NH_4)_3PO_4 + 10\ Ca(NO_3)_2 \quad 10.30$$

The hydrochloric acid acidulation process was originally developed by Israel Mining Industries at Haifa as a production-scale process to phosphoric acid [58], so is commonly referred to as the Haifa or IMI process. It was adopted to such an extent that the licensed capacity in 1965 stood at 50,000 tonnes of P$_2$O$_5$ equivalent per day. Trade hydrogen chloride or hydrochloric acid is normally two to three times the price of sulfuric acid. Hence, this process is primarily of interest to integrated chemical operations having a surplus of hydrogen chloride. This might arise, for example, from the operation of a chlorinated solvents plant producing dichloromethane, or from the manufacture of toluene diisocyanate (Eqs. 10.31–10.33), since the markets for anhydrous hydrogen chloride have traditionally been limited.

$$2\ Cl_2 + CH_4 \rightarrow CH_2Cl_2 + 2\ HCl\ (also\ CH_3Cl,\ CH_2Cl_2,\ CCl_4) \quad 10.31$$

$$CH_3C_6H_3(NH_2 \cdot Cl)NHCOCl + COCl_2 \xrightarrow{heat} \quad 10.32$$
$$CH_3C_6H_3(NHCOCl)_2 + 2\ HCl$$

$$CH_3C_6H_3(NHCOCl)_2 \xrightarrow{heat} CH_3C_6H_3(NCO)_2 + 2\ HCl \quad 10.33$$
$$\text{toluene diisocyanate}$$

Acidulation of fluorapatite with hydrochloric acid produces phosphoric acid in a similar manner to sulfuric acid acidulation, except that in this case, the coproduct calcium chloride is very water soluble (Eq. 10.34).

$$CaF_2 \cdot 3Ca_3(PO_4)_2 + 20\ HCl \rightarrow 6\ H_3PO_4 + 10\ CaCl_2 + 2\ HF \quad 10.34$$

Phosphoric acid recovery from HCl acidulation of phosphate rock, therefore, is not a simple solid–liquid separation as it is from sulfuric acid acidulation;

instead, it requires a solvent extraction to obtain the product free of calcium chloride. Suitable solvents must be polar to permit good solution of phosphoric acid, and have low water solubility to minimize solvent losses. Israel Mining Industries prefers C_4 to C_5 alcohols, though some producers prefer these to be mixed with a water-immiscible solvent such as kerosene. Others prefer trialkylphosphates such as tri-n-butylphosphate or tri-2-ethylhexylphosphate, which have similar physical properties [59]. It should be mentioned here that it is possible to use solvent extraction with sulfuric acid acidulation by using solvent discrimination to obtain a purified phosphoric acid extract [60].

10.5.1. Product Recovery by Solvent Extraction

To establish extraction efficiencies for a substance A one first needs to know, or experimentally determine, the partition coefficient, K_p. This constant relates the concentration of solute A in the extraction solvent to the concentration of the solute in the extracted phase, at equilibrium (Eq. 10.35).

$$K_p = \frac{[A]_{solvent}}{[A]_{water}}$$
                                                                                        10.35

This holds true at any given temperature as long as the distributing species, solute A, exists in the same form in both phases. Otherwise, a more complex "distribution ratio" must be employed. For increased refinement of extraction efficiencies, it may be necessary to employ activities, rather than concentrations in some cases [61].

Let us illustrate the meaning of partition coefficients with the simple extraction system carbon tetrachloride/water. Using ammonia or iodine as the solute, the partition coefficient K_p at 25°C is 0.0042 and 55, respectively [62]. With polar ammonia as the solute, this would move primarily to the water phase on extraction, hence the low K_p. With nonpolar iodine, the higher concentration would accumulate in the nonpolar carbon tetrachloride phase.

FIGURE 10.6 Emission control devices used with the vapor from a phosphoric acid evaporator to decrease fluoride emissions. The water used to condense steam by direct contact with cold water also serves to control fluoride vapor losses quite efficiently. (From Slack [55], page 747, by permission, Marcel Dekker, Inc.)

FIGURE 10.7 Arrangement for the physical components of a mixer settler, such as that used for solvent extraction processes. Units may be separated or combined, in either case yielding discrete solvent and raffinate (extracted) phases with equilibrium, or near equilibrium concentrations of solute.

Obviously, iodine would be much more readily extracted from water than ammonia when using carbon tetrachloride as the extractant.

Other important features in assessing industrial extractions are whether one or more extraction stages are to be used. This will depend in part on the partition coefficient of the system of interest, and on the flow pattern of the solvent and extracted phase (the raffinate) to be used, if more than one extraction stage is employed. Single-stage extraction requires a mixer to bring about intimate contact between the two phases, followed by a settler which allows phase separation and a means for independent removal of the two phases (Fig. 10.7).

For a solute with a relatively poor partition coefficient, one of the two arrangements of multistage extraction, crosscurrent or countercurrent may be selected to improve efficiencies (Fig. 10.8). To decide which would be the advantageous choice, assume we have a 1,000 L/hr aqueous waste stream containing 0.02 M benzoic acid which is to be extracted with 300 L/hr of benzene for benzoic acid recovery. For the extractant benzene, $K_p = 4.3$ at

FIGURE 10.8 Possible arrangements of mixer settlers and process streams for multistage extraction processes. (a) Crosscurrent arrangement. (b) Countercurrent arrangement.

20°C. As a simplification, let us assume complete immiscibility of benzene and water. To accomplish this extraction in a single-stage system, one can set up an equation to calculate the concentration of benzoic acid in each of the streams produced. To do this, the value for the concentration of benzoic acid in the first extract $[BzOH]_E$ is obtained from the known equilibrium which must exist between the two streams of Eq. 10.36. This value is substituted into Eq. 10.37, which defines the balance of concentrations of benzoic acid which must exist between the incoming and outgoing streams.

$$[BzOH]_E = K_p[BzOH]_R, \quad \text{from } K_p = \frac{[BzOH]_E}{[BzOH]_R} \qquad 10.36$$

$$\overset{\text{\textit{Feed Streams}}}{S[BzOH]_S + F[BzOH]_F} = \overset{\text{\textit{Product Streams}}}{E1[BzOH]_{E1} + R1[BzOH]_{R1}} \qquad 10.37$$

where S, F, E_1 and R_1 represent *volumes* of solvent, aqueous benzoic acid feed stream, the first extract and the first raffinate, respectively. The $[BzOH]$ term with its various subscripts represents the *concentration* of benzoic acid present, in moles per liter, in the subscripted stream.

Knowing this, one can derive Eq. 10.38, which allows calculation of the concentration of benzoic acid in the first raffinate, $[BzOH]_{R1}$.

$$[BzOH]R_1 = \frac{[BzOH]_F + \left(\frac{S}{F}\right)[BzOH]_S}{1 + \left(\frac{S}{F}\right)K_p} \qquad 10.38$$

$$\text{For single-stage extraction, } \frac{S}{F} = \frac{E_1}{R_1}$$

By substitution of known values into this expression, the concentration is found to be 9×10^{-3} M. Thus, one extraction stage as described decreases the benzoic acid concentration in the effluent to just under one-half the original value. Similarly, by substituting this concentration value and K_p into Eq. 10.36, the concentration of benzoic acid obtained in the benzene extract is found to be 0.04 M, a ratio corresponding to the benzoic acid partition factor between these two liquids. The same equation may also be used to calculate the single-stage extraction parameters for any other solute of interest, provided that K_p for the system is known, or is determined experimentally.

In a similar manner, one can derive an equation to calculate the benzoic acid in the raffinate for a three-stage, crosscurrent system (Eq. 10.39; cf. ref. [63]).

$$[BzOH]R_3 = \frac{[BzOH]_F}{\left[1 + \left(\frac{S}{F}\right)K_p\right]^n} \qquad 10.39$$

where, for three stages, $n = 3$.

If fresh extracting solvent volumes for each stage differ, the denominator requires expansion to individual terms for each solvent volume (Eq. 10.40).

$$\left[1 + \left(\frac{S_1}{F}\right)K_p\right]\left[1 + \left(\frac{S_2}{F}\right)K_p\right]\left[1 + \left(\frac{S_3}{F}\right)K_p\right]\cdots\left[1 + \left(\frac{S_i}{F}\right)K_p\right] \qquad 10.40$$

TABLE 10.7 Calculated Molar Benzoic Acid Concentrations in Water after Extraction With Benzene[a]

Volume of benzene (L)	Single-stage extraction	Three-stage crosscurrent	Countercurrent	
			Three-stage	Four-stage
100	0.014	0.013	0.012	0.010
300	0.009	0.007	0.003	0.001
600	0.006	0.003	0.0007	4×10^{-5}
1,200	0.003	0.001	0.0001	8×10^{-7}

[a]Calculated for a 1,000 L aqueous wastewater stream after extraction with benzene in various modes at 20°C. Initial benzoic acid concentration in water is 0.02 mol/L, and $K_p = 4.3$.

Assuming that 100 L volumes of benzene are used for each of the three extraction stages in a system of this kind, the benzoic acid concentration in the third-stage raffinate is found to be 0.007 M (Eq. 10.41), or about one-third that of the incoming stream.

$$[BzOH]_R = \frac{0.02 \, mol/L}{\left[1 + \frac{100}{1000}(4.3)\right]^3} = 0.007 \, mol/L \qquad 10.41$$

Carrying this approach to three-stage, countercurrent extraction and solving the pairs of equations for each of the three stages of the process one can derive Eq. 10.42 which allows the determination of the benzoic acid concentration in raffinate 3, the aqueous effluent stream.

$$[BzOH]_R = \frac{[BzOH]_F}{\left[1 + \left(\frac{S}{F} \cdot K_p\right)\right]\left[1 + \left(\frac{S}{F} \cdot K_p\right)^2\right]} \qquad 10.42$$

Substituting into this expression, one obtains a value of 0.003 mol/L for the benzoic acid concentration in the final raffinate, one-seventh that of the incoming stream. Comparing the calculated results for the three modes of extraction demonstrates the clear superiority of the extraction efficiency of countercurrent contacting method and provides the justification for the frequent use of this arrangement for industrial processes. The additional advantage gained by any of the extraction modes as the volume of solvent is increased or with one additional stage of countercurrent contacting is shown by the calculated data of Table 10.7.

10.5.2. Haifa (or IMI) Phosphoric Acid Process Details

A slight excess of the theoretical requirement of hydrochloric acid, based on the BPL level ($Ca_3(PO_4)_2$ concentration) in the rock, is used for acidulation to avoid the formation of monocalcium phosphate ($CaHPO_4$) which is insoluble and hence would result in phosphate losses. The acidulation reaction

FIGURE 10.9 Flowsheet for the Haifa (Israel Mining Industries) hydrochloric acid acidulation route to phosphoric acid. Digestion for a short time is followed by solvent extraction in polyvinyl-chloride mixer settlers, which tolerate the corrosive conditions. (Reprinted from *Chemical Engineering (N.Y.)* [64] by special permission from the journal and McGrawHill Inc.)

(Eq. 10.34) is more rapid with hydrochloric acid than with sulfuric acid, and yields soluble products from both the calcium and phosphate. Thus, there is neither a need to heat the mixture to speed up the reaction, nor are there any crystal form problems requiring temperature stabilization.

About 10% of the raw rock, comprising silica and the like, does not dissolve. However, this material is readily removed by settling the dilute slurry in a thickener, sometimes aided by coagulants (Fig. 6.6). The clarified solution from the thickeners is then extracted with a C_4, C_5 alcohol mixture or trialkylphosphate in a series of three or more mixer settlers producing a solvent phase rich in phosphoric acid, and a raffinate of calcium chloride brine freed of phosphate. Presence of the calcium chloride salt in the aqueous phase undoubtedly assists in driving the phosphoric acid transfer to the organic phase by making the aqueous phase more polar.

To recover the phosphoric acid, the solvent phase is back-extracted with pure water in a further mixer settler (Fig. 10.9). In the absence of dissolved calcium chloride the aqueous phase now extracts the bulk of the phosphoric acid from the solvent, mixed with some residual hydrochloric acid and solvent. Distillation of the extracted stream, usually in three separate evaporators, enables simultaneous recovery of solvent and unused hydrochloric acid, which are both recycled, and phosphoric acid of appropriate concentrations for sale [65].

The product acid from this process is often purer than that obtained from sulfuric acid acidulation, and in addition this process is claimed to recover 98–99% of the available P_2O_5 in the rock, as opposed to the 94–95%

experience of the most efficient sulfuric acid processes. Refined and food grades of phosphoric acid may be obtained after chemical treatment, stripping, and if necessary purification by anion and cation exchange.

An important economic feature of Haifa process operation is the need to maintain good solvent recovery from the brine raffinate as well as from the washed waste solids and the extracted acid. With care, it has been found possible to maintain solvent losses at not more than about 5 kg/tonne P_2O_5 equivalent produced [66].

10.5.3. Haifa Process By-products and Waste Disposal

Fluoride containment still has to be considered in the operation of this process, as with sulfuric acid acidulation, since the acid treatment similarly mobilizes chemically bound fluoride. Problems from this source are minimized by prior calcination of the feed phosphate rock, as has been practiced for the early Japanese experimental operations using this process. In any case, fluoride tends to follow the aqueous phase throughout to end up as a component of the calcium chloride raffinate solution. From this stream, fluoride may either be recovered for sale or removed from the calcium chloride stream and discarded by methods which have been discussed (Section 10.4.4.).

The washed insoluble slurry from acid digestion is generally discarded to landfill, or as a slurry to the sea if the plant operates on ocean frontage.

The brine contains about 2 tonnes of calcium chloride for each tonne of phosphoric acid (100% basis) produced. It may be sold for various uses, such as highway deicing, as a low-temperature heat transfer fluid, or as a heat sink, as has been outlined in more detail with the Solvay process (Section 10.4). If not sold, in the absence of ocean frontage, the waste brine may pose a disposal problem. Deep well disposal into a brine aquifer may be used for inland locations, but geological and watertable factors do not always make this possible locally. However, the smaller waste load compared to the gypsum

TABLE 10.8 Phosphoric Acid Production by Selected Countries[a]

	Thousands of metric tonnes of P_2O_5 equivalent[b]			
	1976	1980	1990	2000
Canada	520	808	402	
China	1,420[c]	2,310	3,040	6,330
France	–	203	626	381
Italy	426[d]	324	497	
U.K.	482	759	93	
U.S.A.	7,210	9,840	10,294	13,143
W. Germany		1,595	165	128[e]

[a]Sources of data include *U.N. Statistical Yearbooks* [3], and *Chemical and Engineering News* [18]. Includes both furnace acid and wet process products.

[b]To obtain production as 100% H_3PO_4, entry shown should be multiplied by 1.38 (i.e., $196 \div 142$).

[c]*Statistical Yearbook of China*, [69]. Unclear whether via H_3PO_4 or from phosphate rock.

[d]Specified as 82% H_3PO_4.

[e]Reunited Germany, 1998.

of the Dorr process and the easier disposal of the calcium chloride brine to the sea are factors which keep the Haifa process of interest to further process development [67], and to phosphoric acid producers [68].

10.6. MAJOR PRODUCERS AND USERS OF PHOSPHORIC ACID

Available production figures for many of the large world producers of phosphoric acid are given in Table 10.8. This product does show a more generally increasing trend with time, since fertilizer demands on phosphoric acid output have generally climbed.

End uses of phosphoric acid are highly dependent on the process used to produce it. About 875,000 tonnes (as P_2O_5) of furnace phosphoric acid was produced in the U.S. in 1984, about 7% of the total U.S. production [70]. Nearly half of this furnace acid is still dedicated to the production of sodium, potassium, and calcium phosphates for use in detergents and commands an 80% or so price premium over fertilizer grade, wet process acid. A further 5% is destined as a food-grade product, which is mostly used as salts in foods, bakery goods, and soft drinks. Some of this grade is also used as a supplement in animal feeds. These are relatively all small-scale uses.

Most of the phosphoric acid production in the U.S., about 85%, is consumed by fertilizer manufacturers, mostly for preparation of ammonium phosphates and triple superphosphate (Chap. 11). This consumption picture may be slightly distorted since the U.S. is also a substantial exporter of phosphoric acid.

REVIEW QUESTIONS

1. (a) Calculate the theoretical quantity of phosphate rock (fluorapatite), silica, and coke required by the electric furnace process to produce 1 tonne of elemental phosphorus.
    (b) Give reasonable explanations for the discrepancies between the theoretical requirements, as calculated in part (a), and the amounts of each of these raw materials used in practice.
2. What quantity of elemental phosphorus and water would theoretically be required using the furnace acid process to produce 1 tonne of 100% equivalent phosphoric acid?
3. A company is considering construction of a phosphoric acid plant employing Haifa process technology, which is to use its own by-product hydrochloric acid for acidulation. Present-day cost of sulfuric acid is $110/tonne.
    (a) Assuming quantitative reactions on fluorapatite in both cases and no solvent losses, what "breakeven" value would the company have to assign to its by-product hydrochloric acid (100% basis) to compete with Dorr process operation using sulfuric acid?
    (b) If the company were able to apply a by-product credit of $40/tonne for Haifa-coproduced calcium chloride, and was not able to sell the

Dorr coproduced gypsum, what hydrochloric acid "breakeven" value should be assigned? (That is, what HCl price would now still be competitive with Dorr acid under these circumstances?)

4. A manufacturer has to choose between by-product anhydrous hydrochloric acid and 100% sulfuric acid available at $50 and $40 per tonne, respectively, for acidulation of fluorapatite ($CaF_2 \cdot 3Ca_3(PO_4)_2$) in order to produce phosphoric acid.

(a) What would be the respective acid costs for each of these alternatives per tonne of 100% equivalent phosphoric acid, assuming 100% conversions and yields?

(b) If there was a market for by-product calcium chloride at $20/tonne, and no market for gypsum, by what dollar value per tonne of 100% phosphoric would the costs of the hydrochloric acid-based process be decreased by the sale of the by-product?

5. A company has to make the decision between using Haifa technology and Dorr technology to build a new phosphoric acid plant. It has access to an extensive source of by-product hydrochloric acid at $50/tonne (100% basis).

(a) Assuming quantitative reactions in both cases and no solvent losses, at what price would sulfuric acid have to be available for the company to be in a "breakeven" raw material cost situation?

(b) If the company were able to obtain a price of $30/tonne for the by-product Haifa calcium chloride and was not able to sell the gypsum if it operated the Dorr process, what would be the "breakeven" price of sulfuric acid in this case? In other words, what would the sulfuric acid price have to be in this case to give an equivalent cost phosphoric acid product?

6. (a) Calculate the masses of gypsum ($CaSO_4 \cdot 2H_2O$) and calcium chloride which would theoretically be obtained from fluorapatite using the Dorr and the Haifa processes, respectively, per tonne of phosphoric acid produced (100% basis).

(b) How would the nature of the by-product produced in each case affect the environmentally acceptable disposal options that a phosphoric acid producer has, when disposal is land based and when disposal can take place to sea?

7. The industrial hygiene limit for fluoride-containing dusts is set at 2.5 mg/m^3, and for hydrogen as 3 ppm (Section 10.4.4).

(a) Can both of these limits be expressed as ppm (meaning by volume)? If so, how? If not, why not?

(b) What would 3 ppm (by volume) hydrogen fluoride be equivalent to in units of mg/m^3?

8. Assuming complete immiscibility of toluene and water, the partition coefficient K_p for diethylamine is 0.48 at 18°C, 0.63 at 25°C, and 0.90 at 30°C.

(a) Which temperature would give the optimum water extraction of diethylamine from toluene?

(b) Using single stage extraction of 0.080 M diethylamine in 100 L of toluene with 5.0 L of water at 25°C, what molar concentrations of diethylamine would result in each of the two solvents?

(c) Name any three variables which could improve the extraction of diethylamine by water over that obtained in 8(b), and indicate the direction of the change in the variable necessary to do this.

9. The distribution coefficient of phosphoric acid in n-butanol and water increased from 0.0516 to 0.177 wt.% when the initial concentration of phosphoric acid in water was raised from 4.55 to 25.9 wt.% (I.N. Shokin *et al., Zhur Prikl, Khin*, **35**, 190–191, (1962)).

(a) What would be the distribution coefficients for each of these two examples when the more common molar concentration basis is used? Densities, etc., are: 4.55 wt.%, 1.025 kg/L, 0.470 M; 25.9 wt.%, 1.147 kg/L, 2.925 M.

(b) Which initial concentration was extracted more efficiently? Explain.

FURTHER READING

Y.C. Athanassiaclis, "Air Pollution Aspects of Phosphorus and its Compounds," NTIS PB-188–073. Washington, DC, 1970.

G.L. Bridger, C.B. Drees, and A.H. Roy, New Process for Production of Purified Phosphoric Acid and/or Fertilizer Grade Dicalcium Phosphate from Various Grades of Phosphatic Materials. *Ind. Eng. Chem. Process Des. Dev.* **20**, 416 (1981).

M.S. Silverstein, G.F. Nordblum, C.W. Dittrich, and J.J. Jakabein, Stable Red Phosphorus, *Ind. Eng. Chem.* **40**, 301 (1948).

S. Skolnik, G. Tarbutton, and W.E. Bergman, Conversion of Liquid White Phosphorus to Red Phosphorus. *J. Am. Chem. Soc.* **68**, 2310 (1946).

L.K. Thompson, S.S. Sidhu, and B.A. Roberts, Fluoride Accumulations in the Soil and Vegetation in the Vicinity of a Phosphorus Plant. *Environ. Pollut.* **18**(3), 221 (1979).

E.N. Walsh, ed., "Phosphorus Chemistry: Developments in American Science." American Chemical Society, Washington, DC, 1992.

REFERENCES

1. J.R. Van Wazer, ed., "Phosphorus and Its Compounds," Vol. 2, pp. 957, 1025, 1158. Interscience, New York, 1961.
2. E.J. Griffith, A. Beeton, J.M. Spencer, and D.T. Mitchell, eds., "Environmental Phosphorus Handbook," p. 24. Wiley, New York, 1973.
3. "United Nations Statistical Yearbook 2001," 48th ed. United Nations, New York, 2003, and earlier editions.
4. S.M. Jasinsky, Phosphate Rock 2001. In "Minerals Yearbook. US Geol. Survey." Available: http://minerals.usgs.gov/minerals/pubs/commodity/phosphate_rock/index.html#myb; & "Phosphate," *Miner. Bull.* MR 160. Energy Mines and Resources Canada, Ottawa, 1976.
5. D.W. Leyshon and R.T. Schneider, "The Evaluation of Phosphate Rock," 2nd Chemical Congress of the North American Continent, Las Vegas, NV, Abstr., FERT 26 (1980).
6. "Mineral Facts and Problems," 1980 ed., Bull. No. 671, p. 663. U.S. Bureau of Mines, Washington, DC, 1980.
7. J.A. Kent, ed., "Riegel's Industrial Chemistry," p. 652. Reinhold, New York, 1962.

8. F.S. Taylor, "A History of Industrial Chemistry," Reprint ed., p. 203. Arno Press, New York, 1972.

9. P. Ellwood, *Chem. Eng.* **72**, 54 (1965).

10. "Kirk-Othmer Encyclopedia of Chemical Technology," 2nd ed., Vol. 15, pp. 257–276. Interscience, New York, 1968.

11. R.C. Weast, ed., "Handbook of Chemistry and Physics," 51st ed., p. B-117. Chem. Rubber Publ. Co., Cleveland, OH, 1970.

12. D.R. Idler, G.L. Fletcher, and D.F. Addison, "Effects of Yellow Phosphorus in the Canadian Environment." National Research Council of Canada, Ottawa, 1981.

13. R. Thompson, ed., "Industrial Inorganic Chemicals: Production and Uses," pp. 373–390. The Royal Society of Chemistry, Cambridge, UK, 1995.

14. Super Problems Belabor Super Plant, *Can. Chem. Process.* **55**(10), 64, Oct. (1971).

15. F. Leder, W.C. Park, P.-W Chang *et al.*, New Process for Technical Grade Phosphoric Acid. *Ind. Eng. Chem. Proc. Des. Dev.* **24**, 688–697 (1985).

16. F.A. Lowenheim and M.K. Moran, "Faith, Keyes and Clark's Industrial Chemicals," 4th ed., p. 641. Wiley-Interscience, New York, 1975.

17. "Kirk-Othmer Encyclopedia of Chemical Technology," 4th ed., Vol. 18, pp. 733–735. Wiley, New York, 1996.

18. Production: Some Improvement, *Chem. Eng. News*, **82**(27), 49–60, July 4 (2004), and same feature, June issues of earlier years.

19. R.J. Lewis, Sr., Sax's Dangerous Properties of Industrial Materials, 8th ed., Vol. 3, pp. 2791–2792. Van Nostrand Reinhold, New York, 1992.

20. D. Rose and J.R. Marier, "Environmental Fluoride." National Research Council of Canada, Ottawa, 1977.

21. J.C. Barber and T.D. Farr, *Chem. Eng. Prog.* **66**(11), 56, Nov. (1970).

22. R.D. Willis, W.D. Ellenson, and T.L. Conner, Monitoring...of Particulate Matter Near a Large Phosphorus Production Facility, *J. Air Waste Manag. Assoc.* **51**(8), 1142–1166 (2001).

23. J.C. Barber, *Chem. Eng. Prog.* **65**(6), 70, June (1969).

24. D.R. Idler, *Chem. Can.* **21**(11), 16, Dec. (1969).

25. J.C. Barber, Incorporation of Phosphorus-containing Water (Phossy Water) in...Fertilizers. Abstr. of Papers of the *Amer. Chem. Soc.* **220**, 21-FERT, Part 1, Aug. 20 (2000).

26. W.J. Schipper, A. Klapwijk, B. Potjer *et al.*, Phosphate Recycling in the Phosphorus Industry, *Environ. Technol.* **22**(11), 1337–1345, Nov. (2001).

27. S.V. Panchenko,...Analysis of the Energy-saving Potential in...Phosphorus Production, *Theor. Foundations Chem. Eng.* **38**(5), 538–544, Sept.–Oct. (2004).

28. J.Q. Hardesty and L.B. Hein, in Reigel's "Handbook of Industrial Chemistry" (J.A. Kent, ed.), 7th ed., p. 537. Van Nostrand-Reinhold, Toronto, 1974.

29. G.S.G. Beveridge and R.G. Hill, Phosphoric Acid, *Chem. Process. Eng.* (N.Y.) **49**(7), 61, July (1968).

30. J.F. Coykendall, E.F. Spencer and O.H. York, *J. Air. Pollut. Control Assoc.* **18**(5), 315 (1968).

30a. R.N. Shreve and j.A.Brink, Jr., "Chemical Process Industries," 4thed., pp. 253, 256, McGraw-Hill, New York, 1977.

31. Acid's value enhanced, *Chem. Eng.* (N.Y.). **71**(1), 26, Jan. 6 (1964).

32. G. Kleinman, *Chem. Eng. Prog.* **74**(11), 37, Nov. (1978).

33. J.F. McCullough, Phosphoric Acid Purification, *Chem. Eng.* (N.Y.) **83**, 101, Dec. 6 (1976).

34. G.S.G. Beveridge and R.G. Hill, Phosphoric Acid, *Chem. Process. Eng.* (N.Y.) **49**(8), 63, Aug. (1968).

35. S.V. Dorozhkin,...Wet-process Phosphoric Acid Production. 2. Kinetics and Mechanism of CaSO4.0.5H2O...Crystallization..., *Ind Eng. Chem. Res.* **36**, 467–73, Feb. (1997).

36. S.V. Dorozhkin,...Kinetics of...Phosphate Rock Dissolution, *Ind. Eng. Chem. Res.* **35**, 4328–35, Nov. (1996).

37. S.V. Dorozhkin, Ecological Principles of Wet-process Phosphoric Acid..., *J. Chem. Technol. Biotechnol.* **71**, 227–233 (1998)

38. TVA Displays Energy Saving, *Chem. Eng. News*, **56**(46), 32, Nov. 13 (1978).

39. P.L. Olivier, Jr., *Chem. Eng. Prog.* **74**(11), 55, Nov. (1978).

40. H. El-Shall, E.A. Abdel-Aal, and B.M. Moudgil, Effect of Surfactants on Phosphogypsum Crystallization and Filtration...During Wet-process Phosphoric Acid Production, *Sep. Sci. Technol.* **35**(3), 395–410 (2000).

41. C. Earl, A. Davister, and F. Thirion, High Strength Phosphoric Acid Process. *Chem. Eng. Prog.* **82**, 34–37, Oct. (1986).

42. G.L. Bridger and A.H. Roy, *Chem. Eng. Prog.* **74**(11), 62, Nov. (1978).

43. SO$_2$ to Process Phosphate Rock, *Chem. Eng. News* **54**, 32, Sept. 6 (1976).

44. J. Bergdorf and R. Fischer, *Chem. Eng. Prog.* **75**(11), 41, Nov. (1978).

45. M. Feki, M. Fourati, M.M. Chaabouni, and H.F. Ayedi, Purification of Wet Process Phosphoric Acid by Solvent Extraction. *Can. J. Chem. Eng.* **72**, 939–944 (1994).

46. W.E. Rushton, *Chem. Eng. Prog.* **74**(11), 52, Nov. (1978).

47. A.P. Kouloheris, *Chem. Eng.* (*N.Y.*). **87**, 82, Aug. 11 (1980).

48. Uranium Recovery from Phosphoric Acid, *Chem. Eng. News*, **55**(42), 17, Oct. 17 (1977); **57**(42), 10, Oct. 15 (1979).

49. J.P. Bolivar, R. GarciaTenorio, and M. GarciaLeon,...Fractionation of Natural Radioactivity in [Wet Process]...Phosphoric Acid..., *Radioanal. Nucl. Chem.-Lett.* **214**(2), 77–88, Sept. 16 (1996).

50. Patty's "Industrial Hygiene and Toxicology," 3rd ed., Vol. 2B, pp. 2950–2951, G.D. Clayton and F.E. Clayton. eds. Wiley-Interscience, New York, 1981.

51. A.W. Frazier, J.R. Lehr, and E.F. Dillard, *Environ. Sci. Technol.* **11**(10), 1007 (1977).

52. C. Djololian and D. Billaud, *Chem. Eng. Prog.* **74**(11), 117, Nov. (1969).

53. P.S. O'Neill, Calcium Fluoride Production in a phosphoric Acid Plant, *Ind. Eng. Chem., Prod. Res. Dev.* **19**, 250 (1980).

54. D. Kitchen and W.J. Skinner, *J. Appl. Chem. Biotechnol.* **21**(2), 53, 56, Feb. (1971).

55. A.V. Slack, ed., "Phosphoric Acid," Vol. 1, Part I, p. 275, and Part II, p. 747. Dekker, New York, 1968.

56. R. Lutz, Preparation of Phosphoric-acid Waste Gypsum for...Building-materials, *Zem.-Kalk-Gips* **47**(12), 690, Dec. (1994).

57. A.J. Poole, D.J. Allington, A.J. Baxter *et al.*, The...Radioactivity of Phosphate Ore...Products Discharged into the Eastern Irish Sea..., *Sci. Total Environ.* **173**(1–6), 137–149, Dec. 1 (1995).

58. R. Blumberg, in "Solvent Extraction Reviews," (Y. Marcus, ed.), Vol. 1, p. 93. Dekker, New York, 1971; R. Blumberg, D. Gonen, and D. Meyer, in "Recent Advances in Liquid–Liquid Extraction" (C. Hanson, ed.), p. 93. Pergamon, Oxford, 1971.

59. Dow to Test Phosphoric Acid Process, *Chem. Eng. News*, **41**(41), 35, Oct. 14 (1963).

60. I. Raz, *Chem. Eng.* (*N.Y.*). **83**, 52, June 10 (1974).

61. R.M. Diamond and D.G. Tuck, in Progress in Inorganic Chemistry, Vol. 2, F.A. Cotton, ed., p. 109. Interscience, New York, 1960.

62. R.H. Perry, ed., "Chemical Engineers Handbook," pp. 14–45. McGraw-Hill, New York, 1969.

63. D.F. Rudd, G.J. Powers, and J.J. Siirola, "Process Synthesis," p. 141. Prentice-Hall, Englewood Cliffs, NJ, 1973.

64. Hydrochloric-based Route to Pure Phosphoric Acid, *Chem. Eng.* (*N.Y.*). **69**(52), 34, Dec. 24 (1962).

65. A. Baniel, R. Blumberg, A. Alon *et al.*, The I.M.I Phosphoric Acid Process, *Chem. Eng. Prog.* **58**(11), 100, Nov. (1962).

66. Israelis Pioneer New Route, *Chem. Eng.* (*N.Y.*) **69**(12), 88, June 11 (1962).

67. A.E.A. Eid, A.A. Hanna, N.S. Youssef *et al.*,...Solvent Extraction of Phosphoric Acid from Phosphate Ores-HCl Digestion, *Asian J. Chem.* **11**(1, 2) 213–224, and 595–605 (1999).

68. New Life Seen for Phosphoric Acid Process. *Chem. Eng. Prog.* **89**, 17, Jan. (1993).

69. "Statistical Yearbook of China." Hong Kong, 1981.

70. Recovery Slows for Large Volume Acids, *Chem. Eng. News*, **62**(24), 10–12, June 11 (1984).

11
AMMONIA, NITRIC ACID AND THEIR DERIVATIVES

...whoever could make two ears of corn, or two blades of grass, to grow upon a spot of ground where only one grew before, would...do more essential service to this country, than the whole race of politicians put together.
—Jonathan Swift (1667–1795)

It has long been known...as...volatile alkali, hartshorn, spirit of sal ammoniac...at present generally by the name of ammonia.
—John Dalton, 1810

11.1. AMMONIA

11.1.1. Background

Early samples of ammonia were obtained from incidental bacterial action on the urea present in urine (Eq. 11.1), or by the dry distillation of protein-containing substances such as bone, horns, and hides.

$$H_2N-CO-NH_2 + H_2O \xrightarrow{\text{bacteria}} CO_2 + 2\,NH_3 \qquad 11.1$$

Small amounts were also obtained as a by-product from the manufacture of coke or coal gas from coal. This source still provides about 1% of current American ammonia [1] (Table 11.1). Today most ammonia is produced by the direct combination of nitrogen and hydrogen.

Other synthetic processes to ammonia have been tested and found to be impractical for large scale commercial exploitation. For example, the French Serpek process for purifying alumina also formed aluminum nitride as a by-product from the heating of bauxite in nitrogen [5] (Eq. 11.2).

$$Al_2O_3 + N_2 + 3\,C \xrightarrow{1,600°C} 2\,AlN + 3\,CO \qquad 11.2$$

321

TABLE 11.1   Production of Synthetic Ammonia by Selected Areas

	Thousands of tonnes[a]			
	1970	**1980**	**1990**	**2000**
Canada	1,220	2,556	3,602	4,888
China	–	12,150	21,251	33,637
France	1,618	2,085	1,478	1,516
W. Germany	–	2,040	1,616	2,530[b]
Italy	1,277	1,714	1,482	n/a
Japan	3,261	2,573	1,831	1,715
U.S.A.	12,541	17,262	15,390	1,490
U.S.S.R.	7,638	16,732	19,990	n/a
World	(62,500)			132,400

[a]Production *capacities* given in parentheses. Data compiled from Kirk-Othmer [2], *Chemical and Engineering News* [3], *Canadian Chemical Processing* [4], and data supplied by the Verband der Chemischen Industrie, e.V. (VCI).
[b]After reunification.

The aluminum nitride could then be hydrolyzed with water to yield ammonia and aluminum hydroxide (Eq. 11.3).

$$AlN + 3\ H_2O \rightarrow NH_3 + Al(OH)_3 \qquad\qquad 11.3$$

The aluminum hydroxide could subsequently be converted to alumina (Al_2O_3) in pure form for aluminum production.

In 1898 Frank and Caro in Germany developed the cyanamide route, in which very high temperatures were used to fix elemental nitrogen to obtain calcium cyanamide from calcium carbide (Eq. 11.4). Hydrolysis of calcium cyanamide then produced ammonia and calcium carbonate (Eq. 11.5). For many years the cyanamide produced predominantly went into ammonia synthesis.

$$CaC_2 + N_2 \xrightarrow{\ 900-1,000°C\ } CaNCN + C \qquad\qquad 11.4$$

$$CaNCN + H_2O \rightarrow 2\ NH_3 + CaCO_3 \qquad\qquad 11.5$$

Direct synthesis from the elements has now displaced cyanamide hydrolysis to produce ammonia.

Ammonia was probably first synthesized directly from its elements in the laboratory by Dobereiner in 1823, as a secondary result of experiments designed to study the direct combination of gaseous hydrogen and oxygen [6]. However, it was not until the early 1900s, as Nernst developed the theory by which equilibrium concentrations could be estimated from thermochemical calculations, and Haber proved the direct reaction of nitrogen and hydrogen on a laboratory scale that there was any promise of adoption of this process on an industrial scale developed. An early bench scale reactor produced about 80 g of ammonia/hour. Later, in collaboration with Badische Anilin-und Soda-Fabrik (BASF), this was rapidly scaled up to plants operating in Germany, which produced 27 tonne/day in 1913 and 500 tonne/day by 1918. Haber in 1918 and Bosch in 1931, earned Nobel prizes for their respective contributions to this area of chemistry, and have jointly lent their names to the process, which originated as a result of their work. Today there are many single-train ammonia plants worldwide operating at capacities of 900 tonne/day, and some as large as 1650 tonne/day [2].

11.1.2. Principles of Ammonia Synthesis: The Haber or Haber–Bosch Process

The ammonia synthesis reaction is represented by Eq. 11.6, which establishes the stoichiometry of 2 mol (volumes) of ammonia produced for each 4 mol (volumes) of reacting gases.

$$N_{2(g)} + 3 H_{2(g)} \rightleftharpoons 2 NH_{3(g)} \quad \Delta H = -92.0 \text{ kJ } (-22.0 \text{ kcal}) \qquad 11.6$$

The principle of Le Chatelier states that when a change is placed on a system in equilibrium this will shift so as to reduce the effect of the change (quantities, pressure, temperature, etc.). Table 11.2 shows that, as predicted by the Le Chatelier principle, an increase in the pressure on this system increases the equilibrium ammonia concentration.

The ammonia synthesis reaction is also faster at higher temperatures – equilibrium conditions are reached more rapidly. At the same time, the reaction is exothermic to the extent of 46 kJ mol^{-1} (about 54 kJ mol^{-1} at normal conversion temperatures), so that as the synthesis temperature is raised the ammonia equilibrium is shifted to the left, again in accordance with Le Chatelier's principle (Table 11.2). Thus, while raising the synthesis temperature does give a faster reaction rate, allowing a greater volume of production to be obtained in the same size reactor; it also displaces the equilibrium of the synthesis (Eq. 11.6) to the left, giving a smaller potential conversion of nitrogen to ammonia. Fortunately, a substantial increase in the pressure on the system substantially improves the equilibrium conversion to ammonia to make the whole exercise practical. For these reasons the conditions used by most synthetic ammonia processes tend to cluster around 100–300 atm and

TABLE 11.2 Equilibrium Percent Concentrations of Ammonia at Various Temperatures and Pressures[a]

Temperature, °C	Absolute pressure (atm)						
	1	10	50	100	300	600	1000
200	15.3	50.7	74.4	81.5	89.9	95.4	98.3
300	2.2	14.7	39.4	52.0	71.0	84.2	92.6
400	0.4	3.9	15.3	25.1	47.0	65.2	79.8
500	–	1.2	5.6	10.6	26.4	42.2	57.5
600	–	0.5	2.3	4.5	13.8	23.1	31.4
700	–	–	1.1	2.2	7.3	11.5	12.9

[a]From reacting a gas mixture of 1 mole nitrogen to 3 mole hydrogen under the indicated conditions. Compiled from the data of Haber *et al.* [7], and Larson [8]. Nielson gives more recent and more detailed data for conditions of commercial interest [9].

400–500°C in an attempt to maximize conversions and conversion rates while moderating compression costs. The highest pressure Claude and the Casale ammonia synthesis processes, operate in France at 900 atm and in Italy at 750 atm, respectively. These are among the highest pressure industrial processes. The Claude process relies on small inside diameter (10-cm) bored nickel chrome ingots for containment of the enormous pressures and in return gains conversions to ammonia of 40% or better. More conventional converters are 7 to 10 m high by 3 m in outside diameter and use 20-cm-thick alloy steel walls for pressure containment to give ammonia conversions of 10–20%.

Successful ammonia conversion required discovery of a catalyst, which would promote a sufficiently rapid reaction at 100–300 atm and 400–500°C to utilize the moderately favorable equilibrium obtained under these conditions. Without this, higher temperatures would be required to obtain sufficiently rapid rates, and the less favorable equilibrium at higher temperatures would necessitate higher pressures as well, in order to obtain an economic conversion to ammonia. The original synthesis experiments were conducted with an osmium catalyst. Haber later discovered that reduced magnetic iron oxide (Fe_3O_4) was much more effective, and that its activity could be further enhanced by the presence of the promoters alumina (Al_2O_3; 3%) and potassium oxide (K_2O; 1%), probably from the introduction of iron lattice defects. Iron with various proprietary variations still forms the basis of all ammonia catalyst systems today.

Another measure of the efficiency of ammonia conversion is the "space velocity" which may be used. Space velocity refers to the volume of reactants fed to a reactor per hour, divided by the volume of the reactor. For liquid reaction streams this relationship is straightforward. For gases, however, the space velocity is defined as being the volume of gases corrected to 0°C and 760 mm Hg (1 atm) passing through the reactor (or catalyst) volume/hour. This amounts to a measure of the gas–catalyst contact time for heterogeneous reaction (Eq. 11.7).

$$\text{space velocity} = \frac{\text{volumetric feed rate}}{\text{volume of the reactor}} \qquad 11.7$$

$$\text{space velocity} = \frac{m^3 \text{ gas(es) hr}^{-1}}{m^3 \text{ catalyst volume}}$$

For a doubly promoted magnetic iron oxide catalyst, working on a nitrogen: hydrogen mole ratio of 1:3 at a pressure of 100 atm and 450°C, and a space velocity of 5,000(hr^{-1}), approximately 13–15% ammonia could be expected [10]. Much lower conversions are obtained from singly promoted or nonpromoted iron.

A space velocity of 5,000 hr^{-1} (corrected to 0°C, 760 mm) corresponds to a gas exchange rate in the catalyst space of 132.4 m^3 of gas/cubic meter of catalyst space/hour at operating conditions (450°C, 100 atm; Eq. 11.8), or an actual gas–catalyst contact time of about 27 sec. Increasing the space velocity simultaneously decreases the gas–catalyst contact time and the percentage of nitrogen and hydrogen converted to ammonia.

Space Velocity Relationships:

$$\text{space velocity} \propto \frac{1}{\text{contact time}} \qquad\qquad 11.8$$

Thus, $5,000\,\mathrm{hr}^{-1}$ space velocity is equivalent to a contact time of $1/(5,000\,\mathrm{hr}^{-1})$, or $2 \times 10^{-4}\,\mathrm{hr}$, or 0.72 ($2 \times 10^{-4}\,\mathrm{hr} \times 3,600\,\mathrm{sec/hr}$) sec contact time, if gases were at $0°C$ ($273\,\mathrm{K}$) and 1 atm.

At constant space velocity then,

$$\text{contact time} \propto \text{pressure}.$$

Thus changing only the pressure to 100 atmospheres raises the contact time to $100 \times 0.72\,\mathrm{sec}$, or $72\,\mathrm{sec}$.

$$\text{contact time also} \propto \frac{1}{\text{absolute temperature}}$$

Thus, simultaneously raising the temperature has the effect of decreasing the actual contact time to

$$72\,\mathrm{sec} \times \frac{1}{(723\,\mathrm{K}/273\,\mathrm{K})} = 22.19\,\mathrm{sec} \text{ or } 27.2\,\mathrm{sec} \text{ contact}$$

time, under converter operating conditions.

Gas changes per hour under converter operating conditions, therefore, will be:

$(3,600\,\mathrm{sec/hr})/(27.19\,\mathrm{sec/change}) = 132.4$ changes per hour.

Doubling the space velocity of a converter may well drop the ammonia conversion percentage to about 10%, which may appear to be a change in the wrong direction. However, if 10% of twice the original volume of reacting gases passing through the converter are converted, this still means that the volume of ammonia recovered is roughly one-third more or $1000\,\mathrm{m}^3/\mathrm{hr}$ NH_3 (10% of 10, 000 m^3) rather than $750\,\mathrm{m}^3/\mathrm{hr}$ NH_3 (15% of 5000 m^3) as obtained at $5,000\,\mathrm{hr}^{-1}$ space velocity. Hence, high space velocities of $10,000$–$20,000\,\mathrm{hr}^{-1}$ or more are common for commercial ammonia converters.

For this combination of reasons, most large ammonia plants have to recycle a large proportion of the converter exit gases in the form of unreacted nitrogen and hydrogen. Thus, a significant component of the engineering of a modern ammonia synthesis unit is concerned with pressure and heat conservation to enable this recycle to be conducted as economically as possible.

11.1.3. Feedstocks for Ammonia Synthesis by Air Distillation

Nitrogen for ammonia synthesis may be obtained either by separation from air by liquid air distillation, or by consumption of the oxygen of air by the burning of a fuel in an air-restricted oxidative process, which leaves a nitrogen residue.

The separation of nitrogen from liquid air is done by cryogenic, or low-temperature, technology. Initial steps involve dust removal, usually by filtration, and removal of the normal carbon dioxide present in air (about

370 ppm), either by scrubbing with an organic base such as monoethanolamine (MEA) or with aqueous alkali (Eqs. 11.9–11.13).

$$CO_2 + H_2O \rightarrow H_2CO_3 \qquad\qquad 11.9$$

MEA absorption:

$$H_2CO_3 + H_2NCH_2CH_2OH \xrightarrow[\text{ambient temp.}]{\text{high pressure}} HCO_3^- + H_3\overset{+}{N}CH_2CH_2OH$$

$$11.10$$

Regeneration (separate unit):

$$HCO_3^- + H_3\overset{+}{N}CH_2CH_2OH \xrightarrow[\text{heat}]{\text{low pressure}} H_2NCH_2CH_2OH + H_2O + CO_2$$

$$11.11$$

Alkali absorption:

$$H_2CO_3 + NaOH \rightarrow NaHCO_3 + H_2O \qquad 11.12$$

$$H_2CO_3 + 2\,NaOH \rightarrow Na_2CO_3 + 2\,H_2O \qquad 11.13$$

Purification is an important preliminary step, particularly in operating areas near a refinery or petrochemical complex, since accumulation of combustible dusts or condensed vapors such as methane or acetylene in cryogenic lines containing liquid oxygen introduces serious operating risks. And, of course, the temperatures in the region of $-200°C$ required to liquefy air are sufficiently low to cause carbon dioxide to solidify, which can plug operating components. After purification, the air is normally dried to as low a dew point as feasible using towers packed with an adsorbent such as activated alumina, a molecular sieve (e.g., Union Carbide type 13X), or silica gel. This helps to avoid problems from water vapor. Two or more drying towers operated on swing cycles provide continuous water adsorption capacity from one tower, while the other is regenerated by heating the adsorbent under reduced pressure.

Liquefaction of the purified air is accomplished using the Joule–Thompson effect, which is the cooling effect obtained from a compressed gas when it is allowed to expand. By using this expansion-cooling effect repetitively, and by employing the chilled expanded gas to prechill the compressed gas before expansion, air may be liquefied by employing compression pressures of only about 10 atm (about 150 psig, Fig. 11.1). It is not possible to accomplish

FIGURE 11.1   Simplified flowsheet for the separation of nitrogen from liquid air.

liquefaction of air by direct compression alone at temperatures above the critical temperatures of the component gases, 126.1 K ($-147.1°C$) for nitrogen and 154.4 K ($-118.8°C$) for oxygen. Even at these temperatures, pressures of 33.5 and 49.7 atm, respectively, would be required for condensation. However, the technique of repeated compression, then external cooling of the compressed gases to well below their critical temperatures makes expansion-cooling in a well-insulated system sufficient to liquefy a part of both gases as they approach their atmospheric pressure boiling points, which are 74.8 K ($-195.8°C$) and 90.2 K ($-183.0°C$) for nitrogen and oxygen, respectively.

On fractionation (fractional distillation) of liquid air, nitrogen distills first, as it is the lower boiling constituent. The oxygen fraction is frequently vented, unless a local petrochemical or metallurgical application can use it. Liquid air distillations are commonly conducted at a pressure of about 5 atm, which both raises the column operating temperatures and increases the difference in boiling points for the component gases to be separated, improving separation efficiency. This is particularly useful if argon (0.93% in air) recovery is practiced. With a boiling point of 87.5 K ($-185.7°C$), argon would be lost with the oxygen stream without these precautions. Incidentally, the close boiling points of argon and oxygen represent physical evidence of the monoatomic versus diatomic existence of these gases; argon with an atomic and molecular weight of 39.95, and oxygen with an atomic weight of 16.00 and molecular weight 32.00.

When liquid air distillation is used as the source of nitrogen, the hydrogen also required for ammonia synthesis is obtained from a variety of sources. Some is obtained as the coproduct from the electrolytic production of chlorine and caustic soda (Chap. 8), some from refinery sources as a by-product of cracking processes or olefin synthesis, some from the water–gas reaction, and some is produced specifically for the purpose by the electrolysis of alkaline water (e.g., by Cominco, Trail, Eq. 11.14).

$$2 \text{ H}_2\text{O} + \underset{\substack{\text{ca. 5 molar} \\ \text{or 26\%}}}{\text{KOH}} \xrightarrow{\text{electrol.}} 2 \text{ H}_2 + \text{O}_2 + \text{KOH} \qquad 11.14$$

Despite the apparently straightforward nature of these procedures to the feedstocks for ammonia synthesis and the "free" source of nitrogen from the air, this combination of approaches has become feasible only for relatively small ammonia plants of around 100 tonne/day, or for special process situations where abundant hydrogen is available. Today, ammonia plants of capacities of 900 tonne/day are common and a few as large as 1650 tonne/day are now operating worldwide. Thus it can be seen that the process sequence described above is not a major contributor to world ammonia production.

11.1.4. Ammonia Feedstocks by Reforming and Secondary Reforming

Hydrogen on a sufficient scale for ammonia synthesis is accessible from coal (indirectly), petroleum, natural gas, and water, and all of these sources are used to some extent. Even biomass has recently been tested [11]. In the original Haber–Bosch process, for example, coke derived from coal was used as the source of hydrogen via the water gas reaction. To do this, initially

■■■ **TABLE 11.3 Relationship Between the Thermally-induced Color of a Substance and its Temperature[a]**

Color of emission	Temperature (°C)	Applications
Dark red	585–870	Al melts at 660°, soft glass at ca. 700°, potter decorating kilns, reformer tubes
Cherry red	880–980	Ag melts at 962°, fired red clay products
Orange red	990–1,090	Au melts at 1064°, Cu at 1083°
Yellow	1,100–1,220	Lime kiln, cement kiln operation
White	1,230–1,280	Lime kiln, cement kiln operation
Brilliant white	1,290–1,520	Ni melts at 1453°, firing of porcelain, spark plug and power line insulators
Dazzling white	1,540–1,800	Fe melts at 1535°, firing temp. of silica bricks and fire brick

[a]Temperature relations selected from Lange [12].

the coke is burned in air until it reaches red heat (Table 11.3), and then the air feed stream is replaced by steam (Eqs. 11.15 and 11.16).

Combustion phase:

$$C + air \rightarrow CO_2 + 4\ N_2 + heat\ (air\ is\ approx.\ O_2{:}4\ N_2) \qquad 11.15$$

Water gas reaction:

$$C + H_2O \xrightarrow[\text{ca. } 1000°C]{} CO + H_2\ (positive\ \Delta H) \qquad 11.16$$

The hot coke in an endothermic reaction produces mainly carbon monoxide and hydrogen, and cools down. The process is periodically switched to combustion air to regain the high coke temperatures necessary for the water splitting reaction. Since carbon monoxide is relatively difficult to separate from hydrogen, and also because it is possible to obtain a further mole of hydrogen from each mole of carbon monoxide, the whole water gas stream is normally put through a shift conversion step (Eq. 11.17).

Shift conversion:

$$CO_{(g)} + H_2O_{(g)} \xrightarrow[\sim 500°C]{FeO + Cr_2O_3} CO_{2(g)} + H_{2(g)} \qquad 11.17$$
$$\Delta H = -41.2\ kJ\ (-9.8\ kcal)$$

Here, the carbon monoxide is converted to carbon dioxide at more moderate temperatures in the presence of a chromium-promoted iron catalyst. Carbon dioxide is readily separated from the hydrogen by absorption in water under pressure (Eq. 11.18).

$$CO_{2(g)} + H_2O_{(\ell)} \xrightarrow{25\ atm} H_2CO_{3(\ell)}\ \ \Delta H = -38.5\ kJ\ (-9.2\ kcal) \qquad 11.18$$

This leaves a hydrogen stream ready for ammonia synthesis, after a final cleanup. The nitrogen required is normally taken from a part of the gas stream obtained from the heating, coke combustion phase of the water–gas reaction sequence, which is depleted in oxygen. Carbon dioxide removal from the nitrogen feed component is achieved in the same manner as from the hydrogen.

However, where natural gas supplies are reasonable in cost, an alternative reforming and secondary reforming sequence is preferred for ammonia feedstock production [13]. Coal-based ammonia requires nearly twice as much capital investment as a plant based on natural gas, and environmental problems are often greater with coal (see also Section 11.1.7 on new developments). Processing of natural gas via reforming and secondary reforming amounts to combustion of methane in the presence of steam with a deficiency of air to give a mixture of hydrogen, nitrogen, and carbon dioxide.

Natural gas reforming can use any kind of petroleum feedstock, not just natural gas, although gas is favored because it has the highest hydrogen content and is the easiest to clean prior to use. Desulfurized natural gas only requires passage through a bed of activated alumina or bauxite for purification by absorption. Use of either gas oil or a residual fuel oil for synthesis of gas generation requires more stringent precautions to remove sulfur since it can permanently poison (decrease the activity) the ammonia synthesis catalyst system [14]. Copper, arsenic, and phosphorus are also serious catalyst poisons.

Regardless of the source of the petroleum, the chemistry to obtain ammonia feedstocks is similar. Since methane is of dominant importance this is used as an example to describe the steps required. Initially methane is mixed with steam and passed into heat resistant nickel-chromium-iron alloy tubes containing a supported nickel catalyst. The tubes are heated externally by a further portion of methane consumed as fuel (Fig. 11.2, Eq. 11.19).

Primary steam reforming:

$$CH_{4(g)} + H_2O_{(g)} \xrightarrow[\text{400 psi, 800°C}]{\text{Ni catalyst}} CO_{(g)} + 3\,H_{2(g)} \qquad 11.19$$

$$\Delta H = +208\ kJ\ (+49.7\ kcal)$$

FIGURE 11.2 Outline of the main components of an ammonia synthesis plant using reforming and secondary reforming as the principal sources of hydrogen. Electrolysis of water is used to supplement this.

This reaction requires continued external combustion of methane to maintain the red heat (Table 11.3) of the reformer tubes, since this is an endothermic reaction. It produces most of the hydrogen required.

Air, methane, and any further hydrogen available are then added to the exit gas stream from the steam reformer in the appropriate proportions to provide all of the eventual nitrogen requirement for ammonia synthesis (Eq. 11.20).

Secondary air reforming:

$$CH_{4(g)} + air_{(g)} \xrightarrow[950°C]{Ni\ catalyst} CO_{(g)} + 2\ H_{2(g)} + 2\ N_{2(g)} \qquad 11.20$$

Air is approx. $1/2\ O_2 : 2\ N_2$. $\Delta H = +35.6$ kJ ($+8.5$ kcal)

This reaction is also endothermic and requires external fuel combustion to maintain the high temperatures. About one-third of the methane consumed up to this stage is used to heat the primary steam, and secondary air reformers.

As with the water gas product of the Haber–Bosch route to hydrogen, each mole of carbon monoxide produced by the two reforming reactions may be converted to a mole of hydrogen via shift conversion (Eq. 11.17). The coproduced carbon dioxide is also more readily removed than carbon monoxide from the product gases, hydrogen and nitrogen (Eq. 11.18). To accomplish this, water is sprayed into the very hot gas mixture emerging from the air reformer to bring the temperature down to the 400°C range for shift conversion, which also provides the steam necessary for this step. Since this step is exothermic it is usually carried out in two separate units to permit better temperature control of the process. Also, once started, it does not require external heat input. At a space velocity of about 100, carbon monoxide conversion to carbon dioxide is achieved to give a residual carbon monoxide concentration of less than 1% [15].

The proportions of the two reforming reactions and shift conversion are so controlled that the gas mixture obtained contains nitrogen and hydrogen in the mole ratio (volume ratio) of 1:3. However, this mixture still contains 20–30% carbon dioxide resulting from the shift conversion reaction and traces of unconverted carbon monoxide. Carbon dioxide can yield carbonates and carbamates in the ammonia synthesis cycle, which are undesirable because they can deposit in the piping. In addition oxygen, and any of its compounds such as carbon monoxide, water, etc., are also ammonia catalyst poisons [13]. Consequently they must be removed.

Carbon dioxide can be removed by scrubbing the pressurized gas mixture with water (Eq. 11.19). In an ordinary scrubber, 21.6 volumes of gas are absorbed per volume of water at 25 atm, 12°C. However, energy costs for moving the large volumes of water are significant. Also it has been found that 3–5% hydrogen loss occurs at sufficiently high pressures for efficient carbon dioxide removal. To avoid this, more favorable gas liquid equilibria are employed. Systems based on monoethanolamine (e.g., the Girbotol process) are capable of decreasing the carbon dioxide content to less than 50 ppm [16] (Eqs. 11.9–11.11). The potassium carbonate–bicarbonate cycle also permits the absorbing base to be recycled, but can only decrease the carbon dioxide content to 500–1000 ppm (Eqs. 11.21 and 11.22).

Absorption:

$$CO_2 + H_2O + K_2CO_3 \rightarrow 2\,KHCO_3 \qquad\qquad 11.21$$

Regeneration:

$$2\,KHCO_3 \xrightarrow[120°C]{} K_3CO_4 + CO_3 + H_2O \qquad\qquad 11.22$$

Sodium carbonate is a less costly absorbent, but also less efficient. Other acid gas removal techniques use physical absorption of carbon dioxide in organic solvents, such as the dimethyl ether of polyethylene glycol, or methanol (the Selexol or Rectisol processes). For effective absorption these require high gas pressures. Regeneration of solvent is done by pressure letdown plus some air stripping (sparging of air through the solvent).

The traces of residual carbon dioxide and any remaining carbon monoxide are then treated catalytically with hydrogen in a methanation step to remove the last traces of oxygen-containing components (Eqs. 11.23 and 11.24).

Methanation:

$$CO + 3\,H_2 \xrightarrow[315°C]{Ni\ catalyst} CH_4 + H_2O \qquad\qquad 11.23$$

$$CO_2 + 4\,H_2 \rightarrow CH_4 + 2\,H_2O \qquad\qquad 11.24$$

If, however, the ammonia plant is obtaining nitrogen from an air separation plant, carbon monoxide and carbon dioxide (boiling points $-205°C$ and sublimation temperature $-78.5°C$, respectively) may be condensed and removed by scrubbing with liquid nitrogen.

11.1.5. Ammonia Synthesis

The feedstock requirements for ammonia synthesis are the same, whether they are obtained from air separation plus supplementary hydrogen or from a reforming sequence. Nitrogen and hydrogen gas are required in a mole ratio (at ordinary pressures close to a volume ratio) of 1:3, and are raised to very high pressures in the 100–900 atm (10,130–91,200 kPa; 1,500–13,500 psi) range, depending on the synthesis system employed [17]. In selecting the synthesis pressure to be used, operators will weigh the following considerations: at low synthesis pressures, compression energy costs will be lower but there will be lower conversions (more recycle cost); by contrast, at high synthesis pressures, compression costs will be higher but higher conversions will be possible.

Early in the development of ammonia synthesis technology it became apparent that iron and high carbon steels became brittle and underwent intergranular corrosion on exposure to high pressures of hydrogen or ammonia, particularly at high temperatures. This caused a modification of the structure of the steel by a little understood process, and made vessels unsafe to use at high pressures after a short time. This process, termed "hydrogen embrittlement," is now understood to involve adsorption of hydrogen onto the steel, followed by ionization (Eq. 11.25).

$$2\,H_{2(g)} \xrightarrow{\text{adsorption}} 2\,H_{2(ads)} \xrightarrow{\text{ionization}} \underset{(+4e^-)}{4\,H^+} \xrightarrow{C_{(Fe)}} CH_{4(Fe)} \rightarrow CH_{4(g)}$$

$$11.25$$

The hydrogen ions formed diffuse into intergranular carbon and react with it, forming methane bubbles. These internally stress the steel, making it much less ductile and causing intergranular cracks.

Since we have gained a better understanding of hydrogen embrittlement phenomena, the design of the pressure shell of all ammonia converters has incorporated features to avoid weakening by this kind of attack [18]. All converters now use the incoming nitrogen/hydrogen stream to keep the inner wall of the pressure shell cool by passing this entering stream immediately inside the pressure shell, thus eliminating the heat factor contributing to embrittlement (e.g., Fig. 11.3). Methanation of any adsorbed or absorbed hydrogen does not occur when alloy steels incorporating titanium, vanadium, tungsten, chromium, or molybdenum are used, since these metals form very stable carbides. Instrumentation has also been developed to monitor continuously hydrogen penetration rates, which adds a useful advance warning. Wrapped converter designs are used where only the inner layer of about a 1.2-cm-thick plate is made leak tight, and the outer multilayers of tightly wrapped 0.6-cm plate are deliberately perforated with small holes to harmlessly release any hydrogen diffused through the initial layer to provide another safeguard.

Ammonia conversion takes place at 400–500°C. Most of the input gas heating is provided by indirect heat exchange with already converted exit gases via an exchanger inside the pressure shell of the converter (Fig. 11.3). During the fraction of a second contact time, the feed gas stream forms a gas

Quench gas

Pressure shell

Reaction cartridge

Gas inlet H_2, N_2

Radial flow catalyst bed

Heat exchanger

Gas outlet H_2, N_2, NH_3

Bypass gases

FIGURE 11.3 Diagram showing design features of a large-scale ammonia synthesis converter. The quench and bypass inlets for feed gas inputs, by foregoing heat exchange before entering the catalyst beds, provide one means of temperature control on reacting gases. The fraction of "inerts" in the feed gas stream is another (see text).

TABLE 11.4 Equilibrium Ammonia Vapor Concentrations in a 1:3 Nitrogen:Hydrogen Gas Mixture Above Liquid Ammonia at 15°C[a]

Pressure (atm)	Approximate equilibrium ammonia concentrations (%)
50	18
100	9
300	6
600	4
1,000	3

[a]Data from Kirk-Othmer [19].

mixture which contains 10–20% ammonia in residual nitrogen and hydrogen. In this process, the pressure drops slightly from both flow resistance through the catalyst beds, and the slight gas volume decrease incurred by the reaction itself. Catalyst life of up to 15 yr can be achieved with careful feed and recycle gas treatment.

Ammonia is condensed from this gas mixture by cooling the gases at converter exit pressures. This practice has been adopted for several reasons. First, it conserves compressor capacity and recompression energy requirements once the liquefied ammonia has been separated, by maintaining the unreacted nitrogen and hydrogen at nearly the inlet pressure required for recycle to the converter again. Second, ammonia (boiling point −33°C at 1 atm pressure) is far easier to condense out of the gas mixture under high pressures than at atmospheric pressure. For processes operating at pressures of 400 atm or above, ordinary process cooling water at 10–20°C is sufficient to condense the ammonia. It is only at pressures below this that refrigeration may be necessary (Table 11.4). The crude liquid ammonia product is then fed to a high-pressure gas separator, and, following this, to a low-pressure gas separator, and then to product storage spheres or shipping units (Fig. 11.4).

Ammonia may also be recovered from converter exit gases or from the gas separators by absorption into water when ammonia solutions are to be

FIGURE 11.4 Simplified flowsheet of the gas recycle and product recovery systems for ammonia synthesis, from the gas mixture leaving the converter.

marketed. Over 600 volumes of ammonia are absorbed per volume of water at 25°C, 1 atm. Solutions containing 28–30% NH_3 by weight are obtained, which is concentrated ammonium hydroxide (Eq. 11.26).

$$NH_3 + H_2O \rightleftharpoons NH_4OH \qquad\qquad 11.26$$

Before the uncondensed and unreacted gases can be recycled from ammonia recovery, the inerts, mainly argon and methane, have to be bled from the system. Otherwise their concentrations build up, which can dilute the active synthesis gases in the system even to the point of blowout. "Blowout" is obtained when the heat of reaction of low residual concentrations of active gases is inadequate to maintain converter catalyst bed temperatures, and the synthesis reaction fails. Dilution is at times intentionally used to control excessive converter temperatures, for example, when fresh, highly active catalyst has been placed. The purge rate used ultimately determines the concentration of inerts in the system. For instance, if the purge rate corresponds to 1% of the rate of addition of make-up gas, and the make-up gas contains 0.1% inerts, then the steady-state concentration of inerts in the converter feed would be 100 × 0.1%, or 10%. At this purge rate, 0.9 mole of nitrogen and hydrogen are discarded for every 100 moles of nitrogen and hydrogen added. At a purge rate of 10% of the make-up rate, the steady-state concentration of inerts in the converter drops to 1%, but 9.9 moles of nitrogen and hydrogen are discarded for each 100 moles added. The hydrogen purged is not a complete loss since it may be recovered, or used along with the methane in the purge gas to fuel the reformers. However, burned hydrogen is not as productive as converted hydrogen. This conflict of objectives between raising the concentration of inerts fed to the converter versus raising the loss rate of the feed gases as reformer fuel represents another case for process optimization.

Occasionally, an ammonia plant practices argon recovery by distillation of liquefied purge gases, particularly if air separation is used to provide at least a part of the nitrogen [20]. But when ammonia synthesis is associated with air separation, very often most of the argon and methane are removed in the liquid nitrogen scrub for carbon dioxide cleanup, giving very low concentrations of inerts in the make-up gas. Recovered argon is used as an illuminating gas for "neon" (decorative) lighting, as an inert filler for filament light bulbs, as an inert blanketing gas for the shielded welding of reactive metals, and as a filler gas for thermopane windows.

A very slight deviation from a 1:3 nitrogen to hydrogen mole ratio in the make-up gases fed to a converter can gradually build up to a significant difference from this ratio in the converter feed, because of the multiple recycle. The desired ratio is directly stabilized to some extent by the purge rate, and may also be compensated for at regular intervals by adjusting the reformer gas feed rates to control the proportion of nitrogen to hydrogen in the make-up gas.

Liquid ammonia from the low-pressure separator will give an analysis of 99.5+% ammonia. This product is stored in refrigerated, insulated storage tanks. A large volume of ammonia storage capacity is necessary because its major market – fertilizer production – is seasonal. For shipment, the ammonia is fed from storage to insulated tank trucks, tank cars, or barges for delivery.

Markets and production facilities in the southwestern U.S. are so well developed that in this area an extensive pipeline network over 4,000 km in extent is used for anhydrous ammonia deliveries of up to 4,500 tonne/day. Smaller quantities are also shipped as ammonia solutions (ammonium hydroxide), which contain 28–30% ammonia by weight, or in pressurized cylinders.

11.1.6. Major Uses of Ammonia

As noted above, the main use for ammonia is for fertilizer-related applications, which represented some 63% of the U.S. ammonia sales in 1968, climbed to about 80% of the total by 1973, and to 85% by 1990, where it has since remained (Table 11.5). The major proportion of this increase is for direct application of ammonia to the soil, either as the gas flowing under pressure to tractor-drawn knives which discharge it just under the soil surface, or as a 28–30% ammonia solution for smaller agricultural operations. The incentives for this large increase in use of ammonia as a fertilizer are its high nitrogen content, ca. 82%, and the strong adsorption by organic matter and clays in the soil which promote efficient retention. Ammonium nitrate, another popular nitrogen component of fertilizers, consumes ammonia both to make the nitric acid required (Section 11.2) and to react with this to form the salt. Fertilizer nitrogen from ammonium nitrate, however, is losing favor to urea, both in the U.S.A. and in Canada. Ammonium phosphate, ammonium sulfate, and other variations are also popular nitrogen components of fertilizers.

TABLE 11.5 Trends in the Use Pattern for Ammonia in the United States since 1968[a]

Use	Percent of total				
	1968	1973	1983	1990	2000
Fertilizer:					
Ammonia and aqua ammonia	6.2	27.1	25	33.3	32.7
Ammonium nitrate	21.6[b]	18.9		15.1	15.5
Nitric acid	25.0	–	20	15.5	16.5
Urea	10.1	13.9	20	16.1	14.8
Other	–	20.5	15[d]	4.9[e]	5.6
Industrial:					
Plastics, fiber monomers[c]	18.9	8.7	10		
Explosives	[b]	4.0	5	15	15
Miscellaneous	17.8	6.9	5[f]		
Total	99.6	100	100	100	100

[a]Compiled from data of Kirk-Othmer [2], Matasa and Tonca [14], and issues of *Chemical and Engineering News*, the *Chemical Marketing Reporter*, and *Statistical Abstract of the United States*.
[b]Includes fertilizer and explosives applications.
[c]Includes amines produced but not destined for polymer applications.
[d]As ammonium phosphate.
[e]Ammonium sulfate.
[f]Includes refrigerants, 2%, and water purification, froth flotation, and other important applications for small amounts of ammonia.

Smaller scale uses of ammonia are mostly indirect, such as the manufacture of explosives via nitric acid or using the new ammonium nitrate/fuel oil or metallized ammonium nitrate slurry explosive systems. Some ammonia is consumed in the preparation of monoamines and diamines as monomers for nylon polymers, in the manufacture of acrylonitrile for Acrilan, and melamine (via urea) for the production of melamine-formaldehyde resins (Chaps. 21 and 23). The low cost of ammonia and its favorable compression–liquefaction properties make it ideal as the working refrigerant in large-scale freezing systems such as for frozen food processing and storage and artificial ice rinks. However, its significant toxicity and flammability make it of only limited use in household refrigerators in favor of the much more costly but less toxic and nonflammable hydrochlorofluorocarbons (HCFCs) and related derivatives. Ammonia is also employed in small-scale applications such as the cation in some sulfite-based wood pulping processes, and as a means of stabilizing residual chlorine in municipal water treatment.

11.1.7. New Developments in Ammonia Synthesis

Even with the large recent price increases for petroleum-based feedstocks, natural gas is still likely to remain a significant, if not the dominant source of the hydrogen required for ammonia synthesis for the foreseeable future [21]. Production facilities that use natural gas as a hydrogen source have focused on boosting energy efficiency which may be achieved by using cryogenic methods [22].

Modifications using waste heat recycle and the more extensive use of heat exchange to simultaneously cool product streams and warm feed streams together with other improvements have reduced the energy cost for ammonia production from 35.2 to 28.7 million Btu/tonne, an 18% saving [23]. A new type of ammonia converter which uses intermediate indirect cooling, rather than quench gas cooling of the reacting gas stream (Fig. 11.3), gives a higher concentration of ammonia in the converter exit gas. This development has parallels with the newest contact sulfuric acid technology. Adoption of two-stage centrifugal compressors driven by steam turbines, in place of reciprocating compressors, achieves further economies [24]. Cost saving through computer regulation of the nitrogen:hydrogen synthesis gas ratio has been found to be sufficient to write off the installation cost in 1 yr [25]. Faster start-ups and planned maintenance of new ammonia plants procedures can contribute to higher actual operating capacities. The causes of shutdowns have been interpreted to minimize this source of inefficiency [26]. Recommendations have also been made to minimize turnaround time (time to get back in production again) in the event of a shutdown [27].

Hydrogen production efficiency has been raised by reducing the steam to carbon ratio in the steam reforming step from 4.5 to about 3, and by improving the carbon dioxide removal system [28]. Skillful control of the concentration of inerts in the recycle gas decreases both feed gas (nitrogen and hydrogen) loss as well as the loss of residual ammonia present in the purge gas if this is burned. In addition, if efficient control of purge ratios is coupled with recovery of the ammonia and hydrogen from the purge gas prior to combustion, hydrogen utilization efficiencies can be raised from the usual 92–95% to about 99.5% [29].

It has been proposed that more water be used to provide the hydrogen requirement [30] (Eqs. 11.27–11.29).

$$2\ CaO + 2\ I_2 \xrightarrow{\text{ca. }100°C} 2\ CaI_2 + O_2 \qquad\qquad 11.27$$

$$CaI_2 + H_2O \xrightarrow{500°C} CaO + 2\ HI \qquad\qquad 11.28$$

$$2\ HI \xrightarrow{300-700°C} H_2 + I_2 \qquad\qquad 11.29$$

This indirect route to cracking water requires 20% less heat to produce hydrogen over existing carbon-based cracking processes. Alternatively, water may be electrolyzed to provide the dominant source of hydrogen using electrolyzers built to operate at 400 atm pressure. If liquid nitrogen were used as the nitrogen source, compressors would not be necessary provided that low-cost power is available for electrolytic hydrogen production. This provides an interesting alternative process sequence to ammonia.

The original Haber–Bosch process relied on coal or coke to produce hydrogen has since been largely replaced by natural gas or petroleum-based technologies. However, this source of hydrogen is still important for 10% of global ammonia production, and may increase as petroleum and natural gas increase in cost [31, 32]. The dominant sequences use coal gasification as the first stage, followed by shift conversion and gas cleanup to obtain ammonia feedstocks (e.g., Eq. 11.30).

$$\text{Coal gasification to} \rightarrow \left\{\begin{array}{ll} CO & 31\% \\ H_2O_{(g)} & 31\% \\ H_2 & 24\% \\ CO_2 & 13\% \\ H_2 & 1\% \end{array}\right\} \begin{array}{l} \text{1. shift conversion} \\ \text{2. clean up} \\ \xrightarrow{\hspace{2cm}} \\ \text{3. } N_2 \text{ wash,} \\ \quad \text{and addition} \end{array} \quad \begin{array}{l} \text{feedstocks} \\ N_2{:}3\ H_2 \end{array} \quad 11.30$$

The methanol-based Rectisol process is favored for gas cleanup, because it depends on physical absorption at high pressures, and permits the recovery of separated hydrogen sulfide and carbon dioxide streams on solvent regeneration. These recoveries are more important with coal feedstocks because of the higher sulfur gas content in the gasifier product. Mass balance predictions certainly appear to make this alternative worth considering (Eq. 11.31), although the increased technological complexity over well-tested natural gas reforming sequences and greater emission control problems, tend to be important negative factors which may restrict future growth.

$$\left\{\begin{array}{l} \text{Coal} \quad\quad 2250\ t/day \\ 2600^*\ \text{tonnes/day} \\ 4.2\%\ S \end{array}\right. \xrightarrow{\hspace{1cm}} \left\{\begin{array}{l} \text{gasified,} \\ \text{cleaned,} \\ N_2\ \text{added} \end{array}\right. \xrightarrow{\hspace{1cm}} \left\{\begin{array}{l} NH_3,\ 1500\ t/day \\ S,\ 107\ t/day \\ \text{wet ash, } 550\ t/day \end{array}\right. \quad 11.31$$

(* 350 t/day burned
for steam, flue gas
scrubbed for SO_2 removal)

Solar ammonia production has been discovered to occur naturally in desert areas via exposure of nitrogen plus water vapor to catalytic sand grain surfaces, and has been confirmed by model experiments in the laboratory [33]. The details of this photochemical water splitting reaction are being further investigated for potential industrial applications (Eq. 11.32).

$$2\,N_2 + 2\,H_2O \xrightarrow{h\nu} 4\,NH_3 + 3\,O_2 \qquad\qquad 11.32$$

Indirect solar energy utilization via the surface-to-depth temperature differential existing in warm oceans has also received consideration. This could be exploited using large-scale floating facilities. Both solar ammonia projects are experimental prospects at present.

11.1.8. Environmental Concerns of Ammonia Production

A recent analysis of the emissions and their discharge rates allows gaseous pollutant emission factors per tonne of ammonia produced to be estimated for relatively old and newer production facilities (Table 11.6).

Sulfur dioxide emissions are low when natural gas feedstocks are used in old or new facilities, as sulfur compounds are removed at the wellhead

TABLE 11.6   Gas Emission Rates from U.S. Ammonia Production Facilities[a]

Source of emission	Emission rate, kg/tonne of ammonia produced			
	SO_2	NO_2	CO	NH_3
Old Plants[b]				
Natural gas cleaning	0.05–0.7	–	–	–
Reformer	0.03–0.3	0.6	–	–
Carbon dioxide removal	–	–	0.03	–
Copper liquor scrubbing	–	–	91.5	3.2
Ammonia synthesis	–	–	–	1.6
Ammonia loading	–	–	–	0.5
Total	0.08–1.0	0.6	91.5	5.3
Modern Plants[c]				
Natural gas cleaning	0.05–0.7	–	–	–
Reformer	0.03–0.3	0.5	–	–
Carbon dioxide removal	–	–	0.03	–
Copper liquor scrubbing	–	–	–	–
Ammonia synthesis	–	–	–	1.6
Ammonia loading	–	–	–	0.2
Total	0.08–1.0	0.5	0.03	1.8
Weighted average emission rates, old and modern facilities (kg/tonne NH_3 produced)	0.4	0.6	6.0	1.3

[a]Selected from summary tables of the National Research Council [1].
[b]Plants using ammoniacal cuprous solutions for carbon monoxide removal.
[c]Plants using methanation for residual carbon dioxide and carbon monoxide removal; ammonia-synthesis purge gas is burned as fuel.

(see Chap. 9). Any traces of sulfur dioxide present in the reformer product must be removed because sulfur is one of the permanent catalyst poisons. Sulfur dioxide is normally removed by adsorption, and the sulfur dioxide discharge occurs when the adsorptive bed is heated for regeneration. The mass rate of emission with natural gas feedstocks is small, of the order of 60 kg SO_2/day for a 1,000 tonne/day ammonia plant. Hence, there has not been a need to change this purging practice for new plants. Use of fuel oil or coal for hydrogen generation will produce more associated sulfur dioxide on reforming and therefore will require integral sulfur dioxide containment [34].

Nitrogen dioxide discharges result from the oxidation of atmospheric nitrogen during combustion of reformer fuel. This occurs to some extent whenever hot nitrogen and oxygen contact heated metal surfaces, with a contribution in ammonia plants from the combustion of traces of residual ammonia still present in purge gases when burned. Nitrogen oxidation also occurs on a much larger scale from the operation of fossil-fueled power stations. The improvement observed for modern ammonia plants results from increased reformer energy efficiency which requires less fuel to be burned per unit of hydrogen, and better control of the ammonia content of purge gases. These measures plus a decrease in the excess air used for reformer combustion improve energy efficiency and decrease NO_x emissions.

Ammonia plants which use aqueous ammoniacal cuprous chloride or formate ion for final stage scrubbing to remove traces of carbon monoxide have experienced significant carbon monoxide losses when the scrubber liquor is regenerated. This can amount to about 60 tonnes of carbon monoxide/day for a 600 tonne/day ammonia plant. Relatively slight losses occurred on regeneration of the scrubbing liquor of the carbon dioxide scrubbing circuit because of the very high selectivity for carbon dioxide of the solutions used for this step. More recently built ammonia plants avoid the major carbon monoxide discharge source altogether by employing methanation (Eqs. 11.23 and 11.24) to remove the last traces of both carbon monoxide and carbon dioxide. They experience losses of only about 30 kg of carbon monoxide/day for a 1,000 tonne/day facility from traces captured in the carbon dioxide scrubber.

Ammonia losses from the synthesis stage primarily arise from purge gas combustion and leakage. Little change is seen in the loss rate from this area with newly adopted technology. However, as more facilities adopt purge gas ammonia and hydrogen recovery systems the loss rate from this operating area should decrease [22]. Improved loading techniques, including provision of ammonia line vapor recompression and continuous control of storage tank vapors have resulted in the decreased losses. Realistic guidelines of daily average ammonia discharge rates of 1 kg/tonne (design) and 1.5 kg/tonne (operating) have recently been set for one jurisdiction which are compatible with the loss rates outlined above [35].

Water impacts from ammonia producing facilities can occur through thermal loading from the discharge of large volumes of cooling water, or from the discharge of dilute ammonia solutions which may also contain organics from recovered process condensates. Lagooning, forced-air cooling, and increased use of heat exchange of hot exit gases to preheat entering raw materials are among the remedies for thermal impacts (Chap. 3). Steam stripping may be used to recover ammonia from dilute aqueous waste streams. However, this is a relatively expensive, energy-intensive procedure for the small quantities likely to be recovered. A better expedient might be to use these streams, which typically contain 400–2,000 ppm ammonia, to simultaneously irrigate and fertilize local croplands. Permissible and desirable criteria for ammonia in public water supplies have been set at 0.5 and 0.01 mgL^{-1}, respectively [36] which limit the acceptable volumes of such waste streams.

Noise problems in large ammonia plants from the movement of large gas volumes and the operation of compressors mainly affect operating employees but have to meet certain local noise criteria. Procedures to locate and localize or dampen (attenuate) noise sources which produce sound levels in excess of 90 dBA have been described [37].

11.2. NITRIC ACID

11.2.1. Background

Nitric acid (*aqua fortis*) was known and its chemistry practiced in the Middle Ages. It was obtained by heating hydrated copper sulfate or sulfuric acid with sodium nitrate (saltpeter or niter) and cooling the vapors generated to obtain a solution of nitric acid [5] (Eq. 11.33).

$$\underset{\text{``niter''}}{\text{NaNO3}} + H_2SO_4 \xrightarrow{\text{ca. 200°C}} \underset{\text{``niter cake''}}{NaHSO_4} + HNO_3 \qquad 11.33$$

While it is theoretically possible to obtain 2 moles of nitric acid for each mole of sulfuric acid used, 900°C temperature required makes this stoichiometry impractical. However, the residual acid value of the niter cake could be used to make hydrochloric acid from salt (Eq. 11.34).

$$\underset{\text{``niter cake''}}{NaHSO_4} + NaCl \rightarrow \underset{\text{``salt cake''}}{Na_2SO_4} + HCl \qquad 11.34$$

For nitrogen fertilizers and for the nitric acid requirement for the manufacture of explosives in the 19th century, natural sodium (or potassium) nitrate provided the only source. India produced some 30,000 tonne/year by 1861 and Chile, in 1870, was exporting some 90,000 tonnes annually which gradually climbed to about 1.4 million tonne/year, in both cases as sodium nitrate.

Present-day nitric acid production is almost entirely via the oxidation of ammonia and absorption of the oxidation products in water. The chemistry of this process was proven experimentally by Kuhlmann in 1839, but had to wait for the development of an economical route to ammonia before it could become commercially significant [38]. Ostwald, working in Germany in about 1900, reexamined and extended Kuhlmann's data and established the proper conditions required for the ammonia oxidation step. Very shortly after this plants operating on these principles were constructed both in Germany and the U.S. Since 1980, 3–5 million metric tonnes of nitric acid (100% basis) have been produced annually in Germany, and 7–9 million tonnes in the U.S. (Table 11.7).

TABLE 11.7 **Production of Nitric Acid (100% Basis) by Selected Countries**[a]

	Thousands of metric tonnes				
	1965	1970	1980	1990	2000
Australia	25	126	180	—	—
Belgium	616	718	1,167		
Canada	344	503	713	965	1,074
Finland	5	261	422	—	—
France	2,319	2,575			
W. Germany	2,599	3,254	3,173	1,880[b]	5,372[b]
Italy	946	1,038	1,011	1,040	1,227
Japan	246	474	577	—	—
Sweden	190	270	346	—	—
U.S.A.	4,444	6,897	8,102	7,033	8,797
Other	3,526	4,565	14,009	—	—
World	15,260	20,681	29,700		

[a]Compiled from Kirk-Othmer [2], *Chemical and Engineering News* [3], and *Statistics Canada* [39], plus data supplied by the Verband der Chemischen Industrie e.V. (VCI). U.S. production for 1960 was 30,071,000 tonnes.

[b]After reunification.

11.2.2. Nitric Acid by Ammonia Oxidation: Chemistry and Theory

Modern nitric acid production uses catalytic oxidation of ammonia in air followed by absorption of the oxidation products in water to yield nitric acid. Unlike the reactions for ammonia feedstock preparation, all of the reaction steps in this sequence are exothermic.

Ammonia is oxidized using a mixture of ammonia gas (9–11%) in air which is passed through multiple layers of fine platinum–rhodium alloy gauze (Eq. 11.35).

$$4\ NH_{3(g)} + 5\ O_{2(g)} \xrightarrow[\text{ca. } 900°C]{\text{Pt/Rh}} \underset{\text{nitric oxide}}{4\ NO_{(g)}} + 6\ H_2O_{(g)} \qquad 11.35$$

$$\Delta H = -907\ kJ\ (-217\ kcal)$$

While this equation represents a close approximation of the stoichiometry of the process, the detailed chemistry is complex and poorly understood. The exotherm of the oxidation causes a temperature rise of about 70°C for each 1% of ammonia in the mix. Prewarming of one or both component streams is used to keep the alloy gauze at close to the optimum 900°C [40]. Operating at a gauze temperature of 500°C produces mostly unreactive nitrous oxide as the oxidation product (Eq. 11.36) which would be lost to acid production.

$$2\ NH_{3(g)} + 2\ O_{2(g)} \xrightarrow[\text{slow}]{500°C} \underset{\substack{\text{nitrous} \\ \text{oxide}}}{N_2O} + 3\ H_2O \qquad 11.36$$

$$\Delta H = -276\ kJ\ (-66\ kcal)$$

This provides the motivation to maintain gauze temperatures at about 900°C.

It may seem to be a backward step to have prepared ammonia from nitrogen and hydrogen, to then burn the ammonia to obtain nitric oxide. However, if one considers the endothermic nature of the reaction between nitrogen and oxygen (Eq. 11.37), it becomes apparent why nitric oxide yields, even at carbon arc temperatures of around 3,000°C, are only a few volume percent.

$$N_{2(g)} + O_{2(g)} \rightarrow 2\ NO_{(g)} \qquad \Delta H = +90.3\ kJ\ (+21.6\ kcal) \qquad 11.37$$

Using ammonia as the starting material for nitric oxide preparation bypasses the enormous energy requirement for elemental nitrogen bond dissociation, and in so doing achieves the product of interest via a net exothermic (thermodynamically favorable) process.

Nitric oxide obtained from ammonia combustion is then further oxidized to nitrogen dioxide in a less exothermic step (Eq. 11.38).

$$\begin{array}{c} 2\ NO_{(g)} \quad + \quad O_{2(g)} \xrightarrow{\text{slow}} 2\ NO_{2(g)} \qquad \Delta H = -113\ kJ\ (-27\ kcal)\quad 11.38 \\ \text{colorless} \qquad\qquad\qquad \text{brown} \\ \text{m.p.} -164°C \\ \text{b.p.} -152°C \end{array}$$

The nitrogen dioxide product participates in a rapid equilibrium to dinitrogen tetroxide (Eq. 11.39).

$$\begin{array}{c} 2\ NO_{2(g)} \quad \rightleftharpoons \quad N_2O_{4(g)} \qquad \Delta H = -57.4\ kJ\ (-13.7\ kcal) \qquad 11.39 \\ \text{brown} \qquad\qquad \text{colorless} \\ \text{m.p.} -11.2°C \\ \text{b.p. } 21.2°C \end{array}$$

At 100°C this equilibrium lies at approximately 90% NO_2, 10% N_2O_4, whereas at 21°C only about 0.1% NO_2 is present [41]. This equilibrium may be important for acid production since there is evidence that dinitrogen tetroxide is the molecular species that reacts with water for acid formation.

The final step in acid formation is absorption of nitrogen dioxide in water, the net result of which is given by Eq. 11.40.

$$3\ NO_{2(g)} + H_2O_{(\ell)} \rightarrow 2\ HNO_{3(aq)} + NO_{(g)} \quad \Delta H = -139\ kJ\ (-33.1\ kcal)\ \ 11.40$$

This is not as straightforward a transformation as it may at first appear since it involves both chemical combinations with water and a net redox reaction for the nitrogen (4+) in nitrogen dioxide to nitrogen (5+) in nitric acid. One nitrogen dioxide thus takes up two electrons (is reduced) to become nitrogen (2+) in nitric oxide to balance this oxidation. The actual chemistry involved is probably better represented as, first of all, a formation of both nitric and nitrous acids (Eq. 11.41).

$$2\ NO_2 + 3\ H_2O \rightarrow HNO_3 + HNO_2 \qquad\qquad 11.41$$

The nitrous acid, being unstable in the presence of any strong mineral acid, disproportionates to yield further nitric acid and nitric oxide (Eq. 11.42).

$$3\ HNO_2 \rightarrow HNO_3 + 2\ NO + H_2O \qquad\qquad 11.42$$

Not all of the nitrogen in the ammonia feed ends up as nitric acid. Some of the ammonia reacts with oxygen of the air to yield elemental nitrogen from ammonia nitrogen, and some reacts with nitric oxide causing loss of potential product from both ammonia and oxidant consumption (Eqs. 11.43 and 11.44).

$$4\,NH_{3(g)} + 3\,O_{2(g)} \rightarrow 2\,N_{2(g)} + 6\,H_2O_{(g)} \quad \Delta H = -1267\,kJ\,(-303\,kcal) \quad 11.43$$

$$4\,NH_{3(g)} + 6\,NO_{(g)} \rightarrow 5\,N_{2(g)} + 6\,H_2O_{(g)} \quad \Delta H = -1806\,kJ\,(-432\,kcal) \quad 11.44$$

Reactions (Eqs. 11.43 and 11.44) are both highly favored thermodynamically, even relative to the desired reaction, Eq. 11.35. The platinum gauze catalytic surface raises the proportion of nitric oxide obtained and decreases nitrogen formation by these pathways, probably by accelerating the rate of the desired reaction without affecting the rates of the others. Use of multiple platinum–rhodium or platinum–palladium alloy gauzes further raises the ammonia conversion to 99% (for 8.3% NH_3 air; gauze temperature 930°C), over the 97.5% obtained using platinum alone, and decreases the metal erosion rate.

From a process design standpoint, an important aspect of the ammonia oxidation reaction, Eq. 11.35, is that since 9 mol react to produce 10 mol of product there is little volume change from this step. Therefore, the equilibrium of this reaction is not much affected by a change in operating pressure. However, an increase in pressure can push more material through the same sized equipment in the same time, which lowers the capital cost of the plant per unit of product. If the pressure increase is moderate, say, 5–10 atm, the metal walls of the process vessels used do not have to be significantly thicker to contain it. This modification also slightly increases the concentration of the nitric acid product from the 50–55% obtained when operating at atmospheric pressure, to 57–65% when under pressure.

Increasing the pressure of the feed gases to the ammonia converter also effectively increases the space velocity through this unit, which slightly decreases the conversion efficiency for ammonia oxidation. But this reaction is so rapid, being essentially complete in 3×10^{-4} sec at 750°C that a few additional layers of catalytic gauze almost completely compensate for the reduced conversion.

The nitric oxide oxidation and water absorption reactions (Eqs. 11.38 and 11.40) are both much slower than the ammonia oxidation reaction (Eq. 11.35), and involve a significant volume decrease on reaction, 3 mol (volumes) to 2, and 3 mol to 1, respectively. Thus the absorbers, where the bulk of these reactions occur, must be large to provide sufficient residence time, and cooled to favor the equilibria in the desired direction. Raising the pressure in the absorbers achieves a significant improvement in performance, in accord with Le Chatelier's principle, because of the volume decrease observed for the absorber reactions. For an increase to 8 atm from 1 atm, the rate of the very slow nitric oxide reoxidation reaction (Eq. 11.45) is accelerated by a factor of the cube of this pressure increase, or 512 times [42].

$$Rate = k_1 (P_{NO})^2 (P_{O_2}) - k_2 (P_{NO_2})^2 \quad 11.45$$

For a process absorber which does not normally operate very close to equilibrium, the rate of the reverse reaction (the second term in this rate equation) can be ignored. The pressure increase also boosts the rate of nitrogen dioxide absorption, once formed, in water. In practice this pressure increase means

FIGURE 11.5   Production of nitric acid by oxidation of ammonia.

that the absorber volume required per daily tonne of nitric acid production capacity can be decreased from about 35 m³ to about 0.5 m³ [43].

11.2.3. Process Description

To put these principles into practice, liquid ammonia is first vaporized by indirect heating with steam, and then filtered to reduce risk of catalyst contamination. This produces an ammonia gas stream at about 8 atm pressure without requiring mechanical compression. An air stream is separately compressed to about the same pressure, preheated to 200–300°C, and filtered prior to mixing with the ammonia (about 10%) gas stream immediately before conversion. This mixture is passed through the red hot platinum–rhodium gauze to produce a hot gas mixture of nitric oxide and water vapor plus the unreacted nitrogen and oxygen components of air (Fig. 11.5), with a yield efficiency (selectivity) under these conditions of about 95%.

Hot converter gas products are cooled in a waste heat boiler prior to absorption, to give simultaneous production of steam and a product gas stream at more moderate temperatures ready for catalyst recovery. The high-pressure nitric acid process, as described here, experiences a gauze metal loss rate of 250–500 mg/tonne of acid produced. This can be kept to the lower end of the range by efficient filtration. If operated at atmospheric pressure, catalyst loss rate is about 50 mg/tonne. After filtration the gas mixture is quickly chilled by heat exchange with process water to condense some dilute nitric acid from the product gases. The dilute nitric acid condensate is also filtered for further catalyst recovery and then trickled into the top of the absorber to raise the acid concentration. The gas stream still contains nitric oxide, but with the elapse of time and having been cooled now also contains nitrogen dioxide and dinitrogen tetroxide (Eqs. 11.38 and 11.39) plus the residual air. This stream is fed into the absorber at an intermediate level.

In the absorbers nitric oxide is oxidized to nitrogen dioxide at the same time as nitrogen dioxide is absorbed in water to give nitric acid plus further nitric oxide. This is not lost but is being continuously reoxidized to nitrogen dioxide like the primary nitric oxide. Cooling water keeps the temperatures down to maintain favorable oxidation and solution equilibria. As the concentration of acid builds up to equilibrium levels in the downward flow of absorption water some dissolved (but unreacted) nitrogen dioxide may also be present, which gives the acid a brown color. This nitrogen dioxide is stripped from the product acid by purging the last few plates of the absorber with "bleach air," to give a colorless to pale yellow product acid containing 60–62% by weight HNO_3 as the bottom product of the absorber. The bleach air, at the same time as stripping excess dissolved nitrogen dioxide from the product, also assists in maintaining an oxygen excess in the middle and upper reaches of the absorber, which favors the formation of nitrogen dioxide from nitric oxide. Residual gases are vented from the top of the absorber via an expander turbine, which recovers energy from the vented gas sufficient to provide about 40% of the power required for driving the inlet air compressor mounted on a common shaft.

11.2.4. Nitric Acid Concentrations and Markets

The ammonia oxidation product is sold in several technical and commercial market grades ranging from about 50% (by weight) HNO_3, up to what is referred to by chemists as ordinary concentrated nitric acid (68%), which is the maximum boiling point or azeotropic composition (Table 11.8). These are the concentrations obtained either directly from the various types of nitric acid plant or obtained by raising the concentration of the acid plant product by simple distillation. The concentrated product ranges in color from water white, through yellow to brown, as the dissolved nitrogen dioxide content

TABLE 11.8 Industrial Market Grades of Nitric Acid

Degrees Baumé[a]	Density (g/mL at 20°C)	Nitric acid concentration (%)
36	1.330	52.3
38	1.3475	56.5
40	1.3743	61.4
42	1.4014	67.2, ordinary conc. nitric
–	1.4521	80.0
–	1.4826	90.0, or higher, is fuming nitric
–	1.4932	95.0
–	1.5492	100.0, anhydrous nitric, unstable[b]

[a]°Bé, an arbitrary specific gravity scale base on density, where specific gravity = 145/(145 − °Bé reading).

[b]Decomposes to NO_2, NO, and water above the freezing point. Hence, fuming grades are more useful at somewhat lower concentrations.

gradually rises in the product with time from the operation of the slow equilibrium reaction, Eq. 11.46.

$$4\,HNO_3 \rightleftharpoons 2\,H_2O + 4\,NO_2 + O_2 \hspace{4em} 11.46$$

These grades or concentrations are used in the production of ammonium nitrate, which at present consumes some 65–75% of the total nitric acid made [44], as well as for many other minor uses. Some of these are, for example, adipic acid production (5–7%), military and industrial explosives (3–5%), isocyanates for polyurethane manufacture (1–2%), nitrobenzene (1–2%), and potassium nitrate preparation from the chloride (about 1%). Many smaller scale uses consume the remaining 10–15% of the acid produced.

For forcing of difficult nitrations and other purposes, which require greater than 68% by weight nitric acid, special techniques have to be used since this is the maximum concentration available by simple distillation. Nitric acid concentrations above 86% consist of concentrated nitric acid plus dissolved nitrogen dioxide. These concentrations are dark brown in color, and tend to lose the dissolved nitrogen dioxide relatively rapidly if left open to the air. For this reason these grades are collectively referred to as "fuming nitric acid" and require considerable caution in their use because of their substantially greater reactivities. The highest concentration of commercial fuming nitric acid contains 94.5–95.5% HNO_3.

One commercial method to obtain nitric acid concentrations above 68% uses concentrated sulfuric acid to dehydrate the azeotropic composition. Hot nitric acid vapor is passed upward against concentrated sulfuric acid, which moved downward (countercurrent) in a tower packed with chemical stoneware to obtain 90+% HNO_3 and a diluted sulfuric acid stream (Fig. 11.6). If this process is practiced on only a small scale, the sulfuric acid may be reconcentrated by addition of oleum, and a portion of the buildup of sulfuric acid in this circuit may be used to make up a commercial nitrating mixture with some of the fuming nitric acid made. Larger scale operation requires the use of a sulfuric acid boiler and a large heat input to reconcentrate the dehydration acid. There is also a noticeable sulfate contamination of the nitric acid product from this process.

Another dehydration procedure employs 72% aqueous magnesium nitrate $(Mg(NO_3)_2 \cdot 4\,H_2O)$ as a dehydrating agent, which gets around some of these difficulties [45]. As in sulfuric acid dehydration, hot nitric acid vapor is fed

FIGURE 11.6 Dehydration of the nitric acid from an ammonia oxidation plant to fuming grades using concentrated sulfuric acid.

countercurrently to hot, molten 72% $MgNO_3$, this time in a multiple-plate column of stainless steel, to achieve similar results. But here the anion of the dehydrating agent and the nitric acid are the same, so there is no sulfate contamination. Also the diluted magnesium nitrate product (approx. $Mg(NO_2)_2 \cdot 6H_2O$; m.p. 89°C) from the bottom of the column requires less heat for reconcentration than diluted sulfuric acid.

11.2.5. Nitric Acid Process Variants and New Developments

Absorption trains of early ammonia oxidation processes to nitric acid were constructed of chemical stoneware or acid-proof brick, which restricted acid production to near ambient atmospheric pressure because of the low strength of the structural materials. The discovery that Duriron (silicon-iron) or high chrome stainless steels could tolerate these corrosive conditions well allowed the adoption of pressure absorption. This measure markedly decreased the size of the absorbers required and reduced nitrogen oxide stack losses. Pressure operation was easiest to achieve by compression of the feed gases at the front end of the process. In this way improved acid production is obtained at comparable capital costs per unit of product by operation at atmospheric pressure.

However, an attendant cost of pressure operation throughout the process is nitric oxide yields of only 90–95% from the ammonia combustion stage compared to yields of 97–98% at atmospheric pressure. For this reason atmospheric pressure processes are still viable and continue to operate. To obtain the greater ammonia oxidation efficiencies possible at near atmospheric pressure and to retain the more efficient absorption at elevated pressures many European producers use split-pressure processes [46]. This combination is particularly valuable when ammonia costs are high. However, the ammonia saving has to be balanced against capital costs of split-pressure plants 1.5 to 2 times that of the other two alternatives because of the need for compressors constructed from exotic metals.

A development in catalyst support systems in which half of the 5–10% rhodium–platinum alloy gauzes were replaced by nonnoble metal supports or by ordinary metal catalysts gave cost economies without adversely affecting operating efficiency [47]. More recently, ammonia oxidation in a two-bed system (Pt gauzes followed by monolithic oxide layers) gave nearly the same ammonia conversion while reducing platinum losses by 50% [48].

Technologies for concentrating nitric acid have been developed to obtain a much higher concentration nitric oxide, nitrogen dioxide gas stream from the acid condenser at a vapor pressure of 2–3 atm [41]. Absorption of this gas in cold concentrated (68%) nitric acid yields a superazeotropic product containing 80% or more nitric acid. Distillation of this produces an overhead stream of 96–99% nitric acid, requiring only six theoretical plates for separation from a bottom stream of the azeotropic composition (Fig. 11.7). Brief operating details of this Espindesa process are available [49].

11.2.6. Emission Control Features

The chief environmental problem of nitric acid plant operation is a discharge of residual nitric oxide (NO) and nitrogen dioxide (NO_2) from the vent stack

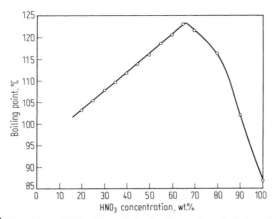

FIGURE 11.7   Plot of the boiling point of nitric acid versus concentration.

of the absorber. For a high-pressure process of the 1960s without the emission controls described here, the total concentration of nitrogen oxides discharged was typically about 0.3% by volume [33]. A detailed breakdown is given in Table 11.9. Nitric acid plants are not the only source of nitrogen oxides (NO_x) but without abatement can correspond to a relatively large point source of emissions, unlike the more diffuse discharge resulting from automobiles (Table 3.1). The primary need to regulate NO_x discharges arises from the implication of nitrogen oxides in photochemical air pollution problems (Chap. 2), although the toxic effects of nitrogen dioxide itself may be felt from its presence in ambient air during photochemical smog episodes.

Nitrogen dioxide is a brown irritating gas, which exists in equilibrium with dinitrogen tetroxide at ordinary ambient conditions (Eq. 11.39). It is dangerous to humans at concentrations above 50 ppm [52], and at high concentrations (1,000–2,000 ppm) in air it can cause loss of consciousness and convulsions. An industrial hygiene standard of 5 ppm for an 8-hr workday has been set. Even though nitric oxide itself is not frequently a problem it does indirectly contribute to atmospheric nitrogen dioxide loadings because any nitric oxide present in vent gases is rapidly oxidized to nitrogen dioxide

TABLE 11.9   Typical Composition of Nitric Acid Plant Vent Gases, Prior to Expansion (at About 30°C, 7.3 Atm)[a]

	Percent, without controls	Percent, with methane catalytic reduction
Nitric oxide	0.05–0.20	0.01–0.20
Nitrogen dioxide	0.05–0.20	
Oxygen	3.0	0.5–2.0
Water	0.6–0.7	2.5–5.0
Nitrogen	96+	92–93.5
Nitric acid	<0.0	<0.0
CO_2	0.0	0–2.0

[a]Compiled from the data of Lund [36], Gillespie *et al.* [50], and Boyars [51].

FIGURE 11.8 One emission control system for nitric acid plant tail gas cleanup via catalytic reduction.

on exposure to air (Eq. 11.38). And nitrogen dioxide, if present in vent gases above about 100 ppm (by volume), gives a visible brown discharge plume from the vent stack.

First-generation emission control focused on catalytic reduction of nitrogen oxides using methane (natural gas) or hydrogen (Fig. 11.8). To achieve this, tail gases from the absorber vent are preheated to about 400°C, and then blended with the appropriate proportion of methane before passage over platinum or palladium catalytic surfaces [33]. The concentration of nitrogen oxides can be decreased by about 90% (Table 11.9), from about 0.3% to 0.01–0.2% (by volume) depending on conditions, from the reactions (Eqs. 11.47–11.49).

$$CH_4 + 2\ O_2 \xrightarrow{\text{cat.}} CO_2 + 2\ H_2O \qquad\qquad 11.47$$

$$\underset{\text{brown}}{CH_4 + 4\ NO_2} \rightarrow \underset{\text{colorless}}{4\ NO} + CO_2 + 2\ H_2O \qquad\qquad 11.48$$

$$CH_4 + 4\ NO \rightarrow 2\ N_2 + CO_2 + 2\ H_2O \qquad\qquad 11.49$$

Sufficient methane is required to first consume the excess oxygen normally present in the vent gases, which is used for reoxidation of nitric oxide to nitrogen dioxide in the absorbers. At the same time, nitrogen dioxide is rapidly reduced to nitric oxide, which removes the brown color from the vent stack discharge, a "decolorization" level of control. However, decolorization to nitric oxide does nothing to decrease the ultimate nitrogen dioxide contributed to the ambient air from vent gases since NO is eventually oxidized back to NO_2 (Eq. 11.38). It also requires sufficient additional methane to achieve this step. This is the only way in which true "abatement" is achieved, which requires additional capital and fuel cost [46] (Fig. 11.8). Fortunately, some of these additional costs may be recovered from additional steam and from increased energy capture by the tail gas expander turbine from the hot, rather than ambient, temperature exhaust gas discharge.

Regulatory requirements for nitric acid plant discharges in the U.S. now stand at 1.5 kg of NO_x (as NO_2 equivalent) per tonne of acid, which equates to about 230 ppm NO_x, and the discharge must be colorless (opacity of 10% or less) for new plants. Older plants may be allowed a maximum of 400 ppm (by volume) NO_x [53]. For European jurisdictions, the requirements vary but generally lie near those required for older U.S. plants. In the U.S.S.R., perhaps because of either very large single-train facilities or because of several fertilizer works in close proximity had set a requirement of 0.55 kg NO_x (calculated as NO_2) discharge rate/tonne of acid.

With careful operation the methane catalytic reduction route is capable of meeting the more stringent regulatory requirements. Capital costs dictate the maximum economic size of absorbing towers so that early designs were only sufficient to achieve 0.1–0.3% NO_x in the tail gases. New plants with larger absorbers and employing refrigeration chilled water systems for absorbers are able to reach tail gas NO_x concentrations of <200 ppm directly. Use of booster compressors to permit absorption at high pressures has also been found to pay off in improved absorption efficiencies, at least for very large plants.

For efficient raw material utilization as well as for fuel conservation (e.g., methane), discharge controls integral with the process absorbers have obvious advantages. Nitric oxide and nitrogen dioxide may be reduced by ammonia using alumina-supported platinum catalysts without the need to consume the residual oxygen first [54] (Eqs. 11.50 and 11.51).

$$6\ NO + 4\ NH_3 \rightarrow 5\ N_2 + 6\ H_2O \qquad\qquad 11.50$$

$$6\ NO_2 + 8\ NH_3 \rightarrow 7\ N_2 + 12\ H_2O \qquad\qquad 11.51$$

This has been shown to meet stringent regulatory requirements with a processing train very similar to that used for methane reduction (Fig. 11.7).

Chemical absorption in scrubbers, which employ any of water, nitric acid, alkaline solutions, or solutions of urea in water can reduce tail gas NO_x concentrations to below 200 ppm [55]. Physical adsorption on molecular sieves is also a feasible control method. Activated carbon cannot be used because of oxidation hazards. Chemical and physical adsorption were improved if both nitric oxide and nitrogen dioxide were present in the tail gas rather than just nitric oxide, perhaps from the formation of dinitrogen trioxide (N_2O_3). An advantage of any absorption or adsorption system is that the NO_x collected is recovered in some form.

Comparative summaries of abatement measures have been published [49, 50].

11.3. COMMERCIAL AMMONIUM NITRATE

11.3.1. Background

Production of ammonium nitrate is the largest single end use for nitric acid and has ranked 10th to 12th in volume of all inorganic chemicals produced in both Canada and the Unites States over the last 20 years (Table 11.10). This interest has been stimulated by the demand for ammonium nitrate as a fertilizer component and as an ingredient in explosives. Its simplicity of production and low cost, about $170/tonne in the U.S. in 1995, add to its product appeal. Fertilizer value is derived both from the high total nitrogen content (~33%) split between fast acting nitrate nitrogen and the slower acting ammonium nitrogen, plus the ease of marketing and handling a solid product.

Explosives applications have been varied from blends with nitroglycerin to make this safer for use in mines, to admixtures with smokeless powders and

TABLE 11.10 Production of Ammonium Nitrate and Urea in Selected Countries[a]

	Thousands of tonnes				
	1970	1975	1980	1990	2000
Canada: NH₄NO₃	777	875	873	1,031	1,110
Urea	367	329	1,274	2,490	3,887
Japan: NH₄NO₃	37	115	123	8	18
Urea	2,463	3,162	1,770	776	641
U.S.A.: NH₄NO₃	5,938	6,430	7,792	6,425	8,244
Urea	2,829	3,446	7,103	7,368	15,322

[a]Compiled from Statistics Canada, cat.no. 46–002 [55a]; FAO UN [56], Japan Economic Yearbooks [57], data supplied by Verband der Chemischen Industrie e.V. (VCI), and *Chemical and Engineering News* [3].

nitro compounds such as trinitrotoluene, to the more recent developments of blends of low-density ammonium nitrate prills with fuel oil (so-called ANFO explosives), or in water slurries with metal powder boosters to produce a low-cost "heaving" power explosive [58]. All explosives applications rely on the large volume of gas released on vigorous detonation of the ammonium nitrate component (Eq. 11.52).

$$2\ NH_4NO_{3(s)} \rightarrow 2\ N_{2(g)} + 4\ H_2O_{(g)} + O_{2(g)} \qquad 11.52$$

The oxygen is consumed by the fuel oil or a readily oxidizable metal such as aluminum or magnesium present to contribute further heat and power to the detonation process. Means to improve the normal handling of ammonium nitrate to decrease the explosion risk under ordinary shipping and storage conditions have been proposed [59].

A small-scale but important use of ammonium nitrate is for the preparation of inhalation grade nitrous oxide for use as an anaesthetic. This is achieved by heating pure ammonium nitrate to about 200–250°C (Eq. 11.53), which gives nitrous oxide virtually free of nitric oxide or

$$NH_4NO_3 \rightarrow N_2O + 2\ H_2O \qquad 11.53$$

nitrogen dioxide. This product is also known as "laughing gas," and is used in dentistry for short term anaesthesia for extraction of molars, etc.

11.3.2. Production of Ammonium Nitrate

Early production involved neutralization of concentrated aqueous ammonium hydroxide with nitric acid, followed by evaporation of much of the water and crystallization of the product. High capital, labor, and energy costs of this batch process have led to it being superseded by the Fauser continuous process or variations of this. The Fauser process contacts ammonia gas with concentrated nitric acid and uses the heat of reaction to evaporate a part of the original water

FIGURE 11.9 Ammonium nitrate by the Stengel process. Intermediate evaporation before prilling or granulation would be employed for the production of a high-density product.

[14]. The Stengel process uses sufficiently high preheat temperatures for ammonia gas and a nitric acid solution such that complete evaporation of the residual water occurs when the heat of reaction is added to this [16].

In the Stengel process ammonia gas preheated to 145°C is blended with about 60% nitric acid preheated to 165°C under pressure in a packed stainless steel reactor (Fig. 11.9). This has now become a routine procedure, yet the mixing of a hot strong oxidant with a hot oxidizable material is a process that is fundamentally unnerving to a chemist. Doing this provides sufficient additional heat from the neutralization reaction to nearly completely evaporate the residual water [60] (Eq. 11.54).

$$\text{NH}_{3(g)} + \text{HNO}_{3(aq)} \rightarrow \text{NH}_4\text{NO}_{3(aq)} \quad \Delta H = -86.1 \text{ kJ } (-20.6 \text{ kcal}) \quad 11.54$$

The steam produced is removed from the molten ammonium nitrate (m.p. 170°C) via the vortex finder of a cyclone separator, and the 99+% molten salt proceeds to either a cooled stainless steel belt to produce a flaked product (Fig. 11.10), or to a 30-m-high prilling tower where droplets of melt fall through a countercurrent dry air stream to produce shot-sized prills (beads) of ammonium nitrate. A hot concentrated solution of ammonium nitrate is explosively sensitized by traces of acid so care is taken to add sufficient ammonia to the wet melt to keep the pH above about 5.

FIGURE 11.10 Steel belt cooler for the preparation of a flaked ammonium nitrate (or other) product.

The product is hygroscopic so that it is often coated with an anticaking compound such as clay or diatomaceous earth before being placed in bulk storage in large moisture-tight silos. Bulk shipment is by covered gondola rail cars, and in smaller quantities in vapor tight bags of polyethylene or waxed paper.

The process itself and environmental aspects of this have been analyzed for efficiency of operation and air and water pollution control measures [61]. Membrane electrolysis coupled to electrodialysis has been studied for wastewater treatment [62].

11.4. PRODUCTION OF UREA

The first synthesis of urea was conducted by Wohler in 1828 via the heating of ammonium cyanate (Eq. 11.55).

$$NH_4OCN \rightarrow H_2NCONH_2 \qquad 11.55$$

This preparation also provided a landmark by being the first time that any compound produced by living things had been synthesized in the laboratory. This event heralded the expansion of study of organic compounds, which had formerly been almost a taboo, to emerge today with a vast array of products and services which are based on this branch of chemistry.

Production of urea on a commercial scale did not assume much significance until about 1930. Early processes were based on calcium cyanamide hydrolysis in the presence of a base (Eqs. 11.56 and 11.57).

$$CaNCN + 2\,H_2O \rightarrow Ca(OH)_2 + H_2NCN \qquad 11.56$$

$$H_2NCN + H_2O \xrightarrow{OH^-} \underset{urea}{H_2NCONH_2} \qquad 11.57$$

Since about 1950, most urea producing units have been based on ammonia–carbon dioxide feedstocks passed through high-pressure equipment, hence the close association with ammonia plants [63]. Over the last 20 years in North America urea production volumes have grown faster than ammonium nitrate for the supply of fertilizer nitrogen (Table 11.10).

Urea from ammonia and carbon dioxide in the various process sequences operated requires, first of all, reaction of these feedstocks under a pressure of 100–200 atm to form ammonium carbamate (Eq. 11.58).

$$NH_{3(g)} + CO_{2(g)} \rightarrow \underset{\substack{carbamic\\acid}}{H_2NCOOH} \xrightarrow{NH_3} \underset{\substack{ammonium\\carbamate}}{H_2NCOONH_{4(s)}} \qquad 11.58$$

$$\Delta H_{25°} = -159\ kJ\ (-38\ kcal)$$

This thermodynamically favorable step of the process is followed by an endothermic thermal decomposition of ammonium carbamate, in a concentrated solution, to give a 50–60% conversion to urea (Eq. 11.59).

$$H_2NCOONH_4 \rightarrow H_2NCONH_2 + H_2O \quad \Delta H = +31.4\ kJ\ (+7.5\ kcal) \qquad 11.59$$

FIGURE 11.11   Production of urea by a simple, one-pass reactor-separator system.

A "once-through" modification of a urea process is shown in Fig. 11.11. Recycle of unconverted ammonium carbamate and of ammonia and carbon dioxide formed in part from ammonium carbamate decomposition (Eq. 11.60) is practiced in other variations on this theme [64].

$$H_2NCOONH_4 + H_2O \rightarrow (NH_4)_2CO_3 \rightarrow 2\, NH_3 + CO_2 + H_2O \qquad 11.60$$

This assists in minimizing the environmental impacts of operation of urea plants. However, despite these recycle measures there is still a small ammonia loss rate, estimated to average 0.6 kg/tonne of urea produced [1]. This is still well below one published guideline of 2.7 kg NH_3/tonne of urea [32].

The principal end use of urea is to provide combined nitrogen for solid fertilizer formulations. Some urea is also converted to biuret ($H_2NCONHCONH_2$, also called carbamoyl urea) and to a sulfur derivative, which are sold as Kedlor [65] and Urasil [66] cattle feed supplements, respectively. Urea is also used to a lesser extent in the manufacture of plastics components, such as melamine, and as a component of urea-formaldehyde resins used as adhesives. It is also used as a component of foam-in-place formulations as a rigid insulating material [67], and as an ingredient in noncorrosive aircraft deicing solutions [67a].

11.5. SYNTHETIC FERTILIZERS

Six of the 15 largest volume inorganic chemicals produced in the U.S. in 1995 directly or indirectly form part of all the various chemical fertilizer formulations produced (Table 1.4). Such is the importance and significance of the fertilizer market in terms of the overall volume of activity of the chemical process industries.

The value of the chemical fertilizer product, of course, lies in the significantly enhanced agricultural productivity possible with their use. This advantage is more easily afforded by the large highly mechanized farming operations of the Western world, so that the per capita consumption in these countries runs from 20 to about 100 kg per capita per year as compared to usage rates in the neighborhood of 1 kg per capita per year for countries such as Bangladesh, mainland China, India, and Pakistan. While the per capita fertilizer consumption for these latter countries is small, the effectiveness of the chemical utilization to improved yields of foodstuffs and other agricultural production, kilogram for kilogram, is probably greater. The importance of chemical fertilizer applications in assisting the stabilization of all major agricultural economies should not be underestimated.

11.5.1. Fertilizer Composition

Plant growth and development requires a total of nine major nutrients or macronutrients to proceed normally [68]. Three of these, carbon, hydrogen, and oxygen, are obtained from air (as carbon dioxide) and soil (as water). Another three, calcium, magnesium, and sulfur, are present to a sufficient extent in most ordinary soils. And another three, fixed nitrogen, phosphorus, and potassium, are present to a small extent in most soils but are rapidly depleted from the soil when most of an agricultural plant is removed from the soil for harvesting. Returning the unused portions of plants to the soil can serve to return some of these elements. But removal of nutrients from whole plant cropping and from soil bacterial processes, which causes loss of nitrogen in volatile forms such as ammonia or elemental nitrogen, requires regular addition to the soil to maintain fertility. Compost, manure, and specially treated sewage sludges comprise "organic" methods to return these elements in low analysis forms. However, supplementation of these methods by the use of chemical fertilizers with higher concentrations of active ingredients can achieve significantly improved yield benefits.

The major constituents or macronutrients present in most commercial fertilizers are nitrogen, which contributes to early plant development and greening; phosphorus, which assists with early growth and seed or fruit development; and potassium, which is used by plants for the production of cellulose and starches. For the fertilizer analysis the three principal ingredients are specified in order as follows: nitrogen, as % N; phosphorus, as % P_2O_5 (equivalent), and potassium, as % K_2O (equivalent) as three numbers in sequence separated only by hyphens. Thus, 15–15–15, one of the popular high analysis fertilizers, would contain 15% N, 15% P_2O_5, and 15% K_2O. High analysis fertilizers offer simplicity and low cost shipping and handling. With the development of agricultural machinery for efficient distribution of these materials, more concentrated fertilizers have occupied for a long time an important place in the range of fertilizer products marketed (e.g., Table 11.11). Some of the other common commercial fertilizer designations are 0–20–20, 2–12–16, 3–12–12, 5–20–20, etc.

In addition to the conventional listing of major constituents, a commercial blend may be formulated to contain one or more of the secondary constituents, calcium, magnesium, or sulfur, which are used to correct local soil deficiencies. These are normally specified in a separate listing. The importance of these, too, cannot be overlooked as, for example, fertilization of a sulfur deficient soil with soluble sulfate has given over 1100% rapeseed (canola) crop improvement [69].

There are also the trace elements boron, copper, iron, manganese, zinc, molybdenum, and occasionally chlorine, which are important for some plant

TABLE 11.11   Commercial Mixed Fertilizers Shipped in Canada in 1969[a]

Analysis	Thousand tonnes	Analysis	Thousand tonnes
0–20–20	29.0	5–10–15	22.0
2–12–10	4.2	5–20–10	43.8
2–12–12	11.6	5–20–20	102.6
2–16–6	1.4	6–12–8	5.7
3–15–6	3.3	6–12–12	86.1
4–12–10	2.2	10–10–10	129.4

[a]Data from Canadian Chemical Processing [4].

metabolic pathways and, which may be deficient in the soils of some areas [70]. These are added as required to special fertilizer formulations made for the purpose, and are also specified separately from the major constituents.

In addition to the chemically important constituents many fertilizers have fillers and/or conditioners added. Fillers can include sand, gypsum, or ground phosphate rock (which is only very slightly water soluble, hence *not* part of P_2O_5 specified), and assist in even distribution of the fertilizer over small areas, such as in home gardens. Sand may also serve a conditioning function by assisting in the breakup of clays, but to be effective requires higher application rates than are normally used for fertilizers. Other soil conditioners, such as dried peat moss, specially processed wood waste, some types of shredded and expanded synthetic polymers (e.g., Krylium), and expanded mica (vermiculite), are used to improve soil structure. Since the conditioners have a much lower density than the active chemical constituents or the fillers, they are usually packaged and used separately from these.

11.5.2. Formulation of Major Active Constituents

Appropriate ingredients for the nitrogen component include ammonia gas itself, which contains 82.5% N, the highest feasible nitrogen content available in a commercially useful compound. But this source is restricted to specialized application methods. Ammonia solutions, which are essentially concentrated ammonium hydroxide of about 25% N, still require specialized equipment and can add only nitrogen. Urea (46.6% N) is also used, either as the neat solid or as an aqueous solution.

For applications of a combination of major nutrients, solid forms are normally preferred for lower cost of shipment and convenient broadcasting in croplands. Thus urea and ammonium nitrate (35% N) serve as useful nitrogen-containing constituents in solid fertilizer blends. In the form of ammonium sulfate (21.2% N) nitrogen addition also adds soluble sulfate as a separately specified secondary nutrient. Of course, nitrogen may also be added in forms chemically combined with other major constituents such as in diammonium phosphate (($NH_4)_2HPO_4$) or potassium nitrate, for example. A few plants, such as peas, clover, peanuts, and alfalfa, have nodules which contain nitrogen-fixing bacteria attached to their roots and so are able to utilize elemental nitrogen from the air to provide their nitrogen requirements.

Crop rotation is already widely practiced to use these natural ways to maintain nitrogen fertility. Genetic engineering experiments may increase the number of species of plants that are able to do this, and in this way decrease the agricultural demand for synthetic fertilizer nitrogen [71].

Finely ground phosphate rock ($Ca_3(PO_4)_2$) is occasionally added to fertilizer formulations as a diluent or filler. However, because phosphate rock has only a very limited water solubility its action as a phosphate nutrient is small and very slow. Hence, this ingredient is not allowed to be included in the % P_2O_5 analysis appearing on fertilizer packaging. Before use as a filler, phosphate rock is normally defluorinated by heating with silica and steam to decrease the risk of soil contamination by fluoride (Chap. 10).

A more soluble form of phosphate ($Ca(H_2PO_4)_2$), variously called "superphosphate" (of lime), monocalcium phosphate, or calcium dihydrogen phosphate, is made by the acidulation of phosphate rock with about 65% sulfuric acid (Eq. 11.61).

$$CaF_2 \cdot 3Ca_3(PO_4)_2 + 7\ H_2SO_4 + 14\ \text{(or more)}\ H_2O \rightarrow$$

$$
\begin{array}{lll}
3\ Ca(H_2PO_4)_2 + & 7\ CaSO_4 \cdot 2\ H_2O + & 2\ HF \uparrow \\
\text{calcium} & \text{gypsum} & \text{collected} \quad\quad 11.61 \\
\text{dihydrogen} & & \text{in scrubbers} \\
\text{phosphate} & &
\end{array}
$$

The thick slurry of calcium dihydrogen phosphate and gypsum obtained is piled outside for 8–10 weeks to complete the reaction (to "cure") after which it is milled to produce a granular product, and then bagged for market. Thus the product, superphosphate, contains the inert gypsum component as well as the fertilizer active calcium dihydrogen phosphate, which is some 1,000 times more soluble than calcium phosphate. The theoretical content of active material for the composition resulting from the reaction of Eq. 11.61 is 21.7% P_2O_5 equivalence, just a shade more than the actual 18–21% P_2O_5 range of the commercial product. The presence of silica or iron oxide in the original phosphate rock, plus the formation of the monohydrate of calcium dihydrogen phosphate ($Ca(H_2PO_4)_2 \cdot H_2O$) or some dicalcium hydrogen phosphate ($CaHPO_4$), all serve to cause these slightly lower than theoretical fertilizer values.

As with the nitrogen component, various other alternatives are available for the phosphate rock acidulation step. For instance, if acidulation is conducted with nitric, instead of sulfuric acid, calcium nitrate is formed instead of gypsum (Eq. 11.62).

$$CaF_2 \cdot 3Ca_3(PO_4)_2 + 14\ HNO_3 + 3\ H_2O \rightarrow$$
$$3\ Ca(H_2PO_4)_2 \cdot H_2O + 7\ Ca(NO_3)_2 + 2\ HF \uparrow \qquad 11.62$$

This produces both P_2O_5 and nitrogen value from the single step. The mixture may be dried and used as obtained, or it may be partially or fully ammoniated to raise the nitrogen analysis still further. Full ammoniation would correspond to the stoichiometry of Eq. 11.63.

$$3\ Ca(H_2PO_4)_2 \cdot H_2O + 7\ Ca(NO_3)_2 + 12\ NH_4OH \rightarrow$$
$$3\ Ca(NH_4)_4(PO_4)_2 + 7\ Ca(NO_3)_2 + 12\ H_2O \qquad 11.64$$

Of course the ammoniation may also be carried out independently of the use of nitric acid if only the slower acting ammonia nitrogen rather than a mixed ammonia – nitrate nitrogen contribution in the formulation is wanted.

Another option to contribute phosphate to a fertilizer formulation is to use phosphoric instead of sulfuric acid for initial acidulation of phosphate rock. In this way the only solid product of acidulation is the soluble calcium dihydrogen phosphate free of gypsum diluent product with about three times the "available P_2O_5" of superphosphate (Eq. 11.64).

$$CaF_2 \cdot 3Ca_3(PO_4)_2 + 14 \; H_3PO_4 + 10 \; H_2O \; \rightarrow \qquad\qquad 11.64$$
$$10 \; Ca(H_2PO_4)_2 \cdot H_2O + 2 \; HF \uparrow$$

This is the reason for the commercial reference to this material as "triple superphosphate". This product too may be ammoniated, or mixed with other nitrogen-containing constituents such as urea or ammonium nitrate nitrogen to provide both nitrogen and phosphorus for plant growth.

Elemental phosphorus itself has been tested as a fertilizer in its more stable red allotropic form but was not found to be commercially useful, presumably because of low availability of phosphorus [72].

Potassium chloride, "muriate of potash," is the principal component used as such to supply the potassium macronutrient in fertilizer formulations. Some crops with a low chloride tolerance, such as tobacco, may require the chloride to be replaced by nitrate, carbonate, or sulfate anion to be useful (e.g., Eq. 11.65).

$$KCl + HNO_3 \rightarrow KNO_3 + HCl \uparrow \qquad\qquad 11.65$$

11.5.3. Environmental Aspects of Fertilizer Production and Use

Emissions during the production phase of most fertilizer constituents have been discussed, along with the details of the processes used to produce the chemical. However, the production of superphosphate and triple superphosphate involves procedures with potential emission problems which differ from those of phosphoric acid production. There is the potential loss of fluoride, primarily as calcium fluoride in the dusts from the primary grinding of phosphate rock, or from milling of the cured product. Hydrogen fluoride, silicon tetrafluoride, and fluosilicic acid vapors are also formed when acid is added to the phosphate rock. Dusts may be controlled in cyclones, for which 80–95% mass collection efficiencies for a particle size range of 40% $< 10 \, \mu m$ has been found [33]. Vapors may be controlled by dry methods involving adsorption on chalk or limestone followed by capture in a cyclone, which gives 95% control [73]. Or methods employing water scrubbers may be used, the most efficient of which achieve 98% fluoride vapor containment [74]. Of course, while scrubbing with water or aqueous solutions can effectively clean waste gases of contaminants, it also changes a potential air pollutant into a potential water pollutant high in fluoride, which must be treated prior to discharge (see Chap. 10).

Use of chemical fertilizers on agricultural land introduces the risk of contamination of surface waters from the leaching of soluble constituents from the soil after application (see Appendix 1). Important precautions to minimize this are the use of good farming practices, application of fertilizer at no more than the recommended rate, and maintaining soil structure through measures such as plowing in a fallow crop. Use of chemical fertilizers at no more than the recommended rate gives the added advantage of providing the best crop return for the fertilizer investment. Maintenance of soil structure helps by providing an insoluble matrix on which complexation assists to hold the soluble fertilizer constituents in place, and minimizes erosion losses of the soil itself. This may be achieved by the use of compost, manure, or specially treated sewage sludge, but these components of a soil management program must also be used carefully to avoid water or soil pollution problems from their use.

Long-term concerns have been expressed that soil applications of fixed nitrogen in various forms provide more substrate for ammonia and nitrous oxide losses to the air via bacterial processes [75] (see also Further Reading). These sources of fixed nitrogen contribute about one-third of the total atmospheric loading of reactive nitrogen (the remainder arises from natural sources), are in turn thought to contribute to depletion of the ozone layer through photochemical reactions.

REVIEW QUESTIONS

1. The Farmers Cooperative ammonia plant has two sources of hydrogen available, 25 tonne/day via a chlor-alkali facility and a further 60 tonne/day from operation of its own reforming and shift conversion processes.

   (a) Assuming there are no losses, calculate the daily production of ammonia in tonnes.

   (b) A part of the ammonia produced is sold directly. One hundred tonnes of ammonia/day is consumed by their nitric acid plant, which operates with 95% yield (selectivity) on ammonia. Calculate the tonnes of ammonium nitrate which would be available if the total nitric acid production was used in its manufacture.

2. (a) An ammonia plant uses steam and air reforming of methane together with shift conversion to provide the nitrogen and hydrogen feed for ammonia conversion. Assuming that air comprises oxygen and nitrogen in the mole ratio 1:4 and ideal gas relations, calculate the correct multiples of the three reforming reactions, which would give rise to a gross nitrogen to hydrogen mole ratio of 1:3 with most efficient use of methane (i.e., no wasted carbon monoxide).

   (b) If the plant in part (a) produces 600 tonnes of ammonia/day (minimum economic size for new ammonia plants), how many tonnes of carbon dioxide would be discharged daily?

3. The inert component (Ar, CH_4) of ammonia plant makeup gas is 1.5% by volume. What purge rate of the recycle gases is necessary, if there is a 10% conversion to ammonia, for the concentration of inerts in the feed gases to the converter to be kept to 15% of the total volume?

4. Assuming ideal gas behavior what would be the gas–catalyst contact time (seconds) for a space velocity of $5000 \, hr^{-1}$ and a gas mixture at:
   (a) 0°C and 1 atm pressure?
   (b) 0°C and 200 atm pressure?
   (c) 450°C and 200 atm pressure?

5. The tail gas nitrogen dioxide concentration from the stack of a nitric acid plant measures 0.20% by volume at 25°C and $1.013 \times 10^5 \, N \, m^{-2}$ (newtons/square meter, i.e., a pressure equivalent to 1 atm).
   (a) To what concentration, in mg/m^3 at the same temperature, would this correspond?
   (b) In the absorbers, where the pressure is $7.091 \times 10^5 \, N \, m^{-2}$ at an operating exit temperature of 25°C, what would the nitrogen dioxide concentration be, expressed in mg/m^3?

6. (a) For a 225 tonne/day nitric acid (100% basis) process using ammonia as the feedstock, assume that all of the nitric oxide oxidation to nitrogen dioxide and the nitrogen dioxide hydration reactions occur in the absorbers. If the combustion gas stream and water for absorption enter the absorber at 25°C and the product acid also leaves at 25°C, how many liters of absorber cooling water would be required per day to keep the temperature rise of this cooling stream to 10°C?
   (b) What approximate proportion of the total nitric acid plant cooling water requirements (i.e., fraction of the total heat removal required) does your answer to part (a) represent?

7. Calculate and give the fertilizer analyses as they would be commercially cited for the following:
   (a) Pure potassium nitrate
   (b) Fully ammoniated triple superphosphate $[Ca\{(NH_4)_2PO_4\}_2]$
   (c) Pure potassium sulfate
   (d) A 30:70 mixture by weight of potassium chloride and fully ammoniated triple superphosphate, respectively

8. (a) What percentage of sand by weight would have to be blended with ammonium nitrate to quote a fertilizer analysis of 15–0–0 on the package?
   (b) What would be the fertilizer analysis that should be quoted on the package for a mixture of 40% by weight ammonium nitrate, 30% potassium chloride, and 30% triple superphosphate $(Ca(H_2PO_4)_2)$?

9. (a) What is the relative cost/tonne of fertilizer nitrogen when added via ammonia at \$210.00/tonne, ammonium nitrate at \$150.00/tonne, or urea, H_2NCONH_2, at \$200.00/tonne?
   (b) Briefly comment on other factors, which might be important, apart from lowest cost per tonne of added nitrogen, in the selection of these three sources of N in the formulation of a blended fertilizer.

10. Calculate and give the fertilizer analyses as they would be commercially cited for the following:
    (a) Superphosphate made by nitric acid acidulation of phosphate rock $(CaF_2 \cdot Ca_3(PO_4)_2)$

(b) Ingredient of part (a), blended with 25% by weight potassium chloride

(c) Pure ammonium phosphate

(d) A 75:25 mixture of ammonium phosphate and potassium chloride

FURTHER READING

G. Bird, D. Limb, and A. Pandit, Flowsheet simulation models crude distillation and ammonia production, *Oil & Gas J.* **88**, 43–47, July 23, (1990).

R.W. Howarth, E.W. Boyer, W.J. Pabich *et al.* Nitrogen use in the U.S. from 1961–2000 and potential future trends, *AMBIO* **31**(2), 88–96, Mar. (2002).

Klaas van Egmond, T. Bresser, and L. Bouwman, The European nitrogen case, *AMBIO* **31**(2), 72–78, Mar. (2002).

R.M. Harrison and H.A. McCartney, Some measurements of ambient air pollution arising from the manufacture of nitric acid and ammonium nitrate fertilizer. *Atmos. Envir.* **13**, 1105 (1979).

K.V. Reddy and A. Husain, Vapor-liquid equilibrium relationship for ammonia in presence of other gases. *Ind. Eng. Chem. Process Res. Dev.* **19**, 580 (1980).

Rosen, M.A., . . . assessment of thermodynamic efficiencies and losses for natural gas-based . . . processes for hydrogen, ammonia and methanol, *Energy Convers. and Manag.* 37, 359–367, Mar. (1996).

REFERENCES

1. Subcommittee on Ammonia, Committee on Medical and Biologic Effects of Environmental Pollutants, and National Research Council. "*Ammonia*," University Park Press, Baltimore, 1979.
2. J.R. LeBlanc, Jr., S. Madhavan, and R.E. Porter, Ammonia, In: "Kirk-Othmer Encyclopedia of Chemical Technology," 3rd ed., vol. 2, p. 470–516, John Wiley & Sons, New York, 1978.
3. Facts & Figures for the Chemical Industry, *Chem. Eng. News,* **83**(28), 41–76, July 11 (2005), and earlier issues.
4. *Can. Chem. Process.* Fertilizers: at rock bottom **54**(9), 55, Sept. (1970).
5. T.H. Chilton, "Strong Water; Nitric Acid: Sources, Methods of Manufacture, and Uses," MIT, Cambridge, MA, 1968.
6. J.W. Dobereiner, *Ann. Chim. (Paris).* **24**(2), 91 (1823); cited by Kirk-Othmer, 2nd ed. Vol. 2, p. 268. Wiley, New York, 1968.
7. F. Haber, S. Tamaru, and C. Ponnaz, *Z. Elektrochem.* **21**(89), 128, and 191 (1915).
8. A.T. Larson, The ammonia equilibrium at high pressures, *J. Amer. Chem. Soc.* **46**, 367–372 (1924), and earlier refs.
9. A. Nielson, "An Investigation on Promoted Iron Catalyst for the Synthesis of Ammonia," 3rd ed. J. Gjellerups Forlag, Denmark, 1968; cited by Kirk-Othmer [2].
10. P.H. Emmett and J.T. Kummer, *Indus. Engin. Chem.* **35**, 677 (1943).
11. S. Czernik, R. French, C. Feik *et al.*, Hydrogen by catalytic steam reforming of byproducts from biomass thermoconversion, *Indus. Engin. Chem. Res.* **41**, 4209–4215, Aug. 21 (2002).
12. N.A. Lange, ed., "Handbook of Chemistry," 10th ed., p. 915. McGraw-Hill, Toronto, 1969.
13. Methane reforming to stay. *Chem. Eng. News.* **59**(35), 39, Aug. 31 (1981).
14. C. Matasa and E. Tonca, "Basic Nitrogen Compounds, Chemistry, Technology, Applications," Chem. Publ. Co., New York, 1973.
15. F.A. Lowenheim and M.K. Moran, Faith, Keyes and Clark's Industrial Chemicals, 4th ed. Wiley, New York, 1975.
16. C.R. Gagliardi, D.D. Smith, and S.I. Wang, Strategies to improve MEA CO_2-removal at ammonia plant. *Oil Gas J.* 44–49, Mar. 6 (1989).
17. R.N. Shreve and J.A. Brink, Jr., "Chemical Process Industries," 4th ed. McGraw-Hill, New York, 1977.

18. M.R. Louthan, Jr., *Process Indus. Corros.* p. 126 (1975).
19. R.E. Kirk and D.F. Othmer, eds., "Encyclopedia of Chemical Technology" Wiley, New York, 1941.
20. Argon recovery plant. *Chem. Eng. News, 58(4), 32, Jan. 28 (1980).*
21. Methane reforming to stay. *Chem. Eng. News* 59(35), 39, Aug. 31 (1981).
22. M. Henderson and M. Gandhi, Cryogenic methods to improve ammonia production, *Hydroc. Process. (Intern. Ed.).* 80(10), 100–102, Oct. (2001).
23. Ammonia plants more efficient. *Can. Chem. Process.* 64(10), 10, Oct. (1980).
24. M. Lauzon, *Can. Chem. Process.* 65(3), 42, May (1981).
25. L.C. Daigre, III and G.R. Nieman, … Computer control of ammonia plants, *Chem. Engin. Prog.* 70(2), 50, Feb. (1974).
26. G.P. Williams and J.G. Sawyer, … What causes ammonia plant shutdowns, *Chem. Engin. Prog.* 70(2), 45, Feb. 1974.
27. J.G. Sawyer and G.P. Williams, Ammonia… Turndown efficiency of single train plants, *Chem. Engin. Prog.* 70(2), 62, Feb. 1974.
28. Ammonia unit, *Chem. Eng. News*, 58(33), 24, Aug.18 (1980).
29. Monsanto sells hydrogen-recovery. *Chem. Eng. News*, 58(18), 8, May 5 (1980).
30. Japanese Technologists, *Can. Chem. Process.* 59(5), 8, May (1975).
31. L.J. Buividas, Coal to ammonia - its status, *Chem. Engin. Prog.* 77, 44–53, May 1981.
32. T.A. Czuppon, S.A. Knez, J.M. Rovner *et al.*, Ammonia, In: "Kirk-Othmer Encyclopedia of Chemical Technology", 4th ed. Vol. 2, p. 638–691, John Wiley & Sons, New York, 1992.
33. Desert sands catalyze ammonia formation. *Chem. Eng. News* 56(46), 7, Nov. 13 (1978).
34. TVA ammonia-from-coal project, *Chem. Eng. News* 57(33), 27, June 4 (1979).
35. "Guidelines for Limiting Contaminant Emissions to the Atmosphere from Fertilizer Plants and Related Industries in Alberta," Standards and Approvals Division, Alberta Dept. of the Environment, Edmonton, 1976.
36. H.F. Lund, ed., "Industrial Pollution Control Handbook," McGraw-Hill, Toronto, 1971.
37. T. Dear, Ammonia plant operations … gas handling noise control, *Chem. Engin. Prog.* 70(2), 65, Feb. 1974.
38. C.F. Kuhlmann, *Justus Liebigs Ann. Chem.* 29, 272 (1839), and reference cited therein.
39. Statistics Canada, "Catalog 46-002," Supply and Services Canada, Ottawa, 1982.
40. R. Thompson, ed., "The Modern Inorganic Chemicals Industry," p. 221. Chemical Society, London, 1977.
41. F.A. Cotton and G. Wilkinson, "Advanced Inorganic Chemistry; A Comprehensive Text," 3rd ed. p. 357. Wiley, New York, 1972.
42. M. Bodenstein, Z. *Electrochem.* 34, 183 (1918).
43. R.M. Stephenson, "Introduction to the Chemical Process Industries," p. 136. Reinhold, New York, 1966.
44. M.C. Manderson, Demand/supply outlook for nitric acid, *Chem. Engin. Prog.* 68(4), 57, April 1972.
45. D.J. Newman and L.A. Klein, Recent developments in nitric acid manufacture, *Chem. Engin. Prog.* 68(4), 62, April 1972.
46. New generation of nitric acid plants, *Chem. Engin. News* 54(52), 33, Dec. 20 (1976).
47. Technology Newsletter, *Chem. Week* 108(13), Apr. 1 (1970).
48. V.A. Sadykov, L.A. Isupova, I.A. Zolotarskii *et al.*, Oxide catalysts for ammonia oxidation in nitric acid production, *Appl. Catal. A-Gen.* 204(1), 59–87, Nov. 6 (2000).
49. L.M. Marzo and J.M. Marzo, Concentrating nitric acid by surpassing an azeotrope, *Chem. Engin. (N.Y.)* 87, 54, Nov. 3, 1980.
50. G.R. Gillespie, A.A. Boyum, and M.F. Collins, Catalytic purification of tail gas, *Chem. Engin. Prog.* 68(4), 72, April, 1972.
51. C. Boyars, Reducing explosion sensitivity of ammonium nitrate fertilizer, *Indus. Engin. Chem., Prod. Res. Dev.* 15(4), 308–309, 1976.
52. F.A. Patty, ed., "Industrial Hygiene and Toxicology," 2nd ed. Vol. 2, p. 918. Wiley, New York, 1963.
53. W. Frietag and M.W. Packbier, *Ammonia Plant Saf.* 20, 11 (1978).
54. M. Yamaguchi, K. Matsushita, and K. Takami, Remove NOx from HNO_3 tail gas, *Hydrocarbon Process.* 55(8), 101–106, Aug. 1976.
55. C.G. Swanson, Jr., J.V. Prusa, T.M. Hellman, and D.E. Elliott, *Pollut. Engin.* 10(10), 52, Oct. (1978).

55a. Industrial Chemicals and Synthetic Resins, Statistics Canada, cat. no. 46-002, December 2000, and earlier years.

56. To ... "United Nations, Rome, 1975." "FAO Statistical Database. Available at http://apps. fao.org/page/form?collection = Fertilizers&Domain = Means&servlet = 1&language = EN

57. "*Japan Economic Yearbook 1981/82*," p. 69. The Oriental Economist, Tokyo, 1981, and earlier issues.

58. I. Dunstan, *Chem. Brit.* 7(2), 62, Feb. (1971).

59. C. Boyars, Reducing explosion sensitivity of ammonium nitrate fertilizer, *Indus. Engin. Chem., Prod. Res. Dev.* 15(4), 308–309 (1976).

60. R. Moniotte, R. Pouilliart, and F. Vanhecke, Make ammonium nitrate with net export steam, *Hydrocarbon Process.* 64(5), 109–113, May 1985.

61. M. Bryson, Experimental design boosts production yields, *Chem. Engin. (N.Y.).* 102, 155 (1995).

62. E. Gain, S. Laborie, P. Viers *et al.*, Ammonium nitrate wastewater treatment by...membrane electrolysis and electrodialysis, *J. Appl. Electrochem.* 32(9), 969–975, Sept. (2002).

63. Two ammonia plants revamped to improve urea production, *Oil & Gas J.* 97(19), 56–58, May 10 (1999).

64. E. Otsuka, S. Inoue, and T. Jojima, What's new in urea technology, *Hydrocarbon Process.* 55(11), 160–164, Nov. 1976.

65. Dow feed-grade Biuret, *Chem. Eng. News* 49(25), 23, June 21 (1971).

66. A new liquid cattle feed supplement, *Can. Chem. Process.* 56(10), 4, Oct. (1972).

67. Product profile urea, *Can. Chem. Process.* 61(10), 50, Oct. (1977).

67a. Ritter, S., Aircraft Deicers, *Chem. Eng. News* 79(1), 30, Jan. 1, 2001.

68. C.J. Pratt, Chemical fertilizers, *Sci. Amer.* 212(6), 62, June 1965. Pratt ref., Chemical Fertilizers

69. Sulphur response, *Chem. Can.* 24(7), 5, Summer (1972).

70. R.J.P. Williams, Inorganic elements in biology and medicine, *Chem. Brit.* 15(10), 506 (1979).

71. Nitrogen fixation research advances, *Chem. Eng. News* 58(49), 29, Dec. 8 (1980).

72. H.P. Rotbaum and W. Kitt, *New Zeal. J. Sci.* 7, 67 (1964).

73. K. Karbe, *Chem. Engin. (London).* 221, 268 (1968).

74. N.L. Nemerow, "Industrial Water Pollution, Origins, Characteristics, and Treatment," p. 613. Addison-Wesley, Don Mills, Ontario, 1978.

75. Fertilizer may deplete ozone layer, *Chem. Eng. News* 56(40), 6, Oct. 2 (1978).

Aerial view of the layout of the Alcan aluminum smelter soon after it started production at Kitimat, British Columbia. Dock facilities for alumina, and aluminum shipments are in the foreground, and alumina storage silos are visible at the left end of the site.

Close-up of the pneumatic conveying facilities used in the operation of the dry scrubber for plant-wide control and recycling of volatile flourides.

12

ALUMINUM AND COMPOUNDS

Science got the beautiful metal aluminum out of the clay, which ignorance trod under foot.
                    —*John Cunningham Geikie, 1976*

12.1. HISTORICAL BACKGROUND

Aluminum is a metal that came of age in the twentieth century. Its volume of production has grown from about 1/100th that of copper, lead, and zinc prior to 1900, to two to four times the volume of these more common nonferrous metals during the last century (Table 12.1). H.C. Oersted, in 1825 (Table 12.1) in Denmark, was the first to isolate aluminum in impure form by the reduction of aluminum chloride with potassium amalgam. Two years later in Germany, F. Wohler obtained higher purity metal and fully described its properties. H.E. St.-Claire Deville put aluminum production into commercial practice in France by 1845 using sodium fusion to reduce aluminum chloride (Eq. 12.1).

$$AlCl_3 + 3\ Na \rightarrow Al + 3\ NaCl \text{ (conducted as a melt)} \qquad 12.1$$

By 1852 the metal sold commercially for about \$545 per pound. An elaborately enameled fan (now housed at the Smithsonian, Washington, DC) was made from aluminum for the Paris Exposition of 1867 while it was still classed as a precious metal, and it is said that Napoleon III had a tea service made from it. Process improvements, such as cryolite fluxing of the melt and lower sodium costs, brought the price of the metal down to U.S.\$8 per pound by 1886, which removed its status appeal.

Paul Heroult in France and Charles Hall in the U.S.A. independently devised electrolytic methods to reduce aluminum oxide to the metal in the early part of 1886 and first made this possible on a large scale. In their procedure, the essence of which is still used today, alumina (Al_2O_3) was dissolved in a bath of molten cryolite (Na_3AlF_6). Passage of a direct current through this bath, via an electrolytic process (not electrothermal, as used for

■■■■ **TABLE 12.1   Growth Rates of World Production of Some Nonferrous Metals Relative to 1900 Levels**[a]

Year	Aluminum		Copper		Lead		Zinc	
	Thousand tonnes	Multiple of 1900	Thousand tonnes	Multiple of 1900	Thousand tonnes	Multiple of 1900	Thousand tones	Multiple of 1900
1900	5.7	1	499	1	877	1	479	1
1960	4,670	819	4,400	8.82	2,630	3.00	3,070	6.41
1970	9,666	1,700	6,227	12.5	3,299	3.76	4,905	10.2
1980	15,368	2,700	6,194	12.4	2,957	3.37	4,354	9.09
1990	21,846	3,830[b]	9,680	19.4[b]	5,950	6.78[b]	7,180	15.0[b]
2000	24,400	4,280	11,080	22.2	3,340	3.81	4,330	9.04

[a]Mass data in thousands of metric tonnes compiled from Kirk-Othmer [1] and *Minerals Yearbooks* [2].
[b]2000 production as a multiple of 1960 values was aluminum, 5.22; copper, 2.52; lead, 1.27; and zinc, 1.41.

phosphorus), produced a layer of molten aluminum on the bottom of the cell. This procedure raised the purity and decreased the price of the metal. The ease of availability to the metal coupled with its lightness and corrosion resistance led to a rapid increase in consumption (Table 12.1).

Continued growth in the production of aluminum is unlikely to be hampered by a shortage of mineral since the earth's crust consists of about 8% aluminum, chiefly as aluminosilicates. Even though aluminum is the most abundant metallic crustal element, bauxite ores suitable for aluminum recovery only occur in more limited areas where natural leaching processes have concentrated the aluminum-containing minerals. Since the free metal is chemically reactive it is never found in nature in this form.

With the rapid acceptance of the Hall–Heroult electrolytic method of aluminum production, facilities using this process have tended to be constructed in areas with abundant, low-cost electric power (Table 12.2). In addition, to minimize shipping costs it is usual to process bauxite ore to purified alumina at the mine site for shipment to smelters (Table 12.2). Thus, aluminum smelters are sited so as to provide easy shipping access for economical alumina delivery from the mine. The production of aluminum from bauxite logically can be considered in two steps: the first, production of high-purity alumina from the working of natural bauxite deposits, and the second, electrolytic reduction of alumina to the metal.

12.2. ALUMINA FROM BAUXITE: THE BAYER PROCESS

Bauxite, the principal ore used for aluminum smelting, is named after Les Baux, Provence, the village where the first deposits were discovered. Bauxite contains hydrated alumina equivalent to as much as 40–60% Al_2O_3, and is free of the other siliceous materials leached out over time. However it still contains 10–30% iron oxide, and some silica and other impurities, making it unsuitable for direct electrolysis. The first commercial-scale recovery of alumina from bauxite was devised by Henri Deville, but by 1900 this was largely

TABLE 12.2 Annual Production of Aluminum by the World's Major Producers, in Thousands of Metric Tonnes and Percent of World[a]

	1945		1965		1970	1980		1990	2000	
	Mass	%	Mass	%	Mass	Mass	%	Mass	Mass	%
Canada	196	22.1	762	11.3	964	1,068	6.9	1,567	2,373	9.7
France	37.2	4.2	340	5.1	380	432	2.8	325	441	1.8
Japan	5.4	6.1	323	4.8	733	1,091	7.1	51	7	<0.1
Norway	7.0	0.8	277	4.1	530	651	4.2	867	1,026	4.2
U.S.A.	449	50.6	2,498	37.1	3,607	4,654	30.3	4,121	3,668	15.0
U.S.S.R.	86.3	9.7	1,279	19.0	1,098	1,787	11.6	–	3,245[b]	13.3
W. Germany	20.0	2.3	234[c]	3.5	308[c]	731[c]	4.8	720[c]	644[d]	2.6
Other	37.3	4.2	1,014	15.1	2,046	4,954	32.3	–	13,004	53.3
World	887.0	100.0	6,727	100.0	69,666	15,368	100.0	21,846	24,400	100.0

[a]List includes all countries whose annual primary aluminum production exceeded 400,000 tonnes in 1980. Compiled and calculated from data of *Minerals Yearbooks* [2].
[b]Russia after perestroika.
[c]Totals for Germany after partition.
[d]Germany after reunification.

replaced by the more economical process based on caustic extraction developed by Bayer in Austria.

Alumina recovery from bauxite by extraction with sodium hydroxide, now referred to as the Bayer process, relies on the amphoterism of aluminum for its success (Fig. 12.1). Details of the alumina extraction procedure

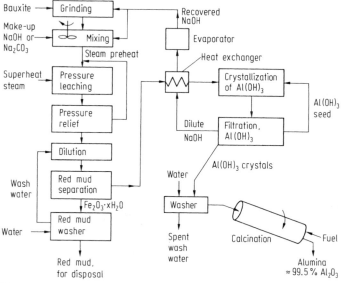

FIGURE 12.1 Production of alumina from bauxite via the Bayer process.

required depend on the form of hydrated alumina, which occurs in the bauxite being processed. Control and simulation of the Bayer process has recently been reviewed [3].

Initially the coarse ore is mechanically reduced to a finely divided form and stirred with the requisite concentration of aqueous base prior to pressure leaching. The alumina from trihydrate bauxite ($Al_2O_3 \cdot 3H_2O$; natural bayerite or gibbsite, $Al(OH)_3$) is relatively easy to dissolve using 15–20% aqueous sodium hydroxide, under pressure, at temperatures of 120–140°C (Eq. 12.2).

$$Al(OH)_3 + NaOH \rightarrow \underset{\text{soluble}}{AlO \cdot ONa} + 2\,H_2O \qquad\qquad 12.2$$

If the content of monohydrate alumina ($Al_2O_2 \cdot H_2O$; e.g., diaspore, $AlO(OH)$) in the ore is significant, more concentrated sodium hydroxide (20–30%) and temperatures of 200–250°C and up to 35 atm are required to effectively leach out the alumina content (Eq. 12.3).

$$AlO \cdot OH + NaOH \rightarrow AlO \cdot ONa + H_2O \qquad\qquad 12.3$$

The more severe conditions will also extract trihydrate alumina from bauxite so the deciding factor for the digestion conditions required is the presence of a significant concentration of monohydrate alumina content. Fortunately, trihydrate alumina is the dominant aluminum species present in the major world deposits of Africa, Australia, the Caribbean, Central and South America, and the U.S.A. European bauxites are mainly the monohydrate.

Iron oxide, clays, and most other impurities do not dissolve under alumina digestion conditions and are quite finely divided (1–10 μm). By adding wash water to decrease the viscosity, it is possible to decant the still hot sodium aluminate solution from the slowly precipitating red muds. These muds that are red from the high iron content are then washed with water to minimize losses of alumina and sodium hydroxide on disposal of the muds. By using this wash water for dilution of the next digester output, alumina and base are not lost from the Bayer circuit. Newer facilities use pressure filtration for both removal and washing of red muds.

Aluminum hydroxide crystals are obtained from the supernatant sodium aluminate solution after a final filtration by ensuring a $Na_2O : Al_2O_3$ ratio of 1.5–1.8:1, by controlling the solution temperature to about 60°C, and by seeding with crystals from an earlier crystallization. Even with seeding the crystallization process requires 2–3 days. However, sufficiently large crystals are obtained by the reverse of the solution process (Eq. 12.4) to enable product recovery by filtration.

$$AlO \cdot ONa + 2\,H_2O \rightarrow Al(OH)_3 + NaOH \qquad\qquad 12.4$$

The aluminum hydroxide product is washed with water and then dried while still on the filters. Evaporation of most of the water from the filtrates allows return of the sodium hydroxide solution to be recycled to the initial grinding and pressure leaching circuits for reuse at appropriate concentrations.

Calcination of the aluminum hydroxide at about 1200°C in either fluidized bed or rotary calciners finally yields alumina of 99.5% purity (Eq. 12.5).

$$2\,Al(OH)_3 \rightarrow Al_2O_3 + 3\,H_2O \qquad\qquad 12.5$$

The chief contaminant is 0.3–0.5% sodium oxide, which fortunately does not affect electrolysis, with <0.05% calcium oxide, <0.025% of silica or iron oxide, and <0.02% of any other metallic oxide [4]. Apart from metal production, some of this high temperature alumina is used for the manufacture of synthetic abrasives and refractory materials. Activated alumina destined for adsorptive uses is produced in the same way, except that more moderate calcining temperatures of about 500°C are employed, which produces a highly porous product with excellent surface activity. The volume of alumina from the world's major producers is listed in Table 12.3. Australia has been the largest producer for many years (Table 12.3).

Gallium is a trace contaminant of chemical and commercial interest, which is present to the extent of 60–200 μg/g in Bayer process liquors. Situated in the same group as aluminum in the periodic table, gallium's very similar chemical properties cause it to be carried through the alumina purification and electrolytic steps along with the aluminum. With a melting point of 30°C and boiling point of 2403°C it is the element with the longest liquid interval. Production is on the scale of 50,000 kg/year, mostly in Australia, Germany, and Russia [2]. It is used in electrooptical devices, such as LEDs, in the GaAs used for the microchannel plates that guide and amplify photons in night vision goggles and sights, and for the integrated circuits in one of the fastest digital signal processors made. Improvement of recovery methods has been examined [6].

It is also possible to obtain alumina from clays or bauxite via leaching with sulfuric acid, a technique, which is especially useful with high silica or high iron oxide bauxites (Eq. 12.6).

TABLE 12.3 Major World Producers of Alumina (Production in Thousands of Metric Tonnes)

	1972	1980	1985	1990	2000
Australia	3,068	7,246	8,792	11,200	15,037
Canada	1,149	1,202	1,020	1,090	1,023
France	1,274	1,173	734	606	200
Jamaica	2,087	2,456	1,513	2,870	3,600
Japan	1,644	1,936	978	481	369
Surinam	1,378	1,316	1,000[b]	1,530	–
U.S.A.	6,114	6,810	3,456	5,230	4,790
U.S.S.R.	2,300	2,700	3,500	5,900	2,850[c]
W. Germany	916	1,608	1,657	922	700[d]
Yugoslavia	135	1,635	1,138	1,090	370
Others	3,535	3,709	7,820	11,681	20,361
World total	23,600	33,426	31,617	42,600	49,300

[a]Includes all those countries producing more than a million tonnes of alumina annually by 1980. Compiled from data in *Minerals Yearbooks* [2] World Mineral Statistics [5].
[b]Estimate.
[c]Russian Federation.
[d]Germany.

$$\text{clays} + H_2SO_4 \rightarrow Al_2(SO_4)_3 + \text{water insoluble residue} \qquad 12.6$$
$$\text{soluble}$$

Crystals of pure aluminum sulfate hydrate ($Al_2(SO_4)_3 \cdot 18H_2O$) are obtained from the leach solution, which may then be calcined to yield alumina (Eq. 12.7).

$$Al_2(SO_4)_3 \cdot 8\,H_2O \rightarrow Al_2O_3 + 3\,SO_3 + 18\,H_2O \qquad 12.7$$

Cryolite ($AlF_3 3NaF$) is the electrolyte used, which is the major component of the electrolytic bath. This is a rare natural mineral originally found only in Gothaab, Greenland. Most present aluminum smelters use synthetic cryolite prepared from alumina and hydrogen fluoride in the presence of sodium hydroxide (Eqs. 12.8 and 12.9).

$$CaF_2 + H_2SO_4 \rightarrow 2\,HF + CaSO_4 \qquad 12.8$$
$$\text{fluorspar}$$

$$6\,HF + \underset{\text{as briquets}}{Al(OH)_3} + 3\,NaOH \rightarrow Na_3AlF_6 + 6\,H_2O \qquad 12.9$$

This produces a cell electrolyte with characteristics identical to the natural mineral.

12.3. ALUMINUM BY THE ELECTROLYSIS OF ALUMINA

Reduction of alumina to aluminum is an electrolytic process like the dominant method for chlorine and caustic production from sodium chloride. This means that there is a direct link between the mass of aluminum produced and the electricity consumed, as with chlorine and caustic soda. The required ionic mobility to accomplish this could theoretically be provided by melting alumina, of melting point 2,050°C. But this temperature is so high (above the useful range of most refractory lining materials) that this is not technically feasible. However, mobility can also be provided by dissolving the alumina in an ionizing solvent, although not water, in which alumina is insoluble. Alumina is, however, sufficiently soluble in molten cryolite (Na_3AlF_6, melting point 1006°C) to enable electrolysis to be carried out. In practice, addition of a few percent of calcium fluoride and/or aluminum fluoride depresses the melting point of the electrolyte some 4–5°C for each 1% of additive [1], which enables electrolysis at ca. 950°C, temperatures technically easier to achieve (Table 12.4).

While the aluminum reduction process is primarily electrolytic, much of the power required is consumed through cell resistance, which keeps the electrolyte in a fluid state.

Alumina electrolysis cells consist of a rectangular steel shell lined with a 25–35 cm layer of baked and rammed dense carbon, which provides both chemical resistance and the cathode contact with the electrolyte via steel bus bars imbedded in the carbon. Normal lining life is 4–6 years, after which it is replaced as large preformed slabs. Once a reduction pot has been started the bulk of the cathode current to the carbon lining is via the pool of newly formed molten aluminum in the bottom of the cell (Fig. 12.2).

TABLE 12.4 Typical Melt Composition for Electrolytic Reduction of Alumina[a]

Component	Range (%)
Cryolite, Na_3AlF_6	80–85
Fluorspar, CaF_2	5–7
Aluminum fluoride	5–7
Alumina, Al_2O_3	2–8

[a]Data from Kirk-Othmer [1]. Normally alumina reduction pots are operated on the acid side (i.e., a net AlF_3:NaF mole ratio of 1.2 to 1.5:1).

Anode elements are commonly prebaked low ash carbon blocks, since any ash residue ends up in the electrolyte. These are electrically connected to copper or aluminum bus bars (heavy electrical conductors) suspended over the cell, which also provide mechanical support and a means for vertical adjustment of the anode elements. An anode variant is the Soderberg paste option, which uses powdered petroleum coke formed into a paste with hard pitch. Electrical contact is established and mechanical adjustment provided by using specially shaped steel pins (Fig. 12.3). As the baked portion of this anode (Fig. 12.3) is gradually consumed, the paste approaches the molten electrolyte and the volatile components in the paste vaporize to leave a hard baked working anode element. Either type of anode element is consumed at the rate of 1–2 cm/day during normal operation, requiring periodic vertical adjustment to maintain an anode–aluminum metal pool spacing of about 5 cm.

To start operation of a pot, anodes are lowered to the lining, the solid electrolyte and alumina components are placed in it, and then the power is turned on. Initial melting of the pot charge is by resistance heating. A typical

FIGURE 12.2 Cross-sectional diagram of a typical aluminum reduction pot using prebaked carbon anodes. The refractory brick lining and the frozen crust beside and on top of the molten electrolyte provide thermal insulation, which raises overall energy efficiencies. (Courtesy of Aluminum Co. of America.)

Anode buses

Anode rod

Steel anode stud

Gases,
to effluent collection system

Anode casing

Gas collecting skirt

Molten electrolyte

Crust

Alumina

Floor-level

Fluid paste

Partially baked anode paste

Burner

Fully baked carbon

Molten aluminum cathode

Cathode bus   Rammed carbon   Carbon block lining   Steel cradle   Steel cathode collector bar

FIGURE 12.3 Diagram of an aluminum reduction pot showing the arrangement of the principal components of a vertical stud Soderberg paste anode [7]. (Reprinted courtesy of Alcan Smelters and Chemicals Ltd.)

cell consumes from 50,000 to 200,000 A at an operating voltage of 4.5–5.5. As the charge melts the anode elements are gradually raised to the normal 5-cm spacing, and electrolysis begins.

At the cathode, aluminum ions pick up electrons to form elemental aluminum (Eq. 12.10), which being denser than the molten cryolite, gradually accumulates as a pool beneath the electrolyte melt layer and rests on the carbon lining.

$$Al^{3+} + 3\ e^- \rightarrow Al^0 \qquad\qquad 12.10$$

For a 125,000 ampere cell the metal production capacity is 810–910 kg/day. Normally, alumina additions are made to each cell at about four hourly intervals. However, if the alumina consumption in any one cell exceeds the addition rate and the alumina content of the electrolyte drops to about 2% alumina or less, there is a sudden rise in the cell operating voltage to about 50 V. This is caused by polarization of the anode from the near stationary film of carbon monoxide and carbon dioxide on its surface [1]. This situation, called the "anode effect," is detected via a 50–60 V light wired in parallel to each cell in a multicell potline. Normal cell operating voltages are too low to cause bulb incandescence, but when an anode effect occurs the light of the affected pot comes on to alert the operators to correct the situation.

To correct the anode effect, fresh powdered alumina prewarmed on the electrolyte crust of the cell is broken into the molten electrolyte with a remotely controlled pneumatic hammer. The anode polarizing gas was once dispelled by the rapid flow of combustion gases from the burning of a green wood pole poked through the crust. Now, however, air lances are used to accomplish this task, referred to as "poling."

During the alumina-deficient situation signaled by an anode effect light, fluorine, as well as oxygen deposition starts to occur at the anodes. At the same time the higher anode–cathode voltage differential may also cause arcing to occur across the anode–aluminum pool gap. These severe conditions cause the abnormally present fluorine to react with the hot carbon, forming carbon tetrafluoride and smaller amount of hexafluoroethane (Eqs. 12.11 and 12.12).

$$2 F_2 + C \text{ (hot)} \rightarrow CF_4 \qquad\qquad 12.11$$

$$3 F_2 + 2 C \text{ (hot)} \rightarrow C_2F_6 \qquad\qquad 12.12$$

These fluorocarbons and carbon monoxide combine to stabilize the polarization of the anode elements by formation of a gas envelope around them [8]. It is these occurrences that cause the rapid rise of the operating voltage of a cell is confirmed by it being temporarily alleviated by stirring the electrolyte, which momentarily clears the gas envelope from the anode surface.

At about daily intervals, as sufficient aluminum accumulates in each cell, a stout pipe attached to an evacuating vessel is broken through the crust on the cell and dipped into the pool of molten aluminum. The pressure in the receiving vessel is decreased, which pulls the melt into the container, a process, which is repeated for each cell down the potline, to collect the molten aluminum in a central holding furnace ready for fabrication.

At the same time as aluminum is reduced at the cathode, oxygen ions migrate to the suspended anode elements and are oxidized to atomic and possibly some molecular oxygen (Eq. 12.13).

$$O^{2-} - 2 e^- \rightarrow 2 [O]; (2[O] \rightarrow O_2) \qquad\qquad 12.13$$

Both forms of oxygen are generated in close contact with a hot carbon element, so it immediately reacts to form carbon monoxide and carbon dioxide (Eq. 12.14–12.16).

$$C \text{ (hot)} + [O] \rightarrow CO \qquad\qquad 12.14$$

$$CO \text{ (hot)} + [O] \rightarrow CO_2 \qquad\qquad 12.15$$

$$C \text{ (hot)} + O_2 \rightarrow CO_2 \qquad\qquad 12.16$$

However, as the carbon dioxide moves away from the anode region it contacts hot aluminum (and possibly sodium) vapor and fine liquid metal droplets (fog), which immediately oxidize to form the corresponding metal oxide and carbon monoxide (Eqs. 12.17 and 12.18).

$$3 CO_2 + 2 Al \rightarrow 3 CO + Al_2O_3 \qquad\qquad 12.17$$

$$CO_2 + 2 Na \rightarrow CO + Na_2O \qquad\qquad 12.18$$

Thus, the gas mixture collected from the skirted anode area consists of carbon monoxide (from incomplete initial oxidation, plus reduced carbon dioxide), and carbon dioxide in a 1:2 to 1:3 weight ratio. The toxic carbon monoxide in these hot gas mixtures normally burns spontaneously as it escapes from the confining crust. For the vertical Soderburg pots, where the gases are captured by the anode skirt, a small burner is used to convert the carbon monoxide content to carbon dioxide at each cell gas exit before the effluent gas mixture proceeds to treatment systems.

The overall electrochemical process is closely represented by Eq. 12.19.

$$Al_2O_3 + 3/2\ C \rightarrow 2\ Al + 3/2\ CO_2 \qquad \Delta H = +1096\ kJ\ (+262\ kcal) \quad 12.19$$

The large endotherm of the reaction is supplied electrochemically during the electrolysis. Substitution of this endotherm in the Gibbs–Helmholtz equation gives us the theoretical equilibrium voltage required for deposition of aluminum, which is 1.7 V (see Section 8.4). It is thought that there is no electrochemical assistance from carbon oxidation by the anode product; hence this single electrochemical potential is the same as the standard reduction potential for aluminum, which is -1.706 V [9]. The actual operating potential required is significantly higher than this, 4.5–5.5 V, which is needed to overcome electrolyte resistivity (Table 12.5) over the 5-cm electrode gap, polarization and other factors, as well as (Table 12.5) to move the process from an equilibrium to a producing situation. Thus, voltage efficiencies range from 24 to about 38% with the bulk of production taking place at the upper end of this range. This is much lower than any of the common chloralkali electrolysis cells and is primarily the result of overall cell resistance, which appears in the form of heat to keep the cell contents molten. The smaller surface to volume ratio for the larger electrolysis cells (up to 260 kA) confers some improvement in voltage efficiency, which improves overall energy efficiencies compared to that of smaller cells.

Like all electrochemical processes the quantity of product obtained for a given current flow/time period is related through Faraday's law (i.e., that in an electrochemical reaction 1 g equivalent weight of substance is deposited on the passage of 96,494 C (1 Faraday) of electricity). Since each aluminum ion,

TABLE 12.5 Electrochemical Parameters of Importance to Hall–Heroult Electrolysis of Alumina[a]

Reduction potentials (V)	
Theory	1.706
Normal operating	4.5–5.5 (to 7 occasionally)
Anode effect, at approximately 2% Al_2O_3	30–50
Operating capacities (kA)	50–260
Anode characteristics	
For prebaked: Resistivity	5–6 ohm/cm
Current density	6.5–13 kA/m^2
For Soderberg: Resistivity	6.5–7.8 ohm/cm
Current density	6.5–9 kA/m^2
Electrode spacing	ca. 5 cm
Electrolyte bath resistivity, at 950°C	0.50 ohm cm
Current efficiency, 100 kA cells	ca. 90%
Normal energy efficiency	34–38%
Typical anode gas composition[b]	85% CO_2, 15% CO

[a]Calculated and compiled from Kirk-Othmer [1, 4], and *The Chemistry of Aluminum* [10].

[b]Gas composition is given as experienced for 90% current efficiencies. At 80% current efficiency the gas composition will be approximately 75% CO_2, 25% CO, i.e., the CO content is inversely proportional to current efficiencies.

to be deposited, requires three electrons (Eq. 12.10), 1 g equivalent weight of aluminum is one-third of its atomic weight (26.9815 g/mole ÷ (3 electrons/mole)) or 8.994 g. Thus, 8.994 g of aluminum would be expected to be deposited in the cell, for each Faraday of electrons. In practice only about 90% of this amount is collected. This is the result of such processes as the reaction of anode carbon dioxide with already reduced aluminum or sodium, which returns these cell "products" to the electrolyte as oxides. It has been found that the addition of aluminum fluoride to the electrolyte tends to decrease the concentration of reduced species in the electrolyte, and in this way helps to raise current efficiencies.

Combining the voltage and current efficiency factors, the overall *energy efficiency* of the cell is usually about 34% (0.90×0.38) although during periods of high voltage operation this can drop as low as 22%. For aluminum production this translates to a power consumption of about 17,600 kWh/tonne. Power and other quantitative aspects of aluminum production, such as anode consumption rates are given in Table 12.6. For quantitative calculations the molar relationships given in Eq. 12.20 have been used, which quite closely reflect the actual operating experience.

$$Al_2O_3 + 2\ C \rightarrow 2\ Al + CO + CO_2 \qquad\qquad 12.20$$

Purity of the direct Hall–Heroult product is 99.5–99.7% aluminum, adequate for most uses requiring the pure metal, and for alloys. The chief impurities are iron, 0.06–0.10%, and silicon, 0.04–0.30%, with a total for all other impurities seldom exceeding 0.10% [1]. Analysis of both the aluminum and the electrolyte of each cell are normally conducted on at least a weekly basis. Among the benefits obtained from these results is advance warning when the carbon lining of a pot has worn through to the steel shell ahead of schedule, which is seen as an increase in the iron content of the aluminum. This cell is

TABLE 12.6 Materials and Energy Required to Produce One Tonne of Commercial Aluminum[a]

Alumina consumption	
Theory	1889.5 kg
Practice	1900–1901 kg
Carbon anode material	
Carbon, theory[b]	445 kg
Carbon, practice	450–550 kg
Pitch, Soderberg	100–200 kg
Electrolyte	
Cryolite	30–70 kg
Aluminum fluoride	ca. 40 kg
Power	15,000–17,600 kWh
Labor	13–31 man-hours

[a]Calculated from data of Kirk-Othmer [1,4], *The Chemistry of Aluminum* [10], Patterson [11], and OECD [12].
[b]Assuming that this is pure carbon, as is nearly true with prebaked anodes, and the stoichiometry of Eq. 12.19.

then taken out of production and repaired before aluminum and electrolyte break through the outer steel shell.

Electrolytic or fractional crystallization methods are employed to produce 99.99% pure aluminum. Electrolytic methods employ a melted aluminum–copper alloy (about 75:25) as the bottom layer, overlain by a high content of a barium salt (either chloride or fluoride) to produce a high-density electrolyte, topped by a layer of the electrolytically refined aluminum as it forms. By passage of a direct current through the cell floor as the anode and suspended carbon cathodes there is a gradual accumulation of refined aluminum in the top layer. Voltages, and hence voltage efficiencies, are similar to those required for primary aluminum production. Zone refining is used for 99.999% or higher aluminum purities. Recycling of scrap aluminum at present accounts for some 20% of aluminum produced. This requires less than 5% of the energy required to produce the metal from alumina [13].

12.4. NEW DEVELOPMENTS IN ALUMINUM PRODUCTION

Despite the world's abundant reserves of bauxite the deposits are not, in many cases, in the major aluminum producing areas. This factor has stimulated the extensive testing of alternative mineral sources of aluminum [11]. High-alumina clays, shales, and other alumina-containing minerals have all been tested as alumina sources, and some of these alternatives, such as nepheline syenite $((Na,K)(Al,Si)_2O_4)$ in the former U.S.S.R., have actually been used. Tests in the U.S.A. have been performed on hydrochloric or nitric acid extraction of alumina from kaolin clays (30–35% Al_2O_3). Alumina recovery from the clays removed from coal (ca. 28% Al_2O_3) during washing and from coal ash, if economically feasible, has the dual advantage of providing a resource from large-scale waste materials. French and Canadian ventures have studied alumina extraction from clays and shales using sulfuric and hydrochloric acids [14]. Recent developments in alumina processing technology have been reviewed [15].

Sumitomo Chemical technology consists of a number of small changes to conventional Hall–Heroult electrolysis cells, which combine to cut power consumption to some 14,000 kWh/tonne from the normal experience of 16,000 to 18,000 kWh/tonne [16]. In addition to reduced power consumption, better emission control, extended cell life, and decreased labor requirements are also achieved by these changes. Better external cell insulation, changes in what is basically a Soderberg anode design, and better fume containment by cell skirt construction modifications provide these improvements.

Advances in alumina reduction technology have been surveyed [17], and alternative pathways to aluminum examined [18]. Aluminum can be obtained by metallic, carbon, or electrolytic reduction of appropriate compounds. The earliest processes successfully used metallic reduction via potassium amalgam or sodium fusion to produce aluminum from aluminum chloride (Eqs. 12.21 and 12.22).

$$AlCl_3 + 3\ K_{(Hg)} \rightarrow Al + 3\ KCl + Hg \qquad\qquad 12.21$$

$$AlCl_3 + 3\ Na \rightarrow Al + 3\ NaCl \qquad\qquad 12.22$$

As Hall–Heroult technology became available these processes were displaced because of higher costs.

Reduction of alumina with carbon is fraught with problems, not the least of which is the very high temperatures required (about 2000°C) and the formation of quite stable aluminum carbide (Eq. 12.23).

$$2\ Al_2O_3 + 9\ C \rightarrow Al_4C_3 + 6\ CO \qquad 12.23$$

Metallic aluminum is sufficiently active that it reduces carbon dioxide to carbon monoxide (Eq. 12.17). This means that the favorable aspects of a solid–gas, alumina–carbon monoxide reduction cannot be utilized. However, good yields have been obtained experimentally by carrying out the reduction in two steps. The first step requires formation of the carbide (Eq. 12.23). Following this, the temperature of the system is raised another 100–200°C to obtain aluminum by the reaction of aluminum carbide with alumina (Eq. 12.24).

$$Al_4C_3 + Al_2O_3 \rightarrow 6\ Al + 3\ CO \qquad 12.24$$

At around 2,100°C, the temperature required for the second step, aluminum (melting point 660°C, boiling point 2,467°C) itself has sufficient volatility to cause metal loss and migration problems.

Carbothermic reduction in the presence of an alloying element, such as copper, iron, or silicon, to decrease aluminum vapor pressures decreases volatility problems but requires a further stage to recover aluminum from the alloy product. It may be selectively dissolved from the alloy with a more volatile metal, such as mercury, lead, or zinc, and then the aluminum recovered by distillation. Or, the tendency for aluminum halides to form more volatile monohalides at high temperatures, which revert to the trihalides at lower temperatures (Eq. 12.25) may be employed.

$$AlCl_3 + \underset{\text{in alloy}}{2\ Al} \underset{800°C}{\overset{1,300°C}{\rightleftharpoons}} 3\ AlCl \qquad 12.25$$

In this procedure, extensively tested by Alcan at one time, dissolution of aluminum from the alloy using aluminum trichloride at about 1,300°C produces the volatile aluminum subhalide [19]. Cooling the subhalide moderately returns a pool of molten product aluminum, and aluminum trichloride (boiling point approximately 183°C), which serves as a working fluid and can be recycled. Despite problems with the aluminum recovery methods there remains an interest in nonelectrolytic aluminum production, such as that offered by carbothermic processes [20].

Recent technology for electrolytic aluminum production employing aluminum chloride has also been of interest because of the about 30% power savings possible [21]. Since aluminum chloride melts at much lower temperatures and forms a much more fluid melt than the standard Hall–Heroult electrolyte matrix, much higher voltage efficiencies are possible. However, sublimation and control problems limit the utility of direct, one-component electrolytic methods. The essence of this idea is employed in the process, developed by C. Toth of Alcoa, which has the additional advantage of enabling clay sources of alumina to be tapped [22] (Fig. 12.4).

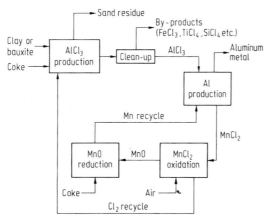

FIGURE 12.4 Simplified flowsheet of new Alcoa aluminum chloride-based route to aluminum. (Information from *Canadian Chemical Processing* [21], and *Chemical and Engineering News* [23].)

While technical details of this process have not been released, production probably involves the sequence of reactions given in Eqs. 12.26–12.29.

$$2\ Al_2O_3 + 6\ Cl_2 + 3\ C \rightarrow 4\ AlCl_3 + 3\ CO_2 \qquad\qquad 12.26$$

$$2\ AlCl_3 + 3\ Mn \xrightarrow[\text{anodes}]{\text{electrolysis}} 2\ Al + 3\ MnCl_2 \qquad\qquad 12.27$$

$$2\ MnCl_2 + O_2\ (\text{from air}) \rightarrow 2\ MnO + 2\ Cl_2\ (\text{Deacon related}) \qquad 12.28$$

$$2\ MnO + C \rightarrow Mn + CO_2 \qquad\qquad 12.29$$

Prototype results with this process were sufficiently attractive for Alcoa to have started production in 1976 using this technology on a scale of approximately 13,500 tonnes/year [16].

12.5. EMISSION CONTROL PRACTICES

12.5.1. Environmental Impacts of Bayer Alumina Production

The mining of bauxite for an alumina product has the same kind and degree of environmental impact as imposed by any strip mining operation. These effects may be minimized by spoils replacement and aesthetic contouring, as well as by the replanting of hardy grasses and shrubs, where appropriate.

Processing of bauxite to produce alumina produces volumes of "red muds," which contain principally water, iron oxide, silica, and the oxides of titanium, chromium, vanadium, and aluminum. The solids in this mixture eventually settle to a relatively high solids content sludge, so that a moderately sized holding pond may be used for many years.

However, as land and raw material costs rise, alumina producers are driven to seek ways to decrease the volume or eliminate this red mud waste. Alcoa has devised the "Alcoa-Bayer" process to recover the formerly discarded alumina fraction as combined alumina precipitated with silica and sodium hydroxide [24]. By adding limestone and sodium carbonate and calcining the mixture to a clinker, it is possible to return the alumina values to the main Bayer circuit. Decreasing the alumina content of the red muds substantially decreases the volume of red mud per tonne of alumina product.

Other tests have shown that the residual Al^{3+} and Fe^{3+} ions present in red muds can be made into inexpensive flocculants for sewage treatments, which are also particularly efficient at removing phosphate [25]. When processed to a finely divided dry powder, this can be used as a pigment, or can yield aluminum, iron, and titanium by reduction metallurgy [25]. In these ways salable commodities may be obtained from a process waste material.

More recently dried, pelletized red mud that is essentially iron oxide, has been reduced to steel via an electric prereduction and smelting process [26]. This technology could be particularly useful in bauxite processing areas where conventional iron ores are low grade, scarce, or nonexistent.

12.5.2. Aluminum Smelter Emission Control

Emission problems of conventional aluminum smelters center on fluoride losses, which preabatement (before ca. 1972) was at the rate of some 21 kg/tonne of aluminum produced (Table 12.7). The bulk of this fluoride loss occurred from the operating electrolytic cells, and two-thirds or more of this was gaseous fluoride. The chief constituents of the fluoride discharge were cryolite (Na_3AlF_6), aluminum fluoride, calcium fluoride, chiolite ($Na_5Al_3F_{14}$), silicon tetrafluoride, and hydrogen fluoride. A rough guide as to mass discharge rates may be obtained from their consumption rates per unit of aluminum produced (Table 12.7). Most of these substances are lost as fumes

TABLE 12.7 Preabatement and Postabatement Atmospheric Fluoride Emission Rates in Primary Aluminum Smelters Operating in the U.S.A. and Canada[a]

	Kilograms of fluoride emitted per tonne aluminum		
	Uncontrolled potlines[b]	Partial abatement	Efficient abatement
Prebaked anodes			
Gaseous	14	13	–
Particulate	9	8	–
Total	23	21	1.7
Range	12–33	5–21	0.6–2.3
Vertical stud Soderburg anodes			
Gaseous	18	13	3.4
Particulate	2–9	2–4	ca. 1
Total	20	21	4.4
Range	18–27	16–22.5	1.1–4.3

[a]Compiled from Brewer *et al.* [7], Iverson [27], Rose and Marier [28], and OECD [12, 29].
[b]In 1973 represented only 3% of primary aluminum producing facilities in the U.S.A., Nemerow [24].

or vapors, except for hydrogen fluoride gas, which results from the reaction of traces of moisture present in alumina added to the cell (Eq. 12.30).

$$3 \text{ H}_2\text{O} + 2 \text{ AlF}_3 \rightarrow \text{Al}_2\text{O}_3 + 6 \text{ HF} \qquad 12.30$$

This mode of fluoride loss is the reason that alumina shipment and transfers are conducted with minimum exposure to air to avoid moisture adsorption. In addition to losses of these substances, carbon tetrafluoride and hexafluoroethane, which are stable and relatively nontoxic, are known to form during an anode effect [7] (Eqs. 12.11 and 12.12). The possible greenhouse effects have been examined [30, 31], and practical abatement measures proposed [32].

While reactive and toxic fluoride emissions have undoubtedly been the impact area of greatest concern, a number of other gases, some arising from the carbon anode baking plant, are also significant. These include carbon monoxide, sulfur dioxide, hydrogen sulfide, nitrogen oxides (NO and NO_2), carbonyl sulfide (COS), and carbon disulfide. In a smelter employing Soderberg anodes, in which the pitch component of the carbon paste is largely vaporized during the *in situ* baking, there will also be formation of hydrocarbon vapors and smokes, including polynuclear aromatic hydrocarbons. The concentration of hydrocarbon vapors collected by the hoods can run to about 3%, but may be decreased to about 0.1% on passage through the carbon monoxide burner [33]. Loss estimates in kg/tonne of aluminum produced are: sulfur dioxide, 1–30; particulates (including alumina, soot, etc.) 5–10; and hydrocarbon vapors (Soderberg cells only), from 0.25–2 [24].

There are also problems dealing with fluoride-contaminated discharge water from wet scrubbers, and the drainage of precipitation from areas where spent pot linings are discarded, since fluoride constituents of electrolyte leach from these. Careful disposal of spent pot linings can minimize this problem. The problem may be eliminated by fluoride recovery followed by recycle of the carbon content.

Fluorides represent the discharge component of greatest hazard to plant life, and to nearby grazing and predator animals [34]. The injury threshold to plants from sulfur dioxide is about 0.1 ppm (100 ppb) whereas the equivalent value for fluoride to fluoride-sensitive plants is <1 ppb [26]. Thus, since fluoride emission has represented the largest mass loss to the atmosphere, this has a significant potential for biotic impact [35].

12.5.3. Smelter Emission Control Strategies

Aluminum smelter control systems may be considered in two groupings, the primary control system, which includes the hoods and ductwork for each cell, and the secondary control system, which provides general ventilation of the buildings, and houses the electrolytic cells [23]. Thus, the primary system collects the bulk of the fume and vapor discharge from a cell during periods when the electrolyte crust of the cell is intact. Scheduled additions of alumina, removal of aluminum, and periodic electrolyte analysis, as well as the occasional "anode effect" all require breaking the electrolyte crust. The secondary collection system is designed to handle losses during these periods [23]. This system consists of a series of forced circulation or convective roof exhaust vents together with floor level inlet vents in the sidewalls of the building, which help to minimize the occupational exposure of cell room operating staff [36].

The primary collection system, where the bulk of the cell fluoride loss is captured, may simply discharge to a tall stack [24]. This aids in dispersal, which may be sufficient for a small smelter. But it neither decreases mass discharge rates nor allows recovery of fluoride for reuse. In some cases, a stack has been found to cost more to operate than abatement methods, which employ fluoride capture.

Wet scrubbers have been in use the longest for cleanup of primary system gases before discharge, mainly because of the high affinity of hydrogen fluoride for water. Most wet scrubber configurations can easily achieve >98% mass containment for *gaseous* fluorides but are less efficient with particulates. Some are only able to capture 90% of particulate mass ($PM_{>10\,\mu m}$), and achieve efficiencies of little better than 10% on the particle fraction less than 5 fm in diameter ($PM_{<5\mu m}$) [24]. The best of these systems use a wet scrubber followed by a wet electrostatic precipitator and have been found to be 98–99% efficient for collection of both gaseous and particulate fluorides [27]. However, the stimulus for the development of alternative primary control systems is the need for a method to regenerate fluorides for electrolyte makeup from spent scrubber liquor (Eqs. 12.14 and 12.15). This can be relatively straightforward if a sieve plate column is used for scrubbing, since this can directly yield a spent scrubber liquor containing 4% hydrogen fluoride [27] (Fig. 12.5).

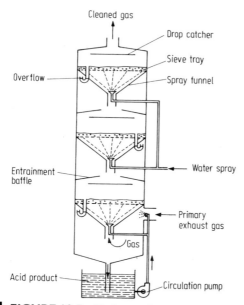

FIGURE 12.5 Hydrogen fluoride removal from smelter primary gas collection system via a sieve tray gas scrubber (only a section shown). Note the recycle of acid product for initial hydrogen fluoride absorption from the exhaust gases to maximize the concentration of recovered acid. (Adapted from Ball and Dawson [33].)

Various dry scrubbing configurations have been applied to primary control systems for fluoride, such as passage of the gases through coated filters or fluidized beds, or injection of alumina into the waste gas stream followed by recovery of the alumina with its adsorbed fluorides. The dry filter is more efficient than most wet scrubber arrangements for capture of fluoride particulates, but the early versions were only about 90% efficient for gaseous fluoride [23]. Both the fluidized alumina bed (Fig. 12.6) and the alumina injection dry scrubbing systems (Fig. 12.7) use the high adsorptive power of γ-Alumina to control both gaseous and particulate fluoride as efficiently as good wet scrubbers, 98+% [37, 38]. Gamma alumina is made at about 500°C, as mentioned earlier, and has a very high specific surface area of $60 + \mathrm{m^2/g}$ [39]. Ordinary α-alumina is unsuitable. The efficiency of the dry methods have the further advantage that simply heating the spent alumina removes adsorbed hydrocarbons, which are burned; and at the same time forms aluminum fluoride (Eq. 12.31).

$$6 \; Al_2O_3 \cdot HF \rightarrow 2 \; AlF_3 + 5 \; Al_2O_3 + 3 \; H_2O \qquad\qquad 12.31$$

In this way the captured fluoride is returned to the cells at the time alumina is added. These methods are used by several large North American aluminum smelters [40].

The incentive for improvement of secondary control systems to decrease gaseous and particulate emissions from the ventilation to the air of the pot line buildings has been twofold. With older facilities where gas collection efficiencies for primary control systems have been 40–60%, a relatively heavy loading of emissions is left to the wall ventilators and roof exhaust fans of secondary control to disperse [24]. Under these conditions, the mass discharge rate of pollutants from the secondary control system is large enough to also require treatment to obtain any substantial reduction in total discharge rates. Sizable equipment is required to handle the large volumes of air moved through the secondary collection system, which substantially raises the costs

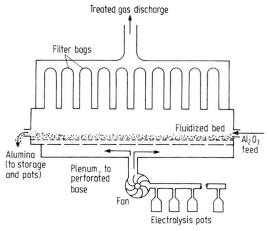

FIGURE 12.6   Typical arrangement of an alumina fluidized bed for primary emission control in electrolytic smelting of aluminum.

FIGURE 12.7 Typical arrangement of dry scrubber components, as used in aluminum smelter primary emission control. Bucket elevator is used to raise spent, fluoridated alumina to bin used to feed alumina to operating pots. Captured fluoride is directly recycled. (From Brewer *et al.* [7], reprinted courtesy of Alcan Smelters and Chemicals Ltd.)

of secondary treatment [27]. This provides a strong stimulus for improvement of the primary control systems, even by retrofitting older facilities.

By increasing the proportion of the cell superstructure, which is hooded, as well as by enclosing and improving the timing of the crust breaking/ alumina addition operations have combined to improve primary collection efficiency to about 95% for new facilities [32]. Cell operations such as electrolyte analyses and periodic aluminum removal require the opening of hooded access doors, which makes 95% containment about the practical limit for primary control. These measures comprise a part of the technology being promoted by Sumitomo [41]. As primary control systems are improved, the need for secondary control is decreased. However, there will always be a ventilating requirement, if for no other reason than to shed the enormous heat load from operating pots.

Liquid effluents from the operation of wet scrubbers and collected precipitation runoff from operating and waste disposal sites require treatment prior to discharge. Fluoride removal is achieved using lime addition to a one-, or two-stage effluent lagoon system (Chap. 10). Hydrocarbons and phenols, mainly from scrubbers operating on the gases from Soderberg cells, have proved to be more of a problem [24]. Traces of fluorides and other salts present in *treated* waste water can be safely discharged into seawater of normal salinity [42], but require prior hydrocarbon removal. Discharge of fresh wastewater into the denser marine environment should be to the bottom of the receiving waters to promote active vertical mixing. Otherwise stratification of the waste waters into top layers will occur and decrease dispersal rates [12].

Discarded pot linings still contain 10–15% absorbed fluorides, which could cause an environmental hazard. However, increased costs for electrolyte fluoride replacement have stimulated more smelters to crush used pot linings and recover the fluoride content by extraction with aqueous base. The filtered extract is then used for the preparation of synthetic cryolite. If spent pot linings are discarded to a landfill, any drainage from the site has to be treated. Landfill disposal also has a risk of occasional fires from spontaneous ignition of lining material.

12.6. PROPERTIES AND USES OF ALUMINUM

Special properties of the metal itself have contributed to the rapid rise in the applications and the volume of production of aluminum. Its natural corrosion resistance due to a firmly bound chemically inert oxide film, contributes to its utility for many uses. The construction industry uses aluminum for window frames, exterior doors and their fittings, decorative and insulating panels, siding, roofing, etc. Anodizing is an anodic electrical passivation of the alumina surface in chromic or sulfuric acid, which is used to enhance corrosion resistance for some construction applications as well as to apply color to some smaller articles [43]. Some polluted atmospheres will cause pitting or a poorly bound powdery white coating on aluminum articles exposed to these conditions, but otherwise no coating finish is required. Mercury should not be allowed to come into contact with an aluminum surface since amalgamation destroys the protective ability of the surface-oxidized aluminum. This is the reason for the strict ban on air freight shipment of mercury.

Aluminum, with a density of $2.70 \, \text{g/cm}^3$, is the third lightest of all metals, which are permanent (stable) in air, after magnesium ($1.74 \, \text{g/cm}^3$) and beryllium ($1.85 \, \text{g/cm}^3$). This property as well as its relatively low cost adds to its advantages in construction-related uses, and contributes to it being a valued component in the transportation industry. Many components of aircraft, the load carrying components (rather than running gear) of unit rail cars and rapid transit systems, truck cabs and bodies, and some engine components are made of aluminum or alloys. In the automotive industry the applications are more extensive, including engine components, wheels, trim, decorative components, and transmission housings. Light weight is also an important rationale for the choice of Al/Mg or Al/Mg/Si alloys in many types of small boats for use in fresh- or salt-water.

Aluminum is an excellent electrical and thermal conductor, the latter contributing to its use in some engine parts. The intrinsic electrical conductivity (i.e., for a test piece of the same dimensions) is $2.655 \, \mu\Omega/\text{cm}$ or 63% of that of copper and 60% of that of silver. But, for an equivalent *weight* of conductor per unit length, aluminum is slightly more than twice as good a conductor as copper and even better than this relative to silver, the reason for its use in long-distance high-tension transmission lines. Cables for this purpose consist of a high tensile strength steel wire core surrounded by a number of aluminum conductors. Electrical connections to aluminum require special techniques because of the nonconductivity of the aluminum oxide surface coating complicated by the plasticity and high coefficient of thermal expansion of the metal itself. These properties have led to problems in some aluminum house wiring installations.

Aluminum's good thermal conductivity (and nontoxicity) has led to extensive use in domestic cookware as well as in industrial heat exchangers, where its very good chemical resistance (e.g., to hydrogen peroxide and nitric acid) is also valued. For cryogenic applications, such as components of liquid air distillation plants, the conductivity, low-temperature toughness, and high reflectivity are also important properties. Related to cryogenic uses, aluminum also has a very low diffusivity (ease of gas transmission through intact metal), less than half that of steel, a property valuable in the construction of double-walled evacuated vessels.

Compatibility with most foodstuffs plus its light weight and relatively high strength have led to extensive use of aluminum in the food packaging industry. Aluminum cans are used for soft drinks and beer, an active area for metal recycling. They are also used for aerosol, meat, prepared pudding, and fish cans. Improved metal properties and packaging technologies have reduced the weight of aluminum per can to one-third of that used 30 years ago. Heavy foil is used for frozen food packaging, which doubles as a preparation container as well. Lighter foil is used for all types of packaging either on its own or in laminates for foodstuffs, tobacco, and confectionery products, and for other materials where its attractiveness, light weight, conformability, and low gas transmission characteristics prove useful.

12.7. PREPARATION AND USES OF ALUMINUM COMPOUNDS

Alumina itself, the product of the Bayer process, is important for many industrial applications other than for the production of aluminum. The red mud wastes that comprise the impurities removed from bauxite during alumina preparation are employed in steel making, the manufacture of red pigments, and for the production of chemicals used for water and sewage treatment [44] (see Section 12.5.1).

By heating the washed aluminum hydroxide (alumina hydrate) to a temperature of approximately 500°C most of the water is removed without altering the ionic arrangement, leaving a highly porous, surface active product. Activated alumina obtained in this way is an invaluable industrial dehydrating agent and a useful catalyst support in the petrochemical industry. The large surface area of some 40–65 m^2/g contributes to the high intrinsic activity and effectiveness of the catalyst. Alumina–silica composite supports are also of catalytic value.

Calcination of washed aluminum hydroxide at 1,100–1,300°C causes shrinkage of the particles as they dehydrate and yields a denser, harder product, α-alumina, useful as a constituent of abrasives. The low porosity and high temperature stability (up to approximately 1,800°C) of this product also make it useful for refractories either as bricks or as a loose fill. Ceramics with an alumina content of 85–95% and fired for up to 2 days at 1,450°C possess superior strength, hardness, and electrical and thermal insulating properties. Typical uses of these materials include spark plug insulators and textile guides as well as demanding applications in the chemical industry. Composites of α-alumina fiber with a metal are showing great promise in achieving from four to six times improved stiffness, and two to four times increased fatigue resistance over the unreinforced metal [45].

Aluminum sulfate ($Al_2(SO_4)_3$), or alum, is invaluable for water treatment (Chap. 5). It is also extensively used in papermaking and to a smaller extent in the leather, drug and cosmetic, and dye industries [46]. Strictly speaking an "alum" is a double salt of a monovalent metal sulfate and a trivalent metal sulfate crystallized with 24 waters (i.e., $(M^+)_2SO_4 \cdot (M^{3+})_2(SO_4)_3 \cdot 24H_2O$). The monovalent cation is usually sodium, potassium, or ammonium, and the trivalent cation is usually aluminum. However, since most traditional applications for alums value the trivalent ion, aluminum sulfate hydrate itself has gradually become known in industry as alum, or "filter alum." It is made by stirring bauxite with hot sulfuric acid (Eq. 12.32), settling, and then treating with a reducing agent, such as sodium sulfide to convert any ferric salts to the less colored ferrous equivalents.

$$Al_2O_3 \ (+Fe_2O_3, \ SiO_2, \ etc.) + H_2SO_4 \rightarrow Al_2(SO_4)_3 + residue \qquad 12.32$$
bauxite

The decanted solution is then concentrated hot, and allowed to cool producing crystals of the octadecahydrate, $Al_2(SO_4)_3 \cdot 18H_2O$, which are filtered off. Partial dehydration of the crystals during drying increases the aluminum analysis of the commercial product to approximately an $Al_2(SO_4)_3 \cdot 14H_2O$ stoichiometry. Alumina, rather than bauxite, may be used as the source of the aluminum ion for a higher purity product for applications, such as dyeing. Alum production provides a potentially profitable outlet for large quantities of by-product sulfuric acid, particularly if there is a local market for alum solutions to save shipping costs.

Aluminum chloride is another commercially important aluminum compound valued as a Lewis acid catalyst in many types of petrochemical reactions and for other process functions. Dry chlorine gas is passed beneath the surface of molten clean factory scrap (e.g., off cuts from door and window manufacturing) (Eq. 12.33).

$$\underset{\text{m.p. } 660°C}{2 \ Al} + 3 \ Cl_2 \rightarrow \underset{\substack{\text{m.p. } 190°C \\ \text{(at 2.5 atm)}}}{2 \ AlCl_3} \qquad 12.33$$

The melt is hot enough to sublime (i.e., $>178°C$) the aluminum chloride out of the melt as it forms, and the product is continuously collected on a water-cooled, scraped-surface heat exchanger. Commercial alkylation reactions are the largest single use for anhydrous aluminum chloride, in particular the alkylation of benzene to ethylbenzene enroute to styrene (Chap. 19). Aluminum chloride made by this route is only slightly more expensive (at $1.80/kg) than aluminum because of the large molar quantity of relatively inexpensive chlorine (23c/kg) added to the relatively expensive aluminum (ca. $1.60/kg) [47]. Liquid (solution in water; specific gravity 1.28) aluminum chloride is a by-product of the manufacture of the anhydrous material. The solution may also be produced by treating alumina with hydrochloric acid (Eq. 12.34).

$$Al_2O_3 + 6\ HCl \rightarrow 2\ AlCl_3 + 3\ H_2O \qquad\qquad 12.34$$

This product is employed in the preparation of antiperspirants and some types of pigments.

Virtually equivalent purity anhydrous aluminum chloride (ca. 99.8%) is also available as a side stream from Alcoa's new aluminum chloride route to the metal [48]. This process require 30% less energy than either of the alternate procedures described above, and gives a fine-grain, free flowing product from a fluid bed reactor.

REVIEW QUESTIONS

1. (a) What mass of sodium would be required theoretically to produce 1.00 tonne of aluminum metal from aluminum chloride?
   (b) How would the theoretical and actual number of Faradays compare for electrochemical reduction of sodium, then use of sodium for Al^{3+} reduction to 1 tonne of aluminum versus direct electrochemical reduction of Al^{3+}?
   (c) How would the theoretical and actual power (kWh/tonne) requirements compare for the two options of part (b)?
2. What is the cause of the "anode effect" as it relates to an aluminum electrolytic cell, and how is this problem corrected?
3. Outline the options that an alumina production facility has to convert the iron-rich "red muds" (a potential waste disposal problem) to salable products. Give equations for their production and describe their applications.
4. What is a "dry scrubber" and how is this device used to help control fluoride loss during the electrolytic production of aluminum?
5. Describe, with equations, the primary means by which aqueous fluoride loss is controlled for aluminum and for phosphorus production facilities.
6. One mole of aluminum can react with 0.1 M sodium hydroxide in water under the right conditions to produce 1.50 moles of hydrogen (J., and E.R. Andersen, *Globe and Mail*, Nov. 25, 2003, p.A8).
   (a) What mass of hydrogen would be expected from complete reaction of 1.00 kg of aluminum with water?
   (b) How many kJ of energy could theoretically be recovered by burning this mass of hydrogen in air? Assume gaseous hydrogen at 20°C, 1.0 atm; oxygen at 20°C, 0.20 atm.
   (c) How does the energy recovered from the mass of hydrogen obtained by step (a) compare with the approximate energy cost of electrolyzing alumina in cryolite to produce 1.00 kg of aluminum?
7. Reaction of aluminum with water and salt can generate hydrogen to drive a fuel cell. (T. Troczynske, A Chaklater, E. Czech, University of

British Columbia, Quirks and Quarks, CBC, Nov. 13, 2004).
In a fuel cell, hydrogen is reacted with oxygen to produce water and electricity at about 0.9 V (theoretical emf is 1.229 from H_2 oxidation, O_2 reduction.

(a) How many kWh of electricity could be available from this fuel cell on consumption of the hydrogen from reaction of 1.00 kg of aluminum?

(b) From first principles, how many kWh of electricity would be required to drive a Hall–Heroult cell operating at 5.00 V to produce 1 kg of aluminum?

(c) What mass of hydrogen could theoretically be produced using 15 kWh of electricity to drive a water-splitting electrolysis cell that operates at 1.5 V?

FURTHER READING

Aluminum and Health, *Search* (Australia), **20**(6), 180–185, Nov/Dec (1989). (A balanced debate by J.F. Ashton and R.S. Laura, The risks of dietary aluminum; C.L. Masters, Alzheimer's link still unproven; and N. Palmer, Temper concern with caution).

A.R. Burkin, "Production of Aluminium and Alumina." Society of Chemical Industry & J. Wiley, Chichester, West Sussex, 1987.

J.P. Hager, "Extraction and Processing for the Treatment and Minimization of Wastes." Proc. Intern. Symp., Minerals, Metals & Materials Soc., Warrendale, PA, 1993.

L.D. Hart and E. Lense, "Alumina Chemicals: Science and Technology Handbook." American Ceramic Society, Westerville, OH, 1990.

C. Misra, "Industrial Alumina Chemicals." American Chemical Society, Washington, DC, 1986.

Surinam: 62 Years of Bauxite Development, 12 Years of Alumina, Aluminum Production, *Eng. Mining J.* **178**, 75 (1977).

E. Vike, Air-pollutant Dispersal Patterns and Vegetation Damage in the Vicinity of Three Aluminium Smelters in Norway, *Sci. Total Environ.* **236**(1–3), 75–90, Sept. 15 (1999).

L.Wang, K.A. Fields, and A.H. Chen, Arsenic removal from drinking water by ion exchange and activated alumina plants, English Internet Resource Computer File, U.S. E.P.A.: National Risk Management Research Laboratory, Cincinnati, OH, 2000.

REFERENCES

1. "Kirk-Othmer Encyclopedia of Chemical Technology," 2nd ed., Vol. 1, p. 929. Wiley, New York, 1963.

2. "2001 Minerals Yearbook," Vol. 1. U.S. Geological Survey. Available: http://minerals.usgs.gov/minerals/pubs/commodity/aluminum/050494.pdf ; and earlier years.

3. L. Sidrak, . . . Simulation and Control of the Bayer Process. A Review, *Indus. Eng. Chem. Res.* **40**(4), 1146–1156, Feb. 21 (2001).

4. "Kirk-Othmer Encyclopedia of Chemical Technology," 3rd ed., Vol. 2, p. 129. Wiley, New York, 1978.

5. "World Mineral Statistics, 1975–1979." Institute of Geological Sciences, H.M. Stationery Office, London, 1981 and 1972–1976 issue.

6. M.E. Afonso de Magalhaes and M. Tubino, Recovering Gallium from Residual Bayer Process Liquor, *JOM* **43**, 37–39, June (1991).

7. R.C. Brewer, J.B. Brodie, L.D. Kornder *et al.*, "Environmental Effects of Emissions from the Alcan Smelter at Kitimat, B.C." Ministry of Environment, Province of British Columbia, 1979.

8. "McGraw-Hill Encyclopedia of Science and Technology," Vol. 1, p. 326. McGraw-Hill, New York, 1977.

9. R.C. Weast, ed., "Handbook of Chemistry and Physics," 51st ed., p. D-111. Chem. Rubber Publ. Co., Cleveland, Ohio, 1970.

10. "The Chemistry of Aluminum." Aluminum Co. Can. Ltd., Montreal, 1970.

11. S.H. Patterson, *Am. Sci.* **65**, 345 (1977).

12. Organization for Economic Cooperation and Development, "Pollution Control Costs in the Primary Aluminum Industry." OECD, Paris, 1977.

13. R.D. Peterson, Issues in the Melting and Reclamation of Aluminum Scrap, *JOM* **47**, 27–29, Feb. (1995).

14. Nonbauxite Aluminum Technology, *Chem. Eng. News*, **52**, 19, Dec. 16 (1974).

15. J.V. Thompson, Alumina: Simple Chemistry: Complex Plants. *Eng. Min. J.* **196**, 42–49, Feb. (1995).

16. M.K. McAbee, *Chem. Eng. News*, **53**(31), 19, Aug. 4 (1975).

17. M.V. Chaubal, J.L Anjier, B.J. Welch *et al.*, Light Metals 1994: Advances in Aluminum Production, *JOM* **46**, 14–23, Aug. (1994).

18. B.J. Welch, Aluminum Production Paths in the New Millennium, *JOM* **51**(5), 24–28, May (1999).

19. Recovery of Aluminum, *Can. Chem. Process.* **51**(3), 75 (1967).

20. New Aluminum Smelting Process, *J. Met.* **29**, 6, Nov. (1977).

21. Alcan's Work, *Can. Chem. Proc.* **57**(3), 52, Feb. (1973).

22. Aluminum Process, *Chem. Eng. News*, **57**, 23, Apr. 30 (1979).

23. New Processes Promise Lower Cost Aluminum, *Chem. Eng. News*, **51**, 11, Feb. 26 (1973).

24. N.L. Nemerow, "Industrial Water Pollution, Origins, Characteristics and Treatment," p. 502 Addison-Wesley, New York, 1977.

25. L. Piga, F. Pochetti, and L. Stoppa, Recovering Metals from Red Mud [from]... Alumina Production, *JOM* **45**, 54–59, Nov. (1993).

26. Red Mud Converted to Steel, *Chem. Eng. News*, **57**(9) 26, Feb. 26 (1979).

27. R.E. Iverson, *J. Met.* **25**, 19, Jan. (1973).

28. D. Rose and J.R. Marier, "Environmental Fluoride." National Research Council of Canada, Ottawa, 1977.

29. Organization for Economic Cooperation and Development, "Air Pollution by Fluorine Compounds from Primary Aluminum Smelting," OECD, Paris, 1973.

30. R.E. Weston, Possible Greenhouse Effects of Tetrafluoromethane and Carbon Dioxide... from Aluminum Production, *Atmos. Environ.* (Oxford) **30**(16), 29012–2910, Aug. (1996).

31. I.K. Larin, A.G. Gushchin, and B.N. Maksimov, Assessment of Greenhouse Potentials of CF_4, C_2F_6, and C_3F_8, *Chem. Phys. Rep.* **16**(12), 2111–2121 (1997).

32. M.J. Gibbs and C. Jacobs, Reducing PFC Emissions From Primary Aluminum Production in the United States, *Light Met. Age.* **54**, 26–34 Feb. (1996).

33. D.F. Ball and P.R. Dawson, Air Pollution from Aluminium Smelters *Chem. Process. Eng.* (N.Y.) **52**(6), 49, June (1971).

34. C.E. Carlson, W.E. Bousfield, and M.D. McGregor, *Fluoride*, **10**(1), 14, Jan. (1977).

35. M.B. Hocking, D. Hocking, and T.A. Smyth, Fluoride distribution... processes about an industrial point source, *Water, Air, Soil Pollut.* **14**, 133–157, (1980).

36. F. Akbar-Khanzadeh, Exposure To Whom It May Concern: Particulates and Fluorides and Respiratory Health of Workers in an Aluminum Production Potroom with Limited Control Measures. *Am. Ind. Hyg. Assoc. J.* **56**, 1008–1015 (1995).

37. A. S. Reid, G.J. Gurnon, W.D. Lamb, and K.F. Denning, Effect of Design and operating variables on scrubbing efficiencies in a dry scrubbing system for V.S. pots. *109th Am. Inst. Met. Eng. Ann. Meet.* New York, p. 759–781 (1980).

38. D. Rush, J.C. Russell, and R.E. Iversen, *J. Air Pollut. Control Assoc.* **23**(2), 98 (1973).

39. J. Miller and D.F. Nasmith, Operation of Alcan/ASV Dry Scrubbing Units with Various aluminas. *102nd Am. Inst. Met. Eng. Ann. Meet.* (A.V. Clack, ed.), New York, pp. 191–208 1973.

40. C.C. Cook, G.R. Swany, and J.W. Colpitts, *J. Air Pollut. Control Assoc.* **21**(8), 479 (1971).

41. Aluminum Cell Modifications Cut Energy Use, *Chem. Eng. News*, **53**(31) 19, Aug. 4 (1975).

42. R.M. Harbo, F.T. McComas, and J.A.J. Thompson, *J. Fish. Res. Board Can.* **31**, 1151 (1974).
43. A. Jenny and W. Lewis, "The Anodic Oxidation of Aluminum and Its Alloys." C. Griffin, London, 1950.
44. D.M. Samuel, "Industrial Chemistry: Inorganic," 2nd ed. Royal Institute of Chemistry, London, 1970.
45. Alumina Fibers used to Strengthen Metals, *Chem. Eng. News*, **58**(26), 24, June 30 (1980).
46. Aluminum Sulfate, *Can. Chem. Process.* **60**(9), 130, Sept. (1976).
47. "Chemical Marketing Reporter," **246**(24), p. 26, June 30. Schnell Publishing Company, New York, 1995.
48. R. Wood, *Process. Eng. (London)*. **57**(9), 9, Sept. (1976).

13

ORE ENRICHMENT AND SMELTING OF COPPER

Chawke may not beare the price of Cheese, nor copper be currant to goe for paiment.
—Thomas Nashe, 1590

The vessel, though her masts be firm,
Beneath her copper bears a worm.
—Henry D. Thoreau (1817–1862)

13.1. EARLY DEVELOPMENT

Copper rivals gold as one of the oldest metals employed by humans. Its first use about 10,000 years ago was stimulated by the natural occurrence of the metal in lumps or leaves in exposed rock formations, so-called "native copper." These exposures enabled simple tools to be fashioned directly by hammering and heat working of these fragments, and in so doing, signaled the end of the Stone Age. The natural occurrence of elemental copper is a feature of its low standard reduction potential of +0.158 V, significantly below hydrogen in the electromotive series and hence relatively easily deposited in elemental form as a consequence of normal geologic processes. This contrasts with the much more easily oxidized aluminum that lies above hydrogen in the electromotive series. With a reduction potential of −1.71 V, it is never found in elemental form in nature.

More purposeful recovery of the metal by mining occurred as early as 3800 B.C. from the dating of early workings discovered in the Sinai. Mines operating in Cyprus around 3000 B.C. were later taken over by the Roman Empire, and the metal product was called *cyprium*, later simplified to *cuprum*, the origin of the Latin name still used for the metal. Falun Copper Mine (Sweden) deserves a mention as one of the longest continuously operating mines in the world, having been worked for a thousand years before closing in 1992 [1]. At some times, centuries ago, it produces two-thirds of the world's copper.

The incentive for these early metal mines probably came from the experiments of early man who first relied upon hammering to shape native copper lumps. Subsequently this was elaborated to heat working and hammering, eventually leading to a need to melt and cast the metal for more fabrication flexibility [2]. It must have first been discovered that coal or charcoal firing was necessary to obtain sufficiently high temperatures to melt the native metal, which is not possible with even a large, hot wood fire. At some time later came the discovery that high grade copper ores, which did not look like copper, could be reduced to the metal [3]. How the knowledge of the requirements to melt metal was extended to the intentional reduction of ore by heat in the presence of carbon is still open to speculation. The reddish sheen of the polished metal and its ease of working undoubtedly provided impetus to develop fabrication methods for decorative art works, weapons, and tools.

From these early beginnings world copper production reached about 8,100 tonnes/year by 1750, 19,000 by 1860, and 50,000 by 1880 [4]. Production of the eighteenth and early nineteenth centuries was dominated by smelters in Mansfeld, Germany, and Swansea, Wales. Swansea accounted for some three-quarters of world production in 1860. However, the importance of both centers to smelting today is slight to nonexistent (Table 13.1). The Rio Tinto Co. in Spain dominated by 1880, producing about 25,000 tonnes of the metal per year. They still have a substantial output, though production levels now fluctuate. By 1900, the Anaconda Company, Montana, was producing around 50,000 tonnes/year. All of the production development described above relied on relatively rich (for copper) grades of ore containing 2–5% copper [7].

Much larger scale mining and smelting of formerly substandard ores (ores having too low a content of copper for smelting) became feasible following the development of froth flotation methods for ore enrichment in around 1920. World production reached levels of 1 million metric tonnes by 1912, and climbed to 2.3 million tonnes for both 1940 and 1950, stimulated by the growing demand and accessibility to lower grade reserves provided by the new enrichment methods. Froth flotation enrichment has also decreased the dependence of smelter siting on ore location, since the value of ore concentrates of 25–30% copper could justify substantial shipping distances for metal recovery. This is why some countries are large scale ore producers only (e.g., Papua New Guinea and the Philippines); some produce mostly ore and some metal (e.g., Australia and Canada) many countries produce similar amounts of ore and metal, and a few (e.g., Japan and Germany) produce metal mostly from imported concentrates (Table 13.1). The high value of copper encourages recycling; one-third of the current metal consumption is from this source. Nearly 60% of the virgin metal produced is ultimately reused.

13.2. MINING OF ORES AND BENEFICIATION

Copper occasionally occurs naturally in elemental forms (e.g., native copper deposits of the "thumb" of Michigan, near Lake Superior, and at Afton Mine, near Kamloops, British Columbia). However, the bulk of the commercial

TABLE 13.1 Major World Producers of Copper Concentrates and Primary Smelted Copper[a]

	Thousands of metric tonnes, copper content						
	1960		1970	1980		1990	2000
	Ore	Metal	Metal	Ore	Metal	Metal	Metal
Australia	111	72	111	217	171	192	315
Canada	399	361	453	710	463	476	544
Chile	536	502	658	1,068	953	1,210	1,461
China	n/a	n/a	100	200	200	359	1,020
Japan	89	188	513	53	861	893	1,331
Peru	182	164	177	365	350	196	340
Poland	11	22	69	346	364	331	508
South Africa	46	46	149	215	181	176	173
U.S.A.	980	1,119	1,489	1,168	1,008	1,160	1,090[b]
U.S.S.R.	n/a	n/a	925	900	905	1,370	580[c]
West Germany	2	62	80	1	160	180	275[d]
Yugoslavia	33	36	108	134	110	174	45
Zaire	302	302	386	459	426	346	–
Zambia	550	568	683	596	617	384	358
Other	347	220	364	945	−37[e]	2,233	911
World totals	3,640[f]	3,670[f]	6,320	7,890	6,828	9,680	9,030

[a]Data for 1975 and earlier compiled from *U.N. Statistical Yearbooks* [5], and more recent data from *Minerals Yearbooks*, [6]. The interfacing years may not entirely correspond.

[b]For 1999. Year 2000 data withheld.

[c]Russian Federation.

[d]Reunified Germany.

[e]Negative value due to rounding and the inclusion of undifferentiated production (e.g., for Poland).

[f]Excludes China, Czechoslovakia, Hungary, North Korea, and the U.S.S.R.

metal is produced from sulfide or oxide minerals. Common sulfide minerals are chalcopyrite ($Cu_2S \cdot Fe_2S_3$; 34.5%Cu), chalcocite (Cu_2S; 79.8%Cu), covellite (CuS; 66.4% Cu), and hornite ($Cu_2S \cdot CuS \cdot FeS$; 55.6%Cu), with many variations possible, both among these minerals and in admixtures with other sulfides. Sulfides of arsenic, antimony, gold, mercury, silver, and zinc are commonly found with copper ores because of the similarity of some chemical properties. The oxide minerals, which are less common than the sulfides, occur as cuprite (red copper oxide, Cu_2O; 88.8% Cu), tenorite (black copper oxide CuO; 79.8% Cu), malachite (basic, or green copper carbonate, $CuCO_3 \cdot Cu(OH)_2$; 57.3% Cu), and azurite (2 $CuCO_3 \cdot Cu(OH)_2$; 55.3% Cu), also with variations.

In the Americas, the minerals that contain copper occur chiefly in porphyry form (i.e., they are igneous in origin and are widely dispersed in the rock). In Africa and Russia, however, porphyry ores account for less than 20% of the total. In all areas copper *ores* have a relatively low copper analysis, in the range of 0.3–4% Cu. Many mines operate today on ores of 1% or lower copper content. Thus, a significant consideration for economical metal production is low-cost breaking up and removal of igneous rock, and recovery of the copper bearing mineral from this.

Open pit mining is used wherever feasible to minimize ore recovery costs. For instance, in the U.S.A. in 1970, open pit mines accounted for 84% of mine output (e.g., by Kennecott Copper Corporation, Salt Lake City) and only 16% was recovered from underground mining. By the use of low-cost explosives such as ammonium nitrate/fuel oil and large-scale loading and haulage units, mining costs to withdraw the ore are kept to less than 25% of the cost of the refined metal. Overburden containing little or no metal value is removed to a spoils disposal area, and the ore with some preselection proceeds to the concentrating circuit.

To access the mineral of value in the mined lumps of rock first requires size reduction. Crushing in two stages in gyratory cone (Fig. 13.1) or jaw crushers, decreases the initial, random lumps to first 15–22 cm average diameters, and then to 2–5 cm (walnut-sized) fragments. The walnut size lumps are then fed to rod or ball mills (Fig. 13.2) for final wet grinding to 65–200 mesh. A "100-mesh" sieve has a nominal 100 openings per lineal inch, equivalent to 10,000 openings per square inch. Thus, a particle passing through a 100-mesh sieve will have an average diameter of 0.01 in., less the sieve wire thickness of 0.0043 in. or about 0.0057 in. (ca. 0.149 μm). Thus, the 65–200 mesh range equates to particle diameters in the 74–230 μm (0.0029–0.0090 in.) range, averaging 150 μm (0.006 in.). A larger target particle size range than this decreases grinding costs, but also decreases mineral recovery from the rock by not exposing all copper minerals in a particle, which is required for efficient flotation. Aiming at a smaller size range than this ensures that most of the mineral is liberated from the associated rock, which increases the selectivity of

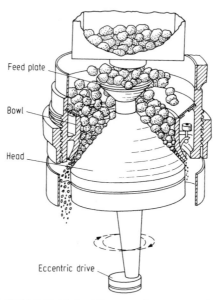

Feed plate

Bowl

Head

Eccentric drive

FIGURE 13.1   Diagram of a cone crusher. (Courtesy of Inco Ltd., Toronto.)

FIGURE 13.2 Cross section of a ball mill used for the wet grinding of copper ore preparatory to froth flotation.

separation in flotation, but raises grinding costs and may cause some loss of nonfloated mineral-containing fines from the flotation circuit [8]. For these reasons the ground ore in water produced by the ball or rod mill is sorted in a cyclone (or similar) classifier into a fines stream, which is fed to the flotation circuit, and a coarse stream, which is returned to the grinding circuit for further size reduction.

13.2.1. Ore Beneficiation by Froth Flotation

Froth flotation is used to raise the low mineral concentrations in ores to concentrations that can be more economically processed. A concentration of 25–30% is suitable for economical smelting of copper. The froth flotation technique was originally developed in about 1910 to raise the copper concentrations of the strip-mined ores of Bingham Canyon, near Salt Lake City [9], and was further perfected for the differential separation of lead, zinc, and iron sulfides at Trail, B.C., at about the same time [10]. Flotation technologies are now widely used for separations such as the beneficiation of low grade Florida phosphate ores from 30–40% to 60–70% concentrations of calcium phosphate (BPL), and the separation of about 98% potassium chloride from sylvinite, a natural mixture of potassium and sodium chlorides. It is also used for bitumen separation from tar sand, removal of slate from coal, and removal of ink from repulped paper stock preparatory to the manufacture of recycled paper stock. More details of these separations are discussed in the relevant chapters.

In general, flotation separation relies on the attachment of properly prepared mineral surfaces to air bubbles in water. Surfaces which are easily wetted by water, such as ordinary ground rock, are not attracted to bubbles and thus sink. Elemental sulfur, graphite (and other forms of elemental carbon, such as the pigment in printers' inks), and talc (minerals with layer lattice type structures) have hydrophobic surfaces, which are attracted to the surface of an air bubble and, if small enough, may be lifted to the surface of the water and separated from the water as a froth layer.

This distinction in behavior in aerated water is primarily dependant on the surface activity, or more precisely wettability, natural or induced, of the material under study. Three types of behavior are observed [11] (Fig. 13.3). If a solid surface is hydrophilic, when a bubble is pushed against it the contact angle θ, is zero (nonexistent) or very small and the surface is relatively easily

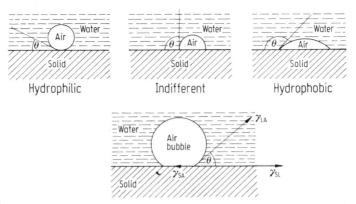

FIGURE 13.3 Behavior at the air–water interface of three types of solid surfaces and the relationship of the three interfacial tensions involved.

water wetted. If the bubble is released from a holder it will not stick to the surface, nor will small particles of the same material stick to a bubble. Finely ground material of this type, such as rock gangue, will sink.

If the test surface is neither readily wetted nor hydrophobic (i.e., it is indifferent to water), the contact angle is about 90° and poor differentiation of the unprepared surface can be expected. However, the behavior of wetted and indifferent surfaces can often be modified by treatment with surface active agents to obtain either a sink or a float behavior, as desired.

If the solid surface is hydrophobic it will tend to develop contact with the air of a bubble pressed against it to give distinct contact angles. For example, the contact angle for paraffin wax is 104°. In this situation there is a significant force component perpendicular to the solid surface to γLA, the liquid–air interface of the force vector diagram (Fig. 13.3), and hence there exists a significant force retaining the bubble to the solid surface. Small particles of this type of material will readily cling to small air bubbles and may be raised to a froth in a flotation cell.

The specific reduction in energy, WSA, from particle adhesion to, rather than release from a bubble by interaction of the three relevant interfaces is given by Eq. 13.1.

$$WSA = \gamma SL + \gamma LA - \gamma SA \qquad 13.1$$

This equation is credited to Dupré and S, L, and A represent the solid, liquid, and air surfaces, respectively (Fig. 13.3) [11]. Combining this equation with the Young equation, 13.2,

$$\gamma SA = \gamma SL + \gamma LA \cos \theta, \qquad 13.2$$

which relates the equality of the components of the three interfacial tensions involved, allows simplification of the Dupré equation to the form of Eq. 13.3.

$$WSA = \gamma LA(1 - \cos \theta) \qquad 13.3$$

For water γLA is readily determined to be about 70 dynes/cm^2. For a solution of a surfactant in water it is lower than for pure water and relatively constant.

The contact angle θ may thus be shown to be a significant variable factor in the work required to obtain particle adhesion. However, it is known that several other factors such as particle and bubble size ranges, the nature, age, and degree of oxidation of the mineral age since grinding of the particle, and the pH with its influence on particle-bubble interaction via neutralization of electrostatic charges on particles, also influence particle separations [12]. Hydrodynamic effects, which occur in agitated mineral suspensions are also of importance [13].

Fine particles of many sulfide minerals including sulfur itself, and sulfides of copper, are naturally hydrophobic. If they are subjected to flotation soon after grinding they will tend to stick to bubbles and to be raised to a froth. Silica and gangue are wetted by water and will tend to sink.

The natural differentiation of valuable mineral from gangue in this way is very good, but the "sticking" of sulfide minerals to air bubbles can be improved by the addition of a collector to the water phase. Typical collectors for sulfide minerals are xanthates ($ROC(S)S^- \ Na^+$) (e.g., potassium ethyl xanthate, $CH_3CH_2OC(S)S^- \ K^+$) or dithiophosphates (($RO)_2P(S)S^- \ Na^+$) (e.g., sodium diethyldithiophosphate, $(CH_3CH_2O)_2P(S)S^- \ Na^+$, and more recently N-butanoylphenylhydroxylamine [14]). These all have in common an oil-attractive (or hydrophobic) group and a sulfide-attractive group. When added to the pulp of ground-up ore in water at the rate of 10–45 g/tonne of ore processed, just before flotation, collectors increase natural water repellency and air avidity of particles of the desired mineral. These techniques increase the fraction of copper sulfide minerals recovered from the gangue and the rejection of gangue in the floated material.

If a bubble carrying air-avid mineral particles reaches the surface of the flotation unit and then bursts, the raised mineral particles will sink again. Bubble stability is maximized by addition of a foam stabilizer or frother, which assists in generating a sufficiently stable foam layer on the flotation unit to enable the foam plus associated mineral to be skimmed from the water phase. The frother also puts an oily phase on the surface of each bubble as it forms, which helps the mineral-gangue differentiation function of the collector. Typical frothers are oily materials of no more than slight water solubility such as pine oil (a mixture of terpenes) or a long chain (C5 or higher) alcohol such as 1-pentanol, and are used at the rate of 20–45 g/tonne of ore.

In addition to the two primary flotation agents there are also a number of modifying and conditioning reagents, which control a variety of ore and flotation circuit variables [13]. For instance, lime is a depressant for pyrite (FeS_2) in the flotation separation of copper sulfides. Lime would normally be added to the grinding circuit of a copper concentrator some time before flotation is carried out, for oxidation of the pyrite and the pH regulating effect to take hold. Sodium carbonate and sodium hydroxide are also common pH regulating agents. Molybdenite (MoS_2), a valuable by-product of copper ore processing, is first floated with copper sulfides then depressed, to the underflow of a separate flotation cell, by the addition of colloidal dextrin to the pulp feed. There are also additives, such as sodium cyanide, which are used occasionally to cause depression of some minerals, as well as dispersants or deflocculants to avoid nonselective aggregation of particles of mineral value with particles of gangue [15].

Separation of copper sulfide minerals from the gangue is normally carried on in a series of large rectangular steel tanks of capacities from 2 to $28\,m^3$ equipped with stirrers and compressed air inlets (Fig. 13.4). These provide vigorous agitation and air blowing through the properly conditioned ground ore pulp. The froth collected from a series of rougher cells is then passed to a set of cleaner cells to further raise the copper sulfide concentrations (Fig. 13.5). The gangue-rich underflow from the rougher cells is passed through a series of scavenger cells to collect any residual copper sulfide values not picked up by the rougher cells. Dried flotation concentrates contain from 20 to 30% copper (frequently 25–28%). The concentration process recovers 80–90% of the metal content of the ore feed. The discarded gangue may still contain ca. 0.05% copper, which may or may not be processed further to recover additional copper values (Section 13.5.2).

Copper (or metal) recovery percentages of the flotation circuits of a copper mine/mill are extremely important. For example, it is estimated that for a plant processing 50,000 tonnes of ore containing 1% copper per day, increasing the copper recovery from 86 to 87% of that contained in the ore could increase revenues by approximately $4 million annually. This revenue increase, estimated from a value of $1.10/kg allowed for the copper content of the concentrate when the metal was selling at $2.25/kg, would be obtained for the same overhead costs; hence the importance of seemingly minor flotation variables. Flotation reagents in the parts per million range are placed in the wet pulp prior to flotation at a cost of 10–35 cents/tonne of ore. Recent developments have been reviewed [16,17].

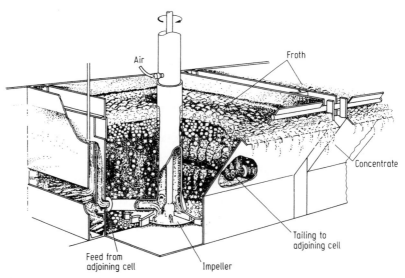

FIGURE 13.4   A cutaway drawing of a froth flotation unit for the concentration of copper ores with a detail photomicrograph showing mineral-loaded bubbles. (Courtesy of Inco Ltd., Toronto.)

FIGURE 13.5 Flowsheet of copper sulfide ore pathway, mined ore to dried concentrate ready for smelting. In some operations there may be additional cells and the thickener, such as a molybdenite (MoS_2) recovery circuit or the like.

13.3. SMELTING AND REFINING OF CONCENTRATES

Much of the copper smelted today still employs procedures, which may be divided into four more or less discrete steps, very similar to those developed by the early Welsh copper smelting industry [3]. However, the chemistry involved is much better understood today.

Roasting involves heating the concentrate in the presence of air to temperatures of 750–800°C in a multiple hearth, or fluidized bed roaster (Fig. 13.6). The charge remains more or less solid throughout. This completes the drying of the concentrate and removes a part of the sulfur from both the copper and iron present, producing oxides from some of the iron (Eqs. 13.4–13.7).

$$2\,CuS \rightarrow Cu_2S + S \qquad\qquad 13.4$$

$$FeS_2 \rightarrow FeS + S \text{ (pyrite pyrolysis)} \qquad\qquad 13.5$$

$$S + O_2 \rightarrow SO_2 \qquad\qquad 13.6$$

$$2\,FeS + 3\,O_2 \rightarrow 2\,FeO + 2\,SO_2 \qquad\qquad 13.7$$

Some external fuel, usually oil or gas, is required initially to raise the roaster temperature to the desired range for these processes to occur. Less is required later since the sulfur combustion is highly exothermic and provides some of the heat. If the concentrate contains more than 25% sulfur and is dry, sufficient heat is evolved from the sulfur combustion to maintain the roaster

FIGURE 13.6 Diagrams of multiple hearth type and fluidized bed roasters, such as used to decrease the sulfur content of copper sulfide ores. (Courtesy of Inco Ltd., Toronto.)

temperature once started. This mode of operation is referred to as "autogenous roasting."

Some sulfur trioxide also forms in the roaster, though this is not the objective of the process. This forms small amounts of ferrous sulfate in the roaster output (Eq. 13.8).

$$SO_3 + FeO \rightarrow FeSO_4 \qquad\qquad 13.8$$

Many modern roasting operations now collect and use the sulfur dioxide present in roaster exit gases to produce sulfuric acid. Fluidized bed roasters are favored since they can give 15% sulfur dioxide concentrations, compared to the 3–6% range expected from earlier designs.

Roasting is followed by reverberatory smelting, carried out on the still solid, hot roaster output. Silica, as a fluxing agent, and slag from a previous converter stage are added to the roasted concentrate and then external heat is applied by the combustion of powdered coal, oil, or natural gas. The flames and hot gases are reflected down from the ceiling of the furnace onto the mixed charge (hence, "reverberatory" furnace). This stage of smelting, where the charge temperature is gradually raised from about 600 to 1,000°C, accomplishes removal of more of the sulfur, by burning. Much of the iron is also removed as it combines with the fluxing agent to form a lighter ferrous silicate slag, which floats on top of the copper melt, as it forms (Eqs. 13.9–13.13).

$$Cu_2O + FeS \rightarrow Cu_2S + FeO \qquad 13.9$$

$$2\ Cu + FeS \rightarrow Cu_2S + Fe \qquad 13.10$$

$$2\ Fe + O_2 \rightarrow 2\ FeO \qquad 13.11$$

$$Cu_2S + 2\ CuO \rightarrow 4\ Cu + SO_2 \qquad 13.12$$

$$FeO + SiO_2 \rightarrow FeSiO_3\ (slag) \qquad 13.13$$

The transfer of a part of the iron-bound sulfur to copper, which is an objective of this stage, is the result of the change in the relative heats of formation of cuprous and ferrous sulfides at these high temperatures (e.g., at 1,300°C, $\Delta H = -174$ and 34.4 kJ, respectively) relative to those at 25°C ($\Delta H = -79.5$ and 95.1 kJ, respectively).

The product of the reverberatory smelting stage is matte copper, containing primarily copper metal and cuprous sulfide. The analysis will range from 15% (particularly with nonconcentrated ores) to 50% copper [18]. Ideally, the copper concentration in matte copper is kept at 40–45% Cu to maintain the matte as an effective collector of precious metals, and to keep the copper loss in the slag to a minimum. Insufficient sulfur (as Cu_2S, melting point 1,100°C) in the matte also makes it difficult to keep the charge molten (melting point of Cu, 1,083°C; which is depressed by the eutectic mixture of the matte copper).

In converting, the third stage of copper smelting, air is blown through the mass of molten matte copper to complete the oxidation of sulfides, since by this stage the bulk of the iron has been removed (Fig. 13.7) (Eqs. 13.14 and 13.15).

FIGURE 13.7 Diagram of a side blown converter used to separate an iron-rich ferro–silicate slag from a sulfur-rich matte copper. (Courtesy of Inco Ltd., Toronto.)

$$2\ Cu_2S + 3\ O_2 \rightarrow 2\ Cu_2O + 2\ SO_2 \qquad\qquad 13.14$$

$$Cu_2S + 2\ Cu_2O \rightarrow 6\ Cu + SO_2 \qquad\qquad 13.15$$

Any residual iron present is oxidized and forms a slag layer with an additional small portion of added silica (Eq. 13.13). This slag layer also captures any traces of oxidized arsenic and antimony, if present, that have not already been volatilized by the blowing. The less volatile metal oxides remaining are lithophilic ("rock-loving"), and thus tend to be absorbed by the slag layer. For the same reason, some oxidized copper also enters this slag layer, but is recaptured when the slag is used as a part of the charge for subsequent reverberatory smelting. Any precious metals present, such as silver, platinum, and gold, are relatively stable toward oxidation and remain with the copper. The product of converting is termed "blister copper," from the pock marks on the surface formed by the escape of sulfur dioxide bubbles from the metal as it cools and solidifies. It is virtually free of sulfur, but contains from 0.6 to 0.9% oxygen (ca. 5.4–8.1% Cu_2O) from the vigorous oxidation.

Fire refining, the final stage of copper smelting, is used to reduce the small percentage of cuprous oxide present in the blister copper back to elemental copper. It is often carried out in the converter itself, once the air flow has been cut off and while the charge is still molten. Originally fire refining was accomplished by the addition of coke and green wood poles, which were firmly pushed under the surface of the melt to bring about reduction of cuprous oxide by the carbon of both additives. Vaporization of the moisture and partial combustion of the green wood also vigorously agitated the melt via the subsurface release of steam, carbon dioxide, carbon monoxide, and hydrogen. The hydrogen contributed to the reducing action of the carbon (Eqs. 13.16–3.18).

$$2\ Cu_2O + C \rightarrow 4\ Cu + CO_2 \qquad\qquad 13.16$$

$$Cu_2O + CO \rightarrow 2\ Cu + CO_2 \qquad\qquad 13.17$$

$$Cu_2O + H_2 \rightarrow 2\ Cu + H_2O \qquad\qquad 13.18$$

"Poling" is now usually carried out by passing natural gas or ammonia into the melt by water (or gas) cooled, hollow steel lances. Fire-refined copper, the product of this smelting stage, gives a copper analysis of 99% or better. Fire-refined copper can be used directly for about 15% of copper uses, including some alloys whose properties are not seriously affected by small amounts of impurities. Metal properties are seriously impaired for some uses, however, by the presence of traces of impurities, which makes further refining necessary. Phosphorus, arsenic, aluminum, iron, or antimony in particular decrease electrical conductivity (e.g., some 6% loss is experienced with as little as 50 ppm phosphorus [7]). For these reasons as well as from an interest in the recovery of the precious metal values which are often found in the fire-refined copper, the fire refined copper is further purified by electrorefining.

13.3.1. Electrorefining of Smelted Copper

Electrolytic purification of smelted copper removes contaminants which adversely affect electrical conductivity, malleability, and other properties. It also permits recovery of the precious metal content of the fire-refined product.

Anodes for electrorefining must first be cast from the smelted metal, with a uniform thickness across each anode as well as from one anode to the next. Anodes which are too thick give too much scrap anode material for recasting when the spent anodes are changed. Anodes which are too thin or irregular in thickness can cause pieces of metal to drop off in the electrolytic cells resulting in electrical shorts and decreased current efficiencies.

Anode slabs are connected in parallel to heavy bus bars (high current conductors) and placed in the electrolytic cells. These are long rectangular vessels of wood or cement lined with lead or plastic (Fig. 13.8). Interleaved between the thick anode castings of fire-refined metal to be purified are thin starting sheets of elecrolytically pure copper, which are connected to the cathode (negative) bus bar. The electrolyte consists of about 20% sulfuric acid (ca. 2 molar), which also contains 20–50 g/L (0.3–0.8 molar) cupric ion. Traces, up to a combined total of 20–30 g/L arsenic, antimony, bismuth, and nickel gradually accumulate in the electrolyte as a result of the dissolution of these electrolytically more active metals (Table 13.2) from the anodes. A low concentration of chloride (ca. 0.02 g/L) supplied by sodium chloride or hydrochloric acid is also maintained in the electrolyte to ensure that silver is precipitated on the bottom of the cell as finely divided silver chloride, rather than being deposited with the copper on the cathodes. As electrolysis proceeds copper is dissolved from the anodes, and is simultaneously deposited from the electrolyte onto the cathodes causing a movement of metal between the electrodes (Eqs. 13.19 and 13.20).

$$\text{Anode reaction: } Cu^0 - 2\ e^- \rightarrow Cu^{2+}{}_{(aq)} \qquad 13.19$$

$$\text{Cathode reaction: } Cu^{2+}{}_{(aq)} + 2\ e^- \rightarrow Cu^0 \qquad 13.20$$

So there is no overall electrochemical reaction, hence, the theoretical (equilibrium) voltage required is zero. The applied voltage of about 0.2 V is required for the process solely to drive the electrolytic purification in the desired direction. Indirect heating of the electrolyte to about 60°C decreases the solution viscosity, which helps to maintain a high production rate at these low operating voltages. Current densities of about 240 A/m² are normal,

FIGURE 13.8 Diagram of a section of an electrolytic cell for the purification of copper.

TABLE 13.2 Standard Reduction Potentials for a Series of Elements of Importance to Electrolytic Purification of Copper[a]

Ion Reduced	Standard reduction potential (V)
Ca^{2+}	−2.76
Mg^{2+}	−2.38
Al^{3+}	−1.71
Fe^{2+}	−0.41
Ni^{2+}	−0.23
Pb^{2+}	−0.126
H^+	0.0
Sb^{3+}	+0.145
As^{3+}	+0.234
Bi^{3+}	+0.320
Cu^{2+}	+0.158
Te^{4+}	+0.63
Ag^+	+0.80
Pd^{2+}	+0.83
Pt^{2+}	ca.+1.2
Au^{3+}	+1.42
Au^+	+1.68

[a]Arranged in order from the most, to the least active elements.

although up to $350\,A/m^2$ is successfully employed with periodic current reversals (PCR) [11]. Current efficiencies for copper deposition are 90–95%, tending toward the lower end of this range with PCR, so that the overall power requirement for copper purification by electrolysis is relatively low. Power consumption of several cell types in commercial electrorefining service is 176–220 kWh/tonne of copper cathode deposited.

During the electrolysis the metals present in the anode, which are of lower electrochemical activity than copper, chiefly silver, tellurium, gold, and traces of platinum, are precipitated as slimes on the floor of the electrolysis cells as they are released from the anodes (Table 13.3). Any mercury, which is frequently present in the ore, is mostly vaporized in the hot smelting stages prior to electrorefining. Lead, which is dissolved from the anode as Pb^{2+}, is immediately precipitated as the insoluble sulfate. Nickel and many other metals more electrochemically active than copper, the sulfates of which are soluble, accumulate in the electrolyte. The accumulated slime sludges are periodically pumped from the floor of the cells and dried prior to recovery. By a complex series of steps which include sulfuric acid acidification, roasting, copper electrowinning, silver cementation, fusions, and further electrowinning operations a precious metal alloy called Doré metal analyzing 8–9% gold and 85–90% silver is recovered from the residual copper, selenium, tellurium, and lead. The value of the precious metal content recovered from fire-refined copper by electrorefining is sufficient to pay for much of the cost of the electrolytic step.

During normal cell operation the anode contaminants that are more electrochemically active than copper (and less than hydrogen), mainly arsenic,

TABLE 13.3   Analytical Ranges of Products of the Copper Refining Industry, U.S.A. and Japan[a]

	Anode (fire-refined) copper	Electro-refined copper	Dry, unprocessed anode slime
Copper, %	99.0–99.6	99.95	20–40
Oxygen, %	0.1–0.3	0.02–0.04	–
Gold, g/tonne	3–140	0.07–0.3	1,700–10,000
Silver, g/tonne	70–3,400	2–20	34,000–300,000
S, As, Sb, or Bi, %	0.003–0.2	0.00001–0.0002	0.5–5.0
Lead, %	0.01–0.15	0.0002–0.001	2–15
Nickel, %	0.01–0.10	0.0001–0.002	0.1–2
Selenium, %	0.01–0.06	0.0003–0.001	1–20
Tellurium, %	0.001–0.02	0.0001–0.001	0.5–8

[a]Compiled from data of Hofman [7] and Kirk-othmer [18].

antimony, and bismuth, gradually accumulate in the electrolyte. These are kept from depositing on the refined copper cathodes partly by maintaining a much higher concentration of cupric ion in the electrolyte. Fortunately for the electrorefining process, cuprous oxide is usually the main impurity of fire-refined copper. Since each mole of the cuprous oxide present in the anodes and exposed to the electrolyte puts a mole of cupric ion into solution by an auto-redox reaction (Eq. 13.21), without electrochemically removing a like amount onto the cathodes, the concentration of copper ion in the electrolyte rises more rapidly than the concentration of the other contaminants.

$$Cu_2O + 2\ H^+ \rightarrow Cu^{2+} + CuO + H_2O \qquad 13.21$$

Incidentally, the elemental copper formed by Eq. 13.21 is the chief source of the copper present in anode slimes. Electrolyte is continually recirculated through the electrolysis cells to minimize the buildup of impurities. When the copper content of the electrolyte reaches 40–50 g/L (from the initial 12–18 g/L), it must be purified before reuse. The concentrations of copper, arsenic, antimony, bismuth, and nickel must all be substantially decreased. A small portion (ca. 2–3% of the total volume, per day) of the electrolyte stream is continuously bled from the main electrolyte stream for this purpose.

Most of the excess copper present in spent electrolyte is electrolytically removed from solution in liberator cells, which are very similar to the electrolysis cells used for electrowinning. These have a pure copper cathode, on which copper is deposited (Eq. 13.20), and a lead anode, rather than the copper anode used in electrorefining. The lead anode reforms sulfuric acid from the hydrogen ions released from water, instead of contributing dissolved lead to the electrolyte (Eq. 13.22).

$$H_2O - 2\ e^- \rightarrow 2\ H^+ + 1/2\ O_2 \qquad 13.22$$

Antimony and bismuth tend to precipitate on the bottom of the cell toward the later stages of this electrolysis, and arsenic reacts with the acid present under electrolytic conditions to form highly toxic arsine gas (Eq. 13.23).

$$As + 3\ H^+ + 3\ e^- \rightarrow \quad AsH_3\ (-0.54\ V) \qquad\qquad 13.23$$
$$\text{arsine, b.p. } -55°C$$

For this reason, liberator cells are normally hooded and well vented to avoid problems from the evolution of these gases. The "liberated" electrolyte is then further treated to remove nickel, either by crystallization as nickel sulfate or by dialysis, before it is returned to the electrolysis circuit for reuse.

Vigorous air agitation has been found to boost cathode purities when electrorefining at very high (ca. $226\ A/m^2$) current densities and suggestions have been made to generally improve electrorefining and electrowinning operations [19, 20].

13.3.2. Fabrication and End Uses

For uses of pure copper or for critical alloys the 99.95% copper cathodes obtained from electrolytic refining have to be washed, dried, and remelted before being cast into stock shapes useful for fabrication. The cast ingots can then be rolled into sheets (or foil), or hot or cold drawn into wire or pipe. Working the metal in these ways hardens and strengthens it, that is it raises the tensile yield point from $30,000-40,000\ lb/in.^2$ to $60,000-70,000\ lb/in.^2$ [7]. With prolonged work this process can result in cracks or fractures in the metal. The metal may be annealed (made malleable again) by heating to a temperature of $500-700°C$ followed by quenching (rapid chilling). The annealed stock may then be worked again and sold either in a work-hardened or annealed state, depending on the requirements of the application.

The high electrical conductivity of copper, $0.586 \times 10^6\ mho\cdot cm$ at $20°C$, or about 95% of that of the most conducting metal, silver, has led to the large-scale use of copper in wiring and electrical machinery of all types. More than half of the metal produced in Japan, the U.S.A., and western Europe is used in electrical or electronic applications [21]. Even though aluminum possesses only some 62% of the conductivity of copper, because of its light weight and relatively lower cost it has become competitive with copper for some electrical applications.

The relative thermal conductivities of the metals quite closely parallel electrical conductivities, so that again copper, with a conductivity of $0.934\ cal/cm^2/cm/sec/°C$ at $20°C$ ($3.98\ W/cm/K$ at $300\ K$) emerges as the best low-cost metal for efficient heat transfer. This property has led to its employment in industrial heat exchangers as well as in food processing and domestic cookware. Cookware is usually lined with stainless steel to prevent the possibility of copper uptake by food. Copper is an essential trace element in human nutrition and the average American intake has been established to be 2–5 mg daily [22]. However, it has caused mild toxic symptoms from ingestion of citrus juices that have been in long-term contact with, or cooked in, copper vessels [23]. Copper and its salts have a relatively low mammalian toxicity, although instances of acute copper poisoning have been described (from 1–12 g of copper sulfate). Chronic toxicity is known to lead to liver cirrhosis and nerve, brain, and kidney damage (Wilson's disease).

Excellent resistance to corrosion both promotes the use of copper in some of the applications mentioned above and favors its use as an expensive, and long-lived roofing material. This property has also led to the use of copper in high-quality piping for gutters and drains.

Its malleability and corrosion resistance are factors in the extensive use of copper as a coinage metal, although undoubtedly here its beauty is an important contributing factor. The attractive red color is also the stimulus for its use in many types of decorative interior fittings.

About one-quarter of the copper produced goes directly into copper alloys, mostly as fire refined, plus some electrolytically refined. The two principal alloy groups are the brasses, which comprise copper alloyed primarily with zinc, and the bronzes which are mainly copper/tin alloys. Both of these alloy groups are stronger than copper, one motivation for their use, but there are other reasons which favor their choice. The brasses are less expensive than pure copper, hence their employment in such uses as cartridge cases for ammunition. While the brasses are less corrosion resistant than copper this property is still sufficiently good for brasses to be used in automobile radiators where its good heat transfer capabilities and higher strength and lower cost favor this choice. The zinc content of brasses ranges from 5%, in gilding metal, through to 30% and 40% in cartridge brass and Muntz metal, respectively, which are the brass alloys containing the highest proportion of zinc. Some brasses also contain 1–3% lead which improves machinability.

The tin content of bronzes is lower, ranging from 1.25 to 8% in the phosphorbronzes, which also contain 0.25% phosphorus, and confer higher strength and better corrosion resistance than pure copper. These alloys are therefore ideally suited to marine exposure in such applications as ships' propellers, exterior fittings, and water lines carrying saltwater, etc. A low coefficient of friction between bronze and steel and the superior strength of bronze over copper have led to its application as a bearing metal and as the casting metal for gears [24].

Two special copper alloys, beryllium copper and nickel silver, although only used on a small scale, contribute valuable metal properties for special uses. Two percent beryllium added to copper gives greater fatigue resistance to the metal and confers a nonsparking (on impact) quality to tools made of this alloy, important for impact tools in flammable or explosive atmospheres. Nickel silver, a copper/nickel/zinc alloy with an appearance very like silver, is important as the strong base metal for silver-plated tableware.

A small amount of copper is consumed for copper salts such as copper sulfate, which are important as agricultural fungicides as a spray or a dust. Copper salts are occasionally used to correct local soil deficiencies. Copper sulfate itself is also an excellent algicide in swimming pools and fish rearing ponds.

13.4. EMISSION CONTROL PRACTICES

13.4.1. Mining and Concentration

Problems at the mining stage center around overburden and waste rock disposal and the potential contamination of surface water courses from disposal of mine drainage waters. Waste rock disposal in mined-out areas is used wherever feasible but because the crushed material does not pack as densely as the bedrock from which it was derived, some surface disposal is also required. By skillful placement and profiling followed by plantings to encourage reestablishment of flora, dry land surface disposal can be accomplished in an aesthetically acceptable manner.

Percolate from such dry land disposal sites, and mine drainage waters, often contain traces of dissolved copper and other metals such as iron (Table 13.4). It also usually has a low pH. These impurities all arise from bacterial action on the sulfidic minerals present (e.g., Eq. 13.24).

$$2\ CuS + 3\ O_2 + 2\ H_2O \xrightarrow[\text{bacteria}]{} 2\ Cu^{2+} + 2\ H_2SO_4 \qquad 13.24$$

Copper may be recovered from such waste streams by cementation, or by cementation plus liming, which can reclaim the metal and decrease the toxicity risks on stream discharge (Eqs. 13.25–13.28).

Cementation:

$$Cu^{2+} + Fe \rightarrow Fe^{2+} + Cu \quad \text{(cement copper)} \qquad 13.25$$

Liming:

$$Cu^{2+} + Ca(OH)_2 \rightarrow Cu(OH)_2 \downarrow + Ca^{2+} \qquad 13.26$$

$$Fe^{2+} + Ca(OH)_2 \rightarrow Fe(OH)_2 \downarrow + Ca^{2+} \qquad 13.27$$

$$2\ Fe^{2+} + O_2 + 2\ H_2O + Ca(OH)_2 \rightarrow 2\ Fe(OH)_3 \downarrow + Ca^{2+} \qquad 13.28$$

Cementation involves passing the solution containing copper ions over a large surface area of scrap iron, followed by settling. Finely divided copper is precipitated. Subsequent liming removes most of the iron as the hydroxide, and raises the pH to more normal values before discharge. Four-stage solvent extraction in the presence of proprietary complexing agents has also been tested and found to be capable of decreasing aqueous Cu^{2+} concentrations from about 400 to about 6 mg/L [25].

A large fraction of the water used for froth flotation may be recirculated by decanting a clear supernatant liquor from a settling lagoon, or from a

TABLE 13.4   Partial Analysis of Typical Copper Mine and Smelter Waters[a]

	Ion concentrations (mg/L)					
	Cu^{2+}	Fe^{2+}	Fe, total	Zn^{2+}	H$_2$SO$_4$	pH
Mine waters:						
Typical	970	8,350	10,500	4,000	–	1.4
S. American	120	2,000	4,000	1,000	–	1.4
Canadian	156	–	176	328	<10	3.8
Typical wash water:	100	>5	10	24	100	2.8
Electrolytic	240	–	3	<1	760	2.0
Copper refinery	446	–	4.5	43	1,810	1.5

[a]Typical data from Barthel [25].

thickener overflow. However, froth flotation is sensitive to low concentrations of surface active agents. This means that it cannot operate as a closed circuit since buildup of impurities could affect flotation efficiencies. Traces of metals in the effluent flotation water, some suspended and some dissolved, are captured prior to discharge by liming and settling as outlined for mine drainage waters or tailings percolate/leachate.

Other types of water effluent problems can arise from froth flotation operations. For example, with multimineral flotation when separate concentrates are required for copper, lead, and zinc, copper would normally be floated first, then lead, and finally zinc sulfide. This separation sequence requires addition of sodium cyanide as a depressant. The waste water stream from this process can contain 50 mg/L or more of dissolved cyanide which has to be detoxified before discharge. Calcium hypochlorite has been found to be effective to neutralize the cyanide by oxidation to cyanate (Eq. 13.29).

$$2 \, NaCN + Ca(ClO)_2 \rightarrow 2 \, NaCNO + CaCl_2 \qquad 13.29$$

When treatment is undertaken on a large scale, however, chlorine addition has been found to be both more convenient and more economical (Eqs. 13.30 and 13.31).

$$Cl_2 + H_2O \rightarrow HCl + HClO \qquad 13.30$$

$$2 \, Ca(OH)_2 + 2 \, HCl + 2 \, HClO \rightarrow CaCl_2 + Ca(ClO)_2 + 2 \, H_2O \qquad 13.31$$

If the chlorine addition is undertaken prior to pH reduction with lime the reaction sequence is thought to involve intermediate formation of cyanogen chloride which is decomposed to cyanate on addition of base (Eqs. 13.32 and 13.33).

$$NaCN + Cl_2 \rightarrow CNCl + NaCl \qquad 13.32$$

$$2 \, CNCl + 2 \, Ca(OH)_2 \rightarrow Ca(CNO)_2 + CaCl_2 + 2 \, H_2O \qquad 13.33$$

This waste treatment sequence can reduce feed values of 68 mg/L cyanide and 42 mg/L copper to concentrations of about 0.1 mg/L for both contaminants [26]. This result requires 3.4 kg chlorine/kg cyanide in the effluent, and a 15–30 min residence time for destruction of the cyanide only. Three to four times this residence time is required for neutralization of copper cyanide. Ideally, sufficient oxidant should be added to convert cyanate to carbon dioxide and nitrogen to avoid the possibility of cyanate reversion back to cyanide after discharge (e.g., Eq. 13.34).

$$2 \, NaCN + 5 \, NaOCl + H_2O \rightarrow N_2 + 5 \, NaCl + 2 \, NaHCO_3 \qquad 13.34$$

Ozonation has also been found to be effective for the neutralization of cyanide in waste waters [27].

Disposal of the waste pulverized gangue may be to underground or strip mined areas. However, excess disposal volumes can be a problem. Dry land disposal followed by aesthetic contouring and reestablishment of plant life may be used. This method requires monitoring and control measures for any dissolved metals in leachate from the tailings. Drainage directly to a small lake or a water course is not permitted by pollution control agencies (e.g. [28]).

Direct tailings disposal to large lakes and ocean inlets has been practiced, but is rarely acceptable now [29].

Solution mining can generate complex recovery problems, particularly with in-place leaching methods. When mining has been completed the fractured deposit requires extensive flushing to remove residues of leaching solution. Deep well injection or various procedures for concentrating solution are used for ultimate disposal of waste solutions from flushing operations.

13.4.2. Smelter Operations

Copper smelting is associated with dust production from materials handling and grinding, and sulfur dioxide from sulfides. A tonne of sulfur dioxide gets dispersed for every tonne of copper obtained from sulfidic ores. In the early days of the industry the sulfur dioxide was discharged from stacks to dissipate in the air around a smelter. The severe effect on local air quality prompted mixing of the smelter gases with large volumes of air before discharge to decrease ground level sulfur dioxide concentrations. Stack heights gradually increased over the years to exploit dilution effects for larger volumes of sulfur dioxide, culminating in the construction of the world's tallest stack, 380 m high, by the International Nickel Company (Inco) at Sudbury for this purpose, in 1970. At the same time, some smelters adopted sulfur dioxide containment and conversion to liquid SO_2 or sulfuric acid to decrease discharge problems. These by-products also provided an economic process credit. Containment and conversion measures have now been widely adopted by the industry, as the limited atmospheric dilution capacity has been recognized [30]. The U.S. Environmental Protection Agency now requires 95% or better of the sulfur in processed ores to be contained during smelting.

Crushing and dry grinding stages as well as final rotary drying are likely to contribute the most to dust losses. Ore grinding wet avoids dust problems for this stage, and improves control of the final particle size. For all processing stages the proportion of copper present, about 10% in roaster dusts, 25% in reverberatory furnace dusts, and 45–55% in converter dusts, is sufficiently high to provide material recovery and emission control incentives for capture.

Simple gravity settling of dust has for many years provided about 84% mass containment success for copper blast furnace discharges [7]. Sometimes dust settling and gas carrying functions are combined in a single large cross-sectional area (slow gas flow) balloon flue (Fig. 13.9). Tighter emission control requirements as well as an interest in cleaner acid plant operation, however, have required employment of electrostatic precipitation and/or scrubbers for smelter flue gas treatment [15]. Electrostatic precipitation controls fumes as well as larger particles with its control efficiency of about 98% by mass of 0.1–100 μm diameter particles [31]. Fumes form from sublimation, or from condensation and solidification of such components as oxides of antimony, arsenic, lead, zinc, and sulfates. Water, sulfuric acid, and mercury may also have been adsorbed onto the fume particles. Dusts containing these elements are treated for recovery of the components of interest (e.g., mercury by sulfatization [32, 33]). Baghouses are also efficient for dust control, which requires additional blower energy to force gases through the

FIGURE 13.9 Operating details of (a) a gravity settling chamber and (b) a section of a balloon-flue for removal of the coarser dusts from copper smelter operation.

filter fabric. The bag fabric has to be compatible with acid gases and flue gas temperatures have to be kept within the working range of the fabric.

The problem with sulfur dioxide containment is that many smelter processing units produce sulfur dioxide concentrations of 1–2% whereas the minimum economic concentration for sulfur dioxide conversion processes (e.g., for sulfuric acid or sulfur production) is 3.5–4%. Thus, the economic solution to containment requires either modification of the smelter process to obtain higher sulfur dioxide concentrations, or capture of sulfur dioxide at relatively low concentrations followed by some means of regeneration of a high concentration of sulfur dioxide (Chap. 3). Throw-away approaches to sulfur dioxide containment, lime or limestone wet scrubbing, are worthy of consideration by small-scale smelters (e.g., Eqs. 13.35–13.37).

$$Ca(OH)_2 + H_2O + SO_2 \rightarrow CaSO_3 + 2\,H_2O \qquad 13.35$$

$$CaSO_3 + 1/2\,O_2 + 2\,H_2O \rightarrow CaSO_4 \cdot 2\,H_2O \qquad 13.36$$

$$CaCO_3 + H_2O + SO_2 \rightarrow CaSO_3 + CO_2 + H_2O \qquad 13.37$$

A switch from chain grate or multiple hearth type roasters to fluidized bed types with improved air to hot sulfide ore particle contact and less entrained air can raise the sulfur dioxide from 1 to 12% in the roaster "off-gases." If the sulfide content of the ore is high enough, a fluid bed roaster can accomplish autogenous (self-heating) roasting. The need for added external fuel for roasters using low sulfide ores dilutes the sulfur dioxide to less than economic concentrations, which has led to the abandonment of roasting as a separate smelter operation under these conditions [18].

Continuous, or flash smelting processes using oxygen-enriched air are being employed to combine two or three smelting stages, roasting, reverberatory smelting and/or converting, into one [34]. This gives sulfur removal with good control, and high sulfur dioxide concentrations in the exit gases. Use of a top blown rotary converter (TBRC) streamlines copper production by merging smelting stages and by allowing the use of oxygen-enriched air for blowing [35]. Oxygen-enriched air makes it possible to obtain flue gases containing 75–80% sulfur dioxide, sufficiently high to practice direct sulfur dioxide recovery, if desired [36]. Use of oxygen-enriched air in ordinary converters is not possible because of heat damage to tuyeres (air nozzles). These developments have been adopted by several smelters [37, 38].

Smelter acid is not as pure as the acid produced from sulfur combustion, so it fetches a lower price. Nevertheless, this product is quite suitable for uses such as fertilizer phosphate production, which gives a by-product credit to the process. Smelters which exercise this choice produce about 4 tonnes of sulfuric acid for each tonne of copper [36]. Smelter sources contributed about 6% of the sulfuric acid produced in the U.S.A. in 1965 and more than 60% of the Canadian total for 1976 [39].

Sulfur dioxide capture and subsequent reduction to a sulfur product has also been tested. The American Smelting and Refining Company's process involves combustion of the cooled and cleaned smelter gases with methane (Fig. 13.10; Eq. 13.38).

$$4\ SO_2 + 2\ CH_4 + air \xrightarrow{1,250°C} S + H_2S + COS + SO_2 + 2\ CO_2 + \qquad 13.38$$
$$3\ H_2O + 4\ N_2$$

Carbonyl sulfide and some of the residual sulfur dioxide are then reacted at 450°C in the presence of a bauxite catalyst to raise the yield of sulfur (Eq. 13.39).

$$2\ COS + SO_2 \rightarrow 2\ CO_2 + 3\ S \qquad 13.39$$

After electrostatic precipitation of the products of these two steps, the remaining 2:1 ratio of hydrogen sulfide and sulfur dioxide are reacted in the presence of a Claus-type catalyst to add to the sulfur yield (Chap. 3; Eq. 13.40). Sulfur recoveries are 90–95%.

FIGURE 13.10   Flowsheet of the American Smelting and Mining Company's process for conversion of sulfur dioxide to sulfur, adapted from U.S. Bureau of Mines [40].

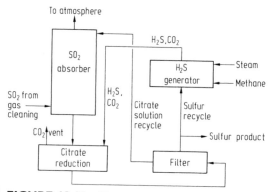

FIGURE 13.11 The citrate process for sulfur dioxide capture and conversion to elemental sulfur developed by the U.S. Bureau of Mines. Details available from U.S. Bureau of Mines [40].

$$2\,H_2S + SO_2 \rightarrow 3\,S + 2\,H_2O \qquad\qquad 13.40$$

The U.S. Bureau of Mines has tested a pilot-scale sulfur dioxide to sulfur conversion process which involves initial absorption of sulfur dioxide in an aqueous solution of citric acid, $HOC(CH_2CO_2H)_2CO_2H$, and sodium citrate (Fig. 13.11). This solution is then contacted with hydrogen sulfide to reduce the sulfur dioxide to elemental sulfur, which is then readily filtered from the resulting slurry (Eqs. 13.41–13.43).

$$SO_2 + H_2O \rightarrow H_2SO_3 \qquad\qquad 13.41$$
$$H_2SO_3 + H_3Cit \rightarrow H_2SO_3 \cdot H_3Cit \qquad\qquad 13.42$$
$$H_2SO_3 \cdot H_3Cit + 2\,H_2S \rightarrow 3\,S + H_3Cit + 3\,H_2O \qquad\qquad 13.43$$

The filtrate is returned to the absorber for reuse. A part of the product sulfur is reacted with methane and steam to produce the hydrogen sulfide required for reduction of the citrate complex (Eq. 13.44).

$$4\,S + CH_4 + 2\,H_2O \rightarrow 4\,H_2S + CO_2 \qquad\qquad 13.44$$

This is over 90% efficient for reverberatory furnace flue gases containing as little as 1.5% sulfur dioxide.

Emission standards for several smaller scale, but more hazardous discharges of copper smelters, such as arsenic and lead compounds have recently been set by the U.S. EPA [41].

13.5. HYDROMETALLURGICAL COPPER RECOVERY

13.5.1. Basic Principles

Hydrometallurgy refers to the application of aqueous solutions for metal recovery from ores, and has been practiced for copper recovery for many years. The original impetus for solution methods for copper extraction before the development of froth flotation technology was the existence of large ore bodies of low copper content which were uneconomic to work using conventional smelting methods. Hydrometallurgy also holds a complementary position to concentrator technologies for its ability to scavenge copper from concentrator tailings, since froth flotation is poor at floating copper oxides which are readily extracted by the appropriate aqueous solutions [42]. Most recently, hydrometallurgical copper recovery methods have been developed to process concentrates of more complex ores with the objective of eliminating the sulfur dioxide discharge problems of conventional smelting. Thus, while all hydrometallurgical methods essentially have a low air pollution potential, some recently developed methods use chemistry which yields elemental sulfur directly from the sulfur minerals present in the ores, to give a cleaner process overall.

Hydrometallurgical copper recovery can be conveniently considered in two stages: the leaching stage, in which the various forms of copper in the ore are placed into an aqueous solution, and the recovery stage, where dissolved copper is recovered as solid, nearly pure copper metal ready for fabrication or final smelting.

The most common solvent is 5–10% sulfuric acid in water. Diluted smelter acid can provide this at low cost. This can dissolve a number of copper minerals, not just the oxide (Eqs. 13.45–13.48).

$$CuO + H_2SO_4 \rightarrow CuSO_4 + H_2O \qquad\qquad 13.45$$

$$Cu_2O + 2\ H_2SO_4 + 1/2\ O_2 \rightarrow 2\ CuSO_4 + 2\ H_2O \qquad\qquad 13.46$$

$$CuSiO_3 \cdot 2\ H_2O + H_2SO_4 \rightarrow CuSO_4 + SiO_2 + 3\ H_2O \qquad\qquad 13.47$$

$$Cu_2(OH)_2CO_3 + 2\ H_2SO_4 \rightarrow 2\ CuSO_4 + CO_2 + 3\ H_2O \qquad\qquad 13.48$$

These reactions, in particular carbonate solvation, deplete the sulfuric acid content of the extraction solution. Later metal recovery from the solutions, however, returns an equivalent acid strength to the water. Also, *Thiobacillus ferro-oxidans* in the presence of air converts sulfide minerals to sulfates, which contributes to the acid content of the solvent when metal is recovered from the solution.

Metal may be extracted by leaching in place, by heap or percolation leaching, or by leaching with mechanical agitation in separate tanks [43]. Leaching in place requires prior fracturing of the ore body by explosives to provide cracks for percolation of leach solution. Copper-containing percolate is recovered for recirculation and eventual metal recovery by pumping this solution from lower galleries of the in-place leaching zone. This method may be used for low-grade zones of an ore body that is loosened in place, or for spent ores which have been returned to the mine cavity for disposal.

Heap and percolation leaching are used with low-grade ores which are crushed to an average 1-cm particle diameter for this purpose. Heap percolation is practiced on a large-scale, 100,000-tonne by 35-m-high flattened top ore heaps, which are constructed so as to allow even distribution of liquor and air throughout the heap. Percolation leaching is conducted in a series of vats containing the crushed ore through which leaching solution is pumped to gradually put the metal into solution.

Agitation leaching is used with very finely divided ores which may have already been partially processed and would pack too tightly for percolation methods. The finely ground pulp is placed with a sufficient volume of leach solution to obtain a very fluid mass, in vats equipped with an air-lift (a "pachuca," after its first use in Pachuca, Mexico) or mechanical agitation to maintain the whole mass in suspension [15]. Extraction is orders of magnitude faster than either in-place or heap/percolation leaching methods. The supernatant liquor containing the dissolved metal values is collected as the overflow from a thickener. Once the copper and any other metals of interest have been obtained in solution, the copper may be recovered by a variety of methods.

Cementation involves precipitation of the copper by passage of the leach solution over scrap iron (Eq. 13.25). This requires about 1 hr of contact time and produces a finely divided form of the metal analyzing ca. 90% Cu (dry) by replacement of the dissolved copper by iron. In theory this should only require 0.879 kg of iron for each kilogram of copper obtained (Eq. 13.25). In practice there is an iron consumption of 1.3–3 kg/kg copper [44]. At times as much as 10% of U.S. production has been from cement copper.

Copper may also be recovered from leach solutions electrolytically. Electrowinning requires the use of an insoluble anode such as hard lead, comparable to the liberator cell used for liquor purification in copper electrorefining. Consequently, there *are* net electrochemical reactions involved in electrowinning (Eqs. 13.20 and 13.22), as opposed to the situation with electrorefining, so that about 1.7 V are required for this step. This results in a much higher electrical power consumption of about 2.8 kWh/kg copper for electrowinning, compared to about 0.2 kWh/kg for electrorefining.

13.5.2. Hydrometallurgy Developments

More selective, cleaner copper recovery from leach solutions may be obtained by solvent extraction using a copper-selective complexer, such as an hydroxyoxime (Fig. 13.12) in the organic phase. Two mole of this complexer to each cupric ion confers a marked preference of the copper (II) species for the organic phase, in which it is captured from a dilute aqueous solution [46]. The metal complex is stabilized by the formation of two six-membered rings which include the cupric ion. The metal-rich organic phase is separated from the aqueous scavenging solution. Then the copper may be back-extracted to a fresh aqueous acid phase ready for electrowinning of the copper [47] (Fig. 13.13). The copper in the primary leach solutions may be pre-concentrated by ion exchange or adsorption/desorption from activated carbon [48].

The Sherritt–Cominco hydrometallurgical process makes the method applicable to chalcopyrite ($CuFeS_2$) ores [49]. After thermal rearrangement and reduction of the concentrate with hydrogen it is leached first with sulfuric acid to dissolve out the iron [50] (Eq. 13.49).

$$FeS + H_2SO_4 \rightarrow FeSO_4 + H_2S \qquad 13.49$$

The hydrogen sulfide is recovered for processing to sulfur and the iron-containing solution is separated for precipitation of iron. The copper, now as Cu_5FeS_4, remains undissolved in the concentrate up to this stage. It is separately dissolved out of the concentrate with sulfuric acid under oxidizing conditions (Eq. 13.50).

FIGURE 13.12 Structure of the two forms of the copper-selective complexer, LIX65N, and the kerosene-soluble copper complex formed on contact of the *anti* form with cupric ion. From Bailes *et al.* [45].

$$Cu_5FeS_4 + 6\ H_2SO_4 + 3\ O_2 \rightarrow 5\ CuSO_4 + FeSO_4 + 4\ S + 6\ H_2O \quad\quad 13.50$$

Metallic copper is then obtained either by electrowinning or hydrogen reduction. At least 98% copper recovery, gold and silver capture equivalent to conventional smelting, and recovery of 60–90% of the sulfur in elemental form has been demonstrated [51].

Cymet's use of aqueous ferric chloride leaching solution has also been proven to work with chalcopyritic ores [52].

Ammonia complexation has also been employed, primarily for nickel-copper separations but also for copper recovery, when this is from native copper or copper oxide ores [15]. The dissolving process is thought to involve formation of soluble cuprous ammonium carbonate [53] (Eqs. 13.51 and 13.52).

$$Cu + Cu(NH_3)_4CO_3 \rightarrow Cu_2(NH_3)_4CO_3 \quad\quad 13.51$$

$$2\ Cu_2(NH_3)_4CO_3 + 4\ (NH_4)_2CO_3 + O_2 \rightarrow \quad\quad 13.52$$
$$4\ Cu(NH_3)_4CO_3 + 4\ H_2O + 2\ CO_2$$

FIGURE 13.13 Countercurrent solvent extraction, using a copper-selective complexer in the organic phase used to raise the concentration and purity of copper ion solutions fed to electrowinning.

Interesting developments relate to combinations of hydrometallurgical with other processes. Flotoextraction combines froth flotation and solvent extraction to give improved metal recovery discrimination, and decontamination of the residual extractant from the recycle water [54]. In different example, 98% leaching of copper from the ore, and simultaneous electrolytic deposition of metal were combined on a laboratory scale in a single electrochemical cell. An "electroslurried" product of 99.9% copper was obtained [55]. The evolution of these and related technologies have been evaluated [56, 57].

REVIEW QUESTIONS

1. (a) Explain the meaning of the term "contact angle" and its relevance to the beneficiation of low-grade copper ores.
(b) Why are good recoveries obtained with froth flotation of sulfide ores of copper, while the oxides are poorly reclaimed?
(c) What influence(s) does the particle size of the ore have on the efficiency of copper recovery from froth flotation, and how does time since grinding affect this?

2. (a) From first principles, what is the theoretical power required (kWh/ tonne) for the electrolytic production of 1.00 tonne of each of the following: chlorine, sodium metal, aluminum, and electrorefined copper?
(b) What is the actual power required for 1.00 tonne of electrorefined copper and for 1.00 tonne of copper recovered by electrowinning. Why is there a difference?

3. (a) Why does the pH of the copper-containing electrolyte drop during electrowinning?
(b) How does this pH change help the hydrometallurgical component of the process?

4. (a) What mass of iron is required theoretically to recover 1.00 tonne of "cement copper?"
(b) What factors contribute to the higher than theoretical amount of iron required for cementation in practice, and how do these affect the process?

5. Compare and contrast the niches served in copper recovery and production by froth flotation, hot smelting, hydrometallurgical extraction, electrowinning, and cementation. Include consideration of environmental factors.

6. Toxic sulfide ion (S^{2-}), present in the waste waters from froth flotation of sulfide ores or from refinery operations, may be neutralized by the addition of the stoichiometric (not excess) amount of chlorine to oxidize sulfide to sulfate.
(a) Give the equations which describe the chemistry of this neutralization.
(b) What rate of chlorine addition (g/hr) would be required to neutralize an alkaline waste stream of 35 L/hr containing 1,200 mg/L sulfide?

(c) What problems, apart from excessive reagent cost, might the operator of such a neutralization plant face on the addition of a large excess of chlorine to the sulfide-containing waste stream?

7. With a 2.5:1 mole ratio of sodium hypochlorite to aqueous sodium cyanide it is possible to convert the toxic cyanide ion to harmless nitrogen gas.

(a) Write the balanced equation for this process.

(b) What amounts of sodium hydroxide and chlorine would be required to neutralize 100 L of effluent containing 30 mg/L sodium cyanide, in this way?

(c) If one were tempted to carry out a similar neutralization using hypochlorous acid made by dissolving chlorine gas in water, what serious exposure risk could arise? If this were to be adopted, what precaution(s) must be taken to avoid this hazard?

FURTHER READING

A. Casali, G., Gonzalez, H. Agusto *et al.*, Dynamic Simulator of a Rougher Flotation Circuit for a Copper Sulphide Ore. *Miner. Eng.* 15(4), 253–262, Apr. (2002).

A.S. Ek and I. Renberg, Heavy Metal Pollution and Lake Acidity Changes Caused by 1000 Years of Copper Mining at Falun, Central Sweden. *J. Paleolimnology* 26(1), 89–107 (2001). (Or see: *AMBIO* 30(2), 96–103, Mar. (2001).)

S.M. Hong, J.P. Candelone, M. Soutif *et al.*, A Reconstruction of Changes in Copper Production and Copper Emissions to the Atmosphere During the Past 7000 Years. *Sci. Total Envir.* 188(2–3), 183–193, Oct. 11 (1996).

L. Landner and R. Reuther, "Metals in Society and the Environment: A Critical Review of Current Knowledge on Fluxes, Speciation, Bioavailability and Risk for Adverse Effects of Copper, Chromium, Nickel and Zinc." Kluwer Academic, Dordrecht, 2004.

J. Leja, "Surface Chemistry of Froth Flotation." Plenum, New York, 1981.

REFERENCES

1. The Falun Copper Mine Foundation, Falun, Sweden, 1999.
2. H.H. Coghlan, "Notes on the Prehistoric Metallurgy of Copper and Bronze in the Old World." University Press, Oxford, 1951.
3. F.S. Taylor, "A History of Industrial Chemistry," p. 37. Arno Press, New York, 1972.
4. R. Chadwick, *Chem. Br.* **17**(8), 369 (1981).
5. "United Nations Statistical Yearbook 1979/80." United Nations, New York, 1981, and earlier issues.
6. "USGS Minerals Yearbook 2001." Available: http://minerals.usgs.gov/minerals/pubs/commodity/myb/and earlier issues.
7. H.O. Hofman, "Metallurgy of Copper." McGraw-Hill, New York, 1914.
8. J. Leja, *J. Chem. Educ.* **49**(3), 157 (1972).
9. D.W. Fuerstenau, ed., "Froth Flotation," 50th Aniversary Volume. American Institute of Mining, Metallurgical, and Petroleum Engineers, New York, 1962.
10. C.J.S. Warrington and R.V.V. Nichols, "A History of Chemistry in Canada," p. 15. Pitman and Sons, Toronto, 1949.
11. "Kirk-Othmer Encyclopedia of Chemical Technology," 3rd ed., Vol. 6, p. 819. Wiley, New York, 1979.

12. W. Tolley, D. Kotlyar, and R. VanWagoner, Fundamental Electrochemical Studies of Sulfide Mineral Flotation. *Miner. Eng.* **9**(6), 603–637, June (1996).

13. V.I. Klassen and V.A. Mokrousov, "An Introduction to the Theory of Flotation" (translated by J. Leja and G.W. Poling), 2nd ed. Butterworth, London, 1963.

14. R. Natarajan and I. Nirdosh, N-arylhydroxamic Acids as Mineral Collectors for Ore-beneficiation. *Can. J. Chem. Eng.* **79**(6), 941–945, Dec. (2001).

15. J.R. Boldt, Jr. and P. Queneau, "The Winning of Nickel." Longmans of Canada, Ltd., Toronto, 1967.

16. R.R. Klimpel, Technological Trends in Flotation Chemistry. *Min. Eng. (Littleton, Colo.)* **47**, 933–942 (1995).

17. G.C. Allan and J.T. Woodcock, A Review of the Flotation of Native Gold and Electrum [copper alloys]. *Miner. Eng.* **14**(9), 931–962, Sept. (2001).

18. "Kirk-Othmer Encyclopedia of Chemical Technology," 2nd ed., Vol. 6, p. 131. Wiley, New York, 1965.

19. W.W. Harvey, M.R. Randlett, and K.I.Bangerskis, *J. Met.* **30**, 32, July (1978).

20. A.S. Gendron, R.R. Matthews, and W.C. Wilson, *Can. Min. Metall. Bull.* **70B**, 166, Aug. (1977).

21. W. Gluschke, J. Shaw, and B. Varon, "Copper: The Next Fifteen Years," p. 30. D. Reidel (The United Nations), Dordrecht (The Netherlands), 1979.

22. G.D. Clayton and F.E. Clayton, eds., "Patty's Industrial Hygiene and Toxicology," 3rd ed., Vol. 2A, p. 1620. Wiley, New York, 1981.

23. "McGraw-Hill Encyclopedia of Science and Technology," Vol. 3, p. 641. McGraw-Hill, New York, 1982.

24. R.C. Weast, ed., "Handbook of Chemistry and Physics," 51st ed., p. F-16. Chem. Rubber Publ. Co., Cleveland, OH, 1970.

25. G. Barthel, *J. Met.* **30**, 7, July (1978).

26. A.G. Eccles, *Can. Mining Metall. Bull.* **70**(785), 141, Sept. (1977).

27. G.I. Mathieu, "Ozonation for Destruction of Cyanide in Canadian Gold Mill Effluents: A Preliminary Evaluation" Canmet Rep. 77–11. Dept. of Supply and Services, Ottawa, 1977.

28. W.A. Triggs and A.M. Laird, *Mining Mag.* **124**(6), 438, June (1971).

29. M.J.R. Clark and T.O. Morrison, "Impact of the Westmin Resources Ltd. Mining Operation on Buttle Lake and the Campbell River Watershed." Waste Management Branch, Ministry of the Environment, Victoria, B.C., 1982.

30. Inco Entering Acid Production, *Can. Chem. Process.* **65**(2), 10, Mar. 27 (1981).

31. P.L. Magill, F.R. Holden, and C. Ackley, eds., "Air Pollution Handbook." McGraw-Hill, New York, 1956.

32. J. Rastas, E. Nyholm, and J. Kangas, *Eng. Min. J.* **172**, 123 (1971).

33. J. Kangas, E. Nyholm, and J. Rastas, *Chem. Eng.* (*N.Y.*). **78E**, 55 (1971).

34. Copper Smelting is Going Continuous, *Can. Chem. Process.* **60**(4), 21, Apr. (1976).

35. Copper Refining Via Rotary Smelting, *Can. Chem. Process.* **60**(1), 6, Jan. (1976).

36. T.J. Chessell, *Can. Chem. Process.* **63**, 21, Mar. 21 (1979).

37. New Copper Recipe, *Dravo Rev.*, Fall, 2 (1978).

38. Afton, New Canadian Copper Mine. *World Min.* **31**(4), 42, Apr. (1978).

39. P.D. Nolan, *Can. Chem. Process.* **61**(4), 35, Apr. (1977).

40. "Control of Sulfur Oxide Emissions in Copper, Lead, and Zinc Smelting," Inf. Circ. 8527. U.S. Bureau of Mines, Washington, DC, 1971.

41. "U.S. Environmental Protection Agency, National Emission Standards for Hazardous Air Pollutants for Primary Copper Smelters." U.S. EPA: Protection Agency, Research Triangle Park, NC, 2001.

42. N. Arbiter and A.W. Fletcher, Copper Hydrometallurgy—Evolution and Milestones. *Min. Eng.* **48**, 118–123, Feb. (1994).

43. A.M. Johnson, Solution Mining. *Min. Eng.* **44**, 452, May (1992).

44. Copper Industry Uses Much Scrap Iron, *Environ. Sci. Technol.* **7**(2), 100 (1973).

45. P.J. Bailes, C. Hanson, and M.A. Hughes, *Chem. Eng.* (*N.Y.*). **83C**, 86, Aug. 30 (1976).

46. A.M. Sastre and J. Szymanowski,...Extraction Properties of Hydroxyoximes. A Review. *Solv. Extraction Ion Exchange*, **22**(5), 737–759 (2004).

47. M.C. Kuhn, *Min. Eng. (Littleton, Colo.).* **29**, 79 (1977).
48. F. Habashi, *Chem. Eng. News*, **60**(6), 46, Feb. 8 (1982).
49. S–C Process Offers Unique Steps, *Can. Chem. Process.* **63**, 25, Mar. 21 (1979).
50. D.E.G. Maschmeyer, P. Kawalka, E.F.G. Milner *et al., J. Met.* **30**, 27, July (1978).
51. D.E.G. Maschmeyer, E.F.G. Milner, and B.M. Parekh, *Can. Min. Metall. Bull.* **71**, 131, Feb. (1978).
52. Cyprus Reveals Details, *Eng. Min. J.* **178**, 27, Nov. (1977).
53. B.D. Pandey, V. Kumar, D. Bagehi *et al.*, Extraction of Nickel and Copper from the Ammoniacal Leach Solution of Sea Nodules by LIX 64N. *Ind. Eng. Chem. Res.* **28**, 1664–1669, Nov. (1989).
54. I.A. Dibrov, N.N. Voronin, and A.A. Klemyatov, Froth Flotoextraction…Method of Metal Separation from Aqueous Solutions. *Int. J. Miner. Process.* **54**(1), 45–58, June (1998).
55. Lower energy Process for Copper Refining, *Chem. Eng. News*, **58**(23), 7, June 9 (1980).
56. J.B. Hiskey, Technical Innovations Spur Resurgence of Copper Solution Mining. *Min. Eng. (Littleton, Colo.).* **38**, 1036–1039 (1986).
57. N. Arbiter and A.W. Fletcher, Copper Hydrometallurgy—Evolution and Milestones. *Min. Eng. (Littleton, Colo.).* **46**, 118–123 (1994).

14
PRODUCTION OF IRON AND STEEL

You should hammer your iron when it is hot.
　　　　—Publilius Syrus, ca. 42 B.C.

Strike the iron whilst it is hot.
　　　　—Francois Rabelais (1494–1553)

Strike now, or else the iron cools.
　　　　—William Shakespeare (1564–1616)

14.1. EARLY HISTORY AND DEVELOPMENT

Iron, at 5% of the earth's crust, is the most abundant metal after aluminum and the fourth most abundant element in the surface rocks of the earth. Iron is also the most generally useful metal for the manufacture of machinery, vehicles, construction, etc. The wide use of iron itself and a wide range of steels (iron alloys) combine to make the tonnage of iron produced each year easily exceed the combined annual production of all the other metals used by our society.

Iron, like aluminum, is relatively easily oxidized and so is seldom found in nature in metallic form, except in recent meteorites. In fact, sideritic (iron-containing) meteorites were probably the source of the first iron worked by early people. It is also possible that lumps of iron-rich ores placed around early, very hot campfires were reduced to a spongy type of iron mixed with slag. Slag could then have been beaten out from these lumps of iron and simple tools fashioned in the same way. Since meteoritic iron would not be as common as lumps of native copper, historians believe that the Iron Age followed the Copper and Bronze ages in prehistoric metallurgical development, following the Stone Age [1]. Those parts of the world without any significant surface exposures of copper minerals could have directly moved from the Stone to the Iron Age. The use of iron was certainly known to the Egyptians dating back to 3–4000 B.C. for tools and simple jewellery [2]. Renovation, and operation of the smelting furnaces of the Roman period of Les Martys (France) has recently been reproduced [3].

The methods used by the early small-scale smelters were probably simpler versions of the methods used in Europe before about 1750. A furnace built of a fire-resisting clay and fired by charcoal was raised to smelting temperatures by a pair of continuously operated foot bellows. Selected ore of 40% or more iron was then added to the furnace in lumps, a few at a time, and the heating continued for several hours. The furnace was then allowed to cool, the front broken open, and the lumps of spongy iron mixed with slag removed. Hammering was used to drive out most of the slag and to shape the residual iron. This reduction method produced a metallic iron low in carbon and malleable enough to be shaped by beating.

The development of the forerunners of the modern blast furnace is believed to have occurred in Belgium in about 1400. These were still fired by charcoal and produced several hundred kilograms every 24 hr, but also required water power to provide a strong air blast. These larger blast furnaces only gradually displaced the early smaller scale operations.

The charcoal required for firing and reduction processes took enormous quantities of wood and rapidly depleted forest stands near smelters in Europe and North America [4, 5]. This led to many attempts to fire blast furnaces with coal with uniformly poor results. The sulfur content of the coal produced a high-sulfur iron which made it brittle when hot, an undesirable property. Gradually it was learned that prior coking of the coal could be used to obtain an equivalent quality of iron as could be obtained using wood charcoal. Only subsequently it was learned that sulfur in the coal produced the brittle iron. As the size of blast furnaces became larger, the steam engine became available to provide power for the air blast.

From these early developments the iron and steel industry has grown to a world production of iron ore, based on iron content, exceeded 600 million tonnes in 2000 (Table 14.1). While the major producers of ore are generally also major producers of pig iron, this is by no means universally true. Sweden, Australia, Brazil, Canada, and India each mine and process more iron ore than the pig iron they produce. Japan and Germany currently import nearly all of their iron ore. The U.S. depends on imported ores for about one-third of its requirements, which has been fairly steady for some time.

Steel derived from new pig iron produced from ores and from recycled steel has better properties than iron for most uses. World steel production currently hovers around 1.33 times world pig iron production, a reflection of the recycled metal content. Again, we have some countries, like China and the U.S.A., whose national steel consumption exceeds even their very large scale of production (Table 14.2).

France, the U.K., and Germany each produce slightly more than they consume. And Japan, even with a very large per capita consumption, produces almost twice its annual requirements.

14.2. REDUCTION OF IRON ORES

14.2.1. Direct Reduction

A number of naturally occurring minerals contain iron and some are rich enough in iron to be classed as ores. In North America, these are the oxides

TABLE 14.1   Major World Producers of Pig Iron[a]

	Thousands of metric tonnes				
	1960	1970	1980	1990	2000
Australia	2,922	5,769	7,407	6,188	7,000
Brazil	1,750	4,296	13,116	21,141	28,553
Canada	4,025	8,424	11,182	7,344	10,034
China	27,500	22,000	39,604	62,380	131,010
India	4,260	7,118	8,718	12,000	27,761
Japan	12,341	69,714	89,130	80,229	81,071
Sweden	1,627	2,842	2,596	2,851	3,146
Turkey	–	1,034	2,072	32,829	300
U.K.	–	17,677	6,316	12,320	9,900
U.S.S.R.	46,800	85,933	107,283	110,166	46,820[c]
U.S.A.	62,202	85,141	63,748	49,668	43,300
West Germany	25,739	33,897	34,055	29,585	30,400[d]
Other	69,434	114,666	134,869	126,355	195,705
World Total	258,600	439,800	520,096	553,056	615,000

[a]Data from *U.N. Statistical Yearbooks* [6] and *U.S. Geological Survey Minerals Yearbook 2001* [7]. World total for 1950 was 112,900 thousand metric tonnes.
[b]For 2000, includes small amounts of direct reduced iron.
[c]Data for Russian Federation.
[d]Data for the unified Germany.

magnetite (Fe_3O_4, 72.4% iron), hematite (Fe_2O_3, 70.0% iron), and limonite, a hydrated iron oxide ($2Fe_2O_3 \cdot 3H_2O$, 59.8% iron). In Europe, siderite ($FeCO_3$, 48.2% iron) is an important ore. The sulfides pyrite (FeS_2) and chalcopyrite ($CuFeS_2$) are more important for their sulfur content than for their iron. While the iron content of natural deposits of these minerals is never

TABLE 14.2   Steel Production by Some Major Steel Producing Countries[a]

	Thousands of metric tonnes					
	1950	1960	1970	1980	1990	2000
China	540	18,460	18,000	39,996	41,936	128,500
France	8,652	17,284	23,773	23,176	19,304	20,980
Italy	2,362	8,231	17,277	26,501	25,647	26,544
Japan	4,848	22,144	93,322	111,395	110,339	106,444
U.K.	16,555	24,680	28,316	11,277	17,841	15,000
U.S.S.R.	27,000	65,318	115,889	147,941	162,326	59,000[b]
U.S.A.	87,848	90,092	119,309	101,456	89,726	90,084
West Germany	12,121	34,110	45,040	43,838	38,433	46,000[c]
Other	26,074	65,597	132,974	196 404	267,389	356,448
World	186,000	345,916	593,900	701,984	772,941	849,000

[a]Countries producing more than 20 million tonnes of steel per year, from *U.N. Statistical Yearbooks* [6] and *U.S. Geological Survey Minerals Yearbook, 2001* [7].
[b]Data for 2000 for Russian Federation.
[c]Data for 2000 for unified Germany.

quite as high as the calculated percentages, many commercial ores contain 50–60% iron and some are as high as 66–68% [8].

Ore in a hard rock matrix is mined from underground or surface deposits by drilling and blasting followed by recovery with power shovels. Many deposits are loosely consolidated, enabling direct recovery using power shovels by open cut, or open pit operation. Ore is also frequently beneficiated (cleaned of extraneous material) at the mine site to improve the grade. This may involve any of rough sorting, crushing, screening, magnetic separation (magnetites only), washing (for clay or sand removal), and drying, all designed to improve the grade of the ore prior to shipping or smelting.

Direct reduction processes similar to prehistoric reduction methods are still used on a small scale to produce spongy and powdered iron products. These involve contact of a solid or gaseous reductant at temperatures of 1,000–1,200°C, with a rich magnetite or hematite ore to obtain reduction without melting [9]. Powdered coal or coke, natural gas, or hydrogen may be used. Direct reduction gives an iron product low in carbon that can be used for steelmaking directly, without requiring prior carbon removal. For this reason the direct reduction product is sometimes called "synthetic scrap," since this type of iron can replace part or all of the scrap used in steelmaking. Also the very high porosities and specific surface areas of these forms of iron are valuable for catalyst applications.

Feed composition and processing constraints required by direct reduction methods have led to blast furnace reduction of iron ores as the dominant method used to obtain metallic iron. Direct reduction methods remain in favor where there is abundant natural gas but negligible coal reserves, such as in Iran, Qatar, and Venezuela.

14.2.2. Blast Furnace Reduction of Iron Ores

Reduction of iron ore to the metal in a blast furnace requires the reduced metal to be melted in the process, the key physical difference between direct reduction and blast furnace reduction. Two to three tonnes of iron ore are required for each tonne of iron produced. The main impurity in iron ore is generally silica. To avoid contaminating the iron with silica, limestone or dolomite ($CaCO_3 + MgCO_3$) is added with the ore to combine with the silica and provide a lithophilic phase to dissolve any other acidic oxide impurities (Eqs. 14.1–14.3).

$$CaCO_3 \rightarrow CaO + CO_2 \tag{14.1}$$

$$CaO + SiO_2 \rightarrow CaSiO_3 \text{ (slag)} \tag{14.2}$$

$$CaO + Al_2O_3 \rightarrow Ca(AlO_2)_2 \tag{14.3}$$

The impurities form a separate molten slag phase which can be easily separated from the molten iron. About 550–1,000 kg of coke is also used per tonne of iron produced, to provide both the heat and the reducing agents necessary for the formation of pig iron.

The blast furnace consists of a large, almost cylindrical stack up to 10 m in diameter and 70 m high, which is tapered to a larger diameter at the

FIGURE 14.1 Diagram of the operating details of a blast furnace for the production of pig iron.

bottom to avoid bridging (jamming) of ore as it moves down the stack (Fig. 14.1). The outer shell is constructed of heavy (ca. 2-cm-thick) riveted or welded steel plates, which is then lined with silica-containing (upper region) or carbon (hearth area) refractory bricks to insulate and make the structure resistant to the intense heat. This unit functions chemically as a heterogeneous, countercurrent reactor in which lumps of the solid reactants are charged in at the top using a skip cart on rails, and the firing air is blown in at the bottom. The limestone and coke are added in alternating layers, in lump form to permit movement of gases through the charges. A double bell arrangement at the top of the furnace prevents gas loss from the furnace during the charging process by only opening one bell at a time during the addition steps. The firing air is preheated to about 850°C in regenerative stoves and then blown through water-cooled steel tuyères (air nozzles) at the hearth area of the furnace.

Initial warming of a newly lined blast furnace is conducted gradually during several days to avoid any unnecessary thermal stresses. Fueled first by wood or oil and then by coke, the approach to normal operating temperature is gauged from peepholes through the tuyères. As these temperatures are reached, ore and limestone are blended with the coke additions through the double-bell arrangement at the top of the furnace. The heat, provided by primary combustion in the oxidizing zone near the hot air blast of the tuyères, is sufficient to cause carbon dioxide and water vapor (from the air) reduction (Eqs. 14.4–14.6).

Primary combustion and reductant formation:

$$C + O_2 \rightarrow CO_2 \qquad\qquad 14.4$$

$$CO_2 + C \text{ (white hot)} \rightarrow 2\ CO \qquad\qquad 14.5$$

$$H_2O + C \text{ (white hot)} \rightarrow CO + H_2 \qquad\qquad 14.6$$

Carbon monoxide is the primary reducing agent of the iron oxides in the ore, assisted by carbon and to a small extent hydrogen (Eqs. 14.7–14.12).

Primary reducing reactions:

$$\text{ca. } 500°C\ \ 3\ Fe_2O_3 + CO \rightarrow 2\ Fe_3O_4 + CO_2 \qquad\qquad 14.7$$

$$\text{ca. } 850°C\ \ Fe_3O_4 + CO \rightarrow 3\ FeO + CO_2 \qquad\qquad 14.8$$

$$\text{ca. } 900\ \ FeO + CO \rightarrow Fe + CO_2 \qquad\qquad 14.9$$

$$\text{ca. } 900\ \ Fe_2O_3 + 3\ H_2 \rightarrow 2\ Fe + 3\ H_2O \qquad\qquad 14.10$$

The solid–solid or liquid–solid reactions between ferric and ferrous oxides and carbon also occur at temperatures above about 500°C [8].

$$Fe_2O_3 + C \rightarrow 2\ FeO + CO \qquad\qquad 14.11$$

$$FeO + C \rightarrow Fe + CO \qquad\qquad 14.12$$

The iron, with its melting point depressed to about 1,100°C by dissolved carbon (m.p. of pure iron 1,535°C), drips through the charge and collects on the hearth. Being the denser fluid product of the blast furnace, it forms a lower phase as it accumulates on the hearth. Molten pig iron is periodically withdrawn from this accumulation by drilling or knocking out a clay plug placed in a lower tap hole in the side of the hearth.

The slagging reactions (Eqs. 14.1–14.3) start to occur as the charge moves down the furnace and the limestone or dolomite flux reaches temperatures of about 900°C or more. As carbon dioxide loss occurs, the lime (and magnesia, if present) starts to react with and dissolve other oxide impurities of the charge into the slag layer. It is also possible to reduce silica and phosphate by the same kinds of reactions used to reduce iron which, in the process, contaminates the pig iron (Eqs. 14.13 and 14.14).

$$SiO_2 + 2\ C \rightarrow Si + 2\ CO \qquad\qquad 14.13$$

$$P_2O_5 + 5\ C \rightarrow 2\ P + 5\ CO \qquad\qquad 14.14$$

The presence of the basic slagging ingredients minimizes but does not completely prevent these reductions from occurring. The molten slag, which is less dense than iron, floats on top of the iron layer and is periodically withdrawn through an upper tap hole in the hearth of the blast furnace. About 0.5–1.5 tonnes of slag are produced for each tonne of pig iron, the low end of the range being obtained with high-grade ores while using a coke with a low-ash content. Thus utilization of this by-product of pig iron production is a significant part of normal blast furnace operation. Slag is used for road aggregate (ca. 50–60% of the total), in crushed form as a stone component of asphalt, and in granulated form (by running the melt into water) for use as an ingredient in the manufacture of cement (ca. 10–20% of the total). It is also used to a small extent (ca. 5%) for the preparation of mineral wool

■■■ **TABLE 14.3 Approximate Composition Ranges of the Products of Blast Furnace Operation**[a]

Iron		Slag		Exit gases	
Component	% by wt.	Component	% by wt.	Component	% by wt.
Iron	90–94	Lime ($CaCO_3$)	30–48	Nitrogen	58–60
Carbon	3–4.5	Silica	28–42	Carbon monoxide	27–30
Silicon	1–2.5	Alumina	10–20	Carbon dioxide	8–14
Phosphorus	0.04–2.5	Magnesia (MgO)	2–14	Hydrogen	0.4–3
Manganese	0.4–0.7	Sulfur	1–3	Water vapor	0.4–1.0
Sulfur	0.02–0.2	Iron oxide (FeO)	0.3–2		
		Manganese oxide (MnO)	0.2–1.5		

[a]Compiled from data of Dearden [2] and Kirk-Othmer [8].

("rock wool") insulation, and has been examined for removal of phosphate from aqueous solutions and for flue gas desulfurization [10, 11]. The approximate chemical composition of blast furnace slag is given in Table 14.3.

The carbon monoxide produced in the furnace is only partially consumed on its passage through the granular charge (Table 14.3). Capture of this important fuel gas for energy recovery, by the double-bell closure at the top of the furnace, is only a relatively recent (ca. 1840) innovation in the history of iron making. After collection from the upper part of the furnace, the exit gas is passed to a dust catcher for cleaning. It is then used partly to fuel three or four regenerative stoves and partly for other purposes. The regenerative stoves are designed to have a large thermal mass from an extensive internal firebrick checker-work of narrow passages, and are first heated by burning blast furnace gas. When the temperature of the brick checker-work of one stove has been raised sufficiently, the blast furnace gas is switched to fuel the burner of another stove. At this point the preheated stove is used to preheat the air blast being fed to the tuyères in the hearth area of the blast furnace. By a fuel gas and air switching arrangement like this, blast air is continuously preheated using the fuel value of the furnace exit gases. This decreases the coke consumption to much less than that required before gas collection was practiced.

Originally the molten iron product was drained from the lower notch on the hearth to a series of shallow trenches shaped in sand. As the molten iron moved along each long trench, it moved through a series of short smaller trenches and filled roughly rectangular hollows at the end of each of these. Each of these formed a 20–25 lb ingot of iron on cooling. This product became known as "pig iron" by the appearance parallels between this method of forming ingots to a sow suckling her young. Today pig iron ingots are made on continuous pig-casting machines, which consist of an endless chain with metal molds into which the molten iron is ladled. As the molds with molten metal move through a water spray, the melt is quickly chilled, allowing pig production at 2 tonne/minute or more. A lime or similar wash is sprayed into the insides of the emptied molds as they return to the ladle for refilling to prevent the pig iron from sticking to the molds.

Blast furnace production of iron allows the hot, newly reduced product to trickle through the bed of heated coke to the hearth. Since carbon is somewhat soluble in molten iron, pig iron usually contains from 3 to 4.5% carbon. It also contains smaller percentages of other reduced elements such as silicon, phosphorus, manganese, etc., generated by the same reducing processes that yielded the iron (Table 14.3). Primarily from the effect of the high-carbon content on the iron crystal structure, the blast furnace product is brittle, hard, and possesses relatively low-tensile strength. Hence the crude pig iron product of the blast furnace is not much used in this form.

After remelting in a cupola furnace and some minor composition adjustments, pig iron is used to make various types of cast irons with improved properties. Molten gray cast iron, for instance, is poured into molds to make automotive engine blocks. The hardness and graphite content of this material contribute to its desirable wear resistance. A large variety of cast irons are important in iron metallurgy. Properties may be adjusted by microstructural constituents and variation of crystal structures. Corrosion resistance of some cast irons makes these useful for the fabrication of certain types of chemical reactors. The low cost of the raw material means that reactor walls may be made sufficiently thick so that some corrosion can be tolerated. Construction of low speed flywheels from cast iron and as ballast in ships depend on low cost and high density. However, most pig iron moves directly to one or another of the various steelmaking processes, either still in molten state or in ingot form.

A number of advantages are obtained from pig iron production in very large furnaces. One of these is the lowered coke consumption from the proportionately lower furnace heat losses, which results from the smaller surface to volume ratio. This factor, together with more extensive use of richer, beneficiated ores which also may be partially prefluxed, and use of a more uniform size range for all charges (by prior agglomeration), plus additions of oxygen and steam to the blast air have combined to nearly halve the coke consumption per tonne of pig iron produced in American practice since 1950 [12]. The addition of steam to the blast helps to control the blast temperature as well as providing additional reducing gas at the bosh zone via Eq. 14.6 [9]. Occasionally fuel oil or natural gas is used to supplement combustion and the reducing capacity of a blast furnace. Recent energy balance data incorporating many of these variations are given by Kirk-Othmer [13].

An idea of the scale of operation of a typical modern intermediate size blast furnace incorporating some of these innovations may be gained from consulting Table 14.4. The largest blast furnaces now operating in the U.S.A. and Japan are capable of producing 10,000 tonnes of pig iron/day, more than 16 times this capacity. The total U.S. iron production in 1979, of some 80 million tonnes, was produced by 168 large and small furnaces, mostly operating in the Pittsburgh and Chicago areas.

14.3. THE MAKING OF MILD AND CARBON STEELS

Pig irons contain 6–8% of total impurities which give the product brittle, low-strength characteristics. Broadly speaking, pig irons may all be classified into

TABLE 14.4 Approximate Material Balance Data for a Modern Blast Furnace Producing 600 Tonnes of Pig Iron/Day[a]

Raw materials		Product and by-products	
Beneficiated iron ore (50–60% Fe)	1,000–1,200 tonnes	Pig iron	600 tonnes
Hard coke	300–600 tonnes	Slag	300–400 tonnes
Limestone (or dolomite)	200–300 tonnes	Dust	60–80 tonnes
Bast air, at >800 °C and	2,000–2,400 tonnes,	Exhaust	2,800–3,400 tonnes,
550 kPa (80 psi)	$1.5–1.9 \times 10^6 \, m^3$ [b]	gases	$2.2–2.6 \times 10^6 \, m^3$ [b]
Water, as steam	20–40 tonnes		
Water, cooling	14,000–18,000 tonnes		
Electricity	2,400–28,600 kWh		

[a]Data compiled from Dearden [2], McGannon [9], *Encyclopedia of Science and Technology* [12], and Kirk-Othmer [13].
[b]Volumes at standard conditions, i.e., 15°C and 1 atm.

two groups: those that are low in phosphorus and those that are high (Table 14.5). Mild steels and high-carbon steels are made by reducing the concentrations of the impurities present in pig irons and controlling the final concentration of carbon. Steels are considerably stronger materials and in addition may have other specific desirable properties contributed to them by control of composition and by heating, or other types of subsequent treatments. To achieve this improvement in properties, more than 90% of all pig iron produced moves directly to various steelmaking procedures. Over 90% of all steels made, the "common steels" of Table 14.5, are carbon steels. Thus, the bulk of the following discussion is centered on this important group of products. Mild steels, the product used for the vast majority of structural and engineering purposes, contain 0.15–0.25% carbon. High-carbon steels are defined as steels containing 0.6–0.7% carbon, and possess important heat treating qualities which can permit them to hold a sharp "edge." Mild steels are not usually heat treated.

The processes used in making common steels may be classified on a chemical basis using the composition of the refractory linings used for the vessels in which they are made, and the primary composition of the slag. The

TABLE 14.5 Percentage of Impurities in Representative Pig Irons from the Two Main Types of Ores, Compared to the Compositions of Common Steels[a]

Element	Pig iron		Common steels (%)
	No. 1 (%)	No. 2 (%)	
Carbon	3.5–4.0	3.0–3.6	0.1–0.74
Silicon	2.0–2.5	0.6–1.0	0.01–0.21
Sulfur	<0.040	0.08–0.10	0.028–0.055
Phosphorus	<0.040	1.8–2.5	0.005–0.054
Manganese	0.75	1.0–2.5	0.35–0.84

[a]Compositions in percent by weight compiled from Allen [14] and Bashworth [15].

processes used may be further subdivided according to whether they are based on pneumatic, open hearth, or electric furnace technologies.

Acidic steelmaking employs silica (SiO_2) refractory linings in the equipment used, and generates silica-rich slags for removal of impurities via additions of sand. Silica, being a nonmetallic oxide, allows removal of carbon, manganese, and silicon during the metallurgical steps but cannot remove sulfur or phosphorus. Thus, pig irons rich in sulfur or phosphorus cannot be satisfactorily processed under these conditions. Basic steelmaking procedures use magnesia (MgO) refractory linings and limestone ($CaCO_3$) or dolomite ($CaCO_3 \cdot MgCO_3$) added to the charge as a slag former. Equipment with a basic lining is not only capable of removing carbon, manganese, and silicon, but can also remove phosphorus and sulfur by capturing their oxides in the basic slag. Addition of basic slag formers to an acid-lined vessel or vice versa is never practiced because lining life is drastically shortened by direct reaction.

The technological classification of steelmaking methods is based on the system used to effect removal of pig iron impurities. All pneumatic methods rely on the heat content of the molten charge plus the heat of the refining reactions to provide the heat required by the process. Purification is effected by blowing air or oxygen through or onto the molten charge. Open hearth methods rely on the combustion of an external fuel, usually natural gas or oil, to provide the heat required for the process. Stoves with their regenerative heating capabilities (as with blast furnace gas) are an important component for open hearth operation. Electric furnace methods derive their heat requirements from electric arc or resistance sources, which permit control of the inert atmosphere above the melting independently of the heating process. This capability makes an important contribution to the production of special alloy steels.

14.3.1. Pneumatic Steelmaking: The Bessemer Process

The Bessemer process was the earliest of the pneumatic methods, and was devised and patented in England in 1855 by Henry Bessemer [16]. Originally prompted by an interest in improving the quality and strength of cast iron gun barrels to obtain greater range from lighter guns, this process was developed from the perception by Bessemer that the carbon in molten iron could be burned out by air blowing. It was anticipated that this method could raise the speed and decrease the cost of the existing crucible method of steelmaking.

Bessemer steel is made in a Bessemer converter, a pear-shaped vessel with a double bottom lined with silica, and capable of being tilted on its axis. A typical converter can hold 15–25 tonnes of metal at one time (Fig. 14.2). It is charged with molten pig iron while the converter is lying on its side. The air pressure to the perforated inside false bottom of the unit is then turned on and the vessel gradually raised to the upright position so that the air is sparged through the molten pig iron via the inner perforated floor of the converter. Blowing is continued for a period of 15–20 min until the operator, from observing the color and luminosity change of the flame at the mouth of the converter, judges the refining process to be complete. Carbon, as well as

FIGURE 14.2 Cut-away view of the main operating details of a Bessemer converter during a "blow." It is filled and emptied by tilting onto its side, while the air flow is off.

silicon, manganese, and some iron are removed, largely via reactions in Eqs. 14.15–14.19.

$$2\,Fe + O_2 \rightarrow 2\,FeO \qquad\qquad 14.15$$

$$2\,Mn + O_2 \rightarrow 2\,MnO \qquad\qquad 14.16$$

$$Si + O_2 \rightarrow SiO_2 \qquad\qquad 14.17$$

$$2\,C + O_2 \rightarrow 2\,CO \qquad\qquad 14.18$$

$$C + O_2 \rightarrow CO_2 \qquad\qquad 14.19$$

The involatile products of these reactions form an acidic slag ($SiO_2.FeO$; $SiO_2.MnO$, etc.), which forms a separate lighter phase on top of the molten iron. These exothermic reactions occurring during the blow are sufficient to raise the temperature of the converted metal from about 1,350 to around 1,600°C or higher, adequate to keep the residual iron molten. This temperature rise is necessary for handling since the melting point of the iron phase rises as the concentration of impurities is decreased by the conversion process.

The Bessemer process is fast; one converter is capable of producing nearly a tonne of steel/minute. This compares to 10–14 days for smaller quantities per unit via the crucible methods used at that time. The product is of acceptable quality for some uses, at least when the pig iron is derived from low phosphorus ores and is smelted in blast furnaces using low sulfur cokes. However, with high phosphorus pig irons the decarbonized product is "cold short" (brittle at ambient temperatures) which makes it less than useful for the majority of applications of the steel [16]. Another difficulty is that if the ore or the smelting process produces a pig iron, which is high in sulfur (e.g., from sulfur-containing cokes) the Bessemer product is "hot short," or brittle when hot. This would make hot shaping of the product impossible, adding to steelmaking difficulties with this product.

The speed of Bessemer steelmaking is both an advantage and a disadvantage, in that there is no time during the process for product sampling, analysis, and composition adjustment. Thus, much of the product quality depends on the skill and judgment of the operator. Overblowing, which results in total carbon removal plus some iron oxide and iron nitride formation, may be partially corrected by blending the blown product with a small amount of hot melt pig iron and mixing well [9]. Other minor adjustments may be made in the ladle receiving the Bessemer product. But removal of the nitrogen is never easy, even with strong deoxidizers such as aluminum, and yet this is essential to obtain good stamping quality steels (e.g., steels destined for shaping into auto or appliance body panels). The combination of these stringent material requirements and the short composition adjustment time thus has restricted production of acid Bessemer steels (produced in converters with acidic linings) to the low carbon, mild steels. However, steelmaking using these technological elements is still being practiced on a small scale as a method for producing low-cost steels.

14.3.2. The Basic Bessemer, or Thomas Process

The problem of phosphorus or sulfur removal from pig irons high in these elements was solved by the use of a basic refractory lining, such as magnesia (MgO) or lime (CaO), in the converter. When this was accompanied by the addition of limestone or dolomite with the molten charge of pig iron, phosphorus and sulfur constituents were removed in the basic slag layer that was formed. These improvements were developed in London in 1878 by the cousins S.G. Thomas and P.G. Thomas, and were adopted by the U.K. steel industry for the processing of moderate phosphorus content ores. The steelmakers of Lorraine, Germany, enthusiastically adopted the Thomas process to achieve phosphorus removal and improved steel quality from their high-phosphorus ores [14].

Phosphorus removal in this process is thought to occur primarily as the oxide, whereas elemental sulfur is thought to form a slag constituent by direct replacement of the oxygen of calcium oxide (Eqs. 14.1 and 14.20–14.22).

$$4\,P + 5\,O_2 \rightarrow 2\,P_2O_5 \qquad\qquad 14.20$$

$$P_2O_5 + 4\,CaO \rightarrow 4\,CaO\cdot P_2O_5 \text{ (the slag)} \qquad 14.21$$

$$2\,S_{(Fe)} + 2\,CaO_{(slag)} \rightarrow 2\,CaS_{(slag)} + O_{2(Fe)} \qquad 14.22$$

An important attraction of the acidic Bessemer and Thomas (basic Bessemer) processes is their rapid steelmaking capability, about a tonne a minute per converter. However, both processes tend to cause some nitriding, which is the absorption of nitrogen from the blowing air by the hot metal (Table 14.6). Nitrides in a finished steel increase its susceptibility to strain aging or embrittlement under stress, a feature which makes these products unsuitable for any cold metal shaping or finishing operations, such as wire drawing. This disadvantage and the inability to handle large quantities of scrap metal or cold iron charges have decreased their relative importance to 10% or less of the total [16]. There has been limited adoption of variants of methods that use oxygen-enriched air, oxygen and steam, or oxygen and carbon dioxide to decrease the nitriding tendency. Pure oxygen introduces a very severe thermal

TABLE 14.6 Nitrogen Content of Blast Furnace Metal (Pig Iron) and various steels[a]

Iron product	Range of nitrogen content (%)
Pig iron	0.002–0.006
Bessemer steels	0.010–0.020
Duplex steels[b]	0.005–0.008
Open hearth steels	0.004–0.007

[a]Compiled from McGannon [9] and Allen [14].
[b]See text for details.

stress on a converter. However, use of a pneumatic process as a preliminary to an open hearth furnace (duplex steelmaking) or an electric furnace (triplex steelmaking) decreases the nitrogen content of the product and improves steelmaking flexibility.

14.3.3. The Open Hearth Process

Improved furnace combustion efficiency was the primary objective of C.W. Siemens, Germany, who conducted most of his combustion trials in Britain. He applied regenerative methods to recover waste heat from a combustion process by passing the combustion gases through a chamber containing an open, firebrick checkerwork, much like the stoves of a blast furnace. When the checkerwork reached nearly the combustion gas temperature, the hot gas flow was diverted to another chamber of firebrick checkerwork to heat this. At the same time the air flow to the combustion process was preheated by passing it through the chamber of hot checkerwork. This air preheat enabled much higher ultimate temperatures to be reached, and on swing cycles improved the efficiency of the combustion process. In this way continuous heat recovery from the spent combustion gases was obtained. By 1857, these developments alone gave fuel savings of some 70–80% for industries such as glass making [16].

These combustion developments were first tested for steelmaking in 1863 in France, under license from Siemens, by a father and son team, E. Martin and P. Martin. Siemens himself tested the furnace for this purpose in Wales, 3 years later, and both operations were pronounced a success. These are now often referred to as the Siemens–Martin process after their joint development.

The open hearth furnace uses the Siemens regenerative principle to raise maximum combustion temperatures to about 1,650°C and possesses a heating capability independent of the heat content or impurity burning reactions of the iron charge. These features enable the charge placed in the furnace to be any ratio of scrap to hot metal up to 100% scrap, since the furnace can melt this. Usual practice, however, is to use a roughly 50:50 mix of scrap to melt [17].

The steelmaking components of an open hearth furnace consist of a rectangular heating zone with a floor containing a long shallow depression of 200–500 tonnes capacity into which the charge is placed (Fig. 14.3). Gas or

FIGURE 14.3 Operating details of an open hearth (Siemens–Martin) furnace for steel production.

oil burners at each end are provided with preheated combustion air via hot checkerwork, and the hot gases are deflected down onto the charge by a convex-down, reverberatory roof. Thus a combination of radiant heating and hot gas contact with the charge transfers sufficient heat to melt it. To protect the structural components of the depression holding the charge from the high temperature, and to enable removal of phosphorus and sulfur impurities from the charge, the depression is lined with a magnesia (MgO) brick refractory.

Steelmaking begins by placing the solid components of the charge, the limestone, any ore requirement, and rusty scrap steel into the depression, cold. The scrap is preferably rusty to provide oxygen which will assist in impurity removal. Some iron ore (e.g., Fe_2O_3) may also be added for the same reason. The fuel, usually natural gas, coal gas, or oil, is then ignited at the burners and the initial charge is heated until it just starts to melt. During this step some oxidation also inevitably occurs as a result of the oxidizing atmosphere of the furnace. At these temperatures carbon dioxide chemically oxidizes iron, and causes formation of some additional iron oxides. Once the initial solid components are melted, the blast furnace melt at 1,300–1,350°C is added, causing initiation of the main carbon removing reaction (Eq. 14.23).

$$2\ C_{(Fe)} + O_{2(Fe)} \rightarrow CO_{(gas)} \qquad\qquad 14.23$$

Oxygen, from the scrap, or the ore, or dissolved in the molten iron from air, reacts with the carbon present in the blast furnace melt to form carbon monoxide. As carbon monoxide bubbles up through the charge it stirs the fluid mass and aids in the rise of silica, manganese oxide, phosphorus oxides, etc., to form a slag layer. While these reactions are going on, limestone decarbonization also occurs and the carbon dioxide formed further contributes to the carbon monoxide formed by its conversion with dissolved carbon (Eqs. 14.1, 14.24).

$$CO_{2(Fe)} + C_{(Fe)} \rightarrow 2\ CO_{(gas)} \qquad\qquad 14.24$$

Thus, the slag formed on the open hearth charge is quite frothy. It is pushed off the pool of molten steel beneath it with a type of hoe, once or twice during the process. Blast furnace metal melts at about 1,130°C, whereas pure iron, more closely representative of steels, melts at 1,535°C. For this reason, as carbon removal from the blast furnace melt proceeds, the temperature of the charge has to be raised to avoid solidification of the charge. Since the bulk of heat transfer in the open hearth process occurs via direct contact of hot combustion gases with the surface of the charge, it is important that the frothy slag layer not be allowed to get too thick or heating may become ineffective.

When a satisfactory composition of the charge is obtained, checked by analysis, the product is tapped from a hole in the base of the hearth depression and cast into large ingots ready for final shaping. Open hearth processing is quite slow, 6–12 hr/batch; about half of this for charging. However, this does provide time for sampling, analysis, and adjustment of the carbon content by additions of further scrap or hot blast furnace metal. Some alloys of up to 10% of another metal are also possible in an open hearth process. Thus greater product flexibility is possible than with the pneumatic steelmaking processes. Material balances of the various steelmaking processes are given in Table 14.7.

Open hearth processes employing silica refractory linings are also known [9]. However, since this type of lining cannot remove phosphorus or sulfur these are usually reserved for open hearth operation using only scrap iron charges, or fed by high-grade iron ores free of these elements. The independent heating capability of an open hearth furnace allows steels to be made locally in these ways.

14.3.4. Electric Furnace Steel

Early development of practical furnaces heated by an electric arc at around 1876 again owed much of their development to the contributions of

TABLE 14.7 Raw Materials Required to Make One Tonne of Ingot by Various Steelmaking Processes[a]

Raw material	Basic Bessemer process	Hot metal, fixed open hearth furnace	Electric arc process		Basic (LD) oxygen furnace
			Common steels	Special alloy steels	
Iron, kg:					
Molten furnace	1,050–1,150	550	nil	nil	850–1,025
Cold pig	nil	slight	some[c]	0–75	nil
Scrap iron (or steel), kg:	0–80	525	1,075	1,000–1,100	100–300
Oxidizing ore or millscale,[d] kg:	low to 60	100	12.5	2–18	7–75
Fluxes, as limestone etc., kg:	110–180	200	55	25–75	60–80
Finishing alloys:	ca.10	14	10	various	6–7
Oxygen, m³ (15°C 1 atm)	nil	varies	150 ft³	varies	1,700–2,000
Primary fuel, kJ (Btu):	nil	4.2–5.3 × 10⁶	nil	nil	nil
As natural or coal gas		(4–5 × 10⁶)			
Electricity, kWh	nil	nil	500–550	650–750	nil

[a]Adapted from data in Allen [14].
[b]Amounts actually consumed in the process are given in kilograms, unless otherwise stated.
[c]Some used occasionally.
[d]Millscale is the iron oxide (mostly Fe_3O_4) "crust" removed from steel billets during rolling operations.

C.W. Siemens. Again it was others who first utilized these principles in steel processing. Stassano succeeded in melting steel scrap in 1899 in Italy, using this furnace design. He used an arc struck between closely spaced carbon electrodes, but this method was unable to reduce iron ores. The designs of Paul Heroult working in France at about the same time were more successful, and became the forerunners of electric furnace steelmaking processes used to the present day [2].

Heroult used experience from his electrolytic method for reduction of alumina and abandoned the indirect arc electric furnace designs, in which the arc was struck between opposed carbon electrodes. Instead he used direct arc designs which still employed the electrodes to carry power to the heating zone of the furnace. But the heating effect was obtained by two arcs, one struck from each of the two electrodes and the surface of the metal. The metal itself served as the power conductor between the electrodes, thus producing two arcs from the same current flow (Fig. 14.4). The double or multiple arcs in direct contact with the metal provided by this design gave much improved heat transfer efficiency and electrical efficiency per mass of metal melted, compared to the earlier indirect arc designs. This heating principle forms the design basis of all modern steelmaking arc furnaces [18].

Even though electric power is more expensive than natural gas, oil, or coal gas as a source of energy, its higher efficiency in the arc furnace makes up for this. It is estimated that an open hearth furnace, even with regenerators, is only able to utilize 15–20% of the heat energy in the fuel supplied to it, whereas the parallel value for the electric furnace is of the order of 55–60% [2].

Electric furnace steelmaking, like the open hearth process, has a means of charge heating independent of a hot metal charge or the combustion of impurities. Since 1900, it has been primarily used for the remelting of clean scrap, particularly of valuable alloys. But since about 1950, electric furnace steelmaking has been much more widely adopted because it permits control of both the heat input *and* the atmosphere above the charge. It does not need an oxidizing atmosphere as required for the pneumatic and open hearth processes. An appropriate atmosphere can be provided for the type of steel being made, which is particularly useful for some of the alloy steels. Special versions

FIGURE 14.4 Schematic design of a direct arc electric furnace for steelmaking. The whole unit may be tilted to the side for emptying the finished steel melt.

of arc furnaces permit melting under a partial vacuum for products destined for demanding applications when small amounts of dissolved gases in the steel could adversely affect properties.

For an electric furnace with basic refractory, a small amount of limestone is charged initially to provide the basic slag-forming components. This is followed by scrap steel for simple refabrication, or scrap steel plus alloying metal(s) for special alloys. If carbon steels are being made, a small amount of ore plus scrap to provide oxygen and solid or molten pig iron will be added. Electric furnaces producing specialized alloys may have only 1–5 tonnes capacity, although some as large as 150 tonnes are in common use. On completion of charging the arcs are struck and heating begins. A reducing atmosphere of carbon monoxide and carbon dioxide is soon produced in the vapor space above the charge. A period of 2–3 hr is required for the charge to melt completely, after which samples can be taken to check for composition.

If the carbon content is still too high for carbon steel or mild steel preparation, further oxygen can be added using a water-cooled lance to discharge the gas about 1–1.5 m above the surface of the melt while the arc is off. The initial slag containing silica, MnO, FeO, P_2O_5, and calcium sulfide in a lime base, called an "oxidizing slag," is skimmed off. Then just before pouring the steel it is "killed" (deoxidized) by addition of one or more star-shaped pieces of aluminum to the melt, depending on the oxygen content determined by analysis. The easily oxidized aluminum combines with the oxygen of any residual iron oxides forming further slag and helping to decrease the incidence of imperfections in castings made from the melt (Eq. 14.25).

$$2\ Al + FeO \rightarrow Al_2O_3 + 3\ Fe \qquad\qquad 14.25$$

For the preparation of alloy steels such as chromium, vanadium, tungsten, or manganese, most of the alloying metal is in place at the start of the melt. To prevent oxidation of the alloying metal a reducing slag of burned lime (CaO) and powdered carbon is used. This provides both the protection of the reducing atmosphere of carbon monoxide and carbon dioxide in the vapor space above the metal, as well as the protective layer on the metal. On completion of melting a sample is taken and analyzed to determine whether any final adjustment of alloying metal concentration is necessary before pouring.

The stainless steels require an electric furnace for their preparation. The corrosion resistance of this class of alloy, conferred by the presence of about 18% chromium plus 8% nickel in the iron, is a valuable feature of its use in industrial vessels, the food industry, and for interior, and exterior fittings.

The stainless steels are only one example of the present day versatility of the electric furnace in steel making. The electric furnace has a high flexibility to accept solid, molten, or a mix of these types of charges, to operate with acidic or basic linings depending on need, and to provide an oxidizing, reducing, or neutral (inert) atmosphere. The combined appeal of these attributes plus its ability to operate under clean conditions, and under a partial vacuum for steels destined for critical service, have led to the rapid adoption of electric furnace methods as a component of steelmaking practice worldwide [19]. They have also helped to revitalize at least one steelworks that had been floundering while using older technology [20].

Electric furnaces which employ induction heating have also been used sporadically for steelmaking. In this procedure high-frequency electric power of 1,000–2,000 Hz is passed through a water-cooled copper coil (conductors are actually heavy walled copper pipe) surrounding a nonmagnetic, nonconducting crucible made entirely of refractory. In this way, the charge placed in the crucible serves as both the "core" and the single turn secondary "winding" of a step-down transformer. The power flow and eddy currents generated in the charge bring about rapid melting and permit composition adjustments to be made. Electrical consumption is about the same as for an electric arc furnace, 660–880 kWh/tonne of steel, but the capital cost of an induction furnace averages two to three times that of an arc furnace, mainly for the motor-alternator and condensers required to provide the high-frequency AC current. However, for steel requiring clean conditions and for specialty steels, induction furnaces fill an important need [14].

14.3.5. Oxygen Steelmaking: The LD or BOS Process

The use of essentially pure oxygen for the oxidation of iron impurities became a commercial possibility as oxygen became available on a tonnage scale in the 1930s. Early development of this process, originally called the "Linzer Dusenverfahren," took place in Linz and Donawitz in Austria during the 1930s and 1940s [16]. Either the original process name or the places of development led to the identification of this development as the LD process. The first two oxygen converter plants using this principle became operational in Austria in 1952 and 1953 [21]. By 1970, 40% of the world's steel was produced by this process, and by 2000, this had grown to more than 60%. The LD process, also referred to as basic oxygen steelmaking (BOS, also refers to the product, basic oxygen steel), is carried out in a basic oxygen furnace (BOF). The vessel used for these processes is similar to the Bessemer converter, but lined with a basic refractory. From a water-cooled nozzle located in the neck of the vessel, a stream of oxygen is directed onto the surface of the mostly molten blast furnace metal charge. Impurities in the hot metal combine rapidly with the high-oxygen content atmosphere. The resulting oxides then combine with lime, or limestone which is also added to the charge, to form a lighter slag layer to permit removal of the impurities.

Oxygen steelmaking is rapid, like the older pneumatic processes. Tap to tap times are normally in the range of 18–20 min and the product is not subject to nitriding since nitrogen contact is avoided. Also, because the heat production from impurity oxidation is roughly equivalent to Bessemer expectations and there is no heat loss to nitrogen of air, it is possible to incorporate up to 30% by weight of scrap steel or high grade iron ores into the 70%, or so of molten blast furnace iron which is the usual charge in the LD process [17]. This ability to accept a proportion of solid charge with the blast furnace melt contributes flexibility to the process. To do all this requires some $65\,m^3$ of oxygen per tonne of steel (about $2100\,ft^3$/tonne, at $16°C$, 1 atm), a net raw material cost to the process. But in return for this, the time from charging to tapping the steel product for the larger LD furnaces of 200 and 300 tonnes capacity is still only about twice as long as for the smaller LD units, about 40 min [22]. In comparison to the 10- to 12-hr turnaround time for a single charge to a 300-tonne capacity open hearth furnace, the cycle time for an oxygen steelmaking unit relates very favorably. Improvements aided by modeling experiments [23], the introduction of computer controls [24], and other measures further raised the attractiveness of oxygen steelmaking. As a result these methods are now used to produce about one-half of the steel of the U.K. and the U.S., about 70% in Germany, and about 80% of the total in Japan.

14.5. PROPERTIES AND USES OF IRON AND STEELS

The wrought iron of historical importance was produced by reducing the iron ore without melting the iron component and then beating out most of the ferrous silicate slag. This product was virtually pure iron containing traces of carbon (0.02–0.04%) and strings of residual slag (1.0–2.0% $FeSiO_3$). The low carbon content was a result of reduction without fusion and contributed to the good malleability, ductility, and the associated good cold working properties that are synonymous with wrought iron. Nineteenth century iron works could form sheets and bars by cold rolling of blooms (balls of reduced iron from which much of the slag has been removed by mechanical hammering) of wrought iron. Cold rolling hardens and generally strengthens the iron, a process called work-hardening, which is a desirable procedure for some end uses (Tables 14.8 and 14.9). If, however, the iron shapes are required to be ductile for subsequent shaping, etc., the cold rolled shapes must first be annealed, i.e., heated to a high temperature and then allowed to cool slowly, to resoften the metal to continue cold shaping.

Wrought iron is still produced on a small scale by pouring molten Bessemer iron into slag that is molten but below the melting point of the iron, and by other

TABLE 14.8   Percentage Composition of Various Forms of Iron and Carbon Steels[a]

Element	A Ingot iron	B Wrought iron	C Gray cast iron	D Carbon steels, SAE 1020	E Ultra-high-strength steel, 300-M
Iron	99.9+	97.5[b]	94.3	99.1	94.0
Carbon			3.4	0.2	0.43
Silicon			1.8	0.25	1.6
Manganese			0.5	0.45	0.80
Nickel					1.85
Chromium					0.85
Molybdenum					0.38
Vanadium					0.08

[a]Compiled and recalculated from data of Perry [25] and Lange [26].
[b]Residue consists of slag.

TABLE 14.9 Selected Properties of Various Forms of Iron and Carbon Steels[a]

Material type[b] and condition	Yield[c] strength (MPa)	Tensile[c] strength (MPa)	Elongation[d] in 5 cm (%)	Hardness[e] Brinell	Density (g/cm³)	Melting point (°C)
A Annealed	130	260	45	67	7.86	1,530
Hot rolled	200	310	26	90		
B Hot rolled	210	330	30	100	7.70	1,510
C As cast	–	170	0.5	180	7.20	1,230
D Annealed	260	450	30	130	7.86	1,515
Hot rolled	290	470	32	135		
Hardened[f]	430	620	25	179		
E Hardened[g]	1,650	2,000	10	535	7.84	1,504

[a]Compiled and calculated from data of Perry [25] and Lange [26].
[b]Labeled compositions are those given in Table 14.8.
[c]Megapascals convertible to pounds/square inch by multiplying by 145.
   Yield strengths correspond to the stresses causing permanent deformation of the metal (i.e., beyond the elastic limit). Tensile strengths correspond to stresses at the point of ultimate metal failure.
[d]A measure of ductility.
[e]The Brinell hardness number is based on the indentation of the surface under test by a ball of specific diameter, usually 10 mm, when pressed into the test surface. It is usually expressed in kilograms/square millimeter, thus small numbers represent low hardness.
[f]Hardened by water quench followed by 540°C temperature.
[g]Hardened by oil quench followed by 320°C temperature.

similar methods. The spongy mass which forms is then mechanically treated to form a wrought iron very similar in properties to the earlier hand-worked material. Residual slag stringers left in the wrought iron contribute to the toughness of the iron in this form, and also make a small contribution to corrosion resistance. The ductility, weldability, and toughness of wrought iron continue to make it valuable for many of the same uses as the original hand-produced material, for example, pipe, boiler tubes, rivets, heavy chains and hooks, and ornamental iron work.

14.4.1. Iron–Carbon Alloys: Cast Iron and Steels

One of the most important characteristics of iron products, which enable the spectrum of properties to be achieved is the concentration of carbon present. Ordinary steels are iron–carbon alloys, which are simply referred to as steel, and are so important that this alloy comprises more than 98% of all iron alloys produced. In most iron–carbon alloys, the carbon is present as iron carbide, Fe_3C, also called cementite. Since the carbon content of cementite is only 6.69%, a small change in the carbon content of an iron causes a large change in the concentration of the cementite present in the iron. Cementite is soluble in molten iron, one of the reasons why carbon is accumulated in the product of the blast furnace process for reduction of iron ores. This is an advantage, since the melting point of the iron–cementite mixture is depressed to about 1,100°C, permitting accumulation of the product in the base of the furnace at temperatures considerably below the 1,535°C, m.p. of pure iron. At the same time it is a disadvantage since, in the process, rather more cementite is dissolved in the molten iron than is wanted for most uses.

If the percentage of carbon in molten iron is 1.7% or less, which amounts to 25.4% or less cementite, it remains dissolved in the iron when solidified by quick cooling (e.g., when quenched in water) to form only one type of crystal. The crystals of this solid solution have cementite evenly distributed in the crystals of γ-iron and yields an extremely hard, brittle product called martensite. If the iron containing from about 0.82–1.7% carbon is allowed to cool more slowly, cementite can separate from the iron and the result is an intercrystalline structure of cementite and pearlite. The formation of pearlite, which consists of alternate layers of ferrite (pure iron) and cementite, contributes significantly softer, more malleable, and tougher properties to the iron than that obtained from a quenched product. As the amount of carbon in the iron is brought down below about 0.82%, slow cooling allows ferrite plus pearlite (interlayered ferrite and cementite) phases to separate, again giving a relatively soft, tough product. More detail concerning these phase changes and their effect on properties of carbon steels is available from more extensive accounts [9].

Iron–carbon alloys are commonly referred to as carbon steels or simply steel, and are classified in accordance with their carbon content. Thus, mild steels normally contain about 0.25% carbon, medium steels about 0.45% carbon, and high carbon steels about 0.7–1.7% carbon. Both medium and high carbon steels are commonly tempered (heat treated) to optimize their properties although any of the carbon steels are amenable to property modification by heat treating [27]. In fact, that is the general definition of a steel: an iron–carbon alloy that contains less than 1.7% carbon and is amenable to hardening by quenching (Tables 14.8 and 14.9). As a rule, the strength and hardenability of a steel is raised at the same time as the toughness and ductility are decreased, with an increase in the carbon content of a steel. A user classification system has been proposed [28].

To obtain optimum properties from heat treating, the finished steel article is first hardened by heating to red heat followed by quenching by dipping in

TABLE 14.10 Tempering (Draw Down) Temperatures for Some Common Uses of Carbon Steels[a]

End use	Reheat temperature (°C)	Color of oxide layer
Razor blades, twist drills	232	Pale straw yellow
Axes, taps, and dies	255	Brown-yellow
Cutlery	277	Purple
Watch springs	288	Bright blue
Chisels	300	Dark blue

[a]Compiled from Dearden [2], Oberg et al. [29], and other sources.

water, brine, or oil – any of which may also be cold or warm depending on the degree of hardness desired (Table 14.10).

The martensite structure obtained in the iron by this treatment is very hard and a tool at this stage will hold a cutting edge well, but it will also be rather brittle and in this condition can fracture quite easily. It will also be under severe internal stress because of the rapid cooling rate, which for a small article will tend to increase its susceptibility to fracture under shock. A large part may actually crack as a result of the quenching alone. For this reason, large parts will be oil or air quenched to reduce this risk. To gain the desired degree of toughness, the quenched article is tempered (or "drawn down") by reheating to a particular controlled temperature according to the end use desired in an air or oil bath, and then allowed to cool slowly. This combined treatment gives a cutting product or spring the desired combination of hardness (sharp edge retention) and toughness (resistance to shattering under sudden physical shock).

If a carbon steel is required in a ductile state either for intermediate cold forming (e.g., auto body panels) or machining, or in its final use form, then it is annealed. To accomplish this, the finished part is heated to above the iron-phase transition temperature, about 650°C, and then allowed to cool very slowly. Annealing not only accomplishes a softening and ductility-improving effect on the steel (Table 14.9), but also achieves stress relief. Large cold-formed or welded industrial vessels or ships' components will have stresses induced in the structure from the fabrication methods used, which may be removed by annealing. In this way the service life and reliability of the product may be greatly improved.

When the carbon content of the iron rises to above 1.7% (>25.4% cementite), it is classed as a cast iron. At this concentration the proportion of cementite becomes so high that it is no longer completely soluble even in solid γ-iron and forms a separate cementite phase on hardening from the melt. Cast iron may be made by mixing molten pig iron with molten steel scrap, or by melting steel scrap in direct contact with coke in a cupola furnace (virtually a small version of a blast furnace). It will ordinarily contain 2.0–4.0% carbon, only a little less than pig iron, together with 0.5–3.0% silicon, 0.5–1.0% manganese, and traces of phosphorus and sulfur. The hard brittle nature of many cast irons, hot or cold, makes these impossible to shape by rolling or forging (hammering while hot) the solidified metal, so that casting, followed by machining if required, are the usual methods of fabrication. Fortunately the presence of the cementite and other impurities depresses the melting point of the iron to the 1,100–1,200°C range making both the melting process and casting easier than it would be for pure iron.

It is possible to produce a moderately tough cast iron by special alloying techniques, and by the control of the microstructure through the selection of the appropriate cooling cycle. And some versions of cast iron may be converted to malleable iron by annealing for a period of several days. The long anneal causes dissociation of the cementite (Fe_3C), the cause of the hardness, into ferrite (iron) and spherical zones of graphite making the product relatively soft and significantly tougher than the material as originally cast. With these special irons, primary fabrication by casting can be followed by other shaping techniques.

14.4.2. Alloy Steels

Mild steels and high carbon steels may be fabricated and heat treated in ways to achieve many desirable combinations of properties. However, plain carbon steels suffer several serious drawbacks in the areas of strength, toughness, hardness, and in corrosion resistance. For example, heat treating of very large machine parts made of carbon steels can only achieve shallow surface hardening. Also, heat treating of large parts introduces severe internal stresses, which can cause distortion from the originally machined shape. This distortion may be acceptable or compensated for in small parts, but with large parts, it can cause fitting problems and a tendency to crack. With a situation such as this, a drastic quench cannot be used and therefore, optimum hardness of the carbon steel is not achievable. Alloying with elements other than carbon can enable achievement of hardenability, a term reserved to mean hardness at depth, with a less severe (such as an air) quench. Addition of molybdenum, for example, not only achieves hardenability but at the same time contributes other desirable properties (Table 14.11).

Examples of the properties achievable with steels for several alloying elements are given in Table 14.11. Many hundreds of alloy steels are available using the elements listed and others to tailor-make a steel to suit a particular application. For example, many nickel–chromium combinations confer stainless properties (resistance to corrosion) with more than 20 different commercial variations available for specific objectives. The stainless steels probably represent the largest single class of alloy steels. Aluminum may be added in trace amounts to tie up oxygen and avoid the deleterious properties induced in a steel by the presence of small amounts of ferrous oxide. While ferrous sulfide introduces brittleness in the steelmaking hot working impossible, manganese sulfide does not interfere in this way. So addition of a small percentage of manganese can tie up any residual traces of sulfur to permit forging operations without introducing defects. Boron is a relatively recently discovered alloying element, which contributes low-cost hardenability with a few hundredths of a percent. Coating developments with hard carbides such as titanium promise to increase the life of cutting tools made with high-speed tool steels [30].

14.5. EMISSION CONTROL IN IRON AND STEEL PRODUCTION

For an industry which involves large-scale handling of finely divided solid materials, emission control of large-scale combustion processes for coking, etc., has to be a priority. Containment is necessary for maintenance of public health in associated built-up areas, for industrial hygiene requirements of plant operators, for better product yields from the raw materials entering the processes used, and for aesthetic reasons.

Many technological changes introduced to the iron and steel industry to improve operating efficiency also reduced levels of emission per tonne of

■ **TABLE 14.11 Compositions and Applications of Some Common Examples of Alloy Steels**[a]

Alloy steels	Percentages of alloying elements	Properties achieved	Typical uses
Chromium	Cr 0.6–1.2, C 0.2–0.5	Improved strength, toughness, heat resistance	Improved castings, stainless steels
	Cr 16–25, Ni 3.5–20	Combination confers stainlessness, heat resistance, strength	Stainless steels: chemical, medical, and food processing equipment
	Cr > 12%, Ni > 8%	As above, plus hardenability	Stainless cutlery
Copper	Cu 0.10, C ca. 0.2	Improved corrosion resistance	Copper-bearing steels, bridges, exterior steelwork
Nickel	Ni 6–9, C 0.2–0.5	Low-temperature toughness	Snowmobile parts, ice-breakers, armor plate (as 6.5% Ni)
	Ni 30–45%	Low coefficient of thermal expansion, i.e., 0.8×10^{-6} as compared to 10 to 20×10^{-6} for most other steels	Invar: graded seals to glass, length standards (surveying, etc.), compensated pendulums and balance wheels
	Ni 60, Cr 16	Heating and electrical resistance units	Nichrome heating elements
Manganese	Mn 10–15, C ca. 1	Tough, acquires hardness on repeated impact	Rock crusher jaws, power shovel bucket teeth, burglar-proof safes
Molybdenum	Mo 5–9	Hardenability, high-temperature strength	High-speed tools, steam boilers and pipes, jet engine turbines
Silicon	Si 3–5	Low hysteresis and eddy current losses, increased magnetic permeability	Electric power transformer plates, motor parts
	Si 13–15, C 0.8	Superior hardness, acid resistance	For example, Duriron (14.5% Si), industrial valves, piping
Tungsten	W 5, C 0.5	Assists in retaining hardness while hot	Tungsten steels, twist drill bits
	W 17, Cr 10, V 0.3, C 0.7	Same as above, remains hard and strong even when hot	High-speed steel, high-speed cutting, and machine tools
Vanadium	V 0.2–0.5, + Cr ca.1	High strength, toughness, heat treatability	Gears, driveshafts, axles, ball bearings, mechanics' tools

[a]Compiled from data of McGannon [9], Perry [25], and Lange [26].

product. For instance, blast furnace top gas containment introduced to conserve the carbon monoxide for preheating of blast air also decreased particulate and gaseous pollutant discharges from blast furnace operations. Until recently blast furnace operation allowed some intentional gas bleeding (for pressure relief), accidental gas losses containing up to $35 \, \text{g/m}^3$ ($15 \, \text{grains/ft}^3$) particulate, and gaseous pollutants [31]. These operations are now regulated by passage through at least two levels of control devices before discharge. In these ways the industry has progressed from no containment to 90–98+% containment to its own and society's mutual benefit. Similar progress has occurred with the BOP (basic oxygen process) for steelmaking. Since the air and water emission control have many overlapping aspects, these aspects will be treated by the emission control area, rather than the process.

14.5.1. Air Pollution Control

Discharges to the air from facilities for producing iron and steel may be conveniently considered in sequence, from the raw material preparation area to the fabrication of finished products. Thus, coke plants and sintering plants used to prepare the blast furnace fuel and ore components are considered first.

The coke ovens used to prepare a dense, desulfurized metallurgical coke from coking grade coals for blast furnace feed can produce substantial particulate and polycyclic aromatic hydrocarbon (PAHs) discharges in the absence of control measures. A battery of coke ovens used to process 4,500 tonnes of coking coal per day to yield about 3,200 tonnes of coke was estimated to generate a particulate discharge of about 0.1% of the mass of coal processed, about 4,500 kg/day [32]. Emission factors for PAHs were strongly correlated with auxiliary fuel used. Average total-PAH emission factors for the auxiliary fuels coal, heavy oil, and electric arc furnace were $4,050 \, \mu g/kg$ coal, $5,750 \, \mu g/L$ oil, $2,620 \, \mu g/kWh$, respectively [33]. Carcinogenic benzo[a]pyrene emission factors for gas plus particulate phases were 2.0 g/kg coal [as cited], $2.4 \, \mu g/L$ oil and $1.4 \, \mu g/kWh$ for the same fuel categories, respectively. These emissions occur at all stages of coke preparation, including charging, pushing, and quenching stages of the operation as well as from leaks in the ovens themselves, thus making control measures difficult. No estimates of current feasible particulate or PAH emission control efficiencies are available.

Coking produces a blast furnace coke feed substantially free of sulfur. However, the gaseous product, coke oven gas, has a sulfur gas content of $900–1,100 \, g/m^3$ (at 15°C, 1 atm) [31]. This is mainly hydrogen sulfide, which may be removed either by the vacuum carbonate or Stretford processes. The sulfur gas removal efficiency of the Koppers Company's vacuum carbonate process is about 90%, which produces sulfuric acid, whereas the Stretford process can achieve 99% containment to a sulfur product (Chaps. 3 and 9). The choice of desulfurization process depends on the efficiency required and the sulfur product desired. Condensible hydrocarbons such as benzene (and other aromatics) and phenols have always been recovered by condensation, etc. [34].

Sintering machines prepare large nodules (lumps) from beneficiated iron ore fines, and iron oxide containing dusts recycled from particulate emission control equipment. It is estimated that uncontrolled operation would discharge particulate at the rate of about 0.3% of the mass of sinter produced, or about 2,700 kg from a machine producing 900 tonnes of sinter per day. Cyclones can decrease the particulate emission to about one-quarter of these levels. It is also possible to use the sintering machine as a roaster to enable sulfur removal from sulfur-containing iron ores. This produces a more amenable ore, but it also produces sulfur dioxide in the waste gas stream. No emission-rate data for sulfur dioxide in sinter plant exhaust gas is available, since this has not normally been recovered. However, a mathematical model which enables estimation has been described [13].

Particle loading of the top gases leaving a blast furnace is very dependent on the particle size distribution of the raw materials charged into the furnace. Without elimination of fines from the ore or flux, dust loadings of about $2 \, g/m^3$, and loss rates of about 200 kg/tonne of pig iron are obtained. Prescreening roughly halves this loss rate and improves blast furnace operation by providing a freer flow of gases (cf. Table 14.4). Prior briquetting, pelletizing, or nodulizing of the raw materials before charging can reduce this to 15–20 kg/tonne of pig iron produced [35].

All the measures described above are precombustion or prereduction methodologies. In addition to these dust *avoidance* techniques, modern blast furnaces also usually require three stages of gas cleaning for the top gases before use as a fuel. A dust catcher first captures the coarse particulate (see Fig. 14.1) and then either low and high energy scrubbers, or a low energy scrubber followed by electrostatic precipitation removes the remaining mid-size and aerosol particles to give particulate loss rates of 1–18 kg/tonne of pig iron [32]. Magnetic filtration has been proposed as a safer method than electrostatic precipitation to clean up combustible blast furnace gases to avoid explosion risks [36].

All steelmaking procedures produce exhaust gases having a moderate particle loading before treatment, but none as severe as from blast furnace operation. Dust loadings of $0.5–1.4 \, g/m^3$ are experienced during the charging phase (loading ore, flux, scrap, etc.) of an open hearth furnace [34]. Higher dust losses of $4.6–7.0 \, g/m^3$ are experienced when oxygen lancing (e.g., basic oxygen process) in an open hearth operation. Originally, open hearth furnaces were operated without emission controls. However, a wet scrubber, or electrostatic precipitation, which gives 90–97% particulate control, is now used [17, 34]. Wet scrubber waste water is cleaned by chemical coagulation and settling followed by pH neutralization [37]. Costs can be reduced by blending of low and high pH scrubber waste liquors.

For Bessemer, electric arc, and basic oxygen furnaces pollutant containment is practiced via a hood with connecting ductwork placed over the opening of the vessel during active steelmaking phases [38, 39]. The collected gases are cleaned by electrostatic precipitation, scrubbers, or with a baghouse [31]. Particulate collection efficiencies of 80–90% are obtained by the first two methods, and 98–99% for baghouses.

14.5.2. Water Pollution Control

Preparation of blast furnace coke involves the heating of metallurgical coal to 1,000–1,100°C in the absence of air in a battery of refractory brick-lined coke ovens. This is referred to as the "by-product coke plant" from the association of by-product recovery with coke formation. The coal charge is heated until all of the volatile matter has been vaporized and pyrolysis is complete, a process which takes 16–24 hr. The residual lumps of coke, still hot, are then pushed out of the oven through a quenching shower of water and into a rail car for final shipment. About 700 kg of coke plus a number of volatile products are recovered from each tonne of metallurgical coal heated. More details on the coking process itself are available [40].

TABLE 14.12 Typical Characteristics of the Waste Water Stream from the Ammonia Still of By-Product Coke Ovens[a]

Parameter	Significance (ppm by weight)
5 day BOD, 20°C	3,974
Suspended solids:	
Volatile	153
Total	356
Nitrogen:	
As ammonia, NH_3	187
Organic and NH_3	281
Phenol	2,057
Cyanide	110
pH	8.9

[a]Data selected from the American Iron and Steel Institute [41].

About $42 \, m^3$/tonne coke (10,000 U.S. gal/tonne) of water is used in a coking operation, mostly for indirect cooling of the volatiles stream to collect coal tar, water, and other condensibles. The cooling water is only heated, which has to be considered on discharge. However, the aqueous condensate stream also contains ammonia, sulfides, hydrocarbons, phenols, etc., condensed from the coal volatiles (Tables 14.12 and 14.13) [41, 42]. This stream requires treatment for chemical recovery and for emission control before discharge.

About $2 \, m^3$ of water per tonne of coke (ca. 500 U.S. gal/ton) is used for quenching purposes, about a third of which is vaporized in the coke cooling process. The vaporized fraction is discharged as steam. The residual water contains suspended coke fines ("coke breeze") plus low concentrations of cyanides and phenols, and requires treatment before discharge. Coke fines are

TABLE 14.13 Approximate Coke and Volatile Product Yields from One Tonne of Metallurgical Coal[a]

Solids and condensed liquids (kg)		Gaseous products (kg)	
Coke, with coke breeze	750	Ammonia	3.5
Light oil	0.5	Hydrogen sulfide	3
Heavy creosote	4.5	Carbon dioxide	9
Pitches	24	Nitrogen	4
Benzene	7.5	Hydrogen	16
Toluene	3.0	Carbon monoxide	45
Xylene	0.8	Methane	65
Naphthalene oil	4.0	Ethane	5.5
Anthracene oil	7.0	Ethylene	10
Other organics	1.3	Other gases	5
Aqueous liquor	60	Total gases	143.5[b]

[a]Data selected and calculated from that of Kent [42].
[b]About $305 \, m^3$ at 15°C, 1 atm.

removed by settling and then blended into a mixture of ore and limestone for briquetting as blast furnace charge, or are used as fuel for sintering. Recycling the supernatant water minimizes the volume that requires treatment. A European Research Project examined the emission control value of preheating techniques and dry cooling of the coke, which produced less air and water pollution and other advantages plus some offsetting disadvantages [43].

The aqueous condensate stream is the most complex to deal with. Phase separation produces a water phase and an oily tarry phase, each of which has to be treated separately. Phenol is kept in the aqueous phase during ammonia recovery by converting it to the less volatile, and more soluble sodium salt (Eq. 14.26).

$$PhOH + NaOH \rightarrow NaOPh + H_2O \hspace{3em} 14.26$$

Ammonia is then removed from the aqueous phase by distillation and either recovered by scrubbing with sulfuric acid to produce ammonium sulfate, which is sold as a fertilizer constituent, or it may be discharged to air. Economical reacidification of the aqueous ammonia recovery residue with flue gas and subsequent distillation allows recovery of both phenol and sodium carbonate (Eq. 14.27, 14.28).

$$CO_2 + H_2O \rightarrow H_2CO_3 \hspace{3em} 14.27$$

$$2\ NaOPh + H_2CO_3 \rightarrow Na_2CO_3 + PhOH \hspace{3em} 14.28$$

Cyanide may be detoxified to cyanate by oxidation with chlorine or hypochlorite (Chap. 15). Or the oxidation may be conducted by air in the presence of sulfur dioxide and copper ion at pH 9 or 10 [44]. Finally BOD reduction to 80–90% of the original values will be achieved via either biological waste treatment or alkaline chlorination methods before discharge.

The separated organic phase of coke oven gas is commonly used as fuel. Otherwise it will be put through a sequence of washes and stills for chemical recovery, or segregated into portions destined for appropriate end use areas [42]. Details of the recoverable products are given in Table 14.13. About one-third of the coke oven gas is burned to provide the heat for coal carbonization and the rest is usually used on site for other energy needs.

Blast furnace water is mainly required for water jacket cooling of the critical tuyères and bosh areas. The primary concern on discharge of this water is to minimize thermal loadings. Water is also frequently used for scrubbing of blast furnace top gases, and less often for fracturing (breaking up) of hot slag. These uses require the water to be settled and chemically neutralized before discharge.

Steelmaking requires 12–15 m^3/tonne (ca. 3,000–3,700 U.S. gal/tonne) of cooling water and 80–100 L/tonne (ca. 20–25 U.S. gal/tonne) for scrubbing of exhaust gases [37]. Cooling water disposal requires consideration of possible thermal effects. Treatment of scrubber effluent is similar to that used with blast furnace scrubber effluent, except that steelmaking waste streams are usually acidic whereas blast furnace effluent is normally alkaline. After particle removal, pH neutralization can be by stream blending, if the two processes are operated close to each other.

FIGURE 14.5 Schematic diagram of a scale pit showing how simultaneous oil and scale removal is achieved from steel rolling mill waste waters.

Water effluent problems of hot and cold steel rolling mills differ from those of other areas of the complex. The primary water requirement here is for cooling of the heavy iron or steel rolls used for shaping the steel. However, this cooling function is by direct spray onto the rollers. Thus, the water picks up small particles of iron, scale (mostly iron oxide), and lubricating oil in its passage across the face of the rollers, which are retained in suspension as the water reaches mill collection channels. Some additional water is used in high-pressure jets to flush scale from a steel sheet as it forms and is collected in the same wastewater system. A large tank, called a "scale pit," serves the combined functions of scale settling and oil creaming to simultaneously remove most of these contaminants before discharge of the water stream (Fig. 14.5). A system of paddles on a slowly moving endless chain is used to move settled scale out of the scale pit, up an inclined plane to drain off most of the fluid, and thence to a collection for incorporation into blast furnace feed [34]. Creamed oil collected on the water surface of the scale pit is removed by a suction pump through a skimmer floating on the fluid surface. If the cleaned exit water from the scale pit still does not meet local effluent guidelines it can be treated further by chemical or biological methods before discharge.

14.5.3. Recycle of Scrap, Precautions

Scrap steel is extensively recycled without incident, as mentioned in the introductory comments. However, before it is accepted into a steel remelting yard it should be checked for any stray radioactive material. Of the 21 known accidents of steel contamination with cobalt 60, cesium 137, or other radioactive materials, eight occurred in the 1992–1993 period [45]. Checking is simple and inexpensive compared with the potentially serious outcome and possibly costly recovery from accidental radioactivity put into steel.

Less critically, galvanizing prolongs the life of iron or steel products which are exposed to corrosive environments, but adds a difficulty by requiring "dezincing" before recycling [46, 47].

REVIEW QUESTIONS

1. (a) What are the usual benefits of large scale contributed by very large blast furnaces as compared to smaller units?

(b) What additional benefit is also obtained in this case which reduces the raw material requirements per unit of pig iron product, and how does this happen?

2. What would be the theoretical coke (take as 100% C) required for only the reduction reactions of 1,000 tonnes of iron ore if the ore is:
(a) 100% by weight Fe_2O_3?
(b) 50% Fe_2O_3, 30% SiO_2, 20% $Ca_3(PO_4)_2$?
(c) How do your answers compare to the actual requirements?

3. An 8-tonne charge of molten iron in an electric furnace contains 0.20% FeO.
(a) What mass of aluminum should be added to be just sufficient to deoxidize this melt (reduce the ferrous oxide to iron)?
(b) How would the aluminum oxide produce be removed from the iron?

4. Compare the types of impurities which may be removed most efficiently during steel processing when using acidic (SiO_2-based) vessel linings and when using basic (MgO-based) linings.

5. (a) Outline the precombustion emission control strategies, which may be employed in blast furnace operations to decrease particulate emissions.
(b) What postcombustion measures are taken to further decrease particulate emissions?
(c) What precombustion and postcombustion measures may be employed to decrease gaseous air pollutants?

6. (a) How do the presence of phosphorus, sulfur, and oxygen affect the properties of iron or steel?
(b) How do these elements get into the product?
(c) What measures may be taken to substantially prevent their presence from developing during melt processing?

7. What would be the theoretical oxygen demand of a wastewater stream from an integrated steel mill which contains:
(a) 200 ppm (by wt.) ammonia (pH about 4)?
(b) 900 ppm phenol (C_6H_5OH)?
(c) Give an alternative method to effectively treat each of the above waste streams that does not involve long-term aeration in a biopond.

8. Name the principal pneumatic steelmaking processes in order of their chronological development, and briefly outline the advantages and disadvantages (if any) achieved by each process.

FURTHER READING

I.L. Bell, "Principles of the Manufacture of Iron and Steel, with Some Notes on the Economic Conditions of Their Production" G. Routledge, New York, 1884.

A.W. Cramb, "The Making, Shaping, and Treating of Steel," AISE (Association of Iron and Steel Engineers) Steel Foundation, Pittsburgh, PA, 1998.

H.B.H. Cooper and W.J. Green, Energy consumption requirements for air pollution control at an iron foundry, *J. Air Pollut. Control Assoc.* 28, 545 (1978).

G. Krause and M. Grossmann, "Principles of Heat Treatment of Steel." American Society for Metals, Metals Park, OH, 1980.

D. Marchand, Possible improvement to dust collection in electric steel plants and summary of all planned and existing systems in the Federal Republic of Germany. *Ironmaking Steelmaking* 3(4), 221 (Discussion p. 230) (1976).

N.T. Stephens, J.M. Hughes, B.W. Owen *et al.*, Emissions control and ambient air quality at a secondary steel production facility, *J. Air Pollut. Contr. Assoc.* 27(1), 61, Jan. 1977.

REFERENCES

1. J. Newton Friend, "Iron in Antiquity," p. 27. Griffin, London, 1926.
2. J. Dearden, "Iron and Steel Today," 2nd ed., p. 19. Oxford University Press, Oxford, 1956.
3. C. Jarrier, P. Andrieux, C. Domergue *et al.*, . . . iron by direct reduction: Attempts to reproduce antique processing, *Rev. Metallurg.-cahier Inform. Tech.* 94(5), 691, May (1997).
4. W.K.V. Gale, "Iron and Steel." Longmans, Green & Co., London, 1969.
5. J.B. Pearse, "A Concise History of the Iron Manufacture of the American Colonies up to the Revolution and of Pennsylvania until the Present Time." Burt Franklin, New York, 1970, repr. of 1876 ed.
6. "United Nations Statistical Yearbook 1993," 40th ed. United Nations, New York, 1995; and earlier editions.
7. M.D. Fenton, Iron and Steel, in U.S. Geological Survey Minerals Yearbook, 2001. Available: http://minerals.usgs.gov/minerals/pubs/commodity/iron_&_steel/festmyb01.pdf
8. "Kirk-Othmer, Encyclopedia of Chemical Technology," Vol. 8. Interscience, New York, 1952.
9. H.E. McGannon, ed., "The Making, Shaping and Treating of Steel," 9th ed. United States Steel, Pittsburgh, 1971.
10. E. Oguz, Removal of phosphate from aqueous solution with blast furnace slag, *J. Hazardous Mater.* 114(1–3), 131–137, Oct. 18 (2004).
11. C.F. Liu and S.M. Shih, Iron blast furnace slag/hydrated lime sorbents for flue gas desulfurization, *Envir. Sci. & Technol.* 38(16), 4451–4456, Aug. 15 (2004).
12. "McGraw-Hill Encyclopedia of Science and Technology," Vol. 13, p. 98. McGraw-Hill, New York, 1971.
13. "Kirk-Othmer, Encyclopedia of Chemical Technology," 3rd ed. Vol. 13, p. 743. Wiley-Interscience, New York, 1981.
14. J.A. Allen, "Studies in Innovation in the Steel and Chemical Industries." A.M. Kelley, New York, 1968.
15. G.R. Bashworth, "The Manufacture of Iron and Steel," 3rd ed. Chapman & Hall, London, 1964, cited by Allen [10].
16. W.K.V. Gale, "The British Iron and Steel Industry." David and Charles, Newton Abbot, 1967.
17. C.S. Russell and W.J. Vaughan, "Steel Production: Processes and Residuals." Johns Hopkins University Press, Baltimore, 1976.
18. "Kirk-Othmer, Encyclopedia of Chemical Technology," 2nd ed. Vol. 18, p. 715. Interscience, New York, 1969.
19. G.E. Wittur, "Primary Iron and Steel in Canada." Department of Energy, Mines and Resources, Ottawa, 1968.
20. G. Krause, Arisen from the ruins: The comeback of Georgsmarienhutte, *Stahl Und Eisen* 123(12), 131–134, Dec. 17 (2003).
21. D. Ameling, Role and effects of oxygen steelmaking – Remarks [at] the . . . "4th European Oxygen Steelmaking Conference," *Stahl Und Eisen* 123(4), 41–44, Apr. 15 (2003).
22. M. Finniston, *Chem. Ind. (London).* 501, June 19, (1976).
23. M.S. Lee, S.L. O'Rourke, and N.A. Molloy, Oscillatory flow in the steelmaking vessel, *Scand. J. Metallurgy,* 32(5), 281–288, Oct. (2003).
24. I.J. Cox, R.W. Lewis, R.S. Ransing *et al.*, Application of neural computing in basic oxygen steelmaking, *J. Mater. Proc. Technol.* 120(1–3), 310–315, Jan. 15 (2002).
25. R.H. Perry, ed., "Chemical Engineers Handbook," 4th ed. McGraw-Hill, New York, 1969.

26. N.A. Lange, ed., "Lange's Handbook of Chemistry," 10th ed. McGraw-Hill, New York, 1969.

27. Potentials for more high-strength, low-alloy steels in autos, *Met. Prog.* **109**(2), 26, Feb. (1976).

28. J.T. Sponzilli, C.H. Sperry, and J.L. Lytell, Jr., *Met. Prog.* **109**(2), 32, Feb. (1976).

29. E. Oberg, F.D. Jones, and H.L. Horton, "Machinery's Handbook," 20th ed. Industrial Press, New York, 1975.

30. R.F. Bunshah and A.H. Shabaik, *Res./Dev.* **26**(6), 46, June (1975).

31. J. Szekely, ed., "The Steel Industry and the Environment." Dekker, New York, 1973.

32. "Handbook of Environmental Control," Vol. 1. CRC Press, Boca Raton, FL, 1972.

33. H.H. Yang, S.O. Lai, L.T. Hsieh, *et al.*, Profiles of PAH emission from steel and iron industries, *Chemosphere*, **48**(10), 1061–1074, Sept. (2002).

34. H.F. Lund, ed., "Industrial Pollution Control Handbook." McGraw-Hill, New York, 1971.

35. Waste iron-bearing fumes, *Chem. Eng. News*, **55**(16), 17, Apr. 18 (1977).

36. Magnetic filter may cut pollution, *Chem. Ecology*, p.11, July (1980).

37. N.L. Nemerow, "Industrial Water Pollution, Origins, Characteristics, Treatment." Addison-Wesley, Reading, MA, 1978.

38. T. Cesta and L.M. Wrona., Optimization of BOF air emission control systems. *Iron Steel Eng.* **72**, 23–31, July (1995).

39. J.A.T. Jones, Interactions between electric arc furnace operation and environmental control. *Iron Steel Eng.* **72**, 37–46, Dec. (1995).

40. G.T. Austin, "Shreve's Chemical Process Industries," 5th ed., pp. 70–88. McGraw-Hill, New York, 1984.

41. "Annual Statistical Report." American Iron and Steel Institute, New York, 1949, cited by Nemerow [37].

42. J.A. Kent, ed., "Riegel's Handbook of Industrial Chemistry," 7th ed. Van Nostrand-Reinhold, New York, 1974.

43. M.A. Diez, R. Alvarez, M Sirgado *et al.*, Preheating techniques to manufacture metallurgical coke, *ISIJ International*, **31**(5), 449–457 (1991).

44. Process destroys cyanide, *Can. Chem. Process.* **66**(6), 8, Sept. 10 (1982).

45. A. LaMastra, The changing face of radioactivity in steel. *Iron Steel Eng.* **72**, 44–45, July (1995).

46. J. Wijenberg and J. Droog, Dezincing of . . . steel scrap in hot caustic soda, *Steel Res.* **70**(6), 227–232, June (1999).

47. M.N.C. Ijomah and A.I. Ijomah, Chemical recycling of galvanized steel scrap, *Indian J. Chem. Technol.* **10**(2), 159–165, Mar. (2003).

15

PRODUCTION OF PULP AND PAPER

...thou hast caused printing to be used and, contrary to the king his crown, and dignity, thou hast built a paper mill.
—*William Shakespeare, 1591*

15.1. BACKGROUND AND DISTRIBUTION OF THE INDUSTRY

The production of wood pulp and paper is a primary industry in the sense of its utilization of wood as its chief raw material. Pulp and paper is also a secondary industry in the sense that it consumes large quantities of bulk inorganic chemicals such as chlorine, sodium hydroxide, sodium sulfate, etc., produced by the primary (commodity) chemical industry. This business area consumes close to 10% of the inorganic chemicals produced in the countries in which it is dominant. Thus, it not only contributes directly to the economy of these areas, but also indirectly by its purchases of chemicals.

Development of a large-scale pulp and paper industry is favored by the availability of an extensive domestic wood supply (e.g., Brazil, Canada, Finland, Sweden, the U.S.A., and the Russian Federation) (Table 15.1). The rate of growth of China's pulp and paper production has exceeded 100% per decade for the last 20 years. Japan, among the large scale pulp producers, relies heavily on imported pulp logs and pulp for her pulp and paper industry. Norway, with a large forest potential, has not exploited this resource to the same extent as other nations. Brazil, a relatively recent pulp producer, is committed to the "farming" of wood on a continuous harvest basis using fast growing varieties of trees. The extent of exploitation of the available wood supply by all the major pulp producing countries has aroused concern about the inability of natural processes to continue to provide wood at the rate that it is being cut. In most cases, this has resulted in better organized care and harvesting of wood, and increased encouragement of renewal by more effective artificial reforestation.

TABLE 15.1   Major World Producers of Wood Pulp, in Thousands of Metric Tonnes[a,b]

	1960	1970	Chemical 1980	Mechanical 1980	1990	2000
Brazil	453	811	3,256	105	4,844	6,473
Canada	10,397	16,609	4,765	8,625	16,466	20,921
China		1,220	6,450	375	17,057	35,439
Finland	3,700	6,471	4,350	1,569	8,777	13,509
France	1,139	1,787	4,891	261	7,049	10,006
Japan	3,524	8,801	15,414	2,674	28,088	31,828
Norway	1,645	2,182	784	589	1,819	2,300
Sweden	5,611	8,142	4,648	1,534	8,419	10,786
U.S.A.	22,966	39,304	52,601	4,238	71,965	81,529
U.S.S.R.	3,213	6,679	7,379	1,354	10,718	5,310[c]
West Germany	1,452	1,732	6,974	606	11,873	18,182[d]
Other	5,480	11,034	33,293	4,213	51,975	87,763
World	59,580	104,772	144,805	26,143	239,050	324,046

[a]Includes totals of both chemical and mechanical pulp where undifferentiated. Total undifferentiated world pulp for 1950 was 34,620 thousand metric tonnes. Pulp weight specified on an air dry basis (i.e., 10% moisture.) Data compiled from Reference Tables [1] and *U.N. Statistical Yearbooks* [2].
[b]Newsprint.
[c]For the Russian Federation.
[d]Data for 2000 is for unified Germany.

Financially, an active pulp and paper industry provides a significant fraction of the gross national product of the countries where this is a major business area. In Canada, this amounts to some $18 billion annually. The value of pulp and paper exports is of substantial importance to a national economy, because of its ability to earn international exchange. This factor is particularly significant to countries, which produce pulp and paper on a very large scale and have a relatively small population and hence consumption (e.g., Canada, Finland, Norway, and Sweden (Table 15.2)).

The largest producers, U.S.A. and Canada, are also the largest consumers, at least on a per capita basis (Table 15.3). The United States alone consumes close to half of the total world production. The United States and Japan, the runner-up after the U.S.A. on a consumption rate by country basis, all show a trend of increasing annual newsprint consumption of 10–60% over the 21-year period. Denmark, the Netherlands, Sweden, and the U.K., all lie in the mid range for per capita newsprint consumption. Among the smaller scale newsprint users, per capita consumption is much less for China, India, Iraq, and Ethiopia, and very much less for Malawi and Uganda.

We may have originally learned how to make paper by careful observation of the construction of wasp nests, as the builders macerated plant fibers for this purpose. Early papermaking was accomplished one sheet at a time, by hand, by methods, which originated in China at about 105 A.D. From that time till about 1760, papermaking remained a batch operation, single sheets

TABLE 15.2   Growth in Exports of Wood Pulp by Some Countries, in Thousands of Metric Tonnes[a]

	1950	1960	1970	1979	2000
Brazil	0		40	582	2,917
Canada	1,675	2,360	5,063	7,090	11,653
Finland	1,056	1,587	2,057	1,865	1,681
Norway	546	802	981	573	408
South Africa	0	74	278	529	284
Sweden	2,091	2,931	3,762	3,519	2,975
U.S.A.	87	1,036	2,808	2,662	5,583
U.S.S.R.	n/a	257	448	680	1,619[b]
Other	243	632	1,410	2,673	–
World Total	5,698	9,679	16,847	20,173	36,562

[a]Data from Reference Tables [1].
[b]For the Russian Federation.

being prepared by dipping a mold (or screen) into a vat holding a suspension of fiber in water. After draining and pressing of the excess water, the sheets were dried individually in the open air, making this an expensive process. N.L. Robert, in 1761, took the first step toward developing a machine for papermaking on a continuous basis [3]. It was not until 1822 that a Fourdrinier paper machine, in France, was first operated as a continuous

TABLE 15.3   Annual Consumption of Newsprint by Selected Countries in Thousands of Metric Tonnes (Mg), and per Capita[a]

	1970		1980		1991	
	Total (Mg)	Per capita (kg)	Total (Mg)	Per capita (kg)	Total (Mg)	Per capita (kg)
U.S.A.	8,924	43.6	10,673	46.9	12,274	48.6
Canada	657	30.8	918	38.1	416	15.4
Australia	449	35.9	557	37.9	561	32.4
Denmark	149	30.2	153	29.8	212	41.3
Netherlands	379	29.1	458	32.4	465	30.9
Sweden	343	42.7	296	35.6	414	48.1
U.K.	1,544	27.8	1,381	24.5	1,851	32.2
Japan	1,973	18.9	2,703	23.1	3,821	30.8
West Germany	1,077	17.7	1,394	22.6	–	–
France	606	11.9	629	11.7	729	12.8
U.S.S.R.	928	3.8	1,604	4.1	1,324	4.7
China	571	0.74	528	0.53	768	0.66
India	182	0.33	310	0.45	450	0.52
Afghanistan	–	–	0.1	0.006	–	–
Ethiopia	0.90	0.036	2	0.05	2.3	0.045
Other	3,773.1	–	–	–	–	–
Total, world mean	21,563	5.86	–	–	–	–

[a]Data from *U.N. Statistical Yearbooks* [2].

papermaking machine. This substantially raised the uniformity and reduced the cost of the product.

A suspension of pulp in water is the common raw material for virtually all types of papermaking. The pulp is mostly cellulose and is derived from wood fibers separated from whole wood. Other sources such as cotton and linen rags (as offcuts from clothing manufacture), which are virtually 100% cellulose are also used to produce small quantities of the highest quality papers. Where wood may be scarce, straw, cornstalks, hemp, jute, or bagasse (sugar cane fiber) may also be employed as raw materials. Papermaking from bagasse has to cope with carbon particles left on the stems when cane field foliage is burned just prior to sugar harvest, and with the high proportion of pith. Esparto grass, a tough dryland cover, and bamboo are also used. The very rapid growth of bamboo allows "cropping" for papermaking and with prior crushing between steel rollers the tough internodes may also be used. Modern chemical pulping processes are also capable of pulping annual plants, which permits the use of a wide range of easily renewable raw materials. For instance, kenaf, a cannabis-free variant of hemp, is used for high-quality pulp production by Australia, China, and Thailand, using much less intensive processes than required for wood. A recent proposal to the Tasmanian government aimed to produce 100,000 tonnes of pulp per year from 15,000 ha dedicated to the growing of kenaf. But by far the dominant raw material used for papermaking worldwide is wood, so the detailed discussion which follows centers on the methods used with this source of pulp.

15.2. WOOD COMPOSITION AND MORPHOLOGY

The woody part of a tree provides both mechanical support to hold the tree upright for optimum exposure of the leaves to sunlight and air, and serves as a conduit to carry water and trace nutrients from the roots to the leaves and photosynthetic products to where needed. Hollow and interconnected fibers composed mostly of cellulose and oriented along the axis of the tree provide both of these functions (Fig. 15.1).

Cellulose, the main component of the conduit fibers, is a polymer of β-D-glucose (Fig. 15.2). The initial step in its formation is the photosynthetic reaction, which produces glucose from water and carbon dioxide (Eq. 15.1).

$$6\ H_2O\ +\ 6\ CO_2\ \xrightarrow[\text{sunlight}]{\text{chlorophyll}}\ \underset{\text{glucose}}{C_6H_{12}O_6}\ +\ 6\ O_2 \qquad 15.1$$

Glucose in solution is carried in the sap to provide energy for plant metabolism, by the reverse of the photosynthetic reaction, and building blocks to the cambium layer of the wood, where the new growth occurs. The cambium layer comprises the interface between the wood and the bark (the "sapwood"). This is where new fibers form by processes, which deposit insoluble cellulose from the soluble glucose and cellobiose, a disaccharide of glucose. Rings of fibers, which form in the spring during a period of rapid growth are relatively thin walled and have a large lumen (bore or duct). These are collectively referred to as springwood. In summer, when growth is slowed,

FIGURE 15.1 Enlarged cubic section of a typical softwood. (From Kirk-Othmer [4], reprinted courtesy of John Wiley & Sons, Inc.)

fibers are formed with thicker walls and smaller lumens, which are collectively referred to as summerwood. It is these differential growth features, which give rise to the characteristic annual rings of wood, and which also allow long term historical climatic variations to be estimated from cross sections of old specimens. The core of the trunk or limb inside the cambium growth layer is dead wood and is termed heartwood.

Cellulose is formed by elimination of water from adjacent β-anomers of the two cyclic forms of glucose (Eq. 15.2), hence it is a "dehydro" polymer.

$$n\ C_6H_{12}O_6 \rightarrow (C_6H_{10}O_5)_n + n\ H_2O \qquad n = \text{about } 10,000 \qquad 15.2$$

Formation of cellulose from the β-anomers is what gives the polymer its characteristic linear form, which favors strong intermolecular hydrogen bonding between adjacent polymer chains. With some 10,000 glucose units (molecular weight about 1.5 million) per chain, individual molecules can be up to 0.012 mm in length [5]. This high molecular weight and the close intermolecular association makes fibers composed of cellulose both strong and insoluble in water. Both factors also introduce significant resistance to hydrolysis because of the relative inaccessibility of the acetal links between adjacent glucose units. The hydrolytic stability, strength, and insolubility are necessary features of the structural role of cellulose in the plant as well as being important for the papermaking process and for the properties of paper.

Chemically cellulose is a two-dimensional high molecular weight polymer with no conjugation, hence it is white (colorless) in color. It has no taste or

α-D-glucose
36%

D-glucose
<1%

β-D-glucose
64%

Starch, poly (α-D-glucose)

Cellulose, poly (β-D-glucose)

FIGURE 15.2 Linear and stereochemical representations of the cyclic forms of glucose, and their relationship to the two regular dehydropolymers of glucose.

odor and is insoluble and relatively stable in water or aqueous alkali. However, it is soluble in high concentrations of aqueous mineral acids, and may be regenerated to lower molecular weight fragments of the solid on dilution with water. Cellulose does not melt on heating and is relatively stable to heat, not decomposing until it reaches 260–270°C. It is also relatively stable toward oxidation. This set of properties is highly suitable for a papermaking raw material. Chemically speaking, cellulose is the same from one species to the next, although there are morphological differences between the cellulosic fibers from different species.

Lignin is an important component of wood, which functions primarily as an interfiber bonding agent. It is a three-dimensional, cross-linked polymer of heterogeneous structure composed of predominantly n-propylbenzene units joined by a variety of ether (C–O–C) and carbon–carbon (C–C) links (Fig. 15.3). Lignin is also biosynthetically derived from glucose, like cellulose,

FIGURE 15.3 Structure and identity of decomposition fragments and an approximation of the structure of a softwood (spruce) lignin. (Reprinted from Bolker [5, p. 579], courtesy of Marcel Dekker, Inc.)

although less directly [6]. The high-aromatic–phenolic content in the polymer makes lignins all more or less brown in color, and they are relatively readily oxidized. The molecular weight of natural lignins, *in situ*, is not well defined. It probably consists of polymers possessing a range of molecular weights. Molecular weights of isolated lignins mostly range from 1,000 to 20,000, depending on the origin of the lignin and the method used for determination.

Lignin is relatively stable to most aqueous mineral acids but is soluble in hot aqueous base and hot aqueous bisulfite ion (HSO_3^-). It has a low-softening temperature and melting point (e.g., native spruce lignin softens at 80–90°C (wet), 120°C (dry), and melts at 140–150°C). Higher temperatures than this cause carbonization. The lignins from different species of tree, and even from the same species growing in different locations are significantly different chemically from one another, particularly when comparing softwoods and hardwoods.

Less dominant components of wood are the hemicelluloses, so called because at first it was thought that this component consisted of lower molecular weight biosynthetic intermediates en route to cellulose. However, the hemicelluloses are soluble in aqueous alkali, which was one of the early means for differentiating them from cellulose. This property is also a clue to the amorphous nature of hemicelluloses, which are branched polysaccharides of a variety of sugars, all of lower molecular weight than cellulose. None has a simple trivial name. The name of the particular hemicellulose is derived from the sugar monomer from which the polymer is composed. For example, xylose forms xylans, galactose forms galactans, mannose forms mannans, etc. As with lignin, the chemical structure and composition of the hemicellulosic fraction of wood also differs from species to species in woods, grasses, and annual plants. With hardwoods the xylans predominate and with softwoods the glucomannans are usually dominant. Isolation of a particular hemicellulosic fraction from wood also modifies it somewhat, so that it is difficult to determine what the original composition of the hemicellulose was. Thus the structure of hemicelluloses remains rather poorly defined, much like lignins.

The amorphous nature and lower molecular weight of hemicellulose makes it an important component of papermaking, by its ability to improve interfiber bonding. Inevitably there is some hemicellulose loss from the pulp, particularly with some chemical pulping processes, because hemicellulose is less stable to acid and alkali than cellulose. The sum of the cellulose and hemicellulose components in wood or pulp is collectively referred to as holocellulose. Some types of chemical pulps (the so-called "dissolving pulps") are destined for end uses requiring pure cellulose such as for the manufacture of cellophane or cellulose acetate. For these applications the presence of hemicelluloses causes problems with filtration, etc. Thus, when pulping to produce a dissolving pulp, the process is modified to remove as much of the hemicellulose as possible.

In addition to these dominant components of dry wood, there is a group of miscellaneous nonpolymeric, or very low molecular weight polymeric materials, which are soluble in cold water and are relatively easily separated from the wood. These are collectively referred to as extractives and comprise of 3–8% by weight of dry wood. Despite their lower importance in paper, the

TABLE 15.4 Typical Percentage Compositions of Dry Woods and Cotton, on an Extractive-free Basis[a]

Constituent	Softwood (e.g., fir)	Hardwood (e.g., aspen)	Cotton
Holocellulose			
Cellulose	42	42–48	99
Hemicelluloses	27	27	1
Total holocellulose	69	75	100
Lignin	29	15–20	0
Pectin, starch, ash, etc.	2	4	0
Total	100	100	100

[a]Analysis after thorough extraction with water and nonpolar organic solvent and drying. Compiled from data of Kirk-Othmer [4] and Walker [5].

extractives can cause processing problems with some species of wood. For example, the presence or formation of fatty acids from some species of wood during pulping can cause foaming problems, either during pulping or on waste stream disposal, or both. Fatty acids and the resin acids formed from oxidized terpenes can also cause equipment corrosion problems.

The proportions of these wood components vary from species to species, and even with specimens of the same species grown under different conditions. However, in general, hardwoods (from deciduous trees) contain more cellulose and less lignin than softwoods (coniferous trees; Table 15.4). Also keep in mind that the chemical nature of the components of wood will also differ significantly from one species to another, which will affect the details of the optimum pulping method for any particular species. The objective of all pulping processes is to produce a fiber suspension in water suitable for papermaking from a wood raw material. Fiber separation may be obtained by largely physical work, or by chemical treatment, or by a combination of these methods.

15.3. PREPARATION OF WOOD FOR PULPING

After wood has been harvested employing extensive mechanization, and brought in to the pulp mill by water, rail or road, it has to be prepared for pulping. Bark removal, or debarking, is a preliminary step common to all pulping processes. Leaving the bark in place as the wood is pulped could slightly increase the fiber yield, at least for some woods and some pulping processes. However, for most species of wood it would contribute an undesirable highly colored, nonfibrous constituent to the pulp, the reason for its removal.

With smaller logs a drum debarker is commonly used. This consists of a steel cylinder 3–4 m in diameter and 15–25 m long, placed on a slightly sloping axis, and capable of slow rotation on its axis. It is fitted with log-tumbling baffles inside and has longitudinal slots cut into the periphery to allow bark fragments to fall out or be flushed out with water. Slow rotation of the drum containing a load of logs removes the bark by a combination of abrasion and impact attrition, after which the debarked logs are removed from the outlet gate. Drum debarkers may be operated dry, producing a dry bark by-product of high fuel value to the pulp mill, but sometimes slowing the debarking step as a result. Or they may be operated wet, which gives faster debarking, particularly with hot water and cold or frozen logs. However, this results in a bark of lower fuel value that requires a drying step before combustion, and also produces a wastewater stream, which requires treatment.

The bark of very large trees is usually removed in an in-line debarker. One horizontal log at a time is moved slowly, in an axial direction, while at the same time being rotated slowly on its axis. At the same time rapidly rotating slack chains (chain flails) beat the bark off, or a water jet operating at 1,000 psi (lb/in.^{2}) or higher and directed to the wood/bark interface (a "hydraulic debarker") literally blasts the bark off. Another version of in-line debarker moves the log axially, without rotation, while a large stout ring fitted with spring-loaded knives pointing radially inward is revolved around the log in a lathe-like fashion, which cuts the bark off as the log moves through.

When the bark is removed from a cord 128 ft^3(3.625 m^3) of wood, about 120 ft^3(3.5 m^3) of debarked solid wood is normally obtained, a yield of 90–95%. Larger logs can give larger yields of wood. The volume measure used by the pulp and paper industry is the cunit (c unit), meaning 100 ft^3 (equivalent to 2.83 m^3) of bark-free wood (i.e., the raw material suitable for papermaking).

Once the bark is removed, small logs, which are to be reduced to a pulp suspension in water by grinding (one form of mechanical pulping) are floated in water channels directly to the stone groundwood operation. Very large logs, which cannot be accommodated by the stone groundwood equipment directly will be sawn to manageable sizes of blocks before grinding. While stone grinding of wood uses short lengths of whole logs, or sawn rectangular blocks of wood for processing (see later section), all of the other methods of pulp production use a chipped wood feed.

Chipping, the process of reducing a log to chips of about $2.5 \times 2.5 \times 0.5$ cm thick is a common preliminary to all pulping methods except stone groundwood. Wood chips are more convenient and uniform for solids transport within the mill complex by conveyor belt or pneumatic delivery systems than are whole logs. Also a chip format is more amenable to direct physical conversion to fibers because of the much easier and more uniform penetration of heat and moisture through the thin sections of wood.

A chipper consists of a 2–4 m diameter heavy steel disk fitted with 5 or 6 radially mounted steel knives, each of which protrude about 2.5 cm (the chip length desired) from one face of the disk. A spout aimed about 30° off the axis of rotation, directs the log to be chipped toward the rotating disk, end first, and at the ideal angle to minimize the rotational energy required to drive the chipper. In the chipping process, the angular feed also produces a splayed angular edge cut to the chip, usually with some small splits, which optimize the ease of liquor penetration when pulped. As the chips are cut, slots through the steel disk ahead of each of the knives allow the chips to pass straight through the disk, aided by blowing vanes attached to the opposite face of the chipper from the knives.

A chipper of the type described, driven by a 500 hp (ca. 800 kW) electric motor is capable of reducing a 0.6 m diameter × 8 m log to chips in less than 30 s and could produce something like 50 tonnes of chips per hour. Greater production rates are obtained from more recent models by using 2,000–2,500 hp (1,600–2,000 kW) drive motors.

It has been proposed that chipping could be conducted in two stages to obtain chips with less compression damage. Whatever method of chipping is used uniformity of size is important to the yield and the strength of the pulp produced. Any fines have to be screened out and cooked under different conditions than regular sized chips to decrease loss of cellulose. Any oversize material has to be further processed for use. Blade sharpness, speed of disk rotation, weight of log(s) in the spout, and whether the butt or the top of the log is fed in first, all affect chip consistency and quality.

15.4. MECHANICAL PULPING

The pulps produced, which rely solely on physical work on the log to reduce the wood to a fiber suspension in water are collectively referred to as mechanical pulps. The pulp produced consists of a mixture of all of the insoluble constituents present in the original wood, less a small amount that is degraded during the pulping step, which is lost in soluble form in the water phase. Thus the yield, or the weight of pulp obtained from the weight of wood pulped is normally about 95%. Low-cost papers are made from this pulp. Light-colored woods of favorable species of trees such as white spruce give a light-colored pulp. The whiteness and strength of the low-cost papers made from mechanical pulps may both be improved by blending some fully bleached chemical pulp into this before papermaking.

Relatively coarse mechanical pulps, with only 10–15% incorporation of bleached chemical pulp to provide sufficient strength, are used in the preparation of textured hanging stock (wallpapers). Newsprint, which is used for newspapers, inexpensive magazines, paperback books, and other low-cost publishing requirements employs a blend of 15–35% chemical pulp stocks into the mechanical pulp, to obtain an economically thin sheet of sufficient strength to avoid breakage in modern high-speed printing presses. And about 10% or so of mechanical pulps, particularly of stone groundwood types, may be blended into the pulps used for some fine papers to improve the smoothness and opacity.

15.4.1. Stone Groundwood

Whole debarked logs or sawn wood blocks with their axes parallel to the axis of the stone (Fig. 15.4) are pressed against a large grindstone rotating in a pit of water, to produce a suspension of 3–4% groundwood pulp in water. The axis of the log is kept parallel to the axis of the stone to obtain as long a fiber

Three-pocket hydraulic grinder

Hydraulic magazine grinder

FIGURE 15.4 Grindstone and log feed configurations used for the production of stone groundwood pulps from whole logs or blocks.

from the wood as possible and at the same time to minimize the grinding power requirement. Friction between the log and the stone produces heat, which softens the lignin and allows cellulose fibers to be torn out with minimum fiber damage. A system of showers with flow control and a pit water level below the base of the stone are often used to obtain better regulation of stone and wood temperatures [7].

Stone dimensions range from 1.8–2 m in diameter and from 1–1.5 m wide. They are made of synthetic grits such as aluminum oxide or silicon carbide, embedded in a ceramic matrix. To drive the stone at the required 250–300 rpm (peripheral speeds of 25–30 m/s), requires 4,000–5,000 hp (ca. 3,200–4,000 kW) electric motors. Grinding pressure, or the force with which the logs are pressed to the stone, is the means used to regulate both stone speeds and the motor power requirement. The mechanical power required is 1,600–1,750 hp hr/tonne of pulp for newsprint grades, and is the chief production cost. For hanging stock, the coarsest pulps, the power requirement is slightly less than this, and for book papers, slightly more.

The last step in stone groundwood production is a screening of the fiber suspension in water to remove shives (fiber bundles or small splinters) from the stock. If paper is made from stock containing shives it will have low-strength properties in the area around each shive because of poor interfiber cohesion. This makes the sheet susceptible to tearing either on the papermaking machine itself or during subsequent printing. The separated shives are usually fed in a water suspension to a refiner for defiberization and then added back to the main groundwood pulp stream.

15.4.2. Chip Refiner Groundwood

It became possible to produce a mechanical pulp from chips when the process of refiner mechanical pulping was developed to accomplish this during the 1950s. This convenient method of producing mechanical pulp was rapidly adopted as a complement to stone grinding by many newsprint mills, but did not, however, completely displace stone grinding as a mechanical pulping procedure because of the differing complementary qualities of the pulp produced by the two methods.

The refiner machine used to accomplish this consists of two 1-m diameter stout steel disks fitted with welded hardened steel bars on one face of each (Fig. 15.5). The barred faces of the two disks face each other, and the spacing between the bars of the facing disks is adjusted to control the fineness of the fiber suspension produced. One disk may be stationary and the other rotated, or they may be counterrotating (rotated in opposite directions) in either case by powerful electric drives of up to 10,000 hp (ca. 8,000 kW).

Wet chips with some additional water are fed to the center of the space between the two refiner disks while they are rotating. The adjacent faces of the disks are each slightly concave so that the chips are able to drop into the wider space existing between the disks, near their centers. Friction between the chips and the disks produces heat and steam, which softens the lignin and starts the defiberization process. As the chip fragments get smaller they are carried by the continuing flow of water, fresh chips, and steam toward the periphery of the disks, where the spacing between the disks is less, and the bars are more numerous. (Fig. 15.5). Chips are rapidly reduced to a separated fiber suspension in water, as the fiber fragments reach the periphery of the disks.

FIGURE 15.5 Operating details of a refiner used to reduce a wood chip feed to pulp.

Peripheral speeds of refiner disks may be as high as 160 m/s, significantly greater than the usual peripheral speeds of stone grinders [4]. Controls of the operating parameters of chip refiners and the use of statistical modelling for optimum operation have been analyzed [8, 9].

Paper produced from this refiner pulp is generally stronger, from the longer average fiber lengths produced, than paper produced from stone grinding of the same wood. However, it also has a higher surface roughness and lower opacity than stone groundwood newsprint, although these factors are affected by refiner adjustments, which increase or decrease the power requirement to obtain the pulp (Table 15.5). Chip refining has a somewhat higher

TABLE 15.5 Comparison of Properties of Papers Produced from Spruce Stone-, or Refiner Groundwood Pulps at Similar and Different Work Inputs[a,b]

	Stone groundwood	Refiner groundwoods	
		Moderate work	Higher work
Power, hp days/air dry tonne	76	75.3	89
Can. standard freeness[b] (mL)	115	164	101
Burst factor	12.1	12.6	18.2
Tear factor	56.7	87.8	86.0
Breaking length, m	2,920	2,680	3,200
Brightness, GE or Elrepho	61.0	59.0	60.5
Opacity, %	95.0	92.5	95.0

[a]Properties selected from Britt [7].

[b]Related Terminology:

Canadian standard freeness (CSF) is a measure of the ease of water flow through a standard perforated plate, specified in milliliter, from a one liter sample of pulp at 20°C and 0.3% consistency (weight fraction of pulp in the suspension). Drained water from the sample passes into a dual outlet funnel, one outlet higher than the other. The amount of water collected from the upper "overflow" outlet, in milliliter, is recorded as the CSF. The rate of water removal on the paper machine of a "free" pulp will be faster than the rate of water removal from a "slow" (slow draining) pulp.

Burst factor. The pressure developed by a fluid, in pounds per square inch, required to split open a sample of the paper when it covers a circular hole 1.20 in. in diameter and is protected by a rubber diaphragm, is defined as burst (lb/in.^{2} or g/cm^2). When burst is divided by the basis weight (e.g., in g/m^2), it is called burst factor.

Tear corresponds to the tensile force in grams required to continue a tear that has already been started in a single sheet of paper. Corresponds to the "internal tear resistance" as opposed to "edge tear resistance," which is frequently a much higher value. Tear, divided by basis weight, is the tear factor.

Breaking length is a hypothetical value calculated from tensile strength and material density measurements, to give a relative measure of the length of a strip of paper, which is just self supporting, when it is held vertically. Generally speaking, the higher this value the higher the tensile strength of the sample.

Brightness is a measure of overall reflectivity of a paper to white light, or whiteness. It is based on a scale of 100 for pure magnesium oxide and 8 for carbon black, when the reflectance of a beam of light, of wavelength 457 nm, is compared at angle of 45° from the axis of the incident light.

Opacity is a measure of sheet "see-through." It is obtained by dividing the reflectance of the sheet when backed by a black body by the reflectance of the same sheet when backed by a white body having an effective reflectance of 89% (absolute), and expressed as percentage.

power cost for pulp production, say, about 80–100 hp days per air dry tonne for spruce, as opposed to about 80 to 90 when stone grinding. These are all reasons for the continuance of stone grinding as a component of the mechanical pulping operations in many newsprint mills, although more recent developments are causing some mills to phase out stone grinding. It is also possible to produce a useful pulp from planer shavings or sawdust by refiner pulping, options, which are not possible with stone grinding. While it is not possible to make a useful paper from sawdust pulp alone, it can be blended with more conventional pulps to produce a lower cost paper sheet to meet a printer's specifications.

15.4.3. Thermomechanical Pulp

As more experience was gained with refiner mechanical pulping, it was recognized that higher temperatures decreased the power input required per tonne of pulp produced. Carrying this realization a step further, the idea of subjecting chips to a short period of preheating before refining was developed with the expectation that this might both decrease refining power and produce a longer fibered pulp, which could produce a stronger paper than the product from conventional refining. These were the benefits achieved by the first commercial thermomechanical installations, which began operation in the early 1970s.

Chips, shavings, or sawdust are first washed to remove any grit or other potentially damaging contaminants, and then digested with steam at 120–130°C, 170–315 kPa (25–45 psi) for 2–3 min [10]. While still under pressure, and with only a 60–75% water content (much less than for conventional chip refining), the fragmented wood is refined and the pressure then released, producing a fluffed up, bulky pulp. After addition of a small amount of water and a further stage of refining at atmospheric pressure a pulp is obtained, which has longer average fiber lengths, higher strength (from less damaged fiber), and a lower fines content than either stone groundwood or chip refiner pulps.

The favorable strength properties of thermomechanical pulp (TMP) allow it to be used to replace part or all of the chemical pulp component in a newsprint furnish, the replacement rate ranging from 2 to 4% for every 1% reduction in chemical pulp content. Since the yield from wood for TMP, about 90%, is nearly twice that of chemical pulps, chemical pulp replacement by TMP in newsprint gives a significant saving in wood requirements for newsprint production. The saving in wood far surpasses the additional refining energy requirement for thermomechanical pulping. For coniferous wood TMP requires about 2,000–2,500 kWh/tonne, about 1,000 more than stone grinding at about 1,200–1,650 kWh per air dry tonne. A further advantage of TMP over the chemical pulps is that it provides easier effluent treatment. Possibilities for further energy reduction for various refining processes, including TMP and chemithermomechanical pulp (CTMP), have been examined [11].

The exposure of the pulp to heat and air in TMP processing can cause some darkening of TMPs. The use of an aqueous sulfite pretreatment before TMP processing helps to reduce the loss of brightness on refining. This variation of TMP processing is still referred to as a mechanical pulping process since refining (mechanical work) is still the primary process used to pulp the chips. It is referred to as CTMP to distinguish it from ordinary TMP [12].

15.5. CHEMICAL PULPING PROCESSES

Processes, which involve the use of chemicals for either softening or removal of much or all of the lignin to assist the process of fiber separation from the wood are called chemical pulping processes. The simpler processes of pulping use a chemical presoak of whole (small) logs or chips prior to the use of conventional mechanical pulping methods of defiberization to produce a fiber suspension in water. These presoak methods are usually reserved for the mechanical pulping of hardwoods and are referred to as chemimechanical pulping, not as true chemical pulping processes.

At the other extreme, the more rigorous chemical pulping methods remove all or nearly all of the lignin from wood chips leaving cellulose fiber as intact as possible. In this case a fiber suspension in water is obtained directly, with little or no mechanical work required for fiber separation. Despite the aim of chemical pulping methods to soften and dissolve lignin alone without affecting the cellulose, there is inevitably some cellulose chain shortening and end group conversion to carboxyl that also occurs. Nevertheless, since 1930, chemical pulping techniques have been developed that can produce satisfactory pulps for papermaking from any species of tree as well as from grasses and annual plants. Thus, a much wider range of raw materials can be handled by chemical pulping methods than is possible by straight mechanical pulping procedures.

15.5.1. Chemimechanical Pulping

Direct grinding or refining of hardwoods (from deciduous trees) tends to give pulps with a short average fiber length, which would produce a low-strength sheet on a paper machine. This is particularly true with the mechanical pulping of *hard* woods from deciduous trees since the fibers in place are shorter and finer, and since there is a greater tendency for fibers to be fragmented, instead of whole, as they are torn out from the denser wood structures.

However, if the wood is given a presoak in a hot solution of sodium hydroxide it is possible to obtain a more useful pulp with less fiber damage. Temperatures may be near ambient up to 80°C, depending on the species to be pulped and treatment time. Small amounts of sulfur dioxide may be added to the liquor to decrease the tendency of the wood to darken. Whole logs or blocks require a 4–10 hr presoak, whereas chips only need 10–15 min. Pulp yields of 85–90% are obtained with properties similar to those expected for groundwood pulps generally, except usually darker in color. Chemical pretreatment also substantially decreases the power input required to reduce the wood to fibers. Blends of softwood and hardwood chips have also been found to be amenable to chemical pretreatment before mechanical pulping to yield useful pulps.

15.5.2. Semichemical Pulping

The essence of semichemical pulping methods is that chips are subjected to a more intensive chemical treatment than a simple presoak but not as severe as for chemical pulping, to cause partial delignification. Intermediate intensity chemical treatment is achieved by decreased ratios of chemicals to wood, decreased cooking times and/or temperatures, and the use of chemicals, which pulp at approximately neutral pH conditions. Following the cook, the chips require much lower power inputs per tonne of pulp for refining than required for straight mechanical pulping, because of the chemical softening and partial solubilization of the lignin. Lignin solubilization during the pulping process lowers the yield of pulp to 65–80%, less than that given by mechanical or chemimechanical methods.

At least half a dozen variants of semichemical pulping have been practiced. Some of these, such as neutral sulfite semichemical (NSSC) pulping, softwood bisulfite high-yield pulping, and softwood sulfate pulping for linerboard production, are still common [4]. Among these, NSSC pulping has a number of differentiating features of interest. Hence this procedure, which is primarily of value for the pulping of hardwoods, will be discussed here as an example of semichemical pulping.

The digestion liquor used for NSSC pulping consists of a solution of sodium sulfite and sodium carbonate in water, but at the proportions and the pH used is actually a mixture of sodium sulfite and sodium bicarbonate. The make-up sodium carbonate is either purchased or made from aqueous solutions of low-grade sodium hydroxide by contacting this with the carbon dioxide in flue gases (Eqs. 15.3 and 15.4).

$$CO_2 + H_2O \rightleftharpoons H_2CO_3 \qquad\qquad 15.3$$

$$H_2CO_3 + 2\,NaOH \rightleftharpoons Na_2CO_3 + 2\,H_2O \qquad\qquad 15.4$$

Elemental sulfur is burned for the first stage of sodium sulfite preparation. Repeated contact of the sulfur dioxide formed with the aqueous sodium carbonate in a gas–liquid absorption unit (sulfiting tower, Fig. 15.6) produces the approximate proportion of sodium sulfite and sodium bicarbonate (about 1 mol:2 mol) required (Eqs. 15.5 and 15.6).

$$SO_2 + H_2O \rightleftharpoons H_2SO_3 \qquad\qquad 15.5$$

$$H_2SO_3 + 2\,Na_2CO_3 \rightleftharpoons Na_2SO_3 + 2\,NaHCO_3 \qquad\qquad 15.6$$

Liquor recycle to the sulfiting tower is continued until the pH of the underflow reaches 8.5, at which time a portion of this stream will be bled off for use as pulping digestion liquor. Under these conditions the proportions of salts present in solution are about 82% Na_2SO_3, 4% $NaHSO_3$, and 14% $NaHCO_3$. For some species of wood or for different pulping objectives liquor of about pH 8 may be used. This is made by more prolonged contact between the sulfur dioxide and sodium carbonate solutions (Eq. 15.7).

$$H_2SO_3 + 3\,Na_2CO_3 + SO_2 \rightarrow 2\,Na_2SO_3 + 2\,NaHCO_3 + CO_2 \quad 15.7$$

FIGURE 15.6 Preparation of neutral sulfite semichemical pulping liquor via countercurrent contacting of a sulfur dioxide gas stream with aqueous sodium carbonate.

Care is taken to ensure that the digestion liquor does not become too acid, otherwise cellulose degradation during the pulping step can become more significant.

For NSSC pulp production the effective chemical concentrations are about 120 g/L of sodium sulfite and about 20 g/L of sodium bicarbonate. Chips to be pulped are cooked in this two buffer system at temperatures of 140–170°C for 1–4 hr depending on the type of wood being pulped and the pulping objective. High-grade pulps destined for bleaching will be pulped at ratios of sodium sulfite to wood of about 15–20:100 by weight, significantly higher than is used for coarser pulps. These also demand higher temperatures and/or longer times, and give lower yields of finished pulp because of the greater delignification achieved. Coarser pulps destined for board production (cardboard, fluting medium, or box liners) are subjected to milder conditions and shorter times. This produces higher yields and retains individual fiber stiffness, a property important for cushioning products such as corrugating medium. The dual buffer system exerts its pulping action by sulfonation of lignin present in the chips and probably by reduction as well [7]. Both processes give soluble lower molecular weight lignin fragments from the original insoluble high-molecular weight lignin.

The cooked pulp is filtered from the pulping liquor, washed, and then refined to complete the defiberization. A high yield of pulp is obtained, which still contains some lignin. The lignin content and the dark color of high-yield NSSC pulps makes this product generally unsuitable for high-grade papers. The lighter color of lower yield grades minimizes bleaching costs for incorporation of this product into intermediate or coarser paper grades. There is also no significant odor problem associated with the pulping step, as there is with some of the other chemically based pulping processes.

A problem with NSSC pulping is the difficulty of either recycle or disposal of the waste (spent) pulping liquor. It can be concentrated and burned but this produces sodium sulfate, not sulfite suitable for reuse. If there is a kraft pulping operation near the semichemical pulping operation, the sodium sulfate may provide both sodium ion and sulfate make-up chemical for the kraft mill. If the kraft mill is three or more times the scale of the NSSC mill rate, the kraft process is capable of profitably using all of the spent liquor of the latter. It is possible to obtain sulfite from sulfate by a complex series of steps but this adds to the cost making recycle less attractive.

Alternatively the spent NSSC liquor may simply be discharged. Untreated, this liquor has a very high biochemical oxygen demand (BOD) and several toxic constituents so that disposal to freshwater has not been practiced for many years. Disposal in the open ocean, at depth, in areas of vigorous currents could probably be accepted. However, mills fronting on the sea are usually on sheltered inlets or fjords where mixing and dilution would be poor so they have the same restrictions on discharges as mills operating on freshwater. Details of other possibilities are discussed with emission control aspects of sulfite pulping.

15.5.3. Chemical Pulping

Wood chips supply the fiber raw material for chemical pulping operations, as for most chemimechanical and semichemical methods. In chemical pulping, the treatment with chemicals is sufficiently severe to dissolve all or much of the lignin originally present in the wood. The chemicals employed operate at low or high pH (not near neutral), with higher ratios of chemicals to wood, and the cooking conditions are usually more severe than for the processes discussed thus far. Thus, the pulps obtained require little or no mechanical work to obtain a fiber suspension in water from the original wood.

Photomicrographs of chemical pulps show that there is very little fiber damage. Consequently papers from chemical pulps, particularly when made from kraft chemical pulps, are strong. Since there is not only lignin loss, but for some chemical pulping processes hemicellulose as well, the yields of pulp from straight chemical processes are in the 40–60% range, much lower than that achieved by the basic mechanical procedures. Chemical pulps are more or less readily bleached to high brightness (whiteness) levels because they consist of almost pure cellulose. The brightness and the strength properties are both much more stable than for mechanical (newsprint-type) pulps. However, these papers are also more expensive because of the increased wood, chemical, and energy cost required for their manufacture. The various chemical pulping processes used to produce these pulps may be conveniently classified according to the pH of the digestion liquor used.

15.5.4. Acid Sulfite Pulping

After the discovery of the softening action of sulfurous acid on wood by Benjamin Tilghman in the U.S.A. in about 1857, the idea was developed experimentally as a chemical method for wood pulping by 1867 [7]. It was adopted on a commercial scale in the next few years, both in the U.S.A. and elsewhere, and rapidly became the dominant chemical pulping method until about 1937, when the volume of kraft pulp produced first exceeded the volume of sulfite pulp.

Digestion liquor for acid sulfite pulping consists of a solution of sulfur dioxide in water containing calcium ion. Sulfur dioxide is obtained by sulfur combustion or by the roasting of pyrite (FeS_2). The lime component is added to the liquor by passage of the cold sulfur dioxide solution down one or more Jenssen towers packed with limestone, in a manner very similar to the sulfiting towers used to prepare NSSC pulping liquor (Eqs. 15.8 and 15.9).

$$SO_2 + H_2O \rightleftharpoons H_2SO_3 \qquad\qquad 15.8$$

$$2\,H_2SO_3 + CaCO_3 \rightleftharpoons Ca(HSO_3)_2 + H_2O + CO_2 \qquad 15.9$$

Since calcium sulfite is not very soluble in water, the last stage(s) of liquor preparation involve absorption of sulfur dioxide into the cold liquor under pressure. This eventually gives a ratio of about 1 mole of calcium hydrogen sulfite (calcium bisulfite) to a total of 4–5 mole of (sulfur dioxide + sulfurous acid). At this stage the pulping liquor is an aqueous solution, which contains 6.5–9% dissolved sulfur dioxide in all forms.

The total dissolved (sulfur dioxide + sulfurous acid) is described as "free SO_2," and the sulfur dioxide tied up in calcium hydrogen sulfite as "combined SO_2" (Eq. 15.10).

$$\begin{array}{ccc} SO_2, H_2SO_3 & & Ca(HSO_3)_2 \\ \text{"Free } SO_2\text{"} & & \text{"Combined } SO_2\text{"} \\ \text{Mole ratio:} \quad 4-5 & : & 1 \end{array} \qquad 15.10$$

However, the effective sulfur dioxide accessible for pulping is a composite of these two forms of sulfur dioxide since the "available SO_2" = "free SO_2" + 1/2 "combined SO_2." The reason for this definition is that, as sulfur dioxide is consumed or vented from the digestion vessel during pulping, the equilibrium represented by Eq. 15.11 is displaced to the right.

$$Ca(HSO_3)_2 \rightarrow CaSO_3 + H_2O + SO_2 \qquad 15.11$$

As this occurs one-half of the combined SO_2 is released to the solution to also become available for pulping.

Conditions required for sulfite pulping vary with the species of wood being pulped, the degree of delignification desired, and the chemical to wood ratio used. Normal range is 8–12 hr at 110–145°C, after a 2-hr liquor penetration period. Typical pulping liquor to wood ratios on the West Coast of North America are 182,000 L (50,000 U.S. gal) of liquor containing 8.80% total SO_2 and 1.10% combined SO_2 at 70°C, to 36 tonnes of bone dry wood. It is not necessary that the chips be dry for digestion. But the moisture content of the chips must be known, to have a consistent ratio of active chemicals to the dry weight of wood to obtain optimum pulping.

The heat and low pH conditions cause hydrolytic fragmentation and sulfonation of the lignin in the chips, solubilizing a large fraction of it to a light-colored, easily bleached pulp. After washing and screening and even without bleaching it has a brightness (whiteness) compatible with direct blending into mechanical pulps for newsprint production. With two or three bleaching stages high brightness bond papers may be produced from these pulps. However, the less than optimum paper strength properties and expensive chemical recovery procedures (see Further Reading Section) have made acid sulfite pulping much less important in recent years. It is still of value for the manufacture of specialty pulps for uses such as tissues for its softer fibers, and for some printing papers [13]. It is also used to produce dissolving pulp for conversion to cellulose derivatives such as viscose rayon and some thermoplastic materials because it is relatively free of hemicellulose. Yield advantages and the simpler bleaching process than required for kraft pulps promise a continued place for sulfite-based pulping processes for these applications [14]. It has also been possible to profitably produce vanillin, a valuable chemical intermediate and flavor constituent from the lignin fragments present in sulfite (or kraft) waste liquor [15]. At one time the Ontario Paper Company, operating at Thorold, Ontario, produced 60% of the world supply of vanillin from this source.

15.5.5. Other Sulfite-based Pulping Processes

Acid sulfite pulping, originally developed around low-cost calcium ion, suffers from a number of disadvantages. Among these, calcium sulfite is virtually insoluble, which makes digestion liquor preparation difficult; chemical recovery is complex making it uneconomic; and disposal of the waste liquor is difficult (Section 15.5.2). These disadvantages have prompted most modern sulfite mills to switch from calcium-based to magnesium-, sodium-, or ammonium-based systems, which circumvent many of these difficulties.

With magnesium-, sodium-, or ammonium-based sytems, the bisulfite and sulfite salts are all soluble at all proportions in the presence of sulfurous acid. Even magnesium sulfite, with a solubility of about 1.25 g/100 mL, cold, is about 160 times as soluble as calcium sulfite at the same temperature and its solubility *increases* with temperature. So liquor preparation with these sulfite salts is easier, whether for acid sulfite, bisulfite, or NSSC pulping conditions, and even for experimental tests under alkaline conditions. For ammonium-based systems, ammonium hydroxide is contacted with a sulfur dioxide gas stream for liquor preparation. Magnesium-based systems use a magnesium hydroxide slurry to contact the sulfur dioxide gas stream. Sodium-based systems normally employ sodium carbonate lumps in a sulfiting tower, in a method similar to that used for NSSC liquor preparation. Sodium hydroxide may also be used if available at low cost.

Ammonium, magnesium, and sodium-based sulfite pulping systems also offer a versatility of operating pH not possible with calcium. However, sodium ion is the only one useful throughout the whole pH range even to monosulfite and alkaline sulfite pulping liquor preparation (Table 15.6). With sodium carbonate as the base it is possible to use carbonate to sulfurous acid ratios ranging from 1:3 where the active pulping species are bisulfite (HSO_3^-) and sulfurous acid (acid sulfite pulping), to 1:2 where the principal pulping species is bisulfite, such as is used in bisulfite pulping. At the 1:1 to 2:1 range of ratios of carbonate to sulfurous acid, sulfite and bicarbonate are the active pulping agents, as in NSSC pulping. And when one reaches the 3:1 to 4:1 range of ratios, sulfite and carbonate are the dominant anions present, important in the experimental alkaline sulfite pulping method (Table 15.7). Analogous results may be achieved with ammonium and magnesium ions, at least with pHs up to the bisulfite range.

TABLE 15.6 Chemical Pulping Processes Classified by Operating pH, and Pulp End Uses

Pulping process	Liquor pH at 20°C	Base options	Appropriate end uses
Acid sulfite	1–2	Ca^{2+}, Mg^{2+} Na^+, NH_4^+	1. Chemical conversion to derivatives of cellulose (e.g., viscous rayon, cellulose acetate) 2. Filler pulps for paper requiring little strength (e.g., writing papers. Provides bulk, cushioning properties)
Bisulfite	3–5	Mg^{2+}, Na^+, NH_4^+	1. Unbleached—in newsprint (25–30%, rest groundwood) 2. Bleached—fine papers (up to 100%)
Neutral sulfite semichemical[a]	6–8	Na^+, NH_4^+	1. Fluting board—provides stiffness, crush resistance to corrugated cardboard 2. Coarse wrapping papers
Monosulfite	8–10	Na^+, NH_4^+	Same as for neutral sulfite
Alkaline sulfite (experimental)	11–13[b]	Na^+	1. Unbleached—box liner-board other packaging materials 2. Bleached—high-quality papers
Soda	11–13	Na^+	Outmoded for chemical pulping; see text
Kraft	11–13	Na^+, (Na_2S)	1. Unbleached—packaging materials: boxboard, sack, and bag papers 2. Semibleached—newsprint (25–35%) 3. Bleached—high quality papers of all types
Prehydrolysis kraft	11–13	Na^+, (Na_2S)	Chemical conversion (e.g., cellulose acetate)

[a]Not generally used as a full chemical pulping process, but may also be used for the production of fine papers. Details of this process are included here to demonstrate the pulping versatility of sulfite ion over the full range of pH conditions.

[b]Only at the laboratory stage at present, but see *Can. Chem. Proc.* [16].

TABLE 15.7 Ratios of Sodium Carbonate to Sulfurous Acid[a] Required for Sulfite Pulping at Various pHs

Process	Required mole ratio of		Active pulping entities present
	Na_2CO_3 :	H_2SO_3	
Acid sulfite, pH ca. 1–2	1	3 (and excess)	H_2SO_3, HSO_3^-
Bisulfite, pH ca. 3–5	1	2	HSO_3^-
NSSC[b], pH ca. 6–8	1 3	1 2	SO_3^{2-} SO_3^{2-}, HCO_3^-
NSSC, and monosulfite, pH ca. 8–10	2	1	SO_3^{2-}, $2HCO_3^-$
Alkaline sulfite[c], pH ca. 11–13	3 4	1 1	SO_3^{2-}, $2CO_3^{2-}$ SO_3^{2-}, $3CO_3^{2-}$

[a]The existence of sulfurous acid (H_2SO_3) is somewhat hypothetical, but is a convenient concept for the considerations required here. Sulfite salts are common.

[b]Neutral sulfite semichemical pulping.

[c]Experimental at present, see *Can. Chem. Proc.* [16].

15.5.6. Alkaline Pulping: the Soda Process

Soda pulping involves the pulping of chips by cooking with aqueous sodium hydroxide at 160–170°C, and was originally devised by Watt and Burgess in England in 1851. It required long cooking times and gave poor yields of a dark pulp, which still had a significant lignin content, even with hardwoods, which gave the most favorable results. Also the strength properties of soda pulps are second rate. Any "soda" mills still operating generally employ a small amount of sulfide in the pulping liquor, which is not strictly soda pulping. This modification bears more than a superficial resemblance to kraft pulping. Or a small proportion of anthraquinone is used as a redox catalyst to improve performance (e.g., [17, 18]). Thus, at present virtually all alkaline pulping, worldwide, is conducted by the kraft process.

15.5.7. Alkaline Pulping: Background of the Kraft Process

The kraft, or sulfate process is today the preeminent chemical pulping process. In Canada alone, 80% of all the chemical pulp produced is by this method, and worldwide some 85% of the total is via this route. The current prominence of this pulping procedure warrants a rather more complete discussion of the details than devoted to the other chemical pulping methods.

In this process, wood chips are cooked in a pulping liquor, which consists of a solution of sodium hydroxide and sodium sulfide in water (so-called "white liquor"). The details were originally developed by Dahl, Germany in 1879 [19]. The presence of the sodium sulfide accelerates the pulping rate so that kraft pulping requires a shorter cooking time than soda pulping. The sulfide also stabilizes the cellulose, so that the pulp suffers minimal fiber damage and hence makes a strong paper sheet. This high-strength pulp product is the origin of the name given to the process, "kraft" being German for "strength." Factors, which contribute to the greater strength of kraft pulps include the presence of sodium sulfide, the shorter cooking time, and the lower ratio of chemicals to wood required to obtaining effective pulping. Kraft pulping can also handle hardwoods or softwoods, even of resinous or pitchy species, since this liquor composition dissolves these resins. Resinous woods are more difficult to handle by any of the acidic sulfite pulping processes.

A feature of the kraft process vital to continued success is its integral, well-tested chemical recovery system. The digestion liquor for each batch of chips to be pulped is mainly obtained from the chemicals recovered from the spent liquor of previous digestions, and has approximately the composition given in Table 15.8. For kraft pulping it is usual to specify all of the components of the digestion liquor on a "Na_2O equivalent" basis. This puts all the active constituents on the same sodium ion content basis. Thus, the actual concentration of sodium hydroxide present for a 73 g/L, Na_2O equivalent is given by Eq. 15.12.

$$Na_2O + H_2O \rightarrow 2\,NaOH \qquad\qquad 15.12$$

73 g/L (Na_2O equiv.) $\div$ 62 g/mol (Na_2O) $\times$ 40 g/mol (NaOH) $\times$ 2 mol/mol

Thus, NaOH from 73 g/L Na_2O = 94.2 g/L NaOH

▮▮▮   **TABLE 15.8   Typical Analysis of White Liquor Used for Kraft Pulping**[a]

| | Concentrations, as | | |
Component	Na$_2$O equivalent (g/L)	Molarity (mol/L)	Function
NaOH	73.0	2.34	Primary pulping agent
Na$_2$S	31.4	0.51	Accelerates pulping, raises yield of cellulose and hemicellulose by protective action
Na$_2$CO$_3$	18.2	0.29	Negligible. Presence incidental as a result of chemical recovery system
Na$_2$SO$_4$, Na$_2$SO$_3$, Na$_2$S$_2$O$_3$	ca. 1.0	–	None. Traces present incidentally

[a]Data from Kirk-Othmer [4] and Britt [7]. Liquor with the composition given here will normally be diluted to a net sodium hydroxide concentration of about 50 g/L Na$_2$O.

By a similar calculation using the stoichiometry of Eq. 15.13 the actual concentration of sodium sulfide present works out to 39.5 g/L.

$$Na_2O + H_2S \rightarrow Na_2S + H_2O \qquad 15.13$$

For pulping purposes, the "effective alkali" present is higher than the actual sodium hydroxide concentration in the white liquor because of the existence of an hydrolytic equilibrium between sodium sulfide and sodium hydrogen sulfide (Eq. 15.14).

$$Na_2S + H_2O \rightarrow NaOH + NaHS \qquad 15.14$$

Thus, the effective alkali comprises the concentration of sodium hydroxide present, plus one-half of the concentration of sodium sulfide present, when both are expressed as Na$_2$O. For the typical analysis of Table 15.8, this would be about 88–89 g/L Na$_2$O. Specified in molar terms, since the molar concentration of sodium hydroxide derived from sodium sulfide is equivalent to the initial molar concentration of sodium sulfide (Eq. 15.14), the effective alkali concentration of the white liquor of Table 15.8 would be 2.85 M.

"Sulfidity" is also an important parameter of the liquor composition for kraft pulping. It refers to the concentration of sodium sulfide divided by the concentrations of sodium sulfide plus sodium hydroxide the "active alkali," as opposed to "effective alkali," all expressed in terms of their Na$_2$O equivalents. Equation 15.15 gives the method of determining sulfidity, expressed as a percentage.

$$\% \text{ Sulfidity} = \frac{[Na_2S]}{([Na_2S] + [NaOH]) \times 100} \qquad 15.15$$

The desirable range for sulfidity is 15–35%. Values lower than this tend to give incomplete pulping and weaker pulps, akin to the properties of pulp produced by the soda process. Higher values tend to give sluggish delignification, and also tend to raise potential sulfur losses in both the pulping and chemical recovery segments of kraft pulping.

15.5.8. Details of the Kraft Process

The cook is carried out in a digester, which is a steel or stainless steel cylindrical pressure vessel about 3 m in diameter and 16 m high (the height of a five-story building), fitted with external heat exchangers. For a digester this size, 50–60 tonnes of chips per batch (containing about 50% moisture) are loaded from the top, while steam is blown into the opening of the digester. Steam blowing displaces much of the air from the digester both during the loading process and on white liquor addition. It also aids in liquor penetration into the chip from its combined air displacement, warming, and wetting actions on the wood. Any knots or uncooked chips screened from a previous digestion will also be added at this stage. At the time of white liquor addition, sufficient "weak black liquor" is also added to bring the content of $NaOH + Na_2S$ present in the digestion liquor to the concentration range of about 50 g/L Na_2O equivalent, and to a liquor to wood ratio of about 16 kg of effective alkali for each 100 kg of dry wood. The heavy steel lid is bolted down and heating begun by pumping liquor from the digester, through the external steam-heated heat exchangers, and back to the digester. This takes about 1.5 hr to bring the temperature up to 170°C and 620 kPa (kilopascals, i.e., 6 atm pressure, 90 psig), after which it is held at this temperature for a further 1.5 hr to complete the digestion. Chemical to wood ratios, digestion times, and temperatures will be varied somewhat depending on the species of wood being pulped, and the degree of delignification desired for the product.

After the digestion is completed a valve at the base of the digester is opened while it is still at operating temperature and pressure, causing a very rapid discharge of the contents of the digester into a blow tank. The fast pressure drop experienced by the pulp on discharge blows the fiber bundles apart, aiding in the production of a pulp from the chips.

Cooked pulp plus weak black liquor, which is the spent white liquor which has picked up lignin and some hydrolyzed cellulose and hemicellulose, etc., proceeds to screens which serve to remove any uncooked fragments and knots from the pulp. The bulk of the weak black liquor is separated from the cooked pulp in a decker, basically a filter that captures the pulp on a screen and removes the bulk of the liquor by suction. The pulp is then rinsed in washers, units very like drum filters, equipped with two or three freshwater fed shower tubes, which run parallel to the axis of the drum and, which rinse off any residual weak black liquor from the pulp. At this stage the separated kraft process product is a fairly clean pulp, though still dark brown in color from residual lignin, etc. (this is the raw material for "kraft" paper bags and building papers). The weak black liquor filtrate contains all of the spent chemical constituents of the original white liquor plus the components of the wood, which were dissolved during the course of the pulping process.

The stages just described, from the point of loading chips into a cylindrical digestion vessel, comprise *batch* pulping of wood. To maintain a continuous stream of cooked pulp for subsequent stages requires, say, six to ten of these digesters, operating in parallel in a staggered loading sequence, with each discharging five to seven loads of 12–15 tonnes (about one-quarter of the weight of wet wood charged to each digester) of cooked pulp from each load, per 24 hr. It is also possible to conduct all the stages described to this point continuously in a single, more complex digestion unit, which performs all these functions in turn on a mass of chips and, which moves steadily through the unit. One commercially available version is the Kamyr continuous digester (Fig. 15.7). Many pulp mills use a combination of batch and continuous digesters to obtain maximum flexibility in terms of the types of wood being pulped, and to obtain the desired degree of pulp delignification required for the end use objectives of the pulp.

15.5.9. Kraft Cyclic Chemical Recovery

There are three objectives of chemical recovery from the weak black liquor separated from the pulp after kraft pulping. Recovery and regeneration of most of the chemicals employed in the original digestion liquor are primary objectives. A further objective is to generate a significant fraction of the steam requirement of the pulp mill from combustion of the dissolved lignin residues and other organic constituents (comprising nearly half the original dry weight of the wood) contained in the weak black liquor. Burning the organic con-

FIGURE 15.7 Schematic of the operating details of a Kamyr continuous digester.

stituents in the spent liquor also accomplishes the third objective, which is to avoid of the significant polluting potential of untreated discharge of spent pulping liquor. An alternative to recovery, treatment of the spent liquor by artificial aeration, would be a substantial net cost to the process rather than a net chemicals and energy credit.

Initially the weak black liquor is oxidized, either by sparging air through a tank of the liquor, or by causing thin films of black liquor to fall through an air chamber. The objective of this step is to decrease the volatility and losses of foul-smelling volatile sulfur compounds, which would otherwise occur during subsequent evaporation or combustion steps (Eqs. 15.16–15.18).
Volatile sulfide loss:

$$H_2O + CO_2 \text{ (from combustion)} \rightarrow H_2CO_3 \qquad 15.16$$

$$Na_2S + H_2CO_3 \rightarrow Na_2CO_3 + H_2S \uparrow \text{ (b.p.} -60°C) \qquad 15.17$$

$$2\,NaSCH_3 + H_2CO_3 \rightarrow Na_2CO_3 + \underset{\text{methyl mercaptan}}{CH_3SH} \text{ (b.p. }6°C) \qquad 15.18$$

By oxidizing sodium sulfide to sodium thiosulfate and methyl mercaptan (methanethiol) to dimethyldisulfide, the first component is made virtually involatile, and the volatility of the second is greatly decreased (Eqs. 15.19 and 15.20).

Black liquor oxidation:

$$2\,Na_2S + 2\,O_2 + H_2O \rightarrow Na_2S_2O_3 + 2\,NaOH \text{ (complex process)} \quad 15.19$$
$$\Delta H = -900 \text{ kJ } (-215 \text{ kcal})$$

$$2\,CH_3SNa + 1/2\,O_2 + H_2O \rightarrow \underset{\text{(b.p. }116°C)}{CH_3SSCH_3} + 2\,NaOH \qquad 15.20$$

After oxidation of the weak black liquor, of about 15–18% total solids content by weight, it is concentrated. Concentration before combustion is necessary to increase the energy recovery from combustion of the organic constituents. Otherwise a large proportion of the energy derived from combustion would be consumed to evaporate the high-water content. Evaporation to 50–55% solids is usually carried out in three to six stages of multiple-effect evaporators. By operating under reduced pressure, and by using the steam obtained from the first and each of the subsequent stages to heat the liquor of the following stage, 5–5.5 kg of water removal can be achieved for each kilogram of steam used for heating the first stage in a sextuple-effect evaporator.

The last stage of black liquor evaporation, from 50 to 55% solids to about 65–70% solids, is by direct contact of the black liquor with hot flue gases from the recovery boiler (Fig. 15.8). Increased surface area for evaporation and increased evaporation rates are obtained by slowly rotating parallel disks, which are half immersed in the liquor and half exposed to the hot flue gases. In this manner direct contact evaporation also accomplishes a moderate scrubbing effect on the flue gases, at the same time consuming the bulk of any residual sodium hydroxide (Eqs. 15.21 and 15.22).

$$2\,NaOH + CO_2 \rightarrow Na_2CO_3 + H_2O \qquad 15.21$$

$$2\,NaOH + SO_2 \rightarrow Na_2SO_3 + H_2O \qquad 15.22$$

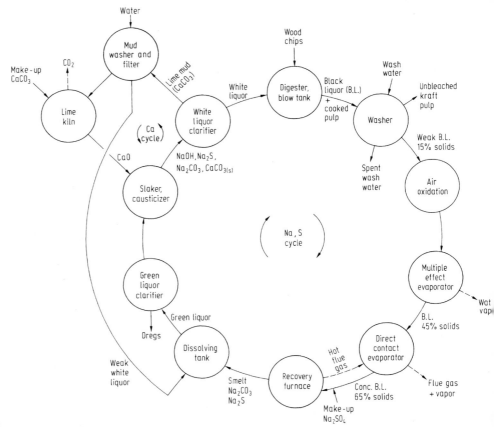

FIGURE 15.8 Cyclic process for recovery of chemicals and energy from the spent pulping liquor the kraft process.

This process also transfers heat from the flue gases to the black liquor so that it leaves the last stage of evaporation hot. At this point, it is referred to as strong black liquor. Addition of make-up chemicals, either sodium sulfate ("salt cake") or sulfur as necessary is carried out at this stage to replace any losses of sodium and sulfur occurring elsewhere in the process. In fact, "sulfate process" is synonymous with "kraft process" because the primary make-up chemical requirement of the kraft process is sodium sulfate, although sodium sulfate is not an active constituent of kraft pulping liquor.

The objectives of black liquor combustion in the furnace of the recovery boiler are complex. These include combustion of the organic components present in the strong black liquor to generate heat and produce steam, and to recover the inorganic chemicals originally present in the black liquor, in *reduced* form (i.e., as sodium sulfide rather than sodium sulfate). Preheated (about 120°C) strong black liquor is first sprayed into the very hot, drying zone of the recovery furnace (Fig. 15.9). The residual water in the black liquor instantaneously flashes off, producing a char, part of which adheres to the walls of the furnace. As the thickness of the layers of carbonized char on the walls builds up, pieces fall off and land on the hearth. At the hearth,

FIGURE 15.9 Diagram of the main components of a kraft recovery boiler, for strong black liquor combustion. (Reprinted from *Canadian Chemical Processing*, November 1970, courtesy of Southam Business Publications Ltd.)

conditions are chemically reducing, maintained by providing inadequate air for total combustion in this zone, and temperatures of about 1,000°C prevail. These temperatures provide the energy required for the endothermic reducing reactions. Here, sodium sulfate and sodium thiosulfate, produced at the black liquor oxidation stage, are reduced to sodium sulfide, and any remaining sodium hydroxide is converted to sodium carbonate (Eqs. 15.23–15.28).

Examples of recovery furnace reducing zone reactions:

$$2\,C + O_2 \rightarrow 2\,CO \qquad\qquad 15.23$$

$$C + H_2O \rightarrow CO + H_2 \qquad\qquad 15.24$$

$$Na_2SO_4 + 4\,CO \rightarrow Na_2S + 4\,CO_2 \qquad\qquad 15.25$$

$$Na_2SO_4 + 2\,C \rightarrow Na_2S + 2\,CO_2 \qquad\qquad 15.26$$

$$Na_2S_2O_3 + 3\,CO + 2\,NaOH \rightarrow 2\,Na_2S + 4\,CO_2 + H_2O \qquad 15.27$$

$$Na_2CO_3 \rightarrow Na_2O + CO_2 \text{ (negligible, Chap.7)} \qquad 15.28$$

The hearth is sloping so that as the sodium sulfide and sodium carbonate (melting points 950 and 851°C) accumulate in the char on the hearth the mixture, called smelt (literally a molten rock), trickles out. The smelt is immediately dissolved in lime mud wash water for the first stage in fresh liquor preparation. Dissolving of the molten smelt in water is a noisy procedure from the large temperature differential between the smelt and the water, and from the exothermic nature of the solution process as well.

At the upper parts of the recovery furnace secondary air is admitted to obtain an oxidizing atmosphere, and the temperatures are still sufficiently high that any residual reduced compounds present in the gases are oxidized (Eqs. 15.29–15.32).

Recovery furnace oxidizing zone reactions:

$$2\ CO\ +\ O_2\ \rightarrow\ 2\ CO_2 \tag{15.29}$$

$$2\ H_2\ +\ O_2\ \rightarrow\ 2\ H_2O\ \text{(minor)} \tag{15.30}$$

$$2\ H_2S\ +\ 3\ O_2\ \rightarrow\ 2\ SO_2\ +\ 2\ H_2O\ \text{(traces)} \tag{15.31}$$

$$CH_3SH\ +\ 3\ O_2\ \rightarrow\ SO_2\ +\ CO_2\ +\ 2\ H_2O\ \text{(traces)} \tag{15.32}$$

As a result, gas temperatures rise to the neighborhood of 1250°C in this region of the furnace, from which heat is recovered via steam superheating tubes and water tubes to generate high- and low-pressure steam. After heat recovery the flue gases are first passed through the direct contact evaporator, for additional heat utilization in the last stage of concentrating black liquor, and through electrostatic precipitators, for particulate removal, before delivery to the stack for discharge. The electrostatic precipitator catch is usually returned to the black liquor stream for recycling.

When smelt is dissolved in lime mud wash water the resulting solution is referred to as "green liquor," from the color imparted to it from suspended inorganic insoluble matter, or "dregs," which are present. The dregs, which consist mostly of carbon plus other inorganic insolubles such as iron, are settled out to leave a supernatant clarified green liquor, which is essentially a solution of sodium carbonate and sodium sulfide in water.

For white liquor preparation, clarified green liquor is fed to a causticizer, which forms sodium hydroxide on the addition of slaked lime, leaving the sodium sulfide unchanged (Eq. 15.33).

$$Na_2CO_3\ +\ Ca(OH)_2\ +\ Na_2S\ \rightarrow\ 2\ NaOH\ +\ CaCO_3\downarrow\ +\ Na_2S \tag{15.33}$$

The insoluble calcium carbonate (lime mud) is settled out from this mixture, filtered, and rinsed with water, which is then subsequently used for dissolving smelt. The supernatant liquor is now a solution of sodium hydroxide, sodium sulfide, and a small amount of unreacted sodium carbonate of nearly the correct composition required for a new batch of white liquor. Calcium carbonate in the lime mud is recycled by first calcining to decarbonate, followed by slaking to regenerate the calcium hydroxide required for a subsequent causticization (Eqs. 15.34 and 15.35).

$$CaCO_3\ \xrightarrow{1,200°C}\ CaO\ +\ CO_2\uparrow \tag{15.34}$$

$$CaO\ +\ H_2O\ \rightarrow\ Ca(OH)_2 \tag{15.35}$$

Several important by-products are recovered from kraft pulping operations, among them kraft turpentine, tall oil, and dimethyl sulfoxide. Kraft turpentine is condensed from digester relief gas while pulping coniferous species. Pine, for example, yields about 10 kg/tonne of pulp. Kraft turpentine contains a number of diterpenes such as α- and β-pinene, camphene, dipentene, etc., similar to the composition of the turpentine obtained from the dry distillation of pine wood [20] (e.g., the naval stores industry). Crude kraft turpentine also contains water, volatile acids, and reduced sulfides such as methyl mercaptan and hydrogen sulfide in addition to the terpenes. By careful washing and fractional distillation of this crude product several useful terpenes are recovered for use as constituents in fragrances, and as precursors for drug synthesis.

Tall oil, derived from the Swedish "tallolja" meaning pine oil, is recovered from the black liquor of softwood pulping. It is taken out at an intermediate stage of the multiple-effect evaporation when the liquor contains about 30% total solids, after it is allowed to stand [21]. The soaps (sodium salts of fatty acids present) are insoluble, cream to the top of the vessel, and are skimmed off. The residual black liquor is returned to the evaporators to continue chemical recovery. The soap yield, which can range from 10 to 200 kg/tonne of pulp (or even higher for pine), is then acidified and the free fatty acids and resin acids obtained are separated by distillation. The fatty acids recovered consist mainly of oleic and linoleic acids and are employed in soap manufacture and as the drying oil components of paints and varnishes [22] (Chap. 19). Resin acids consist of terpene acids such as abietic acid and its positional and reductive variants, and are mainly employed in paper sizing.

Dimethyl sulfoxide $(CH_3)_2SO$; b.p. $189°C$) is made from dimethyl sulfide recovered by condensation from kraft digester relief gases. This useful polar, and water-miscible industrial solvent is produced by air oxidation of the sulfide in the presence of catalytic concentrations of nitrogen oxides.

15.6. BLEACHING OF WOOD PULPS

Mechanical and chemical pulps are normally more-or-less brown as initially obtained. The color is due in part to residual lignin or chemically modified lignin and miscellaneous color bodies present in the pulp. The objective of bleaching is to obtain the maximum increase in brightness possible consistent with the pulp end use, with a minimum yield loss and bleaching cost. Brightness is a term applied to the whiteness of a pulp, determined from a small paper sample or "handsheet" prepared manually from a pulp sample. It is measured in a spectrophotometer at an angle of $45°$ to the incident beam of blue light of 457 nm. It is based on an arbitrary scale of 8 for the brightness of carbon black, and 100 for the brightness of pure magnesium oxide. On this basis, brightnesses of about 55 to 65 are common for unbleached mechanical pulps, which may be raised to the 65–75 range by bleaching. Unbleached sulfite pulps will have a brightness of 60–70, and kraft pulps of 40–45. Both may be bleached to brightnesses in the 80–85 range, the brightness of ordinary bond papers, with the appropriate number of bleaching stages. Very high brightnesses of 85–90 are achievable with chemical pulps using one or two additional bleaching stages. Since the bleaching procedures for the high-lignin content mechanical and chemimechanical pulps differ from those used with chemical pulps, which have a low-lignin content, the two groups are discussed separately. Neutral sulfite semichemical pulps are not usually bleached, since their uses do not require this.

15.6.1. Bleaching of Mechanical and Chemimechanical Pulps

Mechanical pulps, since they comprise the whole wood changed into a pulp suspension in water, have approximately the same lignin content as the wood from which they were derived and an as-pulped brightness usually in the

50–65 range, depending on species pulped and process. Either a chemically oxidizing or a chemically reducing system, or sometimes both, is used with these pulps to gain a 10–15 point improvement in brightness. The bleaching methods used are selected to minimize attack of lignin, since the objective of these is to decolorize lignin without significant solubilization in order to maximize the yield of bleached pulp obtained. Brightening is the term used to reflect these objectives.

Oxidizing systems are based on peroxide, and may use solutions of either sodium peroxide or hydrogen peroxide in water, usually with added disodium ethylenediaminetetraacetic acid (Na_2 EDTA) to suppress metal ion-catalyzed spontaneous decomposition of the peroxide. Capital P, is the symbol used to designate a peroxide stage in bleaching. The active decolorizing agent is probably hydroperoxide anion, HOO^-, formed by an equilibrium between hydrogen peroxide and sodium hydroxide (Eqs. 15.36 and 15.37). It is also possible that peracetate is involved [23].

$$H_2O_2 + NaOH \rightarrow HOO^- + Na^+ + H_2O \qquad 15.36$$

$$H_2O + Na_2O_2 \rightarrow HOO^- + 2\,Na^+ + OH^- \qquad 15.37$$

Peroxide-induced lignin decolorization is achieved by hydroperoxide ion attack of the carbonyl carbon of the quinones and ketones, which form a part of long conjugated chains (and contribute to the color) in the lignin. In this way the conjugation is interrupted, which reduces the intensity and wavelength of the contributing color [24] (e.g., Eqs. 15.38 and 15.39).

By this means peripheral, color-inducing functionalities are oxidized and decolorized over a bleaching period of 2–3 hr while the bulk of the lignin macromolecules remain intact (insoluble). Thus a brightness improvement of 8–10 points is possible while the yield loss is kept low. Unfortunately peroxide-based bleaching systems tend to be expensive relative to other alternatives for groundwoods, and hence this method is focused on pulps destined for tissue papers, paper napkins, and some specialty papers, rather than for newsprint applications. However, the increased resistance to color reversion of peroxide-bleached pulps tends to offset the increased cost of this step.

Sodium bisulfite ($NaHSO_3$) was once used for reductive bleaching of mechanical pulps, but its action is slow and ineffective, so, its use has been discontinued. Present reductive bleach systems for groundwoods employ either zinc or sodium hydrosulfite (ZnS_2O_4; or $Na_2S_2O_4$) dissolved in water.

Zinc hydrosulfite, developed in the mid-1950s, was inexpensively made at the mill site by reacting powdered zinc with common pulp mill chemicals (Eq. 15.40).

$$Zn + 2 H_2SO_3 \rightarrow ZnS_2O_4 + 2 H_2O \qquad 15.40$$

This powerful reducing agent produces brightness improvements of 10–12 points, if care is taken to exclude oxygen. Any oxygen present wastes a part of the active reducing capacity from the very rapid (probably only diffusion limited) reaction of zinc hydrosulfite with oxygen (Eq. 15.41).

$$2 ZnS_2O_4 + 2 O_2 + 2 H_2O \rightarrow Zn(HSO_4)_2 + Zn(HSO_3)_2 \qquad 15.41$$

However, the disposal of spent groundwood bleach liquor from zinc hydrosulfite has posed zinc toxicity problems for shellfish, so that mills have now switched to sodium hydrosulfite. This is equally effective (although more expensive) and avoids this problem. Sodium hydrosulfite is made at the mill from purchased sodium borohydride (Eq. 15.42).

$$NaBH_4 + 8 NaHSO_3 \rightarrow NaBO_2 + 4 Na_2S_2O_4 + 6 H_2O \qquad 15.42$$

If the two groundwood bleaching methods are to be used in sequence, a sulfur dioxide solution will be used as a stop bath for the pulp, after the peroxide stage, to ensure that any residual oxidants are kept out of the hydrosulfite stage. Combination bleaching will usually be conducted in this order, not the reverse, to obtain optimum results. Brightness improvements of 16–18 points are possible, better than with either bleach system alone but not quite an additive total. The residual lignin present in these pulps makes the brightness unstable, which may be observed by the significant darkening of a newspaper if it is left exposed to strong sunlight for an afternoon.

15.6.2. Bleaching of Chemical Pulps

High brightness papers are produced from chemical pulps, which are cooked under conditions chosen to leave a low-residual lignin content in the pulp. Bleaching of these pulps involves selective solubilization and washing out of lignin in the early stages, with some yield loss. This is followed by decolorization of any lignin residues present in the cellulose fiber in later stages, which results in a much smaller yield loss. Thus, bleaching of chemical pulps results in a yield loss proportional to the number of bleaching stages.

The ease of bleaching of a chemical pulp is related to its initial brightness (color) and its lignin content. To establish this, a permanganate number (also kappa, or K-number) test, specified as the number of milliliters of 0.10 N potassium permanganate, which is decolorized by 1 g of dry pulp at 25°C in 5 min, is run on the pulp to determine its bleachability. Easy bleaching pulps give permanganate numbers of 6–10 or less, while pulps which give permanganate numbers of 20 or more are generally classified as unbleachable.

The chlorination, or C stage, of bleaching is conducted at a pH of about 2 by a solution of chlorine in water. This relatively rapidly chlorinates and demethoxylates lignin via the action of both dissolved chlorine and of undissociated hypochlorous acid [25] (Eq. 15.43).

$$Cl_2 + H_2O \rightleftharpoons HCl + HOCl \qquad\qquad 15.43$$

A slower oxidation of lignin also occurs, which, in combination with the first two processes, yields more soluble, lower molecular weight fragments of lignin, which later can be washed out of the pulp. Chlorination thus removes much of the lignin, though the pulp does not appear to be substantially brighter at this stage. Cellulose is not chlorinated under these conditions, although undissociated hypochlorous acid causes some oxidative degradation.

A caustic extraction, E stage, uses a solution of about 0.5–0.7% sodium hydroxide in water at 0–60°C to remove solubilized lignin produced by either the C or H (hypochlorite) stage. The high-pH conditions convert any phenolic hydroxyls to their sodium salts, raising their solubility, and converts any residual fatty acids and rosin acids present to salts, aiding in their removal by washing. Caustic extraction also removes some hemicellulose even though this component can be useful for papermaking. Hemicellulose losses are minimized by keeping the caustic solutions dilute. Use of 1–5 M (4–20%) *cold* aqueous sodium hydroxide can maximize hemicellulose removal when required, such as for the preparation of dissolving pulps.

Hypochlorite, H stage, bleaching employs a solution of either calcium or sodium hypochlorite in water, made at the pulp mill by adding chlorine to a suspension of lime or sodium hydroxide in water (Eqs. 15.44 and 15.45).

$$2\ Ca(OH)_2 + 2\ Cl_2 \rightarrow Ca(OCl)_2 + CaCl_2 + 2\ H_2O \qquad 15.44$$

$$2\ NaOH + Cl_2 \rightarrow NaOCl + NaCl + H_2O \qquad 15.45$$

Excess base is used to ensure that the pH is kept at or below 12, to ensure that hypochlorite, OCl^- (and not hypochlorous acid, HOCl), is the bleach active species (see Fig. 15.10), and to stabilize the hypochlorite ion for this purpose. Both the C and H, stages are designed to operate at pHs well clear of neutral to minimize the concentration of undissociated hypochlorous acid and attendant cellulose degradation. Hypochlorite ion is a less powerful oxidant than chlorine and hence more suitable for later bleach stages. Hypochlorite oxidizes and dissolves the chlorinated residues from the chlorination stage, the so-called

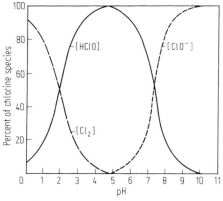

FIGURE 15.10 Variation with pH of proportions of chlorine, hypochlorous acid, and hypochlorite ion in 0.05 M chlorine water at 25°C. (Data from Nevell [25].)

"refractory material" remaining after the E stage. Some cellulose degradation is also experienced here.

A solution of chlorine dioxide, ClO_2, 12 g/L in water is the active agent in a D bleaching stage. Good bleaching action is obtained with less loss in pulp viscosity (i.e., less cellulose degradation) than from the other procedures just described. Since solutions of chlorine dioxide are unstable this is made as needed, right at the mill site, by treating an aqueous solution of sodium chlorate with sulfur dioxide (Eq. 15.46).

$$2\ NaClO_3\ +\ SO_2\ \rightarrow\ 2\ ClO_2\ +\ Na_2SO_4 \qquad 15.46$$

The by-product sodium sulfate from this process can provide the pulping chemical make-up for a kraft pulp mill. Chlorine dioxide is a milder oxidizing agent than hypochlorite and may be employed either on its own, as a separate bleach stage, or in combination with chlorine to replace a part of the stronger acting chlorine [26]. Chlorine dioxide may be economically produced at the same time as chlorine by employing sodium chloride or hydrochloric acid in the generators (Eqs. 15.47 and 15.48).

$$2\ H_2SO_4\ +\ 2\ NaClO_3\ +\ 2\ NaCl\ \rightarrow\ 2\ ClO_2\ +\ Cl_2\ +\ 2\ Na_2SO_4\ +\ 2\ H_2O \quad 15.47$$

$$4\ HCl\ +\ 2\ NaClO_3\ \rightarrow\ 2\ ClO_2\ +\ Cl_2\ +\ 2\ NaCl\ (recycled)\ +\ 2\ H_2O \quad 15.48$$

Combination usage, as C_D (excess chlorine) or D_C (excess chlorine dioxide) stages, gives pulps of higher strength than obtained using chlorine alone in the first stage. These modifications also produce much less chlorinated material in the aqueous effluent and greatly facilitate chemical recovery of bleach plant effluent, if desired [27].

A peroxide stage, P stage, is occasionally used as a polishing bleach for chemical pulps, using the conditions discussed. Peroxide has also been tested for initial bleaching of softwood kraft pulps after an acid pretreatment. A significant reduction in waste loads is obtained but at high cost [28].

Oxygen bleaching, an O stage, has become a part of the bleach sequence of many pulp mills. Delignification and decolorization are effected by treating the chemical pulp at medium-to-high consistencies (10–15%, up to 30% pulp by weight) in a dilute solution of sodium hydroxide, with up to 5 atm pressure of oxygen [29]. Addition of a small amount of magnesium or manganese salt stabilizes the cellulose against severe degradation. Strong incentives for use of an oxygen bleaching step are that it can replace a C E, or C E H combination, and it produces a decreased waste stream of lower toxicity. It does, however, require investment in rather different, pressurized bleach plant equipment to put into practice.

Easy bleaching (low permanganate number) sulfite pulps can be bleached to a brightness of 80, or so with a single H stage, and to 85 or higher with a C E H sequence. Three to five bleaching stages (e.g., C E H, C E H D, and C E H E D sequences) are required for the sulfite pulps that are harder to bleach.

To contrast with sulfite pulps, dark kraft pulps can only be bleached to the 40–50 brightness range with a single hypochlorite stage. For semibleached grades of brightness, the 70–75 range, at least three or four bleach stages (e.g., C E D, C E H D) were required. A typical progression of pulp, chemicals, and water for a semibleached kraft pulp is given in Fig. 15.11. To obtain high or

FIGURE 15.11 Progression through a typical C E H bleach sequence starting with 500 B.D. (bone dry) tonnes of unbleached kraft pulp.

"super" brightness kraft pulps of 85–90 brightness levels, five to seven stages of bleaching were required. Sequences such as C E D E D, C E H D E D, C E H E D P, and C E H E D E D P were commonly used, which required a period of 18–24 hr for passage of the pulp through the bleach plant.

The heavy dependence on chlorine required by the bleach sequences used in the 1980s and earlier that have just been described were found to give very high BOD aqueous waste streams. They also contained a number of chlorinated organics, including phenols, many of which were persistent and quite toxic to aquatic life [30, 31]. Removal methods or waste treatment to destroy, these compounds were difficult to operate and expensive. As a result, since the 1990s there has been a pronounced move away from the use of chlorine towards the application of other more benign agents for bleaching of kraft pulps [32, 33]. The substitutions adopted replaced chlorine in water and hypochlorite-based bleaching systems by options such as variants of the O stage previously described [34], and have led to an intensive examination of the limits to the parameters, which control this process [35–37]. These variants have collectively become known as Elemental Chlorine Free (ECF) processes (Table 15.9). Other variations eliminated the use of chlorine in all forms, including chlorine dioxide, and became known as Totally Chlorine

▮▮ **TABLE 15.9 A Typical Kraft Bleach Sequence of the 1990s to Decrease Chlorinated Organics and Oxygen Demand of Wastewaters**[a,b]

Symbol	Bleach step details
W_1	Open stage water wash, effluent sewered
D_0	First chlorine dioxide stage, acidic. ("D_1" already used, hence D_0). Typically 1.8–2% ClO_2 % by wt. on "bone dry" pulp, about half the total used.
W_2	Using filtrates from other bleach stages (e.g., W_4 and W_5).
E_{op}	Caustic extraction of "chlorinated organics" after ClO_2 treatment. Reinforced with high-pressure oxygen and hydrogen peroxide in a 6–8 min, 20–25 psi upflow preretention tube
W_3	As before (e.g., W_4 first, then W_5)
D_1	Second ClO_2 stage, acidic. Typically 1–1.5% applied, based on pulp lignin content, for a final brightness of 78–80% ISO.
W_4	As before (e.g., W_6)
E_{2P}	Second extraction, alkaline. Also reinforced with peroxide for additional brightness to 82–85% ISO
W_5	e.g., hot water first, then clean "white water"
D_2	An optional third ClO_2 stage, acidic. About 0.25% applied, to take brightness to 88–89% ISO
W_6	Clean "white water"

[a]Personal communication Flater [38].

[b]Symbol sequence would be $W\ D_0\ E_{op}\ D_1\ E_p\ D_2$. Late stage wash filtrates are reused for earlier wash stages to minimize water usage. Selection of wash water used takes recognition of the alternation of pH with bleach stages, so called "jump stage" countercurrent washing.

Free (TCF) processes. The ECF processes substantially eliminated the formation of chlorinated organics in kraft bleach effluent. Both ECF and TCF processes also reduce the BOD and COD of bleach effluent significantly, since the same process, which oxidized the lignin fragments in the pulp also produced largely oxidized products from these. This gave a lower overall oxygen demand from the effluent, which made it much less expensive to treat.

15.7. MARKET PULP AND PAPERMAKING

Bleached or unbleached chemical pulps may progress along either of two paths. They may be fed as a dilute water suspension onto a fairly coarse wire mesh screen to form a thick pulp mat, which is then dried to about 10% moisture (roughly in equilibrium with atmospheric moisture). The mat is then cut into sheets of about 0.6 × 1.2 m, wrapped, and baled for shipment as 232 kg (500 lb) units of market pulp. Or they may be fed in diluted form directly onto the Fourdrinier wire of a paper machine, with the intention of producing directly one of the wide variety of trade papers. The sheets produced for market pulp are deliberately made thick and loosely consolidated to facilitate repulping by stirring with water at the destination point, for eventual papermaking. Mechanical pulps are much less color stable and cannot be dried or repulped without a loss in sheet properties. For these reasons they are nearly always converted to the finished product, newsprint or paper "boards" of various types, at the pulping site.

With any pulp source, eventual papermaking requires the appropriate single or a blend of pulps to be prepared into a furnish, which is a dilute pulp suspension (about 2%) in water ready to be fed to a paper machine. Before papermaking, chemical pulps may be beaten (mechanically worked in units similar to refiners) to a varying extent depending on the properties desired in the finished sheet. A typical furnish composition for a newsprint might be a blend of 60–70% groundwood (or mechanical) pulp, and 30–40% sulfite, semibleached, or bleached kraft pulps to provide strength. Bag papers and board, such as linerboard for corrugated board manufacture would normally use 100% unbleached kraft pulp, hence the term "kraft paper." Board destined for use as the flutes in corrugating medium would normally use a higher yield stiff pulp fiber such as that produced by the NSSC process. Bond or book papers would use fully bleached sulfite, rag, or kraft pulps, or blends of these.

The furnish is fed to the headbox of a paper machine (Fig. 15.12), about the length of a city block. Sufficient additional fresh water (white water) is added to bring the pulp concentration in the headbox down to about 0.5%. White water, the water removed from the fibers as the paper sheet is formed on the machine, is continuously recycled. This low-concentration of pulp in water is necessary to obtain an even distribution of fiber (a smooth "formation") as the sheet is formed. The diluted furnish is fed through an adjustable horizontal slot (the slice) in the headbox onto an endless wire screen (the Fourdrinier wire), which moves rapidly away from the headbox. Direct drain-

FIGURE 15.12 Principal components of a Fourdrinier paper machine. (From Kirk-Othmer [4], courtesy of Beloit Corporation.)

age plus suction boxes draw off more than 90% of the water while the sheet is still on the wire, enough that the delicate 20% solids sheet may be passed on to the press section of the paper machine, where the sheet is now supported on felts. By pressing the damp sheet into the drying felts the pulp content is nearly doubled, giving the sheet enough strength to carry it through the drying section. Here, 30–50 steam heated rolls rapidly evaporate the remaining water to give a sheet containing only 5–8% moisture as it emerges from the dryer. The last function of the paper machine carried out on the still warm paper sheet is to smooth it, performed as the sheet passes through the calender stack. Here, heavy smooth steel rollers decrease the surface roughness and eliminate as much as possible any residual rough, and smooth "sidedness" of the sheet as it formed on the Fourdrinier wire.

Any adjustment of the final paper roll width, from the 7 to 10 m width produced by the typical paper machine, is carried out as a separate step on a slitter-rewinder, using the 20 tonne rolls of paper produced by the paper machine. The final step is a heavy paper wrapping, which is applied to protect the contents from physical damage during shipment and to control the moisture content of the product.

Paper is an extraordinarily interesting commodity because of the wide variations possible in many of its properties. While superficially it appears to be uniform in two dimensions, it is actually significantly anisotropic. Even the best papers are usually sided, that is they have a top side, which is smoother and a wire side, which is rougher. An exception is papers produced on twin wire machines from which both sides of the sheet will be slightly rough. A newsprint will usually have a higher proportion of chemical fibers on the wire side than the top side of the sheet. The sheet will have a machine direction and cross direction to it, an anisotropy feature, which is under some control by the paper machine operator. More of the fiber will be aligned closer to the machine direction than the cross direction, by slightly stretching the damp sheet as it is dried. This fiber alignment gives tensile and folding strengths to the sheet, which are greater in the machine direction than in the cross direction. This property is also important in printing, for instance, since books lie better for use with the pages having the machine direction up and down the page rather than across the page. There are many other properties under the control of the papermaker such as density, thickness, porosity, smoothness, stiffness, strength, etc., as well as specialized measurements, which are used to quantify each of these properties, which are beyond the scope of the present discussion.

15.8. PULPING EMISSION CONTROL MEASURES

Each of the pulping methods—mechanical, chemimechanical (presoak), semi-chemical, and the kraft and sulfite chemical processes—produce different types of wastes, so that emission control measures for each should be considered separately. Since the mechanical and kraft chemically based pulping methods dominate the world markets (Table 15.1), the discussion will focus on these two processes. Resource use and pollution control problems of the pulp and paper industry have been surveyed [39], and the details compared with those of competitive materials [40, 41].

Wastes generated from wood preparation for pulping do not generally pose a major problem. Hydraulic debarker effluent amounting to some $14\,m^3$/tonne (3,700 U.S. gal/tonne) is relatively clean, containing only about 10 ppm suspended solids as wood fines and grit and about 5 mg/L BOD. Primary clarification is required to decrease the suspended solids, while the BOD level is regarded as inconsequential. Toxic rosin acids flushed from the bark/wood interface of some species are handled by dilution with other mill effluents. Many mills use drum, or in-line dry debarking systems, particularly when there is a limited water supply. The bark itself is normally burned in a bark-fueled (hog fuel) boiler with energy recovery as steam.

15.8.1. Effluents of Mechanical Pulping

The nature of mechanical pulping processes means that they do not directly present any air emission problems. Problems are, however, possible at the point of electric power generation, if this is from combustion sources. Pulp mills often use on-site thermal power generation. Control of these potential emission sources is discussed later.

There is, however, a large water requirement, estimated to be about $21\,m^3$/tonne (5,500 U.S. gal/tonne) of newsprint. This is used for log or block transport, wet grinding operations, and sheet preparation on the paper machine itself. The types and approximate levels of impurities present in the wastewater stream are given in Table 15.10.

The philosophy of pulp mill emission control is to recover as much fiber as possible before employing other treatment measures. Recovery is obtained by sedimentation and/or flotation methods. Sedimentation may be conducted in a pond, which is periodically drained and the settled material removed (as a semibatch process). Or a clarifier (which operates in a similar manner to a thickener) may be used for continuous operation. The sediment collected contains 90–95% water plus fiber and grit solids. For disposal the waste is dewatered to 50–60% solids in a filter or centrifuge and then burned, which accomplishes both disposal and can permit energy recovery.

TABLE 15.10 Typical Characteristics of the Wastewater from Groundwood Pulping[a]

Component	Value (ppm)
Total solids	1160
Suspended solids	600
Ash of above	60
Dissolved solids	560
Ash of above	240
BOD, 5 day	250

[a]Data selected from Nemerov [39].

Flotation methods allow recovery of fibers from water, which are raised to the surface on bubbles as a thick mat, which forms on the top of an aeration cell. This mat, which is virtually free of grit, is recycled to the pulp working stream. Slime growth in the aeration cell is discouraged by strong alternation of conditions such as pH, temperature, or chlorination level. Organomercury slime control agents were once used for this purpose.

After grit removal and fiber recovery, wastewater streams from mechanical pulping are treated in aeration lagoons to decrease the BOD. A 40–75% decrease of the BOD in 4 days and a 90% or better decrease in 7 days can be achieved using a surface to volume ratio of about 1 ha (1–2 acres) per million gallons per day of wastewater flow [42]. Floating mechanical aerators, rather than activated sludge treatment, are the optimum solution for BOD reduction. The large volume and good mixing provided by this system are tolerant of the shock loadings, and phosphate and nitrogen nutrient requirements are less than for alternatives. Activated sludge treatment has been examined for thermomechanical pulping effluents [43].

15.8.2. Kraft Pulping Air Pollution Control

Air pollution problems center around control of the formation and discharge of reduced sulfur compounds, which cause severe odor problems (Table 15.11), and the loss of particulates. Hydrogen sulfide and methyl mercaptan are more readily lost from aqueous solution because of their low-boiling points and high Henry's law constants. Hydrogen sulfide, however, is relatively easily dispersed in air, because of its more similar density, whereas the others tend to hug the ground and settle in depressions.

The reduced sulfur compounds, methyl mercaptan and dimethylsulfide, arise from the action of hydrosulfide and methyl sulfide anions on the methoxyl groups present in the lignin of wood (Eqs. 15.49–15.51).

▮▮▮ TABLE 15.11 Reduced Sulfides Contributing to the Odor Control Problems in Kraft Pulping[a]

Compound	Boiling point (°C)	Henry's law constant at 38°C[b]	Vapor density rel. to air, as 1.0	Odor threshold (ppb)	Industrial hygiene standard (ppm)[c]
Hydrogen sulfide, H_2S	−62	700	1.2	5	10
Methyl mercaptan, CH_3SH (methanethiol)	6	180	1.7	5	0.5
Dimethylsulfide, CH_3SCH_3	38	50	2.2	ca. 10	–
Dimethyldisulfide, CH_3SSCH_3	116	30	3.3	ca. 100	–

[a]Compiled from data of Sarkanen [44], Karnovski [45], and Smet [46].

[b]Equilibrium mole fraction in the gas phase divided by the mole fraction in a water phase at 38°C. If base is present the equilibria of the first two compounds will affect this ratio. H_2S, $k_1 = 2.1 \times 10^{-7}$, $k_2 \leq 10-14$; CH_3SH, $k = 4.3 \times 10^{-11}$ for aqueous solutions at 100°C.

[c]Time weighted average.

$$HS^- + lignin-O-CH_3 \rightarrow lignin-O^- + CH_3SH \qquad 15.49$$

$$CH_3SH + NaOH \rightarrow CH_3S^- + Na^+ + H_2O \qquad 15.50$$

$$CH_3S^- + lignin-O-CH_3 \rightarrow lignin-O^- + CH_3SCH_3 \qquad 15.51$$

In this way, about 3% of the total sulfide loading to the digester of some 80–100 kg/air dry tonne of pulp is converted to methyl mercaptan and dimethylsulfide. High sulfidities, high-pulping temperatures, long cooks, the time, or pulping of hardwoods are all factors, which tend to produce proportionately more of these sulfides [44]. The much higher methoxyl content of hardwood lignins nearly doubles the amount of volatile sulfides generated relative to that expected from softwoods. About 6 to 8% of the residual sulfide in the liquor becomes bound to lignin, mostly in soluble but involatile forms. The residue is converted to various forms of oxidized, dissolved sulfur.

Escape of the volatile sulfides can occur when excess digester pressure is vented, which occurs at intervals during the digestion, at the blow, when the contents of the digester are released to the blow tank, and during the last stage of black liquor evaporation in the direct contact evaporator, when there is rapid passage of a gas stream past the hot concentrated liquor. Details are given in Table 15.12. Control measures focus on sulfide containment from the major loss points, the direct contact evaporator and digester relief and blow gases. Black liquor oxidation (Fig. 15.8), decreases the volatile sulfide losses on direct contact evaporation. For optimum benefit the time interval between oxidation and evaporation should be kept short [47]. Otherwise sugar-based reductants still present in the liquor can cause reformation of volatile sulfides on standing. Indirect black liquor evaporation for all stages has been advocated by some Swedish mills.

Control of digester relief and blow gases is obtained by chilling, which removes a "foul condensate." The uncondensed gas fraction is moved on to a gas accumulator, which smooths out the highly variable gas flow rates. The accumulator is vented at a steady rate through a shower of weak black liquor,

TABLE 15.12 Total Reduced Sulfide Emissions from the Chemical Recovery Operations of a Kraft Pulp Mill, in kg S per Air Dry Tonne of Pulp[a]

Source	Total reduced sulfide emissions (kg S/tonne)	
	No controls	With controls
Direct contact evaporator	7–10	0.05–1
Recovery furnace	0.1–1	0.05–1
Digester and evaporators, noncondensible gases	1.0–1.5	0.0–0.3
Black liquor oxidation	0.05–0.2	0.05–0.2
Pulp washer hoods	0.05–0.2	0.05–0.1
Dissolving tank vents	0.05–0.2	0.05–0.1
Lime kilns	0.05–1	0.05–0.1
Totals	8.3–14.1	0.03–2.8

[a]Compiled from data of Sarkanen et al. [44].

which has some reserve alkalinity and, which captures at least hydrogen sulfide and methyl mercaptan. Any residual dimethylsulfide and dimethyl-disulfide, which are not captured are destroyed by venting these into the combustion air stream of the lime kiln. Novel scrubbing techniques have been described [48].

Sulfide emissions may of course be bypassed altogether by a change of the pulping process to avoid the use of sulfide [49]. Both the old soda process and nitric acid pulping suffer from the inferior strength properties of the pulps produced. Use of anthraquinone with basically a soda pulping procedure could put this alternative on a more competitive basis, but has not been adopted [50]. So this is not a practical option.

Particulate emissions can amount to as much as 0.25 tonne/tonne of pulp produced, particularly for an older pulp mill operating without controls (Table 15.13). Electrostatic precipitators, wet scrubbers, or occasionally both can achieve particulate containment of 95% or better [49]. The precipitator catch is returned to the black liquor stream for chemical recovery from the captured inorganic chemicals. Fumes lost from the dissolving tank vent are captured by demister pads or small low-energy scrubbers, which return the collected material to the green liquor circuit. Wet scrubbers are used for lime kiln dust containment. The waste liquor is used for slaking or other recycle functions in the lime circuit (Chap. 7), helping to decrease make-up chemical requirements and water emissions.

15.8.3. Kraft Pulping Water Pollution Control

The principal wastewater producing sectors of a large fully bleached kraft mill are summarized in Table 15.14, which gives ranges of values to cover normal operating variations as well as to cover a range of ages of pulp mills. Hot wastewater streams are rarely a problem because of blending with other wastewater streams. The total volumes of water consumed are so large, $20-80 \, m^3$/tonne of unbleached kraft pulp and $100-220 \, m^3$/tonne for fully bleached kraft pulp, that the blended wastewater stream is negligibly higher in temperature than the source water.

TABLE 15.13 Particulate Emission Loads from the Chemical Recovery Operations of a Kraft Pulp Mill, per Air Dry Tonne of Pulp[a]

Potential source	Particulate composition	Particulate discharge (kg/tonne pulp)	
		No controls	**With controls**
Recovery furnace	Na_2SO_4, Na_2CO_3, $NaCl$[b], etc.	100–200	1–10
Lime kiln	CaO dust	10–20	0.4–2
Dissolving tank vent	Na_2S, Na_2CO_3, Na_2SO_4, $NaCl$[b]	3–4	0.4–1
Total, kg/tonne pulp		113–224	1.8–13

[a]Air dry tonne is defined as pulp with 10% moisture content, which is below the normal equilibrium moisture content of pulp exposed to ambient air.

[b]Salt build-up will particularly affect coastal mills, which use tidewater transport and storage of logs, particularly when good particulate emission control is in place (see text).

Kraft process effluents, which could affect surface water quality parameters, are those which are high in suspended and dissolved solids, highly acidic or alkaline, and those which have a high BOD. Other parameters of concern include intensely colored streams, some of which have a foaming tendency, and streams that are toxic to fish, all caused by dissolved constituents. Useful reviews of wastewater treatment for the pulp and paper industry have been published [52]. Keep in mind that the chemical recovery part of kraft pulping has already served to decrease potential effluent loadings of about 1,000 kg of BOD and 200 kg of dissolved inorganic chemicals per tonne of pulp, which would otherwise have to be treated. This is far greater than the total pollutant loadings of Table 15.14.

The need for treatment depends on the degree to which the water quality parameters of the effluent stream differ from those of the receiving body of water and on the total effluent volume of any stream, which does show significant differences. A low-volume stream, which shows a significant deviation from normal in one or more water quality parameters may be adequately diluted on blending with $200 \, m^3$ of process water to no longer pose a problem. However, discharge regulations are usually based on the mass of pollutant discharge per tonne of pulp, not on concentration. Also it is frequently easier to devise efficient treatment measures to improve the quality of differentiated streams than of a blended one.

Using these criteria, the main problem streams from kraft pulping are those from the hydraulic debarker (if used), the pulping and screening area, the causticization section of chemical recovery, and both bleach plant streams. Chemical recovery process wastewaters are relatively benign, as long as relief gas or evaporator foul condensates are kept separated from them [53]. Paper or pulp machine waters are quite clean, as long as specialty papers involving use of other chemicals, dyes, pigments, or fillers are not being produced.

A clarifier can be used to remove 90–95% of settleable material from waste streams containing suspended solids. The sludges obtained are dewatered by filtration or centrifugation to 40–50% solids or higher and then burned. Incombustible sludges from the lime circuit clarifiers can be filtered and recycled to the lime kiln for burning.

pH deviations from normal can be corrected by blending streams of opposing pHs as far as possible, *after* other required treatment steps. Final adjustment may be necessary through reagent addition, using 90+% sulfuric acid for correction of excess alkalinity, and lime or 50% sodium hydroxide for correction of excess acidity. An automated operating system has been described [51].

Reduction of the BOD is desirable, particularly for waste streams where the loading is 3–4 kg/tonne of pulp, or more. In concentration terms, BODs of 75–300 ppm (mg O_2/L) are common in kraft pulping effluents [39]. Biochemical waste treatment in activated sludge plants or artificially aerated waste lagoons, after adjustment of the water conditions to a pH in the 6.5–9 range, the temperature to 12–32°C, and addition of ammonium phosphate to provide nitrogen and phosphorus nutrients, accomplishes accelerated BOD reduction. Many of the bacteria, which are involved in this process have recently been isolated and identified [54]. Use of oxygen, rather than air, for

TABLE 15.14 Aqueous Effluent Sources, Characteristics, and Volumes per Tonne of Pulp (TP) from a Fully Bleached Kraft Pulp Mill[a]

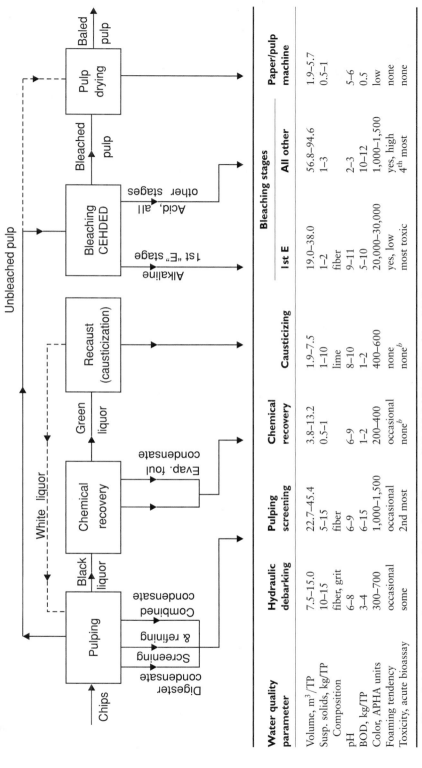

Water quality parameter	Hydraulic debarking	Pulping screening	Chemical recovery	Causticizing	Bleaching stages		Paper/pulp machine
					1st E	All other	
Volume, m³/TP	7.5–15.0	22.7–45.4	3.8–13.2	1.9–7.5	19.0–38.0	56.8–94.6	1.9–5.7
Susp. solids, kg/TP	10–15	5–15	0.5–1	1–10	1–2	1–3	0.5–1
Composition	fiber, grit	fiber		lime	fiber		
pH	6–8	6–9	6–9	8–10	9–11	2–3	5–6
BOD, kg/TP	3–4	6–15	1–2	1–2	5–10	10–12	0.5
Color, APHA units	300–700	1,000–1,500	200–400	400–600	20,000–30,000	1,000–1,500	low
Foaming tendency	occasional	occasional	occasional	none[b]	yes, low	yes, high	none
Toxicity, acute bioassay	some	2nd most	none[b]	none[b]	most toxic	4th most	none

[a]Compiled from personal contacts and data from Nemerov [39], McCubbin [42], and Lund [51].
[b]Effluents marked as "none" are only nontoxic if no condensates are added to these streams.

treatment has also been suggested (e.g., UNOX) [55]. The polishing of secondary treated pulp mill effluent by passage through a constructed wetland has been found to decrease BOD and suspended solids by a further 55 and 70% [56].

The intense brown color of the effluent from the first caustic extraction or E stage, may be dealt with in a number of ways [57]. Addition of lime $(Ca(OH)_2)$ can decrease the intensity of the color by precipitating a lime/organics complex [58]. However, the mixed organic-inorganic sludge poses disposal problems. Activated carbon has also been used for color adsorption, but carbon regeneration is difficult. It is, however, possible to operate using mill generated carbon [59]. Soil infiltration has been found to be effective [60], but its applicability depends on the climate, soil type, and land available at the mill site. Reverse osmosis has also been suggested [61].

The toxicity of effluent may be from sulfides originating from digester relief gas or black liquor evaporation, from unsaturated fatty acids released from wood during pulping [62], or from chlorinated compounds produced in bleaching steps [63]. The sulfides from foul condensates may be detoxified by air or steam stripping of the condensate, and capture of the volatiles released in a black liquor scrubber, or by combustion in the recovery boiler. Normal biodegradation is sufficient to deal with the toxicants released from wood, but chlorinated phenols are more resistant to this process [64]. The foaming tendencies of these effluents may be used to generate a head of foam in which the toxicants are concentrated, analogous to froth flotation for mineral separations. Separation of the foam removes these toxic compounds [65]. Use of lime addition for color removal also decreases toxicities via adsorption of these compounds onto the sludges as they separate. Toxic constituents may also be removed by adsorption onto activated carbon.

Another approach is to change the pulping technology. Use of oxygen, or air pulping [66], and/or bleaching could avoid many of the problems outlined

FIGURE 15.13 Proposed scheme for a foam production and separation device for the removal of both toxicants and foamable constituents from kraft process wastewater streams. (From Leach et al. [64], reprinted courtesy of Wheatlands Journals.)

[67]. It has been found possible to recycle bleach plant effluents as well as spent liquor from pulping, in the process regaining the chemicals required for pulping and bleaching [27].

Waste streams with a tendency to produce foam can be controlled by adding a commercial antifoaming agent but many of these are also toxic, which could upset the operation of a biobasin or contribute a toxicant to the effluent. Foams can be generated in a treatment plant, then skimmed and burned simultaneously destroying foam and toxicants (Fig. 15.13). Mills operating in cold climates find that a 0.5- to 1-m head of foam on a biopond can actually be an aid to maintaining optimum pond temperatures for continued BOD reduction under winter operating conditions. Without this assistance, floating plastic foam mats may be used to keep pond temperatures high enough during cold periods to maintain biochemical and biological activity.

15.9. ENVIRONMENTAL ASPECTS OF PAPERMAKING AND PAPER RECYCLING

Wastewater streams from paper machines while operating entirely on wood pulp do not pose serious waste loads on discharge. But when filled, coated, sized, or colored papers are being produced additives accumulated in the white water have to be substantially removed before water discharge. The volume of wastewater from this source that is required to be treated is minimized by operating the paper machine using 90% or higher direct water recycle. This measure also achieves a substantial direct saving of fillers, dyes, or the expensive titanium dioxide-based coatings by capture of these constituents into the paper sheet as it is formed. Treatment measures for the rejected fraction of recycled water, necessary to avoid buildup of resinous components, dissolved solids, or the establishment of microbiological growths, involve settling for waste streams containing primarily suspended solids, as already described. Color may be removed by activated carbon adsorption or one of the other color removal techniques already described under wastewater emission control.

Fiber reuse by repulping of printers' offcuts, computer and office paper wastes, as well as the paper component of municipal garbage is also an excellent way to decrease the overall impact of papermaking operations. Wood requirements are greatly reduced, though not eliminated, since it is usual to blend virgin fiber with recycle stock [68]. Also energy and chemical utilization are substantially reduced, as are the waste loadings to air or water. However, recycling does not eliminate external impacts since significant waste loads to water are generated from the repulping, de-inking, washing, and simple bleach stages required [69].

Waste paper stock destined for recycling is normally segregated into grades according to whether the original pulp was produced by chemical or mechanical methods. The separate stocks are then repulped (dispersed) in a hot dilute solution of sodium hydroxide, sodium phosphate, or sodium silicate in water, and cleaned via a series of raggers, cyclones, and screens. Self adhesive labels and tapes pose a perennial problem in recycle pulps that a recently developed test method may help to solve [70]. Standard printer's inks are removed by thorough dispersal and washing [71]. For the recently introduced thermoplastic inks used by laser printers, and some fax machines and photocopiers that use the same printing system, the polymeric pigments are present in the repulped stock as larger, hydrophobic particles, so are not removed by conventional wash de-inking. Froth flotation techniques are used to remove of the ink from these pulps via a pigment-rich froth.

After a further screening and wash, the stock is bleached with one or two stages of sodium peroxide, hypochlorite, or a chlorine dioxide-rich mix of chlorine and chlorine dioxide. On completion of these steps there is perhaps a 15–20% mass loss in reprocessing of chemical pulps and a 20–25% loss from groundwood pulps [72]. The product pulp may be incorporated into pulps prepared from new fiber, at varying proportions, to produce a sheet with little difference in final properties from a sheet produced entirely from new fiber [73]. Fine papers and newsprint may be made from the cleaner grades of recycle fiber; liner boards, corrugating medium, and asphalt shingle stock are prepared from the less clean grades.

REVIEW QUESTIONS

1. Biosynthesis of cellulose involves photosynthetic conversion of carbon dioxide to glucose, followed by dehydropolymerization of β-D-glucose to cellulose (polymerization accompanied by loss of water). What mass of carbon dioxide would theoretically be utilized by a tree to produce 1 tonne (1000 kg) of cellulose?

2. How do starch and cellulose differ stereochemically and functionally in the plant from one another, and why are these distinctions important in the utilization of cellulose rather than starch for papermaking?

3. (a) What weight of sodium hydroxide would be required to produce 10,000 L of solution containing 53 g of sodium carbonate per liter, after contacting with flue gas (for CO_2 content)?
   (b) How much sulfur would have to be burned to provide the sulfur dioxide for absorption into 10,000 L of aqueous sodium carbonate (53 g/L) to produce a solution containing equal molar concentrations of sodium bicarbonate ($NaHCO_3$) and sodium sulfite (Na_2SO_3)? Assume no volume change on absorption.

4. A typical analysis for kraft white liquor for pulping is given below in terms of the sodium oxide equivalence. This puts all the dissolved components on the same sodium ion content basis. Using this information, calculate the actual molarities present of the three dissolved components in this sample of kraft white liquor.

Na_2O equivalent content (g/L)	
NaOH	68
Na_2S	25
Na_2CO_3	16

5. The lime kiln in the causticization circuit of the chemical recovery operation for a large kraft mill produces 100 tonnes (1 tonne = 1000 kg) of burned lime (CaO) daily.

$$CaCO_3 \rightarrow CaO + CO_2$$

(a) How many cubic meters of CO_2 at 760 mm and 10°C would be discharged into the air around this operation daily, as a result of this reaction? (Use ideal gas relations.)

(b) Given the heats of formation of $CaCO_{3(c)}$, $CaO_{(c)}$, and $CO_{2(g)}$ as -1209.9, -635.5, and -393.5 kJ/mol, what gross daily heat requirement would be required theoretically to supply the endotherm of this reaction?

(c) Assuming 25% overall efficiency of heat transfer to the reaction from the combustion process, and given that the thermal value of the No. 2 fuel oil used is 35,000 kJ/L, what would be the daily fuel oil requirement for this level of production?

6. Clarified green liquor from the chemical recovery circuit of a kraft pulp mill has a density of 1.21 g/mL at 20°C and contains 20% sodium carbonate and 5% sodium sulfide, both specified by weight. On causticization with a 10% excess of lime (Ca(OH)$_2$) 90% of the sodium carbonate is converted to sodium hydroxide.

(a) What would be the percentage *and* the molar concentrations of sodium hydroxide, sodium sulfide, and sodium carbonate in the clarified white liquor obtained? Assume no volume or solution density changes.

(b) What would be the "effective alkali" concentration, as g/L of Na_2O equivalent, in this white liquor?

(c) What would be the sulfidity of this solution?

(d) Does this white liquor require any adjustment (make-up) in terms of concentration and/or chemical composition to be suitable for chip digesting in the Kamyr kraft method of pulping? If so, specify details.

7. The objective for emission of bivalent sulfur compounds (H_2S, CH_3SH, $(CH_3)_2S$, etc.) from an existing kraft pulp mill in British Columbia has been set on the basis of 3.74 lb of sulfur equivalent per ton of air dry pulp produced per day.

(a) If this limit was reached solely from methyl mercaptan (CH_3SH) for a mill producing 600 tonnes of kraft pulp per day, what mass of CH_3SH could be expected?

(b) What volume of air, in cubic meters, would be required to dilute the mass discharged to the point where it would be undetectable by the nose (0°C, 760 mm Hg)? The olfactory threshold of methyl mercaptan is 40 ppb by volume.

(c) Briefly outline any three control measures that could be taken to minimize emission of reduced sulfides from a kraft pulping operation.

8. (a) About 6.4% by weight chlorine in water is used in the first stage of bleaching of kraft pulps. What initial percentage by weight and molarity of chlorine and hypochlorous acid would result if half the

added chlorine reacted with water and half remained as dissolved chlorine?

(b) What masses (kg) of calcium hydroxide and chlorine would theoretically be required to make up 10,000 L of solution, 1.30% by weight in calcium hypochlorite, suitable for hypochlorite bleaching of pulp?

(c) What would be the molarity of 1.30% by weight calcium hypochlorite?

9. (a) What are the two main categories of air pollutants that are potential problem areas for a kraft pulp mill?

(b) Briefly, how may these be controlled?

10. Assume complete biodegradation of a tonne of waste paper in a landfill according to the following equation:

$$(C_6H_{10}O_5)_n \xrightarrow[\text{H}_2\text{O, hydrolases}]{\text{Hydrolysis}} n\ C_6H_{12}O_6 \xrightarrow[\text{methanogenic bacteria}]{\text{Anaerobic decomposition}} 3n\ CH_4 + 3n\ CO_2$$

What would be the masses of glucose, methane, and carbon dioxide produced if 100% hydrolysis occurred, followed by 90% anaerobic decomposition?

FURTHER READING

C.J. Biermann, "Handbook of Pulping and Papermaking," 2nd ed. Academic Press, San Diego, 1996.

B.N. Brogdon and C.C. Walker, Fundamental Advances and Innovations. In "The Pulp and Paper Industry." American Institute of Chemical Engineers, New York, 1999.

J.P. Casey, ed., "Pulp and Paper: Chemistry and Chemical Technology," 3rd ed., Vols. 1–4. Wiley, New York, 1979–1981.

J.J. Garceau, S.N. Lo, and L. Marchildon, A Cost Evaluation of Alternative Sulphite Spent Liquor Strategies. *Pulp Pap. Can.* 77(10), T174 (1976).

J. Gullichsen and C.-J. Fogelholm, "Chemical Pulping." Fapet Oy; [Atlanta, Ga. : TAPPI press] Helsinki, Finland, 1999.

"Kirk-Othmer Encyclopedia of Chemical Technology," 3rd ed., Vol. 19. Wiley, New York, 1982.

W.F. Sinclair, "Controlling Pollution from Canadian Pulp and Paper Manufacturers: A Federal Perspective." Environment Canada, Ottawa, 1990.

J. Sundholm, "Mechanical pulping." Published by the Finnish Paper Engineers' Association and TAPPI, Helsinki [Atlanta], 1998.

REFERENCES

1. "Reference Tables," 35th ed. Canadian Pulp and Paper Association, Montreal, 1981.

2. "U.N. Statistical Yearbook 2001," 47th ed. United Nations, New York, 2003, and earlier editions.

3. R.H. Clapperton, "The Papermaking Machine." Pergamon, Toronto, 1967.

4. "Kirk-Othmer Encyclopedia of Chemical Technology," 2nd ed., Vol. 16, p. 680. Wiley, New York, 1968.

5. H.I. Bolker, "Natural and Synthetic Polymers: An Introduction." Dekker, New York, 1974.

6. K.V. Sarkanen and C.H. Ludwig, eds., "Lignins, Occurrence, Formation, Structure, Reactions." Wiley-Interscience, Toronto, 1971.

7. K.W. Britt, ed., "Handbook of Pulp and Paper Technology," 2nd ed. Van Nostrand-Reinhold, New York, 1970.

8. J. Hill, K.Saarinen, and R. Stenos, *Pulp Pap. Can.* **94**(6), 43–47 (1993).

9. G. Broderick, R. Lanouette, and J.L. Valade, *Can. J. Chem. Eng.* **75**(1), 79–87, Feb. (1997).

10. J.C. Gauss and D.J. Wachowiak, *TAPPI J.* **62**(9), 27, Sept. (1979).

11. V.P. Puri and H.G. Higgins, *APPITA* **37**(6), 496–501 (1984).

12. J.A. Wright, M.J. Sabourin, and W.S. Dvorak, Laboratory Results of TMP and CTMP Trials, *TAPPI J.*, **78**(1), 91–96, Jan. (1995).

13. T. Fossum, *TAPPI J.* **65**(10), 69, Oct. (1982).

14. A. Wong, *TAPPI J.* **65**(10), 11, Oct. (1982).

15. M.B. Hocking, Vanillin: Synthetic Flavoring from Spent Sulfite Liquor. *J Chem. Educ.* **74**(9), 1055–1059 (1997).

16. Alkaline Sulfite May Challenge Kraft, *Can. Chem. Process.* **54**(11), 56, Oct. (1970).

17. M. Kleen, J. Sjoberg, O. Dahlman *et al.*, *Nordic Pulp. Pap. Res. J.* **17**(3), 357–363 (2002).

18. M.B. Hocking, H.I. Bolker, and B.I. Fleming, *Can. J. Chem.* **58**(18), 1983–1992 (1980).

19. J.N. McGovern and R.R. Fuller, 1984: Centennial of Kraft Pulping Patent. *TAPPI J.* **67**(11), 48–49 (1984).

20. J. Drew and G.D. Pylant, Jr., *TAPPI J.* **49**(10), 430, Oct. (1966).

21. "Crude Tall Oil Recovery," Tech. Sect. Proc., 47th Annu. Meet. Can. Pulp and Paper Assoc., Montreal, 1961.

22. Tall Oil Fatty Acids, *Chem. Eng. News*, **57**(40), 10, Oct. 10 (1979).

23. M.B. Hocking and J.P. Crow, On the Mechanism of Alkaline Hydrogen Peroxide Oxidation of the Lignin Model *p*-Hydroxyacetophenone. *Can. J. Chem.* **72**, 1137–1142 (1994).

24. K. Kratzl, P.K. Claus, A. Hruschka, and F.W. Vierhapper, *Cell. Chem. Technol.* **12**, 445 (1978).

25. T.P. Nevell, Chlorine in the Textile and Paper Industries. *Chem. Indus. (London)*. 253–256, Mar. 15 (1975).

26. A. Teder and D. Tormund, *TAPPI J.* **61**(12), 59, Dec. (1978).

27. H. Rapson, C.B. Anderson, and D. Reeve, *Pulp Pap. Mag. Can.* **78**(6), T137, June (1977).

28. M. Ruhanen and H.S. Dugal, *TAPPI J.* **65**(9), 107, Sept. (1982).

29. L. Nasman and G. Annergren, *TAPPI J.* **63**(4), 105, Apr. (1980).

30. C. Leuenberger, W. Giger, R. Coney *et al.*, Persistent Chemicals in Pulp Mill Effluents, . . . Behavior in an Activated Sludge Plant. *Water Resource*, **19**(7), 885 (1985).

31. R.H. Voss, Chlorinated Neutral Organics in Biologically Treated Bleached Kraft Mill Effluent. *Environ. Sci. Technol.* **17**(9), 530 (1983).

32. E.M. Kirschner, Closing the Loop. *Chem. Eng. News*, **75**, 15–17, Sept. 29 (1997).

33. K.R. Solomon, Chlorine in the Bleaching of Pulp and Paper. *Pure. Appl. Chem.* **68**(9), 1721 (1996).

34. B.S. Gevert, S.F. Lohmander, and O.H. Samuelson, Interactions Between Metal Compounds During Oxygen Bleaching of Kraft Pulp. *TAPPI J.* **80**, 177–187, July (1997).

35. O. Samuelson and U. Ojteg, Behavior of Calcium, Magnesium and Manganese Compounds During Oxygen Bleaching of Kraft Pulps. *J. Wood Chem. Technol.* **15**(3), 303–328 (1995).

36. L.G. Akim, J.L. Colodette, and D.S. Argyropoulos, Factors Limiting Oxygen Delignification of Kraft Pulp. *Can. J. Chem.* **79**, 201–210 (2001).

37. S.-Li. Chen and L.A Lucia, . . . Mechanism of Oxygen Delignification of Kraft Pulps, II. Application of Surfactants. *Cell. Chem. Technol.* **36**(5–6), 495–505 (2002).

38. D. Flater Norske Canada and B.C. Crofton, Personal Communication, June (2004).

39. N.L. Nemerov, "Industrial Water Pollution," 439. Addison-Wesley, Reading, MA (1978).

40. M.B. Hocking, Paper Versus Polystyrene, a Complex Choice. *Science*, **251**, 504–505 (1991).

41. M.B. Hocking, Relative Merits of Polystyrene Foam and Paper in Hot Drink Cups: Implications for Packaging. *Environ. Manage.* **15**(6), 731–747 (1991).

42. N. McCubbin, "The Basic Technology of the Pulp and Paper Industry and Its Environmental Protection Practices." EPS 6-EP-83-1. Environment Canada, Ottawa, Oct., 1983.

43. S.N. Lo, H.C. Lavallee, R.S. Rowbottom *et al.*, Activated Sludge Treatment of TMP Mill Effluents. *TAPPI J.* **77**(11), 167–178, Nov. (1994).

44. K.V. Sarkanen, B.F. Hrutfiord, L.N. Johanson *et al.*, *TAPPI J.* **53**(5), 766, May (1970).

45. M.A. Karnovski, Odor Generation in the Kraft Process. *J. Chem. Educ.* **52**, 490 (1975).

46. E. Smet, P. Lens, and H. Van Langenhove, Treatment of Waste Gases Contaminated with Odorous Sulfur Compounds. *Crit. Rev. Environ. Sci. Technol.* **28**, 89–117 (1998).
47. J.M. Bentvelzen, W.T. McKean, and J.S. Gratzl, *TAPPI J.* **59**(1), 130, Jan. (1976).
48. A.J. Teller and L.L. Bebchick, *Pulp Pap. Mag. Can.*, **81**(12), T358, Dec. (1980).
49. V.R. Parthasarathy, R.C. Grygotis, K.W. Wahoske *et al.*, A Sulfur-free, Chlorine-free Alternative to Kraft Pulping, *TAPPI J.* **79**(6), 189–198, June (1996).
50. M.B. Hocking, H.I. Bolker, and B.I. Fleming, ... Anthraquinone-catalysed Alkaline Pulping via ... Modelling and ESR Measurements. *Can. J. Chem.* **58**, 1983 (1980).
51. H.F. Lund, ed., "Industrial Pollution Control Handbook," 18–1. McGraw-Hill, New York, 1971.
52. "Proceedings of Seminars on Water Pollution Abatement Technology in the Pulp and Paper Industry," Rep. EPS 3-WP-76-4. Environment Canada and the Canadian Pulp and Paper Association, Ottawa, 1976.
53. B.R. Blackwell, W.B. Mackay, F.E. Murray *et al.*, *TAPPI J.* **62**(10), 33, Oct. (1979).
54. R.R. Fulthorpe, S.N. Liss, and D.G. Allen, Characterization of Bacteria Isolated from a Bleached Kraft Pulp-mill Wastewater Treatment System. *Can. J. Microbiol.* **39**(1), 13–24 (1993).
55. M.V. Nelson, *TAPPI J.* **63**(3), 61, Mar. (1980).
56. K. Hatano, D.J. Frederick, and J.A. Moore, Microbial Ecology of Constructed Wetlands Used for Treating Pulp-mill Wastewater. *Water Sci. Technol.* **29**(4), 233–239 (1994); and *ibid.* pp. 241–247.
57. R.J. Rush and E.E. Shannon, "Review of Color Removal Technology in the Pulp and Paper Industry," Rep. EPS 3-WP-76-5. Environment Canada, Ottawa, 1976.
58. D.J. Bennett, C.W. Dence, F.L. Kung *et al.*, *TAPPI J.* **54**, 2019 (1971).
59. New Scrubbing and Effluent Treatment Processes Use Activated Carbon, *Chem. Can.* **28**(11), 19, Dec. (1976).
60. J.C. Mueller, "B.C. Guidelines." B.C. Research, Vancouver, Feb., 1981.
61. H. Lundahl and I. Mansson, *TAPPI J.* **63**(4), 97, Apr. (1980).
62. J.M. Leach and A.N. Thakore, *Prog. Water Technol.* **9**, 787 (1977).
63. A.N. Thakore and A.C. Oehlschlager, *Can. J. Chem.* **55**, 3298 (1977).
64. J.M. Leach, J.C. Mueller, and C.C. Walden, *Process Biochem.* **10**(1), 7 (1976).
65. K.S. Ng, J.C. Mueller, and C.C. Walden, *Can. J. Chem. Eng.* **55**, 439 (1977).
66. C. de Choudens and P. Monzie, *Pulp Pap. Mag. Can.* **78**(6), T118, June (1977).
67. J. Luo and P.K. Christensen, Oxygen Delignification with Magnesium as Base: A Good Solution for Sulfite Pulp. *TAPPI J.* **75**(6), 183–187, June (1992).
68. O.Y. Mansour, M.A. Nassar, and M. El Meliegy, Recycling of Waste Papers and Their Reuse. *Cell. Chem. Technol.* **35**(1), 135–149, Jan.–Apr. (2001).
69. G. Kaufman, Activated Sludge Treatment of De-inking Wastewater at Fraser Inc., Thorold. *Pulp Pap. Can.* **86**(2), T36–T41, Feb. (1985).
70. O. Heise, B.J. Cao, J., Dehm *et al.*, A New Stickies Test Method—Statistically Sound and User Friendly. *TAPPI J.* **82**(2), 143–151, Feb. (1999).
71. L.D. Ferguson, Deinking Chemistry: Part 1. *TAPPI J.* **57**(7), 75–83, July, 1992; Part 2. *Ibid.* (8), 49–58, Aug. (1992).
72. D.W. Duncan, *TAPPI J.* **62**(7), 31, July (1979).
73. R.C. Howard and W. Bichard, The Basic Effects of Recycling on Pulp Properties. *J. Pulp Pap. Sci.* **18**(4), J151–J159, July (1992).

16

FERMENTATION AND OTHER MICROBIOLOGICAL PROCESSES

Ale, man, ale's the stuff to drink
For fellows whom it hurts to think.
                    —*Alfred E. Housman, 1896*

16.1. GENERAL MICROBIOLOGICAL PRINCIPLES

Fermentation, strictly speaking, refers to the process of anaerobic breakdown or fragmentation of organic compounds by the metabolic processes of micro-organisms. However, fermentation can more generally relate to the chemical changes of a substrate accomplished by selected microorganisms or extracts of microorganisms, to yield a useful product. This less specific definition includes microbiological processes carried out under anaerobic (fermentative, or in the absence of air), aerobic (respiratory), and enzymatic (via extracts) conditions.

Microbes of one kind or another have been used for centuries for the preparation of a variety of foods and beverages. Each fermentation product probably arose from what was originally an accidental discovery, for example, the fermentation of grape juice on storage or mold growth on curds (milk solids). These would have been recognized as changes, which increased the variety in their menus and the storage life of more perishable foods, hence the appeal of these products. It is easy to speculate, then, that each accidental discovery led to experiments aimed at being able to reproduce the change in the character of the food intentionally. From these early beginnings, yeasts have been used in the arts of wine, beer, and spirits production, as well as for the *in situ* production of the carbon dioxide used for leavening of bread and other bakery goods. Bacteria have been employed for the preparation of a variety of cheeses, sauerkraut, and yogurt. Molds too, still play a part in the preparation of certain cheeses. However, it is only since about 1900 that fermentation processes have been employed to any significant extent for the production of industrial chemicals, in addition to foods.

Whilst the world of microbiological processing has a long history, in another sense it is as up-to-date as the recent captivating interest in the nano-sciences. In the cells typically used for a microbial transformation, several enzymes (catalysts) are packed into a cell volume of the order of 10s of pL ($1 \text{ pL} \equiv 10^{-12}$ L) for yeasts, and of about 100 pL for bacteria. Since a nanoliter is 10^{-9} L, and microbes are a fraction of this volume, these may be thought of as the ultimate nanoreactors. Appreciation of these aspects of microbial processing could prompt the use of nanotechnology-based probes to further understand, and possibly even to influence, these processes.

The incentive for industrial utilization of microorganisms is based on several favorable factors. The most important of these is that a fermentation process is very often the only feasible way to obtain a complex product, or a specific chemical transformation. Also a chemical change conducted enzymatically (via microorganisms or their enzymes) has a far lower energy barrier than by the same chemical synthetic step. This means that microbiological processes may often be carried out at easily attainable near-ambient conditions. Under milder conditions, the enzymatic processes can be 10^9 to 10^{12} times as rapid as the corresponding chemical route to the same product [1]. Another feature of importance for the production of some complex biological products is that many enzymes are totally specific and stereoselective for the product of interest, unlike many chemical reactions.

Under appropriate conditions the rate of reproduction of yeasts and bacteria is extremely rapid, with doubling rates measured in minutes, so that there is little delay in building up the number of cells to maximize the conversion rate of raw material (Table 16.1).

TABLE 16.1   Key Details of the Four Principal Groups of Microorganisms Useful in Industrial Fermentations[a]

Class, complexity, and size	Reproduction	Occurrence (common uses)
Bacteria		
Simple, single chromosome (prokaryote), usually single-celled organisms, 1–6 μm diameter, spherical (coccus), or rod-like (bacillus) in shape	Asexually, by binary fission (simple cell division)	Air, water, and soil (sour milk, cheeses, yogurt, sauerkraut)
Yeasts		
Two or more chromosomes (eukaryote), single-celled near spherical, 5–12 μm diameter	Usually asexual by "budding," under some conditions sexual (via spore cells)	Air, water, and soil (beer, wine, spirits, bread)
Actinomycetes		
Multicellular organisms intermediate in complexity between bacteria and molds, filamentous bacteria whose mycelium is readily fragmented	See bacteria and molds/fungi	Normally inhabit soil (antibiotics)
Molds, fungi		
Two or more chromosomes (eukaryote), readily visible multicellular filaments	Varies, generally asexually via wind-transmitted spores, also sexually via wind ascospore formation	Air, water, and soil (cheeses, antibiotics)

[a]Compiled from Kent [2], Shreve and Brink [3], and Phaff [4].

Their small size, and thus large surface to volume ratio, is also favorable for the rapid diffusion of substrate into the cell, and the metabolized product out. However, using microorganisms to effect a chemical change introduces a change in the profile of a rate plot for the appearance of product. If all of the reactants are present in the reactor at the start of a conventional chemical batch reaction, an initial maximum rate is observed, which gradually tapers to lower rates. In contrast to this, chemical conversions using microbes usually start relatively slowly while the number of organisms builds up from the initial small number during a reproductive phase (Fig. 16.1). Following this lag phase or induction period, the maximum conversion rate of substrate to product is observed during the interval B–C, when there is little or no further increase in numbers of organisms. The interval C–D corresponds to the period when the concentration of available substrate has decreased sufficiently to limit its availability to organisms, which then slows the rate of metabolic conversion to product.

For large-scale fermentations the proportion of time during which a fermenter is producing under high-rate conditions should be a substantial fraction of the total fermenter operating time to minimize operating costs. That is, the time period B–C, during which maximum production is observed, should be large relative to the total of the other operating intervals of the fermenter, as given by Eq. 16.1.

$$\text{If interval B–C} >> \text{A–B + C–D + D–A,} \qquad 16.1$$

then batch operation is generally favored.

In this expression, interval D–A corresponds to the turnaround time, or time required to empty the fermenter, clean and sterilize it, refill it with fresh sterile substrate, and introduce a "starter" of the appropriate organism. If this situation can be arranged, then batch operation of the fermentation process is commercially attractive.

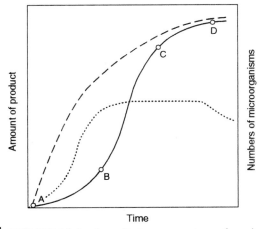

Time

FIGURE 16.1 A qualitative comparison of production curves via a conventional chemical process (––––) and a fermentation process (——) correlated to the propagation and metabolic rates of the microorganism used (. . . .).

If, however, B–C, the interval for maximum production rate, is short relative to the total fermenter operating time, then continuous operation is generally favored (Eq. 16.2).

$$\text{If interval } B–C < A–B + C–D + D–A, \qquad 16.2$$

then continuous operation favored.

Under these circumstances, there is an incentive to develop a continuous, rather that batch operation for the process. Continuous operation is particularly attractive for very large-scale fermentations, and is often staged in several vessels rather than in a single large one. In this way, a high fermentation rate is obtained continuously in one or more of the vessels to give a more rapid overall rate of conversion of substrate to product.

There are problems attached to operation of continuous, rather than batch fermentations. Among these are the potential for genetic variation of the microorganism itself with time, and the difficulty of maintaining ideal conditions for growth. Nevertheless there are a number of commercial continuous fermentations. Baker's yeast, yeast from petroleum or petroleum products, vinegar, and a number of aerobic processes use continuous culture methods. Even for the market-sensitive brewing and wine-making industries, which have traditionally been batch operations, continuous fermentations have been found to be commercially feasible. Wastewater treatment plants also operate in a continuous mode. However, wastewater treatment employs a wide variety of microorganisms rather than a single species to accomplish the effluent improvement, which is a nonspecific conversion (Chap. 5).

The requirements for producing a specific product via a fermentative industrial process are, first and foremost, the need for a culture of a specific microorganism that produces the desired end product. This is usually the dominant product, such as is seen for the anaerobic production of ethanol (ca. 95% selective) from sugar by yeasts. In this situation, the process is termed homofermentative. On the other hand there are also fermentations by a single organism, which produce several products, such as the anaerobic production of butanol (60–70%), acetone (20–30%), and ethanol (about 10%) from sugars by *Clostridium acetobutylicum* [5]. This type of process is called heterofermentative, and may still be commercially useful if all the products are easily recovered and marketable. Homofermentative processes are generally preferred because they avoid the difficulty of marketing fixed ratios of several products from a single process.

Other requirements are that the microbe must be relatively easy to culture and to preserve in a dormant state for reference and stock renewal purposes. The microbe should also be able to be easily maintained in an active state with consistent genetic characteristics. Raw materials appropriate for conversion must be economic for the end product desired (i.e., sugars and starches for beverage alcohol), low-cost sugar waste streams, starches, or cellulosic substrates for fuel alcohol, etc. Acceptable to good yields should be obtained, based on the relative prices of substrate and end product. The fermentation should proceed rapidly under conditions (pH, temperature, pressure, nutrients, etc.) that can be easily provided. Finally, recovery of the products of the fermentation should be easy.

Many of the requirements summarized above are general for any viable industrial process. The key difference here is the participation of the microorganism in the process, the locating, selection, and improvement of which is primarily the province of microbiologists. Cultures of the desired strains are maintained in a dormant state and are supplemented under sterile conditions. Portions of these cultures can then be taken and provided with ideal propagating conditions to build up the numbers of organisms available to provide a "starter" for inoculation to the medium of a full-scale fermenter. The medium, or growth environment, will consist of a solution of the primary substrate to be converted, usually in water. It will also have small amounts of nutrients, such as ammonium phosphate, to provide a source of nitrogen and phosphorus, plus traces of other minerals that may be required for microbial metabolism and reproduction. After pH adjustment and sterilization of the medium to avoid interference from any competing organisms, the temperature will be adjusted to the appropriate range and the starter added.

One remaining consideration which affects industrial fermentations is whether it is to be operated as an aerobic or an anaerobic process. Some microorganisms can only function under aerobic conditions (in a respiratory mode) and are referred to as obligate aerobes. Others, such as the obligate anaerobes, can only function in the absence of air (i.e., in a strict fermentative sense). Many species of obligate anaerobes are so sensitive to air that they are killed on exposure to it. Yet a third group, referred to as facultative microorganisms, is capable of functioning under either anaerobic or aerobic conditions. The metabolic processes and end products of facultative organisms usually differ under anaerobic or aerobic conditions. For example, yeasts in the presence of air ferment simple sugars to carbon dioxide and water. Under anaerobic conditions, they produce carbon dioxide and ethanol.

The type of fermenter used will be determined by whether the fermentation is to be conducted aerobically or anaerobically, as well as other factors. However, even anaerobic fermentations such as those used for alcohol production, can be conducted in open-topped fermenters when located inside a building. To do this, the top of the fermenting medium is protected from contact with air by the blanket of coproduced carbon dioxide, which is heavier than air. However, closed top fermenters are usually preferred for aerobic or anaerobic fermentation to reduce the risk of contamination of the process by unwanted organisms.

16.2. BREWING OF BEER

Brewing is used to refer to the large-, or small-scale manufacture of beer. Beers are beverage products made from a fermented brew of malted cereal(s) and hops that have been produced at least since the dawn of recorded history and probably for more than 8,000 years [6]. Brewing is the sector of the beverage fermentation industry with the highest volume of product and the highest value of annual sales. It is dominated by nations, which have a temperate climate suitable for cereal crops (Table 16.2). Beer production also occupies a larger total fermenter capacity than any other fermentation product,

TABLE 16.2   Annual Production of Beer by the Major World Producers[a]

	Millions of liters				
	1968	1978	1980	1990	2000
Australia	1,408	2,001	2,023	1,939	1,679
Canada	1,506	2,129	2,266	–	–
Czechoslovakia	2,010	2,206	2,393	2,197	1,779[b]
France	1,996	2,278	2,129	–	1,655
Japan	2,428	4,421	4,559	5,053	5,434
Mexico	1,252	2,257	2,688	3,873	5,985
Spain	1,026	1,735[c]	2,061	2,794	2,639
U.K.	5,143	6,642	6,484	7,080	5,891
U.S.S.R.	3,830	6,414	6,133	6,251	5,160[d]
U.S.A.	13,790	18,000[c]	22,777	23,667	24,429[e]
West Germany	7,373	8,792	8,932	9,138	10,690[f]
Other	10,125	16,431	29,241	47,359	n.a.
World	56,682	80,480	91,686	109,351	n.a.

[a] All countries producing more than 10 billion (10^{10}) liters annually in 1978. From Lom [7], *U.N. Statistical Yearbooks* [8], and Kirk-Othmer [9].
[b] For the Czech Republic.
[c] Estimated.
[d] For the Russian Federation.
[e] Calculated from the 48.9% fraction of U.S. production (101.8 million bbl; 1 bbl = 31 U.S. gal) reported to be held by Anheuser Busch in 2002. Available at Website for Standard and Poor. Production in 1996 was 23,349 million L (*U.N. Statistical Yearbook*, [8]).
[f] Reunited Germany.

a gross of over $128,000 \, \mathrm{m}^3$ for the U.K. beer production is thus the most important sector of beverage alcohol production. In the U.K., the only "fermentative sector," which exceeds brewing in the total volume employed is the 2.8 million cubic meters in use for wastewater treatment.

Beer is a perishable product and is largely consumed in the country of origin. Brazil, China, Germany, Kenya, and the U.S. were among the countries that produced more than 6 million liters in the year 2000 (Table 16.2). On a per capita production basis, Australia, Czechoslovakia, and Germany each produced over 140 L/person/year [8]. In contrast to this, the countries with a much lower levels of production average less than 5 L per capita per year.

The manufacture of beer is an anaerobic fermentation, which uses varieties of yeast to produce a beverage of low alcohol content from a solution containing mostly simple sugars in water. All brewery processing is conducted in glass-lined, or epoxy resin-lined steel, or stainless steel to avoid corrosion problems. The exception is the cereal cooker where copper is often used to obtain more efficient heat transfer. Four main types of beer are recognized, based on their alcohol content by volume. Low alcohol or alcohol-free beers have 0.5–1.3% ethanol by volume. Light beers contain 3.5–4% alcohol. Regular beers have about 5%, and the malt liquors have 5.5–6.5% alcohol. In addition to these main lines, specialty beers such as bock are produced at some times of year, which can have an alcohol content as high as sherries.

To outline the process, the aqueous substrate for beer production is made by brewing cereals with hops, which provide the characteristic bitter flavor component. Barley is the chief cereal ingredient, which is first malted to give enzymes required to breakdown starches to simple sugars, and also to contribute an important flavor component. Barley is a fairly expensive source of starch, so it is usually supplemented by the addition of less expensive starch adjuncts such as corn (maize), oats, millet, wheat, or rice. The adjunct contributes to carbohydrates, but makes little contribution to the flavor. Various strains of yeasts selected for their efficient conversion of glucose to ethanol are refined by individual brewers to achieve desired product qualities.

16.2.1. Malting, Mashing, and Fermentation

The barley component of the brew is first malted. This begins with cleaned, graded whole barley grains, which are soaked in cold water until the desired moisture content of 44–48% has been reached. Then the grains are spread out to germinate. During the germination step of 5–8 days, the enzymes diastase and maltase are formed by the sprouting seeds to mobilize the starch reserves in the seed ready for growth. Diastase promotes the hydrolytic conversion of starch, a polysaccharide, into the disaccharide, maltose. This hydrolysis is accomplished relatively easily since starch is a polymer of about 2,000 units of the alpha anomer of glucose, in which the glucosidic links are relatively accessible to water. Maltase in turn rapidly hydrolyzes maltose, a disaccharide of two glucose units, to glucose itself. Once these enzymes have formed in the germinating barley it is heat-killed and dried in a current of hot air. This product, "light malt," is ready for incorporation into a brew. Or, light malt may be cooked or roasted to a varying extent in a kiln to form a medium ("crystal") or a dark ("black malt") to contribute additional color and stronger flavor elements to particular brews. The enzymes of a dark malt may be denatured. Careful control of temperature, humidity, and air flow is required through all malting stages to produce an acceptable product. Malting details, and the integration of these into the brewing process have recently been reviewed [10].

The considerable time and space required for malting, and industry standardization of this process, has led to its virtual takeover by large, centrally located malting companies. Dried malt of various grades is shipped from the central processor to order, as required by individual brewers.

In mashing, the next step in the process, the dried malt and the starch adjunct (malt surrogate, 25–30% of the total starch) are ground up to coarse particles. Some of the ground malt and the whole of the adjunct are then mixed with water and boiled for half an hour in a cereal cooker to convert insoluble to soluble starch (Eqs. 16.3 and 16.4).

$$\begin{array}{ccc} \text{insoluble starch} & \rightarrow & \text{soluble starch} \\ \text{(intact cell membranes)} & & \text{(ruptured cells)} \end{array} \qquad 16.3$$

$$\begin{array}{ccc} (C_6H_{10}O_5)_n & \rightarrow & (C_6H_{10}O_5)_n \\ \text{intracellular} & & \text{extracellular} \end{array} \qquad 16.4$$

Meanwhile, in the mash tun (tub or mixer), the remainder of the malt is suspended in water and warmed to about 50°C. The soluble starch in water from the cereal cooker, while still hot, is then added to the mash tun containing the warmed malt in water to produce a resultant temperature of 70–75°C (Fig. 16.2). At this temperature (not higher or enzymes are destroyed) about half an hour is provided for hydrolysis of the soluble starches to maltose and glucose (Eqs. 16.5 and 16.6).

$$\underset{\text{(extracellular)}}{\text{soluble starches}} \xrightarrow[\substack{\text{amylases,} \\ \text{e.g., diastase}}]{\text{hydrol.}} \text{disaccharide} \xrightarrow[\text{maltase}]{\text{hydrol.}} \underset{\text{(simple sugar)}}{\text{monosaccharide}} \quad 16.5$$

$$\underset{\text{starch}}{(C_6H_{10}O_5)_n} \xrightarrow{n/2\ H_2O} \underset{\text{maltose}}{n/2\ (C_{12}H_{22}O_{11})} \xrightarrow{n/2\ H_2O} \underset{\text{glucose}}{n\ C_6H_{12}O_6} \quad 16.6$$

Under these conditions, the mashing step converts 75–80% of the starches to simple sugars [7]. This step is sometimes referred to as saccharification, literally "sugar formation".

The product of the mashing stage, called a wort, consists of varying proportions of maltose and glucose, flavor constituents from the malt, plus dextrins (incompletely hydrolyzed starches of 30–100 glucose units), and undissolved fragments of grain, malt, and husks (Table 16.3).

The hot wort is first lautered, a coarse filtration step, by passing it into a vessel with a perforated false bottom (lauter tun), which retains any undissolved grain, malt, and husk fragments. Filtered wort is then boiled for a period of 2–3 hr, hops being added after the first hour, usually in several small portions. Hops, which consist of the bracteole or cone of a vine *Humulus lupulus*, provide the characteristic bitter flavor component of a beer. These are contributed by several humulones and related compounds [11, 12] (Eq. 16.7), which are extracted from the hops during this boiling step.

humulone   lupulone   16.7

These extracts are also bacteriostatic (i.e., they discourage bacterial growth), which was one of the original reasons for using hops. The boiling period extracts the hoppy flavor constituents, destroys residual enzymes, coagulates undesirable proteins, and provides a sterile medium for subsequent fermentation. After boiling, the hot wort is settled to remove the coagulated proteins while being kept covered to avoid contamination by unwanted organisms. Subsequently, it is cooled through a plate cooler (via heat exchange to water) to 8–11°C. The product at this stage is referred to as hopped wort.

Before, or immediately after the yeast is added for fermentation, the hopped wort is oxygenated briefly. This helps to start the fermentation process more rapidly once the brew is placed under anaerobic conditions. A period of 6–7 days is required to complete the fermentation. During the

FIGURE 16.2 The principal steps of beer production in a modern, large-scale brewery. (Reprinted from Lom [7], courtesy of Southam Business Publications, Ltd.)

TABLE 16.3 Breakdown of the Principal Carbohydrate Constituents of a Normal Wort for Making Beer^a

Component	Concentration (g/L)
Maltose	57.7
Tri- to nanosaccharides	22.8
Higher carbohydrates	10.8
Total carbohydrates (91.0% of extract)	101.1
Fermentable carbohydrates (69.7% of extract)	77.4

^{a}Data selected from the more detailed listing of Kirk-Othmer [9].

active phase this is accomplished by a yeast count in the fermenting wort of about 12 million/mL, originally added as a starter of yeast cells in a propagating suspension of about 1% of the volume of hopped wort.

For each mole of glucose fermented about 80 kJ of heat is evolved (Eq. 16.8).

$$C_6H_{12}O_6 + H_2O \xrightarrow[\text{(in yeast)}]{\text{zymase}} 2\ CH_3CH_2OH + 2\ CO_2 \qquad 16.8$$

$$\Delta H = -74.5 \text{ to } -93.3 \text{ kJ } (-17.8 \text{ to } -22.3 \text{ kcal})$$

The heat evolution on maltose fermentation can be approximated by multiplying the figure for glucose by two and adding the further 15.9 kJ/mole required for maltose hydrolysis (Eq. 16.9). Fermentation temperatures of 15–18°C are desirable for beer production.

$$C_{12}H_{22}O_{11} + H_2O \xrightarrow{\text{maltase}} 2\ C_6H_{12}O_6 \qquad 16.9$$

$$\Delta H = -15.9 \text{ kJ } (-3.8 \text{ kcal})$$

Above this range some flavor elements are lost with the carbon dioxide bubbles leaving the wort, and head (froth) formation of the finished beer is poor. This is why refrigeration is normally used to maintain fermentation temperatures in this optimum range.

Yeasts used in brewing (and for winemaking and spirits production) are homofermentative, that is, under the anaerobic conditions of use they are about 95% selective for ethanol as the end product of glucose metabolism [13]. Thus, Eqs. 16.8 and 16.9 closely approximate the overall chemical change accomplished by fermentation, even though they do not do justice to the sequence of multiple biochemical steps required. Using these equations, it is possible to calculate the approximate alcohol concentration to be expected on the fermentation of any given initial concentration of fermentable sugars. During the fermentation step, virtually all of the fermentable sugars present, which comprise about two-thirds of the original carbohydrate mashed, are converted to ethanol and carbon dioxide (Table 16.3). Tri- and higher polysaccharides, which are not fermentable by yeast remain as a dissolved carbohydrate fraction in the beer. Some breweries collect the carbon dioxide formed during fermentation and clean it after compression by two or more stages of active carbon adsorption, followed by a water wash. It is then liquefied for storage for later use. In continuous brewing systems, part of the carbon dioxide produced is used to keep the yeast and the wort agitated during the fermentation step.

After fermentation the brew is separated from the yeast by centrifuging, or by decantation through an outlet raised slightly above the bottom of the fermenter, which leaves nearly all of the yeast behind as a "heel" in the fermenter. Each fermentation produces enough yeast to pitch (start) five further brews. After careful checking for strain purity and freedom from mutations, the middle portions of this yeast will be reused, sometimes for as long as 3 months or 8–12 fermentations, before it is arbitrarily discarded. A starter freshly prepared from pure culture stock is then used to reinitiate the whole process. The decanted, fermented brew is then filtered. The opalescent product at this stage is referred to as a "green" beer.

Green beer requires aging at temperatures as near $0°C$ as possible for a period of 2–6 weeks, usually with at least one intermediate decantation plus filtration. During this period, some further protein settling occurs, which helps to clarify the product, and a certain amount of flavor mellowing (maturation) is achieved to produce a more palatable product. The aged product, with an analysis as given in Table 16.4, is now ready for final packaging and marketing.

Finished beer is marketed in two forms, bottled (or canned) and draft. The proportions of each sold varies widely depending on market area. Prior to packaging of either form of beer in North America, the mature product is usually given a final polishing filtration to yield the product of high brilliance (clarity) demanded by the market. In Europe, some opalescence (haze) is generally accepted by customers, so that the final product is not usually filtered before packaging. The product is carbonated to about 0.4% carbon dioxide by weight (2.6–2.9% by volume), using a part of the purified carbon dioxide produced by fermentation [15]. It is then loaded aseptically into clean bottles previously flushed out with carbon dioxide, if possible with the bottling equipment used. Following bottling, the shelf life of the bottled product is improved to about 3–4 months by pasteurizing for 6–10 min at $60°C$. After pasteurization, a simple residual "head" check determines the quality of the seal obtained on capping, and any bottles with imperfect seals are discarded. Optimum product shelf life is obtained under cool, dark storage conditions.

Draft beer is produced from the same brew, but has a lower level of carbonation, 2.2–2.4% carbon dioxide by volume. In Canada, it is packaged in aluminum or stainless steel "half barrels" of 12.5 Imperial gallons capacity, and is not pasteurized. Thus, draft beer has a much shorter shelf life than bottled beer, about 21 days with refrigerated storage. For this reason, a brewer apportions the newly brewed product appropriately to ensure that all that are packaged as draft consumed within this period. The incentive for the draft package lies in the very much lower packaging cost in this form. In return, consumers enjoy a flavor unaffected by pasteurization, and less affected by carbonation. Even for a large-scale brewery, the cost of bottling very often exceeds all the other costs of brewing.

TABLE 16.4 Composition by Weight and Other Properties of a Typical Finished Beer[a]

	Composition[b]	Per 341 mL bottle,[c] approx.
Alcohol content, by weight	3.8–3.9%	11.3 g
Alcohol content, by volume	ca. 5.2%	–
Degree of fermentation	66%	–
Density	1.003–1.007 g/mL	–
Total unfermented carbohydrates	4.0–4.6%	12.6 g
Fermentable sugars (as maltose)	0.90%	2.6 g
Water	92%	270 g
Proteins (as N, ×6.25)	0.30–0.50%	1.3 g
Humulones	ppm range	–
Minerals, ppm		
Calcium (as Ca)	40	15 mg
Phosphorus (as P)	150–250	93 mg
Vitamins, ppb		
Thiamin (vitamin B_1)	20–90	19 μg
Riboflavin (vitamin B_2)	300–1300	270 μg
Pantothenic acid (vitamin B_5)	1000	340 μg
Pyridoxin (vitamin B_6)	400–800	205 μg
Nicotinamide	10,000	3,400 μg
Inositol	29,000	9,900 μg
Biotin (vitamin H)	10	3 μg
Other properties		
pH	4.1–4.4	–
Food energy (as alcohol, carbohydrates, proteins)	1,780–1,970 kJ/L (425–470 kcal/L)	607–672 kJ (145–160 kcal)

[a]Data compiled from Lom [7], Kirk-Othmer [9], and Donaldson and Lampert [14].
[b]By weight, except where otherwise stated.
[c]Equivalent to a 12-ounce bottle.

16.2.2. Product Variety and Quality

Many factors other than packaging influence the quality and type of product emerging from the brewing process. To start with, the water quality and dissolved ion content are important, both for the chemistry of the starch conversion step and for the flavor of the finished beer. Very soft waters may require addition of small amounts of calcium chloride and/or sulfuric acid, or sometimes gypsum ($CaSO_4 \cdot 2H_2O$) to provide sufficient trace nutrients for the yeast and a sufficiently low pH for a good product.

A breakdown of the main ingredients used in brewing is given in Table 16.5. Lager beers have two distinctions. First, they are produced by a bottom-fermenting strain of yeast, *Saccharomyces carlsbergensis*, a variety of *Saccharomyces uvarum*. This variety is named after its discovery and isolation at the Carlsberg Institute, Copenhagen, by E.C. Hansen in about 1870. Bottom-fermenting yeasts settle out or sink in the green beer when fermentation is complete. Two or more additional settlings or rests of the chilled decanted lager beer from the fermentation tank remove much of the yeast as well as other proteins. Hence, the name lager, since "lagern" is German for "to rest".

■ **TABLE 16.5 Major Ingredients and Appropriate Current Costs of Brewing, Per Canadian Barrel and Per Hectoliter of Beer**[a,b]

	Per Canadian barrel	Per hectoliter (100 L)
Brewer's malt	34–37 lb	11–12 kg
Malt adjunct(s)	12–14 lb	4–5 kg
Hops	0.27–1.5 lb	0.090–0.50 kg
Yeast	ca. 1 lb	ca. 0.3 lb
Production costs:		
Labor	$4.50	North American
Power, steam	$4.00	cost estimates, 1988
Bottling	$6.00	

[a]Canadian barrel of beer = 25 Imperial gallons (or 30.023 U.S. gallons); U.S. beer barrel = 31 U.S. gallons; English beer barrel = 36 Imperial gallons; 1 dozen small bottles = 0.90 Imperial gallons; 1 hectoliter (European marketing unit) = 100 L = 22.00 Imperial gallons.
[b]Ranges calculated from data of Shreve and Brink [3] and Kirk-Othmer [9].

Lagers comprise more than 95% of the brew sold in the U.S.A., but just over 40% of the beer sales in Canada, and less than 5% of the beer sold in the U.K. More than 50% of the beer sold in Canada and nearly 90% of that sold in the U.K. is an ale. The basic difference between these brews is that, an ale is produced by a top-fermenting yeast, such as *Saccharomyces cerevisiae,* from which the yeast is skimmed from the top of the beer on completion of fermentation. Fermentation with these yeasts is carried out at 15–20°C, slightly higher than bottom fermentations. Ales are also hopped at a higher rate than lagers, which gives the product its characteristic more bitter, hoppy flavor.

Low alcohol or alcohol-free beers of 0.5–1.3% alcohol may be produced by interruption of the fermentation, by boiling of the beer to distill part of the alcohol, by vacuum distillation, or by alcohol removal by reverse osmosis. The first two methods are the most common [16]. However, these methods result in the loss of desirable aroma substances along with the alcohol. This is remedied with used yeast, which is a reservoir of aroma substances. This is added at 2–10 mg/L to the dealcoholized beer just before filtration to restore the beer aroma.

Other brew variations are pilsner, a light beer of 3.4–3.8% alcohol content and a little higher hop content, originally produced in Pilsen, Bohemia; and bock beer, a heavy, darker, sweet product of lower hop content is produced as a specialty product in the spring by German breweries. Porters and stouts are dark, heavy, strongly flavored English beers produced using more hops and using malts, which have been baked (roasted or black malt), and little or no adjuncts. Porters and stouts, produced by top fermentation, are also sold in Canada but together comprise less than 1% of the market.

The perishability of beer is attributable to its low-alcohol content, which is inadequate to prevent attack by microorganisms, and to the unstable nature of many of its constituents. Thus, bottled and draft beers are at their best immediately after packaging. Optimum keeping qualities are obtained by ensuring minimum oxygen content and aseptic conditions during aging and packaging, by keeping the product under cool conditions to slow any chemical changes, and by storage of the product in the dark or subdued light to avoid the generation of "light struck flavors" [17].

16.2.3. Brewing Emissions and Controls

Air emission problems of breweries are relatively minor since mass emission rates are low and discharges are largely nontoxic. Emission control measures focus on containment of process odors. Water scrubbers are used on brewhouse stacks to eliminate this potential problem area [7]. If spent yeasts and grains are dried on site, the resulting odors are controlled by afterburner pyrolysis before the dryer vent stacks [18].

Wastewater streams with a high BOD and high suspended solids are the main aqueous emission problem areas. It is estimated that the equivalent of about 9 kg of BOD loading results from each 1000 L of beer (90 lb BOD/ 1,000 Imp. gal.) produced in a commercial brewery, with a breakdown as to origin roughly as given in Table 16.6. Beer itself has BODs of 50,000–90,000 mg/L and CODs 20–30% higher, which have to be dealt with in the event of spillage.

The heaviest BOD loadings come from the spent yeast, beer spillage, and trub (combined clarification precipitates) streams. Where the local sewerage regulations permit, the brewery may pay a treatment surcharge to the local authority and discharge the brewery liquid wastes to the municipal sewerage system. The surcharge offsets the higher treatment costs imposed on the sewage treatment plant by the very high BOD wastes. Or the waste streams may be treated to reduce the waste load before they enter the sewerage system.

The spent grains and yeasts, for instance, can be recovered and sold or given away for use as cattle feed or poultry supplements [18, 20]. As long as the feed is used promptly these wastes can be sold wet. This has two advantages. It avoids the formation of another high BOD "screen and press liquor" stream,

TABLE 16.6   Characteristics of Typical Aqueous Brewery Waste Streams[a]

Origin	Biochemical oxygen demand (mg/L)	Suspended solids (mg/L)
Beer spillage	90,000	4,000
Washings from process units	200–7,000	100–2,000
Cleaning solutions	1,000	100
Screen and press liquor[b]	15,000	20,000
Wort trubs[c]	50,000	28,000
Spent yeast	150,000	800
Precipitates from clarification	60,000	100

[a]Data obtained from Beszedits [19].
[b]From the pressing of wet spent grains prior to drying.
[c]Consist of haze, which form either during wort boiling or cooling, and which is removed before fermentation.

which would result from filtration, and the capital and energy costs of drying. If these waste materials cannot be consumed promptly they must be pressed and dried to improve keeping qualities and reduce shipping costs. If no feasible feed markets exist, the dried wastes may be incinerated for disposal.

The beer spillage, washings, etc., have a lower BOD, but still high enough to have a sewage treatment impact. Conventional activated sludge methods are sometimes used to reduce the BOD and suspended solids of these streams to acceptable values [18]. With added nutrient, BOD and suspended solids reductions of 95–97% have been obtained [21]. Rotating biological contactors (RBCs or "biodisk units"), plastic media-filled biotowers [22], and the UNOX process developed by Union Carbide [18] have all been used. A mixture of brewery wastes and domestic sewage of about 1,000 mg/L BOD and 400 mg/L suspended solids, for example, gave a 95% improvement using the UNOX system.

16.3. WINEMAKING

The art of winemaking, like brewing, was developed well before the beginnings of recorded history. It is easy to speculate, again that these skills were acquired by experimentation following the accidental fermentation of stored fruit juices by natural yeasts present on the surface of the fruit. Several theories exist. The theory that grape cultivation spread parallel with culture from East to West is most widely accepted. It is thought that viticulture originated in the lands south of the Caspian Sea, since signs of grape cultivation and winemaking have been discovered in Mesopotamia dating back 4,500 years. Certainly winemaking and aging skills were well known to the Greeks some 2,500 years ago, and gradually spread from there, and probably other centers, in the following centuries [23].

From a chemical standpoint, winemaking is similar to brewing since the essential alcohol-forming step still involves an anaerobic fermentation using yeast. But there are also some significant differences. The key distinction is that winemaking starts with a fruit, usually grapes, which already contain sugars, the fermentation substrate. Thus many of the preliminary steps required in brewing to produce a sugar solution from starch-based raw materials are unnecessary for winemaking. This simplifies the early steps of the winemaking process.

A number of varieties of grapes form the dominant winemaking raw material but many fruits other than the grape, such as apples, loganberries, raspberries, strawberries, etc., are also used. Apart from the motivation to produce an appealing beverage product, there is also an economic value-added incentive to winemaking. To illustrate this, in the mid-1970s grapes sold wholesale at some $190/tonne and retailed fresh at something like $660/tonne. The roughly 600–800 L of wine obtainable from the tonne of grapes (ca. 160 Imp. gal/ton grapes) was worth some $1,600 at that time, an approximation of the profit potential of winemaking. Consequently, half or more of the grapes grown go into winemaking, and over 75% in some countries.

The types of grapes grown depend on the climate and the local microclimate, the soil types of an area, as well as on the type of wine to be produced. Also, the sugar content of the grape will vary significantly with growing conditions, which will affect the type and quality of finished wine produced. These factors affect the development of the wine industry of every aspiring country (Table 16.7). It is possible to ship fruit from a grape-growing to a wine-producing area, such as to supplement a low sugar content of a local juice. However, the perishability and relatively low value of raw grapes discourage this. Thus, winemaking is more or less restricted to areas, which are able to grow their own suitable grapes. This reduces the number of countries engaged in commercial winemaking to about one-third of the number, which produce beer [8].

Another natural factor, which affects the volume of wine produced is the availability of fruit. This factor alone can affect the quality, and cause the volume of wine produced by a country to vary from year to year by a factor of two or more. However, short-term world production variations are less than this (Table 16.7). Beer production shows far smaller annual variations in

TABLE 16.7   Wine Production by a Selection of the Major Wine-producing Countries[a,b]

	Millions of Liters				
	1960	**1970**	**1980**	**1990**	**2000**
Argentina	1,675	1,836	2,300	1,404	1,254
Australia	154	287	414	445	806
Bulgaria	405[c]	409	445	228	210
Chile	485	401	570	398	642
France	6,311	7,540	7,155	6,553	5,754
Greece	290	453	440	353	356
Hungary	296	438	565	547	300
Italy	5,534	6,887	7,900	5,487	5,162
Portugal	1,146	1,150	943	1,137	669
Romania	88[c]	449	895	471	546
South Africa	287	424	630	952	695
Spain	2,126	2,501	4,243	3,969	4,169
U.S.A.	1,033	969	1,729	1,600	2,330
U.S.S.R.	777	2,685	2,940	1,570	290[d]
West Germany	684	1,012	399	949	985[e]
Yugoslavia	335	548	678	517	197
Other	(361)[c]	841	872	1,646	3,224
World total	24,300	30,199	33,921	28,226	27,589

[a]Selected from *U.N. Statistical Yearbooks* [8] and from the Wine Institute (California) [24].

[b]To convert liters to Imperial gallons, divide by 4.55; to U.S. gallons divide by 3.79. A common international unit for wine production volume is the hectoliter, which is 100 L.

[c]Averages for 1961–1965 period; 1960 not available. Not included in the world total figure for 1960, hence the negative entry under "Other".

[d]Russian Federation.

[e]Reunified Germany.

Table 16.8   Wine and Beer Production and Consumption by the Top Six Wine Producers in 2000

	Wine[a]				Beer[b]		
	Production (million L)	Per capita production		Per capita consumption (L)	Production (million L)	Per capita production	
		Liters	Rank			Liters	Rank
France	5,754	97.0	2	58.2	1,655[b]	27.9	4
Italy	5,162	89.6	3	53.4	1,117	19.3	5
Spain	4,169	104.2	1	34.6	2,640	64.8	2
U.S.A.	2,330	8.4	6	7.7	24,429[c,d]	85.7	1
Argentina	1,254	33.9	4	33.7	1,209	32.6	3
Germany	985	11.9	5	23.6	10,690[b]	130.0	6
World	27,589	4.5		6.7	n.a.	n.a.	

[a]Data selected from the Wine Institute (California) [24].
[b]Production data from the *U.N. Statistical Yearbook* [8].
[c]Data for 1998.
[d]Calculated from the 48.9% (101.8 million bbl (of 31 U.S. gal)) market share held by Anheuser Busch for 2002. (See Table 16.2)

production volume than winemaking since its volume of production is less limited by raw material supply. This means that some of the largest producers of wine are also very large producers of beer (Table 16.8).

Both beer and wine markets are dictated to some extent by changing beverage tastes and social customs of the producing area.

16.3.1. Classification of Wines

A wider range of product variety and alcohol content is possible with wine than with beer. The "natural" wines, also called "table" or "dinner" wines, are those in which the alcohol content is derived entirely by natural fermentation. This limits the alcohol content to about 7–14%, by volume. The upper end of this concentration range is only attainable by using special varieties of wine yeasts. However, all have a higher alcohol content than even the malt liquor types of beer.

Most natural wines are sold "still," that is, with no residual carbonation. However, some are carbonated or "sparkling" varieties, which effervesce on opening from the release of dissolved carbon dioxide. Originally, carbonation was accomplished by bottling the wine prior to complete fermentation so that the carbon dioxide produced during a secondary fermentation carbonated the product. However, its difficulty of application and high cost means that it is now little used. Fermentation in large pressure tanks lowers the cost and the risks of this method. More often today the wine is carbonated at the time of bottling using purchased or fermentation carbon dioxide. Sparkling wines are classified according to their alcohol and carbon dioxide content. Thus, "crackling" wines are those carbonated to less than 2 atm pressure and contain 7–13.5% alcohol, by volume. "Sparkling" wines all contain about

7% alcohol, and may be carbonated to a varying extent. "Champagnes" may have an alcohol content in the 7–13.5% range and are carbonated to 2–6 atm pressure, depending on type.

An alcohol content of about 14% is the highest, which can ordinarily be achieved directly by fermentation because at this concentration further sugar conversion by yeast is inhibited. Concentrations of up to about 18% by volume are possible with special strains of wine yeasts. A wine of higher alcohol content, however, is possible by the addition of brandy (distilled from wine) or fermentation alcohol to a natural wine. Wines where the natural alcohol content is supplemented to 14–20% by volume are referred to as "fortified," or dessert wines, and include the varieties commonly known as sherries and ports. Fortification improves the keeping qualities for shipment and export by inhibiting further yeast action. Fortified wines are produced on about the same scale as natural wines.

Natural and fortified wines may be further classified according to their residual sugar content. Dry wines have 1% or lower residual sugar, i.e., as little unfermented sugar as the vintner is able to achieve. This will be mostly unfermentable pentoses [25]. Red natural wines are obtained by fermenting the juice of special varieties of grapes in the presence of the skins to extract some of the color, or sometimes by heating the must (juice for fermentation) in the presence of the skins for extraction before fermentation. These wines tend to be drier than the white varieties, which are fermented in the absence of skins. Sweet wines, which are usually whites, contain from 2.5 to 10% residual sugars. These are produced by starting with an initial sugar content, which is higher than can be completely fermented by the yeast. Occasionally, however, the sweeter character is obtained by a stopped fermentation, which is achieved by inactivating or removing the yeast from the fermentation at the appropriate stage.

16.3.2. Principal Steps of Winemaking

Grapes are harvested in the fall, the exact timing depending on the grape variety, current weather conditions, and grape sugar content. On arrival at the winery, the green stems are separated from the grapes, and the fruit is crushed. The time from picking the fruit to crushing is kept to less than 24 hr whenever possible. During this process the fruit or juice is kept out of contact with air as much as possible to prevent oxidation, which can affect the flavor. For temporary storage, chilled fruit can be kept in a nitrogen or carbon dioxide atmosphere, or the fruit may be treated with sulfur dioxide. For production of white wines the juice ("must") is separated from the skins on crushing. The red grapes for red wines are crushed and kept with the juice for fermentation. Much of the color is extracted from the skins ("pomace") by the alcohol as it forms in the fermenting mixture. This is promoted by carbon dioxide evolution during the fermentation, and by occasionally pressing the thick mat of skins at the top of the fermenter under the surface of the must, or by pumping the fermenting must over the skins.

Large numbers of natural yeasts (and other microorganisms) are found adhering to the fruit in the field. Originally these were the yeasts used to ferment the juice, with somewhat unpredictable results. Now, however, the wild yeasts in the must are inactivated by addition of 50–200 mg/L of "available sulfur dioxide" ($SO_2 + H_2SO_3$) before fermentation. Liquefied sulfur dioxide itself or a 10% solution of sodium or potassium metabisulfite ($K_2S_2O_5$) may be used for this purpose. A 10% metabisulfite solution, equivalent to about 5% available sulfur dioxide (Eqs. 16.10 and 16.11), is a more convenient form for small-scale use.

$$Na_2S_2O_5 + H_2O \rightarrow 2\ NaHSO_3 \qquad\qquad 16.10$$

$$NaHSO_3 + H_2O \rightarrow NaOH + SO_2 \qquad\qquad 16.11$$

Sulfur dioxide sterilization inhibits the activity of the wild yeasts present, serves as a chemically reducing preservative, and lowers the pH of the must. These conditions favor the activity of the acid-tolerant yeasts added by the winemaker over the wild yeasts present. If boiling were used for sterilization in winemaking, as it is in brewing, many of the subtle fruity elements of the product would be lost. Before fermentation, the must is analyzed for sugar and total acid content and the concentrations adjusted with sweeter juices or sucrose, and tartaric or citric acids if necessary [13]. The product at this stage of the process will have an analysis roughly that given in Table 16.9.

Before fermentation, the cooled must is allowed to stand for 2–24 hr for the sterilizing action of the sulfur dioxide to take effect, and to allow a decrease in the sulfur dioxide content by forming combined sulfur dioxide and by vaporization losses. This period also allows heavy sediments to settle (fruit particles, seeds, sand, etc.) from which the juice is separated to give a better preservation of the natural fruitiness of the product.

After this standing period, a starter suspension of the desired yeast at 1–2% of the must volume is added, and fermentation starts at close to 20°C (e.g., Eqs. 16.8 and 16.9). Fermentation temperatures for winemaking are less critical than for beer, so that many winemakers simply allow the process to take its course. However, better control of flavor and potential foaming problems is obtained by controlling white wine fermentations to within 2–3° of 15°C using chilled water. This process takes 1–2 weeks to complete [13]. Some quality wines may use lower temperatures to produce a wine with better varietal character (reflecting the single species of grape fermented), higher residual sugar, and higher volatile acidity.

Red wines are fermented at about 20°C in the presence of the skins for 3–6 days, depending on the intensity of color (anthocyanins) and dry flavor (tannins) desired. The partially fermented must is then decanted and pressed from the skins, and a secondary slower fermentation carried out to the extent required. Acid and glycerin formation are small but important components of both white and red wines. The concentration of these components starts to increase towards the end of the fermentation (Eq. 16.12), when the inhibiting influence of a high concentration of ethanol in the wine starts to inhibit the primary ethanol-forming reactions of the yeast (Eqs. 16.8 and 16.9).

$$2\ C_6H_{12}O_6 + H_2O \xrightarrow{\text{yeast}} CH_3CH_2OH + CH_3COOH + \qquad 16.12$$
$$2\ CH_2OHCHOHCH_2OH + 2\ CO_2$$

TABLE 16.9   Range of Concentrations of the Principal Constituents of a Must and a Dry Wine, in Percent by Weight[a,b]

Component	Must	Wine
Water	70–85	80–90
Total carbohydrates	15–25	0.1–0.3
Glucose	8–13	0.1–0.5
Fructose	7–12	0.05–0.10
Pentoses	0.08–0.20	0.08–0.20
Alcohols		
Ethanol	trace	8–15
Methanol[c]	0.0	0.01–0.02
C_3 and higher	0.0	0.008–0.012
Glycerol	0.0	0.30–1.40
Total organic acids	0.3–1.5	0.3–1.1
d-Tartaric	0.2–1.0	0.1–0.6
l-Malic	0.1–0.8	0.0–0.6
l-Citric	0.01–0.05	0.0–0.05
Succinic	0.0	0.05–0.15
Lactic	0.0	0.1–0.5
Acetic	0.00–0.02	0.03–0.05
pH	<3.3–3.6	
Tannins	0.01–0.10	0.01–0.30
Aldehydes	trace	0.001–0.050
Nitrogenous compounds	0.03–0.17	0.01–0.09
Inorganic compounds	0.3–0.5	0.15–0.40

[a]Selected from the comprehensive listings of Amerine [13] and Amerine *et al.* [26].

[b]Composition of the starting grape is about 80–90% pulp and juice, and 10–20% seeds, skin, and stems [13].

[c]Methanol is not a product of the fermentation, but arises from the hydrolysis of naturally occurring pectins [26].

Completion of fermentation for both types produces a "young" table wine, which requires maturation before it becomes a palatable product.

As soon as the fermentation is stopped sulfur dioxide is added to bring the free SO_2 concentration to 25 ppm for the subsequent clarification steps. This prevents oxidation, assists in biological stabilization, and binds aldehydes in the new wine. The binding of aldehydes (mostly acetaldehyde) improves the taste of the wine.

Clarification is needed after fermentation to remove the spent yeasts in the wine to prevent their autolysis in the product, and to remove any other suspended material (Fig. 16.3). Simple settling and racking (decantation) repeated several times accomplishes much of this. Occasionally fining agents, such as isinglas (a fish protein), gelatin, a bentonite clay, or a highly refined diatomaceous earth may be added to promote coagulation and precipitation of colloidal matter. Prolonged chilling to near-freezing temperature is maintained to promote tartrate crystallization, or ion exchange methods are being employed by some wineries to remove excess tartrates, particularly before the bottling of young red wines [13]. This practice avoids the formation of haze (turbidity) from the crystallization of the residual tartrate in the bottle, which can occur when the wine is chilled or is aged in the bottle. A final polishing

FIGURE 16.3 Flowsheet outlining the sequence of steps required to produce wine on a commercial scale.

step is filtration, used primarily to ensure that no yeast or other microorganism remains in the wine during the subsequent aging steps, and to produce a brilliantly clear product [27]. At this stage the wine will have an analysis in the ranges quoted in Table 16.9.

Aging accomplishes a maturation of many of the flavor components of the young wine to give it a less harsh, more palatable flavor [28]. A marrying of flavors and a development of bouquet (aroma) occur through a variety of chemical steps. Probably the most important of these is esterification, the combining of the acids and alcohols present to form esters. These contribute a more fruity, less harsh flavor and "nose" components than the initial precursors that were present in the young wine (Eq. 16.13).

$$\text{RCOOH} + \text{R}'\text{CH}_2\text{OH} \xrightarrow{\text{H}^+} \text{RCOOCH}_2\text{R}' + \text{H}_2\text{O} \qquad 16.13$$

Ethyl acetate is the most likely ester to form, because of the relatively higher concentrations of ethanol and acetic acid than other ester forming constituents present. It is also the most important for the progress of maturation. Malic to lactic acid conversion is also an important part of the aging process, particularly for red wines (Eq. 16.14).

$$\underset{\text{malic acid}}{\text{HOOC–CHOH–CH}_2\text{–COOH}} \xrightarrow{\text{Lacto-bacillaceae}} \qquad 16.14$$

$$\underset{\text{lactic acid}}{\text{HOOC–CHOH–CH}_3} + \text{CO}_2$$

Some North American white wines are bottled and marketed after only about 4 months aging and two or three rackings. Many are marketed in under a year [13]. Red wines are normally aged a minimum of 24 months, and some as long as 4 to 6 years before bottling. Sherries and ports, the fortified wines, are generally aged 3 years or longer to allow full maturation and smoothness to develop before marketing. During this aging process sherries are partially oxidized with air, and then "baked" at about 50°C for 1–2 months to produce the characteristic sherry flavor. A beverage very like a sherry may also be made from the by-products from beekeeping, with or without supplementation from honey [29].

Prior to bottling, most commercial wines are stabilized and blended. Stabilization improves the keeping qualities of a wine. Filtration through a small pore filter removes yeast cells that might continue to ferment residual sugar when the wine is in the bottle. Filtration also prevents lysis (cell rupture) from contributing intracellular amino acids, etc., to the wine. Pasteurization for 2 min at 86°C, may also be used, either in series with filtration or on its own. This kills, but does not remove microorganisms, although it coagulates some proteins, thereby assisting in their removal. The wine is then gently sparged with carbon dioxide or nitrogen to dispel any dissolved oxygen, which might be present to help prevent oxidation on storage. Residual sulfur dioxide, present at an average level of about 200 mg/L (as combined, and free) in Canadian wines, and at up to twice this concentration in some European varieties, also helps to prevent wine oxidation or secondary fermentation on extended storage [30].

Finally, the products of different batch fermentations based on similar or different grapes are blended to achieve consistent flavor characteristics for a particular brand name of wine. Once the stabilized blend is bottled, particularly when using airtight caps, there is negligible further change on storage. Bottled wine kept in a cool, dark place, is a relatively stable product even on prolonged keeping, which is assisted by the stabilization steps described above. Regulations in Canada and the U.S.A. permit up to 350 mg/L total sulfur dioxide in natural or dessert wines, still below the detectable level for the average palate. However, a trained wine taster can taste and roughly quantify concentrations of free sulfur dioxide higher than about 25 mg/L. Combined sulfur dioxide is present mostly as its bisulfite addition compound with acetaldehyde [25]. In the U.K., sodium or potassium sorbate and ascorbic or erythorbic (isoascorbic) acid addition, are now also permitted to absorb oxygen. These help to maintain a free SO_2 level for a long period so that less sulfur dioxide is needed for good keeping qualities.

Related to product quality control the concentration of asbestos fiber in all types of finished wines is being minimized through better final product sampling and tests, and by the avoidance of asbestos-containing filtering materials [31]. The wide dispersal of asbestos fiber in the atmosphere means that it is found on the surface of all harvested fruit, inevitably putting some fiber into wine musts being processed. Traces of fungicides and insecticides have been detected in wine, most of which has been found to be removed in the yeast cake and lees separated from the wine [32]. Fortunately, distillation of these by-products for alcohol recovery again leaves most of these pesticides in the residue.

16.3.3. Utilization and Disposal of Winery Operating Wastes

Stems and pomace (skins and seeds) together amount to 10–20% of the weight of grapes crushed. These represent a significant disposal problem, particularly for large wineries producing upward of 36 million liters (ca. 8 million Imp. gallons) of wine per year. Some wineries return these directly to the soil of the vineyards, as an organic fertilizer [33]. At one time pomace was dehydrated for use as cattle feed, the seeds being separated out first for oil recovery, but this is now less important because of low nutritional value and storage difficulties [13]. Pomace stored in heaps can start to ferment, and if fed to cattle in this state can be fatal. South Africa has made extensive use of winery by-products; alcohol, tartrates, cattle feed, methane, and fertilizer are being produced from pomace [34]. Many wineries recover the sugars from the pomace of white wines, and the residual sugars plus alcohol from the pomace of red wines by extraction with water. Fermentation of the sugars in the extracts produces ethanol, which is recovered by distillation and used for fortified wines or brandies.

Alcohol may also be recovered from fresh lees (spent yeast) by distillation of the wine that separates with the yeast, and from pomace by steam distillation [35]. Some wineries pressure filter the lees to improve wine recovery and produce a dry yeast cake, which may be dried and sold as cattle feed [13].

Monopotassium tartrate may be recovered from pomace, from lees, or from the virtually pure material obtained from tartrate removal processes [36]. Monopotassium tartrate from this source provides most of the commercially marketed material used in such diverse products as baking powder and cream of tartar.

A poor batch of wine can sometimes be improved by chemical means, and then marketed as a blend. If it is not possible to correct the "off" wine, it may be distilled for recovery of alcohol, which again may be used for wine fortification. The aqueous residue, now significantly lower in BOD, is then discarded. An "off" wine may also be converted to wine vinegar by acetification [37](Eq. 16.15). This process is sufficiently profitable that one California winery has specialized in producing wine vinegar, rather than wine.

$$CH_3CH_2OH + O_2 \xrightarrow[aceti]{Acetobacter} CH_3COOH + H_2O \qquad 16.15$$

With appropriate local circumstances few of the by-products and waste materials of a winery operation need be wasted.

Dual aerated lagoons operated in series have been found to be an effective way of dealing with high BOD aqueous wastes, and with malodorous effluent as long as the TOC was less than 500 mg/L [38]. Better than 95% BOD removal has been obtained for influent BODs of 131–5,000 mg/L when retention times averaged 30 days for the first lagoon and 100 days for the second.

16.4. BEVERAGE SPIRITS

Spirits are a class of alcoholic beverage with an alcohol content of 20% or more, above the range achievable solely by fermentation. Thus, the spirits classification consists of beverages where distillation is needed to raise the alcohol content above that reached by fermentation, and where the distillate forms an essential and significant part of the final product.

To be able to produce beverage spirits requires a knowledge of both fermentation processes and of the art of distillation. Thus, the preparation of spiritous liquors is historically much more recent than the arts of beer or wine production. Probably the earliest successful production of a distilled beverage from wine occurred in Greece or Egypt some 1,800 years ago, most likely by alchemists [39]. The original discoveries did not become widespread because of the secretiveness of the early practitioners. Perhaps a thousand years elapsed before distilled beverages became at all widely available. Synonyms for whiskey, the Gaelic "uisge-beatha," the French "eau de vie," and the Russian "vodka," all came into use in the last 800–900 years. All can be liberally translated as "water of life," which gives some idea of the early recognition of its stimulant effects.

16.4.1. Specifying the Alcohol Content of Spirits

Much of the English-speaking world uses the proof system on a degrees Sikes scale, which originated in the U.K. in 1816, for specifying the alcohol content of beverage spirits. In this system, 100% alcohol is equivalent to 175.2° proof. Proof alcohol, synonymous with 100° proof on this scale, is thus equivalent to 57.1% ((100°/175.2°) × 100%) alcohol by volume. In Canada and the U.K., proof spirit (100° proof) is specified in old statutes on a density basis as being exactly 12/13th the weight of an equal volume of distilled water at 51°F (ca. 10.5°C) [39]. Rumor has it that the early basis for deciding the alcohol content of spirits lay in the test that gunpowder moistened with an equal amount of proof (or higher alcohol content) spirits still ignited when touched with a flint or lighted taper. When moistened with underproof spirits it would not.

Using this proof system, spirits that are 20° overproof, which is the same as saying 120° Sikes or 120 proof, corresponds to a beverage, which contains 68.5% ((120°/175.2°) × 100%) alcohol by volume (Table 16.10). Similarly, spirits that are "70° proof" are 70° Sikes or 30° underproof, and will contain ((70°/175.2°) × 100%), or 40% (39.95%) alcohol by volume.

The most convenient way of determining ethanol concentration is by measuring the density and temperature of the solution and relating this to concentration by the use of appropriate tables. The concentration by volume determined in this way can then be related to the British proof system, or to the Sikes scale for specification on the bottle.

The density system is used in a number of countries to determine alcohol concentration. The United States uses a simpler proof system based on 100% alcohol being equivalent to 200° proof. Fractions of this concentration, by volume, are specified by multiplying the concentration, in percent by volume, by two (Table 16.10). In France, alcohol concentrations in spirits are simply specified as percent alcohol by volume, sometimes referred to as the Gay-Lussac system. Percent alcohol by *weight* is used in Germany.

The excise tax levied on distilled spirits is based on the volume of proof spirits produced, which corresponds to 57.1% ethanol (or ethyl alcohol) by

TABLE 16.10   A Comparison of the Systems Used for Specifying the Alcohol Content of Beverages

French, % alcohol by volume[a]	Approximate density at 15°C[b]	Canadian/ British proof[c]	American proof system	German, % alcohol by weight[a]
100	0.7939	175.2	200	100
97	0.8079	170	194	95.3
91	0.8306	160	182	87.0
86	0.8465	150	172	80.6
80	0.8639	140	160	73.5
74	0.8799	130	148	66.8
69	0.8925	120	138	61.4
63	0.9068	110	126	55.2
57.1	0.9199	100 (proof)[d]	114.2	49.3
51	0.9325	90	102	43.4
50	0.9344	87.5	100 (proof)[e]	42.5
46	0.9419	80	92	38.8
40	0.9519	70	80	33.4
34	0.9604	60	68	28.1
29	0.9666	50	58	23.8
23	0.9730	40	46	18.8
17	0.9791	30	34	13.8
10	0.9866	17.5[f]	20[f]	8.04[f]
5	0.9928	8.8[f]	10[f]	4.00[f]
0	1.0000	0	0	0

[a]System used in France, sometimes referred to as the Gay-Lussac system. In Germany, alcohol concentrations are specified on the basis of percent by weight.

[b]Related to the density of water, also at 15°C. Values will deviate slightly from this with sugar content, and at differing temperatures. Selected and calculated from the data of Lange [40].

[c]Values on this scale are sometimes referred to as "degrees Sikes". Entries are obtained by multiplying the percent alcohol by volume by 1.752; consequently, these are not exact figures. Values are rounded downward for ease of comparison and to keep proof value cited conservative for the percent alcohol given. Values above "proof" are sometimes quoted as "overproof," using (the value minus 100). Values below proof are sometimes quoted as "underproof," using (100 minus the value) to refer to lower alcohol concentrations.

[d]Proof alcohol in the British system, or 57.1% alcohol, by volume.

[e]Proof alcohol in the American system, or 50.0% alcohol by volume.

[f]Correspondence between the different scales becomes inaccurate in this range.

volume in Canada and the U.K., and 50.0% alcohol by volume in the U.S.A. Since the concentration basis of proof alcohol for taxation in the former areas is higher and the Imperial gallon is equivalent to 1.20 U.S. gallons, the alcohol content of one British "proof gallon" is equivalent to the alcohol content of 1.37 U.S. "proof gallons".

16.4.2. Steps in Spirits Production

The initial procedure for spirits production parallel those described for beer or wine in that one starts with either a cereal source for starch to be hydrolyzed to sugar, or an inexpensive direct source of sugar itself. Thus a variety of grains or molasses are common raw materials. Whiskey has been, and still is,

TABLE 16.11   Trends in the U.S. Consumption of Classes of Distilled Beverages (Spirits)[a]

Class of spirits	Percent of U.S. consumption, by class and year					
	1949	1960	1966	1976	1990	2001
Whiskeys[b]	87.8	74.4	68.2	53.5	38.2	28.9
Gin	7.1	9.3	10.5	9.8	8.6	7.4
Vodka	0.0	7.8	10.4	18.0	22.7	25.1
Cordials	2.2	3.8	4.3	6.8	11.1	11.7
(Liqueurs) brandy	1.3	2.6	3.2	3.9	4.9	6.2
Rum	1.3	1.6	2.2	4.2	8.3	11.9
Other	0.3	0.5	1.2	3.8	6.2	8.8
	100	100	100	100	100	100
Total volumes:						
L × 10[6]	641.5	888.3	1,169.2	1,621.8	1,386.0	1,352.8
Wine gallons[c] ×10[6]	169.5	234.7	308.9	428.5	366.2	357.4

[a]Selected from data of *Business Week* [41], Kirk-Othmer [39], and Adams Handbook [42].
[b]Include blends, straights, bonds, Scotch, Canadian, and other types of whiskeys.
[c]The wine gallon is a strictly volume measure; in the U.S.A., equivalent to a U.S. gallon (3.7854 L); in Canada and the U.K., equivalent to an Imperial gallon (4.5460 L).

the dominant product of the distilled beverage class, although there has been an increase in the proportions of vodka, cordials, and brandies in the U.S.A. over the last 25 years (Table 16.11). The discussion of distilled beverage production will be centered on whiskey because of its continued dominance in this class.

The choice of the main cereal component used is based on the lowest cereal cost on a starch content basis, since the sugar ultimately available for fermentation depends on the amount of starch hydrolyzed. Small amounts of a particular cereal, for example, rye for rye whiskey, corn for bourbon, may still be employed to provide the desired flavor element in the final beverage. Rye is used extensively by distillers in the areas where it is plentiful. Corn, wheat, millet (65–68% starch), rice, and potatoes also provide supplementary sources of starch for spirits production, depending on local cost, usage, and availability. Small amounts of hydrolyzed rye are sometimes used to favor initial yeast propagation (about 2% of the total starch input).

Whatever grain is employed, high quality and an accurate knowledge of the water content at the time of purchase are vital since every bushel of grain shipped to a distillery has to be accounted for in terms of the proof gallons produced. For any grain the standard "distillers bushel" is 56 lb. Since this is measured on a mass, and *not* on the usual volume basis, an unsuspected high water content can cost the distiller not only the lost raw material but also the excise tax on the proof gallons not obtained, for which they are still liable based on the quota of grain received. With a cereal start, the preliminary steps are very similar to the first steps of beer production, to obtain a sterile mash ready for inoculation. The mash is cooled to about 20°C. The yeast "starter," which is propagated from a master culture in four to five stages to a volume of about 2–3% of the mash volume to be fermented, is then added. Fermentation temperatures are less critical for spirit production than for beer, since the important flavor elements are less affected by temperature. However, only a few minutes at 50°C is enough to kill the yeast, so sufficient cooling is applied to keep the closed fermenter temperature below about 38°C to avoid this. The higher temperatures complete the fermention in a period of only about 3 days instead of the 6–7 days required for beer. By the time 50% of the sugar in the mash has been converted, there is no further yeast propagation. The remaining sugar conversion is accomplished by the existing yeast population to a "beer" of 7.5–8% alcohol by volume. Beverage spirits are obtained from this intermediate fermentation product by distillation, blending, and aging procedures.

16.4.3. Distillation of "Beers"

The theoretical basis of the success of distillation as a method for enrichment of the alcohol content of aqueous solutions is based on the application of Raoult's law. This law quantitatively relates vapor composition to liquid composition, for mixtures of ideal liquids, A and B, according to Eq. 16.16.

$$P_A = X_A \, P_A^0, \qquad\qquad 16.16$$

where

P_A = partial pressure of the vapor of liquid A, in the vapor mixture;
X_A = the mole fraction of A in the liquid phase;
P_A^0 = the vapor pressure of pure liquid A, at the same temperature.

While few real liquids behave quite like the ideal liquids required for strict adherence to Raoult's law, many mixtures of real liquids, including ethanol/water mixtures, approximately follow this relationship on heating (Fig. 16.4). Water has a vapor pressure of 327 mmHg at 78.3°C.

Ethanol, with a boiling point of 78.3°C, has a vapor pressure of 760 mmHg at this temperature and consequently forms a higher mole fraction in the vapor space above a heated ethanol/water mixture than it does in the liquid phase. Condensation of the alcohol-enriched vapor mixture obtained in this way produces a solution of ethanol in water again, but now enriched in the concentration of ethanol. In a laboratory batch distillation the process described above may be carried out very easily, but this only achieves a limited (by the liquid–vapor composition diagram) improvement in concentration of ethanol obtained with each repetition of the distillation (Fig. 16.5a). Also, as the distillation proceeds, the concentration of alcohol in the distilling vessel becomes depleted. Consequently there is also a gradual depletion in the alcohol concentration obtained in the vapor, and the condensate from this. Despite these problems, many small distilleries still use batch distillation to raise the alcohol concentrations to the requirement of their product [44].

It is also possible to operate a distillation in a continuous mode. When an incoming beer stream is fed continuously to a distilling vessel, and a vapor stream and a distilled (alcohol depleted) liquid stream are continuously removed from the vessel, this is referred to as continuous distillation.

FIGURE 16.4   Liquid–vapor composition diagram for ethanol–water mixtures. (Reprinted from Shreve and Brink [3], courtesy of McGraw-Hill, Inc.)

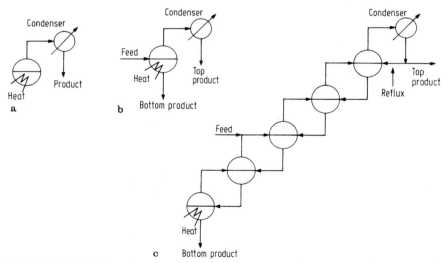

FIGURE 16.5   The relationship of batch (simple) distillation to single and multistage continuous distillation. (Reprinted from Reuben and Burstall [43], courtesy of M.L. Burstall.)

Continuous distillation simplifies quality control, because of the predictable and steady-state alcohol concentrations obtained from both product streams (Fig. 16.5b). The compositions of both the distillate and the residue streams, within the limits dictated by the liquid–vapor composition diagram, will depend primarily on the beer feed rate and the heat input to the system.

With a feed beer stream consisting of 7–8% alcohol, the increase in alcohol concentration, which may be achieved with one stage of distillation under ideal conditions is close to 55% alcohol. This figure may be estimated by drawing a horizontal line from the 8% alcohol point of the liquid composition line to the vapor composition line. To allow more latitude in congener (flavor component) addition, blending, and the alcohol concentrations that are taken for aging, most distillers strive for higher alcohol concentrations than this. This may be achieved either by several repetitions of simple (batch) distillations or by several stages of continuous distillation (Fig. 16.5c).

Continuous distillation is the only way to handle the large scale of a North American distillery, which can ferment over 1,000 bushels of grain per day. This is carried out in a distilling column in which a series of perforated plates or bubble cap plates separate the stages of distillation (Fig. 16.6). Using this arrangement, hot vapors from each stage heat the next higher stage as they partly condense, so that a preheated beers stream and only a single source of heat at the bottom of the column are sufficient to warm all the higher plates. As excess higher boiling fluid builds up on a plate, it overflows into a "downcomer" tube, which carries it down to the next lower plate. Any more volatile, lower boiling alcohol present in the fluid on a plate has a tendency to be vaporized and carried up the distilling column in vapor form as the fluid on the plate becomes heated by the upward moving vapors passing through it. In these ways, the liquid on each plate gradually comes to equilibrium with the hot vapors continuously being forced through the liquid on the plate.

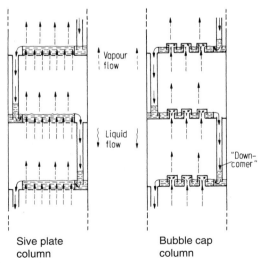

Sive plate
column

Bubble cap
column

FIGURE 16.6 Horizontal sections of a sieve plate (left) and bubble cap (right) distilling columns.

Plate choice is a function of the variables, which the particular distilling column is required to meet. Perforated plate columns are the most efficient for good liquid/vapor contact but are rather sensitive to distillation rates. Too low a distillation rate dries the plates from drainage through the perforations as well as via the downcomer; too high a distillation rate flood the perforated plate column (become overloaded with fluid), and hence lose distillation efficiency. A bubble cap column, in contrast, can handle a wider range of distillation rates for a given column without the drying out or flooding problems of a perforated plate column. But a bubble cap column provides less efficient liquid–vapor contact than a perforated plate column, and also costs more, for a given capacity. Bubble cap plates are usually specified when flexibility in operating rates is essential.

For a spirits producer fermenting over 1,500 bushels/day, the main beer distilling column can be shorter if the separation task is split among several columns. The selective distillation, the aldehyde, and the fusel oil columns can all be significantly smaller than the main whiskey separating column, which has the job of alcohol enrichment for the whole of the beers stream (Fig. 16.7) [45]. The first, and largest column, the beer distilling or whiskey separating column, produces an overhead stream, which contains 55–60% alcohol, together with the remaining volatiles, mainly aldehydes and fusel oils. This stream is the "high wines" stream of the distillery. At one time, with only some subsequent adsorption of impurities on charcoal, this stream comprised the whiskey product of early distilleries, which was sent to aging [46]. The bottom stream from this column is "stillage," which is water from which all the alcohol has been stripped, but which will still contain some unfermented sugars and dextrins.

FIGURE 16.7 A suitable configuration of columns, which can be used by a distillery for spirit production. (Simplified from Kirk-Othmer [45].)

The smaller selective distillation column separates a neutral spirits stream (essentially 95% ethanol, i.e., the azeotropic composition) plus aldehydes (overhead) and fusel oils (bottom product) streams from the high wines stream of the beer column. The last two columns recover high wines streams from the aldehydes and the fusel oil fractions. The more concentrated streams of aldehydes and fusel oils obtained are about 8 and 7% of the volume of the high wine stream, respectively [16].

16.4.4. Distinct Distilled Beverage Products

The aldehydes and fusel oils recoverd from distillation are also referred to as congeners, or flavor components. Most of the aldehyde content is acetaldehyde, although other aldehydes are also present. The fusel oils (from the German "fuselöl" meaning "bad spirits") are thought to arise via yeast fermentation of amino acids present in the original mash. They include amyl, and isoamyl alcohols (about 50%), n-butanol (about 20%), and a residue of fatty acids and several other higher alcohols [47] (Eq. 16.17).

$$CH_3CH_2CH_2CH_2CH_2OH \qquad CH_3CH(CH_3)CH_2CH_2OH \qquad \qquad 16.17$$
$$\text{amyl alcohol} \qquad\qquad \text{iso-amyl alcohol}$$

$$CH_3CH_2CH_2CH_2OH$$
$$\text{n-butanol}$$

Once the congeners are separated from the high wines stream, varying proportions of each may be added back to the neutral spirits fraction as required for the type of product being made. In this way the distiller obtains good control of the whiskey flavor of the final product. Thus, many whiskey products start out as blends of various proportions of neutral spirits, aldehydes ("heads"), and fusel oils ("tails"), together with enough water to bring the alcohol content to the range required by statute. Bourbon-type whiskeys generally have the highest concentration of congeners, followed by Scotch whiskeys, though it can be seen that there is also a difference in the ratio of aldehydes and fusel oils (Table 16.12). Canadian blended whiskeys, in contrast, are generally lower in congeners than either the Bourbon or Scotch counterparts, only some 0.5–1% of the heads and tails streams being added to neutral spirits before aging to produce these.

The newly mixed solution of neutral spirits, congeners, and sufficient water to bring the mixture to 60% alcohol is not a finished whiskey. It still must be matured to obtain a mellowing and smoothing of the initial blend to produce a more palatable product. The aging processes are carried out in new charred oak barrels (required in the United States) that are stored at a constant temperature as near 13–17°C as possible, and a relative humidity of about 65% for a period ranging from 4 to 18 years. If the conditions are cooler than this, the product doesn't mellow well, and if warmer or drier it causes greater than acceptable annual evaporative losses, something the distiller tries to avoid. During the process, water and alcohol diffuse out through the wood, and air diffuses in. The air very gradually oxidizes the aldehydes and other congeners present. Flavoring and coloring is also picked up from

▆▆▆ **TABLE 16.12   Congener Content of Some Common Types of Whiskey**[a,b]

Component	American blended whiskey	Bourbon whiskey	Canadian blended whiskey	Scotch blended whiskey
Fusel oil	83	203	58	143
Total acids (as acetic acid)	30	69	20	15
Esters (as ethyl acetate)	17	56	14	17
Aldehydes (as acetaldehyde)	2.7	6.8	2.9	4.5
Furfural	0.33	0.45	0.11	0.11
Total solids	112	180	97	127
Tannins	21	52	18	8
Total congeners, wt/vol %	0.116	0.292	0.085	0.160

[a]Selected examples from Kirk-Othmer [45].

[b]Grams per 100 L at 100° U.S. proof, density of about 0.914 g/mL, at 20°C. Thus, multiplying the numbers given by 10 gives a close approximation of respective concentrations in mg/L, and by 9.14 to give mg/kg.

the charred lining of the barrels, which at the same time adsorbs a certain amount of the fusel oil component from the mix. Congener content in the product may be evaluated by headspace solid phase microextraction [48]. The complex changes in congener concentrations, which occur during the aging process have been determined [49].

In Canada, used charred oak barrels are generally used. Here, the government requires a distiller to demonstrate production of at least 2.5 proof gallons (2.5 Imp. gal. of 57.1% alcohol by volume) for each bushel of grain mashed. If this is not achieved, the distiller is still taxed on this basis, hence the strong emphasis on pure, dry, high-quality raw materials by the distiller. In practice a distiller can generally produce 3.5–4 proof gallons (Canadian) per bushel, which is still taxed at the same rate but allows some tax margin for error. The government also allows for an aging loss of 3% per year. If the loss rate exceeds this, the distiller is still taxed for product volume it should have had with a 3% per year compound loss rate. This factor, plus product quality are strong incentives for proper aging conditions.

After aging the product is bottled, sometimes after blending the contents of a number of barrels with differing flavor nuances, to yield a whiskey with consistent flavor characteristics. It is easy to see why an 8-, or 12-year old whiskey costs more per bottle than a 4-year-old whiskey, since there could be 10–20% less whiskey to bottle after these aging periods together with the inventory and inflation costs of any stored product. Once bottled, there is very little, if any, change in the product because of the impermeability of the glass. Also stability is not a problem since whiskeys, like spirits generally, have a sufficiently high alcohol content to inhibit or kill any microorganisms, which may be present.

Vodka, presently second in sales volume to whiskeys in the United States, consists of straight neutral spirits (95% alcohol) diluted with pure water to the marketing concentration required. Ideally there should be no residual congeners in a first-class vodka. This is ensured by some producers by passing the neutral spirits through activated charcoal, or by redistillation before water addition. Perhaps the growth in vodka sales can be attributed to the fact that it is a beverage without significant aroma, character, or taste. It readily lends itself to the preparation of many types of mixed drinks, punches, and home-made cordials.

Gin is produced entirely from neutral spirits too (i.e., without the fermentation congeners), but has an aromatic flavor element derived from a final distillation over juniper berries or from added coriander seeds (a member of the carrot family).

Brandies have the special feature that they are prepared by distillation of a wine, generally a grape wine, rather than a "beer". The great variety in flavor and qualities of the various brandies originates from the different types of wines distilled, whether the distillation was batch or continuous, and the aging and blending details.

Rums are a distinctive beverage of this class in that they consist of the distillate from fermented molasses or other solution of sugarcane juice origin. So-called light or heavy rums are obtained depending on the type of sugarcane product that is fermented and the details of the distillation and maturation process used.

Cordials (liqueurs) may be made from any of the distilled spirits, which are blended with various fruit juices, flowers, botanicals, and miscellaneous other flavoring materials [49]. Maceration, percolation, and distillation techniques are all used to place the blend of flavor components into the spirits base used for solution. Over the years many generic names have come into common use to enable recognition of the principal flavor elements, such as Creme de Cacao (chocolate), Creme de Menthe (peppermint), Triple Sec and Cointreau (citrus peel), Anisette (aniseed), etc. Also certain brand names have, by tradition, established a well-known flavor association with them, such as Benedictine (aromatic plant flavors) and Drambuie (whiskey flavor). Considering the wide range of flavors of the cordial class alone, the range of flavor variations possible from the distilled beverage class far exceeds that possible from either of the other two classes of alcoholic beverage. However, a producer of distilled beverages has to face a greater technological and investment input and far closer government regulation than the other two classes to produce a competitive consumer product for market.

16.4.5. Environmental Aspects of Distillery Operations

Similarities of raw materials and many of the procedures of distillery operations to those of brewing and winemaking mean that many of the emission control problems and solutions are similar. Much of the discussion of those sections is also appropriate here.

The higher fermentation temperatures used for whiskey production compared to those used for beer or wine, increase the vaporization rate of organics from the fermenting mash. One distillery had an average ethanol vapor emission rate of $182 \, g/m^3$ of grain processed [50]. Lower average loss rates of ethyl acetate, $0.59 \, g/m^3$, isoamyl alcohol at $0.17 \, g/m^3$, and isobutyl alcohol at $0.051 \, g/m^3$ were also reported, all expressed on the same basis.

Activated charcoal may be used for vapor capture and analysis, or for control and recovery of these organics by heating, if necessary.

Distillery operations produce a stillage or "distillery slops" aqueous waste stream, as well as the usual brewery aqueous wastes. Beer still contains virtually all of the unfermented original ingredients, as well as the fermentation products. Distillery operations, however, since they are primarily interested in the distilled alcohol and congener fractions, are left with residual wastewater from the distillation, which contains 2–7% total solids [51]. The solids consist of mostly starches, dextrins, and unfermented sugars with beer-based spirits, or unfermented sugars with wine-based spirits. There may also be a significant nitrogen content in the stillage when the distillery has fed the yeasts (lees) to the still with the beer or wine stream. Biochemical oxygen demands of 1,500–5,000 mg/L have been cited for stillage in the absence of lees, and in the 7,000–50,000 range for lees stillage [13]. Mass emission rates of 0.60 kg BOD/tonne grapes and 0.90 kg COD/tonne grapes, have been cited for a winery operating with stills.

If the distillery is operating in a large metropolitan center, liquid wastes can generally be discharged directly to their sewer system. With dilution, normal sewage treatment should be able to process it. However, it may require the payment of a surcharge to compensate for additional treatment costs.

Stillage may be used as the organic feed to an anaerobic digestion unit in which bacteria convert the carbon content of the wastes into methane [52]. Combustion of methane rather than oil for a distillery's energy requirements, could save an estimated 116,000 Imp. gallons of oil annually.

If the distillery is operating in an area with available land, stillage may be discharged to an aerated pond for treatment, or it may be used for irrigation [45]. Decomposition of the waste in soil is accelerated by regular disking or light harrowing.

An alternative is to recover dried yeast and dried spent grains from the stillage for sale as cattle feed. BOD removal from "weaker" stillage is more economical. A trickling filter, or a biofiltration plant using two aeration units in series, has demonstrated 60–90% BOD reductions [45].

16.5. INDUSTRIAL ETHYL ALCOHOL

Before 1945, most of the supply of ethyl alcohol for industrial solvent or feedstock uses was derived from fermentation (Table 16.13). Since this time, the reliability and low cost of petrochemical routes to the product caused a rapid displacement of fermentation sources in the U.S. Since 1975, however, subsidies for fermentation alcohol have changed this. Large new fermentation units have been constructed, and distilleries formerly used for spirits production have been converted to industrial alcohol production [56]. Increased costs of American synthetic ethanol have kept its production at two-thirds of the total. The early petrochemical sources were based on the formation and hydrolysis of ethyl sulfate, but in North America, this has been replaced by the direct gas phase hydration of ethylene (Eqs. 16.18–16.20).

TABLE 16.13   Trends in the U.S. Supply of Industrial Ethanol from Fermentation Versus Synthetic Sources^a

	Industrial ethanol (million L of 95%)			Percent by fermentation
	Fermentation^b	Synthetic^c	Total	
1930	364	3.4	367.4	99
1935	325	35	360	90
1940	364	122	486	75
1945	669	223	892	75
1950	232	394	626	37
1955	168	652	820	26
1960	95	939	1,034	10
1965	229^d	1,126	1,455	16
1970	241^d	1,164	1,305	18
1975	316	850	1,166	27
1980	246	574	810	30
1985	140	680	822	17
1990	265	630	895	30
2000	–	–	6,500	–

^{a}Compiled from Eveleigh [52], Kirk-Othmer [54], and *Chemical and Engineering News* [55].
^{b}Fermented from molasses, grain, sulfite liquors, cellulose, etc.
^{c}From ethyl sulfate, and directly from ethylene, the latter predominating since 1970.
^{d}Includes beverage alcohol, therefore is higher than the straight industrial product volume.

$$CH_2 = CH_2 + H_2SO_4 \rightarrow CH_3CH_2OSO_3H \qquad 16.18$$

$$CH_3CH_2OSO_3H + H_2O \rightarrow CH_3CH_2OH + H_2SO_4 \qquad 16.19$$

$$CH_2 = CH_2 + H_2O \xrightarrow{H_3PO_4} CH_3CH_2OH \qquad 16.20$$

Starches, fruit, and sugarcane sugars are used for industrial alcohol production using methods similar to those used for beverage products [57]. Industrial end uses have less stringent raw material quality requirements. For example, a mobile, trailer-mounted fermentation/distillation unit can profitably convert overripe fruit and spoiled starch crops to fuel-grade ethanol for a farmer on a contract basis. Other sources of fermentable sugars, such as waste sulfite liquors, can yield as much as 75 L of ethanol per tonne of wood pulp [58]. Or cellulose may be deliberately hydrolyzed, promoted by heat and hydrochloric (40%, at 20°C) or sulfuric (0.5%, at 130°C) acids in what are known as the Bergius, and the Scholler process, respectively (Eq. 16.21).

$$\underset{\text{cellulose}}{(C_6H_{10}O_5)n} + n\,H_2O \xrightarrow{H^+} \underset{\text{glucose}}{n\,C_6H_{12}O_6} \qquad 16.21$$

Enzymatic hydrolysis of cellulose appears to be more economically attractive than chemical hydrolysis. [59]. Regardless of hydrolysis method, cellulose opens up the possibility to use of all kinds of farm wastes, straw, sawdust, wastepaper, etc., to produce sugars for fermentation ethanol.

All fermentation processes strive to obtain a 6% or higher concentration of ethanol, the minimum at which recovery is economic. The azeotropic composition, 95% alcohol/5% water, is the maximum concentration of ethanol, which can be distilled from the fermentation product. However, this azeotrope is suitable unchanged for some industrial uses.

For the 10% alcohol–gasoline blends, a formula for the "gasohol" fuels for automotive use, as well as for some industrial applications, anhydrous ethanol is required because this is miscible with gasoline and many other nonpolar solvents. The presence of even a trace of water in the alcohol or the hydrocarbon component causes a water-rich phase to separate from the composite fuel.

Small amounts of 95% ethanol may be conveniently dried by addition of the correct amount of quicklime (calcium oxide, Eq. 16.22).

$$CaO + H_2O \rightarrow Ca(OH)_2 \qquad\qquad 16.22$$

No stable hydrate is formed so that a mole of drying agent is required for each mole of water to be removed. The dry product is obtained by filtration.

Azeotropic distillation is the most economical method to dry large volumes of ethanol. A third component, usually benzene, is added to the 95% alcohol azeotropic composition. The boiling point of an azeotrope is always below that of the component with the lowest boiling point because the vapor pressures of each of the components present at any given temperature are additive. Therefore, heating the 95% ethanol benzene solution produces the ternary azeotrope of water, alcohol, and benzene, with a boiling point of 64.6°C, as the overhead product from the first distilling column (Table 16.14). If sufficient benzene is added to remove all of the water from the ethanol/water azeotrope in this way, the bottom product from the distilling column A, is pure, dry ethanol (Fig. 16.8). A slight excess of benzene is not critical to this result. Traces of benzene remain in this alcohol product making it unsuitable as a beverage component. Nor can it be used as a solvent for measurement of UV spectra because of the strong UV absorption pattern of benzene.

TABLE 16.14 Boiling Points and Compositions of Azeotropes Appropriate to the Drying of Ethanol[a]

Components and % by vol. composition	Boiling point (°C)
Water/ethanol/benzene 7.4/18.5/74.1	64.6
Benzene/ethanol 67.6/32.4	67.8
Benzene/water 91.1/8.9	69.4
Ethanol/water 95.6/4.4	78.2
Pure ethanol	78.5
Pure benzene	80.1
Pure water	100.0

[a]Data selected from *Handbook of Chemistry and Physics* [60].

FIGURE 16.8 Drying of 95% ethanol by distillation with benzene. (Reprinted from Perry et al. [61], courtesy of McGraw-Hill, Inc.)

Condensation of the top product from column A, which is the ternary azeotrope, then allows two phases to separate. The upper phase is rich in benzene (84.5%) with only a trace of water present (1.0%). The lower phase of the separator is rich in ethanol and water. The upper phase is returned to an upper plate of the first column to recycle most of the benzene added to this column.

The lower phase from the separator is redistilled in column C, to recover the ternary azeotrope as the overhead product and water-rich ethanol as the bottom product. The ternary azeotrope from column B joins that from column A. A third distilling column, column D, completes the recovery of ethanol as the ethanol/water azeotrope top product of this column, from the water-rich ethanol stream of the second column. The bottom product of the third column is the water originally present in the ethanol/water azeotrope feed to column A, now entirely freed of solvents.

Columns C and D are smaller than A since the volume to be distilled is much less at these stages. The only raw material cost to dry ethanol by this method is for the benzene "makeup" (to replace process losses), which amounts to about 0.2% of the volume of ethanol produced. Some of this is lost in the dried ethanol, as evidenced by the strong benzene absorption seen in the ultraviolet spectra of absolute alcohol. Thus, for an ultraviolet spectrum free of benzene absorption it is better to use 95% ethanol as the solvent, to avoid this interference.

16.6. AEROBIC MICROBIOLOGICAL PROCESSES

Here the term "fermentation" is used in its more general sense to refer to a chemical conversion accomplished by microorganisms, rather than in its strict sense of a microbiological conversion carried out in the absence of air. Most microorganisms are either aerobes or they are facultative, i.e., they are capable of functioning under either aerobic or anaerobic conditions. Therefore, to accomplish fermentations under aerobic conditions gives far greater scope to the types of chemical change possible than if one is limited to anaerobic transformations. The wide range of aerobes and the variety of conditions to which they can adapt also means that the risk of contamination by stray organisms is much greater than with anaerobic fermentations. Also the very need of an air transfer system by aerobic processes provides a greater risk of exposure from foreign organisms present in this air. Nevertheless, development of the technology for aerobic fermentations has substantially increased the number of options for microbiological products. Yeasts can function in an aerobic mode, which is used for bakers' yeast production, or for bread-making. When increased cell mass or leavening are the primary process objectives, aerobic function is valuable. However, the end products of aerobic fermentation of sugar by yeast are carbon dioxide and water, not very useful chemical conversions.

16.6.1. Operating Details of Aerobic Processes

Neither oxygen nor air is very soluble in water so that the rate of exchange of dissolved gases from a medium to and from air rapidly becomes the limiting factor for aerobic conversions. In the laboratory, the rate of air exchange per unit volume of liquid is increased by use of a thin layer of the liquid fermentation medium in the bottom of a gently shaken Erlenmeyer flask. Air exchange between the flask air and the outside air is obtained using a tight, sterile cotton plug, which also serves as a filter barrier to intrusion by outside organisms. These methods function well for preliminary tests and process exploration, but are inappropriate for larger scale operations.

For larger scales, shallow trays made of corrosion-resistant metal or plastic placed in portable trolleys may be used for efficient liquid–air contact. The trolleys are placed in a sterile room with frequent sterile air changes under closely controlled temperature and humidity conditions. Efficient air–liquid contact may also be obtained from a loosely packed, moist, solid medium permeated with air. For instance bran is used as the substrate for enzyme production by *Aspergillus oryzal*. Wine vinegar (acetic acid) is produced by trickling wine (basically ethanol in water) over a short tower of wood chips, which are used both as a support and to provide air to *Acetobacter aceti* [13] (Eq. 16.15). By law, the acetic acid (3–5% in water) in vinegar *must* be produced by fermentation, and not by petrochemical methods.

The most common procedure for large-scale aerobic fermentations is to use deep tanks of a stirred liquid medium. Efficient air exchange is obtained by continuous pumping of sterile air into crossed or coiled perforated pipe placed at the bottom of the tank at rates of up to one volume of air per unit of medium volume per minute [62]. This method promotes aerobic growth throughout the medium, which accomplishes large volume production in a limited space. But it also has problems, which relate to the maintenance of sterility and occasionally from foam formation. Overall, however, submerged aerobic fermentations are the most common method used by industry for aerobic processing.

16.6.2. Aerobic Processes to Single Cell Protein

Without knowing the detailed metabolic pathways, many fermentative products today are obtained via aerobic fermentation processes. Probably the largest scale, single product aerobic fermentations are those used to produce single cell protein (SCP), or yeast biomass from hydrocarbons [63]. Here, an aerated yeast suspension in water, which contains trace nutrients is used to generate yeast biomass from gas oil, a mixture of straight chain hydrocarbons in the C_{12} to C_{20} range (Eq. 16.23).

$$\begin{array}{cccccc} \text{yeast} & + & \text{water} & + & \text{gas oil} & + & \text{air} & \rightarrow & \text{yeast biomass} & & 16.23 \\ & & \text{(with} & & \text{(C_{12} to} & & \text{(with} \\ & & \text{nutrients)} & & \text{C_{20})} & & \text{ammonia)} \end{array}$$

The yeast produced by continuous culture techniques is separated from the liquid medium and solvent washed by centrifugation or filtration techniques. After drying, a protein supplement is obtained, which contains 65–68% protein and is suitable for addition to animal feeds. This protein content compares very favorably with that of dry fish meal, which contains about 65%, and dry skim milk powder with about 32%. The SCP processes have operated on the thousands of tonne/year scale in the U.K., France, and Italy, but regulatory problems with facilities operating on unpurified gas oil feedstocks have caused some shutdowns [64]. Nevertheless, because of a cell mass doubling time of 2.5–3 hr and the efficient carbon conversion to protein of this technology, these developments deserve to be explored further.

It is also possible to convert methane to a mixed cell biomass in a similar type of fermenter [65]. Difficulties with obtaining sufficiently high concentrations of methane in the water phase, and the associated pressure required to keep it there, have led to disfavor with this system.

Methanol, which is relatively easily derived from methane and is also readily purified, avoids some of the contamination problems of the higher, n-paraffin carbon sources. Also methanol is easily put into aqueous solution at any desired concentration. This is the basis of ICI's process to manufacture protein from methanol using *Methylophilus methylotrophus*, on a scale of 30,000–50,000 tonne/year [66] (Fig. 16.9). Recombinant DNA technology was employed to raise the efficiency of methanol conversion by this organism. The dry product trade named Pruteen contains 72% protein and is suitable for animal feed supplementation.

Digestive sensitivity problems have prevented any of these SCP products from being used for human consumption, but desirable features contribute to a continuing interest in these types of processes.

16.6.3. Aerobic Processing to Substituted Hydrocarbons

Aerobic microorganisms are capable of selective oxidation of specific hydrocarbon substrates, which would be less efficient or impossible by direct

FIGURE 16.9 Details of the ICI single cell protein (SCP) process based on methanol. Reprinted from Prave and Faust[67], courtesy of the Royal Society of Chemistry.

chemical means. Strains of *Pseudomonas desmolytica*, for example, can carry out the specific oxidation of p-cymene to cumic acid [68], a process difficult to achieve by ordinary chemical means (Eq. 16.24).

$$16.24$$

p-cymene   cumic acid

Some steroid transformations, which are difficult to achieve by ordinary chemical means may also be accomplished readily by microbiological methods. Salicylic acid may be obtained in a 94% yield, on a weight for weight basis, by the action of *Pseudomonas aeruginosa* on naphthalene [69] (Eq. 16.25).

$$16.25$$

naphthalene   salicylic acid   "ASA"

Acetylation of the product with acetic anhydride or ketene nearly quantitatively converts this to acetylsalicylic acid (ASA, or aspirin), one of the most widely sold, nonprescription analgesics.

16.6.4. Microbiological Processes to Amino Acids

As the pressures on the world's available food supply become greater, the per capita availability of first class (animal) protein will become more limited. This will place an increasing dependence on plant (vegetable) sources of protein for this necessary component of our diets. Most of the first class, animal-derived protein (meat, fish, eggs, milk, cheese, etc.) has a food demand to produce it of 2–20 times the weight of the protein obtained. Thus the amount of arable land required to support the production of a given amount of animal protein is much larger than would be required to supply a similar amount directly from plant sources.

The problem with using plant sources of protein is that they do not provide the proper balance of the amino acids required for human nutrition [70]. Some of the deficient amino acids may be synthesized by the body. But some, the "essential amino acids," must be acquired through the diet for proper human nutrition and growth. Commercial production of amino acids has mainly been for animal feed supplementation. However, cereals supplemented with synthetically produced essential amino acids could improve the nutritional value of plant crops to supply the human protein requirement. In this way a proportion of the crops now grown for animal feed could be diverted to human consumption, and more people could be fed from the same area of arable land than would be possible using animal protein.

At the moment L-glutamic acid, a nonessential amino acid, is produced on the scale of about 340,000 tonne/year, about a quarter of this in Japan. This is the largest volume of any commercially produced amino acid. Common starting materials for fermentation production of glutamic acid are glucose, molasses, corn or wheat glutens, or Steffen's waste liquor (from beet sugar mills) plus trace nutrients, although it is also possible to use purified fractions of n-paraffins (C_{12} to C_{20}) or acetic acid as substrates [57] (Eq. 16.26).

$$\text{glucose or molasses} \rightarrow \underset{\text{glutamic acid}}{\text{HOOCCH}_2\text{CH}_2\text{CH(NH}_2)\text{COOH}} \rightarrow \qquad \text{16.26}$$

$$\underset{\text{zwitterionic form}}{\text{HOOCCH}_2\text{CH}_2\text{CH(}\overset{+}{\text{N}}\text{H}_3)\text{COO}^-}$$

Glutamic acid, as the end product of this fermentation, results from a tinkering of the normal cell metabolism of *Micrococcus glutamicus* in such a way that glutamic acid, as it is formed, "leaks" through the cell membrane and accumulates in the aqueous medium. Treatment of the glutamic acid product with sodium carbonate then yields monosodium glutamate (MSG). The MSG is produced to supply its large market as a flavor enhancer, not as a salt of an essential amino acid. However, the experience gained from the fermentative production of glutamic acid on a commercial scale has paved the way for the commercial production of essential amino acids.

When monosodium glutamate is produced by a petrochemical route (Eq. 16.27), it yields the racemic mixture of D-, and L-glutamic acid. This product is only about half as effective as the fermentative product on a weight basis.

$$\text{CH}_2\text{=CH—CN} \xrightarrow[\text{reaction}]{\text{oxo}} \text{OHCCH}_2\text{CH}_2\text{CN} \xrightarrow[\text{2. NH}_3]{\text{1. HCN}} \qquad \text{16.27}$$

$$\text{NCCH(NH}_2)\text{CH}_2\text{CH}_2\text{CN} \xrightarrow[\text{H}_2\text{O}]{\text{hydrolysis}} \text{HO}_2\text{CCH(NH}_2)\text{CH}_2\text{CH}_2\text{CO}_2\text{H}$$

For this, and other economic reasons the petrochemical route is not as attractive as the fermentative route, so is not practiced commercially.

Of the two essential amino acids, which are generally deficient in cereals, L-lysine and L-methionine, only L-lysine is currently produced by fermentation to any significant extent [53]. Again a carbohydrate source is a suitable substrate for cultures of *Corynebacterium glutamicum* to produce L-lysine (Eq. 16.28).

$$\text{carbohydrate source} \rightarrow \text{H}_2\text{NCH}_2\text{CH}_2\text{CH}_2\text{CH}_2\text{CH(NH}_2)\text{COOH} \rightarrow \qquad 16.28$$
$$\text{lysine}$$

$$\text{H}_2\text{NCH}_2\text{CH}_2\text{CH}_2\text{CH}_2\text{CH(}\overset{+}{\text{N}}\text{H}_3)\text{COO}^-$$
$$\text{zwitterionic form}$$

As with glutamic acid production, some modification of the normal metabolic pathway of *C. glutamicum* imposed by the fermentation conditions, is necessary in order to accumulate lysine in the medium. At present, worldwide production of L-lysine by fermentation routes totals some 70,000 tonnes annually, about 40% of this by Japan. About 9,000 tonnes of racemic lysine (DL mixture) per year is also produced synthetically from petrochemical sources. One interesting synthetic substrate for DL-lysine is caprolactam (or 6-amino-caproic acid), a starting material, which is readily available on a tonnage scale from its use as a nylon monomer.

As yet methionine, the other amino acid in which cereals are deficient, is only produced from pentaerythritol or acrolein by synthetic means, not by fermentation [71], which yields the DL product (e.g., Eq. 16.29).

$$\text{CH}_3\text{SH} + \text{CH}_2{=}\text{CHCHO} \rightarrow \text{CH}_3\text{SCH}_2\text{CH}_2\text{CHO} \xrightarrow[\text{2. (NH}_4)_2\text{CO}_3]{\text{1. NaCN}}$$

$$\underset{|}{\overset{}{\text{RCHNHCOO}}} \overset{- \ +}{\text{NH}_4} \qquad \text{CH}_3\text{SCH}_2\text{CH}_2\text{CH}\!-\!\text{CO}$$
$$\text{CONH}_2 \qquad \qquad \qquad \text{NH}\!-\!\text{CO} \qquad 16.29$$

$$\text{CH}_3\text{SCH}_2\text{CH}_2\text{CHCOOH}$$
$$|$$
$$\text{NH}_2$$
$$\text{DL-methionine}$$

However, it is probably only a matter of time before the production of this and other essential amino acids are produced via biochemical methods. This will come either from a detailed understanding of the metabolic processes of more types of suitable microorganisms or by genetic engineering techniques to generate specific new types of microorganisms with the required synthesis characteristics [72].

Plant genetic experiments presently being undertaken with corn (maize) and with rice may improve the nutritional properties of these cereals. These may help to close the amino acid nutritional gap, which exists between plant sources of proteins and first class proteins [70]. However, this will still not remove the need for synthetic amino acids for their many other uses: (e.g., for chelating agents), as dietary supplements, as building blocks for specialty proteins, and as components of intravenous feeding solutions. All of these uses will continue to provide incentives for large-scale biochemical and chemical synthesis of amino acids. Prospects and applications of amino acids have recently been surveyed [73].

16.6.5. Other Aerobic Fermentation Products

Since the original development and commercialization of penicillin production in about 1943 [74], the antibiotics as a class and the penicillins in particular are still dominant in sales value in the aerobic fermentation products area [75]. The submerged culture methods used to prepare penicillin itself, which sold for $330/g in 1943, were sufficiently improved and refined that by 1970 the price had dropped to about $2/g (Eq. 16.30).

$$
\begin{array}{c}
\text{H} \qquad \text{S} \\
\qquad\;\; \diagup \quad \diagdown \\
\text{RCONH} - \text{C} - \text{CH} \quad \text{C(CH}_3)_2 \\
\quad | \qquad\; | \qquad\quad | \\
\qquad\quad \text{C} - \text{N} - \text{CH(COOH)} \\
\;\; \diagup\!\!\!\!= \\
\text{O}
\end{array}
$$

16.30

$R = PhCH_2 -$	$R = PhOCH_2 -$	$R = PhCH(NH_2) -$
"Penicillin"	Phenoxymethyl	Ampicillin
benzylpenicillin,	penicillin, or	(semisynthetic)
or penicillin G	penicillin V	

Details of production methods and process flowsheets have been given by Shreve and Brink [3]. Despite the decreased cost of penicillin and of the related antibiotics such as streptomycin, neomycin, the tetracyclines, as their production methods were streamlined, the medical benefits of these drugs are so well established that the value of U.S. production of drugs in this class now exceeds a billion dollars annually [75].

Many vitamins too are produced by fermentation methods, although some are also produced by chemical synthesis. These represent another important class of submerged aerobic culture products. The value of prescription vitamins alone that were marketed in the U.S.A. in 1980 exceeded $130 million [75]. In 1959, during the early development of the commercial fermentative product vitamin B_{12}, it sold for $139/g. Through processing improvements, the price of this product is now $8/g. Just the two B vitamins, riboflavin (vitamin B_{-2}), (at $56/kg) and vitamin B_{-12}, now account for U.S. sales of more than $20 million annually.

Citric acid is the largest scale pure organic acid produced by a microbiological process. World production in 1990 was about 550,000 tonnes. It is produced by *Aspergillus niger* from a medium containing sugars, generally low-cost cane or beet molasses. The stoichiometry of this process was originally established by Currie [76] (Eqs. 16.31 and 16.32).

$$C_6H_{12}O_6 + 3/2\ O_2 \rightarrow HOOCCH_2COH(COOH)CH_2COOH + 2\ H_2O$$
glucose $\qquad\qquad\qquad\qquad\qquad\qquad$ citric acid

16.31

$$C_{12}H_{22}O_{11} + H_2O + 3\ O_2 \rightarrow 2\ HOOCCH_2COH(COOH)CH_2COOH + 4\ H_2O$$

sucrose   citric acid

16.32

To obtain citric acid as the metabolic product again requires interference with the normal Krebs tricarboxylic acid cycle in such a way that citric acid metabolism is blocked. Usually this is achieved by careful regulation of concentrations of trace metals available as coenzymes to the various enzyme pathways used by *A. niger,* so that some of these are rendered ineffective (are blocked).

About 40% of the citric acid produced is used as a synthetic flavoring agent, to give tartness to the fruit flavors in soft drinks. It is also a valuable chelating agent, helping to flavor and preserve foods, as well as contributing to the cleaning action of detergents. There is also interest in its use as a component of an aqueous sulfur dioxide scrubbing solution. No chemical synthesis of citric acid yet devised has been able to compete effectively with this microbiological method.

Several other organic acids are also made commercially on a large scale by microbial techniques. Some of the better known examples of these are acetic acid (vinegar) via the metabolism of alcohol in water by *Acetobacter aceti,* and lactic acid (Eq. 16.33), produced by *Streptococcus lactis* fermentation of hexose sugars.

$$CH_3CH(OH)COOH \qquad HOCH_2(CHOH)_4COOH$$

lactic acid   gluconic acid

16.33

$$H_2C = C(COOH) - CH_2COOH$$

itaconic acid

Common sugar sources for lactic acid are molasses, hydrolyzed corn, and whey, a by-product of cheese production. Gluconic and itaconic acids are also microbiological oxidation products, obtained by methods similar to those used for citric acid [2]. Gluconic acid is formed, via simple oxidative fermentation, by *Aspergillus niger* or *Acetobacter suboxydans,* of the glucose aldehyde group.

16.6.6. Soluble, and Immobilized Enzymes

A branch of the fermentation industry has pursued the isolation of enzymes, which are the catalytically active proteins of a cell. In the presence of the appropriate coenzyme, these proteins are the active principals of the cell, which carry out the chemical changes from which the cell gains energy. Rennet, a milk curdling enzyme used in cheese-making, and the amylases, used in food processing and for laundry stain removal, are among the earliest examples of isolated enzymes, which have been used commercially [77]. The essence of the value of enzyme use is that the enormous rate enhancements obtained by microbiological means can be achieved in the absence of the organism, if the enzyme responsible is stable and can be isolated. A third of current enzyme production is also directed toward therapeutic uses. A direct result of these wide applications is an increase in the value of the world enzyme market from $300 million in 1979, $480 million in 1985, to over a billion dollars in 1990.

The advantage of using isolated enzymes rather than whole cells to carry out a chemical change is that the process does not have to cater to the special requirements of living cells. However, enzymes, too, are sensitive entities so that care has to be exercised in the conditions under which they are used in order to maintain catalytic activity. An additional problem is that use of the isolated enzyme is frequently a more expensive proposition, on a single-use basis, than the use of propagated cells. Another disadvantage is the need to recover a costly enzyme from the product, once the desired change is completed. Alternatively, long reaction times may be necessary if a low concentration of an expensive enzyme is used to decrease enzyme costs [78].

The full potential of enzyme use may be realized by fixing or immobilizing the enzyme in some way. The substrate solution is then passed through a bed or tube of the immobilized enzyme, and the product(s) collected as the solution emerges after conversion by the enzyme.

A number of methods are being tested for enzyme immobilization. The method selected depends on the operating details of the enzyme system employed and the nature of the solvent to be used, which is usually water. Enzyme, or inactivated cells, may be encapsulated in a film, or encased in a gel, which is permeable to both the substrate and product, but not to enzyme [77]. Porous glasses or insoluble polymers such as a derivatized cellulose may be used as a support onto which enzyme is adsorbed. Pendant functional groups of a polymer, such as those of the ion-exchange resins, can be used either to ionically bind the enzyme to the resin active sites or to covalently bond the enzyme to the resin [79]. The enzyme may be bonded to a polymer backbone chain using a bifunctional monomer such as glutaraldehyde to react with enzyme sites that do not affect its catalytic activity [80].

Hollow fiber technology may also be usefully wedded to immobilized enzyme technology by making the wall of a hollow fiber out of a gel in which the enzyme is already immobilized. The finished fiber may then be packed in small pieces inside a larger tube to be used as a large surface area for enzyme activity. Or, it may be used as a membrane into which substrate solution permeates from the outside, under a pressure differential. Passage through the tube wall allows contact of the substrate with the gelled enzyme, which converts substrate to product. Then the differential pressure across the wall of the tube would force the product to flow through to the inside of the tube, to be collected at a multitube header.

Immobilized enzymes have already been successfully tested for sugar conversion to ethanol [81]. Conversion of the lactose of whey to glucose and galactose is another confirmed application. Preliminary tests predict 300-day operating cycles, when processing 6,500 L/day (1800 U.S. gal/day) of whey containing 5% lactose [82]. This converts the lactose in a cheese-maker's waste stream into two monosaccharides of higher sweetening power and substantial food processing value.

Enzymes, in both soluble and immobilized forms, are also increasingly being employed in the food processing industry to convert other sugars with low sweetening power into more potent sweeteners. As examples, glucose is isomerized to fructose, which is 2.5 times as sweet, by an immobilized glucose isomerase enzyme in a fixed, or a fluidized bed [83]. Also a high-fructose corn syrup, to be produced from corn using two stages of immobilized enzymes, is likely to soon become a commercial reality. Starch hydrolysis to glucose would be accomplished by an immobilized glucoamylase in the first stage, in 6 min, followed by glucose isomerization to fructose, as already outlined.

Enzymic interconversion of sugars has also been exploited as an additional fresh milk processing step to convert the normal lactose milk sugar, to which some people react unfavorably, into glucose and galactose, which are much better tolerated [82]. Products from this milk can now be enjoyed by those who could only tolerate milk substitutes previously.

Amino acid production too may benefit from this immobilized enzyme technology. Optically active D(-)-a-phenylglycine is one example, which has been confirmed, but more are likely to follow.

The promise of fixed enzymes to process many food and waste products is only just beginning to be realized [84]. Combining this technology with the developing field of synthetic enzyme preparations will lead to a wide range of feasible, clean, and specific chemical conversions never before thought to be possible [85].

REVIEW QUESTIONS

1. In the process of photosynthesis green plants use carbon dioxide and water to produce sugars according to the reaction:

$$H_2O + CO_2 \rightarrow C_{12}H_{22}O_{11} + O_2 \text{ (unbalanced)}$$

(a) What volume of CO_2 at 20°C and 760 mm Hg pressure is used by a plant to make 1 kg of sucrose, $C_{12}H_{22}O_{11}$?
(b) What volume of air is deprived of its normal amount of CO_2 in this process? (Air contains 0.037% CO_2 by volume.) Assume a barometric pressure of 760 mm Hg and a temperature of 20°C, and that photosynthetic CO_2 depletion is complete.

2. Normal brewing involves yeast fermentation of a wort initially containing 12% sugars (as $C_6H_{12}O_6$, 120 g/L solution) down to a beer containing 2.4% sugars plus ethanol, with evolution of carbon dioxide. Assume quantitative sugar conversion by zymase, negligible volume change on conversion, and unit density of starting and product solutions.

(a) What percent concentration of alcohol, weight/liter, should this produce?
(b) What mass of carbon dioxide would be coproduced per liter of wort fermented?
(c) If 0.029 L of carbon dioxide at 5°C and 1 atm pressure are required to carbonate 1 L of beer, what proportion of the carbon dioxide produced is required for carbonation?

3. Winemaking essentially involves fermentation of a must containing sugars and flavor elements to a solution of ethanol plus flavor elements still containing some residual sugar.

(a) If a must containing 185 g/L sugar is fermented to a green wine with a sugar content of 45 g/L, what percent concentration of alcohol by weight should this produce? Assume quantitative conversion, of $C_6H_{12}O_6$, glucose, negligible volume change on conversion and a density of 1.075 g/mL both before and after fermentation.
(b) What would be the technical classification of this wine, if the carbon dioxide produced was allowed to escape?
(c) What mass of carbon dioxide would be produced at the same time per liter of must fermented?
(d) What fraction of the cleaned-up carbon dioxide would be required for reinjection into the wine on bottling to produce a champagne? Bottling conditions are 1.0 L CO_2 added per L of wine, at 5 atm absolute pressure, and 0°C.

4. White vinegar is a solution of acetic acid in water. Knowing only that acetic acid contains C, H, and O atoms in the ratio of 1:2:1, respectively, and that a sample of vinegar gives an analysis of 2.00% carbon by weight, what percentage (by weight) of the sample is actually acetic acid?

5. An American 20° overproof navy rum is to be sold in Canada.
(a) What is the percent alcohol (ethanol) by volume in this product?
(b) What would be the proper designation of this product using the Canadian proof system (which is equivalent to degrees Sikes)?
(c) How should a brandy containing 45.7% alcohol by volume be specified on the Canadian proof system, i.e., degrees Sikes?

6. Corn contains 82% starch (dry weight basis) and an unfermentable residue of corn oil, gluten, fiber, etc. How many litres of 100% ethanol (density 0.789 kg/L) would be obtained from one metric tonne of dry corn, assuming quantitative conversions of starch to fermentable sugars, and sugar fermentation to ethanol?

7. Corn contains 82% starch (dry weight basis) and an unfermentable residue of corn oil, gluten, fiber, etc.
(a) What dry weight mass of corn would be required to produce 1,000 L of 100% ethanol (density 0.789 g/mL) assuming quantitative conversions of starch to fermentable sugars, and sugar fermentation to ethanol?
(b) Ethanol (100%) currently sells at $1.50 Imp. gallon. At what price would corn have to be available, allowing 50% of the price of the ethanol for operating overheads (capital costs, energy, labor, etc.) to enable this to be an economic source of the starch? 1 Imp. gal. = 4.546 L.
(c) At a straight gasoline price of 42¢/L, what would be the price of gasohol at 10% ethanol by volume?
(d) Name any two other potential sources of fermentable sugars for fuel alcohol production, which are likely to be significantly less expensive than corn.

8. At one time a test for the detection of adulteration of tobacco by sugar involved extraction of a weighed dry tobacco sample with hot water, and fermentation of the extract by yeast [P. Hammond, *Chem. Brit.*

28(9): 796–798 (1992)]. Unadulterated tobacco contains 1–2% by weight fermentable sugars. Fermentation of the water extract from a 100-g sample of tobacco gives 6.6 g of pure ethanol and 0.13 mole of carbon dioxide. Assume quantitative fermentation using $C_6H_{12}O_6$.

(a) What is the percent sugar by weight in the tobacco sample from the ethanol recovered?

(b) What is the percent sugar by weight in the tobacco sample calculated from the moles of carbon dioxide collected?

(c) Is the tobacco sample adulterated or not, with consideration of the possibility of experimental error in each of the two results?

9. Vinegar is produced by the oxidative microbiological conversion of a 6.0% by weight solution of ethanol in water to acetic acid in water as the sole carbon-containing product.

(a) What concentration of acetic acid by weight would be present in the product, assuming quantitative conversion of ethanol and unit density for the starting and product solutions?

(b) What volume of air (assume $1/2\ O_2 : 2\ N_2$) at 1 atmosphere and $0°C$ would be required to carry out the conversion of $2.00 \times 10^3\ L$ of the starting solution?

10. (a) What mass of powdered calcium oxide would theoretically be required to dry 100 L of 95.6% alcohol (ethanol) by volume (density 0.814 g/mL, 4.4% water)?

(b) What factors would have to be considered to determine the relative cost of this chemical drying method, versus the costs of azeotropic drying?

FURTHER READING

American Chemical Society, "Wine Production Technology in the United States," ACS Symp. Ser., No. 145. American Chemical Society, Washington, DC, 1981.

E.D. Baxter and P.S. Hughes, "Beer Quality, Safety and Nutritional Aspects," Roy. Soc. Chem. Cambridge, U.K., 2001.

R.G. Board and D. Jones, "Microbial Fermentations: Beverages, Foods and Feeds," Blackwell Science, Osney Mead, Oxford, 1995.

M.E. Bushell, "Computers in Fermentation Technology." Elsevier, Amsterdam, 1988.

I. Goldberg, "Single Cell Protein." Springer-Verlag, New York, 1985.

K. Mosbach, "Immobilized enzymes and cells, Part B." Academic Press, Orlando, FL. 1987

O.P. Ward, "Fermentation Biotechnology: Principles, Processes, and Products." Prentice-Hall, Englewood Cliffs, NJ, 1989.

REFERENCES

1. P. Dunnill, *Chem. Ind. (London).* 204, Apr. 4 (1981).

2. J.A. Kent, ed., "Riegel's Handbook of Industrial Chemistry," 7th ed., p. 156. Van Nostrand-Reinhold, New York, 1974.

3. R.N. Shreve and J.A. Brink, Jr., "Chemical Process Industries," 4th ed. p. 525. McGraw-Hill, New York, 1977.

4. H.J. Phaff, *Sci. Amer.* 245(3), 76, Sept. (1981).

5. A. Mulchandani and B. Volesky, Production of Acetone–Butanol–Ethanol by *Clostridium Acetobutilicum*, *J. Biotechnology*, **34**, 51–60, Apr. 30 (1994).

6. F.W. Salem, "Beer, Its History and Economic Value as a National Beverage," Arno Press, New York, 1972, reprint of 1880 edition.

7. T. Lom, *Can. Chem. Process.* **58**(12), Dec. 14 (1974).

8. Calc. from "United Nations Statistical Yearbook 2001," 48th ed., United Nations, New York, 2003, and earlier years.

9. "Kirk-Othmer Encyclopedia of Chemical Technology," 3rd ed., Vol. 3, p. 692. Wiley, New York, 1978.

10. M. Linko, A. Haikara, A. Ritala *et al.*, Recent Advances in the Malting and Brewing Industry, *J. Biotechnol.* **65**(2–3), 85–98, Oct. 27 (1998).

11. K.E. Behre, The History of Beer Additives in Europe—A Review, *Veg. Hist. Archaeobot.* **8**(1–2), 35–48, June (1999).

12. C. Bamforth, Brewing a Better Beer, *Chem. Brit.* **33**(8), 37–39 Aug. (1997).

13. M.A. Amerine, *Sci. Amer.* **211**(2), 46, Aug. (1964).

14. G. Donaldson and G. Lampert, eds., "The Great Canadian Beer Book." McLelland and Stewart, Toronto, 1975.

15. R.I. Tenney, The brewing industry. In: "Industrial Fermentations," (L.A. Underkofler and R.J. Hickey, eds.), Vol. 1, p. 172. Chem. Publ. Co. New York, 1954.

16. "Kirk-Othmer Encyclopedia of Chemical Technology," 4th ed, vol. 4, pp. 56, 160. Wiley, New York, 1992.

17. A. Vogler and H. Kunkely, Photochemistry and Beer, *J. Chem. Educ.* **59**(1), Jan. 25 (1982).

18. Avoiding an Environmental Hangover, *Can. Chem. Process.* **64**(2), 29, Mar. (1980).

19. S. Beszedits, *Water Pollut. Control.* **115**(10), 10, Oct. (1977).

20. M.L.Westendorf and J.E.Wohlt, Brewing By-products:…Use As Animal Feeds, *Vet. Clin. N. Am-Food A.* **18**(2), 233, July (2002).

21. H.M. Malin, Jr., *Envir. Sci. Technol.* **6**, 504 (1972).

22. J.F. Walker, *Water Pollut. Manag.* p. 174 (1972), cited by *Water Polln Abstr. (U.K.)*, **45**, 2953 (1972).

23. H.W. Allen, "A History of Wine," Horizon Press, New York, 1961.

24. Wine Institute (California) data. Available from: http://www.wineinstitute.org/communications. keyfacts_worldwineproduction02.htm

25. P. Duncan and B. Acton, "Progressive Winemaking," Amateur Winemaker, Andover, Hants, 1967.

26. M.A. Amerine, H.W. Berg, and W.V. Cruess, "The Technology of Winemaking," 3rd ed. Avi Publ. Co., Westport, CT, 1972.

27. L. Trauberman,…Winemaking…Product Stabilization, *Food Engin.* **39**(11), 83, Nov. (1967).

28. V.L. Singleton, Maturation of Wines and Spirits—Comparisons, Facts, and Hypotheses, *Amer. J. Enol. Viticult.* **46**(1), 98–115 (1995).

29. B. Hocking, Mead—A Silk Purse from a Sow's Ear. *Can. Bee J.* **63**(12), 4–6, Dec. 18 (1955).

30. L. Morisset-Blais, *Can. Chem. Process.* **57**(12), 31, Dec. (1973).

31. H.M. Cunningham and R. Pontefract, *Nature (London)*, **232**, 332 (1971).

32. P. Cabras, A. Angioni, V.L. Garau *et al.*, Pesticides in the Distilled Spirits of Wine and Its By-products, *J. Agr. Food Chem.* **45**(6), 2248–2251 June (1997).

33. W.H. Detwiler, *Chem. Eng. (N.Y.)*, **75**, 110, Oct. 21 (1968).

34. A. Agostini, Wynboer **32**(395), 13 (1964); cited by Amerine *et al.* [26].

35. M.L. Silva, A.C. Macedo, and F.X. Malcata, Review: Steam Distilled Spirits from Fermented Grape Pomace, *Food Sci. Technol. Internat.* **6**(4), 285–300, Aug. (2000).

36. A.Vernhet, K. Dupre, L. Boulange-Petermann *et al.*, Composition of Tartrate Precipitates During…Cold Stabilization of…White Wines, *Am. J. Enol. Viticult.* **50**(4), 391–397 (1999); Red wines, *ibid.*, 398–403.

37. G. Fregapane, H. Rubio-Fernandez, and M.D. Salvador, Continuous Production of Wine Vinegar in Bubble Column Reactors…, *Eur. Food Res. Technol.* **216**(1), 63–67, Jan. (2003).

38. J.M. Guillot, V. Desauziers, M. Avesac *et al.*, …. Treatment of Olfactory Pollution Emitted by Wastewater in Wineries…., *Fresenius Envir. Bull.* **9**, 243–250 (2000).

39. "Kirk-Othmer Encyclopedia of Chemical Technology," 2nd ed., Vol. 1, p. 501. Wiley-Interscience, New York, 1963.

40. N.A. Lange, ed., "Lange's Handbook of Chemistry," 10th ed., p. 1183. McGraw-Hill, New York, 1969.

41. Business Week, issues of Feb. 17, 1962 and March 21, 1977. Cited by Kirk-Othmer [45].

42. "Adams Liquor Handbook 2002," p. 9. Adams Publishing Co., New York, 2002.

43. B.G. Reuben and M.L. Burstall, "The Chemical Economy," p. 428. Longman Group Ltd., London, 1973.

44. J.S. Swan and S.M. Burtles, *Chem. Soc. Rev.* **7**(2), 201 (1978).

45. "Kirk-Othmer Encyclopedia of Chemical Technology," 3rd ed., Vol. 3, p. 830. Wiley, New York, 1978.

46. W.F. Rannie, "Canadian Whisky, The Product and the Industry," W.F. Rannie Publishing Co., Lincoln, Ontario, 1976.

47. "McGraw-Hill Encyclopedia of Science and Technology," 5th ed., Vol. 4, p. 346. McGraw-Hill, New York, 1982.

48. W. Wardencki, P. Sowinski, and J. Curylo, . . . Analysis of Volatile Carbonyl Compounds in Spirits . . . , *J. Chromatogr. A.* **984**(1), 89–96, Jan. 10 (2003).

49. P.V. Price, "Penguin Book of Spirits and Liqueurs," Handbook Series. Penguin Books, New York, 1981.

50. R.V. Carter and B. Linsky, *Atmos. Envir.* **8**(1), 57 (1974).

51. N.L. Nemerow, "Industrial Water Pollution, Origins, Characteristics, Treatment," Addison-Wesley, Reading, MA, 1978.

52. Methane Pilot for Distillery, *Can. Chem. Process.* **66**(1), 8, Feb. 19 (1982).

53. D.E. Eveleigh, Microbiological Production of Industrial Chemicals, *Sci. Amer.* **245**(3), 154, Sep. (1981).

54. "Kirk-Othmer Encyclopedia of Chemical Technology," 3rd ed., Vol. 9, p. 338. Wiley, New York, 1980.

55. No Recovery in Sight for Major Solvents, *Chem. Eng. News*, **59**(45), 12, Nov. 9 (1981).

56. National Distillers to Convert Plant for Gasohol, *Chem. Eng. News*, **58**(19), 8, May 12 (1980).

57. W.L. Faith, D.B. Keyes, and R.L. Clark, "Industrial Chemicals," 3rd ed., p. 521–527. Wiley, New York, 1965.

58. C.R. Noller, "Textbook of Organic Chemistry," 3rd ed., p. 121. Saunders, Philadelphia, 1966.

59. W. Worthy, *Chem. Eng. News*, **59**(49), 35, Dec. 7 (1981).

60. R.C. Weast, ed., "Handbook of Chemistry and Physics," 56th ed. CRC Press, Cleveland, OH, 1975.

61. R.H. Perry, C.H. Chilton, and S.D. Kirkpatrick, eds., "Chemical Engineers' Handbook," pp. 13–50. McGraw-Hill, New York, 1969.

62. B. Kristiansen, *Chem. Ind. (London)*, p. 787, Oct. 21 (1978).

63. G. Hamer and I.Y. Hamdan, *Chem. Soc. Rev.* **8**(1), 143 (1979).

64. BP Contests Petroprotein Plant Health Issue, *Chem. Eng. News*, **56**(12), 12, Mar. 20 (1978).

65. J.C. Mueller, *Can. J. Microbiol.* **15**(9), 1047 (1969).

66. ICI's Bugs Come on Stream, *Chem. Brit.* **17**(2), 48, Feb. (1981).

67. P. Prave and U. Faust, *Chem. Brit.* **14**(11), 552 (1978).

68. K. Yamada, S. Horiguchi, and J. Takehashi, *Agric. Biol. Chem.* **29**, 943 (1965).

69. T. Ishikura, H. Nishida, K. Tanno et al., . . . Salicylic Acid from Naphthalene, *Agric. Biol. Chem.* **32**, 12 (1968).

70. M. Hamdy, Nutritional Value of Vegetable Protein, *CHEMTECH.* **4**(10), 616 (1974).

71. R.F. Goldstein and A.L. Waddams, "The Petroleum Chemicals Industry," 3rd ed. p. 345, 407. Spon, London, 1967.

72. D.A. Hopwood, *Sci. Am.* **245**(3), 90, Sep. (1981).

73. "Kirk-Othmer Encyclopedia of Chemical Technology," 4th ed. Vol. 2, pp. 504–579. Wiley, New York, 1992.

74. S. Aldridge, A Landmark Discovery, *Chem. Brit.* **36**(1), 32–34, Jan. (2000).

75. Y. Aharanowitz and G. Cohen, Microbiological Production of Pharmaceuticals, *Sci. Amer.* **245**(3), 140, Sep. (1981).

76. J.N. Currie, *J. Biol. Chem.* **31**, 15 (1917); *Chem. Abstr.* **11**, 2814 (1917).

77. J.L. Meers, Enzymes for the Future, *Chem. Brit.* **12**(4), 115, Apr. (1976).

78. C. Bucke and A. Wiseman, *Chem. Ind. (London)*, p. 234, Apr. 4 (1981).

79. C.J. Suckling, *Chem. Soc. Revs.* **6**(2), 215 (1977).

80. W.R. Vieth and K. Venkatasubramanian, *CHEMTECH.* **4**(1), 47 (1974).

81. Y. Shabtai, S. Chaimovitz, and A. Freeman, Continuous Ethanol Production by Immobilized Yeast Reactor...*Biotechnol. Bioeng.* **38**, 869–876 (1991).

82. R. Greene, Slow Road for Immobilized Enzymes, *Chem. Eng. (N.Y.)*, **85B**, 78, Apr. 10 (1978).

83. Enzymes Now Winning New Applications, *Can. Chem. Process.* **61**(12), 20, Dec. (1977).

84. D. Perlman, *CHEMTECH.* **4**(4), 210 (1974).

85. C.J. Suckling and K.E. Suckling, *Chem. Soc. Rev.* **3**(4), 387 (1974).

Bucket excavator used for the large-scale shifting of surface materials required for the tar sands processing. Syncrude, Fort McMurray, Alberta.

17

PETROLEUM PRODUCTION AND TRANSPORT

Energy is beauty—a Ferrari with an empty tank doesn't run.

—Elsa Peretti (1940)

17.1. PRODUCTION OF CONVENTIONAL PETROLEUM

Early use of petroleum or mineral oil, as distinct from animal or plant oils, was achieved by direct harvesting of the crude product from surface seeps and springs. For example, tar obtained from the Pitch Lake (La Brea) area, Trinidad, has been used for the caulking of ships since the Middle Ages and is still marketed to the extent of about 142,000 tonnes per year [1]. Tar from the Alberta tar sands was used in the 1700s by Cree Indians of the Athabasca river area to seal their canoes, as recorded by Peter Pond. Also a thick bituminous gum was collected from the soil surface near the St. Clair River in Southern Ontario [2], and from Guanoco Lake, Venezuela [3], and these too were marketed for a range of purposes.

It was a small but important step to advance from opportunistic oil recovery from surface seeps to the placement of wells in or near these surface deposits to try to increase production. Probably the first wells in the Western world for oil recovery were drilled during the period 1785–1849 in the Pechelbronn area of France [4]. None of these wells, which ranged in depth from 31 to 72 m, was a prolific producer but they did establish the practice of drilling for oil in Europe.

In North America, "Colonel" Edwin H. Drake drilled a well to about 19 m (60 ft) in depth at Titusville, Pennsylvania, in 1859, where he struck oil. This well initially produced about 20 barrels/day. Some historical accounts of the development of the North American petroleum industry credit this event as the first crude oil production from a well [5]. More comprehensive accounts also describe the activity of James M. Williams in Canada at about the same time, who had already marketed some $680 \, \text{m}^3$ (150,000 gallons) of oil produced from his wells near Black Creek, near Petrolia and Oil Springs, Southern Ontario [6]. The first of Williams's wells, which were placed by digging rather than drilling, produced oil at a depth of about 18 m (59 ft) in late 1857 to early 1858. By 1860 each of Williams's five successful wells were producing about $3.2 \, \text{m}^3$ (20 barrels) of oil a day, netting some $1600 \, \text{m}^3$ (10,000 barrels) for Ontario in that year [2]. American production rapidly made up for the later start with a gross of more than $40,000 \, \text{m}^3$ (250,000 barrels) in the same year.

17.1.1. Modern Exploration and Drilling for Oil and Gas

Prospecting for petroleum (literally "rock oil") draws significantly on the present state of knowledge regarding its genesis. Inorganic, biogenic, and dualistic theories have been proposed, with biogenic origins being generally favored [7]. Biogenic oil formation requires the presence, millions of years ago, of shallow seas having a rich microscopic and/or larger animal life and probably also plant life. For the formation of oil deposits these conditions must have been accompanied by a large sediment source, such as a turbid river. Thus, as organisms died and were trapped in the accumulating sediments, anaerobic bacterial decomposition occurred. As further sediments accumulated the transformation process was assisted by contributions of heat (geothermal and decompositional) and pressure. The pressure generated by accumulating sediment eventually caused expression of the fluid and gaseous decomposition products. These then migrated through a porous deposit to formations of lower pressure, generally upward or sideways (horizontally), until the mobile products reached impermeable strata. If this type of hydrocarbon trapping occurred a petroleum reservoir or deposit was formed (Fig. 17.1). Or there may have been no nonporous barrier between the formation zone and the surface, or a fault line may have subsequently permitted migration in which case a large part of the volatile hydrocarbon content either moved to a fresh entrapment zone or was lost to the atmosphere and surroundings by evaporation and weathering. Losses of this type probably combined with some further bacterial action are what caused, and continue to contribute to surface deposits of viscous, high molecular weight hydrocarbons, such as those that occur at Pitch Lake, Trinidad, and the other surface seeps and springs mentioned earlier.

While the harvesting and tapping of more or less self-evident oil deposits was sufficient to provide the early demand for petroleum products, the incentive soon developed to devise methods to explore areas without any such obvious signs to improve the prospects of success on drilling. In the 1940s and 1950s, exploration geologists looked for large-scale structural "traps" that could be seen as domes or anticlines on the surface. For a new area this initially involved aerial strip photography, with overlap of adjacent strips to allow stereoscopic viewing to discover these features. Maps were then made to locate the important features. Aerial surveys were continued via gravity meter measurements, in which the small gravitational differences in rock distribution are measured, and by magnetometer surveys, used to measure

FIGURE 17.1   Three types of structural traps, which can provide a situation favorable to the formation of oil accumulations. Other types of structural traps, such as salt domes are also recognized.

local variations in the earth's magnetic field. These methods are used to estimate the density and thickness of underlying sedimentary rock formations. More recently, geochemical methods, which rely on determination of the hydrocarbon gas content above surface soils, are also being employed [8]. The presence of a hydrocarbon "halo" at the surface from the diffusion of gases from an underground deposit is taken as promising evidence of an accumulation.

More detailed ground surveys are conducted in the more likely areas revealed from the large-scale methods. The first essential of surface study is geologic mapping, which may include fossil collecting and analysis. This is commonly followed by seismic surveys to try to obtain evidence of an anticlinal or other rock structure, which may hold petroleum. Evidence of the presence of a thick sequence of marine sedimentary rock, including some rich in kerogen, would be a good sign of hydrocarbon formation.

Seismic study may employ mechanical "thumpers," or one or more small explosive charges to set off shock waves, which are transmitted through the rock formations. Placement of 100 or more sensitive geophones (listening devices) in a pattern around the site of the explosive charge allows the travel time and frequency of the shock waves to be recorded digitally. Processing and study of these results, in turn, provide geologists with 3D or 4D pictures of shock wave intensities and times, including reflections from deep-seated formations, to provide a better estimated shape of underlying formations [9, 10].

If the site area still looks promising, test drilling will follow these preliminaries. Subsurface geological information is derived from a detailed study by the rig geologist of the rock cuttings (chips) brought up with the flow of drilling mud from the regular rotary bit. This will be followed particularly closely with a new wildcat well. When cuttings show signs of oil-stained and/ or porous zones, or fossils of special interest, core samples (clean cylinders of rock) are taken using a special circular cutting drilling bit. The high cost of coring keeps this method of obtaining information about underlying strata to a minimum. Cores are stored at the surface in the sequence obtained. Study of these permits determination of formations or strata at this drilling site, which are parallel with those previously collected at other drilling sites. It also enables any fossils present to be identified and from a prior knowledge of these, the approximate age of the formations may be established. Stratum dating is an important feature of core analysis because of the good correspondence between formation age and the geological periods during which many oil deposits were laid down [11]. Core samples can also help to establish whether the underlying strata at the test site are appropriate for petroleum entrapment by basic mineralogy, porosity, and permeability measurements. The presence of vugs (holes) in a limestone formation could signal the possibility of an oil reservoir. Evidence of a porous and permeable zone of the appropriate geological age overlain by an impermeable shale would also be promising.

Drilling success rates (i.e., wells drilled, which show *any* evidence of the presence of oil or gas), are markedly improved with the sophistication of the initial exploration techniques used. Ratios of 1 success in 30 or more holes is about all that can be expected without the application of the methods outlined. However, success ratios of 1 in 5 to 1 in 10 are achievable when using evidence gained from these procedures [12]. Since to drill and complete (bring into production) a typical oil well in an accessible region can cost upwards of half a million dollars, and $10 million or more in frontier regions, it is important to assemble as much favorable evidence as feasible before drilling. For a well to be worth bringing into production it must be capable of a significant initial rate of oil or gas production. Otherwise the cost of the associated surface collection equipment becomes too high in relation to the anticipated value of the product, so the well is capped or abandoned, by filling the drilled hole with cement, depending on perception of future prospects. In new areas, the success rate of discovering producing wells, compared to dry holes or nonproducers, runs to about 1 in 9 or 10 wells completed in North America; hence the high exploration investment required by the oil industry.

Early drilled wells were made by a cable-tool system in which a heavy tool, about 10 cm in diameter by 1.5–2 m long and with a sharpened end, was alternately lifted and then let drop in the hole by a cable and winch system. At intervals the loosened material, suspended in a few centimeters of water at the base of the well, was lifted out with a bailer fastened to the end of the cable system. The bailer consisted of a length of pipe fitted with a weighted valve at the lower end, which opened when the pipe was lowered to the end of the hole, and which reclosed on lifting. Then the percussion drilling could be resumed. This method, which had been used by the Chinese for water and brine wells for more than a thousand years, was effective, simple, inexpensive, and could reach depths as great as 1,100 m [4]. It was virtually the only method used to drill oil wells in the nineteenth century and the predominant method for the first two decades of the twentieth century.

However, cable-tool systems were slow relative to other developing methods, particularly in softer formations. They had a depth limitation too and, using this system, it was not possible to provide an effective safeguard for pressure containment in the event that the drill penetrated formations under high gas pressures.

Rotary drilling gradually became the dominant well drilling method in 1900–1925, and enabled more rapid well completion. It also permitted working to depths as great as 8,000 m but required a greater equipment and labor investment to achieve this. First, a larger derrick or support structure 40 or more meters high is required to enable lifting, lowering, and guidance of the drill string and to act as a temporary storage rack for lengths of drill pipe during bit changes, etc. (Fig. 17.2). The actual drilling is accomplished by special types of bit, about 16 cm in diameter, threaded to the end of 9-m lengths of about 8-cm-diameter special high-strength steel pipe sufficient to reach the bottom of the hole from the surface. Bit and pipe diameters can be varied depending on desired hole diameter and depth of well. A lifting system from the top of the derrick allows control of the weight or force on the bit and

FIGURE 17.2 Schematic diagram of a rotary drilling rig showing hoisting derrick and routing of drilling mud.

hence the torque required to rotate it as the bit cuts into the rock formations being penetrated. Since three 9-m lengths (the working unit kept stacked inside the derrick) of even 12-cm-diameter drill pipe weighs some 730 kg, sufficient pipe to reach a 2,000-m depth would weigh some 54 tonnes. Thus, this control task using massive equipment is a delicate one, assigned to the head driller of each operating crew to ensure optimum drilling rates without breakage of the drill string from the application of excessive vertical pressure on the bit. A rotating table at the base of the derrick, driven by 3,000–4,000 hp of tandem diesel engines, is used to grip and rotate the pipe string from the top of the string at 100–250 rpm [12].

A specially formulated drilling mud pumped down the inside of the pipe string and returning between the pipe and the side of the hole provides lubrication and cooling for the pipe string and drill bit. It also carries rock chips up to the surface for removal from the mud by screening, assists in sealing any porous formations, which are being penetrated, and, most importantly, provides a safety feature in the event that a formation under high gas pressure is reached. The density of the drilling mud used will be deliberately increased by the addition of powdered barytes ($BaSO_4$; density $4.15 \, g/cm^3$), iron oxide, or the like when it is anticipated that a high-pressure formation is likely to be encountered. This can be predicted, for instance, when further developing an oil reservoir with known high formation pressures, which has already been tapped by exploratory wells.

Blowouts or the loss of drilling fluid and sometimes the whole of the drill pipe and derrick as well when unanticipated high pressures are encountered, only rarely occur today. Warning signs, such as a rapid unexplained increase in drilling mud volume, sometimes accompanied by a frothy appearance, are taken seriously and the drill mud exit stream flow rate may be restricted to increase the drilling mud pressures in the hole. Also modern drill rigs are equipped with blowout preventers, massive valves placed immediately under the derrick floor and cemented into the top of the hole, and sometimes additional valves, which may be placed at one or two intermediate depths for deeper wells, and can be closed at a moment's notice on the threat of a blowout. In these ways both the hazards and the significant hydrocarbon and formation pressure losses, which can result from a blowout are generally avoided.

In addition to the cable-tool and rotary methods already described, one further mode of drilling is the turbo drill. This method, refined substantially by the Russians, uses the same derrick, draw works (lifting system), and mud pumps of the rotary drilling method. But instead of imparting the turning force to the bit by rotating the string of pipe from the top of the string like a rotary drill, this system uses a turbo-powered unit located in the drill string at the bottom of the hole and just above the drill bit itself to rotate the bit. The pipe string remains stationary, except for vertical movement during drilling, bit changes, etc. Power is carried to the turbo unit and thence to the bit via the high-pressure mud stream itself, flowing down the inside of the string of pipe. Since the drill string does not need to provide the rotating torque to the drill bit with this method, the pipe stresses are much less. For this reason turbo-drilling is of primary value for the making of very deep holes of 5,000 m or more, and for directional and horizontal drilling [13]. Derrick lifting capability of 300 or more tonnes is required for these rigs.

17.1.2. Petroleum Production

Conventional petroleum accumulations can generally be classified into one of four types of oil reservoirs (Fig. 17.3) [14]). Production from a water drive reservoir involves the movement of petroleum upward into the producing well by displacement of petroleum from lower portions of the reservoir into the producing zone by the hydrostatic pressure of underlying water. Conversely, a gas cap drive reservoir relies on the pressure of a separate natural gas phase within the formation to move oil downward toward a producing well. In some reservoirs containing both oil and gas, however, the gas is dissolved in the oil throughout the formation, and is described as a dissolved gas drive. When this occurs, both components move into the producing well. When the pressure drops during production, gas can separate from the oil either in the reservoir or in the well bore. This can interfere with production. The fourth type of reservoir, where little or no production assistance is derived from the natural pressure of the formation contents, is described as a gravity drainage reservoir. In some petroleum producing situations combinations of these reservoir types may be encountered.

The fraction of oil-in-place recoverable from conventional petroleum reservoirs varies greatly with the reservoir type, oil viscosity, formation pressure, production rate, and finesse employed. The positive displacement aspect of water drive reservoirs generally gives them the highest ultimate petroleum recoveries, up to 70% of the oil-in-place [14]. Estimates of ultimate recoveries possible from gas cap drive and dissolved gas drive types of reservoirs are usually much lower, 25–50% for the former and 10–30% for the latter. Recovery from gravity drainage reservoirs will be at the lower end of the ranges of the two gas drive reservoir types.

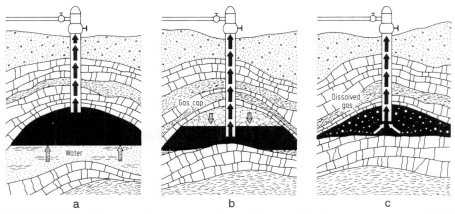

a   b   c

FIGURE 17.3 Diagrammatic representations of three of the four common types of oil reservoir. *Left:* Water drive , *Center:* Gas cap drive, *Right:* Dissolved gas drive. (From Purdy [14], courtesy G.A. Purdy.)

When a new oil formation of one of the first three types is first tapped, formation pressures of up to 20 mPa (3000 lb/in.^{2}) or more may be encountered. This is more than sufficient to force the oil being produced to the surface. Thus, artesian well fashion, the crude oil (plus associated gas) simply requires piping to a temporary holding area or to a wellhead separation manifold. In the early days of the oil industry, when competing interests were all vying for product from the same oil reservoir under little or no government regulation, it became a race to access as many producing wells into the deposit as soon as possible to obtain the maximum share of the product [15]. At the time there was also little application for the coproduced natural gas so that any which did separate from the oil product during production was simply flared at the well site. While these practices may have given one company a larger share of the product than another, they also decreased the ultimate recovery of oil possible from the field to the detriment of all producers. This yield decrease occurred as a net result of the premature loss of formation pressure, an increase in the average viscosity of residual formation oils still in place, and from discontinuities in fluid-filled voids in the rock. All of these production problems were a direct result of the rapid production rates, sometimes complicated by reservoir structural features.

For some newly developed oil fields and for most oil fields when they reach a mature stage of production, formation pressures are inadequate to raise the oil to the surface although they may be sufficient to raise the oil some distance toward the surface of a producing well. In these instances pumping with specially designed versions of deep well lift pumps is used to bring the oil to the surface. When gas is present with the oil, gas separators are placed in the well below the pumping zone to avoid the interference of vapor lock, which could prevent pumping.

17.1.3. Economizing Techniques

Today most oil producing areas have drilling and production regulations in place to put the development of oil fields and oil production on a more orderly and less wasteful basis. For instance, in British Columbia only one well may be drilled per 4 ha (10 acres). Regulations also limit the production rate per well based on a formula, which takes into account the unrestricted production rate for the well on initial testing, estimated reserves present in the particular oil field, and product properties. The regulated production rate of a particular well is controlled by installation of a choke, a plate with a hole of an appropriate size for the required production rate, or by appropriately adjusted valves installed at the wellhead. One estimate of an efficient production rate for a water drive reservoir is that the total annual production of all wells producing from the reservoir be not more than 3–5% of the estimated ultimate yield (not oil-in-place) of the reservoir [16].

Despite conservation measures taken during the early life of mature oil fields, and sometimes even for new oil fields, which have low initial formation pressure, secondary oil recovery by pumping has to be used to initiate or maintain petroleum production. This is one way in which the producing life of an oil field may be extended for some years, at the same time as increasing ultimate yield.

Water flooding, by pumping fresh or salt water plus additives into the formation via one well, while producing displaced oil plus water from surrounding wells, is a tertiary method of improving the ultimate recovery of oil [17]. These methods represent two types of "enhanced oil recovery" (EOR) methods [18].

Some recovery of reservoir pressure and decrease in viscosity of the residual petroleum in the reservoir may be obtained by returning natural gas to the formation. High formation pressures contribute to the solubility of methane, ethane, propane, etc., in the residual petroleum, which brings about the decreased viscosity. Carbon dioxide injection is also used, which accomplishes similar objectives as natural gas return [19], at the same time as offsetting releases of carbon dioxide by fossil fuel combustion (sequestration). Nitrogen has also been used for this purpose [20].

An effective variant of the gas injection EOR procedure is the proposal to inject molasses or some other low-cost fermentable material plus a species of bacterium, such as *Clostridium acetobutylicum* into the formation [21]. Fermentation products, such as acetone, butanol, and lower molecular weight acids are produced. These and the carbon dioxide also produced can augment production by 200–250%.

Direct application of heat via *in situ* combustion or via superheated steam generation at the surface and injection are other effective methods to boost production, either in mature oil fields or in "heavy oil" fields where the petroleum is naturally quite viscous. While both formation heating methods achieve production rate improvements by viscosity reduction, the apparent simplicity of the *in situ* combustion concept is offset by the difficult separation of recovered oil from an aqueous solution containing nitrogen oxides, sulfur oxides, and other polar combustion products. The acids present in the aqueous phase contribute to the stability of the emulsions obtained from the producing wells and are highly corrosive to steel pipes and tanks.

Sometimes there is a need to increase the porosity of the formation in the producing zone, either to initiate production in a new oil field or to stimulate production in a mature one. This may be achieved hydraulically (called "fracturing") by pumping a mixture of water and coarse sand or high strength ceramic particles into the well at a high flow rate and pressure [22]. Under the impetus of pumps driven by 5,000–10,000 hp, the injected high-pressure stream of water and sand opens up cracks (fractures) in the formation extending radially from the well being stimulated. When pumping is stopped the coarse sand jams in the cracks and holds them open against the weight of the overlying rock when the pressure from injected water is removed. The propped open cracks now serve as channels through which oil (or gas) may flow more readily to the producing well.

If the producing zone of a well is in a limestone formation, acidizing techniques may be used to stimulate production [23]. Corrosion-inhibited hydrochloric acid is pumped down the well, which dissolves the limestone with which it comes into contact and in this way generates new channels for oil flow (Eq. 17.1).

$$CaCO_3 + 2\,HCl \rightarrow CaCl_2 + H_2O + CO_2 \qquad 17.1$$
$$\text{insoluble} \qquad\qquad \text{soluble}$$

Explosives may also be used in various ways to augment oil flow. For old or new wells, any accumulated wax or bitumen in the casing at one time was removed and the producing zone enlarged by setting off a carefully placed charge of nitroglycerin in a "torpedo," a special device designed for this purpose. More often today casing cleaning is conducted under better control by using a combination of solvent and scraping tools. A more specialized well servicing unit, the "gun perforator," may be used to boost production of a sluggish producing zone. The unit is lowered to the required level where high powered, steel-piercing shells previously loaded into the device are fired. The paths cut through the casing and into the rock by the high-velocity shells provide new channels for enhanced oil flow. A gun perforator may also be lowered to a higher or lower section of the steel casing of the well, which was not originally brought into production but which was noted as a potential producing zone at the time the well was originally drilled and logged (formation details recorded). Firing the shells at this new perforator position both penetrates the steel casing at this point and provides short flow channels in the rock at the new producing zone to improve production.

Of all the economizing techniques described water flooding is generally considered to be the most effective, particularly in sandstones. Frequently, however, a combination of techniques are used.

17.1.4. Supply Prospects of Conventional Petroleum

World crude oil production and consumption grew rapidly from 1930 to 1970 [24]. From a production of 177 million metric tonnes in 1930, this has grown to 3.65 billion metric tonnes for 2002 (Table 17.1). Half the cumulative total oil produced to 1970 took from 1857 to 1960, a 103-year period, and the other half took only 10 years, from 1960 to 1970 [24]. The doubling time, or the time in years to reach double the initial annual production rate stood at 21 years for 1930–1940, and rapidly decreased to 8.2 years for 1960–1970 (Table 17.2). This trend (Table 17.2) led Hubbert, in 1956, to estimate that world petroleum production would peak at about the year 2000, and that new finds at that time would fail to keep up with the increase in demand [28, 29]. However, since this estimate, the doubling time for the 1970–1980 interval actually increased to a period of 32 years. Significant contributing factors undoubtedly were the oil embargo and price increases by the Organization of Petroleum Exporting Countries (OPEC) in 1974. Global political disturbances since then appear to have caused a brief reversal in production volumes, which seem to verify Hubbert's prediction in the short term. It remains to be seen whether this reversal is sustained as the continuing increase in the price of oil encourages further exploration and development.

Since 1978 Canada, China, Norway and the U.K. have had substantial production increases, while the production of Indonesia, Iran, Iraq, and Saudi Arabia declined. This decline was not the consequence of decreased reserves (Table 17.3).

TABLE 17.1 Annual Petroleum Production Trends by Major World Producers of Crude Oil[a]

	Density (t/m^3)	Millions of metric tonnes[b]					
		1950	1960	1970	1978	1990[c]	2002
Algeria	0.80	<0.1	8.8	49.0	54.0	52.9	77.0
Canada	0.85	3.7	26.0	62.0[d]	64.3[d]	81.9	142.1
China	0.86	—	3.7[e]	23.9	104.1	138.3	169.0
Indonesia	0.85	6.4	20.6	42.6	80.5	74.9	63.0
Iran	0.86	32.3	52.2	191.3	262.8	155.9	168.0
Iraq	0.85	6.5	47.5	76.5	125.6	99.2	100.1
Kuwait	0.86	17.3	81.9	150.6	126.0	54.2	93.4
Libya	0.83	—	—	159.8	95.4	65.8	62.3
Mexico	0.89	10.3	14.4	21.5	60.8	147.9	185.9
Nigeria	0.85	—	0.9	54.2	94.9	88.3	99.3
Norway		0	0	0	17.6[f]	82.1	164.3
Saudi Arabia	0.86	26.9	62.1	188.4	476.3	330.6	433.2
United Arab Emirates	0.85	—	—	37.7	89.6	110.9	112.0
U.K.	0.86	<0.1	ca. 0.1	ca. 0.1	52.9	91.6	122.9
U.S.A.	0.85	270.1	348.0	475.3	429.2	411.8	379.7
U.S.S.R.	0.86	—	147.9	353.0	572.5	570.8	384.2[g]
Venezuela	0.90	78.2	152.4	194.3	113.5	114.4	153.6
Other		33.3	90.3	196.6	284.2	475.3	736.9
World total		485.0	1056.8	2276.8	3086.6	3146.8	3646.9

[a]Data for those countries producing more than 50 million metric tonnes in 1978, from *U.N. Statistical Yearbooks* [25], OECD [26], and *BP Statistical Review of World Energy* [27]. Does not include natural gas or natural gas liquids.

[b]Volume in petroleum industry "barrels" may be obtained by dividing metric tonnes by the appropriate density figure (tonnes/m^3) and then multiplying by 6.2898 bbl/m^3. The standard petroleum industry barrel is equivalent to 0.159 m^3, or 35 (34.97) Imperial gallons, or 42 U.S. gallons.

[c]Includes natural gas liquids.

[d]Includes synthetic crude oil produced from tar sands: 1.63, and 2.70 million tonnes in 1970 and 1978, respectively. Very little was produced from tar sands in 1960, and none in 1950.

[e]Data for 1959.

[f]Production started in Norway in 1971.

[g]Russian Federation.

Their known reserves actually increased, which combined with their reduced production rates, caused all of their ratios of reserves to production to go up. The anomalously high values seen for Iraq and Kuwait are largely the consequence of the 1990 Gulf War and its aftermath.

Worldwide resource life projections of this kind show a similar trend from a value of 29.7 "years" in 1970, 25.2 "years" in 1978, and leveling off at 42.5 and 38.8 "years" in 1992 and 2000. Large-scale consumers like the U.S., Canada, and the European Union, and more recently China have consumptions that are large fractions of their reserves, which has depressed the world petroleum life projections. Conventional petroleum production rates for Canada peaked in 1973, and for the U.S. in 1970 (Fig. 17.4).

Table 17.2   Slowing of the Growth Rate of World Oil Production from Lengthening of Doubling Times[a]

Time interval	Production (10^9 bbl/year)	Doubling time[a] (year)
1930–1940	0.177	21
1940–1950	0.262	11.8
1950–1960	—	8.5
1960–1970	7.9	8.2
1970–1980	17.5	32
1980–1990	23.0	256
1990–2000	23.9	72.4
2000–2002	27.2	[680][b]

[a]Calculated from Hubbert [24] and BP Statistical Review of World Energy [27].

Doubling times given in the years it would take for world production to reach twice the initial production at the rate of increase for the stated period.

[b]Time in years it would take to decrease to half the initial production at the rate of *decrease* for the stated period, i.e., a *halving* time in years.

China, Norway, and the U.K. have been able to increase their petroleum production from either diligent exploration, late resource development, or a combination of these factors (Tables 17.1 and 17.2). Norway and the U.K., from their first discoveries of natural gas and oil in the North Sea in the late 1960s and early 1970s, both increased their domestic petroleum production by a factor of more than 600 from 1970 to 1980. Since then their production volume has more than doubled again.

The large-scale petroleum consumers are exploring ways to supplement the convenient liquid fuel and chemical feedstock aspects of a conventional petroleum resource. Coal could be used in this way since the known reserves of hard coal have an estimated life of more than 200 years. However, extraction of coal is generally more difficult, equipment to consume it directly is more complex, and efficient emission control is more complicated than for petroleum combustion.

Coal may be partially converted to a liquid fuel by anaerobic distillation (Table 14.13), or may yield a higher return of liquid fuel by gasification followed by catalytic hydrogenation using Fischer–Tropsch technology [31]. This technology was developed and operated on a large scale in plants in Germany when the country was cut off from sources of conventional petroleum, and was dismantled at the end of the World War II. The plants operated today by the South African Coal, Oil, and Gas Company (SASOL) were operated for similar political reasons and are also not profitable.

Hydrogenated urban refuse has also been studied as a potential source for liquid fuel, but provides too small a return of liquid fuel to justify the large capital cost [32]. Refuse pyrolysis to produce liquid fuels has been found to be economically feasible by some cities [33]. The mixture of oils produced may be used as a fuel for space heating, though not for vehicles. The volume of municipal waste available is large, but not large enough to provide more than a small fraction of our liquid fuel requirements by these methods. Also the growing increase in efficiency of recycling of the paper and plastics content of

TABLE 17.3 Trends in Petroleum Reserves and the Anticipated Reserve Life in Years for Major Oil Producing Countries[a]

	1970		1978		1992		2002	
	Reserves million (tonnes)	Ratio of reserves to production	Reserves (million tonnes)	Ratio of reserves to production	Reserves (million tonnes)	Ratio of reserves to production	Reserves (million tonnes)	Ratio of reserves to production
Algeria	1,056	22.3	1,309	24.2	1,170	19.1	1,170	15.2
Canada[b]	1,157	19.1	791	12.8	893	10.1	932	6.6
China	—	—	2,738	26.3	3,340	23	2,470	14.8
Indonesia	1,367	32.5	1,070	13.3	740	10.1	675	10.7
Iran	8,204	42.8	6,148	23.4	13,535	74	1,212	73
Iraq	3,873	50.7	4,702	37.4	13,405	568	15,200	152
Kuwait	10,443	76.0	10,184	80.8	25,040	286	13,040	141
Libya	3,959	24.5	3,719	39.0	2,900	42.4	3,990	59
Mexico	779	35.5	3,884	63.8	6,930	45.0	1,783	9.6
Norway	—	—	—	—	1,271	11.1	1,392	8.5
Nigeria	757	14.0	1,678	17.7	2,419	25.1	3,243	32.7
Saudi Arabia	12,041	68.1	15,911	33.4	34,400	81	35,380	72
U. Arab Emirates	2,161	56.9	4,318	48.2	10,707	94	13,220	118
U.K.	132	1590[c]	1,393	26.3	581	5.8	635	4.9
U.S.A.	5,270	11.1	3,801	8.9	3,851	9.9	4,108	10.8
U.S.S.R.	7,930	22.5	7,990	14.0	6,554	16.5	8,108	21.4
Venezuela	2,009	10.4	2,492	22.0	8,998	69	11,130	724
Other	6,562	—	5,559	—	—	—	—	—
World	67,700	29.7	77,687	25.2	134,211	42.5	141,600	38.8

[a]Data from *U.N. Statistical Yearbooks* [25], *BP Statistical Review of World Energy* [27], and Kirk-Othmer [30]. Reserve data may be converted to petroleum industry barrels by dividing metric tonnes by the appropriate density figure (tonnes/m^3, from Table 17.1) and multiplying by 6.2898 bbl/m^3. The petroleum industry barrel is equivalent to 0.159 m^3, ca. 35 (34.972). Imperial gallons or 42 U.S. gallons (exactly).

[b]Does not include synthetic crude oil produced from the tar sands nor the reserves represented by this source.

[c]Anomalously high values for the ratios of reserves to production are obtained here during the early stages of development of a new large petroleum producing area, North Sea oil (see text).

a

b

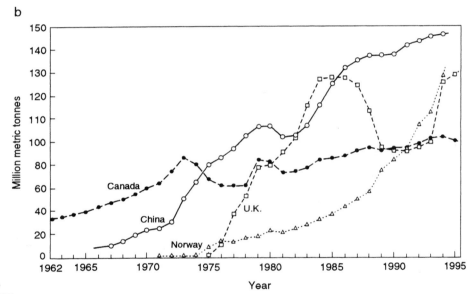

FIGURE 17.4 Trends in annual conventional petroleum production volumes for examples of (a) large producers where this appears to have peaked, and (b) smaller scale, more recent producing countries that have shown dramatic growth.

municipal waste now practiced reduces its potential for liquid fuel production. Thus municipal waste is unlikely to make a significant contribution to the world liquid fuel supply.

Liquid fuel, more specifically ethanol, may also be produced by fermentation or other processing of carbohydrates (see Chap. 16). Brazil has developed this technology to the extent that domestic automobiles are capable of operating entirely on ethanol produced from sugarcane. However, this renewable fuel technology is being employed more as a supplement to conventional automotive fuels as a "gasohol," rather than as a means of replacing conventional petroleum.

Chemical conversion of vegetable oils to general purpose liquid fuels ("biofuels," and "biodiesel") has also been successfully explored [34, 35]. However, the small size of this resource makes it unlikely that this could do more than supplement petroleum-based sources. Probably the more significant developments to extend petroleum-based liquid fuels lie in the recovery of oil from the tar sands, and the pilot plant projects involving oil shale pyrolysis experiments to liquid fuels.

17.2. LIQUID FUEL FROM NONCONVENTIONAL SOURCES

17.2.1. Petroleum Recovery from Tar Sands

Major deposits of tar sands, also called bituminous sands or oil sands, represent considerable hydrocarbon reserves (Table 17.4). The largest of these reserves-in-place occur in (Table 17.4) Canada in the Fort McMurray area of Alberta. While the reserves in place for these deposits are estimated to be of the order of 120×10^9 metric tonnes it is thought that only about one-quarter of this, or about 30×10^9 metric tonnes, is eventually recoverable

TABLE 17.4 Locations and Reserves of the World's Tar Sands Deposits (in Millions of Metric Tonnes)[a]

Deposit location and name	Areal extent (km² or 10³ ha)[b]	Overburden (m)	Reserves in place[c]
Canada, Alberta			
Athabasca	2,323	0–580	99,510
Cold Lake	486	210–790	8,188
Peace River	456	90–430	5,310
Wabasca	714	76–762	5,500
Totals	3,979		118,508
Venezuela			
Oficina-Tremblador	2,323	0–900	31,790
Malagasy			
Bemolanga	39	0–30	278
U.S.A.			
Utah deposits	19.8	0–610	276
California deposits	2.7	0–180	34
New Mexico, Santa Rosa	1.9	0–12	9
Kentucky, Asphalt	2.8	2–15	8
Totals	27.2		327
Albania			
Selenizza	2.1	Shallow	59
Trinidad			
La Brea	0.05	0	10

[a]From the extensive data of Berkovitz and Speight [36] and Bridges [37].
[b]To convert to acres multiply by 2.47.
[c]To convert into units of petroleum industry barrels, since the density of these heavy oils is close to 1.0, multiply the number of tonnes given by 6.290.

[36]. Nevertheless the potential recoverable synthetic crude oil from the tar sands of this one locale is equivalent to about 95% of the estimated reserves in the Middle East.

The second largest tar sand deposits occur in the Oficina-Tremblador area of Venezuela, and have a significant development potential. Smaller deposits also occur in Malagasy, the U.S.A., Albania, and Trinidad, all of which are important for the locales in which they occur. Other oil sand deposits are known, such as the Lower Triassic Bjorne formation of Melville Island in the Canadian Archipelago, but their extent is less well defined.

The tar sands situated in Alberta consist of oil-bearing sandstones. Surface exposures occur in parts of the Athabasca deposit, but much of the deposit lies 100 m or more beneath the surface. Oil present in these Lower Cretaceous sandstone deposits is very viscous, thick, and partially oxidized so that it cannot be recovered by pumping alone. Hence, the term "tar" or "bitumen" describes this heavy oil fraction, which averages about 12% ranging up to as much as 18% of the deposit by weight. Deposits with less than 2–3% bitumen are excluded from the reserve data given in Table 17.4.

The surface exposures and the near surface deposits (<45 m of cover) of the Athabasca region can be strip-mined for bitumen recovery. They amount to about 10% of the Athabasca deposit. The remainder of the Alberta tar sands lie under 75 m or more of overburden. It is uneconomical to surface mine, and too poorly consolidated for underground mining. These deeper deposits are yielding bitumen to the surface via various *in situ* techniques.

To obtain bitumen free of sand from the strip-mined tar sands requires treatment with hot water and steam, together with a small amount of sodium hydroxide, which produces surfactants from naturally occurring organic acids. This process was originally devised by K.A. Clark [38, 39]. In primary separation cells, the hot frothy bitumen rises to the surface and is skimmed off for further processing (Fig. 17.5).

The separated sand sinks to the bottom of the cell and is moved hydraulically to the tailings disposal area. By directing the discharge piping carrying the sand tailings onto the walls of the tailings pond, these are continuously raised and made thicker by the precipitated sand. The tailings pond is used by a hot water process tar sands extraction plant to retain waste waters, which contain about 20% of accumulated mineral fines plus 1–2% bitumen in a very stable suspension [40, 41].

There is also a middle, less well-differentiated fraction from the primary separation cell, which contains a mixture of water, clay fines, and bitumen. This "middlings" fraction is processed through scavenger cells in which more vigorous aeration and flotation techniques recover some further bitumen froth by a process, which has very recently become better understood [42, 43]. This is eventually combined with the bitumen layer from the primary separation cell for further processing. It is the tailings (wastewater stream) from these cells that contribute much of the mineral fines fraction to the tailings pond, which has been found to be very difficult to separate. However, it has recently been discovered that addition of gypsum (calcium sulfate) to the tailings pond reduces the settling time for the solids fraction to a one or two decades, at which time reclamation should be possible [44].

FIGURE 17.5 Schematic diagram of the hot water process for the extraction of bitumen from tar sands. (Reprinted from Hocking [46], with permission.)

The bitumen layer from the two separation processes is a similar density to water and also quite viscous. It contains some occluded water plus a small amount of mineral fines. By diluting this fraction with naphtha, in which the bitumen is soluble, the viscosity is decreased sufficiently to allow cleaner phase separation to take place. Phase separation, accelerated by centrifuging, produces separate streams of the wastewater (plus some mineral fines) and the naphtha solution of bitumen. Flash distillation of the naphtha solution then yields the crude bitumen product and recovers the naphtha for recycle.

Crude bitumen from the Alberta tar sands is a black, very viscous (about 100 centipoise at 38°C) rather intractable material as initially obtained. It is not very useful directly, since the temperature at which it softens is too low to make it useful as an asphalt for roads. Its high viscosity and relatively high pour point (minimum temperature at which it will still just pour) of about 10°C would make it expensive to move it via pipeline to another location for further processing. So, at each extraction plant, the hot crude bitumen obtained from the naphtha recovery unit is fed to a coker. Here, by applying heat at temperatures up to about 500°C, gases and the volatile liquids present are distilled off and the heavier asphaltene fraction (Eq. 17.2), consisting of higher molecular weight polyaromatics, is cracked to more volatile hydrocarbons and coke (Table 17.5; Fig. 17.6).

TABLE 17.5   Approximate Breakdown of Products on the Delayed Coking of Bitumen Recovered from the Alberta Tar Sands by Hot Water Extraction[a]

Component	Weight (%)	Boiling range (°C)	Density (g/cm³)
Gases[b]	8	—	—
Naphtha	12.7	95–190	0.775
Kerosene	15	190–260	0.832
Gas oil + heavy fuel oil	42.1	260–460	0.884
Coke	22.2	>460	—

[a]Compiled from Berkovitz and Speight [36] and Bachman and Stormont [45].
[b]Breakdown given in Table 17.5.

Examples of hypothetical asphaltene skeletons [47]:

ca. $C_{200}H_{249}N_2S_7O_4$ 17.2

Eq. 17.3 gives an example of the kinds of chemical changes occurring during this processing step.

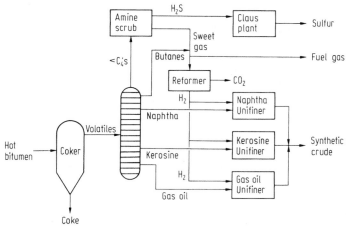

FIGURE 17.6   Flow diagram for the purification and stabilization of the products of bitumen pyrolysis. (Reprinted from Hocking [46], with permission.)

$$C_{42}H_{56} \rightarrow 2\ C_8H_{18} + C_4H_{10} + C_2H_6 + C_2H_4 + 18\ C \qquad 17.3$$

a typical   octane   butane   ethane   ethylene   coke

asphaltene

The gases obtained from volatilization and asphaltene pyrolysis comprise about 8% by weight of the bitumen coked, more than half of which consists of hydrogen and methane (Table 17.6).

The crude gaseous product is a valuable fuel and can also be used for the production of hydrogen. Volatile liquid products amount to some 68% of the bitumen feed, leaving a residue of about 20–22% coke. Overall mass balances and the properties of the products have been published [46].

The coke residue is the result of the hydrogen-deficient stoichiometry of the process. With a proximate analysis (determination of the compounds, moisture, ash, etc., present) of carbon 80%, volatiles 10%, sulfur 6%, and ash 4%, the coke has a fuel value near that of high rank coals. It is burned in the site power plants to provide steam and electrical power for oil sands processing. However, the high sulfur content detracts from its wider utility as a fuel. Any coke in excess of the current fuel requirement is finely powdered and incorporated into the dyke walls to help trap any hydrocarbons present in water seepage through the wall.

The crude liquid product contains alkenes and other unstable components resulting from the pyrolysis reactions and is high in bound sulfur. These interact with air to cause accumulation of gummy material on storage. Also the presence of sulfur compounds is undesirable in fuels. High-pressure hydrogenation (unifining) is used on site to produce more stable alkanes from alkenes, and volatile hydrogen sulfide from the sulfur bound to organic compounds (Eqs. 17.4 and 17.5).

TABLE 17.6 Composition of Gas Stream from Delayed Coker, Before and After Sweetening[a]

Component	Mole fraction (%)	
	Crude	Sweetened
Hydrogen	24.20	31.7
Carbon monoxide ($+N_2$)	0.9	1.7
Carbonyl sulfide	0.1	0.0
Hydrogen sulfide	15.2	0.0
Carbon dioxide	1.2	0.0
Methane	32.9	37.9
Ethane	11.6	13.1
Ethylene	2.3	2.6
Propane	7.6	8.6
Propylene	3.4	3.8
Butane	0.6	0.6
	100.0	100.0

[a]From Bachman and Stormont [45].

$$H(R)C = CH_2 + H_2 \xrightarrow{\text{catalyst}} H_2(R)C-CH_3 \qquad 17.4$$

$$R_2S + 2 H_2 \xrightarrow{\text{high pressure}} 2 RH + H_2S \qquad 17.5$$

The hydrogen sulfide is now easily separated from the liquid hydrocarbon stream by distillation, and is then converted to elemental sulfur, another product of tar sands operations, via the Claus process. The stabilized liquid hydrocarbon stream is the synthetic crude oil product of tar sands extraction plants.

In this way the useful petroleum fractions are recovered from the surface or near surface exposures of tar sand by the two currently operating hot water process extraction plants in Alberta. The production of synthetic crude oil by Alberta tar sand processors has risen from 28 million barrels (ca. 4 million tonnes) in 1978, to 77.3 million barrels (ca. 10.5 million tonnes) in 2003, which now supplies about 13% of Canada's current crude oil requirements [48]. Other processes for bitumen recovery from minable sands, such as preliminary partial sand removal with the help of cold water, followed by direct coking of the whole of the bitumen/solid residue, and solvent extraction methods have both been tested but are apparently not attractive for commercial development [49].

Various measures are being developed for recovery of the 90% of bitumen, which is in deposits too deep for surface mining. Among these experimental *in situ* methods being tested are formation heating using steam assisted gravity drainage (SAGD), *in situ* bitumen combustion [50], or electrical resistance heating [51] to increase bitumen temperatures sufficiently to make it flow and allow recovery by pumping. Hot water flooding with a polymer or sodium hydroxide additive content to enhance the interfacial interactions, in the presence of heat, has also demonstrated some positive results [52]. Mining integrated with one or more of the above methods is also under consideration as an *in situ* recovery method. As these *in situ* combinations are refined to efficient larger scale production, it should become possible to realize the petroleum recovery potential of the deeper deposits of the Alberta tar sands.

17.2.2. Petroleum from the Oil Shales

The world oil shale deposits are distributed among a number of different countries, and represent a considerable potential source of oil (Table 17.7). However, realizing this potential with an oil shale is a significantly different proposition than with the tar sands. First, the organic fraction present in the rock is not an oil or bitumen, but kerogen. Kerogen is a complex, high molecular weight (>3000), three-dimensional polymeric solid precursor to petroleum which is intimately distributed through the rock. It is insoluble in water or in common organic solvents [55]. As an indication of this, less than 1% of the organic content is recovered on extended Soxhlet extraction of oil shale with boiling toluene, and the extract is mostly traces of bitumen not kerogen itself. Thus, in contrast to the tar sands, the organic component of oil shales in place is not an oil. However, on vigorous heating of the shale in the absence of air to temperatures of the order of 500°C, a process called

TABLE 17.7 Estimates of the Potential Oil-in-Place by Country, of the Major Oil Shale Deposits of the World[a]

	Potential oil-in-place ($10^9 m^3$)
Brazil	127
Canada	8.0
China	4.5
Sicily	5.6
Congo	15.9
U.S.A.	350
U.S.S.R.	16.7
Other	3.4
Total	531.1

[a]Corresponds to 1969 estimates from Kirk-Othmer [53]. Data of 1981 give a world total potential of $332.8 \times 10^{12} m^3$, considerably more than quoted here, distributed by grade as follows: 83%, 21–42 liter/tonne shale; 16%, 42–104 liter/tonne; and 1%, 104–417 liter/tonne from Kirk-Othmer [54].

retorting, it may be made to yield from 21 L/tonne shale (5 U.S. gallon/short ton) to occasionally 417 L/tonne (100 U.S. gallon/ton) of a dark viscous oil (Tables 17.8 and 17.9).

A second misnomer of the oil shale resource is the matrix in which the kerogen resides. This is not a true shale but a marlstone composed of dolomite ($CaCO_3 \cdot MgCO_3$; ca. 32%), calcite ($CaCO_3$; 16%), quartz (SiO_2; 15%), illite (a silica clay; 19%), and an albite ($Na_2O \cdot Al_2O_3 \cdot 6SiO_2$; ca. 10%), and smaller amounts of several other minerals [57]. The carbonate constituents are susceptible to thermal decomposition (e.g., dolomite at 600–750°C), so their presence influences the composition of the pyrolysis gases, particularly at higher retorting temperatures (Eq. 17.6; Table 17.10). They also increase the heat required for kerogen pyrolysis.

$$CaCO_3 \cdot MgCO_3 \xrightarrow[>600°C]{} CaO + MgO + 2 CO_2 \qquad 17.6$$

TABLE 17.8 Percent Elemental Composition by Weight of Kerogen and Products from Retorting of Oil Shale[a]

	Raw shale	Kerogen	Retorted shale	Crude oil
Organic carbon	16.5	80.5	4.94	84.68
Hydrogen	2.15	10.3	0.27	11.27
Nitrogen	0.46	2.39	0.28	1.82
Sulfur	0.75	1.04	0.62	0.83
Oxygen	–	5.75	–	–

[a]Data from Atwood [56] and Bozak and Garcia [57].

TABLE 17.9 Properties of Typical Crude Shale Oil, and Composition of Gases from Surface Retorting of Oil Shale[a]

Shale oil composite properties		Typical composition of gases (mol %)	
Gravity, °API[b]	22	Hydrogen	22.4
Pour point (°C)	−1.0	Carbon dioxide	21.4
Carbon (%)	84.7	Carbon monoxide	3.6
Hydrogen (%)	11.3	Methane	15.2
C:H ratio	7.5	Ethane	10.3
Nitrogen,[c] (%)	1.8	Ethylene	5.4
Sulfur (%)	0.8	Propane	4.0
Boiling ranges (%) by weight:		Propylene	3.7
Initial to 204 (°C)	18	Butane	1.6
205–316 (°C)	24	Butenes	2.7
317–482 (°C)	34	C_5 and higher	5.4
483°C and above	24	Hydrogen sulfide	4.3
Total	100		100.0

[a]From Atwood [56].

[b]°API gravity, short for degrees, American Petroleum Institute gravity is a density scale used to relate this property of different crude oils and distillate fractions.

$$°\text{API gravity} = \left\{ \frac{141.5}{\text{specific gravity } 60°/60°F} \right\} - 131.5$$

Thus, water, with specific gravity of 1.00 has an °API of 10, crude oils run the range from about 5 to about 65, lubricating oils run about 26–35, and gasolines run about 60. The less dense the crude or distillate fraction, the higher the °API.

[c]Nitrogen-containing constituents present are mainly pyridines, quinolines, pyrroles, and carbazoles from Yen[52].

The Fischer assay is a standardized laboratory test in which an oil shale sample is retorted at 500°C to determine its oil yield. It provides a measure of the grade of oil shale being processed. Commercial processes such as The Oil Shale Corporation's TOSCO II process give oil recoveries (synthetic crude or syncrude) of up to 100% of the Fischer assay. Oil and gas formation from

TABLE 17.10 Effect of Retorting Temperature of Colorado Oil Shale on Product Distribution[a]

	Distillate hydrocarbon yield (volume percent)[b]		
Retorting conditions	Saturates	Olefins	Aromatics
537°C	18	57	25
649°C	7.5	39.5	53
760°C	0	2.5	97.5
871°C	0	0	100
Simulated *in situ*	41	37	22
Actual *in situ*	59	16	65

[a]Data selected from Kirk-Othmer [54].

[b]Volume percents quoted for product test distillation to a 300°C boiling point.

the kerogen occurs by what is now believed to be two successive first-order processes to give a product distribution of about 70% oil, 10% gas and light oils, and 20% coke, which remains on the solid residue [58] (Eqs. 17.7 and 17.8).

$$\text{kerogen} \xrightarrow{k_1} \text{bitumen} + \text{gas} + \text{coke} \qquad 17.7$$

$$\text{bitumen} \xrightarrow{k_2} \text{oil} + \text{gas} + \text{coke} \qquad 17.8$$

Thus, the observed products arise both directly from kerogen pyrolysis, and indirectly from kerogen via a bitumen intermediate formed from the kerogen. The product distribution is similar to that obtained on the coking of tar sands bitumen. Heat for the pyrolysis is provided from combustion of a part of the kerogen in the shale feed, and sometimes by burning a portion of the gases and/or the residual coke in the pyrolyzed shale (Fig. 17.7, [55]). Higher overall thermal efficiencies are obtained for those processes, which utilize the residual coke as well as pyrolysis gases for shale heating.

The key difference between the methods used for oil recovery from oil shales and that used for tar sands is in the methods used for separation of the organic constituent from the naturally occurring material. Oil shale processing requires the whole of the mined material to be heated to pyrolysis temperatures of 500°C or more, whereas the hot water process for tar sands extraction requires the mined tar sand (plus process water) to be heated to only around 70–80°C. Only the extracted bitumen from the tar sand, some 10–12% of the mined mass, has to be heated to ca. 500°C during the coking step to obtain synthetic crude. Because *all* the oil shale must be heated to pyrolysis temperatures to effect oil recovery, efficient heat transfer and

FIGURE 17.7 Diagram of the components of The Oil Shale Corporation's TOSCO II process for the retorting of mined oil shale for the production of synthetic crude. [From Hall and Yardumian [55], courtesy L. Yardumian.)

reclamation from hot spent shale is crucial to the commercial success of oil recovery from oil shales. This is achieved in the TOSCO II process by recirculating steel or ceramic balls (Fig. 17.7). A number of variations of this process are being tested to determine the most efficient heat transfer methods and to confirm product quality. One recent procedure borrowed from coal gasification technology uses the hot spent shale itself to transfer heat to the incoming pulverized raw shale. This Lurgi-developed process obtains oil recoveries equivalent to 100% of the Fischer assay at thermal efficiencies of 92.7% [59]. Process development in other parts of the world has been reviewed [60, 61].

The crude black, viscous pyrolysis oil product requires an upgrading step to make it suitable as a refinery feedstock. This is accomplished by high-pressure hydrogenation in a manner very similar to the upgrading step used for the coking products in tar sands processing. The influence of overall process conditions on the polycyclic aromatic compounds found in the product has been examined [62], and the supercritical water extract of the pyrolytic product has been characterized [63].

In situ methods are required to economically exploit the deeper oil shale deposits. These involve underground kerogen pyrolysis methods, which permit simultaneous or subsequent recovery of the oil. The most promising methods to date use explosives to fracture the oil shale formation to initially loosen the material sufficiently to permit underground combustion, and flow of hot gases to take place [52]. Air is then injected in the presence of an ignition source located at the top of the loosened rubble. This heats the formation sufficiently to cause kerogen pyrolysis and oil generation. As the oil forms, it trickles to the bottom of the fractured rubble from where it is recovered by conventional pumping.

At the moment high projected prices for the production of oil from oil shale have stalled the construction of commercial-scale oil shale processing plants, although several pilot-scale test plants have been built. One recent venture, that of The Oil Shale Corporation (TOSCO) constructed near Grand Valley, Colorado, was designed to process 100 short tons of shale per day (ca. 36,000 U.S. gallon/day) and was operated close to this scale of production at intervals from 1965 to 1967. Increases in the world price of conventional petroleum have reawakened an interest in the development of U.S. oil shale deposits. Commercial-scale plants capable of producing 1600 m³/day (10,000 barrels/day) to be constructed by 1982, to be enlarged to 8000 m³/day (50,000 barrels/day) by 1990, were announced [64]. Engine and furnace combustion tests of the product have been conducted [65, 66].

17.3. ENVIRONMENTAL ASPECTS OF PETROLEUM PRODUCTION

17.3.1. From the Exploratory Geology Phase

The significance of any environmental impact of petroleum production is affected by the stage of the producing process. Land exploration by aerial surveying, or by the use of aerial or surface-based geophysical techniques such as seismic, gravimetric, magnetic, or electrical methods have relatively little direct effect on the land surface. Disturbances caused by "thumping," or the placement of seismic shots and travel requirements of a field crew will have to be assessed, particularly in tundra, permafrost, or other ecologically sensitive regions. Summer passage of exploration teams in the Arctic tundra can leave thermokarst line scars, which can remain for decades because of the low temperatures, limited diversity of plant species, and short growing season. Impacts on these polar and subpolar "frontier regions" can be minimized, however, through use of proper vehicles, such as hovercraft, helicopters, and vehicles, such as the Rolligon, which use wide, ultralow-pressure pneumatic tires [67]. Timing this activity to coincide with a frozen surface layer can also help to minimize impact.

Offshore exploration has risks of sediment or underwater installation disturbance, but the overlying water column serves to minimize any visual impacts of these activities.

17.3.2. Impacts of Drilling Activities

Wells drilled to test seismic information are placed with little knowledge of what to expect. The greatest environmental risk faced by these operations comes from the threat of a "blowout," an uncontrolled release of drilling fluid mixed with oil, gas, and brine under high pressure when a drill rig penetrates a high-pressure reservoir. The on-site hazards of such an event include not only the mud and debris from the well itself but could involve the forceful ejection of the whole of the drill string (one or more kilometers of heavy pipe). There is also the risk of a wellhead explosion or fire from the release of flammable gases at the surface. Hydrogen sulfide gas, which often occurs in natural gas, may also be released. Fortunately the rotten egg smell of hydrogen sulfide provides a warning, at least for a short time (see Section 2.6.4). These collective risks in the drilling of exploratory wells have led to them being termed "wildcats" by drillers, as recognition of their unpredictable nature.

The ability to prevent a blowout can reduce the risk of injury to personnel and damage to equipment and has favorable environmental consequences. These factors give blowout control a high priority in the drilling phase. Increasing the density of drilling mud and restricting its outflow from the mud circuit are primary blowout control measures. These are accompanied by at least one pair of blowout preventers, which consists of stout large-bore valves, actuated by a pair of hydraulic rams. The upper valve is used for control when a drill string is in place, and has semicircular cutouts in the two faces of the valve corresponding to half the cross-sectional area of the pipe used in the drill string. When activated, the two halves close tightly around the drill string controlling both movement of the pipe itself and formation pressures. A lower blowout preventer without the semicircular cutouts can be used to close off the well entirely when there is no drill string present. In deeper wells, when the size of the well casing (pipe lining) is decreased to a smaller size for the lower reaches of the well, an additional one or more blowout preventers will be installed at the junction. This increases the capability to control very high pressures, which are more likely to be encountered at depth. Blowouts are now rare, but they still do occur occasionally, despite the precautions. Shielding of potential ignition sources reduces the risk of fire from the dissipation of flammable gas captured in the drilling mud, should there be a blowout.

An unsuccessful wildcat is called a "dry hole," and a successful wildcat is a "discovery well". When drilling equipment is removed and the well is brought into production there is very little impact on the surrounding area during the production phase. "Appraisal wells" may be drilled to better determine the size and extent of the initial discovery. These will have a much lower risk of the occurrence of a blowout because the depths and pressures of likely formations will already have been established. "Development wells" may or may not be drilled following completion of the appraisal wells, depending on the results of the appraisal.

Exploratory drilling in tundra or permafrost areas requires greater care because of the much slower natural recovery capability [68]. Wildcat wells in these areas require more precautions for the handling of warm drilling mud, etc., to avoid terrain collapse from the introduction of a thaw-susceptible area into permafrost [69].

Offshore drilling introduces several complicating factors simultaneously. Drilling from a ship is accomplished by radial placement of four (or more) anchors or sometimes by computer-controlled propellers. In rough weather drilling may have to be curtailed. A sea-based drilling platform uses ballasted stabilizing legs extending well into ocean depths not subject to wave action, or to the sea floor. Drilling from these, or from artificial islands constructed from material dredged from the surrounding area, is positionally stable. However, severe weather can introduce risks to any of these situations. Add the problem of floating massive sea ice to northern offshore drilling and one can see why oil exploration in arctic frontier areas is so expensive.

Blowout control methods for offshore drilling are still basically the same as those used on land. However, in the event of the failure of blowout prevention equipment an oil pollution risk is introduced to a much wider area than would be the case on land.

Oil production, particularly in offshore areas, is facing increasing difficulty in meeting the tighter regulations for oil discharges at sea. Water that is received at the surface with the oil, "produced water" is now required to be processed through three or more stages of treatment before discharge to control any losses of oil [70].

17.3.3. Emission Problems of Synthetic Crude from Tar Sands

One of the problem areas of the hot water process for tar sands extraction arises from the clay mineral fines, which comprises from less than 1% to over 15% of the mined material. These mineral fines interfere with efficient bitumen separation in the primary separation cell and require the operation of the backup scavenger cell to maintain bitumen recovery efficiencies. Selective mining could be used to avoid the problem by leaving high fines tar sands in the deposit. However, this practice would raise mining complexity and cost and would waste bitumen present in the fines, so it is not used. As a result of this, the spent water discharged from the scavenger cells still contains much of the mineral fines plus traces of bitumen. This is accumulated in large holding ponds, totaling $30\,km^2$ (about 11 square miles) in extent for the larger of the two older extraction plants, which are operating at Fort McMurray.

Prolonged settling in the holding ponds allows the supernatant water to be recycled into the extraction process. A sludge consisting of about 78% water, 20% mineral fines, and 1–2% bitumen gradually accumulates in the holding pond. Ordinarily, the residual sand after the bitumen has been extracted could be accommodated in the mined-out areas as mining proceeds. But the large volume of sludges makes the gross volume of waste greater than the volume of mined-out areas. An economic method of sludge dewatering could allow the spoils to be accommodated [71]. Spherical agglomeration methods, which have been tested for direct solvent extraction of bitumen from tar sands could be applied to this problem area [72].

Sulfur gases arising during synthetic crude production from the bitumen and from the high-pressure hydrogenation process for synthetic crude stabilization are captured in amine scrubbers, and subsequently converted to sulfur via the Claus process (Chap. 9). Approximately 1500 tonnes of sulfur are produced daily from these sources by the two older hot water extraction plants. Occasionally these control measures have been inadequate to maintain low ambient air sulfur dioxide concentrations, particularly during an inversion episode. The potential of vanadium and nickel recovery from fly ash, which is possible on the scale of 1,600 and 3,900 tonnes/year, respectively, has also been considered.

In situ bitumen recovery from tar sands promises to have a lower environmental impact than surface mining and extraction. Water recovered with the bitumen from steam drive *in situ* tar sands processing requires treatment before reuse or discharge.

17.3.4. Impact Control of Oil Shale Processing

Production of synthetic crude oil from oil shale has several environmental problem areas. These center on a scarcity of water in the operating area (at least in the U.S.), disposal of spent shales, reclamation of disposal areas, and sulfur gas containment. Even though water is not required for extraction, the water needs are substantial for pretreating shale, condensing the crude and upgraded shale oils, and for the operation of any scrubber-based emission control devices. Thus, while the present water supply is adequate for pilot-scale recovery plants, provision of an adequate water supply for large-scale shale oil producing units is a major concern [73].

The "burned" shale is mineralized; it contains little in the way of plant nutrients and has a high salt content so is not easily reclaimed. This is complicated by the dry conditions and the large amounts involved, though because the waste is dry its volume is a closer match to the volume of the mined-out areas. However, an oil shale industry that produces a million barrels of oil per day would produce over 1.25 million tonnes of burned shale each day, so the areas potentially affected by production-scale units are large [74]. However, use of oil shale solid waste for soil stabilization [75] or reclamation activity is not likely to be complicated by health problems from either the raw oil shale rock or from the spent shale [76].

Sulfur gas containment from either oil shale pyrolysis or oil upgrading can be accomplished using the largely existing technology already outlined for tar sands processing. Environmental aspects of oil shale processing have been reviewed [54].

Recovery of shale oil from *in situ* shale pyrolysis processes is likely to have a lower environmental impact than the methods based on surface mining. Small volumes of formation water mixed with the shale oil recovered by pumping from the lower reaches of the fractured deposit will require treatment before being recycled.

17.3.5. Loss Prevention During Petroleum Shipment

The major petroleum producing areas of the world do not coincide very closely with the major consuming areas, so that much ocean shipping activity today involves tanker movement of oil. When the producing area is accessible to the consuming area by a land route, or for the transport of oil to or from an oil port to a refinery, a pipeline is normally used. For smaller quantities, segregated products, or shorter distances, small tanker ships, tank trucks, or rail tank cars are generally used.

Since the volumes of oil shipped long distances are very large, and transportation costs per tonne of oil shipped are lowest when using large ocean tankers or pipelines, more than half the present world's ocean cargo weight is now oil, and has been since the mid-1960s. More than 75% of this oil traffic is now crude oil, which means that very large tankers increase shipping economy [77]. This has led to a rapid escalation in the size of the largest tankers in use, from 5,000 to 15,000 deadweight tonnes (dwt) size range in the l939–1945 period to more than 85 vessels of over 320,000 deadweight tonnes operating in the mid-1980s. More than 80% of the tonnage was vessels of over 72,000 dwt and several were 430,000 dwt.

This escalation in tanker size has decreased vessel maneuverability. The standard World War II tanker of 15,000 dwt had a loaded draft of about 9 m and a length of 170 m, whereas the loaded 430,000 dwt very large crude carrier (VLCC) has a draft of about 27.5 m and a length of over 400 m, both adversely affecting maneuverability. In the event of a serious accident, involving a VLCC the polluting potential is also enormous, but the probable number of accidents for the same volume of oil transported is less because of the smaller number of vessels. It would take some 30 or more World War II vintage tankers to move the same volume of oil as one VLCC. Shipment of oil by double-hulled tankers would decrease the risk of loss for a relatively modest increase in cost.

TABLE 17.11   Estimates of Direct Petroleum Losses to the World's Oceans[a]

Source	Thousands of metric tonnes	Percent of total
Waste oil disposal of industrial and motor oils	745–3300	25–67
Tankers, normal operations, tank cleaning, ballasting	530–725	11–25
Other ships, normal operations, cleaning of bilges	500–635	10–22
Refineries, petrochemical plants	300–405	6–14
Tanker and other ship accidents, other accidental spills	200–300	4–9
Offshore petroleum production, normal operations	100–144	2–5
Range of total	2954–4930[b]	

[a]Compiled from Pryde [78] and Matthews *et al.* [79].
[b]Amounts to 0.09–0.15% of current world production.

These factors combine to result in only about 9% of the world oil loss to the oceans being via accidental tanker spills (Table 17.11). This represents 100,000–270,000 tonnes/year, about one major VLCC accident every 2 years or 10 accidents involving smaller vessels every year. Much larger quantities than this, from 530,000 to 670,000 tonnes per year are estimated to be discharged to the sea as a result of normal tanker operations. Direct discharge to the sea of water used for tank cleaning operations, or used as seawater ballast for vessel stability on the return trip, contributed to this total. Today, most tanker facilities are equipped with shore lagoon storage facilities to accept large volumes of oily ballast and tank cleaning wastewaters, which are cleaned using oil–water separation units before the saltwater is discharged.

Waste oils and used oils are being re-refined either to produce petroleum products or for energy recovery [80, 81]. These measures are decreasing the oil lost to the oceans by poorly used oil disposal practices. The motivation for these actions has come from the increase in value of petroleum and from an increased recognition of the harm done to oceans by large discharges of used oil.

Oil loss may occur from pipeline rupture caused by corrosion, subsidence or land slip, or from seismic disturbances [82]. Regularly scheduled inspections of pipeline rights of way, and efficient containment and cleanup operations can keep these losses small. Diking of any pipeline rights of way alongside a lake, stream, or maritime coast at the time of construction can be a useful precaution. This can be aesthetically blended into natural contours, and seeded for erosion protection.

When an oil spill does occur, a first priority is containment to limit the spread as much as possible, using sandbags or earth dikes on land, or floating booms with underwater skirts on water (Fig. 17.8). Then as much of the oil as possible is picked up by pumping from land-based spills, or by using one of

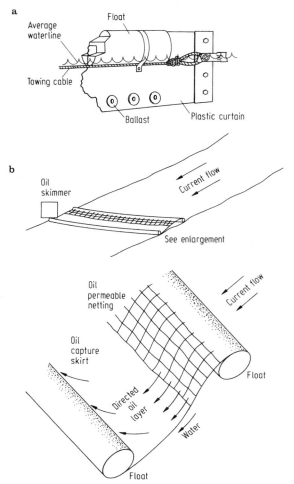

FIGURE 17.8 (a) Important components of an oil boom constructed to contain an oil spill under open water conditions. (b) A boom specifically designed to capture spills in moving water. (Reprinted from Smith [83], courtesy Pollution Engineering.)

the commercially available oil–water separation devices (Fig. 17.9). Final cleanup is by oil sorbents, detergents, and high-pressure water jets. Residues in soils may be biodegraded with the assistance of the correct balance of micronutrients or may be removed to a pyrolysis device and the oil content of the soil burned.

More novel proposals for frontier oil shipment have included dirigibles having a 450-tonne payload. Methane could be used as a lifting gas, which would allow the craft to serve a dual role, movement of oil and natural gas. As another alternative, Boeing officials conducted a feasibility study in 1972 in which 12-engined jet aircraft having a 154-m wingspan and capable of one million kilogram payloads were judged to be a competitive option. Perhaps accident risk perceptions caused these alternatives to be abandoned.

To external pump

Oil slick

To external pump

Weir skimmer

Oil slick

Suction skimmer

To pump and separator

Water pump

Water

Centrifugal skimmer

Collection well

Pump to storage

Rotating belt

Vessel

Oil slick

Scraper

Direction of vessel movement

Submersion skimmer

Squeeze rollers

Oleophilic rope (Oil clings to rope water does not)

Collection well

Tail pulley

Surface skimmer

FIGURE 17.9 Illustrations of the operating principles of different types of recovery equipment for oil spills on water. (From Shell Canada [84], with permission.)

REVIEW QUESTIONS

1. (a) Compare the structures and properties of the material in place for conventional petroleum, "heavy oil," tar sands, and oil shale deposits.
(b) How do the recovery methods and the recovery rates (as a proportion of the resource in place) compare for the hydrocarbon sources of part (a)?
(c) What postproduction measures are necessary for the reclamation of exhausted production areas for the hydrocarbon sources of part (a)?
2. (a) What production practices are used to help maximize the life and the ultimate hydrocarbon recovery from a newly discovered conventional oil or gas field?
(b) Outline the details of any four measures (secondary and tertiary recovery methods) that may be used to maintain production of an aging oil field. Explain how each of these functions.
(c) Why is air not used to help maintain formation pressure during tertiary recovery measures?
3. (a) What is the major constituent of natural gas, and what are the minor constituent(s) that may be present?

(b) Why do these minor constituents have to be removed before pipeline delivery, or use, and how is the wellhead composition cleaned up for these purposes? (See Section 9.5.)

4. Qualitatively rank the energy requirement for the production of conventional petroleum in comparison to the requirement for oil production from the tar sands and from the oil shales. Explain the reasons for this ranking.

5. (a) Define the meaning of °API and explain how it is evaluated.
(b) What general composition information does this quantity convey with respect to petroleum fractions?
(c) What range of °API values could be expected for crude oils, for gasolines, and for lubricating oils?

6. (a) Fully define the petroleum industry volume units: the barrel (bbl) and, in relation to natural gas, the "standard cubic foot".
(b) Make the following unit conversions: 1000 bbl/day of 0.85 kg/dm^3 density, to metric tonnes/day; 30 metric tonnes/day of crude oil of density 0.90 kg/dm^3, to bbl/day; 22,000 ft^3/day of natural gas at 1200 lb/in.^{2} pressure and 20°C to std ft^3/day (state conditions) *and* to m^3/day at 0°C and 1 atm pressure. (Assume ideal gas relations for gas conversions.)

7. (a) Outline precautions that could be taken at the time of construction of land-based facilities for handling oil that could provide a passive way to minimize losses in the event of an accident or spill.
(b) What sequence of steps would help to minimize oil loss and clean-up costs from an oil spill on land?

8. (a) How might ship design and the provision of shore-based facilities for loading and unloading ships help to reduce the losses of oil to oceans?
(b) What sequence of steps and preparedness measures could help to reduce the oil loss, cleanup costs, and potential environmental damage from an oil loss at sea?

FURTHER READING

J.W. Bunger and N.C. Li, eds., "Chemistry of Asphaltenes," Adv. Chem. Ser., No. 195. American Chemical Society, Washington, DC, 1981.

R. L. Busby, "International Petroleum Encyclopedia: 2002," Pennwell Corp., Tulsa, Oklahoma 2002.

J.J. Fitzgerald, "Black Gold With Grit," Gray's Publishing Ltd., Sidney, B.C., 1978.

J.H. Gary, ed., "15th Oil Shale Symposium Proceedings," Colorado School of Mines Press, Golden, 1982.

J. Hupka, J.D. Miller and J. Dreilich, Water-Based Bitumen Recovery from Diluent-Conditioned Oil Sands, *Can J. Chem. Eng.* **82**(5), 978–985, Oct. (2004).

J. Masliyah, Z.J. Zhou, Z.H. Xu *et al.*, Understanding Water-Based Bitumen Extraction from Athabasca tar Sands, *Can J. Chem. Eng.* **82**(4), 628–654, Oct. (2004).

J.G. Speight, "The Chemistry and Technology of Petroleum," 2nd ed. Marcel Decker, New York, 1991.

"World Energy Assessment: Energy and the Challenge of Sustainability," (J. Goldemberg, ed.), U.N. Dev. Program, U.N. Dept. of Econ. and Soc. Affairs, World Energy Council, New York, NY, 2000.

F. Zurcher and M. Thuer, Rapid weathering processes of fuel oil in natural waters. *Environ. Sci. Technol.* **12**, 838, July (1978).

REFERENCES

1. J.F. van Oss, "Chemical Technology: An Encyclopedic Treatment," Vol. 4, p. 34. Barnes and Noble, New York, 1972.
2. R.M. McPherson and R.W. Ford, "A History of the Chemical Industry in Lambton County," Chemical Institute of Canada, Sarnia, 1964; and RWF, ibid., 4th ed. Sarnia, 2000.
3. "Encyclopaedia Brittannica," 15th ed. Vol. 18, p. 712. Macropaedia, Encyclopaedia Brittannica Inc., Chicago, 1974.
4. J.E. Brantly, "History of Oil Well Drilling," Gulf Publishing, Houston, TX, 1971.
5. P.H. Giddens, "The Birth of the Oil Industry," Arno Press, New York, 1972 (reprint edition).
6. "The Canadian Petroleum Industry," Chemical Division, Shell Oil Co. of Canada, Ltd., Ryerson Press, Toronto, p. 31, 1956.
7. C. Barker, "Origin of Petroleum", In "Kirk-Othmer Encyclopedia of Chemical Technology," 4th ed. Vol. 18, p. 346–352. John Wiley & Sons, New York, 1996.
8. Fresh look at geochemistry, *Chem. Eng. News*, **59**(15), 60, April 13 (1981).
9. Norcen Inc., Annual Report, p.13, 1992.
10. R.N. Anderson, Oil Production in the 21st century, *Sci. Am.* **278**(3), 86–91, Mar. (1998).
11. B. Ortiz, E. Diaz, E. Quiroz *et al.*, First Deep Core Taken Provides...Results, Mexico, Proc. SPE Intern. Petrol. Conf. & Exhib. Villahermosa, Mexico, Mar. 5–7 1996, *Soc. Petrol. Engin.* 35323, p. 189–194 (1996).
12. "The Petroleum Handbook," 4th ed. Shell International, London, 1959.
13. R.N. Anderson, Oil Production in the 21st Century, *Sci. Am.* **286**(3), 86–91, Mar. (1998).
14. G.A. Purdy, "Petroleum, Prehistoric to Petrochemicals," McGraw-Hill, New York, 1958.
15. C.J. Campbell and J.H. Laherrère, The End of Cheap Oil, *Sci. Am.* **278**(3), 78–83, Mar. (1998).
16. L.M. Fanning, ed., "Our Oil Resources," 2nd ed., McGraw-Hill, New York, 1950.
17. C. Tielong, S. Zhengyu, F. Song et al., A Pilot Test of Polymer Flooding in....Reservoir, In SPE Reservoir Engineering, *Soc. Petrol. Eng.* **1**(1), 24–29 (1998).
18. K. Borchardt, (Petroleum) Enhanced Oil Recovery, In Kirk-Othmer Encyclopedia of Chem. Technology, 4th ed., Vol. 18, p. 405–427, Wiley & Sons, New York, 1996.
19. Alberta Stimulates Enhanced Oil Recovery, *Can. Chem. Process.* **66**, 4, Nov. (1982).
20. Nitrogen Scheduled for Oil Recovery, *Chem. Eng. News*, **58**(32), 7, Aug. 11 (1980).
21. T. Jack, Using Bacteria to Enhance Oil Recovery, *B.C. Guidelines to Ind. Prog.* (9), 1, Sept. (1980).
22. R.E. Blauer, I.N. Onat, and M.J. Lemieux, Production Data Indicate Reservoir and Fracture Performance....Proc. 1992 Permian Basin Oil and Gas Recovery Conference, Mar. 18–20, 1992, Midland, TX, USA, p. 503–512.
23. K.C. Taylor and H.A. Nasr-El-Din,...Acid Stimulation of SEAWATER injection Wells ..., SPE 58736, SPE Intern. Symp. on Formation Damage, Lafayette, LA, USA, Feb. 23–24,2000. *Synopsis J. Petrol. Technol.* **52**(3), 30–31 (2000).
24. M.K. Hubbert, Can. Min. Metall. Bull. p. 37, July 1973.
25. "United Nations Statistical Yearbook 1979/80," United Nations, New York, 1981, and earlier years.
26. Organization for Economic Cooperation and Development, "IEA Statistics, Oil and Gas Information, 1994," p. 77, OECD, Paris, 1995.
27. BP Statistical Review of World Energy, June 2003. Available: http://www.oilsurvey.com/php/link.php3?CoId=901006
28. M.K. Hubbert, Nuclear Energy and Fossil Fuels, "In Drilling and Production Practice," Am. Petr. Inst. Washington, DC, p.7–25 1956.
29. L.F. Ivanhoe, Updated Hubbert Curves Analyze World Oil Supply, *World Oil*, **217**(11), 91–94, Nov. (1996).
30. "Kirk-Othmer Encyclopedia of Chemical Technology," 4th ed. Vol. 18, p. 471, Wiley, New York, 1995.
31. Liquid Fuels from Coal, *Hydrocarbon Process.* **54**(5), 119, May (1975).
32. Work Intensifies on Biomass, *Chem. Eng. News*, **57**(40), 34, Oct. 1 (1979).

33. M.B. Hocking, A Chemical Input-Output Analysis of Municipal Solid Waste Incineration. *J. Environ. Syst.* **5**(3), 163 (1975).

34. D. Gaskell, Sowing the Seeds of Sustainability, *Chem. Br.* **35**(2), 49–50, Feb. (1998).

35. W.Y. Zhou, S.K. Konar, and D.G.B. Boocock, Ethyl Esters from . . . Ethanolysis of Vegetable oils, *J. Am. Oil Chemists Soc.* **80**(4), 367–371, Apr. (2003).

36. N. Berkovitz and J.G. Speight, *Fuel*, **54**, 138 (1975).

37. L. Bridges, *Sci. Affairs*, **9**(3), 4 (1976).

38. K.A. Clark, *Can. Min. Metall. Bull.* **212**, 1385 (1930).

39. K.A. Clark, *Trans. Can. Inst. Min. Metall. Min. Soc. N.S.* **47**, 257 (1944).

40. M.B. Hocking, Physical Characterization and Microbiological Settling Rate Modification of Aqueous [Waste from] . . . Oil Sands Extraction, *Fuel*, **56**, 334 (1977).

41. M.B. Hocking and G.W. Lee, The Effect of Chemical Agents on Settling Rates of Effluent from Hot Water Extraction of Athabasca Oil Sands. *Fuel.* **56**, 325 (1977).

42. L.L. Schramm, E.N. Stasiuk, H. Yarranton *et al.*, Temperature Effects from the . . . Flotation of Bitumen From Oil Sands . . . , *J. Can. Petrol. Technol.* **42**(8), 55–61, Aug. (2003).

43. L.L. Schramm, E.N. Stasuik, and D. Turner, The Influence of Interfacial tension in the Recovery of Bitumen [from] . . . Athabasca Oil Sands, *Fuel Proc. Technol.* **80**(2), 101–118, Feb 15 (2003).

44. R.L. George, Mining for Oil, *Sci. Am.* **278**(3), 84–85, Mar. (1998).

45. W.A. Bachman and D.H. Stormont, *Oil Gas J.* **65**, 69 (1967).

46. M.B. Hocking, The Chemistry of Oil Recovery from Bituminous Sands, *J. Chem. Educ.* **54**(12), 725 (1977).

47. T. Ignasiak, A.V. Kemp-Jones, and O.P. Strausz, *J. Org. Chem.* **42**, 312 (1977).

48. S. Yui and K.H. Chung, How Sweet it [Synthetic Crude Oil] is, *Can. Chem. News*, **56**(8) 16–19, Sep. (2004).

49. M.A. Carrigy, ed., "The K.A. Clark Volume," Research Council of Alberta, Edmonton, 1963.

50. R.E. McRory, "Oil Sands and Heavy Oils of Alberta," Alberta Energy and Natural Resources, Edmonton, 1982.

51. D.E. Towsonin, The Future of Heavy Crude and Tar Sand, Proc. First Unitar Conference Edmonton, 1979, p. 410 (1979).

52. T.F. Yen, ed., "Shale Oil, Tar Sands and Related Fuel Sources," Adv. Chem. Ser., No. 151. American Chemical Society, Washington, DC, 1976.

53. "Kirk-Othmer Encyclopedia of Chemical Technology," 2nd ed., Vol. 18, p. 1. Wiley, New York, 1969.

54. "Kirk-Othmer Encyclopedia of Chemical Technology," 3rd ed., Vol. 16, p. 333. Wiley, New York, 1981.

55. R.N. Hall and L.H. Yardumian, The Economics of Commercial Shale Production by the Tosco II Process, *61st Annu. Meet., Am. Inst. Chem. Eng.* Los Angeles, 1968 (unpublished).

56. M.T. Atwood, The Production of Oil Shale, *CHEMTECH.* **3**, 617 (1973).

57. R.E. Bozak and M. Garcia, Jr., Chemistry in the Oil Shales, *J. Chem. Educ.* **53**, 154, March (1976).

58. H.C. Stauffer, ed., "Oil Shale, Tar Sands, and Related Materials," ACS Symp. Ser., No. 163. American Chemical Society, Washington, DC, 1981.

59. Shale Oil Processes Ready, *Chem. Eng. News.* **60**(15), 60, April 12 (1982).

60. V. Yefimov, Oil shale processing in Estonia and Russia, *Oil Shale*, **17** (4), 367–385 (2000).

61. V. Yefimov, and T. Purre, Main Factors Influencing Processing of . . . Oil Shale in Vertical Retorts, *Oil Shale*, **13**(2), 123–132 (1996).

62. P.T. Williams and J.M. Nazzal, Polycyclic Aromatic Compounds in Shale oils: Influence of Process Conditions, *Environ. Technol.* **19**(8), 775–787, Aug. (1998).

63. J. Yanik, M. Yuksel, M. Saglam *et al.*, Characterization of . . . Shale Oil Obtained by Pyrolysis and Supercritical Water Extraction, *Fuel.* **74**(1), 46–50, Jan. (1995).

64. Production, Engine Tests Start for Shale Oil, *Chem. Eng. News*, **58**(25), 39, June 23 (1980).

65. M. Abu-Qudais and M.I. Al-Widyan, Performance . . . of a Diesel Engine Operating on Shale Oil, *Energ. Convers. Manag.* **43**(5), 673–682, Mar. (2002).

66. M. Abu-Qudais, Performance . . . of a . . . Furnace Using Non-Petroleum Fuel, *Ener. Conserv. Manag.* **43** (5), 683–691, Mar. (2002).

67. W.E. Rickard, Jr. and J. Brown, *Environ. Conserv.* **1**(1), 55 (1974).

68. J.P. Schumacher, E. Malachosky, D.M. Lantero et al., Minimization and Recycling of Drilling waste on the Alaskan north slope, *J. Pet. Technol.* **43**, 722–729 (1991).
69. H.M. French, *Arctic*, **3**(4), 794 (1980).
70. S. Davies, Recycling, Water Output Limitation Challenge . . . Treatment Methods, *Offshore Int.* **54**, 73–74, Sept. (1994).
71. N.N. Bakshi, R.G. Gillies, and P. Khare, *Environ. Sci. Technol.* **9**(4), 363 (1975).
72. W. Campbell, *Sci. Dimens.* (Ottawa). **8**(1), 10 (1976).
73. D.L. Klass, *CHEMTECH.* **5**, 1, Aug. (1975).
74. R.N. Heistand, Retorted Oil Shale Disposal Research. In "Oil Shale, Tar Sands, and Related Materials" (H.C. Stauffer, ed.), ACS Sympos. Ser., No. 163. American Chemical Society, Washington, DC, 1981.
75. J.P. Turner, Soil Stabilization Using Oil-Shale Solid Waste. *J. Geotech. Eng.* **120**, 646–660 (1994).
76. Shale oil materials pose few health problems, *Chem. Eng. News*, **56**(41), 13, Oct. 9 (1978).
77. "BP Statistical Review of World Energy 1981," British Petroleum Company, London, 1982.
78. L.T. Pryde, "Environmental Chemistry," Cummings, The Philippines, 1973.
79. W.H. Matthews, F.E. Smith, and E.D. Goldberg, eds., "Man's Impact on Terrestrial and Oceanic Ecosystems," M.I.T. Press, Cambridge, MA, 1971.
80. D.J. Skinner, "Preliminary Review of Used Lubricating Oils in Canada," Rep. No. EPS 3-WP-74-4. Environment Canada, Ottawa, 1974.
81. N.J. Weinstein, Re-Refining Schemes Compared, *Hydrocarbon Process.* **53**(12), 74, Dec. (1974).
82. R.O. Van Everdingen, "Potential Interactions Between Pipelines and Terrain in a Northern Environment," Paper No. 8. National Hydrology Research Institute, Ottawa, 1979.
83. M.F. Smith, *Pollut. Eng.* **2**(5), 24, Nov./Dec. (1971).
84. "Perspective on Oil Spills," Shell, Canada, Montreal, 1981.

18

■ PETROLEUM REFINING

*No fewer than ten substances are obtained from
petroleum by the refining process...2nd, gasolene,
used in artificial gas machines.*
　　　　　　　—*Century Magazine, July 1883*

18.1. COMPOSITION OF CONVENTIONAL PETROLEUM

It is commonly thought that crude oil from conventional oil wells is quite
similar in appearance and composition, regardless of the source. This is
not so. The physical characteristics and the composition of crude oils vary
widely with the location of the oil field and the stage of production (new or
mature).

Conventional crude oil ranges from green to brown or black in color
depending on the petroleum type and the mineral matter present and is
composed of a heterogeneous mixture of liquids, solids, and gases. Some
components of the crude oil are dissolved and some are a separate phase.
Water may occur with the petroleum produced as a readily separated
phase or as an emulsion containing as much as 80–90% water. The "pour
point" is a measure of the low-temperature viscosity or flow characteristics of
an oil. It is defined as being 3°C (or 5°F) above the setting temperature
(maximum temperature at which no observable flow occurs) of the oil.
Pour points of some viscous conventional crudes can lie above 5°C, while
the pour points of the less viscous, or lighter crudes can be less than −15°C
(Table 18.1).

The industry standard for bulk measurement for both crude oil and liquid
products was the petroleum industry barrel (abbreviated bbl) but now
the metric tonne and the cubic meter are more commonly used, especially
in international trade. The metric tonne may be converted to the barrel
volume unit by dividing the mass unit by the density (specific gravity) of the

▮▮▮▮ TABLE 18.1   Specifications of Examples of Typical Conventional and Synthetic Crude Oils

Base oil	Conventional crude oils[a]			Synthetic crude oils	
	Paraffin naphthene base (wax bearing)	Intermediate Oil (wax bearing)	Naphthene base oil (wax free)	From tar sands[b]	From oil shales[c]
Density (g/cm³)[d] (specific gravity)	0.781	0.964	0.910	0.84	0.93
Pour point (°C)	<-15	5	<-15	-35	5
Saybolt viscosity at 38°C, seconds[e]	34	4,000	55	34	78
Color	Green	Brown-black	Green	Pale yellow	Black
Sulfur content (%)	0.10	3.84	0.14	0.03	0.72
Distillation:					
1st drop (°C)	34	138	157	–	–
Fractions (%)					
Gasoline + naphtha	45.2	2.9	1.1	30	18
Kerosene	17.7	4.5	0.0	20	24
Gas oil	8.3	10.6	55.5	50	34
Nonviscous lube	9.8	8.6	14.2	–	–
Medium lube	3.8	6.7	4.7	–	–
Viscous lube	0.0 1.0	1.0	11.6	–	–
Residue (%)	14.7	58.4	12.7	–	24
Distillation loss (%)	0.9	1.9	0.2	–	–
Carbon in residue (%)	1.1	18.2	4.5	–	–
Carbon in crude (%)	0.2	10.6	0.6	$(22)^b$	–

[a]Selected from Lane and Garton [1]. Distillate and residue breakdowns are quoted in percent by weight.

[b]Data calculated from that for composite synthetic crude obtained after coking and Unifining of extracted bitumen, from Bachman and Stormont [2]. Before Unifining (hydrogenation) mean density of the composite stream would be somewhat higher, and sulfur content would be about 3%. Proportions of distillate components are approximate; carbon content quoted is the coke residue on pyrolysis of bitumen.

[c]Data calculated from that given for crude Fischer assay oil (the crude pyrolysate) from Atwood [3]. Proportions given on distillation are approximate. The percent residue quoted corresponds to the fraction of the pyrolysate having a boiling point higher than 482°C.

[d]Density is a necessary property for the conversion of volume units such as m³, bbl (barrels), or liter to mass units, such as tonnes. It is also a useful indicator of the composition of the crude oil (see text). °API is a petroleum industry density unit obtained from the specific gravity at 16°C: °API = (141.5/specific gravity) −131.5.

[e]Saybolt universal viscosity, measured by the number of seconds required for a sample to pass through a standard orifice in a Saybolt viscosimeter, usually specified at 100°F.

particular oil being measured and multiplying by 6.2898. For example 1 tonne of Canadian crude, of density 0.85 g/cm³, equates to 7.4 bbl of oil (Eq. 18.1).

$$(1 \text{ tonne} \times 6.2898 \text{ bbl/m}^3) \div (0.85 \text{ tonne/m}^3) = 7.4 \text{ bbl} \qquad 18.1$$

In all bulk measurements of petroleum and especially of crude oil it is necessary to specify the gross concentration of solids and nonpetroleum liquids present. This is normally stated as percent bottom sediment and water (% BS&W), and is determined by centrifuging of a representative sample.

Light petroleum gas is almost always present in solution in conventional crude oil and some wells are brought into production to produce solely natural gas. However, even when the objective is to produce oil this may contain as much as $50 \, m^3/m^3$ (ca. 300 ft^3/bbl) of dissolved gas. High formation pressures help to dissolve the gas in the oil. Gas is usually separated at well gathering stations by the controlled release of gas from solution. Depending on the producing temperature and pressure of the oil, gas may be released from the oil in several stages and flared. Or it may be compressed for pipeline transport or for injection back into the formation when oil, rather than gas, is the product desired.

Formation water is produced along with the crude oil at rates that average about 10% by volume. Much higher volumes are obtained from oil fields, which are using water flooding for enhanced recovery. The water and any dissolved salts are separated from the oil by gravity settling in field tank batteries after degassing, which leaves the crude oil ready for delivery. Shipped oil usually contains less than 0.5% by volume sediment and water. Sediment and salts, mainly sodium chloride, can occur either dissolved in the aqueous phase of the oil, or as a fine particulate suspension in the oil phase. Other entrained insoluble mineral matter, such as sands and silts may also be present to the extent of 15–30 g/tonne oil (5–10 lb/1,000 bbl), although the salt content alone can run as high as 300–1,500 g/tonne (100–500 lb/1,000 bbl). If this mineral matter is not removed before distillation, it can concentrate and accumulate on heat transfer surfaces during distillation and decrease thermal efficiency of the distillation. Thus, removal of nondissolved material has to be considered in the design of refinery process sequences.

The petroleum itself consists of a mixture of straight hydrocarbons and hydrocarbons containing one or more heteroatoms. Even natural gas that is mostly methane also contains small amounts of four or five other low-molecular weight hydrocarbons. The paraffin hydrocarbon component of crude oil has the generic formula C_nH_{2n+2}, and may comprise gases, liquids, or solids depending on their molecular weight (or the value of n). The simplest example of this group is methane, CH_4, the chief constituent of natural gas. It has a boiling point of $-164°C$ and a melting point of $-183°C$. Propane and butane, C_3H_8, and C_4H_{10}, both are gases under ordinary conditions, but with boiling points of -42 and $-1°C$, respectively, are relatively easily liquefied. Butane and the higher (larger carbon number) members of this series occur not only in the straight chain form, referred to as the normal or "n" form, but also in various branched chain structures of the same molecular formula but with different physical and chemical properties (Eq. 18.2).

$$\underset{\substack{n\text{-butane (or butane)}\\ \text{b.p. } -0.5°C}}{CH_3CH_2CH_2CH_3} \qquad \underset{\substack{\text{isobutane}\\ \text{b.p. } -12°C}}{CH_3-\overset{\overset{\textstyle CH_3}{\textstyle |}}{CH}-CH_3} \qquad 18.2$$

Pentane isomers, C_5H_{12}, have boiling points in the range of normal ambient conditions, and represent the approximate borderline between gases and liquids in the paraffin series (Eq. 18.3).

$$CH_3CH_2CH_2CH_2CH_3 \qquad CH_3\overset{\overset{\displaystyle CH_3}{|}}{C}HCH_2CH_3 \qquad CH_3-\overset{\overset{\displaystyle CH_3}{|}}{\underset{\underset{\displaystyle CH_3}{|}}{C}}-CH_3$$

18.3

n-pentane	isopentane	neopentane
b.p. 36°C	b.p. 28°C	b.p. 9.5°C

As the carbon number gets larger in the paraffin series the number of possible structural isomers also gets larger so that hexane (C_6H_{14}), for example, has five structural isomers and heptane (C_7H_{16}) six. All 18 isomers of octane have been isolated or synthesized, as have the 35 isomers of nonane. Beyond this, however, little is established about the natural occurrence of the 75 possible structural isomers of decane ($C_{10}H_{22}$) or the over 4,000 possible isomers with pentadecane ($C_{15}H_{32}$). However, many of the possible paraffin isomers have been found and isolated from exhaustive separations of natural petroleum [4].

Normal octane, C_8H_{18}, with a melting point of $-57°C$ and boiling point of 126°C, lies near the upper end of the liquid paraffinic constituents of gasoline. *n*-Eicosane ($C_{20}H_{42}$; $CH_3(CH_2)_{18}CH_3$), with a melting point of 36°C and a boiling point of >340°C, is the first of the higher (longer carbon chain or larger molecular weight) paraffins, which is isolated in the solid state (a wax) under ordinary conditions. These larger saturated hydrocarbons occur dissolved in the lower molecular weight liquid hydrocarbons, which comprise the bulk of light natural petroleums.

Saturated hydrocarbons also occur in petroleum in cyclic form, and will have generic molecular formulas of the form C_nH_{2n} (if monocyclic). These cycloparaffins are referred to as naphthenes in the petroleum industry and occur primarily as five, six, and seven-membered rings, with and without alkyl substituents. They also occasionally occur as various combinations of two of these ring systems linked or fused together, in which case their molecular formulas become of the form C_nH_{2n-2} (Eq. 18.4).

18.4

cyclohexane
b.p. 81°C

ethylcyclopentane
b.p. 103.5°C

bicyclo[4.3.0]nonane
(hexahydroindane) b.p. 161°C

Aromatic hydrocarbons occur to a varying extent and have a higher ratio of carbon to hydrogen than any of the commonly occurring paraffins or naphthenes. They correspond to a molecular formula of C_nH_{2n-6} if they are mononuclear (single ring only). Benzene, toluene, and cumene are mononuclear examples of this series, and naphthalene is a dinuclear example (Eq. 18.5).

18.5

benzene
b.p. 80°C

toluene
b.p. 111°C

cumene
b.p. 152°C

naphthalene
b.p. 218°C
m.p. 81°C

These and many other aromatics have been isolated from petroleum fractions. The aromatic content of crude oils can vary widely, but an aromatic content of a third or more of the total, as has been noted for some Borneo crudes, is not unusual [5]. The density (or °API) of a crude oil is an indicator of the aromatic content since the high C:H ratio of aromatic components tends to make these the most dense constituents present. This is particularly true with polynuclear aromatic constituents since each additional ring further reduces the hydrogen count by two, increasing the already high C:H ratio and therefore also the density.

While the bulk of the hydrocarbon content of crude oils is represented by the paraffins, naphthenes, and aromatics, small percentages of several other types of compounds are also present. Olefinic hydrocarbons, unsaturated chain compounds having a carbon–carbon double bond and a type formula C_nH_{2n}, also occur in natural petroleum but only to a very small extent since they are relatively unstable. They are mentioned here, however, since substantial amounts of these are formed by some refinery processes, in particular those involving cracking.

A wide variety of compounds that contain a heteroatom as well as carbon and hydrogen also occur in petroleum but generally to a lesser extent. Hydrogen sulfide and a variety of thiols, sulfides, and thiophenes are some examples of the sulfur compounds present [6] (Eq. 18.6).

$$SCH_2CH_2CH_3$$

$$CH_3(CH_2)_5SH \qquad \qquad \qquad \qquad \qquad \qquad \text{—CH}_3 \qquad 18.6$$

l-hexanethiol   l-thiapropyl-benzene   2-methylbenzo-(b) thiophene

The sulfur content of petroleum containing these compounds is usually quite low but it can be as high as 6% of the total. The more stable oxygen derivatives of hydrocarbons, such as the paraffinic acids, ketones, and phenols (Eq. 18.7), also occur in crude oils.

$$\overset{CH_3}{\underset{|}{CH_3(CH_2)_2CHCH_2CO_2H}}$$

3-methylhexanoic acid   fluorenone   p-cresol   18.7

And nitrogen compounds, either on their own or complexed to a transition metal, such as vanadium, also occur in petroleum to a small extent. Pyridine, quinoline, and many other heteroaromatics (Eq. 18.8) have also been found.

pyridine   quinoline   carbazole   benzonitrile   18.8

The first two of these compounds occur at sufficiently high concentrations in some crudes to enable production of them from petroleum. Benzonitrile has also been detected [6]. Much of the trace metal content of petroleums, in particular vanadium and nickel, is present in association with petroporphyrins, which are polycyclic pyrroles closely related in structure to the hemes and chlorophylls. These materials are examples of the more complex nitrogen heterocycles to be found in petroleum. These particular heterocycles with their complexed metal atoms contribute much to our present knowledge of the original biogenesis of the petroleum hydrocarbons [6].

If the proportion of low-molecular weight hydrocarbons (low-carbon number, small molecules) to high-molecular weight hydrocarbons is high, then a light, very fluid (low-viscosity and density) crude is obtained. If, however, high-molecular weight paraffins or polynuclear aromatics (asphaltenes) predominate, then this will give a high viscosity (and density), high pour-point crude, the so-called "heavy crudes." The composition of crude oil produced from a particular oil field will vary somewhat with the stage of production. The proportion of lighter (lower molecular weight) hydrocarbons will generally be higher in the early stages of production of a new oil field. The ratio of the percentage of paraffins to the proportion of naphthenes and aromatics present will be relatively consistent from the different producing wells of a particular oil field but can vary widely from one oil field to another. With all these variables in mind, refineries seldom rely on crude oil from a single oil field for the whole of their production. A selection of different crudes will be employed, depending on price, availability, operational processing equipment, and the proportions of the various refined products that are desired.

18.2. DESALTING AND DISTILLATION

18.2.1. Crude Oil Desalting

Crude oils delivered to the refinery frequently contain substantial quantities of water, silts, sand, extraneous salts, etc. "Desalting" involves removal of most of these impurities before further processing [7]. If not removed they can increase scaling rates (buildup of deposits) or cause blockages of refinery equipment, such as heat exchangers, and increase corrosion. Two desalting methods are in common use, each using several processing units operating in series.

Chemical desalting is accomplished by addition of about 10% water to the oil to be treated, plus sulfuric acid or sodium hydroxide as necessary for crude pH adjustment. Reagents are added just before the oil enters the desalting system charging pump to obtain good mixing [8] (Fig. 18.1). The acid or base addition may be sufficient by itself to cause rapid demulsification after mixing, or a small concentration of a proprietary demulsifier, such as Tretolite (a polyethylenimine) may have to be added with the water to assist the process. The mixture is then heated to 65–180°C (depending on oil viscosity and boiling point) and mixing is assured by passage through a

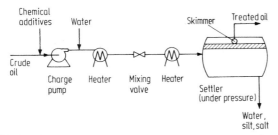

FIGURE 18.1 Flowsheet for the main steps involved in chemical desalting. (From Hydrocarbon Processing [8] with permission.)

mixing valve [8]. It is then held for a time to allow separation of the oil and aqueous phases. Cleaned oil is drawn from the top of the vessel and the aqueous phase containing the salts, sand, silt, and other extraneous material is drawn from the bottom.

Electrostatic desalting is also common [9]. As in chemical desalting, water is added to the oil stream (to dissolve any suspended salts) and the stream passed through a mixing valve into the desalter. Here oil–water separation is mainly accomplished with a high-potential electrostatic field instead of with demulsifying chemicals. The electrostatic field induces rapid coalescence and settling of water droplets together with any other water-associated impurities and "creaming" of the oil phase. The water phase is drawn continuously from the bottom of the desalter vessel and passed to the refinery effluent-treatment plant. The oil phase is skimmed from the top for further processing. Desalting by one or the other or both of these two procedures is now a routine for the refining of most crudes. It is also sometimes used as a secondary clean-up to remove excess salt still remaining in distilled atmospheric gas oil before it is further processed [10]. Desalting techniques have recently been extensively re-examined experimentally [11].

18.2.2. Petroleum Distillation

Distillation of crude oil accomplishes a rough sorting of the main components on the basis of boiling point differences. This was originally carried out on a batch basis, where the crude oil to be distilled was entirely in place in the distilling vessel at the start of heating. With a batch still the components of the crude are obtained in vapor form in sequence as the heat is applied, the lighter (lower boiling point) components first. Midrange and heavier, less volatile constituents are distilled later in the sequence, eventually leaving behind a viscous, high boiling point asphaltic residue in the distilling vessel. Batch distillations are inefficient because of the need for separate steps for filling, heating, stopping, pumping out the residue, and refilling of the distilling vessel again. For this reason, they tend to be economic only for distillations on a scale of less than 300 bbl/day (ca. 12,000 U.S. gallons, 10,000 Imp. gallons, or 45 m³/day). This scale of production was only appropriate for local, very small distillate requirements from a simple refinery, which also happened to be close to a petroleum producing area, not a common situation today.

Even a small modern refinery distills 2,000–10,000 bbl/day and the largest American refineries process 175,000 bbl (27,800 m^3) or more crude oil per day [12]. These are all scales, which require continuous operation for efficient production.

For continuous distillation the crude oil is first heated to 400–550°C while it is continuously flowing through a pipe still, using natural gas, "light ends" (miscellaneous low boiling point hydrocarbons), or fuel oil for fuel. The heated crude is then passed into a fractionating tower near the bottom, the hotter end of the tower. In the fractionating tower, a unit 2–3 m in diameter and 30–40 m high for a large modern refinery, the lower boiling components move up as vapors (Fig. 18.2). As the vapor moves upward past each plate of the column it is forced, via the bubble caps ("bell caps") of that level, to pass through and come to thermal equilibrium with the liquid on that plate or tray (see Fig. 16.6). Hydrocarbons having a boiling point lower than the temperature of the liquid on the plate will continue to move up the column, in vapor form. Components of the heated crude having a boiling point higher

FIGURE 18.2 Partial cut-away view of continuous crude oil distillation via a pipe still and fractionating column. Frequently the heat content of the distillate fractions is employed to preheat the incoming crude oil (not shown here).

or the same as the temperature of the liquid on the plate will tend to condense in the liquid on the plate. As liquids of similar boiling points accumulate on the plate they either overflow into the projecting end of a "downcomer" pipe to the next lower plate, or they are drawn off the plate and the column as one of the product streams of the crude distillation. The lower end of each downcomer is of sufficient length that it dips some distance into the liquid on the lower tray to prevent vapor movement up the downcomer.

In this way propane and butane (C_3 and C_4 hydrocarbons) that are the lowest boiling components, and frequently some water, are collected as vapors from the top of the fractionating column. The vapors pass through a dephlegmator (a partial condenser), which condenses naphtha, gasoline, and water vapor components out of this vapor stream, and allows passage of butane and lighter fractions through as vapors (Table 18.2) [13]. Water is phase-separated from the condensed liquids for removal in a unit outside the fractionating column, and most of the condensed naphtha/gasoline fraction is withdrawn from this as a product stream.

A part of the phase-separated hydrocarbon stream is returned to the top plate of the fractionating column as a reflux for temperature control. This return of condensed light distillate to a top plate of the column is what controls the temperature of the cool end of the fractionating column to achieve the thermal gradient of 12–15°C/m required. The bottom and hot end of the column is kept at about 500°C via heat brought in through the entry of preheated crude oil, already mostly vaporized, and also occasionally by means of a reboiler operating at the bottom of the column indirectly heated by heat exchange fluid. To decrease the thermal gradient per meter of height and sharpen (narrow) the boiling range of each fraction separated from crude

TABLE 18.2 Atmospheric Pressure Boiling Point Ranges, Trends in Properties, and Approximate Composition of Representative Fractions from Crude Oil Distillation[a]

Fraction	Boiling range (°C)	Density (g/cm³)	Viscosity[b] (cSt at 38°C)	Range of n-paraffins[c]	No. of compounds[d]	Sulfur (wt.%)
Gas	<40	0.4–0.6[e]	–	C_1-C_5	7	0.001
Naphtha/gasoline	40–180	0.70	Low	C_6-C_{10}	>100	0.011
Kerosene	180–230	0.79	1.1	C_{11}, C_{12}	>> 40	0.20
Light gas oil	230–320	0.85	3.7	$C_{13}-C_{17}$	>> 12	1.40
Heavy gas oil	320–400	0.90	50	$C_{18}-C_{25}$	>> 10	2.0
Lubricant	400–520	0.91	60	$C_{26}-C_{38}$	>> 7	3
Residue	>520	1.02	Solid[b]	$>C_{38}$	–	4–6

[a]Compiled from Rossini [4], Kirk-Othmer [12], and Guthrie [13]. The values given are approximate. Uncertainty is indicated by significant figures. Trends seen are realistic.

[b]Viscosity in centistokes is equivalent to units of mm²/s. For the lubricant fraction the value would be about 7 cSt at 99°C and for the residue about 13,400 at 99°C.

[c]Carbon numbers quoted are for normal (i.e., straight chain) paraffins. For branched paraffins the carbon numbers will tend to be somewhat higher, and for cyclic paraffins (naphthenes) and aromatics somewhat lower.

[d]Numbers given are preceded by a single carat if the compounds isolated comprised most of the sample of that boiling range, and a double carat if this comprised only a small fraction of the sample.

[e]A range since this is a composite of the densities at 15°C of methane, 0.3; ethane, 0.37; propane, 0.51; isobutane, 0.56; butane, 0.58; and pentane/isopentane, 0.63.

oil, the fractionating column will occasionally be split into two columns. One of these will normally operate over the temperature range of about 40°C to a maximum of about 350–370°C. The other will be used to fractionate all components boiling above about 370°C. This will normally operate under reduced pressure to provide sharper fractions of the higher boiling constituents of the crude, not possible with a single column.

Whether the crude distillation is conducted in single or dual columns, the principle of operation is the same. Hydrocarbon vapors plus a naphtha/gasoline fraction boiling in the 40–180°C range are the top take-off products of the crude still. A kerosine/jet fuel fraction boiling at approximately 180–230°C is taken off a plate further down the column, followed by fractions of light gas oil, b.p. 230–300°C, and heavy gas oil, b.p. 300–400°C, from the appropriate lower plates of the column (Fig. 18.2). The highest boiling distillate (i.e., volatile) fraction from the crude still is the lubricating oil stream, b.p. 400–520°C, which also contains much of the grease and wax yield of the crude oil. The residue or the bottom stream of the crude fractionating tower includes all the crude components not vaporized below about 520°C, and consists of mostly asphalt and suspended petroleum coke. This is the highest distillation temperature normally used for crude distillation because at these temperatures the residual large hydrocarbon constituents begin to decompose or break down.

Since the boiling point ranges of the petroleum cuts (fractions) obtained from the crude distillation column(s) are quite wide, many of these crude streams will be redistilled and/or stripped to sharpen (narrow down) the boiling point ranges. The gas stream that is composed of components boiling below 40°C, is normally redistilled under pressure. Pressure distillation increases the boiling point and condensation temperatures of the components present, which sharpens the separations and enables condensation of many of the separated compounds as liquids. At atmospheric pressure refrigeration would be required to achieve condensation of these components. A separate column is generally used for the final clean-up of each of these components.

The naphtha/gasoline, kerosine, and higher boiling streams will usually be sharpened to narrower boiling ranges than obtained directly from the crude still by heating with steam in a reboiler to strip (remove by vaporization) the excess volatiles. Alternatively, these streams may be redistilled in small columns to separate each of them into two or three separate sharper fractions.

The trends in density, viscosity, sulfur content, etc., that are obtained as one proceeds from the low to higher boiling point fractions are quite informative as to content (Table 18.2). For instance, as the carbon to hydrogen ratio increases there is a corresponding increase in density. Viscosity also increases steadily with an increase in molecular size. The sulfur also tends to occur in the larger molecules from the increase in sulfur content observed with an increase in boiling point. The content of transition metals, such as vanadium, nickel, and iron tends to predominate with the high boiling and residual fractions of crude oil distillation [9]. Up to 250 ppm of vanadium is not uncommon in the residues [14].

Simple distillation, secondary distillation, and stripping processes separate crude petroleum into useful hydrocarbon fractions of similar physical

TABLE 18.3 Some Primary and Secondary Uses of the Gas Stream from Petroleum Refining

Component	Direct uses	Primary products	Secondary products
Methane	Natural gas, heating	Hydrogen	Ammonia, methanol
Ethane	Natural gas, heating	Ethylene	Polyethylene, ethanol, styrene, ethylene glycol
Propane	Bottled gas (liquefied petroleum gas, LPG)	Propylene	Polypropylene, propylene glycol, cumene
Butane	Bottled gas, natural gasoline (winter blends)	Butadiene, polymer gasoline	Synthetic rubber, adiponitrile (nylon polymers)
Pentane	Natural gasoline	1-, and 2-pentenes, cyclopentane	*sec*-Amyl alcohols (oxo process)

properties. Each fraction can contain from 5 to over 100 different compounds (Tables 18.2 and 18.3). But the proportions of the main fractions separated vary widely depending on the origin of the crude oil (Table 18.1). Some crude oils contain as little as 1–2% of a gasoline fraction. The demand for gasoline in Japan and Western Europe is 16–20% of the petroleum refined, and in the U.S., it is about double these figures [15, 16] (Table 18.4). In northern

TABLE 18.4 Principal Uses and Demand Distribution for the Primary LIQUID refinery Products[a]

Refining stream	Components	Direct uses	Demand distribution (wt.%)		
			Japan	U.S.A.	W. Europe
Gases	C_1–C_5	Natural gas, LPG[b] etc	1–2	1–2	1–2
Naphtha/gasoline	C_6–C_{10}	Aviation gasolines, gasoline motor gasolines, light weed control oil, dry cleaning, and metal degreasing solvents	20	42–45	16–21
Kerosene/jet fuels	C_{11},C_{12}	Diesel fuels, jet fuels, illuminating and stove oils, light fuel oils	8–9	9–11	4–8
Light gas oil	C_{13}–C_{17}	Gas turbine fuels, diesel fuels, cracking feedstock, furnace oil	7–8	20–22	33–35
Heavy gas oil	C_{18}–C_{25}	Cracking feedstock, fuel oils	48–50	8–10	35–37
Lubricant fraction	C_{26}–C_{38}	Gear and machinery lubricating oils, cutting and heat-treating oils, lubricating greases, medicinal jelly, paraffin waxes	2–3	1–2	1–2
Residue	$> C_{38}$	Roofing, waterproofing, and paving asphalts, residual fuel oils	–	5–6	2
Refinery consumption and losses			–	20–22	11–13

[a]Compiled from data of Kirk-Othmer [12, 15], Guthrie [13], and Garner [16].
[b]Liquefied petroleum gas (i.e., propane and butane).

countries there is also a seasonal swing in gasoline demand from 27% of refinery output in winter to about 35% in summer. These poor matches between the proportions of constituents actually present in crude oil and the proportions demanded of refineries means that molecular modification of the crude distillation fractions is necessary to achieve a better correspondence of product to demand.

18.3. MOLECULAR MODIFICATION FOR GASOLINE PRODUCTION

The naphtha/gasoline fraction recovered from the crude by distillation is not large enough to supply the demand. Thus, gasoline sold at the pump contains not only desulfurized straight-run gasoline obtained directly from the crude, but also chemically modified crude fractions referred to as cracked gas oil, reformate, polymer gasoline, and "natural" gasoline to supplement this. Natural gasoline, also referred to as casinghead gasoline, is the liquid fraction of C_5 and higher hydrocarbons, which is condensed from natural gas at the well head in a field installation. Natural gasoline is added to the crude oil before primary distillation. Cracked gas oil comprises the C_6-C_{10} hydrocarbon fraction recovered from the partial thermal breakdown of the larger molecules of gas oil. Polymer gasoline is produced by fusing the smaller hydrocarbon molecules, primarily C_3 and C_4 hydrocarbons, to produce C_6-C_8 hydrocarbons for blending into gasoline. Reformate is straight-run gasoline, which is desulfurized and upgraded to a higher octane rating by thermal or catalytic reforming. So by a combination of decreasing the molecular size range of crude oil components containing more than 10 carbon atoms, and increasing the carbon number of the C_3 and C_4 fractions of crude oil, the gasoline yield is increased substantially to better match the demand. The direct separation of the first two gasoline components from the crude oil has already been discussed. Details of production of the other two components follow.

18.3.1. Thermal Cracking of Gas Oils

Cracking is a process in which high temperatures and moderate pressures are used to break down the larger molecules of gas oil into smaller ones suitable for incorporation into gasoline. Cracking processes for the production of gasoline constituents are the most important petroleum modifying processes used by refineries today, with a capacity in North America that exceeds 50% of the current crude distillation capacity. If the thermal decomposition is relatively mild and is conducted on a residual feedstock to decrease its viscosity or to produce a heavy gas oil from it, the process is called viscosity-breaking, or "vis-breaking" for short. If the decomposition is sufficiently severe that coke is the residual product, the process is termed coking. If, however, the feedstock being treated is already in the naphtha/gasoline boiling range and is only subjected to mild decomposition by heat to decrease the boiling point and increase the octane number slightly, then the process is referred to as thermal reforming [15]. Vis-breaking, coking, and thermal reforming are all related to, but less important than the processes, which are used to crack gas oil directly to gasoline fractions.

Gas oil may be cracked by straight thermal cracking or by one of the catalytic cracking processes. Thermal cracking, or the application of heat and pressure alone to modify gas oil refinery streams, was the original method by which the gasoline yield from a crude oil was increased. Today straight thermal cracking is used only in "simple" refineries where the available petroleum modification equipment is modest and where the gasoline requirement is small. Gas oil vapor is heated to 500–600°C under a pressure of about 3.5×10^6 Pa(N/m^2; or about 34 atm) for a thermal contact time of about 1 min. These high temperatures are sufficiently energetic to cause homolysis (rupture to two radicals) of carbon–carbon bonds to form two highly reactive radical fragments (Eq. 18.9).

$$\underset{\underset{H}{|}}{\overset{\overset{H}{|}}{R C}} : \underset{\underset{H}{|}}{\overset{\overset{H}{|}}{C}} CH_2CH_2R' \rightarrow \underset{\underset{H}{|}}{\overset{\overset{H}{|}}{R C}} \cdot + \cdot \underset{\underset{H}{|}}{\overset{\overset{H}{|}}{C}} CH_2CH_2R' \qquad 18.9$$

Either of the radicals can then attack another hydrocarbon molecule, by collision and abstraction of a hydrogen atom from the new molecule to form a new stable species from itself and another radical species from the other molecule (Eq. 18.10).

$$R\dot{C}H_2 + RCH_2CH_2CH_2CH_2CH_2CH_2CH_3 \rightarrow \qquad 18.10$$
$$RCH_3 + RCH_2CH_2CH_2CH_2\dot{C}HCH_2CH_3$$

This process is thermodynamically favorable since a secondary (or internal) radical is more stable than a primary (terminal) one. Or the fragment can undergo β-fission to lose ethylene and form a new smaller primary radical (Eq. 18.11).

$$R'CH_2CH_2\dot{C}H_2 \rightarrow R'\dot{C}H_2 + CH_2=CH_2 \qquad 18.11$$

It is the β-fission reaction that produces much of the ethylene obtained from thermal cracking.

The secondary radical can also abstract a hydrogen atom from another hydrocarbon molecule, but this process will not be as favorable as the process, which produced it since neutralization and formation of radicals will be energetically similar. Also there would be no reaction progress. However, this radical center can undergo β-fission too, forming a smaller hydrocarbon radical and olefin products (Eq. 18.12). This process can continue (Eq. 18.13).

$$RCH_2CH_2CH_2CH_2\dot{C}HCH_2CH_3 \rightarrow \qquad 18.12$$
$$RCH_2CH_2\dot{C}H_2 + CH_2=CHCH_2CH_3$$

$$RCH_2CH_2\dot{C}H_2 \rightarrow R\dot{C}H_2 + CH_2=CH_2 \qquad 18.13$$

Some of the larger olefins formed by these processes are useful gasoline constituents. The ethylene also formed is used for other purposes.

There is another process open to a terminal radical on a long hydrocarbon chain, which can give rise to useful gasoline constituents, which is called "backbiting." The radical on the terminal carbon can bend back on itself to abstract a hydrogen atom from the carbon atom six from the end of the chain, forming a more stable secondary radical from a primary one (Eq. 18.14). Progression of the resulting secondary radical center then gives heptene plus a primary radical fragment (Eq. 18.15).

$$\begin{array}{ccc} & H_2C\begin{smallmatrix}CH_2\\ \end{smallmatrix} & \longrightarrow & H_3C\begin{smallmatrix}CH_2\\ \end{smallmatrix} & & 18.14 \end{array}$$

$$RCH_2CH_2\dot{C}HCH_2CH_2CH_2CH_2CH_2CH_3 \rightarrow \qquad\qquad 18.15$$
$$RCH_2\dot{C}H_2 + CH_2\!=\!CHCH_2CH_2CH_2CH_3$$

Thus, by the occurrence of many variations of processes, such as these, thermal cracking produces a mixture of gaseous and liquid hydrocarbons of paraffinic (saturated) and olefinic (unsaturated) types, plus coke (Eq. 18.16).

$$\begin{array}{l} \text{gas oil} \quad \xrightarrow[\text{34 atm, ca.60sec}]{\text{500–600°C}} \quad \left\{ \begin{array}{l} \text{–gases} \qquad\qquad\qquad 18.16\\ \text{–naphtha/gasoline}\\ \text{–cracked gas oil (heavier fractions)}\\ \text{–coke} \end{array} \right. \end{array}$$

The gasoline component is separated from the product mixture by distillation (Fig. 18.3). Of course, the objective of the cracking process is to produce as much hydrocarbon in the C_5–C_9 range appropriate for gasoline blends as possible (e.g., Eq. 18.17).

$$\begin{array}{ccccc} C_{17}H_{36} & \rightarrow & C_8H_{18} & + & C_9H_{18} \qquad 18.17\\ \text{pentadecane} & & \text{octane} & & \text{nonene (olefin)} \end{array}$$

Larger paraffinic hydrocarbons cannot be cracked into two smaller paraffins because of the hydrogen-deficient nature of cracking stoichiometry. This is why the best that can be done is to obtain a paraffin plus an olefin. This process is a sensitive one. At too high a temperature or too long a contact time, the coke-forming reactions become more significant at the expense of hydrocarbons of the desired size range (e.g., Eq. 18.18).

$$\begin{array}{ccccc} 9\,C_{15}H_{32} & \rightarrow & 16\,C_8H_{18} & + & 7\,C \qquad 18.18\\ \text{pentadecane} & & \text{octane} & & \text{coke} \end{array}$$

Thus care is required to select suitable cracking conditions to minimize the extent of coke formation.

18.3.2. Catalytic Cracking

In about 1940 it was discovered that when a suitable catalyst was used for cracking to gasoline the process was accelerated by several hundred to a thousand times that of straight thermal cracking [15]. This discovery firmly

TABLE 18.5 Trends in the Utilization of Cracking Processes for Petroleum Modification in the U.S.A.[a]

Cracking process	Cracking capacity as % of crude capacity						
	1930	1940	1950	1960	1970	1979	1993
Thermal	42	51	37	16	12	n/a	1.5
Catalytic	0	3	23	36	37	33	35
Hydrocracking	0	0	0	0	5	5	9

[a]Data from Kirk-Othmer [12], Considine [17], and OECD [18].

established the dominance of "cat-cracking" for the production of gasoline from gas oils (Table 18.5).

Catalysts used in this process are of two types. Special acid-washed clays of particle diameters in the 200–400-μm range are used in fluidized bed versions of cracking [19]. In these units the catalyst is kept suspended or "fluffed up" on an upward moving stream of hot gas oil vapors, which ensures continuous exposure of all catalyst faces to the raw material and provides continuous turnover of catalyst. Synthetic catalysts are prepared from a mixture of 85–90% silica and 10–15% alumina, or from synthetic crystalline zeolites (molecular sieves) [19]. They are either used in a small particle size suitable for use in a fluidized bed, or can be formed into 3- to 4-mm diameter pellets appropriate for crackers, which use a moving bed for catalyst cycling.

A common feature of these catalysts is their acidic nature (i.e., they all act as solid phase acids in the hot gas oil vapor stream). Synthetic silica/alumina catalyst composites, for example, have an acidity of 0.25 mEq/g distributed over an active surface area of some $500 \, m^2/g$. This acidity is the key feature that distinguishes catalytic cracking from straight thermal cracking.

Among the various possibilities open to acid catalyst action is that it may donate a proton to an olefin, generating a carbocation. The carbonium ion may simply lose a proton again but this time from an internal position along the chain, which yields an olefin, a stable product (Eq. 18.19).

$$RCH_2CH_2CH = CH_2 \xrightarrow{H^+} RCH_2\overset{+}{CH_2}CHCH_3 \rightarrow \qquad 18.19$$
$$RCH_2CH = CHCH_2 + H^+$$

Or the cationic center might migrate to a position further along the chain which then favors alkyl group migration toward a more stable carbonium ion, rather than olefin formation (Eq. 18.20).

$$RCH_2CH_2\overset{+}{C}HCH_3 \rightarrow RCH_2\overset{+}{C}HCH_2CH_3 \rightarrow RCH_2\overset{+}{C}H-\overset{CH_3}{\underset{|}{C}H_2} \qquad 18.20$$

Migration of a hydride ion from the branching carbon of the primary (terminal) carbonium ion so formed can then produce a relatively stable tertiary carbonium ion. An unchanged olefinic product may be formed from this by loss of a further proton (Eq. 18.21).

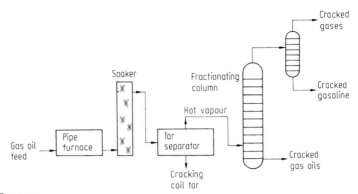

FIGURE 18.3 Tube and tank cracking unit for straight thermal cracking.

$$\underset{\substack{|\\ RCH_2-\overset{+}{\underset{}{C}H}-CH_2}}{\overset{CH_3}{}} \; \rightarrow \; \underset{\substack{|\\ R-CH_2-\underset{+}{C}-CH_3}}{\overset{CH_3}{}} \; \rightarrow \; \underset{\substack{|\\ RCH_2-C=CH_2 + H^+}}{\overset{CH_3}{}}$$

18.21

These kinds of processes lead to the internal olefinic (alkene) and the branched paraffinic (alkene) and other branched olefinic products which are characteristic of catalytic cracking processes.

When a carbonium ion is formed several carbons from the end of a large hydrocarbon molecule, either by withdrawal of a hydride ion by another carbonium ion species in the gas phase or by an electron-deficient area on the surface of the catalyst, it can form a neutral product by one of the processes just outlined. If so, the properties of the product will change but there will be no significant decrease in average molecular weight of the product. Or it can undergo β-fission via either of two possible routes each yielding two smaller fragments, one with a terminal carbonium ion and the other with a terminal olefin (Eq. 18.22)

$$RCH_2CH_2CH_2 \overset{1}{\underline{}} CH_2\overset{+}{C}HCH_2 \overset{2}{\underline{}} CH_2CH_2CH_2R' \qquad 18.22$$

Fission at:

1. $\rightarrow RCH_2CH_2\overset{+}{C}H_2 + CH_2=CHCH_2CH_2CH_2R'$

2. $\rightarrow RCH_2CH_2CH_2CH_2CH=CH_2 + \overset{+}{C}H_2CH_2CH_2R'$

These are the types of processes that give the molecular size reduction activity of catalytic cracking processes.

These acid-catalyzed changes are brought about by passing hot gas oil vapors either through a mechanically circulated bed of the larger pellets or upward through a fluidized bed of more finely divided catalyst particles. The upflow of vapors provides the fluidizing action of the catalyst, hence the units are referred to as "fluidized cat crackers" or FCC units. To avoid loss of

FIGURE 18.4 Main operating components of a fluid catalytic cracking unit. (From [20], courtesy of G.A. Purdy.)

catalyst as the vapors of cracked gas oil leave the catalyst bed, the vapors from fluidized bed crackers are passed through two cyclone separators in series (Fig. 3.4). These retain any entrained solid particles in the vapor stream from the catalyst bed and return these to the bed [20] (Fig. 18.4). The various products formed are then separated from the composite vapor stream on a distilling column.

Overall, catalytic cracking processes achieve a far higher cracking rate than is possible from straight thermal cracking. Furthermore, this is achieved at somewhat lower temperatures and much lower pressures [21] (Table 18.6). A further benefit of modern catalytic cracking units is that a much higher proportion of gasoline-suitable products are obtained, 50–80% of the total. Continuous decoking of the catalyst is necessary at the high cracking rates of these systems in order to maintain catalytic activity. This is accomplished by continuously moving a portion of the catalyst from the catalyst bed to a regenerator, using pneumatic conveying or gravity (Fig. 18.4). Here, the still hot catalyst particles are suspended in a current of air, which forms a second fluidized bed in the regenerator and effectively burns off much of the carbon (Table 18.5). This process also reheats the catalyst particles before they are returned to the endothermic processes going on in the cracker using hot gas oil vapor as the driver. This method of catalyst regeneration allows an easy means to maintain high recycle rates and provides an efficient solid heat transfer medium between the regeneration unit and the cracker. It also represents a substantial petroleum engineering feat since a fluid bed cat cracker processing 100,000 barrels of gas oil per day moves catalyst between the cracker and the

TABLE 18.6 Comparisons of the Conditions Required for Straight Thermal and Catalytic Cracking of Gas Oil to Gasoline[a]

	Cracking conditions	
	Straight thermal	Fluidized bed
Temperature (°C)	450–565	465–540
Pressure[b] N/m²	$1.8-6.2 \times 10^6$	$6.9-14.5 \times 10^4$
(atm)	(14–60)	(0.7–1.4)
Contact time (seconds)[c]	40–300	100–300
Carbon removal method, specifications	Periodic shutdown	0.5–2.6% C on spent catalyst; 0.4–1.6% C on regenerated catalyst

[a]Selected from the data of Kent [19, 21], and converted to appropriate units.
[b]Above atmospheric pressure (i.e., gauge pressure, in pascals) (N/m²).
[c]Estimated from a space velocity range of 0.5–3.0. Space velocity here is defined as (weight of oil feed per hour)/(weight of catalyst in reactor), since catalyst volume is not fixed in the fluid bed mode of operation.

regenerator at the rate of a boxcar load per minute, all without any moving parts [20]. The waste combustion gases from the regenerator exit through two cyclones in series to avoid catalyst loss, followed by heat exchange units for energy recovery as steam finally passing through an electrostatic precipitator for fine particulate emission control. It has been conventional wisdom for refiners to keep the catalyst level in the FCC unit as low as possible to minimize catalyst inventory. However, it has been found that increasing the level results in increased catalyst activity, lower catalyst consumption rates, and lower stack losses [22].

18.3.3. Polymer and Alkylate Gasolines

The volume of polymer gasoline formed by joining together two (or more) light hydrocarbons is only 2–5% of the volume contributed by catalytic cracking. Nevertheless, production of polymer gasoline is an important part of integrated refinery operations. It uses large amounts of the C_3 and C_4 hydrocarbons, which are formed as by-products of cracking processes. Thus, it further increases the ultimate yield of gasoline possible from gas oil and is a valuable high octane component for gasoline blends.

Polymerization is carried out on a $C_3 + C_4$ hydrocarbon stream containing a high proportion of olefins and is usually catalyzed by phosphoric acid on a solid support [23]. At pressures of $2.8 - 8.3 \times 10^6$ N/m² (27–82 atm), higher than that used for cracking, and more moderate temperatures in the 176–224°C range, carbon–carbon bond formation occurs, aided by the acid catalysis and close molecular proximity from the high pressure. With isobutylene (2-methylpropene), for example, initial protonation gives t-butyl carbonium ion), which then adds to a further molecule of isobutylene to give a new branched eight-carbon skeleton containing a different tertiary carbonium ion (Eq. 18.23).

$$2\ CH_3\overset{\underset{|}{CH_3}}{C}\!\!=\!\!CH_2 \;\rightarrow\; CH_3\overset{\underset{|}{CH_3}}{\underset{\underset{CH_3}{|}}{C^+}} \;+\; CH_2\!\!=\!\!\overset{\underset{|}{CH_3}}{C}CH_3 \;\rightarrow$$

$$\tag{18.23}$$

$$CH_3\overset{\underset{|}{CH_3}}{\underset{\underset{CH_3}{|}}{C}}CH_2\overset{\underset{|}{CH_3}}{\underset{+}{C}}CH_3 \;\xrightarrow{-H^+}\; CH_3\overset{\underset{|}{CH_3}}{\underset{\underset{CH_3}{|}}{C}}CH_2\overset{\underset{|}{CH_3}}{C}\!\!=\!\!CH_2$$

Loss of a proton back to the catalyst surface again releases a molecule of 2,4,4-trimethylpentene-1, a branched C_8 olefin, from the catalyst surface. Processes analogous to this can occur not only between two branched four-carbon olefins but also between straight chain olefins, propylene and C_4 olefins, and two propylene molecules. The carbonium ion formed by coupling of any two of these olefin units may also, occasionally, add a third olefin. The composite result of all of these processes is that a stream of C_6, C_7, and C_8 olefins, plus much still unreacted material, is obtained from the predominantly C_3 plus C_4 feed stream (Eq. 18.24).

Polymerization:

$$\text{C3, C4 olefins} \xrightarrow[\text{catalysis}]{H_3PO_4} \underset{\text{(some branched)}}{\text{C6, C7, C8 olefins}} \tag{18.24}$$

Components appropriate for blending into gasoline are separated from this product stream by fractionation.

Polymerization was the first refining process to produce larger molecules appropriate for gasoline. However, a chemically related process called alkylation is of more importance today [15]. This uses concentrated sulfuric acid or hydrofluoric acid as a liquid phase catalyst to form dimers or trimers from C_3 and C_4 olefins reacted with paraffinic hydrocarbons [24]. The yield of the alkylation process based on the olefin feed is about twice of that of polymerization. This factor and that the product has at least as good octane rating as polymer gasoline has made alkylation a dominant contributor to gasoline components.

Alkylation conditions are quite mild. Processes employing concentrated sulfuric acid operate at 2–12°C and $4.1\text{–}4.8 \times 10^6\ N/m^2$ (4.1–4.8 atm), with a gas contact time of 5–30 min. Catalysis with anhydrous hydrogen fluoride requires slightly higher temperatures (25–45°C) and pressures than sulfuric acid, and contact times of 20–40 s. Both catalysts involve initial protonation of the double bond of an olefin. With isobutylene, for example, t-butylcarbonium ion is formed. This can add to another isobutylene in the same manner as occurs during polymerization (Eq. 18.25).

$$(CH_3)_3C^+ + CH_2\!\!=\!\!C(CH_3)_2 \;\rightarrow\; (CH_3)_3C\text{–}CH_2\text{–}\overset{+}{C}(CH_3)_2 \tag{18.25}$$

But rather than lose a proton, as would normally occur under polymerization conditions, the dimeric ion obtained abstracts a hydride ion (H^-) from a neighboring isobutane to give the *saturated* dimer 2,2,4-trimethylpentane, and regeneration of a tertiary butyl carbonium ion (Eq. 18.26).

$$(CH_3)_3C-CH_2\overset{+}{C}(CH_3)_2 + HC(CH_3)_3 \rightarrow \qquad\qquad 18.26$$
$$(CH_3)_3C-CH_2CH(CH_3)_2 + {}^+C(CH_3)_3$$

The regenerated carbonium ion can of course continue the process, a key feature being that under alkylation conditions this active species is formed from saturated alkane, not an olefin as required by polymerization. Different alkenes, such as propylene, 1-butene, or the 2-butenes may also form carbonium ions in a similar manner to the process of Eq. 18.25. However, neither *n*-butane nor *n*-pentane can replace an isoalkane for the hydride transfer since an *n*-alkane is not capable of forming a stabilized carbonium ion. Nevertheless, this is one advantage that the alkylation process has over polymerization as a route to gasoline; it is able to use both light hydrocarbon alkanes (as long as they are branched) and alkenes. Alkylation and polymerization both produce branched products, but the alkylation products are saturated (Table 18.5) whereas the polymerization products are alkenes.

18.3.4. Upgrading of Gasoline Components

Polymer and alkylate gasoline components characteristically have high octane numbers because of the high degree of branching of these products (Table 18.5). Cat-cracked gasoline also has a high octane, 92 or better research octane number (RON), mainly contributed by its high olefin content. In contrast, straight-run gasolines generally give low octane numbers because they are mainly composed of normal alkanes. The octane number of a gasoline blend was improved within limits for many years by the addition of tetraethyl lead or other antiknock compounds. With increasing incentives to reduce lead dispersion into the atmosphere there is a growing trend to upgrade the octane number of a gasoline primarily by raising the octane number of the hydrocarbon feedstock blend supplemented by nonleaded antiknock additives when necessary.

Catalytic thermal reforming, under a variety of trade names, such as Platforming, Ultraforming, Renoforming, etc., and using similar but somewhat milder conditions than catalytic cracking, introduces many octane improving changes into normal (straight chain) alkane streams [25]. For example, naphthenes (cycloalkanes) are dehydrogenated to benzenes, and alkyl naphthenes undergo dehydroisomerization (Eq. 18.27 and 18.28).

$$18.27$$

$$18.28$$

In the same processing step alkanes are also isomerized to branched alkanes (Eq. 18.29).

$$CH_3CH_2CH_2CH_2CH_2CH_3 \rightarrow CH_3-\underset{\underset{CH_3}{|}}{CH}-CH(CH_3)_2 \qquad 18.29$$

Since aromatic compounds, such as benzene, toluene, and branched alkanes all possess higher octane numbers than their precursors, the octane number of the fuel is substantially improved by these changes.

It should be noted that since sulfur is a poison for the platinum-based catalysts used for these changes, the feed for the catalytic reformer has to be essentially sulfur free. Sulfur is removed by passing the feedstock through a cobalt/molybdenum catalyst bed in the presence of hydrogen, which can come from the catalytic reformer. Carbon-bound sulfur is converted to hydrogen sulfide (Eq. 18.30).

$$\underset{\underset{CH_3}{}}{\overset{}{\text{S}}} + 2H_2 \longrightarrow CH_3CH_2CH_2CH_2CH_3 + H_2S \qquad 18.30$$

Hydrogen sulfide is easily separated from the other constituents after hydrogenation by stripping or fractionation. Hydrogen sulfide and mercaptans (thiols) may also be removed from refinery streams by washing with aqueous sodium hydroxide (lye treating, Eq. 18.31, 18.32).

$$2\,NaOH + H_2S \rightarrow Na_2S \;(water\;soluble) + 2\,H_2O \qquad 18.31$$

$$NaOH + RSH \rightarrow NaSR \;(water\;soluble) + H_2O \qquad 18.32$$

Sodium hydroxide may be regenerated, and mercaptan recovered from the alkaline scrubber effluent by steam stripping (Eq. 18.33).

$$NaSR + \underset{(as\;steam)}{H_2O} \rightarrow NaOH + \underset{mercaptan\;oils}{RSH \uparrow} \qquad 18.33$$

Any mercaptans boiling below 80°C are also readily dissolved in alkaline solutions (Eq. 18.32). A common process for the removal of dissolved mercaptans of this kind, especially from catalytically cracked gasoline and liquefied petroleum gas, is the Universal Oil Products (UOP) Merox process that also uses caustic soda for extraction. In this process, however, the mercaptans are then oxidized to disulfides using air assisted by a metal complex catalyst dissolved in the caustic soda (Eq. 18.34).

$$4\,NaSC_2H_5 + O_2 + 2\,H_2O \rightarrow \underset{diethyldisulfide}{2\,C_2H_5SSC_2H_5} + 4\,NaOH \qquad 18.34$$

In this way the sodium hydroxide is regenerated for further use, and the disulfides, which are not soluble in sodium hydroxide, form an oily layer that can be removed. This is one way in which the sulfur content of gasolines is kept below the 0.1% limit required for marketing.

18.3.5. Gasoline Blending

For optimum use of the various refinery fractions in gasoline, they must be blended to obtain the correct volatility and rate of combustion for the engines in which they are to be employed. Winter conditions require more volatile gasolines, which are obtained by increasing the proportions of butane and pentane in the blend. To decrease standing losses (vaporization from the carburation systems, fuel tank, etc.) in summer the proportions of the more volatile constituents are decreased.

Control of the rate of combustion of gasoline vapor is necessary to obtain smooth engine operation over a range of loads and operating speeds. Rough engine operation, or knocking, is caused by too rapid combustion of the gasoline vapor/air mixture in the cylinder. This condition is aggravated when the engine is operating under a heavy load. Rough operation may also be the consequence of preignition (too sensitive ignition), which might in turn be caused by hot carbon deposits on the cylinder head, by an overly hot exhaust valve, or by compression ignition of the fuel vapor–air mixture in the cylinder rather than by the correct ignition timing provided by the spark plug.

The components of commercial gasolines are complex mixtures of hydrocarbons, which are best tested in an operating engine for their resistance to knock, either individually or in already-mixed blends. An octane rating is then assigned to the tested fuel from its resistance to preignition as compared to that of a standard fuel of known octane rating. Octane numbers have been defined for some specific hydrocarbons. Pure 2,2,4-trimethylpentane, for example, burns smoothly in a high compression engine, and is assigned an octane number of 100. n-Heptane that has a strong tendency to knock, is assigned an octane number zero. Octane numbers of other pure hydrocarbons or convenient blends can also be determined experimentally [26]. In general, the octane number of a straight chain hydrocarbon is inversely proportional to molecular weight and directly proportional to the degree of branching or unsaturation (Table 18.7). Rather than using the expensive pure hydrocarbons for test purposes, commercial blends of hydrocarbons of known standardized

TABLE 18.7 Motor Octane Numbers Related to Degree of Branching and Molecular Weight of Some Representative Gasoline Constituents[a]

Component	Branches	Molecular weight	Octane number
n-Butane	0	58.1	89
1-Butene	0	56.1	92
trans-2-Butene	0	56.1	95
Isobutane	1	58.1	97
n-Pentane	1	72.2	63
Isopentane	1	72.2	90
n-Hexane	0	86.2	26
2-Methylhexane	1	100.2	47
3-Methylhexane	1	100.2	55
2,2-Dimethylpentane	2	100.2	96
2,2,3-Trimethylbutane	3	100.2	101

[a]Data selected from Hutson and Logan [26]. Motor octane numbers (MON) are determined under more rigorous test conditions than research octane numbers (RON), and hence are generally somewhat lower than the latter.

octane ratings are used to determine market suitability, and adjustments made if necessary.

If the octane number of the blend is not high enough, it may be raised by increasing the proportion of high octane alkylate, catalytically reformed product, or aromatics in the blend [27]. Or an antiknock compound may be added to accomplish the same thing inexpensively. Use of tetraethyl, or tetramethyl lead with other additives, has been discontinued in Canada and the United States for two decades and are now almost phased out in most other areas. When these additives were in common use in "leaded" gasolines, it took 1.8 mL of tetraethyl lead per Imperial gallon to increase the octane rating of a paraffinic gasoline from 75 up to about 85. However, there is a decreased return from further increases of additive. Doubling the amount added to 3.6 mL/Imperial gallon, the maximum permitted by law for toxicity reasons, only raised the octane number from 85 to about 88. Both tetramethyl and tetraethyl lead function by breaking up (interrupting) the preflame chain reactions leading to combustion, which decreases the tendency of the fuel–air mixture to preignite [28]. The routine levels of tetraethyl lead addition in the U.S. used to be in the 0.48–0.83 mL/L range (1.8–3.15 mL/U.S. gal; 2.2–3.8 mL/Imperial gallon).

In addition to the tetraethyl or tetramethyl lead, both types of antiknock fluids also contained 1,2-dichloroethane and 1,2-dibromoethane (ca. 35% by weight) to react with the lead released on combustion to form lead bromide and lead chloride. These lead halides are volatile at the cylinder combustion temperatures of 800–900°C, and leave the combustion chamber with the exhaust, which prevented the buildup of lead deposits. This was also the final step in the chain of events occurring with the alkylated lead antiknock compounds, which contributed to the widespread dispersal of lead compounds to the air and soil wherever gasoline powered vehicles operated. For this reason, and the toxic exposures during refueling, the alkylated lead addition rate was reduced to not more than 0.5 g of contained lead per U.S. gallon by 1980, even for leaded gasolines [29], and was phased out in the U.S. and Canada by 1985.

By the 1990s MTBE (methyl *t*-butyl ether) became the dominant replacement additive used to raise the octane rating of gasolines in the U.S., and methylcyclopentadienyl manganese tricarbonyl (MMT) in Argentina, Bulgaria, and Canada, among other countries [30]. Fortunately, MTBE is relatively inexpensive (Chap. 19) since much higher addition rates are necessary for an equivalent octane response. However, both replacement antiknock additives have developed usage problems. Methylcyclopentadienyl manganese tricarbonyl has been found in drinking water supplies for some time [31], and environmental issues begin to cloud the use of MMT [30].

A number of other compounds have been tested as antiknock additives. Ferrocene and other related organometallics have shown antiknock activity, but are not as compatible in fuels as the lead alkyls. Many oxygen-containing compounds other than MTBE, such as *t*-butanol, *t*-butyl acetate, and ethanol also have antiknock capabilities [32], and like MTBE require addition at a higher rate to obtain equivalent antiknock activity. The direction to be taken for trouble-free antiknock additives remains to be seen.

The final step in gasoline production is the addition of antioxidants. Mono-, di-, and triolefins, such as are found in catalytically cracked gasoline components, are highly susceptible to gum formation on exposure to air. Contact with air peroxidizes the labile allylic hydrogens of the olefins present to produce free radical centers, which can then initiate polymerization to give the observed gums in the gasoline. However, addition of one, or a combination of *p*-phenylenediamine (PDA), *N*,*N*'di-isobutyl-*p*-phenylenediamine (DBPDA, Eq. 18.35)

$$H_2N-\langle\bigcirc\rangle-NH_2 \quad (CH_3)_2CHCH_2NH-\langle\bigcirc\rangle-N(H)CH_2CH(CH_3)_2$$

PDA   DBPDA

18.35

or 2,6-di-*t*-butylphenol as antioxidants effectively trap radical centers to prevent polymerization [32]. The antioxidant product is a stable free radical, too sluggish to initiate polymerization.

Other minor but important additives, such as corrosion inhibitors, anti-icers, carburetor detergents, and intake valve deposit control additives form part of various gasoline blends. These help to maintain smooth engine operation and long service intervals. Thus even though gasoline appears to be simple material, it is actually a highly complex carefully engineered product of the petroleum refining operation.

18.4. MANUFACTURE OF LUBRICATING OILS

There are four main steps to produce lubricating oil from the appropriate high boiling fractions separated on the crude distillation column. These are vacuum fractionation, to decrease the boiling point range of the fractions used, solvent dewaxing, to lower the pour point (gelling point) of lubricating oils, followed by decolorization and formulation steps.

18.4.1. Vacuum Fractionation

Redistillation of the appropriate lube oil fractions obtained from crude distillation is required to remove the "lighter ends" (lower molecular weight components, such as gas oils), which are poor lubricants, as well as most of the higher boiling point, high-molecular weight asphaltic, and wax constituents from the lube oil base stock. The boiling points of these fractions under ambient conditions are quite close to the temperatures at which cracking and pyrolysis occur. For this reason distillation is conducted under reduced pressures in the range 6–9 kPa (0.06–0.09 atm) [15]. This decreases the boiling points by 50°C or so, sufficient to minimize thermal degradation of the oil on distillation.

Two or three steam ejectors (similar to laboratory water aspirators) operating in series with interstage condensers provide the pressure reduction necessary and are connected to the top end of the vacuum distilling column (Fig. 18.5). Each take-off point on the main vacuum column is also passed

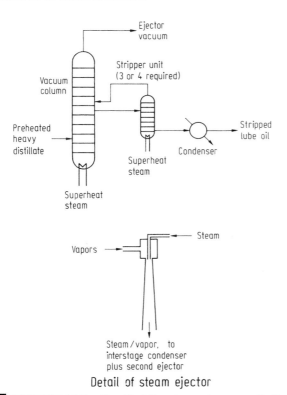

Detail of steam ejector

FIGURE 18.5 Simplified flowsheet of vacuum distillation for lubricating oil refining.

through a stripper unit, a smaller version of the main column, to ensure thorough removal of volatiles before the lubricating oil proceeds to finishing stages. Fractions more volatile than lubricating oils are returned to the column for redistillation.

18.4.2. Solvent Dewaxing

Waxes are mainly paraffinic hydrocarbons in the $C_{20}-C_{30}$ molecular weight range. Their presence in lubricating oils raises the pour point roughly proportionally to the wax content. A lubricating oil with a high pour point can make a cold engine impossible to start under winter conditions. Thus, wax removal is a necessary part of lubricating oil manufacture. It is also one of the most complicated and expensive of lube oil stock preparation steps [33].

Solvents in use for wax removal include propane, which also serves as an autorefrigerant for the process by solvent evaporation, methyl ethyl ketone (MEK)-toluene mixtures, and methyl isobutyl ketone (MIBK). The last two systems are in widest use at present. MEK-benzene was once a common solvent combination for dewaxing but has now been abandoned from benzene toxicity considerations.

For dewaxing with an MEK-toluene mixture, the solvent is mixed with 1 to 4 times its weight of vacuum-fractionated lube oil stock. The intimate mixture is

FIGURE 18.6 Flowsheet of the steps required for solvent dewaxing of a lubricating oil stock.

then chilled to somewhere in the -10 to $-25°C$ range, depending on the degree of dewaxing required (Fig. 18.6). Ammonia (b.p. $-33°C$) or propane (b.p. $-42°C$) evaporation is used as the refrigerant. Scrapers operate continuously on the oil side of the exchanger to prevent the accumulation of adhering wax on the chiller exchange surfaces. The emerging stream from the chillers is a slurry consisting of crystallized wax suspended in a mixture of dewaxed oil and solvent. While the mixture is still cold, the wax is filtered from it and then washed with a small amount of fresh chilled solvent to recover traces of oil still adhering to the wax. The dewaxed oil is recovered from the oil–solvent filtrate in a solvent stripper, which removes the volatile solvent(s) for re-use.

Waxes in the 200–400 molecular weight range covering several melting point ranges, are recovered in finished form after further processing. The 50,000 tonne/year wax market in Canada of the 1970s was distributed as follows: packaging, 32%; tires and molded rubber, 10%; wire and cable insulation materials, 10%; candles, 8%; carbon paper, 8%; other uses, 32% [34].

18.4.3. Lubricating Oil Decolorization

Vacuum fractionation and dewaxing helps to refine lubricating oil stock, but color, caused by olefin and asphaltic residues, still must be removed for long lubricant life. Originally these impurities were removed by the formation of an acid-soluble phase of alkyl hydrogen sulfates and sulfonic acids (Eq. 18.36–18.38),

$$RCH = CH_2 + HOSO_3H \rightarrow RCH(SO_3H)CH_3 \qquad 18.36$$

$$2 H_2SO_4 \rightarrow SO_3 + H_3O^+ + HSO_4^- \qquad 18.37$$

$$C_6H_6 + SO_3 \rightarrow C_6H_5SO_3H \qquad 18.38$$

obtained by treating the lubricating oil stocks with concentrated sulfuric acid. Waste disposal and other problems have led to its gradual abandonment.

Today, clay adsorption or hydrogen treating (catalytic hydrogenation) are used to accomplish decolorization [35]. Passage of oil through a clay coating on a continuous filter is quite effective and is suitable for small refineries.

Larger refineries with a wider product range can justify catalytic hydrogenation facilities. Olefins and aromatics are reduced to their saturated equivalents, paraffins and naphthenes. This method of decolorization decreases the number and extent of conjugated double bonded species and stabilizes the lubricating oil stocks toward oxidation.

The product at this stage is suitable for simple lubricating functions. However, engine lubricants are called on to meet many demands in addition to lubrication, which requires blending and several additives.

18.4.4. Formulation of Lubricating Oils

Paraffinic and naphthenic (cycloparaffinic) stocks may be used for the formulation of lubricating oils, each with favorable characteristics for particular uses. Paraffinic stocks are generally preferred for their superior lubricating power and oxidation resistance. Naphthenic stocks, on the other hand, have naturally lower pour points, i.e, they maintain flow characteristics at lower pour-points than paraffinics (Table 18.8) and are better solvents, features which are more important for applications such as heat transfer, metal working, and fire-resistant hydraulic fluids [33]. Any residual aromatics in the lubricating base stock will have been removed before formulation by solvent extraction, using N-methylpyrrolidone, furfural, or less frequently today, phenol (Eq. 18.39).

$$\text{N-methylpyrrolidone} \qquad \text{furfural} \qquad \text{phenol} \qquad 18.39$$

When oil is used for lubrication of an internal combustion engine its primary function is to reduce the friction of moving parts under a wide range of operating temperatures and loads. At the same time it has to perform efficient heat transfer, deal effectively with accumulating carbon and dust particles, gum deposits, water and corrosive gases, and accomplish all of this without

TABLE 18.8 A Comparison of the Properties of Lubricating Oil Stocks

	Paraffinic	Naphthenic
Density (g/mL)	0.887	0.905
Pour point (°C)	7	−29
Flash point (°C)	246	224
Viscosity (cS)		
38°C	107	120
100°C	11.3	10.0
Viscosity index	100	57

[a]Calculated from the data of Gillespie *et al.* [33].

decomposing in the process [36]. Several major additives to the base oil are used to fulfill these complex functions [37].

Oil-soluble detergents are added to the oil at 0.02–0.2% by weight to keep accumulating impurities dispersed in the oil. Various sulfonates or phenolates are used as their calcium or barium salts to confer oil solubility to the dispersant (e.g., CaX_2 where X is a sulfonate or phenolate ligand Eq. 18.40).

$$CH_3(CH_2)_{11}SO_3^-, \quad \text{wax}\!-\!\!\langle\bigcirc\rangle\!-\!SO_3^-, \quad CH_3(CH_2)_7\!-\!\!\langle\bigcirc\rangle\!-\!O^- \quad 18.40$$

Basic alkyl sulfonates, $Ca(OH)SO_3R$, are also used to contribute acidity neutralizing and dispersing activity to the oil. These detergents (dispersants), help to keep particulate impurities, such as carbon, dust, or metal fines (e.g., lead or other metal oxides and salts) and water or acid droplets suspended in the oil to prevent deposition on or attack of critical moving parts. At the appropriate oil change interval most of the accumulated suspension is removed from the engine with the spent oil.

Mineral oils and in fact most liquids show a decrease in viscosity with an increase in temperature (i.e., an inverse relationship exists between the viscosity of an oil and its temperature). At higher operating temperatures the viscosity of the oil decreases (the oil becomes "thinner"). If this decrease of viscosity with increased temperature is too large, lubrication can become impaired under hot conditions, which can cause excessive engine wear and increase oil consumption. Thus, it is important that the change in viscosity of an oil with temperature be kept within reasonable limits. The ratio of high to low-temperature viscosity of an oil is referred to as its viscosity index, and is an important property of a lubricating oil. The viscosity index is an empirical property determined by measuring the viscosities of an oil at 40°C and at 100°C, and comparing these values with American Society for Testing Materials (ASTM) tables. Under this system Pennsylvania oils that have a relatively small change in viscosity with temperature, are assigned a viscosity index of 100. U.S. Gulf Coast oils, which have a larger change in viscosity with temperature, are assigned a viscosity index of zero.

A few percent of a synthetic polymeric material, such as poly(methyl methacrylate), polyisobutylene, or an ethylene–propylene copolymer added to an oil improves its viscosity index [38, 39]. In this way the viscosity of a low pour point oil that remains fluid at low temperatures, it is still kept acceptably high when it is hot because of the presence of the polymer additive (Fig.18.7) [40]. The additive is thought to function by the individual polymer molecules taking up compact spherical shapes at low temperatures because of the low solvent power of the base oil under these conditions. At higher temperatures the oil is a better solvent for the polymer, the polymer molecules are well solvated and become extended chains, which increases intermolecular entanglements. In these ways the presence of the extended polymer molecules decreases the tendency of the base oil to "thin out" at high temperatures [40] (Fig. 18.8). Viscosity stabilization enables some unsuitable base oils to be used for engine lubrication. These additives also enable the formulation of the multi-grade oils for summer and winter use without requiring seasonal oil changes. Part of the enablement of multi-season use is because some additives

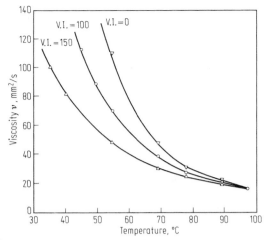

FIGURE 18.7   Relationship of viscosity indices to viscosity change with temperature at different polymer loadings. (From Müller [40].)

are also efficient pour point depressants. These help to prevent any residual wax from crystallizing out at low temperatures so that an engine may still be started under cold conditions.

Rapidly moving engine parts can tend to whip up froth in the lubricant as they operate, and may be aggravated by detergent additives in the oil. Oil foam can cause air to enter the oil circulating pump and cause temporary vapor lock failure of the lubricant circulation system in an engine, leading to possible engine breakdown as a result. Addition of 1–10 ppm of silicone antifoam agents can prevent this without harming lubrication since silicones are also effective lubricants.

Many other additive materials are formulated into lubricating oils. Some are polar compound "oiliness" additives, such as lead soaps, that tend to form a stable film on metal surfaces by plating out on them and maintain lubrication even under extreme pressures. Some contain very finely divided Teflon or graphite particles. Greases are used for lubricating functions where less viscous materials would be forced out of or lost from the friction zone. They consist of a higher concentration of a thickening agent combined with a lubricating oil. Common thickening agents used for greases are the soaps (metal salts of a fatty acid) of aluminum, barium, calcium, lithium, or sodium.

FIGURE 18.8   Mechanism of action of polymeric viscosity index improvers.

Lubricating oils represent only about 2–3% of the volume of crude oil processed. They are a high value component, however. Since about half of the automotive oil sold is potentially recyclable there is a substantial potential for resource conservation and environmental benefit compared to irresponsible disposal. With promotion, this philosophy appears to be gaining acceptance. In Canada in 1974 only about 10% of the potentially reclaimable oil was being recycled, and by 2003 this figure had risen to about 65% [41].

There are several lube oil reclamation options. If only water and sludges are removed in a relatively simple process the reclaimed lubricating oil may be downrated and used as a fuel oil [42]. Full re-refining for use as a lubricant requires six or more stages of upgrading. The higher value of the lubricating oil stock makes this a plausible option [43]. Reclaimed oil may also be processed by delayed coking (cracking) to yield gasoline, gas oil, and petroleum coke [44, 45].

18.4.5. Synthetic Engine Oils

Many special lubricant properties, such as high viscosity indices, very low pour points, high thermal stability, and greater oxidation resistance can be achieved simultaneously by the use of synthetic lubricating fluids [46]. The first synthetic lubricants used alkylated aromatics as the bulk constituent, but these had worse low-temperature viscosity characteristics than refined petroleum stocks and were still susceptible to oxidation. Olefin oligomers (low-molecular weight polymers) developed subsequently were much better in these respects [47]. Blends of olefin oligomer and high-molecular weight esters, such as di-2-ethylhexyl sebacate or trimethylolpropane triheptanoate, are presently favored for automotive use, from cost and performance considerations (Eq. 18.41).

$$[H_3C(CH_2)_3CH(CH_2CH_3)CH_2O_2C(CH_2)_4]_2 \qquad [CH_3(CH_2)_5CO_2CH_2]_3CCH_2CH_3$$
$$\text{di-2-ethyl sebacate} \qquad\qquad \text{trimethylolpropane triheptanoate}$$

18.41

These blends provide reduced wear, better engine cleanliness, longer oil change intervals, slightly improved fuel economy, and lower oil consumption rates than are experienced with conventional petroleum-based lubricants [48]. However, their much greater cost limits their use to the fraction of the automotive market where their advantages outweigh this factor [49].

18.5. FUEL OILS, ASPHALTS, AND PITCHES

The fuel oils include a wide variety of petroleum end uses, from jet fuels at the more volatile low-molecular weight end of the scale, through kerosene, stove oil, and several grades of diesel fuel, to industrial heating fuel and ultimately to asphalt for road cements at the high-molecular weight end. A convenient classification to consider this wide range of products is related to their distillate or residual origin, at the stage of crude distillation.

18.5.1. Distillate Fuel Oils

There is considerable overlap in the boiling ranges of the diesel, jet, kerosene, and light fuel oils represented by the distillate fuel oil class. All, if desulfurized, have the capability of burning relatively cleanly with little more than a trace of ash as a residue. The ash consists of transition metal oxides (mainly vanadium).

Jet fuels are a blend of the higher boiling constituents of the gasoline/naphtha fraction together with a proportion of the C_{11}, C_{12}, and higher components of the kerosene fraction. These are the lowest boiling range of the fuel oil product categories.

Kerosene is the lightest straight fuel oil in the distillate category and has uses, which range from lamp oils, to light stove oils, and diesel fuels for use in Arctic service. Pour points of below $-50°C$ are the attraction of kerosene for low temperature diesel fuel applications. A small addition of lubricating oil stocks is made to diesel fuel in this service.

Conventional diesel fuel is made from crude oil fractions boiling in the range 190–385°C. It is sometimes also referred to as No. 1 fuel oil. Narrower fractions of this boiling range are used depending on the class of diesel fuel being produced. These differ in volatilities and pour points. Diesel fuels have 40–50°C higher flash points than gasoline or jet fuels, which makes them safer [15].

After its required physical properties, the most important specification of a diesel fuel is its cetane number. This is a measure of the ease of autoignition of a diesel fuel, which is critical because this is the means of ignition used in diesel engines. Cetane (*n*-hexadecane, $C_{16}H_{34}$) itself performs very well in a diesel engine, meaning that there is negligible delay between the time of injection of the fuel into the cylinder and fuel combustion. Pure cetane is assigned a cetane number of 100. In contrast to this, α-methylnaphthalene ($CH_3C_{10}H_7$), which performs very poorly (does not readily ignite) in a diesel engine, is assigned a cetane number of 0. Cetane numbers intermediate between these two extremes correspond to the percentage of cetane in a blend of these components. A commercial diesel fuel is composed of a large number of hydrocarbons with physical properties in the right range. It is blended to produce a match of its ignition properties to those of the desired percentage of cetane in a research fuel.

Diesel fuels of cetane number in the 25–40 range are appropriate for operation of slow-speed engines, such as used for the propulsion of ships, and some stationary engines. Medium- and high-speed engines require cetane numbers of about 40, and 50–65, respectively, for smooth operation. Another factor important at low temperatures is that the higher the cetane number the lower the operating temperature at which a diesel engine may be successfully started. In general, the higher cetane number fuels also tend to give smoother engine operation, once it reaches operating temperature, and to reduce the formation of engine deposits and smoke.

The higher boiling No. 2 and No. 3 distillate fuel oils are largely used for space heating (homes and buildings) and can also form part of the fuel composition used in slow-speed diesels. Depending on availability, they are used as the solvent in the formulation of residual fuel oils.

18.5.2. Residual Fuel Oils

As the name implies, these lower cost fuels are formulated from the residues of the distillation and refining processes. Pitchy residues of the crude distillation are dissolved in ("cut with") varying amounts of the residual heavy oil fractions remaining after catalytic cracking, so-called cat-cracked gas oil, or cycle oil (recycle stock). Thus, this group of fuels contains a relatively high concentration of all the less volatile and involatile impurities present in the original crude. These include trace metals, high-molecular weight sulfur-containing compounds, asphalts, coke, and any residual incombustibles, such as salt, silt, etc. A fuel value 5–10% higher than the same volume of distillate fuel helps to offset the disadvantages of the incombustibles [50]. However, residual fuels do require combustion equipment that can cope with their higher viscosities and ignition temperatures and emission control devices to control pollutant losses to the atmosphere.

Bunker A, or No. 4, fuel oil, so-called because of its use in the fueling (bunkering) of ships, is the lightest grade of residual fuel oil. It has the highest proportion of cracked gas oil to pitch, and is designed to stay fluid at ambient temperatures above freezing (a pour point of approximately $-7°C$) without difficulty. Its lower viscosity and hence greater versatility command a price somewhat higher than the other residual fuels. It is an attractive fuel choice for smaller combustion units operating under colder than average conditions, such as space heating, or ships in Arctic service. The use of especially low pour point ($-40°C$) No. 2 light diesel oil is sometimes necessary under these conditions.

Bunker B, or No. 5, fuel oil has a lower proportion of cat-cracked gas oil to pitch than No. 4 fuel oil, but still remains fluid down to storage temperatures of about $10°C$. Combustion equipment must have some provision to heat the fuel prior to atomization, in order to obtain efficient combustion. Outside storage at temperatures below $10°C$ would require insulated tanks fitted for heating to enable the fuel to be pumped.

Bunker C, or No. 6, fuel oil has the lowest proportion of cracked gas oil to pitch of all the residual fuels. Consequently storage tanks for Bunker C require heating even under ordinary ambient conditions to enable the fuel to be pumped and to assist in atomization for burning. Heated storage is not difficult to arrange in return for the lowest in cost and the highest in fuel value among the fuel oils. The fuel value of No. 1 fuel oil is in the range 38,000 to 39,000 kJ/L (135,800–138,800 Btu/U.S. gal) whereas No. 6 fuel oil produces some 42,200–42,600 kJ/L (150,700–152,000 Btu/U.S. gal) on combustion.

An occasional problem with large thermal combustion sources coupled with little air movement is that excessive accumulations of sulfur dioxide may develop in the ambient air while using high sulfur content residual fuels (Section 18.6.1). However, the sulfur content may be decreased by using hydrodesulfurized gas oils as cutting solvents, and by selecting the residual fractions from low sulfur crude oils for residual fuel formulation. The residues from high sulfur content crudes with appropriate properties could then go into the production of pitches and asphalts, where the sulfur content is not subjected to combustion and discharge. A less desirable option for the residues from high sulfur crude oils is to formulate these for marketing outside regulated areas where thermal combustion sources may be smaller and less numerous. It is possible to hydrodesulfurize the formulated residual fuel itself, although the processing costs to do this are much higher than for the desulfurization of distillate fractions [51].

18.5.3. Asphalts and Pitches

The residue from crude distillation in North America is usually processed to an asphalt product as much as possible, since as a component of residual fuel oils it generally fetches a lower price than as asphalt itself. The residual fuel oil market is quite competitive since ships and tankers can choose to load at centers, which offer the lowest cost fuels. Also the price of the heavier grades of bunker fuel used to supply power stations has to be comparable to coal to be competitive.

Asphalts made from different crudes are not always mutually soluble and may require instrumental work before mixing to determine compatibility [52]. Asphalts are marketed in liquid and low-, medium-, and high-melting point solid grades, depending on the intended end use. The melting point and degree of hardness of an asphalt is affected by the completeness of removal of volatile fractions [53]. If the melting point or hardness of a crude separated asphalt product is too low, it is oxidized by blowing air through the heated asphalt until the melting point is raised to the extent desired [54].

Lower melting point asphalts are used for the waterproofing of flat, built-up roofs and the like, where the self-sealing qualities are an attraction. For sloping built-up roofs, higher melting asphalts are required. For asphalt shingles, roll roofing, and similar sheet roofing products, the melting point has to be high enough to give an essentially nonsticky surface after fabrication, under ordinary ambient conditions. Road asphalts normally have to pass a melting point specification, a penetration test ("pen test"), and other requirements. The pen test is the extent to which an asphalt sample at 25°C is penetrated by a steel ball or pin of standard dimensions, while a force of 100 g is applied for 5 seconds. It is an important measure of the suitability of the asphalt for various paving applications. For application as road surfacing small crushed gravel aggregate is blended into hot asphalt, which serves as the binder.

By the variety of processing procedures outlined very little of the crude oil processed is wasted. From the high volatility dissolved gases present, all the way to the nearly involatile asphaltenes, and suspended carbon, etc. present in the original oil are converted to a host of useful products. Any combustible components that are not directly used toward the formulation of a product are used to generate energy for the refinery itself.

18.6. REFINERY EMISSION CONTROL

The incentive to produce useful products from the feedstocks entering a refinery, and recognition of the potential hazards to operating personnel and the local environment in the event of any process losses, have meant that at least good control of process streams has always been an objective of operations. Some aspects of refinery operations require closer attention than others to ensure safe and environmentally acceptable production. For the convenience of discussion the areas of concern have been grouped as they relate to atmospheric emission control, aqueous emission control, and waste disposal practices.

18.6.1. Atmospheric Emission Control

Hydrocarbon vapor losses from refineries are about 5–10% of the total hydrocarbons discharged to the atmosphere, from pollutant inventories [55]. However, refinery losses are highly localized so that control is important to minimize local ambient air concentrations of hydrocarbons. Any volatile sulfides require more stringent monitoring and control measures because of their toxicity and very low odor thresholds.

Losses from pump seals, flanges connecting lengths of pipe, and operating vessels are best controlled by properly scheduled maintenance procedures [56]. Losses of methane or ethane fuels from crude or process stream preheat furnaces using these hydrocarbons as fuels used to occur occasionally from inadequate combustion air. A slight excess of combustion air is now ensured by continuous monitoring of the flue gases with an oxygen analyzer. This measure also contributes more energy recovery from a given quantity of fuel.

Potential hydrocarbon losses from the overpressuring of operating vessels are controlled first by staged computer alerts and/or manual alarms to provide for correction of the condition. If the overpressure exceeds a second set point, pressure relief valves vent the vessel contents to a flare release system. The flare system provides a means of controlled burning of hydrocarbon vapors at a nonhazardous point to avoid fire or explosion risks. Smoke problems from flares are avoided by more efficient designs that use multiple nozzles and low pressure operation to promote clean combustion [57]. Greenhouse gas concerns should more frequently stimulate an interest in energy recovery options from flared hydrocarbons.

Storage tanks for crude oil and refined products are always vented. Those used for more volatile liquids can lose a significant amount of product as vapor, particularly with short cycle times. Apart from venting needed for filling and emptying of tanks, it is also required to prevent overpressuring from vapor inside on a hot day, which could cause tank rupture, or collapse of the tank from vapor condensation and atmospheric pressure acting on the outside of the tank during a cool night. A single storage tank can lose as much as 190,000 L of gasoline per year from venting losses alone [58]. These may be controlled cost effectively by fitting floating roofs to tanks used to store products of high vapor pressures, which eliminates the vapor space responsible for most of the losses from a fixed roof tank [59]. Or low cost hollow polyethylene spheres, 2–3 cm in diameter and two to three layers deep, may be floated on the surface of the liquid inside the tank to decrease losses from the vapor space by 90+%. A slight positive pressure can also help to reduce vapor loss.

Crude oils all contain some sulfur, mostly as thiols, sulfides, and hydrogen sulfide. The sulfur content ranges from about 6% for light Arabian crudes, to 1–2% for California crudes, down to as little as 0.4% for some Canadian Peace River crudes [60]. Loss of volatile sulfides of stored crude oil is also controlled by using floating roof storage tanks.

The hydrogen sulfide content of gaseous hydrocarbon streams is removed during refining by either scrubbing with organic bases (for details see Chap. 7) or by physical absorption (e.g., by the sulfinol process). Liquid process streams in the middle boiling point range require catalytic hydrogenation first (hydrofining or hydrotreating) to convert less volatile sulfides to hydrogen sulfide and paraffinic hydrocarbons (Eq. 18.42).

$$R_2S + 2 H_2 \rightarrow 2 RH + H_2S \qquad\qquad 18.42$$

The more volatile hydrogen sulfide is then separated from the hydrocarbon stream by stripping (sparging) with air or steam. Separation of the sulfur-containing compounds present in higher boiling point fractions and involatile residues also requires preliminary catalytic hydrogenation to convert high boiling sulfides to the more volatile hydrogen sulfide and thiols. Following this they are separated by fractionation or stripping. Sulfur removal from the residue stream is the most difficult and costly since the content of metallic and other impurities slows down or rapidly deactivates hydrogenation catalysts. This is why desulfurization of crude residues is more expensive. The sulfur content of a residual fuel oil may be minimized without having to hydrodesulfurize the distillation residues content by thorough desulfurization of the catalytically cracked gas oil diluent before it is blended with the residue.

Hydrogen sulfide accumulated from desulfurization processes and any on-site sulfur dioxide are converted to elemental sulfur using two or three stages of Claus reactors [61] (Chap. 9). However, Claus plant operation is seldom a profitable operation for a refinery since the recovered sulfur amounts to only a few tonnes per day. However, the improved ambient air quality achieved in the vicinity of the refinery makes this measure worthwhile.

Carbon monoxide is a potential emission problem. Fortunately, it is a valuable industrial fuel and may be burned in a "CO boiler" to recover energy as steam and discharge carbon dioxide. However, it can also arise from the catalyst regenerators of some recent models of catalytic cracking, units which ran at temperatures too low to obtain complete oxidation of carbon to carbon dioxide. The most recent designs of regenerators operate at higher temperatures to achieve complete conversion of carbon to carbon dioxide.

Control of catalyst particle losses from both the cracker and regenerator of fluid catalytic cracking units is achieved by two cyclones operating in series right inside each unit. This is usually followed by an electrostatic precipitator for fine particle control, working on the exhaust side of the catalyst regenerator [62]. The metal content of spent catalysts may be recovered for reuse [63].

Nitric oxide can form to the extent of 270 ppm or more at the hot metal surfaces of the refinery combustion units [64] (Eq. 18.43).

$$N_2 + O_2 \xrightarrow[\text{metal}]{\text{hot}} 2 NO \qquad\qquad 18.43$$

Its formation can be kept to a minimum by keeping the excess air supplied to combustion units to a minimum value for safe complete combustion [56]. Burner designs that produce a more diffuse flame front (large flame volume) achieve lower peak combustion temperatures, which helps to decrease the formation of nitric oxide. Injection of ammonia into the flue gas while it is still hot can decrease NO_x concentrations down to 80–120 ppm, one-third to one-half that of uncontrolled discharges [64]. Measures for NO reduction during operation of fluid catalytic crackers have been evaluated in pilot scale reactors [65].

18.6.2. Aqueous Emission Control

Standards for aqueous effluent from refineries have been set, both in Canada and the United States (Table 18.9). For parameters, such as total suspended solids, oil, grease, and ammonia nitrogen, American standards appear to be more strict than Canadian federal standards. Also provincial standards in Canada are generally more restrictive than the federal standards, perhaps because of the more numerous very large refineries operating in the United States.

Aqueous desalter effluent contains sediments, oil, dissolved salts, and sulfides. It is treated initially by using an American Petroleum Institute (API) separator for residual oil removal and recovery (Fig. 18.9). Any oil droplets larger than about 150 μm (0.15 mm) in diameter are collected as an oil phase, which is routed to the refinery "slop oil" stream to join any other off-specification liquid oil streams for reprocessing. Inclined plastic plates inserted into an oil–water separator of this type substantially increase the speed and

TABLE 18.9   Aqueous Effluent Standards for Petroleum Refining[a]

Effluent component or property	Average of daily values for one month, max. permitted (g/m³ crude)			Maximum values, any one day (g/m³ crude)	
	U.S.A.	Canada[b]	B.C.[b,c]	U.S.A.	Canada[d]
Total suspended solids	2.0	20.6	20[d]	2.4	42.8
BOD	2.0	–	6.6	2.5	–
COD	8.0	–	–	10.0	–
Oil and grease	0.4	8.6	1.7	0.5	21.4
Phenols	0.0060	0.86	0.066	0.012	2.1
Sulfide	0.035	0.29	0.031	0.05	1.4
Ammonia nitrogen	0.51	10.3	1.65	0.68	20.6
Total chromium	0.105	–	0.57	0.124	–
pH	–	6–9.5	6.5–8.5	–	6–9.5

[a]Derived from Water Resources Service [66], Greenwood et al. [67], and Water Pollution Control Directorate [68].

[b]Recalculated from units of pounds per thousand barrels to grams per cubic meter of crude by multiplying by 2.855.

[c]British Columbia Level A standards, test intervals vary from daily to monthly.

[d]Units of mg/liter, not g/m³.

FIGURE 18.9 Operating details of an American Petroleum Institute separator, followed by a dissolved air flotation unit for oil–water separation.

completeness of creaming (rising to the top), and reduces the average droplet size collected to about 60 μm (0.06 mm) [69]. After oil separation the water phase of the desalter is passed through a sour water stripper (Fig. 18.10), where hydrogen sulfide is removed by sparging with low-pressure steam, 140 kPa (ca. 130°C, 20 lb/in.^{2}) [70]. The separated hydrogen sulfide is processed to elemental sulfur in a Claus unit. The sweetened water obtained as the bottom stream from the stripper contains <10 ppm hydrogen sulfide and <50 ppm ammonia [71]. It may be recycled to add to the water for desalting, or to help control losses of volatile compounds [72]. In case of excess water buildup in the desalter circuit, it is bled from this point in the recycle to aerated bio-ponds, etc., for further treatment before discharge.

Sour water (hydrogen sulfide in water) is also obtained as aqueous condensate from the fractionation of the volatile products of refinery operations, such as the fluid catalytic cracker (FCC), hydrotreater, hydrocracker, or coker. In fact, in a refinery, which is endeavoring to minimize water use, sour waters may make up as much as 25% of the total aqueous effluent discharged.

FIGURE 18.10 Details of hydrogen sulfide removal from water in a sour water stripper.

Procedures for hydrogen sulfide (and ammonia) removal can be similar to those used for desalter effluent [69], and have been described in detail [73].

Site drainage waters from surface runoff invariably include some oil picked up from small spills and process leaks. These will also be routed through an API separator first, sometimes followed by a dissolved air separator (Fig. 18.8). Oil separated from this drainage water stream will join the slop oil circuit for recycling. During times of desalter operation when water is short, treated site drainage waters may be used for this purpose. Any water in excess of desalter requirements will proceed to an artificially aerated bio-pond for further BOD reduction before discharge [74].

Phenols are removed from refinery hydrocarbon streams by treating with a solution of sodium hydroxide in water. This produces a waste stream, which is difficult to treat in a bio-pond [75]. If feasible, this stream will be sold (or given) to a neighboring petrochemical plant if it can make use of the phenol content. Phenol is economically recovered from this stream by acidification using flue gases, followed by solvent extraction (Eq. 18.44).

$$\text{NaOPh} + \text{H}_2\text{CO}_3 \rightarrow \text{NaHCO}_3 + \text{PhOH} \qquad 18.44$$

Caustic in water is also frequently used for removal of sulfides from certain liquid hydrocarbon refinery streams (Eq. 18.45 and 18.46).

$$2\,\text{NaOH} + \text{H}_2\text{S} \rightarrow \text{Na}_2\text{S} + 2\,\text{H}_2\text{O} \qquad 18.45$$

$$\text{NaOH} + \text{CH}_3\text{SH} \rightarrow \text{NaSCH}_3 + \text{H}_2\text{O} \qquad 18.46$$

Sodium sulfide, at least, is of value to pulp mills, which use the kraft process for pulping and, if near by, can provide a useful outlet for this waste product. However, if this recycle method is not feasible a "caustic oxidizer" may be used to convert the sodium sulfide to the more environmentally acceptable sodium sulfate, for ultimate disposal by landfill (Eq. 18.47).

$$\text{Na}_2\text{S (aqueous)} + 2\,\text{O}_2 \xrightarrow[\text{heat}]{} \text{Na}_2\text{SO}_4 \qquad 18.47$$

Or, alternatively, aqueous sodium sulfide may be acidified to produce a "sour" water (containing hydrogen sulfide) stream (Eq. 18.48).

$$\text{Na}_2\text{S} + \text{H}_2\text{SO}_4 \rightarrow \text{Na}_2\text{SO}_4 + \text{H}_2\text{S} \qquad 18.48$$

This can then be routed to the sour water stripper for hydrogen sulfide recovery and conversion to sulfur. In this case flue gas contacting, which is the more economical method of acidification, cannot be used since this would liberate highly volatile hydrogen sulfide into the flue gases contacted.

When operating in areas with a restricted water supply, or on more costly municipal water services, significant economies in water usage may be achieved by recycling once-used and warmed cooling water through direct or indirect cooling towers for reuse. Process cooling directly with air, by fan forcing against finned tubes, is also a measure, which can reduce overall refinery water requirements, particularly when operating in cold climates.

The potential impact of inadvertent losses of some constituents of refinery liquid wastes has been reviewed by Cote [76], and a summary of aqueous treatment processes has been presented by Baker [77].

18.6.3. Refinery Waste Disposal Practices

The components of most refinery liquid waste streams are recovered and reused, whenever feasible. Unfortunately, some of these, such as aqueous caustic phenolic or caustic sulfidic wastes, do not lend themselves readily to reuse. Deep well disposal, incineration, or precipitation in some manner and landfilling of the separated solids are the measures used in these instances. Raising the concentration of brine streams by reverse osmosis before discharge can help decrease final disposal costs by decreasing the waste volume [78].

Sludges, such as those accumulated in the monoethanolamine or diethanolamine streams used for hydrogen sulfide removal by scrubbing, comprise another intractable waste material. Sludges may be reprocessed to recover valuable materials [79], or failing this they may be safely discarded by proper incineration. Others use landfilling (dry) or landfarming as disposal methods. Biotreatment or deep well disposal are also used, but to a more limited extent [80].

Spent acids are used where feasible to neutralize alkaline waste streams, or are neutralized with purchased lime, or caustic, and are then routed to the normal effluent treatment system for further cleanup before discharge [61]. Waste solids from refinery operations include such materials as spent clay from decolorization of lubricating oils or waxes, sand used as a catalyst support, filter aid, or filter base, and exhausted Claus catalysts (primarily ferric oxide on alumina). These are disposed of by landfilling by the majority of Canadian refineries. However, some use landfarming for disposal of these materials. One refinery recovers spent Claus catalysts for regeneration into new catalyst.

Landfarming of oily sludges for degradation by soil micro-organisms has been found to be an efficient method of disposal, providing proper conditions are maintained to achieve optimum oil decomposition rates within the soil. Oil degradation rates in outside experimental plots of soil are high (Table 18.10) [81]. Decomposition rates range from about 150 kg oil carbon/hectare/day in the temperate climate of the area of Edmonton, Alberta [79], up to as high as 569 kg oil carbon/hectare/day for landspreading operations in Texas [82]. Experiments that examined the rates of degradation of crude oil spills on land showed that the rate was significantly enhanced by the application of

TABLE 18.10 Decomposition Rates of Crude Oil in the Field, Edmonton, Alberta[a]

Elapsed time (days)	Application rate/oil content (kg/m^2)[b]		
	Light	Medium	Heavy
0	5.5	13.7	24.9
43	3.7	10.7	17.2
340	2.1	8.5	11.5
460	2.0	7.2	11.0
735	1.5	4.2	7.8

[a]Calculated from the data of Toogood [81].
[b]Cultivated to a 30-cm depth.

TABLE 18.11   Percent Composition of Crude Oil after 308 Days in Contact with Soil Plus Various Additives, Swan Hills, Alberta[a]

Crude oil fraction	Percent composition of oil[b]			
	In barrel	Soil alone	Soil plus fertilizer[c]	Soil + bacteria[d] + fertilizer
Asphaltenes, soluble	2.77	1.35	2.01	2.58
Asphaltenes, insoluble	4.19	12.7	20.7	21.3
Saturates	59.7	54.7	39.3	37.9
Aromatics	23.6	17.9	20.9	21.4
NSO's[e], soluble	5.64	6.85	9.06	9.98
NSO's[e], insoluble	5.03	6.42	8.04	7.27

[a]Data selected from that of Cook and Westlake [83, 84]. Little change in composition was noted in these experiments after 12 days, but significant degradation was already evident after 66 days and a further progression of degradation was observed after 433 days, the length of the experiment.

[b]Local crude oil was distributed at the rate of 60 liters per 9 m² plot; results averaged over 4 replicate plots.

[c]Fertilizer (27-27-0) was applied at the rate of 60 g of nitrogen per square meter of plot.

[d]A mixed culture of suitable bacteria was applied at the rate of 10⁶ per square centimeter of plot.

[e]NSO's short for nitrogen-, sulfur-, and oxygen-containing compounds.

fertilizer, and helped somewhat by the addition of a suitable mixed culture of bacteria (Table 18.11).

The methods outlined above, or Biopile methods [85] may also be used for other biodegradable waste products of refinery operation, such as amine sludges and spent oily clay and sand. With only light sludge loadings the land can be returned to agricultural use after one season. However, some areas have regulated against landfarming for sludge disposal (e.g., the State of Minnesota). The Water Quality Programs Committee of the International Joint Commission (U.S.A. and Canada) recommends that groundwater pressure and quality be monitored at different depths, and that routine soil samples be taken for oil content and trace metal analysis from any operating landfarm disposal system. It is also recommended that, when retired from this use, these soils not be used for the growing of crops or grazing because of the risk of contamination from metals or other materials that could have accumulated [86]. Concern with land disposal of petroleum processing wastes has led to the use of these materials as fuel for cement kiln operation [87].

REVIEW QUESTIONS

1. Why is 520°C approximately the upper limit for atmospheric pressure distillation of crude oil? How does this factor relate to the need to use reduced pressure for the refinement of lubricating oils by distillation?
2. (a) Describe the similarities and the differences between straight thermal cracking and catalytic cracking as oil modification procedures.

(b) How do the details of the process parameters affect the types and proportions of the observed products from the two processes?

3. (a) Explain the chemical basis for the inevitable formation of olefins and/or coke (as well as saturated paraffins) from cracking reactions.

(b) Write two plausible equations for cracking reactions to useful products from n-dodecane ($C_{12}H_{26}$), and from n-hexadecane ($C_{16}H_{34}$).

(c) What mass yield of each of 1-heptene and the other expected product would be obtained from one pass of 1,000 kg of n-dodecane through a catalytic cracker for a conversion of 52% and a selectivity of 86%?

(d) Name and explain at least two other products that could have resulted from the 14% of dodecane that does not form the products given in part (c).

4. (a) Explain and give equations to illustrate your explanation of the mode of formation of a high proportion of branched, rather than straight chain products formed during the formation of polymer alkylate for use in gasoline.

(b) What property is contributed by this component of gasoline that is valued by refiners?

(c) Explain the relationship between the typical structures of polymer alkylate and the valuable property that this fraction contributes to gasoline. Equations may assist in interpretation and presentation.

5. (a) Assuming that a statistical distribution of products is obtained from a polymerization process to a gasoline fraction, what proportion of C_6, C_7, and C_8 products would result from the polymerization of 1,000 kg of isobutylene with 1,000 kg of propene?

(b) For a 60% conversion, 80% yield on polymerization of 1,000 kg of isobutylene, what total mass of octanes and octenes would be expected, theoretically?

6. (a) Compare and contrast the conditions required and the type of hydrocarbon produced by polymerization versus alkylation to a gasoline component.

(b) How do the stability, the octane ratings, and the relative volumes of the two product types compare?

7. What is the major acid gas component, and what are the more minor acid gases present in "flue gas" that provide a low cost way to acidify aqueous alkaline waste water streams to assist recovery of pollutants, such as phenols and hydrogen sulfide?

FURTHER READING

Anon. "Equipment Leaks of VOC in Natural Gas Production Industry: Background Information for Proposed Standards." U.S. Environmental Protection Agency, Office of Air, Noise and Radiation, Research Triangle Park, N.C. National Technical Information Service, 1983.

Anon. "Technical Note on the Best Available Technologies to Reduce Emissions of Pollutants into the Air from the Refining Industry: Final Report." Office for Official Publications of the European Communities; Publ. Luxembourg, & Lanham, Md., 1994. UNIPUB [distributor].

C.F. Conaway, "The Petroleum Industry : A Nontechnical Guide." PennWell Publ. Co., Tulsa, Oklahoma, 1999.

R.A. Meyers, "Handbook of Petroleum Refining Processes." McGraw-Hill, New York, 1986.

M.L. Occelli and R. Chianelli, "Hydrotreating Technology for Pollution Control: Catalysts, Catalysis, and Processes." Marcel Dekker, New York, 1996

J.G. Speight, "The Chemistry and Technology of Petroleum." M. Dekker, New York, 1999.

J.G. Speight, "The Desulfurization of Heavy Oils and Residua." M. Dekker, New York, 2000.

REFERENCES

1. E.C. Lane and E.L. Garton, Rep. Invest. U.S., *Bur. Mines*, **RI-3279** (1935), cited by J.A. Kent, ed., "Riegel's Handbook of Industrial Chemistry," p. 406. Van Nostrand-Reinhold, New York, 1974.

2. W.A. Bachman and D.H. Stormont, *Oil Gas J.* **65,** 69 (1967).

3. M.T. Atwood, *CHEMTECH* **3,** 617 (1973).

4. F.D. Rossini, J. *Chem. Educ.* **37**(11), 554 (1960).

5. L.F. Fieser and M. Fieser, "Advanced Organic Chemistry," Reinhold, New York, 1961.

6. G. Eglinton and M.T.J. Murphy, eds., "Organic Geochemistry," pp. 640, 676. Longman-Springer-Verlag, Berlin and London, 1969.

7. Crude Desalting, *Hydrocarbon Process.* **67**(9), 63, Sep. (1988).

8. Chemical Desalting, *Hydrocarbon Process.* **49**(9), 235, Sep. (1970).

9. Electrostatic Desalting, *Hydrocarbon Process.* **49**(9), 237, Sep. (1970).

10. X.D. Ye, Gas Oil Desalting Reduces Chlorides in Crude, *Oil Gas J.* **98**(42), 76–78, Oct. 16 (2000).

11. M. Al-Otaibi, A. Elkamel, and T. Al-Sahhaf *et al.*, Experimental Investigation of Crude Oil Desalting and Dehydration, *Chem. Eng. Commun.* **190**(1), 65–82, Jan. (2003).

12. C.E. Jarnig, Petroleum (Refinery Processes Survey), In: "Kirk-Othmer Encyclopedia of Chemical Technology," 3rd ed., Vol. 17, p. 183. Wiley Interscience, New York, 1982.

13. V.B. Guthrie, ed., "Petroleum Products Handbook," McGraw-Hill, New York, 1960.

14. E.M. van Z. Bakker and J.F. Jaworski, "Effects of Vanadium in the Canadian Environment," National Research Council of Canada, Ottawa, 1980.

15. W.L. Nelson, Petroleum (Refinery Processes), In: "Kirk-Othmer Encyclopedia of Chemical Technology," 2nd ed., Vol. 15, p. 1. Wiley-Interscience, New York, 1968.

16. P.J. Garner, Impact of Automobile on Chemical Industry, *Chem. Ind. (London)*. p. 131, Feb. 16 (1974).

17. D.M. Considine, ed., "Chemical and Process Technology Encyclopedia." McGraw-Hill, New York, 1974.

18. Organization for Economic Cooperation and Development, "IEA Statistics, Oil and Gas Information 1994," p. 81. OECD, Paris, 1995.

19. J.A. Kent, ed., "Riegel's Handbook of Industrial Chemistry," 7th ed. Van Nostrand-Reinhold, New York, 1974.

20. G.A. Purdy, "Petroleum Prehistoric to Petrochemicals," McGraw-Hill, Toronto, 1958.

21. J.A. Kent, ed., "Riegel's Industrial Chemistry," 6th ed. Reinhold, New York, 1962.

22. R.F. Wong, Increasing FCC Regenerator Catalyst Level, *Hydrocarbon Process.* **72**(11), pp. 59–61, Nov. (1993).

23. A. de Klerk, D.J. Engelbrecht, and H. Boikanyo, Oligomerization of...olefins: Effect on...motor-gasoline quality, *Indust. Eng. Chem. Res.* **43**(23), 7449–7455, Nov. 10 (2004).

24. L.F. Albright, *Chem. Eng. (N.Y.)* **73,** 143, Aug. 15 (1966), cited by Kirk-Othmer [12].

25. S.R. Tennison, Recent Advances in Catalytic Reforming, *Chem. Brit.* **17**(11), 536 (1981).

26. T. Hutson, Jr. and R.S. Logan, Estimate Alky Yield and Quality, *Hydrocarbon Process.* **54**(9), 107–110, Sept. (1975).

27. G.C. Ray, J.W. Myers, and D.L. Ripley, Cat Gasoline Improved by Isom, *Hydrocarbon Process.* **53**(1), 141–143, Jan. (1974).

28. P. Polss, What Additives do for Gasoline, *Hydrocarbon Process.* **52**(2), 61, Feb. 1973.

29. EPA Mulls Over Lead-in-Gasoline Rules, *Chem. Eng. News*, **60**(18), 28, May 3 (1982).

30. J. Zayed, Use of MMT in Canadian Gasoline: Health and Environment Issues, *Am. J. Ind. Med*. **39** (4), 426–433, Apr. (2001).

31. F.E. Ahmed, Toxicology and Human Health Effects Following Exposure to Oxygenated or Reformulated Gasoline, *Toxicol. Lett*. 123(2–3), 89–113, Sep. 15 (2001).

32. DOE Endorses Use of Alcohol Fuels, *Chem. Eng. News*, **57**(29), 5, Jul. 16 (1979).

33. B. Gillespie, L.W. Manley, and C.J. Di Perna, *CHEMTECH* **8**, 750 (1978).

34. L. Morriset-Blais, *Can. Chem. Process.* **57**(8), 48, Aug. (1973).

35. E.R. Booser, Lubrication and Lubricants, In: "Kirk-Othmer Encyclopedia of Chemical Technology," 3rd ed., Vol. 14, p. 477. Wiley-Interscience, New York, 1991.

36. G. Smith, Intimate Contact [in engines], *Chem. Brit*. **36**(4), 38–41, Apr. (2000).

37. R. Glyde, Smoothing the Ride, *Chem. Brit*. **33**(7), 39–41, Jul. (1997).

38. Z. Janovic, K. Saric, and K. Sertic-Bionda,...Properties of Some Alkylmethacrylates as Lubricating Oil Viscosity Modifiers, *Chem. Biochem. Eng*. Q **12**(1), 19–24, Mar. (1998).

39. I.C. Liou, R.C.C. Tsiang, J. Wu *et al*.,...a star-shaped poly(ethylene-co-propylene) copolymer as a viscosity index improver...*J. Appl. Polym. Sci*. **83**(9), 1911–1918, Feb. 28 (2002).

40. H.G. Müller, Mechanism of Action of Viscosity Index Improvers, *Tribol. Int*. **11**(3), 189, Jun. (1978).

41. Western Canada Recycling Programs, Why recycle? Available at: http://www.usedoilrecycling.com/html/why5.htm

42. E. Baumgardner, Crankcase Oil Cleaned for Fuel, *Hydrocarbon Process*. **53**(5), 129, May (1974).

43. D. Kress, Thermal Cracking Compared to Re-refining of Used Oil, *Can. Chem. News*, **52**(3), 20–23, Mar. (2000).

44. J. Sternberg, Used Lubricants to Gasoline, *Biocycle*, **33**, 80, May (1992).

45. D. Kress, Thermal Cracking of Used Oil to Produce Distillate Gasoil: A Process Overview, *Can. Chem. News*, **52**(2), 29–32, Feb. (2000).

46. E.E. Klaus and E.J. Tewkesbury, Product Specs for Liquid Lubes, *Hydrocarbon Process*. **53**(12), 67, Dec. (1974).

47. D.B. Barton, J.A. Murphy, and K.W. Gardner, Synthesized Lubricants Provide Exceptional...Passenger Car Performance. *SAE Tech. Pap. Ser*. **780951** (1978).

48. V.V. Grigo'ev, Influence of Molecular Structure on Physicochemical and Service Properties of Synthetic Oils, *Chem. Tech. Fuels Oil*, **35**(2), 94–97, Mar.–Apr. (1999).

49. Synthetic Lubricants Poised for Big Growth, *Chem. Eng. News*, **58**(13), 12, Mar. 31 (1980).

50. R.H. Perry, ed., "Chemical Engineers Handbook," 4th ed., p. 9–6. McGraw-Hill, New York, 1963.

51. F. Tamburrano, Disposal of Heavy oil Residues. *Hydrocarbon Process*. **73**, 79–84, Sep. (1994).

52. A.T. Pauli and J.F. Schabron, Stability and Compatibility Testing of Petroleum and Asphalt, *Amer. Lab*. **35**(19), 8–10, Sep (2003).

53. M. Freemantle, Asphalt, *Chem. Eng. News*, **77**(47), 81, Nov. 22 (1999).

54. Y.T. Qi and F.X. Wang, Study and Evaluation of Aging Performance of Petroleum Asphalts... During Oxygen Absorption...*Petrol. Sci. Technol*. 21(1–2), 283–299 (2003).

55. State of the Environment Reporting, "The State of Canada's Environment," Govt. of Canada, Ottawa, 1991.

56. H.F. Elkin and R.A. Constable, Source Control of Air Emissions, *Hydrocarbon Process*. **51**(10), 113, Oct. (1972).

57. Anon., Almost Unnoticeable Gas Flaring, *Proc. Indust. Canada*, **73**(2), 24, Mar. (1989).

58. P.L. Magill, F.R. Holden, and C. Ackley, eds., "Air Pollution Control Handbook," McGraw-Hill, New York, 1956.

59. Control of Atmospheric Emissions from Petroleum Storage Tanks, *J. Air Polln. Control Assoc.* **21**(5), 263 (1971).

60. C.M. McKinney, Sulfur in Products vs Crude Oil, *Hydrocarbon Process*. **51**(10), 117, Oct. (1972).

61. W.J. Racine, Plant Designed to Protect Environment, *Hydrocarbon Process*. **51**(3), 115, Mar. (1972).

62. C.S. Russell, "Residuals Management In Industry, A Case Study of Petroleum Refining." Johns Hopkins University Press, Baltimore, 1973.

63. G. Berrebi, P. Dufresne, and Y. Jacquier, Recycling of Spent Hydroprocessing Catalysts: EURECAT Technology, *Environ. Prog.* **12**(2), 97–100, May (1993).

64. Pollution Control Process Demonstrated, *Chem. Eng. News*, **56**(8), 23, Feb. 20 (1978).

65. E.A. Efthimiadis, E.F. Iliopoulou, A.A. Lappas *et al.*, NO Reduction Studies in the FCC Process . . . , *Ind. Eng. Chem. Res.* **41**(22), 5401–5409, Oct. 30. (2002).

66. "Pollution Control Objectives for the Chemical and Petroleum Industries of British Columbia." Water Resources Service, Queen's Printer, Victoria, 1977.

67. D.R. Greenwood, G.L. Kingsbury and J.C. Cleland, "A Handbook of Key Federal Regulations and Criteria for Multimedia Environmental Control," Rep. 600/7-79-175. U.S. EPA, Washington, DC, 1979.

68. "Petroleum Refinery Effluent Regulations and Guidelines," Rep. EPS 1-WP-74-1. Water Pollution Control Directorate, Ottawa, 1974.

69. J. Wardley-Smith, "Prevention of Oil Pollution." Graham and Trotman Ltd., London, 1979.

70. R.J. Klett, Treat Sour Water for Profit, *Hydrocarbon Process.* **51**(10), 97, Oct. (1972).

71. Fina Recovers H2S from Wastewaters, *Can. Chem. Process.* **60**(10), 28, Oct. (1976).

72. A. Hasbach, Closed Loop System Recycles VOC's from Refinery Waste Water, *Pollut. Eng.* **24**, 69 (1992).

73. C. Stein, J. Elster, I. Heine *et al.*, Removal of Ammonium Nitrogen and Hydrogen Sulfide from Oil Refinery Waste Water, *Oil Gas-Eur. Mag.* **24**(4), 36–39, Dec. (1998).

74. J.F. Ferrel and D.L. Ford, Select Aerators Carefully, *Hydrocarbon Process.* **51**(10), 101 Oct. (1972).

75. J.C. Hovious, G.T. Waggy, and R.A. Conway, "Identification and Control of Petrochemical Pollutants Inhibitory to Anaerobic Processes," EPA-R2-73-194. Environmental Protection Agency, Washington, DC, 1973.

76. R.P. Cote, "The Effects of Petroleum Refinery Liquid Wastes on Aquatic Life, with Special Emphasis on the Canadian Environment." National Research Council, Ottawa, 1976.

77. D.A. Baker, J. Petroleum Processing Waste, *Water Pollut. Control Fed.* **46**(6), 1298–1301 (1974).

78. T.A. Krug and K.R. Attard, Treating oily waste water with reverse osmosis, *Water Pollut. Control*, **128**(5), 16–18, Oct. (1990).

79. W.J. Hahn, High Temperature Reprocessing of Petroleum Oily Sludges. *SPE Production Facilities*, **9**, 179–182, Aug. (1994).

80. W. Wimberly, To Dispose of Waste Wisely, *Hydrocarbon Process.* **68**(8), 45–49, Aug. (1989).

81. J.A. Toogood, ed., "The Reclamation of Agricultural Soils After Oil Spills," AIP Publ. No. M-77-11. Dept. of Soil Science, Univ. of Alberta, Edmonton, 1977; cited by Beak Consultants Ltd., [86].

82. C.B. Kincannon, "Oily Waste Disposal by Soil Cultivation Process," EPA-R2-72-110. U.S. EPA, Washington, DC, 1972; cited by Beak Consultants Ltd., [86].

83. F.D. Cook and D.W.S. Westlake, "Biodegradability of Northern Crude Oils," Task Force on Northern Oil Development report No. 73-20, Information Canada Cat. No. R72-8373, Ottawa, 1973.

84. F.D. Cook and D.W.S. Westlake, "Microbiological Degradation of Northern Crude Oil," Task Force on Northern Oil Development Report No. 74-1, Information Canada Cat. No. R72-12774, Ottawa, 1974.

85. R. Corbin, . . . Biological Treatment for Hazardous Sludge, *HazMat Management*, **16**(5), 14–15, Oct/Nov. (2004).

86. "Manual for Landspreading of Petroleum Industry Sludges." Rep. No. 81-5A, Beak Consultants Ltd., Petroleum Association for Conservation of the Canadian Environment, Ottawa, 1981.

87. D. Gossman, The Reuse of Petroleum and Petrochemical Waste in Cement Kilns, *Envir. Prog.* **11**(1), 1–6, Feb. (1992).

19

■ PETROCHEMICALS

Take Carbon for example then
What shapely towers it constructs
To house the hopes of men!
                    —*A.M. Sullivan (1896–1980)*

19.1. BACKGROUND

Organic chemicals, such as dyestuffs and derivatives of benzene and hetero-cyclic compounds, were first produced from coal-based raw materials. Since the mid-nineteenth century these products were obtained by the further processing of discrete volatile fractions recovered during the coking of coal. A large fraction of the coke product was destined for the reduction of iron ore to produce pig iron (Chap. 14). Although volatile fractions from the coking of coal, such as benzene, toluene, and the xylenes are still items of commerce, petroleum-based sources for these materials have now become much more important [1]. The attractiveness of petroleum-based feedstocks grew as both the reliability and scale of petroleum production increased. These aspects, plus the gradually increased sophistication of refining methods of the oil industry of the early twentieth century contributed to the increased appeal of petroleum as a resource for chemicals production.

As a consequence of its recent development the petrochemical industry is relatively much younger than the major inorganic chemicals industry. However, one can easily be misled by the classification of products that are termed "petrochemical". Basically a petrochemical is derived directly or indirectly from a petroleum or natural gas fraction. It may be organic, such as ethylene, benzene, or formaldehyde, or it may be inorganic, such as ammonia, nitric acid, and ammonium nitrate (Chap. 11). So a "petrochemical" is not synonymous with an organic chemical, although most petrochemicals are also organic chemicals.

What is the incentive to produce petrochemicals? It is producer accessibility to profit from the additional value-added stages that are obtained from

637

conversion of a refined petroleum or gas fraction into a petrochemical, compared to the value of crude petroleum. Depending on the complexity and number of stages involved, a petrochemical product may command a price that is five or six times the price of the oil or gas used to produce it. If more value-added stages are used to convert a petrochemical commodity into a consumer product, this may fetch a price as high as 20 times that of the oil from which it is derived. In short, oil (or natural gas) is worth far more when converted to chemicals than as fuel.

In considering the scale of production of the top 30 industrial chemicals without regard to source, one may gain an impression of the relative importance of the petrochemical sector to the chemical industry as a whole. Fully 21 of the top 30 chemicals are classified as petrochemicals with a gross annual production in the United States in 1995 of 140 million metric tonnes (Table 19.1) [2]. Only nine are nonpetrochemical inorganic compounds. Nevertheless, the gross volume of production of these nine compounds of 166 million metric tonnes annually is still larger than the annual mass of petrochemicals. The product distribution pattern for Canada and other developed nations is similar. Sulfuric acid is the largest scale inorganic commodity chemical produced in developed nations. In Canada, where the sulfur for sulfuric acid production is derived largely from the sweetening of sour natural gas (Chap. 9), it could be argued that in this case even sulfuric acid should be classified as a petrochemical.

Not only are petrochemicals produced on a vast scale but the diversity of materials represented is substantial. More than 3,000 petrochemicals are in current commercial production. Despite this large scale, however, only 5–6% of current oil and gas production is consumed to produce organic petrochemicals and another 4–5% for inorganic petrochemicals. It seems to be an extravagant use of a key chemical raw material to apply only 10% of it for high value-added products, many of them in recoverable or durable applications, and simply burn the remaining 90% for energy production.

19.2. FIRST PETROCHEMICAL PRODUCTS

Carbon black was the first simple product to be commercially produced from hydrocarbons, and has continues without interruption since 1872. Isopropyl alcohol was the first commercial petrochemical requiring more than one step and continues to be manufactured by virtually the same process as its initiation by Standard Oil in 1920. Details of these early processes follow.

19.2.1. Carbon Black

Carbon black is the largely carbon-containing product of complex structure derived from the low temperature combustion of a hydrocarbon with a deficiency of air. This was initially produced from natural gas (mostly methane). Normal combustion of natural gas with adequate air produces carbon dioxide and water together with a flame temperature of 1,000–1,200°C. However, if the combustion is carried out in a deficiency of air with strong cooling of

TABLE 19.1 The 30 Largest Volume Chemicals and Their Scale of Production in the U.S.A. and Canada in 1995[a]

U.S.A. rank 1995	Petrochemical	Product	Production (million tonnes) U.S.A.	Canada[b]
1		Sulfuric acid	43.25	4.22
2		Nitrogen	30.86	–
3		Oxygen	24.26	–
4	√	Ethylene	21.30	3.13
5		Lime	18.70	–
6	√	Ammonia	16.15	4.66
7		Phosphoric acid	11.88	–
8		Sodium hydroxide	11.88	1.18
9	√	Propylene	11.65	0.72
10		Chlorine	11.38	1.13
11		Sodium carbonate	10.11	–
12	√	Methyl-t-butyl ether	7.99	–
13	√	Ethylene dichloride	7.83	–
14	√	Nitric acid	7.82	0.99
15	√	Ammonium nitrate	7.25	1.03
16	√	Benzene	7.24	0.78
17	√	Urea	7.07	3.33
18	√	Vinyl chloride	6.79	–
19	√	Ethylbenzene	6.20	–
20	√	Styrene	5.17	–
21	√	Methanol	5.12	–
22	√	Carbon dioxide	4.94	–
23	√	Xylene	4.25	–
24	√	Formaldehyde	3.68	–
25	√	Terephthalic acid	3.61	–
26	√	Ethylene oxide	3.46	–
27		Hydrochloric acid	3.32	0.14
28	√	Toluene	3.05	0.28
29	√	p-Xylene	2.88	0.43
30	√	Cumene	2.55	–
Total inorganic chemicals (9)			165.64	
Total petrochemicals (21)			140.00	

[a]Calculated from data of Chem. Eng. News [2].
[b]Some data unavailable because of *Statistics Canada* confidentiality regulations.

the flame, a large proportion of the "combustion" proceeds to yield finely divided carbon (carbon black) and water vapor (Eq. 19.1).

$$CH_4 + O_2 \xrightarrow{\text{ca. } 500°C} C + 2\,H_2O \qquad 19.1$$

The flame was cooled by cold iron bars (channels) placed into the flame and onto which the carbon deposited. The product, called channel black, was periodically scraped from the iron channels for further processing and sale. Yield of carbon from this process was about 1 tonne per 50,000 m^3 (at 1 atm,

15.6°C) of natural gas, or about 5% [3]. The low yields from this process and the increasing concerns about the traces of particulate lost in combustion gases from channel black plants have contributed to their obsolescence.

Better yields of carbon black are obtained by improved separation of the combustion function from the carbon-forming function, as is accomplished in the newer furnace black processes. This approach enables "gas oil" (high boiling point liquid petroleum fractions) or natural gas to be used to produce carbon black. The cooling function from 1,400°C to about 200°C is accomplished by direct water sprays. The product is removed from the gas stream via a combination of cyclone collectors and glass or Teflon fiber bag filters. One tonne of furnace black is obtained from 5,300 to 7,000 m^3 (1 atm; 15.6°C) of natural gas, or 1,400–2,800 L of gas oil corresponding to 50–70% yields. Oil-based furnace black now supplies about 90% of the current carbon black market, although the special features of the product from small-scale processors still contribute some product [4].

"Thermal black" is produced by another variant in which natural gas is cracked to carbon and hydrogen by the direct application of heat via firebrick checker work in a cyclical process (Eq. 19.2) [5].

$$CH_4 \xrightarrow{>870°C} C + 2\,H_2 \qquad\qquad 19.2$$

Elemental analysis of carbon blacks gives a carbon content of 90+%, and in some grades as high as 97%. Although the chemical composition is simple, the structures of carbon black products are complex and vary widely with the feedstock used [6], the process and conditions employed, and minor amounts of feedstock additives that may be used [7]. Since the discovery of the rubber reinforcement properties of carbon blacks in about 1912, this industry has grown to about 400,000 and 1.5 million tonnes/year in Canada and the United States, respectively. It ranks about 35th in volume among large-scale commodity chemicals. About 90% of the oil-based furnace blacks are used in rubber reinforcement, mostly for automobile tires. Printing inks and plastics are much smaller but important pigment uses for carbon blacks.

Other forms of carbon are also of significant industrial interest, among them the various types of activated carbon that are used in many air and water emission control applications. These porous forms of carbon are made from raw materials, such as lignite, coal, wood if it is to be used for liquid phase adsorption, and petroleum residues, coal, or coconut shells if for gas adsorption. Carbon fiber used for reinforcement of composite materials in aerospace and sporting goods applications can sell from $22 to more than $220 per kilogram, a very attractive value added product [8, 9]. Industrial diamonds are used for the cutting and grinding of very hard materials, for optical devices, such as lasers, for heat dissipation, and for diamond coatings to produce enhanced wear resistance. This also represents a small-scale, high-technology carbon product [10].

19.2.2. Isopropanol (Isopropyl Alcohol)

This petrochemical was first produced commercially via basically a scaled-up laboratory procedure. By 1925 its scale of production reached about 68 tonnes/year. By 1995 volumes of about a million tonnes per year were produced in the United States.

The initial step is the absorption of propylene-rich gases into 85% sulfuric acid at temperatures of 20–25°C to form isopropyl hydrogen sulfate via addition of the acid across the double bond (Eq. 19.3). Control of acid temperature and concentration is important to provide selectivity and avoid

$$CH_3-CH=CH_2 + H_2SO_4 \cdot H_2O \rightarrow CH_3C(H)(SO_3H)CH_3 \qquad 19.3$$

absorption of olefins, which require higher concentrations or temperatures to react. Hydrolysis of the intermediate monoester with water in a lead-lined vessel then gives isopropanol and diluted sulfuric acid (Eq. 19.4).

$$CH_3C(H)O(SO_3H)CH_3 + H_2O \text{ (excess)} \rightarrow CH_3CHOHCH_3 + \qquad 19.4$$
$$H_2SO_4 \text{ (dil.)}$$

The heat of reaction, plus additional heat from added steam, drives off the crude isopropanol from the spent sulfuric acid, now about 20% acid. The acid is reconcentrated to 85% by boiling off the excess water, after which it is ready for reuse. Since it plays an active part in the process, the sulfuric acid is a recycled reactant rather than a true catalyst in this instance.

The condensed, crude isopropanol-water solution is concentrated and purified by distillation to the 91% isopropanol–water azeotrope (boiling point 80.4°C, Fig. 19.1).

Any higher alcohols that may have formed in the process from traces of higher olefins present in the propylene feed are absorbed from the azeotrope into mineral oil, in which isopropanol is insoluble. Pure isopropanol is obtained by ternary distillation of the cleaned water azeotrope with the appropriate proportion of added di-isopropyl ether. The ternary azeotrope (Table 19.2) is the top product from the column, and pure isopropanol is removed from the bottom (see Section 16.4 for related information).

About two-thirds of the isopropanol produced is used as an intermediate to form other chemicals, mostly for acetone production. The remainder is used as a solvent in paints and coatings and other chemical processes.

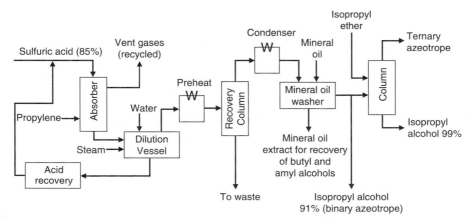

FIGURE 19.1 Simplified flowsheet for isopropanol production from propylene. (Adapted from Lowenheim and Moran [5].)

TABLE 19.2   Parameters for the Drying of 91% Isopropanol via the Ternary Azeotrope

| | Percent composition | | | | |
| | Boiling point (°C) | | | Condensate layer | |
Component	Pure	Azeotrope	Azeotrope	Upper	Lower
Isopropyl ether	67.5		91.0	94.7	1.0
Isopropanol	82.3	61.8	4.0	4.0	5.0
Water	100.0		5.0	1.3	94.0

From the chemically simple process of methane or gas oil to carbon black to the first more complex petrochemical process to isopropanol the scope of this industry has expanded to the production of a truly vast range of large-scale chemicals. Here we will cover outlines of representative samples of the more important of these chemicals. For more detailed information consult the items listed in the Further Reading section at the end of this chapter.

19.3. ALKENE AND AROMATIC PRODUCTS

The majority of the chemical processes of the modern petrochemical industry rely on just three refinery streams, methane (or natural gas), olefins, and aromatics comprising just seven hydrocarbons—methane, ethylene, propylene, the butylenes, benzene, toluene, and the xylenes (dimethyl benzenes). Details of the processes used to isolate naturally occurring methane from raw oil or gas streams have been discussed (Chap. 18). A few of the other hydrocarbons listed also occur to a varying extent in crude oil streams but in concentrations that are inadequate to meet the demand. Thus molecular changes are necessary to convert these hydrocarbon streams. Some of the procedures used to do this, such as polymerization, cyclo-alkylation, and aromatization, produce components of gasoline as has already been described (Section 18.3). Others differ in important respects but still employ similar principles to obtain the desired product, namely sufficiently high temperatures to cause carbon–carbon and/or carbon–hydrogen bond homolytic cleavage, plus suitable conditions (contact time, catalyst, concentration, etc.) to achieve the optimum product yield. Since many refinery operations lead to several of these products simultaneously, computer programs are used to aid in process optimization to produce the currently required ratio of products. The tendency to get several simultaneous products from a single process will be evident from the following discussion of related product groups.

19.3.1. Ethylene (Ethene)

Ethylene, the largest scale petrochemical, is probably also the oldest to have had confirmed preparation in the laboratory. The initial method used, catalytic dehydration of ethanol, was reported as early as 1797 (Eq. 19.5).

$$CH_3CH_2OH \xrightarrow[\text{catalyst}]{300-400°C} CH_2=CH_2 + H_2O \qquad 19.5$$

Although this method is capable of well over 90% yields of ethylene, and may be used when the local availability and price of ethanol (e.g., occasionally fermentation sources, Brazil) are attractive, other methods are usually more important.

Refinery ethylene is usually made by the catalytic cracking of ethane, propane, or a mixed hydrocarbon stream, such as recovered natural gas liquids, naphthas, or gas oil [11]. Cracking conditions are quite severe: 750–900°C and 0.1–0.6 second residence time for a low partial pressure hydrocarbon stream. A number of metal oxide catalysts have recently been evaluated for this purpose [12]. The usual diluent is steam, used at a weight ratio of steam to hydrocarbon of 0.2:1 for ethane feed, to progressively higher ratios with the higher molecular weight hydrocarbons of up to 2.0:1 for gas oil.

With an ethane feed the dominant cracking reaction is to ethylene and hydrogen, together with traces of methane and higher hydrocarbons. The initiation step involves the homolysis of the carbon–carbon bond (Eq. 19.6).

$$CH_3CH_3 \rightarrow 2\ H_3C^\cdot \qquad 19.6$$

Carbon hydrogen bonds, which are stronger, are much less affected. The methyl radicals formed can then abstract hydrogen atoms from ethane in the first of a series of rapid propagation reactions (Eqs. 19.7–19.9).

$$H_3C^\cdot + CH_3CH_3 \rightarrow CH_4 + {^\cdot}CH_2CH_3 \qquad 19.7$$

$${^\cdot}CH_2CH_3 \rightarrow CH_2=CH_2 + H^\cdot \qquad 19.8$$

$$H^\cdot + CH_3CH_3 \rightarrow H_2 + {^\cdot}CH_2CH_3 \qquad 19.9$$

Termination reactions are one of the ways in which the higher hydrocarbons are formed in the cracking of ethane (Eqs. 19.10 and 19.11).

$$2\ {^\cdot}CH_2CH_3 \rightarrow CH_3CH_2CH_2CH_3 \qquad 19.10$$

$$H^\cdot + {^\cdot}CH_2CH_3 \rightarrow CH_3CH_3 \qquad 19.11$$

Propane cracks to both ethylene and methane, and to propylene and hydrogen (Eqs. 19.12 and 19.13) with product proportions, which change with the cracking temperature (Fig. 19.2).

$$CH_3-CH_2-CH_3 \rightarrow CH_2=CH_2 + CH_4 \qquad 19.12$$

$$CH_3-CH_2-CH_3 \rightarrow CH_2=CH-CH_3 + H_2 \qquad 19.13$$

The initial step in the process with propane is still predominantly carbon–carbon bond cleavage, with methyl and ethyl radicals and to lesser extent hydrogen atoms serving in the propagation steps, which lead to the products.

Naphtha, or gas oil, may also be efficiently cracked to ethylene and propylene, although the proportion of ethylene in the product stream is generally lower for these feedstocks than from ethane or propane. However, this flexibility of possible feedstocks to ethylene enables a refinery to process those streams in current excess in order to produce the olefins that it may currently need.

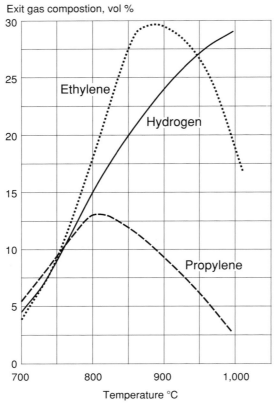

FIGURE 19.2 Changes in the product distribution with temperature on the thermal cracking of propane. (Drawn from the data of Frolich and Wiezevich [13].)

Recovery of ethylene (b.p. −103.7°C) from the mixed hydrocarbon product vapors is accomplished by compression, condensation, and then fractionation at progressively lower temperatures. Ethane (b.p. −88.6°C) is the most difficult constituent to separate from ethylene, but even this is technically feasible.

The single largest use of ethylene is for the production of all types of polyethylene, which consumes about half of the total. Other major uses are for the production of ethylene oxide, ethylene dichloride (1,2-dichloroethane), and ethylbenzene enroute to styrene. The remaining 10–15% of production is distributed among a multiplicity of smaller scale products, including vinyl acetate, ethanol, and acetaldehyde.

19.3.2. Propylene (Propene)

Propylene, or propene by IUPAC (International Union of Pure and Applied Chemistry) nomenclature, is probably the oldest petrochemical feedstock, employed as it was in the early processes to isopropanol. It is produced by the cracking of propane or higher hydrocarbons in the presence of steam under conditions very like those outlined for ethylene. Propane is the feed-stock of choice for a product stream to contain the maximum concentration of propylene. The yield of propylene rises as the cracking severity of the propane is increased, but reaches a maximum before the maximum conversion of ethylene is achieved (Fig. 19.2). Thus, even with use of propane as the feedstock for propylene production, it is common to have more moles of ethylene than propylene product in the exit stream for optimum cracking economics.

Propylene is also recovered as a by-product of other refinery operations, principally from the fluid catalytic cracking (FCC) of gas oils and to a lesser extent from the volatile products of coking, when coking is used. All refinery streams containing recoverable fractions of propylene will be combined into a mixed C_3 stream for propylene separation. Distillation of this combined stream then gives propylene (b.p. $-47.7°C$) as the overhead product and propane (b.p. $-42.1°C$) plus traces of other higher boiling point products as the bottom fraction.

About 45% of the propylene produced is used for the production of isotactic polypropylene. About 10% fractions of the total are consumed for each of acrylonitrile production, the preparation of the "oxo-alcohols," for propylene oxide, and for cumene (isopropylbenzene) production.

19.3.3. The Butylenes (Butenes)

There are four C_4H_8 mono-olefins because of the various possible geometric isomers. These are 1-butene, *cis*-, and *trans*-2-butene, and isobutylene (2-methylpropene). All occur in many catalytically processed refinery streams, and predominantly in the C_4 fraction separated from the products of thermal or fluid catalytic cracking. Some may also be produced in the refinery processes designed to raise the octane rating of gasolines. Including the butanes, which are also present in this stream, we get a total of six compounds present in this C_4's fraction, all boiling in the range -11.7 to $+3.7°C$. This presents a formidable separation task (Table 19.3). Fractional distillation is capable of separating the isobutane, isobutylene, and 1-butene fraction from the n-butane and the 2-butenes. Extractive distillation, using acetonitrile or aqueous acetone, is used to separate the 1-alkenes from isobutane. Extractive distillation with furfural separates the components of the bottom fraction of the distillation. The production of methyl *t*-butyl ether (MTBE), used as an octane improver in gasolines, has been is the single largest consumer of the isobutylene fraction of the butylenes. The next largest market for the butylenes (both 1-butene and the 2-butenes) is for the production of 1,3-butadiene, destined for the production of various types of synthetic rubber.

19.3.4. Benzene, Toluene, and the Xylenes (BTX)

This BTX monoaromatic product stream is usually produced in several different refinery operations. Each of these rank in the top 30 chemicals produced in the United States by volume. They have reached this significance because of their large demand as feedstocks for many of the important

TABLE 19.3 Boiling Points of the Butanes and Butylenes (the "BB Fraction") of the C$_4$ Refinery Stream[a]

Hydrocarbon	B.p. (°C)
2-Methylpropane (isobutane)	−11.7
2-Methylpropene (isobutylene)	−6.9
1-Butene	−6.3
n-Butane	−0.5
trans-2-Butene	+0.88
cis-2-Butene	+3.7

[a]1,3-Butadiene, which may occasionally be present in this fraction, boils at −4.4°C.

plastics. Preparation of these materials is discussed together because several are produced at the same time from many of the processes from which they arise. The chief sources of these aromatics may be conveniently divided into three groups.

In the first group, the production of aromatics is a complementary objective to the refinery processing of gasoline fractions to raise the aromatic content, which evidently links these refining functions. Catalytic reforming processes are used to convert paraffins to naphthenes (cycloparaffins) to be followed by dehydrogenation of naphthenes to aromatics (Chap. 18). Since aromatization of naphthenes is an easier process to accomplish than cycloalkylation, the emphasis in refinery operations is on maximization of the second step in this sequence, when there is an adequate supply of naphthenes. The demand for the aromatics component of gasoline will compete with the feedstock aromatic need from this source.

The second group includes some of the processes used for ethylene production, particularly those using naphtha or gas oil as feedstocks, which may also produce large amounts of BTX. The conditions used (temperature, pressure, feedstock ratios, etc.) to operate these processes provide some flexibility to alter the product distribution slightly in response to the distribution of the demand.

The third group of processes to benzene, toluene, and xylenes, or *p*-xylene is discretionary and depends on the proportion of these products that are available from the first two classes of processes in comparison to the current demand. Benzene consumption is more than twice that of toluene. It may be conveniently produced in almost 90% yields by the catalytic hydrodealkylation of toluene (Eq. 19.14) [14]. Or if both benzene and xylenes are in short supply, toluene may be catalytically transalkylated to mainly benzene and xylenes [15].

$$C_6H_5CH_3 + H_2 \xrightarrow[\text{H}_2,\ \text{catalyst}]{540-800°C} C_6H_6 + CH_4 \qquad 19.14$$

Ethylbenzene, too, may be isomerized to xylenes. Xylenes are normally recovered from a mixed aromatics stream by extraction with sulfolane or a glycol.

If the market for one xylene isomer is greater than that available by separation of the mixture, the low demand isomers may be isomerized to raise the proportion of the desired isomer(s).

About half of the benzene produced as a chemical feedstock is for styrene production, followed by large fractions for phenol and cyclohexane-based products. As much as half of the toluene produced is converted to benzene, depending on the price and demand differential. The largest use of toluene itself is as a component of gasoline. Much smaller amounts are used as a solvent, or in the manufacture of dinitrotoluene and trinitrotoluene for military applications. Xylenes are also used in gasoline formulations and function as octane improvers like toluene. *para*-Xylene and *o*-xylene are the dominant isomers of value as chemical feedstocks, for the production of terephthalic acid (and dimethyl terephthalate) and phthalic anhydride, respectively. Polyester and the synthetic resin markets, in turn, are major consumers of these products. *meta*-Xylene is oxidized on a much smaller scale to produce isophthalic acid, of value in the polyurethane and Nomex aramid (poly(*m*-phenylene isophthalamide)) technologies.

19.4. PRODUCTS FROM METHANE

Ammonia is by far the largest scale chemical product derived from methane. Production of ammonia and its downstream products nitric acid, urea, and ammonium nitrate, which currently rank 6th, 14th, 15th, and 17th in American volume of production, are discussed in Chapter 11. Two other petrochemicals derived from methane, methanol and formaldehyde, currently rank 21st and 24th in volume in the United States at about 4 million tonnes per year each. Details of their production will be outlined here.

Most of the methane recovered, primarily from gas wells, is not used for chemicals, but is simply burned for heat and/or electricity production.

19.4.1. Methanol (Methyl Alcohol)

Methanol is made from a mixture of carbon monoxide and hydrogen. The feed gases required are produced by the reforming of natural gas (Section 9.1). At about 800°C in the presence of a promoted nickel catalyst a gas mixture of close to the correct proportions of carbon monoxide and hydrogen is obtained (Eq. 19.15).

$$3\ CH_4 + CO_2 + \underset{\text{(as steam)}}{2\ H_2O} \longrightarrow 4\ CO + 8\ H_2 \qquad 19.15$$

Similar gas mixtures may also be made by the old water-gas reaction using coal, which gives an approximately 1:1 ratio of carbon monoxide to hydrogen. The hydrogen may be supplemented to the correct proportions from refinery sources or from hydrogen obtained by the electrolysis of water or brine.

The methanol synthesis reaction requires compression of the gas mixture to 200–330 atm, temperatures in the region of 300°C, and a catalyst, such as a chromium oxide-promoted zinc oxide, to result in an approximately 60% conversion to methanol (Eq. 19.16).

$$CO + 2 H_2 \rightarrow CH_3OH \qquad \Delta H = -103 \text{ kJ } (-24.6 \text{ kcal}) \qquad 19.16$$

Methanol is condensed from the exit gases, still at converter pressures, and the unreacted gases are recycled after a partial purge and recompression. As with ammonia production, the purge is required to avoid accumulation of inert gases in the recycle stream. The product under these conditions is about 99% pure methanol plus 1–2% of dimethyl ether and traces of higher alcohols.

The process just described may also be operated at somewhat lower pressures and temperatures than outlined by using some of the more recent copper-based catalysts. Energy saved by doing this achieves some production economies. It is also possible, with an extra mole of hydrogen, to produce methanol from carbon dioxide (Eq. 19.17).

$$CO_2 + 3 H_2 \rightarrow CH_3OH + H_2O \qquad 19.17$$

In this instance distillation in a series of columns is necessary for methanol recovery [5].

Major end uses for methanol are for the production of formaldehyde, about 30%, which is used for the preparation of phenol-formaldehyde resins. About 20% is used for the production of methyl *t*-butyl ether, which is used as an additive alone, and in blends with methanol as a fuel component. Further uses are for the esterification of terephthalic, and acrylic acids, and for acetic acid preparation, about 10% each.

19.4.2. Acetic Acid

The Monsanto process to acetic acid uses a rhodium-iodine-containing catalyst to convert a mixture of methanol and carbon monoxide under mild conditions and ambient pressure to very high (99+% based on methanol) yields of the product (Eq. 19.18) [16]. Several stages of distillation are required to recover the acetic acid (b.p. 118°C) in purities of 99.8% or

$$CH_3OH + CO \xrightarrow[\text{catal.}]{} CH_3COOH \qquad 19.18$$

better. A substantial fraction of acetic acid is also made commercially via the liquid or gas phase oxidation of acetaldehyde or the liquid phase oxidation of butane, in each case using air plus a catalyst.

The largest single use for acetic acid, about half the total, is for the manufacture of vinyl acetate for use in the production of poly(vinyl acetate) (Section 19.10.4). Cellulose acetate destined for fiber spinning consumes a further quarter.

19.4.3. Formaldehyde

Catalytic oxidation of methanol in the presence of a catalyst based on transition metal oxides produces good yields of formaldehyde [17]. Both oxidation and dehydration reactions participate in this process (Eqs. 19.19 and 19.20).

$$CH_3OH + 1/2\ O_2\ \text{(from air)} \rightarrow CH_2O + H_2O \qquad 19.19$$

$$CH_3OH \rightarrow CH_2O + H_2 \qquad 19.20$$

Evidence for the contribution of the second reaction is obtained by the detection of hydrogen in the reactor exit gases. Formaldehyde (b.p. $-21°C$) is recovered from these by fractionation.

Most of the formaldehyde produced is consumed in the production of urea-formaldehyde resins and phenol-formaldehyde resins. These cross-linked polymer products are in turn used in adhesive and laminate applications.

19.5. PRODUCTS FROM ETHYLENE

The various types of polyethylene together consume close to half of the total ethylene produced. Most of the remainder is used to produce the following high volume chemicals.

19.5.1. Ethylene Oxide (EO)

Production of ethylene oxide is the next largest single use of ethylene. It used to be produced via the chlorohydrin process, as propylene oxide still is. However, now it is made by the direct oxidation of ethylene with air in the presence of a catalyst of silver supported on α-alumina (Eq. 19.21).

$$CH_2\!=\!CH_2\ +\ \tfrac{1}{2}\ O_2\ \rightarrow\ \underset{\displaystyle O}{CH_2\text{-}CH_2}\ (+\ \text{some } CO_2,\ H_2O) \qquad 19.21$$

Temperatures in the 270–290°C range, a pressure of about ambient, and a 1-second contact time give conversions to ethylene oxide of about 60%. The ethylene oxide is absorbed in water from the reactor exit gases, and the aqueous solution is then fractionated for ethylene oxide (b.p. 13.5°C) recovery. The nonabsorbed ethylene and the absorption water recovered from the bottom of the ethylene oxide fractionator are both recycled. Recently use of a Cu/Ag alloy catalyst has shown greater selectivity for ethylene oxide in this process [18].

Close to half of the ethylene oxide produced is directly converted to ethylene glycol (1,2-ethanediol) by acid-catalyzed or pressure hydration. Roughly half the ethylene glycol is as an automotive coolant antifreeze and half in polyester (poly(ethylene terephthalate)) fiber production (Chap. 21). Smaller amounts are consumed for alkyd resins production and for the formulation of latex paints.

19.5.2. Ethylene Dichloride (1,2-Dichloroethane) and Vinyl Chloride

Direct reaction of ethylene and chlorine in the presence of a catalyst, in either vapor phase or liquid phase reactors under mild conditions produces a 96+% yield of ethylene dichloride (Eq. 19.22) [19].

$$CH_2 = CH_2 + Cl_2 \rightarrow ClCH_2CH_2Cl \qquad\qquad 19.22$$

Some producers with a good supply of hydrogen chloride available as a by-product from the production of chlorinated solvents may prefer to use the oxychlorination route to ethylene dichloride (Eqs. 19.23 and 19.24).

$$CH_4 + Cl_2 \rightarrow CCl_4 + 4\ HCl \qquad\qquad 19.23$$

Oxychlorination:

$$CH_2 = CH_2 + 2\ HCl + 1/2\ O_2\text{(from air)} \xrightarrow{\quad CuCl_2 \ cat. \quad} \qquad 19.24$$
$$ClCH_2CH_2Cl + H_2O$$

Small amounts of ethylene dichloride (b.p. 83.5°C) are also recovered as a by-product from the direct chlorination of ethane to ethyl chloride (chloroethane).

Whatever the source, 80% or more of the ethylene dichloride produced is directly converted into vinyl chloride [20]. Target conversions are 50% under the usual operating conditions of about 3–4 atm pressure and 500°C to give better than 94% yields of vinyl chloride (Eq. 19.25).

$$ClCH_2CH_2Cl \rightarrow CH_2 = CHCl + HCl \qquad\qquad 19.25$$

Gas exposure times to furnace temperatures are kept short by a direct spray of ethylene dichloride at ambient temperatures onto the exit gases and by an indirect heat exchanger. Hydrogen chloride may be recovered by scrubbing with water, or by some form of fractionation if it is to be fed in anhydrous form to an oxychlorination unit. Many vinyl chloride producers operate an oxychlorination unit solely to convert the by-product hydrogen chloride from the ethylene dichloride cracker to produce additional ethylene dichloride feedstock. Vinyl chloride (b.p. −13.4°C) is almost entirely consumed for the production of poly(vinyl chloride) (PVC).

19.5.3. Ethylbenzene and Styrene (Vinylbenzene)

Ethylbenzene production has a smaller demand for the ethylene stream than the products already described and currently stands 19th in volume of production. It is made by both liquid phase processes under moderate conditions employing aluminum chloride catalysis and by vapor phase processes at 150–250°C and 30–50 atm in the presence of a supported boron trifluoride catalyst (Eq. 19.26).

$$C_6H_6 + CH_2 = CH_2 \xrightarrow{\quad catalyst \quad} C_6H_5CH_2CH_3 \qquad 19.26$$

Ethylbenzene (b.p. 136°C) is recovered from the excess unreacted benzene and the more heavily alkylated material by fractionation. The polyalkylated residues may be partly recycled to take advantage of transalkylation reactions to produce more ethylbenzene, or may be burned as fuel to help drive the process.

Virtually all of the ethylbenzene produced is destined for styrene production, ultimately leading to polystyrene and other polymers. A preheated mixture of ethylbenzene vapor and superheated steam passed over one of several possible dehydrogenation catalysts preheated to 620–640°C crack 30–40% of the ethylbenzene in one pass (Eq. 19.27).

$$C_6H_5CH_2CH_3 \xrightarrow{\text{catalyst}} C_6H_5CH = CH_2 + H_2 \qquad 19.27$$

The exit stream is condensed and, after addition of a polymerization inhibitor (usually sulfur), is successively distilled under reduced pressure to a styrene purity of better than 99.7%. A few ppm of 4-*t*-butylcatechol inhibitor is added to the purified product after which it is ready to be shipped. Industrial yield (selectively) of styrene from ethylbenzene is about 90%.

19.5.4. Vinyl Acetate

Currently ranked 41st in American volume of production, vinyl acetate is made by the vapor phase reaction of ethylene with acetic acid at 170–200°C. Pressures of 5–8 atm and a palladium catalyst supported on carbon are required to give conversions of 10–15% on ethylene and 15–30% on acetic acid. Equation 19.28 is an overall representation of the stoichiometry of the complex process.

$$CH_2 = CH_2 + CH_3COOH + 1/2\ O_2 \rightarrow CH_3COOCH = CH_2 + H_2O$$
$$19.28$$

The reactor exit gases are cleaned up by various washes and the vinyl acetate (b.p. 72.2°C) separated from unreacted starting materials, water, and a small amount of polymeric material by distillation. Industrial yields based on ethylene are 92–95%. The product purity is 99%, to which traces of inhibitor, such as hydroquinone will be added to stabilize it during storage and shipment.

All of the product vinyl acetate is consumed in the production of homopolymer (one monomer) and copolymer products, among them poly(vinyl acetate) latices and resins, poly(vinyl alcohol) by postpolymerization hydrolytic removal of the acetate, and copolymers with vinyl chloride and ethylene.

19.5.5. Ethanol (Ethyl Alcohol)

Production of ethanol by fermentation processes probably represents among the earliest processes operated by humankind. Certainly until the late 1940s it was the dominant process used to produce industrial alcohol (Fig. 19.3). The world market for industrial ethanol in 1991 was about 19×10^9 L. Most of this was for fuel use, Brazil alone consuming about 12×10^9 L, more than half of the total, and the United States a further 3.5×10^9 L. Most of the Brazilian market was supplied by the domestic fermentation of sugar solutions plus distillation, although the proportion produced in this way varies inversely with world sugar prices (Chap. 16). Any Brazilian supply shortfall that may result is supplied from imports of industrial alcohol. In the United States fermentation supplied almost all of the industrial ethanol in 1930. Fermentation sources provided only about a tenth of the total in 1960 but with the stimulus of subsidies had grown to about one-third of a larger total by 1990. The balance of the American industrial ethanol market was supplied from

Ethanol, 10⁶ L

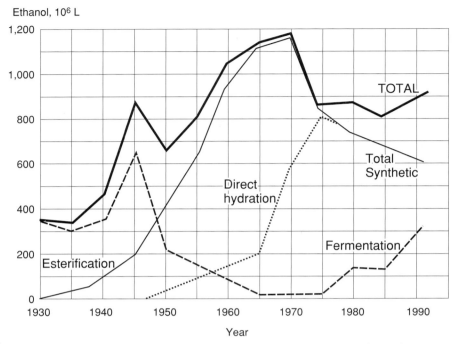

FIGURE 19.3   Volume of production of industrial ethanol in the United States, by process.

petrochemical sources. Today fermentation of sugarcane juice provides most of the world's ethanol, thanks to the massive contribution to this source from Brazil.

The earliest petrochemical process to convert ethylene to ethanol was first practiced in about 1930, and was indirect. In this process ethylene, under pressure, is absorbed by countercurrent passage against sulfuric acid (90–98%) at about 80°C in an absorber to form a mixture of the monoesters, and diesters (Eqs. 19.29 and 19.30).

$$CH_2 = CH_2 + H_2SO_4 \rightarrow CH_3CH_2OSO_3H \qquad 19.29$$

$$2\ CH_2 = CH_2 + H_2SO_4 \rightarrow (CH_3CH_2O)_2SO_2 \qquad 19.30$$

Both the acid concentration and the temperature have to be higher for ester formation from ethylene than for ester formation from propylene because of the less reactive carbon–carbon double bond in ethylene. The monoester is preferred for its easier hydrolysis and its lower tendency to form diethyl ether. After reaction, the mixed esters are hydrolyzed with just sufficient excess water to complete the hydrolysis, which takes about 2 hr at 60–70°C (Eq. 19.31).

$$CH_3CH_2OSO_3H + (CH_3CH_2O)_2SO_2 + 3\ H_2O \rightarrow \qquad 19.31$$
$$3\ CH_3CH_2OH + H_2SO_4\ (dil.)$$

The ethanol product (b.p. 78.3°C) is recovered by distillation and the diluted acid is recovered as the bottom stream of the column. Yields (selectivities) on ethylene are 90–95%. Diluted acid is reconcentrated in several stages, the last stages under reduced pressure, for recycling to the absorption stage. The volume of ethyl alcohol produced in the United States by this process peaked in about 1960. Since the late 1940s it has gradually been supplemented and then largely replaced by the process, which uses direct hydration of ethylene (Fig. 19.3).

The direct hydration process began to be important for American industrial alcohol production in the late 1940s. It has grown from this beginning to produce roughly 70% of the total industrial alcohol in 1975 and certainly has provided most of the supply since then. Like many catalytic petrochemical processes, high pressures (ca. 70 atm) and temperatures of around 300°C are required to convert a 0.6:1 mole ratio of steam:ethylene, to ethanol. In the presence of a supported phosphoric acid catalyst using a space velocity of $1800\,hr^{-1}$ gives conversions of 4.0–4.2% and yields on ethylene of 95+% (Eq. 19.32).

$$CH_2 = CH_2 + H_2O \rightleftharpoons CH_3CH_2OH \quad \Delta H = -40.2 \text{ kJ } (-9.6 \text{ kcal}) \quad 19.32$$

The exit stream is cooled and the condensate separated in a high-pressure separator. After scavenging any remaining product from this stream, the unreacted ethylene is recycled.

Ethanol is recovered from the condensate of the direct hydration process by distillation. Traces of acetaldehyde, which probably arise from the hydration of a small amount of acetylene present in the feed ethylene, are catalytically reduced to ethanol with hydrogen (Eqs. 19.33 and 19.34).

$$HC \equiv CH + H_2O \rightarrow CH_3CHO \qquad 19.33$$

$$CH_3CHO + H_2 \xrightarrow[\text{Ni cat.}]{} CH_3CH_2OH \qquad 19.34$$

Traces of diethyl ether impurity are hydrated back to ethanol using conditions similar to those used for the ethylene hydration reaction.

Ethanol is used in a variety of fuel, solvent, and chemical applications in industry. Chemical applications have dropped sharply in the last 20 years as more direct routes to products, such as acetaldehyde, acetic acid, ethyl acetate, and ethyl chloride, which were formerly made from ethanol, came on stream. The automotive fuel component, as a component to improve octane rating, is anticipated to have the highest potential for growth.

19.6. PRODUCTS FROM PROPYLENE (PROPENE)

Almost half of the propylene produced is used to prepare polypropylene (Chap. 23). Most of the remainder is consumed for production of the following products.

19.6.1. Acrylonitrile (Vinyl Cyanide)

This versatile monomer is produced by the catalytic ammonoxidation of propylene. A mixture of propylene, ammonia, and air in the volume ratio

1:1:10 provides roughly two volumes (moles) of oxygen per volume (mole) of the other reactants, or a 30% excess to that required by the process. The gas mixture at near ambient pressures is given a 10–20 second contact time over a molybdenum-based catalyst on a silica support kept at about 400°C. Better than 70% conversion to acrylonitrile is obtained, together with traces of by-product acetonitrile and hydrogen cyanide (Eq. 19.35).

$$CH_3-CH=CH_2 + NH_3 + 3/2\ O_2 \rightarrow CH_2=CHCN + 3\ H_2O \quad 19.35$$

Reactor exit gases are cooled and washed with water, from which the acrylonitrile is recovered and purified by distillation.

A recently developed one-step ammonoxidation process to acrylonitrile developed by BP Chemicals uses propane instead of propylene feed [21]. Key to the feasibility of this lower cost option was the development of the new catalyst system, which is now at the commercial demonstration stage. Almost all the acrylonitrile production goes into synthetic polymers and copolymers mostly for applications as fibers, some for plastics applications, and a small percentage to elastomer markets (the nitrile rubbers).

19.6.2. Isopropyl Alcohol (Isopropanol)

The indirect, sulfate ester-based process, used to introduce this chapter, was one of the first petrochemical processes (Section 19.3). This method still has a place since its capability of selective absorption of propylene enables it to be used with refinery gas streams to capture low concentrations of propylene, unlike the conditions for alternative processes. The direct hydration of propylene to isopropanol is also possible, but this requires a refinery stream containing a much higher concentration of propylene to be competitive. For this reason both petrochemical processes are still viable in their respective feedstock niches.

With the direct hydration process to ethanol, the Le Chatelier principle would suggest that high pressures and low temperatures would tend to favor product formation. This is also true with propylene. Pressures of 100–250 atm and temperatures of 150–270°C are used because, unfortunately, all known catalysts (supported sulfonic acids, or phosphoric acid) for this process require high temperatures to be effective (Eq. 19.36).

$$CH_3CH=CH_2 + H_2O \rightleftharpoons (CH_3)_2CHOH \quad\quad 19.36$$

$$\Delta H = -210\ kJ\ (-50\ kcal)$$

Isopropanol formation is thought to proceed via an intermediate isopropyl carbocation generated by interaction with water and the acid catalyst, which is then hydrated with water. Recovery is by distillation, to the 91% azeotrope, or by ternary distillation of the azeotrope to 100% isopropanol.

19.6.3. The Oxo Alcohols: *n*-Butanol and Isobutanol (2-Methylpropanol-1)

The oxo reaction is used to produce aldehydes or alcohols containing one additional carbon atom, from the corresponding α-olefin. Propylene, in the presence of a 1:1 mixture of carbon monoxide and hydrogen (synthesis gas) and raised to 100–250 atm and about 170°C, gives 70–80% yields of the initial products butyraldehyde and isobutyraldehyde (Eq. 19.37).

$$CH_3CH{=}CH_2 + CO + H_2 \rightarrow CH_3CH_2CH_2CHO + (CH_3)_2CHCHO$$
$$\phantom{CH_3CH{=}CH_2 + CO + H_2 \rightarrow} 60{-}40\% 40{-}60\%$$

$$19.37$$

An organocobalt derivative, such as cobalt naphthenate, is used as a catalyst. In a second step, the mixed aldehydes are reduced with hydrogen in the presence of a nickel-based catalyst to give *n*-butanol (b.p. 118°C) and iso-butanol (b.p. 108.1°C) (Eq. 19.38).

$$CH_3CH_2CH_2CHO + H_2 \rightarrow CH_3(CH_2)_2CH_2OH + \qquad 19.38$$
$$CH_3CH(CH_3)CH_2CH_2OH$$

The products are recovered and purified by distillation. It should be mentioned that this process is not limited to the oxygenation of propylene, but may be applied to α-olefins generally.

19.6.4. Propylene Oxide (PO)

Two process routes to propylene oxide are commercially practiced; hydroperoxide formation and then use of this to oxidize propylene, and formation of propylene chlorohydrin followed by treatment with a base to form propylene oxide [22, 23]. It has not been possible to produce adequate yields of propylene oxide via the direct oxidation of propylene with air in the manner in which ethylene oxide is now produced, although attempts to come close to this continue [24].

Epoxidation of propylene by a hydroperoxide is commercially viable for those hydroperoxides which, when spent, can easily give another commercial product. One option is to produce *t*-butyl hydroperoxide (plus some *t*-butanol) by the air oxidation of isobutane at 90°C and ca. 30 atm in the presence of molybdenum naphthenate (Arco and Texaco). The separated hydroperoxide is then used to oxidize propylene to propylene oxide with yields on hydroperoxide of over 90% (Eqs. 19.39 and 9.40).

$$(CH_3)_2CHCH_3 + O_2 \rightarrow (CH_3)_3COOH \qquad 19.39$$

$$(CH_3)_3COOH + CH_3{-}CH{=}CH_2 \rightarrow \qquad 19.40$$
$$CH_3{-}\underset{\underset{\displaystyle O}{\diagdown\diagup}}{CH{-}CH_2} + (CH_3)_3COH$$

Coproduct *t*-butanol is produced at the rate of about 2.2 kg per kilogram of propylene, and is converted to methyl *t*-butyl ether for sale as an automotive fuel additive.

Ethylbenzene hydroperoxide from ethylbenzene provides the other viable option to oxidize propylene (Arco, Eq. 19.41 and 19.42).

$$PhCH_2CH_3 + O_2 \rightarrow PhCH(OOH)CH_3 \qquad 19.41$$

$$PhC(OOH)CH_3 \ + \ CH_3CH=CH_2 \ \rightarrow \qquad\qquad 19.42$$

$$CH_3-\underset{\underset{O}{\diagdown\diagup}}{CH-CH_2} \ + \ PhCH(OH)CH_3$$

In this case the spent oxidant is α-methylbenzyl alcohol, which is produced at almost twice the mass per unit of propylene oxide as *t*-butanol. However, easy dehydration of α-methylbenzyl alcohol converts the coproduct to styrene, for which a large market exists (Eq. 19.43).

$$PhCHOHCH_3 \ \rightarrow \ PhCH=CH_2 \ + \ H_2O \qquad\qquad 19.43$$

The chlorohydrin process practiced by Dow Chemical in the United States, weds the chlorine component of chloralkali technology to propylene oxide production. Chlorine added to water produces hypochlorous acid and hydrochloric acid (Eq. 19.44).

$$Cl_2 \ + \ H_2O \ \rightleftharpoons \ HOCl \ + \ HCl \qquad\qquad 19.44$$

Contacting this solution countercurrently with propylene converts some of the propylene to propylene chlorohydrin, which dissolves in the aqueous phase leaving the bottom of the contacting tower. Both possible isomers are formed (Eq. 19.45).

$$CH_3-CH=CH_2 \ + \ HOCl \ \rightarrow \ \underset{90\%}{CH_3-CH(OH)CH_2Cl} \ + \qquad 19.45$$

$$\underset{10\%}{CH_3-CH(Cl)CH_2OH}$$

The rate of reaction of chlorine with propylene is much slower, though 7–8% by weight propylene chloride (1,2-dichloropropane) is also formed. In a second stage the propylene chlorohydrin is contacted with lime slurry in water, which cyclizes the chlorohydrin to the epoxide (Eq. 19.46).

$$2 \ CH_3-CH(OH)CH_2Cl \ + \ Ca(OH)_2 \rightarrow \qquad\qquad 19.46$$
(or other isomer)

$$2 \ CH_3-\underset{\underset{O}{\diagdown\diagup}}{CH-CH_2} + CaCl_2 + 2 \ H_2O$$

To minimize hydrolysis of the propylene oxide (b.p. 34.2°C) as it is formed, it is flashed (rapidly removed) from the reactive lime slurry. Yields of propylene oxide are 75% or better based on propylene. The advantage of the chlorohydrin route to propylene oxide over the two hydroperoxidation processes is that it yields essentially a single product to market. The disadvantage is the large quantities of coproduced aqueous calcium chloride that has to be discarded safely. The small amount of by-product 1,2-dichloropropane may be pyrolyzed to allyl chloride, useful for the preparation of allyl monomers, allyl alcohol, and allylamines. Or it may be blended with 1,3-dichloropropene to produce an effective soil fumigant.

Close to two-thirds of the propylene oxide produced goes to polypropylene glycol and polyester glycols production destined for use in polyurethanes. The polyol diols are made by the base-catalyzed reaction of propylene glycol with propylene oxide (Eq. 19.47). Replacing propylene glycol by glycerol in this process produces polyol triols.

$$CH_3CH(OH)CH_2OH + n\ CH_3-\underset{\underset{O}{\diagdown\diagup}}{CH}-CH_2 \xrightarrow{\ HO^-\ } \qquad 19.47$$

$$HOCH_2CH(CH_3)O[CH_2CH(CH_3)O]_{n-1}CH_2CH(CH_3)OH$$

Most of the rest of the propylene oxide is hydrolyzed in water using a trace of mineral acid to give propylene glycol (1,2-propanediol, b.p. 187°C) used for the manufacture of polyester resins.

19.6.5. Cumene (Isopropylbenzene) and Phenol

About 10% of the propylene produced is consumed for the alkylation of benzene to produce cumene. Use of a supported phosphoric acid catalyst and a high-purity stream of propylene in the presence of a large excess of benzene (to minimize dialkylation and oligomer formation) gives cumene with a 90+% yield based on ethylene, 96+% yield on benzene (Eq. 19.48).

$$C_6H_6 + CH_3-CH{=}CH_2 \rightarrow C_6H_5CH(CH_3)_2 \qquad 19.48$$

Reaction conditions are about 200°C at 20–30 atm pressure. Product cumene (b.p. 152°C) is recovered and purified by distillation.

Almost all of the cumene produced is consumed by the cumene peroxidation process to phenol and acetone. Aspects of this process resemble the peroxidation processes to propylene oxide already described, except that in this case the final products are formed by an intramolecular rearrangement of the same molecule that is peroxidized rather than by a reaction of a peroxidized molecule with a second component, propylene. In the first stage of the cumene to phenol process a suspension of purified cumene in a dilute solution of sodium carbonate in water is heated to about 110°C under slight pressure. Air is blown through this suspension until about 25% of the cumene has formed the hydroperoxide (Eq. 19.49).

$$PhCH(CH_3)_2 + O_2\ (from\ air) \rightarrow PhC(CH_3)_2OOH \qquad 19.49$$

The dissolved sodium carbonate helps to maintain the system to near neutral pH to avoid premature rearrangement of the hydroperoxide. The partially converted cumene from the first stage is carefully concentrated to about 80% cumene hydroperoxide before proceeding to the rearrangement step.

Dilute sulfuric acid at 70–80°C is used to rearrange the cumene hydroperoxide to phenol and acetone, for which the stoichiometry is represented by Eq. 19.50.

$$PhC(CH_3)_2OOH \rightarrow PhOH + (CH_3)_2CO \qquad 19.50$$

The details of the complicated mechanism of this rearrangement were only discovered some time after the startup of commercial facilities (Fig. 19.4). Distillation is used to recover the 90+% yield of the rearrangement products. Unreacted cumene is recycled after careful purification to remove any traces of phenol, which would be a powerful inhibitor of hydroperoxide formation,

$$PhC(CH_3)_2OOH \underset{-H^+}{\overset{H^+}{\rightleftharpoons}} PhC(CH_3)_2OOH_2^+ \rightarrow PhC(CH_3)_2O^+ \xrightarrow{1,2 \text{ shift}}$$

$$PhOC^+(CH_3)_2 \xrightarrow{H_2O} PhOC(CH_3)_2OH_2^+ \underset{H^+}{\overset{-H^+}{\rightleftharpoons}} PhOC(CH_3)_2OH \xrightarrow{H^+}$$

$$PhOH + (CH_3)_2CO$$

FIGURE 19.4 Mechanistic details of the acid catalyzed rearrangement of cumene hydroperoxide to phenol and acetone.

and α-methylstyrene (a by-product), which would reduce the yield of the peroxidation stage.

Any separated α-methylstyrene may be sold as such or may be catalytically reduced with hydrogen back to cumene again.

Early petrochemical routes to phenol involved sulfonation followed by hydrolysis, or catalytic chlorination in the presence of iron followed by drastic hydrolysis (Eqs. 19.51–19.53).

$$C_6H_6 + H_2SO_4 \rightarrow C_6H_5SO_3H, \quad C_6H_5SO_3H \xrightarrow{H_2O} C_6H_5OH + H_2SO_4 \qquad 19.51$$

$$C_6H_6 + Cl_2 \rightarrow HCl + C_6H_5Cl \qquad 19.52$$

$$C_6H_5Cl + H_2O + NaOH \rightarrow C_6H_5OH + NaCl \qquad 19.53$$

Improved yields under less drastic conditions and the decreased waste stream problems of the cumene process to phenol have led to the phasing out of these older processes.

A large fraction of the phenol product, currently ranked 34th in American volume of production, is directed toward the production of several types of phenol-formaldehyde resins of utility as plywood adhesives and as components of laminates (e.g., Arborite). Some is routed to the preparation of *bis*-phenol A enroute to epoxy resin production.

19.7. PRODUCTS FROM ISOBUTYLENE (2-METHYLPROPENE)

Isobutylene itself ranks 38th in volume in the American chemical industry. Its two major uses are for the production of methyl *t*-butyl ether and 1,3-butadiene, which stood 12th and 36th in volume of production in 1995.

19.7.1. Methyl *t*-butyl Ether (MTBE)

Commercial production of methyl *t*-butyl ether began in 1979, shortly after the discovery of its octane-improving capability for motor fuels. Although a higher proportion of this additive was required for equivalent octane enhancement, it was less costly and eliminated the lead particulate discharges associated with the tetraethyl lead previously used (Chap. 18). By 1984 it ranked 49th in American volume of production and jumped to 12th by 1995, an average compound growth of almost 50% per year. It had been predicted that by the year 2000 MTBE production would rank second in volume after ethylene among American organic chemicals, but problems from its contamination of water supplies have reversed its fortunes.

Methyl *t*-butyl ether production is the single largest consumer of butylenes. Liquid phase reaction of methanol with isobutylene in the presence of an acidic ion-exchange resin catalyst at temperatures of below 100°C and moderate pressures give excellent yields of this oxygenated gasoline additive (Eq. 19.54).

$$CH_3OH + CH_2{=}C(CH_3)_2 \rightarrow CH_3OC(CH_3)_3 \qquad 19.54$$

Purification of MTBE (b.p. 55°C) for general solvent use is by distillation. Its use in general solvent applications is still small. However, since MTBE has no secondary or tertiary hydrogens it is very resistant to oxidation and peroxide formation. This makes it an attractive replacement for the more traditional diethyl, and diisopropyl ethers.

19.7.2. 1,3-Butadiene

The other major use for the *n*-butene fraction (both 1-butene and the 2-butenes) of the butylenes is as a feedstock for 1,3-butadiene manufacture, ultimately destined for the production of synthetic rubber. Careful thermal catalytic dehydrogenation in the presence of steam gives a 75–86% yield of 1,3-butadiene from a 25–30% butene conversion per pass (Eq. 19.55).

$$n\ C_4H_8 \xrightarrow{\text{ca. }650°C} CH_2{=}CH{-}CH{=}CH_2 + H_2 \qquad 19.55$$

In 1997 1.7 million metric tonnes of *n*-butenes were consumed in the United States for this purpose, about one-third of that consumed to produce MTBE.

Butane may also be used as the raw material for butadiene production in a different, though related, process. The need to remove four hydrogens requires process operation at lower conversions of 10–11% per pass to maintain reasonable yields, and results in ultimate yields of 1,3-butadiene from this process of about 60%.

19.8. PRODUCTS FROM BENZENE, TOLUENE, AND THE XYLENES (BTX)

The number of derivatives commercially produced from this group of aromatic substrates is extensive, many targeted as monomers for polymerization to the large-volume fibers and plastics. Some of these, the ethylbenzene to styrene sequence and cumene to phenol, have already been discussed in connection with ethylene and propylene derivatives. Details of some further representative examples of major products are outlined here.

19.8.1. Benzene to Cyclohexane

Benzene is the aromatic petrochemical used as such to ultimately produce styrene and phenol, both very large-scale products in their own right. Other major products from benzene use it in hydrogenated form, as cyclohexane.

Adipic acid (48th in volume), for example, is almost entirely derived from cyclohexane (47th) as is much of the production of caprolactam and hexamethylenediamine (1,6-diaminohexane). The production sequences to all of these products will be outlined.

Cyclohexane is made by the catalytic reduction of benzene with hydrogen, usually in a train of several reactors with partial conversion conducted in each. This process uses a nickel or platinum catalyst and requires temperatures of about 200°C, 25–40 atm pressure, and the presence of a cyclohexane recycle diluent to help absorb the exotherm of the hydrogenation (Eq. 19.56).

$$C_6H_{6(g)} + 3\ H_{2(g)} \rightarrow C_6H_{12(g)} \qquad \Delta H = -151\ kJ\ (-36\ kcal) \qquad 19.56$$

Cyclohexane recovery (99+% yields) is by distillation.

19.8.2. Cyclohexane to Adipic Acid (1,6-Hexanedioic Acid)

Oxidation of cyclohexane, usually in two stages, gives adipic acid. In the first stage cyclohexane is oxidized with air in the presence of a cobalt naphthenate catalyst under moderate conditions to give a mixture of cyclohexanol and cyclohexanone (Eq. 19.57, stoichiometry only).

$$2\ C_6H_{12} + O_2 \rightarrow \quad \bigcirc\!\!=\!\!O \quad + \quad \bigcirc\!\!-\!OH \quad + \quad H_2O \qquad 19.57$$

On completion, water is added to the mixture after which it is fractionated. Cyclohexane (b.p. 81°C) containing some benzene is collected from the top of the column, and after hydrogenation of the benzene, is recycled. The cyclohexanol–cyclohexanone mixture consists of approximately equal volumes of cyclohexanol (b.p. 161°C), cyclohexanone (b.p. 156°C), plus a mixture of several esters and ethers. It is collected from the bottom with 80+% yields on cyclohexane. An alternative route to cyclohexanol used by some plants is to catalytically hydrogenate phenol.

The cyclohexanol–cyclohexanone mixture isolated from air oxidation, without separation, is oxidized with 5 volumes of 50% nitric acid plus catalyst at 50–90°C, and pressures only slightly above ambient for 10–30 min, depending on the temperature used (Eq. 19.58).

$$C_6H_{11}OH + \quad \bigcirc\!\!=\!\!O \quad + \quad 4\ HNO_3 \rightarrow \qquad 19.58$$

$$2\ HO_2C(CH_2)_2CO_2H + 2\ H_2O + 2\ N_2O\uparrow$$
$$\text{(approximate stoichiometry)}$$

Adipic acid (m.p. 152°C) is crystallized from the reaction mixture by cooling this to ca. 5°C.

Over 90% of the adipic acid is consumed for the production of nylon for fiber and engineering resin applications. A large fraction of this consumption is direct, as the adipic acid used in the production of nylon 6,6, but a substantial fraction of the adipic acid is further processed to give hexamethylene diamine, the other monomer required. A further small fraction of the adipic acid is converted into di-octyl (di-2-ethylhexyl) or di-hexyl esters for use as plasticizers in flexible grades of PVC, etc., or as a very stable high boiling point component of synthetic motor oils.

19.8.3. Hexamethylenediamine (Hexane-1,6-diamine)

Vapor phase reaction of adipic acid with ammonia at about 400°C and 270–410 atm in the presence of boron phosphate catalyst gives good yields of adiponitrile (Eq. 19.59).

$$HO_2C(CH_2)_4CO_2H + 2\,NH_3 \rightarrow NC(CH_2)_4CN + 4\,H_2O \qquad 19.59$$

Alternatively, adiponitrile may also be made from 1,3-butadiene via chlorination, followed by treatment with sodium cyanide. Catalytic reduction of adiponitrile with hydrogen then produces hexamethylenediamine (b.p. 205°C, Eq. 19.60).

$$NC(CH_2)_4CN + 4\,H_2 \rightarrow H_2N(CH_2)_6NH_2 \qquad 19.60$$

Product recovery is by distillation.

Production of this petrochemical is dedicated to a single end use, the preparation of nylon 6,6.

19.8.4. ε-Caprolactam (2-Oxohexamethyleneimine)

This cyclic amide is of major importance as the single monomer used to produce nylon 6, which rivals nylon 6,6 in utility and in volume of production (Chap. 21). Cyclohexanone is required as the starting material for ε-caprolactam production. The mixture of cyclohexanol and cyclohexanone obtained initially from the air oxidation of cyclohexane for the second stage en route to adipic acid may be used. Subjecting the mixture to dehydrogenation converts the cyclohexanol component to cyclohexanone producing suitable starting material for this process (Eq. 19.61).

$$C_6H_{11}OH \rightarrow C_5H_{10}C{=}O + H_2 \qquad 19.61$$

Alternatively, pure cyclohexanone may be made by the catalytic hydrogenation of phenol in the presence of palladium on charcoal.

Cyclohexanone oxime, an intermediate, is made by reacting aqueous hydroxylamine sulfate with cyclohexanone (Eq. 19.62).

$$\text{(cyclohexyl)}{-}OH \rightarrow \text{(cyclohexylidene)}{=}O + H_2 \qquad 19.62$$

A Beckmann rearrangement of the oxime in hot concentrated sulfuric acid then gives the desired seven-membered cyclic amide, ε-caprolactam. The crude product forms a separate oily phase which is separated from the reaction mixture and purified by distillation under reduced pressure (b.p. 136–138°C at 10 mm; m.p. 72°C). There are also several alternative routes to produce caprolactam [25].

Caprolactam production, like that of hexamethylenediamine, is dedicated to the production of a single product, in this case the various forms of nylon 6,6.

19.8.5. Toluene to Toluenediisocyanate (TDI)

Toluene output at one time used to be almost entirely directed to trinitrotoluene (TNT) production for industrial and military uses. Cheaper industrial explosives, such as various ammonium nitrate formulations have now become available making TNT a less important product. Uses of TNT are now almost entirely confined to military applications. However, the first stage in the preparation of toluenediisocyanate still involves the mono-, and di-nitration of toluene using a mixture of nitric and sulfuric acids, as in the first stages of TNT production. Depending on the nitration strategies employed, one of two isomer mixtures of 2,4-dinitrotoluene and 2,6-dinitrotoluene, an 80:20 ratio and a 65:35 ratio, or the pure 2,4-dinitrotoluene may be obtained (Eq. 19.63) [26].

$$C_6H_5CH_3 + HNO_3/H_2SO_4 \rightarrow \qquad\qquad\qquad 19.63$$
$$2,4\text{-, and } 2,6\text{-PhCH}_3(NO_2)_2 + \text{dil. acids}$$

All three possible toluene nitration outcomes are separately carried through the reduction and phosgenation stages to give the corresponding diisocyanate mixtures, or pure 2,4-toluene diisocyanate. Reduction to the diamines with hydrogen (ca. 50 atm) is conducted catalytically using Raney nickel in the liquid phase at about 90°C, to avoid the considerable explosion risk of gas phase operation (Eq. 19.64).

$$2,4\text{-, or } 2,6\text{-PhCH}_3(NO_2)_2 + 6 H_2 \rightarrow 2,4\text{-, or} \qquad\qquad 19.64$$
$$2,6\text{-PhCH}_3(NH_2)_2 + 4 H_2O$$

Before the catalytic route had been developed, iron in hydrochloric acid was used to reduce the dinitrotoluenes to the diamines.

The second stage, phosgenation of the mixed diamines, or the 2,4-diaminotoluene is conducted in an inert (aromatic, or chlorinated) solvent at 20–50°C. This uses an excess of phosgene in one, or preferably two (for better yields) stages (Eq. 19.65).

$$R(NH_2)_2 + 2 ClCOCl \rightarrow R(NHCOCl)_2 \rightarrow \qquad\qquad 19.65$$
$$R(N{=}C{=}O)_2 + 2 HCl$$

The intermediate carbamoyl chloride is dissociated to the diisocyanate with the aid of second stage heating of the reaction mixture to 100–110°C plus sparging with natural gas or nitrogen to remove the hydrogen chloride as it forms. Material for sale is produced by reduced pressure distillation; pure toluene-2,4-diisocyanate has a b.p. of 120°C at 10 mmHg pressure.

The largest volume product, and the least expensive from this process, is the 80:20 2,4-, to 2,6-toluenediisocyanate blend, which is the composition obtained without any intermediate separations of nitrotoluenes. The 65:35 2,4-, to 2,6-TDI blend, and the pure 100% 2,4-toluenediisocyanate are progressively more expensive and are produced in lower volumes than the 80:20 blend since they each require one nitrotoluene isomer separation step. However, regardless of composition essentially all commercial TDI is consumed in the production of a wide variety of polyurethanes (Chap. 21). Diphenylmethane-4,4′-diisocyanate (MDI), with a somewhat lower volatility than TDI, is also used in some coating, elastomer, and polyurethane fiber applications.

19.8.6. p-Xylene to Dimethyl Terephthalate (Dimethyl Benzene-1,4-dicarboxyate (DMT))

Catalytic oxidation of p-xylene with air is the chief commercial method used to produce terephthalic acid. A solution of p-xylene in acetic acid, together with manganese or cobalt derivative and heavy metal bromides, which serve as cocatalysts, is fed to a continuous reactor, vigorously stirred, and heated to 200°C while under about 25 atm pressure. Air is continuously fed into the reactor at the same time as a small stream of partially reacted solution is removed (Eq. 19.66).

$$C_6H_4p(CH_3)_2 \; + \; 7/2 \; O_2 \; \rightarrow \; C_6H_4p(COOH)_2 \; + \; 2 \; H_2O \qquad 19.66$$

The pressure is released on the exit stream, which simultaneously flashes off much of the excess p-xylene and the acetic acid, and cools the residual solution causing terephthalic acid to crystallize out. Residual acetic acid and p-xylene are removed from the crystals by centrifugation. The crude product acid is then slurried in hot water first for washing, and then for hydrogenation to decolorize any residual traces of colored impurities. It is then recrystallized and dried to give "fiber-grade" material (m.p. >300 °C) in yields of about 90%.

p-Xylene is also oxidized to terephthalic acid by systems, which replace the bromide component of the catalyst system by acetaldehyde or methyl ethyl ketone. These systems function at lower temperatures and pressures, presumably via peroxide-derived radicals.

Terephthalic acid is usually converted directly to the dimethyl ester for polymer synthesis. This is prepared by superficially conventional technology using methanol and a mineral acid under moderate conditions (Eq. 19.67).

$$C_6H_4p(CO_2H)_2 \; + \; 2 \; CH_3OH \; \rightarrow \; C_6H_4p(CO_2CH_3)_2 \; + \; 2 \; H_2O \quad 19.67$$

Proportions and exact conditions required for optimum conversions and yields of diester are proprietary. Recovery of the dimethyl terephthalate product from unreacted starting materials, etc., is by distillation using a series of four or five columns. Polymer-grade material (m.p. 142°C; b.p. 288°C) is obtained by distillation under reduced pressure from the top of the last fractionating column.

Almost all of the dimethyl terephthalate (DMT) output is consumed for the production of poly(ethylene terephthalate), PET, by transesterification with ethylene glycol (ethane-1,2-diol), mostly for fiber, film, and beverage packaging applications.

19.8.7. m-Xylene to Isophthalic Acid

Isophthalic acid is produced from m-xylene by processes analogous to those used for terephthalic acid except that five to six times higher pressures are required. The product (m.p. 342°C) is chiefly used for the preparation of the high thermal stability aramid polymer, Nomex (Dupont), used for electrical insulation in thermally aggressive environments.

19.8.8. *o*-Xylene to Phthalic Anhydride

Air is sufficient to oxidize the methyl groups of *o*-xylene, under the right conditions, like it is with *p*-, or *m*-xylene just described. However, here the similarity ends since commercial *o*-xylene oxidation is a vapor phase process [27]. *ortho*-Xylene vapor, mixed with a large excess of air to ensure operation outside the explosive range, is fed to a reactor containing a supported vanadium pentoxide catalyst and heated to about 550°C. Using about a 0.1-second contact time under these conditions produces exit gases composed of phthalic anhydride, water, and carbon dioxide (Eq. 19.68).

$$C_6H_4O(CH_3)_2 \ + \ 3 \ O_2 \ (\text{from air}) \ \rightarrow \ C_6H_4(CO)_2O \ + \ 3 \ H_2O \qquad 19.68$$

Cooling the exit gases causes condensation and crystallization of phthalic anhydride, which is collected and recrystallized to yield >99.5% pure product (about 75% yield). Phthalic anhydride is also made by catalytic oxidation of naphthalene with air.

About one-half of the phthalic anhydride production is consumed for the preparation of plasticizers, mostly for the various flexible grades of poly(vinyl chloride). The remainder is roughly split between alkyd resin preparation used for many types of surface coatings, and for polyester resin composites with fiberglass reinforcement, the so-called "fiberglass" resins used in boats and other sporting equipment as well as for corrosion-resistant vessels and ducts used in chemical processing, some automotive parts, and as a convenient means of field repair of many of these items.

19.9. ENVIRONMENTAL CONCERNS

The principal objectives of a modern petrochemical complex are good to excellent yields to minimize waste production and reuse or disposal costs, and good containment throughout to minimize losses of raw materials and products to air, water, or soil. Many newer facilities are now able to operate with only make-up water requirements and have no aqueous waste streams (see [28]). Aqueous streams containing volatile organics may be sparged with air to remove these, and the organic vapor–air stream produced used for feed air to a boiler to eliminate air pollution from this source. Bacterial treatment of the residual waste water stream under anaerobic and/or aerobic conditions can do much to consume and remove the residual impurities [29]. By-product organics that are not amenable to recycle or combustion for energy recovery on site may be cleanly consumed for this purpose in cement kilns [30].

Let us use as an example one recent environmental case history relevant to petrochemicals production. Nitrous oxide (N_2O) is implicated as a contributor to stratospheric ozone damage, and is also a potent greenhouse gas (Table 2.7). Recent work indicated that about 10% of the nitrous oxide contributions to the atmosphere was from the world's adipic acid plants, slightly more than the fraction contributed by biomass burning [31, 32]. It has been estimated that about 1 mol of N_2O is produced per mole of adipic acid, or about 0.3 kg of N_2O per kg of adipic acid.

Nitrous oxide control options at various stages of development included thermal catalytic reduction of N_2O in the presence of methane, conversion of N_2O to recoverable NO and use of this to prepare nitric acid, and catalytic dissociation of N_2O to nitrogen and oxygen (Eq. 19.69–19.73) [33, 34].

$$CH_4 + 4\,N_2O \rightarrow 4\,N_2 + CO_2 + 2\,H_2O \qquad 19.69$$

$$CH_4 + 4\,NO \rightarrow 4\,N_2 + CO_2 + 2\,H_2O \qquad 19.70$$

$$N_2O + M \rightarrow N_2 + [O] + M \qquad 19.71$$

$$N_2O + [O] \rightarrow 2\,NO \text{ (to use for nitric acid, Chap.11)} \qquad 19.72$$

$$N_2O + [O] \rightarrow N_2 + O_2 \qquad 19.73$$

It could also be recovered as nitrous oxide for sale, but this has not been economically attractive. The major adipic acid producers, worldwide, have agreed to implement N_2O abatement measures using one or the other of these options by 1998. Feasibility studies of a possible longer range solution have been conducted of an alternate enzymatic route that converts glucose to muconic acid, which could then be catalytically reduced to adipic acid, but this has not as yet been proven on a commercial scale (Eq. 19.74 [35]).

$$C_6H_{12}O_6 \xrightarrow{\text{enzyme}} \underset{cis,\,cis\text{-muconic acid}}{HO_2C\text{–}CH{=}CH\text{–}CH{=}CH\text{–}CO_2H} \xrightarrow{H_2,Pt}$$

$$\underset{\text{adipic acid}}{HO_2C(CH_2)_4CO_2H} \qquad 19.74$$

REVIEW QUESTIONS

1. (a) What would be the mass (kg) of carbon theoretically possible from the processing of $1.00 \times 10^4\,m^3$ (standard conditions) of natural gas (standard conditions are 15.6°C, 1 atm pressure)?
   (b) Compare the percent yields reported for channel black and thermal black in this chapter with the percent yields calculated using the mass ratio determined in part (a).

2. (a) What mass of isopropanol (100% basis) would be expected from the passage of 1,000 kg of propylene through the two-step process, ester formation and hydrolysis, assuming a 52% conversion and 71% selectivity (industrial yield)?
   (b) Assuming the same conversion and yield as in part (a), what mass of propylene would be required to produce 1,000 kg of isopropanol (100% basis)?
   (c) What happens to the unreacted propylene?
   (d) Explain some of the possible causes of the relatively low selectivity.

3. (a) By using plausible reactions of ethyl radicals with ethylene and related reactions, sketch reasonable schemes leading to the various possible C_4 hydrocarbon products from butane, which applies the basic mechanisms to ethylene outlined in Section 19.5.

(b) Sketch plausible propagation and termination reactions for the products observed from the cracking of propane.

4. (a) What pentanol isomer(s) would be possible from the oxo reaction using isobutylene $(CH_2 = C(CH_3)_2)$ as the feed olefin? Outline the equations for the steps involved.

(b) Describe and explain plausible methods, which could be used to isolate and purify the pentanol product(s).

5. (a) Calculate the research (academic) yield and the conversion and selectivity (industrial yield) on ethylene from the following data collected from an autoclave (pressure vessel) experiment on the direct hydration of ethylene.

	Feed to reactor	Recovered from reactor	
Ethylene	280.0 g		268.3 g
Water	300.0 g		292.6 g
Acid catalyst	trace	Ethanol	19.17 g

(b) Based on the research yield, would you authorize construction of a plant to use this process? Explain.

(c) Consider how knowledge of the selectivity of the process might affect your decision given in part 2 (b) and explain your answer.

(d) Outline positive and negative selection features of the direct hydration ethylene process in comparison to the esterification (indirect) route (Section 19.10.5).

6. (a) Calculate the theoretical masses of propylene, chlorine, and calcium hydroxide required to produce 1,000 kg of propylene oxide via the chlorohydrin process, and the mass of calcium chloride (100% basis) that would be coproduced.

(b) What actual masses of starting materials would be required and mass of 12% by weight calcium chloride in water that would be produced for each 1,000 kg of propylene oxide when the yield (selectivity) on each of the starting materials is 75%?

(c) What would be the theoretical, and actual (same selectivity), base requirement for 1,000 kg of propylene oxide if the process was switched to use sodium hydroxide instead of calcium hydroxide as the base? What mass of sodium chloride (100% basis) would also result?

(d) Using the theoretical raw material requirements, at what price would sodium hydroxide have to be available to equal the base cost when slaked lime $(Ca(OH)_2)$ is purchased at \$66 per metric tonnes (1,000 kg).

(e) What environmental factor(s) might influence the selection of the base, apart from equivalent raw material cost?

7. Compare the contact times (in seconds) and the respective space velocities for each of the three following process conditions.

Ethylene oxidation: 1.0-second contact time at 280°C, 1.0 atm
Ethylene hydration: $1800 \, hr^{-1}$ space velocity at 300°C, 70 atm
Propylene ammonoxidation, 20-second contact time at 400°C, 1 atm.

8. (a) Use either a web search engine or a computer-based science database to compile a brief list of the antiknock features and the documented environmental problems of methyl *t*-butyl ether (MTBE) as a gasoline additive.

    (b) Explain any two possible technological fixes, which might reduce or eliminate the environmental problems of this application of MTBE.

9. (a) To avoid the capital cost of an oxychlorination unit, and assuming perfect (100%) conversions and yields, what fraction of the chlorine consumed in the initial addition to ethylene is ultimately present in the vinyl chloride product?

    (b) Explain the details of an alternate method to oxychlorination for conversion of hydrogen chloride back to chlorine.

    (c) How might the hydrogen chloride be used to produce a valuable fertilizer ingredient?

10. (a) Demonstrate by calculation the theoretical mass of *t*-butylhydroperoxide, ethylbenzene hydroperoxide, and cumene hydroperoxide required to oxidize 1 tonne (1,000 kg) of propylene to the epoxide.

    (b) Explain briefly any two other considerations that would be important to consider in the decision regarding the most appropriate oxidant to use.

FURTHER READING

A. Chauvel and G. Lefebvre, "Petrochemical Processes", Vols. 1 and 2, Gulf Publ. Co., Houston, TX, 1989.

W.H. Hedley, Potential Pollutants From Petrochemical Processes, Corporate Author: Monsanto Research Corporation, Technomic Publ. Co., Westport, Conn., 1975.

S. Matar and L.F. Hatch, Chemistry of Petrochemical Processes, Gulf Professional Publ., Boston, 2001.

S. Matar, M.J. Mirbach, and H.A. Tayim, Catalysis in Petrochemical Processes, Kluwer Academic Publishers, Dordrecht & Boston, 1989.

Organisation for Economic Co-operation and Development, "Petrochemical Industry: Energy Aspects of Structural Change", OECD, Paris, 1985.

P.H. Spitz, "Petrochemicals: The Rise of An Industry", Wiley, New York, 1988.

REFERENCES

1. G.T. Austin, "Shreve's Chemical Process Industries", 5th ed., p. 75. McGraw Hill, New York, 1984.

2. Facts and Figures for the Chemical Industry, *Chem. Eng. News*, **74**(26), 38–80, June 24 (1996).

3. W.L. Faith, D.B. Keyes, and R.L. Clark, "Industrial Chemicals", 3rd ed., p. 205, Wiley, New York, 1966.

4. A.M. Thayer, Carbon Black Industry Rattled by Exit of Two Established Producers. *Chem. Eng. News*, **73**(29), 33–40, July 17 (1995).

5. F.A. Lowenheim and M.K. Moran, "Faith Keyes and Clark's Industrial Chemicals", 4th ed., p. 210, Wiley, New York, 1975.

6. M.Srivastava, I.D. Singh, and H. Singh,...Petroleum Based Feedstocks for Carbon Black Production, *Petr. Sci. Technol.* **17**(1–2), 67–80 (1999).

7. G.D. Ulrich, Flame Synthesis of Fine Particles. *Chem. Eng. News*, **62**(32), 22–29, Aug. 6 (1984).

8. M.S. Reich, Demand for Carbon Fibers Rebounds, *Chem. Eng. News*, **72**(48), 23–24, Nov. 28 (1994).

9. P. Beardmore, J.J. Harwood, K.R. Kinsman *et al.*, Fiber-Reinforced Composites: Engineered Structural Materials, *Science*, **208**, 833–840 (1990).

10. P.K. Bachmann and R. Messier, Emerging Technology of Diamond Thin Films. *Chem. Eng. News*, **67**(20), 24–39, May 15 (1989).

11. J. Andryiszyn,...A New Competitive Benchmark for Ethylene Production, *Can. Chem. News*, **52**(9), 27–30, Oct. (2000).

12. A.A. Lemonidou, I.A. Vasalos, E.J. Hirschberg *et al.*, Catalyst Evaluation and Kinetic Study for Ethylene Production. *Ind. Eng. Chem.* Res. **28**, 524–530 (1989).

13. P.K. Frolich and P.J. Wiezevich, Cracking and Polymerization of Low Molecular Weight Hydrocarbons. *Ind. Eng. Chem.* **27**(9), 1055–1062 (1935).

14. D. Netzer and O.J. Ghalayini, Improve Benzene Production from Refinery Sources, *Hydrocarbon Process.* **81**(4), 71, Apr. (2002).

15. A. Kareem, S. Chand, and I.M. Mishra, Disproportionation of Toluene to Produce Benzene and *p*-xylene: A review, *J. Sci. Ind. Res. India*, **60**(4), 319–327, Apr. (2001).

16. T. Kinnunen and K. Laasonen,...Catalytic Carbonylation of Methanol...*J. Molec. Struc. Theochem.* **540**, 91–100, May 4 (2001).

17. A. Faliks, R.A. Yetter, C.A. Floudas *et al.*, Optimal...Catalytic Methanol Conversion to Formaldehyde, *J. Phys. Chem.* A **105**(10), 2099–2105, Mar. 15 (2001).

18. S. Linic, J. Jankowiak and M.A. Barteau,...Bimetallic Ethylene Epoxidation, *J. Catal.* **224**(2), 489–493, Jun. 10 (2004).

19. J.A. Orejas, Model Evaluation for...Direct Chlorination of Ethylene..., *Chem. Eng. Sci.* **56**(2), 513–522, Jan. (2001).

20. M.K. Naqvi, A.K. Kulshreshtha, Vinyl-chloride Manufacture: Technology Trends and an Energy Economic-Perspective, *Polym-Plast Technol.* **34**(2), 213–226 (1995).

21. J. Krieger, Propane Route to Acrylonitrile Holds promise...*Chem. Eng. News*, **74**(39), 18–19, Sept. 23 (1996).

22. S.J. Ainsworth, Propylene Oxide Producers Look for Ways to Counter Sluggish Market. *Chem. Eng. News*, **70**(9), 9–12, Mar. 2 (1992).

23. Arco Expands PO and SM. *Chem. Brit.* **32**(8), 9, Aug. (1996).

24. T. Hayashi, L.-B. Han, S. Tsubota *et al.*,...Propylene Oxide by the Gas-Phase Reaction of Propane and Propene Mixture with Oxygen, *Ind. Eng. Chem.* **34**, 2298–2304 (1995).

25. F.A. Lowenheim and M.K. Moran, "Faith, Keyes and Clark's Industrial Chemicals", 4th ed., p. 201, Wiley-Interscience, New York, 1975.

26. K.J. Saunders, "Organic Polymer Chemistry", 2nd ed., Chapman & Hall, London, 1988.

27. C.R. Dias, M.F. Portela, M. Galan Fereres *et al.*, Selective Oxidation of *o*-Xylene to Phthalic Anhydride, *Catal. Lett.* **43**(1–2), 117–121 (1997).

28. D. Dembicki and K. Tsang, Off the river. *Hazard. Mater. Manag.* **8**(2), 4–9, April–May (1996).

29. D. Dempster, Bacterial Bed Consumes Phenol. *Can. Chem. Proc.* **65**(1), 36–37, Feb. (1981).

30. D. Gossman, The Reuse of Petroleum and Petrochemical Waste in Cement Kilns. *Envir. Prog.* **11**(1), 1–6 Feb. (1992).

31. M.H. Thiemens and W.C. Trogler, Nylon Production: an Unknown Source of Atmospheric Nitrous Oxide. *Science*, **251**, 932–934 (1991).

32. W.R. Cofer III, J.S. Levine, E.L. Winstead *et al.*, New Estimates of Nitrous Oxide Emissions from Biomass Burning, *Nature (London)*, **349**, 689–691 (1991).

33. R.A. Reimer, C.S. Slaten, M. Seapan *et al.*, Abatement of N_2O Emissions Produced in the Adipic Acid Industry. *Envir. Prog.* **13**(2), 134–137, May (1994).

34. Y. Inamori, Y. Kimochi, R. Inamori *et al.*, Control of...N_2O Emissions from Several Industrial Sources...*J. Chem. Eng. Jpn.* **36**(4), 449–457, April (2003).

35. J.W. Frost and K.M. Draths, Sweetening Chemical Manufacture. *Chem. Brit.* **31**(3), 206–210, Mar. (1995).

20

CONDENSATION (STEP-GROWTH) POLYMER THEORY

Mr. McGuire: I want to say just one word to
you...just one word.
Benjamin Braddock: Yes, sir.
Mr. McGuire:...Plastics.
        —The Graduate, Embassy Pictures, 1967

...each primordial Bean
Knew cellulose by heart: Nature alone of Collagen
And Apatite compounded Bone.
                —John Updike, 1968

20.1. BACKGROUND

Natural polymers with a wide variety of fascinating properties have been around for a very long time. Materials such as wood, cotton, hides, wool, muscle, collagen, and starch, all comprise examples of natural polymers, most of which have been used by humans since before historical records were kept, for heat, shelter, tools, food, and other purposes. For all of these applications the natural materials were used in the chemical form in which they originally occurred. It was not until the 19th century when our understanding of the nature of very basic chemical transformations began to develop that technologists began to experiment with the modification of natural polymers to yield products with more versatile properties. By these means the applications possible with natural polymeric materials were gradually expanded beyond those possible with the native materials.

Public awareness of natural rubber began as a scientific curiosity, when it was discovered by Priestley that it could be used to "rub out" pencil marks. This novel capability gave the material its name, but still gave rubber little utility because of its dimensional instability and poor durability. The discovery of vulcanization by Charles Goodyear in 1839 changed all this. Vulcanization, or the heating of natural rubber mixed with small amounts of sulfur, permitted the stabilization of formed rubber shapes and rapidly expanded the uses of this modified natural polymer.

Chemically modified cellulose in the form of cellulose nitrate or nitrocellulose was made and tested for commercial applications in Britain in the 1855–1860 period without much success. The discovery by Hyatt, in 1863, that cellulose nitrate could be plasticized with camphor to give moldability to the blend, made this material much more useful. By 1870, celluloid (plasticized cellulose nitrate) was being produced into a variety of commercial products such as billiard balls, decorative boxes, and combs. Nitrocellulose was also soluble in organic solvents, unlike cellulose, and so could be applied to surfaces in solution to form a coating, as in airplane dopes and automobile lacquers. It could also be solution spun into fibers (synthetic silk) and formed into photographic film, or used as a laminating layer in early auto safety glass. It was also used as an explosive. The hazard introduced to many of these uses of nitrocellulose by its extremely flammable nature resulted in an interest to discover other cellulose derivatives that could still be easily formed, like nitrocellulose, but without its extreme fire hazard.

Cellulose acetate was first prepared in 1865, and was a suitable candidate to replace nitrocellulose. Cellulose triacetate, however, was less soluble in common solvents than nitrocellulose and also was difficult to dye. Lower levels of acetylation gave more tractable, fiber-formable products, which could be more easily dyed, and could be formed into film suitable for photographic use. These were the so-called acetate rayon fibers and plastics, which were also considerably safer to use than nitrocellulose.

Later still, methods were discovered by which cellulose could be dissolved unchanged, and then reprecipitated as fiber or film by subsequent chemical treatment. This process yields viscose rayon fiber and cellophane film from the reconstituted cellulose and is still in use. These are much more useful products than were available from the early acetate rayon.

We learned much from nature with these early attempts to produce useful polymer products based on modified, or reconstituted ("semisynthetic") natural polymers, and many of these processes are still in use today. The first of the purely synthetic commercial polymers came with the small-scale introduction of Bakelite in 1907. This phenol–formaldehyde resin product was developed by Leon Baekeland. It rapidly became a commercial reality with the formation of The General Bakelite Company by Baekeland, and construction of a larger plant at Perth Amboy, New Jersey, in 1910. At about this time styrene was being combined with dienes in the early commercialization of processes to produce synthetic rubber. Polystyrene itself was not a commercial product in Germany until 1930 and in the U.S.A. in 1937. The only other purely synthetic polymers that made a commercial appearance during this early development period were polyvinyl chloride and polyvinyl acetate, both in the early 1920s.

While the relatively young polymer industry was learning much from the modification of natural polymers, production figures were small. As the purely synthetic polymers came on stream, the range of valuable applications for plastic materials rapidly increased. Today, the annual mass of polymeric materials produced is substantially greater than the scale of aluminum production, and rivals pig iron production, both in the U.S.A. and in the world. The lower density of organic polymeric materials means that on a volume-basis production, these products already exceed the combined *volume* of production of these two important fabrication metals. By substituting the low-density, high-strength plastics for metals, corrosion resistance, and substantial weight savings are being realized in the construction of automobiles, buses, aircraft, etc. This, in turn, is resulting in a substantial saving in the energy required to transport people and goods. The modern paints and coatings industry also relies heavily on some sectors of polymer chemistry. A further relevant factor to those who study polymer chemistry could be that it is estimated that one out of every two chemists is employed in some aspect of the polymer industry.

20.2. BASIC POLYMER CONCEPTS

What comprises a polymer? A general definition, which can include natural, modified natural (semisynthetic), and purely synthetic polymers of all types is that a polymer is a large molecule built up of small structural units combined in any conceivable pattern. Staudinger, a major contributor to the early development of polymer theory, set as an arbitrary guideline that a "polymer" was a molecule with a molecular weight of more than 10,000, or that consisted of a total of more than 1,000 carbon atoms. While there are also a number of important polymers with an inorganic backbone, such as silica, the silicones, and phosphonitrilic compounds, where the second criterion would not apply, they would still qualify under the first.

The repeating unit of a polymer, the "-mer" or monomer is the basic building block required for polymer preparation. To be capable of polymerizing it must have two or more "bonding units" (i.e., it must have a functionality or a potential functionality, of two or more).

$$8 \times \cdot A \cdot \rightarrow A \cdot A \cdot A \cdot A \cdot A \cdot A \cdot A \cdot A$$

The "degree of polymerization" is the number of monomer units that the polymer contains, 8 in the example above. This is sometimes expressed as the DP, X_n, or simply X, for short. If the monomer used has a functionality of 2 we normally obtain polymers of long continuous chains, but may get occasional short branches from impurities in the monomer or irregularities in the polymerization process (Eq. 20.1):

20.1

Polymerization of a monomer with a functionality of 3 gives more possible structures. Short or long branches will be numerous, as will be branching branches and links between adjacent chains to form a cross-linked, three-dimensional polymer (Eq. 20.2).

20.2

Functionalities of greater than 3 may be used for some purposes, which will further increase the complexity of the product. Thus, the functionality of the monomer has a significant influence on polymer structure and also on the resultant properties of articles made from that polymer.

Another factor which differentiates polymers arises from the number of different monomers used in their preparation. The functionality discussion above implied the use of a single monomer in the polymer synthesis, which would give a "homopolymer" product, or simply "polymer" for short. A polymer composed of two types of monomers, A and B, would be called a "copolymer." If A and B are only bifunctional, the copolymer possibilities are as follows:

(a) Random copolymer, ~AAABABBAABABBB~, in which the monomer elements are present in a linear polymer in a statistically random order

(b) Alternating copolymer, ~ABABABABABAB~, in which the monomer elements occur in a regular alternating sequence

(c) Block copolymer, ~AAAABBBBBBAAAA~, in which predictable blocks of first one monomer and then the other occur regularly in the polymer chain

(d) Graft copolymer,

```
                              B
                              B
                              B
                              B
                              B
      ~AAAAAAAAAAAAAAAAAAAAAAAAAAAAAAA~
               B
               B
               B
               B
               B
               B
               B
```

in which blocks of one monomer are grafted, or bonded, to a backbone of polymer formed from the other monomer.

There are also the possibilities of using three monomers to produce a terpolymer, and four (or more), to produce a tetrapolymer and so on. Of course if the monomers used to prepare a copolymer have functionalities greater than 2, then the structural complexity of the various possible copolymers obtained also rises.

20.3. POLYMERIZATION PROCESSES

Monomers can be joined by means of two principal methods to form polymers, and these methods are used as the broad basis for classification of synthetic polymers. The first of these, condensation, or step-growth polymerization, involves the use of functional group reactions such as esterification or amide formation to form polymers. When each of the molecules involved has only one functional group then the reaction between a carboxylic acid and an alcohol gives an ester (Eq. 20.3). In this equilibrium reaction water removal will help drive the reaction to the right.

$$CH_3CO_2H + C_2H_5OH \overset{H^+}{\rightleftharpoons} CH_3CO_2C_2H_5 + H_2O \qquad 20.3$$

A functionality of 2 is required in each of the monomer units in order to obtain a linear condensation polymer from esterification reactions. The initial dimeric product still has residual, potentially reactive functionalities (Eq. 20.4).

$$\underset{\text{terephthalic acid}}{HO_2C\text{-}C_6H_4\text{-}CO_2H} + \underset{\text{1,2-ethanediol}}{HOCH_2CH_2OH} \rightleftharpoons \qquad 20.4$$

$$\underset{\text{ethylene terephthalate}}{HO_2C\text{-}C_6H_4\text{-}CO_2CH_2CH_2OH} + H_2O$$

Further reaction and water removal from this system will ultimately give polyethylene terephthalate, well known commercially in fiber form as Dacron, or Terylene, and as a film by the Mylar trade name (Eq. 20.5).

$$n \; HO_2C\text{-}C_6H_4\text{-}CO_2CH_2CH_2OH \rightleftharpoons \qquad 20.5$$

$$\underset{\text{polyethylene terephthalate}}{HO \left[CO\text{-}C_6H_4\text{-}CO_2CH_2CH_2O \right]_n H} + (n-1) \; H_2O$$

Features that generally serve to recognize condensation polymerizations are the loss of a small molecule such as water, methanol, or hydrogen chloride during the polymerization process, and a relatively slow progression toward high molecular weight material, which is aided by the removal of the small molecule from the system.

In contrast with the usually slow progress of condensation polymerization the second major classification, addition, or vinyl-type polymerizations, usually proceed very rapidly, so rapidly that they are referred to as chain reaction polymerizations. This method of producing synthetic polymers uses the potential dual functionality present in a carbon–carbon double bond. The process is initiated by the use of radical or charged initiator species to form new sigma bonds from the carbon–carbon double bonds of the monomer, to link the monomer units (Eq. 20.6).

n CH_2=CHX → $+CH_2-CHX+_n$

e.g., for X = H: n CH_2=CH_2 → $+CH_2-CH_2+_n$,
$$\text{polyethylene, or PE} \qquad 20.6$$

for X = Cl: n CH_2=CH → $+CH_2-CH_2+_n$,
$\qquad\qquad\qquad\qquad |\qquad\qquad\qquad\qquad |$
$\qquad\qquad\qquad\qquad$ Cl $\qquad\qquad\qquad\qquad$ Cl
$$\text{polyvinyl chloride, or PVC}$$

for X = phenyl: n CH_2=CH → $+CH_2-CH+_n$,
$\qquad\qquad\qquad\qquad\quad |\qquad\qquad\qquad\qquad |$
$\qquad\qquad\qquad\qquad\quad$ Ph $\qquad\qquad\qquad\qquad$ Ph
$$\text{polystyrene, or PS}$$

Addition polymerization is usually such a rapid process that only monomer and final polymer chains are present. Very little of the active material in the system is oligomeric, that is, consisting of only a small number of linked monomer units en route to polymer, at any one point in time. Also, in addition polymerization the whole monomer molecule adds to form polymer. No small molecule is lost in the process. Further details of the polymerization processes involved and the structural properties of the products obtained from addition polymerization are discussed in Chaps. 22 and 23. This chapter will focus on the background theory of synthetic polymers, and condensation, or step-growth, polymerization.

20.4. POLYMER MOLECULAR WEIGHTS

With either type of synthetic polymerization we are dealing with a series of completely random events. As a result the molecular weights of the polymer products that are obtained will be within a range, rather than an exact, single molecular weight. Synthetic polymer samples are therefore described as being polydisperse with regard to molecular weights; any particular sample will consist of a composite of chains of many different molecular weights. Some natural polymers are made in a template fashion and can have a single, exact (and characteristic) molecular weight as, for example, some of the proteins. These polymers are described as being monodisperse, in contrast to the usual spread of molecular weights of synthetic polymers.

In work with the molecular weights of the polydisperse synthetic polymers, averages are used a great deal. The number average molecular weight is one of the important averages in common use. Number average molecular weight M_n, $\overline{M}_n$, or $<M_n>$, is defined as:

$$M_n = \frac{w}{\displaystyle\sum_{i=1}^{\infty}} N_i = \frac{\displaystyle\sum_{i=1}^{\infty} N_i M_i}{\displaystyle\sum_{i=1}^{\infty}} N_i \qquad 20.7$$

where w = mass of polymer sample measured
$\quad$ N_i = the number of molecules (moles) of molecular weight i in the sample
$\quad$ M_i = the molecular weight.

Since the numerator of Eq. 20.7 represents the mass of the polymer sample measured, and the denominator the total number of polymer molecules present, the result of this calculation is simply the average molecular weight per polymer molecule in the sample.

The number average molecular weight of a polymer sample is experimentally determined by any method, which determines colligative properties of polymer solutions. Any measurements such as osmotic pressure, boiling point elevation, or freezing point depression, where the result is affected by the *numbers* of molecules present, can be used to give a number average molecular weight result. Experimental details are available [1].

The number average molecular weight and the molecular weight distribution of a synthetic polymer markedly affect the processing and the final properties of articles made from it. To better visualize and understand how a number average result is obtained, let us take a hypothetical sample of polymer containing an equal weight of material of 1,000 molecular weight and 10,000 molecular weight (Fig. 20.1). To have the same weight, or mass, of each molecular weight polymer in the mixture, we must have

$$N_{1,000} = 10 \times N_{10,000}$$

To calculate M_n we need to know only the molar ratio of the two (or more) molecular weight fractions, not the actual numbers of molecules. If we let x be the number of 10,000 molecular weight molecules, then

$$M_n = \frac{10x(1,000 \, \text{g/mol}) + x(10,000 \, \text{g/mol})}{11x} = 1,818 \, \text{g/mol}$$

The number average molecular weight of this mixture is 1,818 g/mol, which rounds off to 1,820 g/mol.

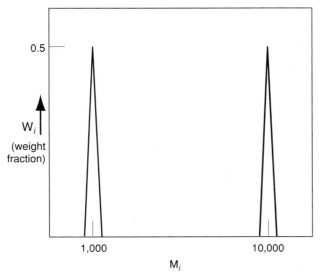

FIGURE 20.1 A representation of a hypothetical polymer sample composed of equal weights of 1,000 and 10,000 molecular weight (M_i) material.

The other very generally useful molecular weight measure in polymer work is the weight average molecular weight M_w, $\overline{M}_w$ or $<M_w>$. It is defined as

$$M_w = \frac{\sum\limits_{i=1}^{\infty} N_i M_i^2}{\sum\limits_{i=1}^{\infty} N_i M_i} \qquad 20.8$$

Weight average molecular weight may be experimentally determined by any method in which molecular size or molar mass is the parameter being measured (e.g., light scattering or ultracentrifugation).

We can calculate weight average molecular weight by methods similar to number average molecular weight. Using the same hypothetical polymer sample used for the calculation of number average molecular weight gives us

$$M_w = \frac{10x(1,000\,\text{g/mol})^2 + x(10,000\,\text{g/mol})^2}{10x(1,000\,\text{g/mol}) + x(10,000\,\text{g/mol})} = 5,500\,\text{g/mol}$$

as the weight average molecular weight, or molar mass, for this sample. This markedly different value results from both the extra molecular weight term present in the numerator and denominator for the definition of M_w, and from the bimodal (not "normal") distribution of molecular weights in the hypothetical sample. However, it is true that, for any ordinary synthetic polymer sample, M_w will always be greater than M_n. The "most probable" molecular weight is defined as the molecular weight of the most numerous molecular weight fraction of a normal distribution of molecular weights in a synthetic polymer sample. (Figure 20.2) gives the relationship of the most probable molecular weight to M_n and M_w for a typical molecular weight distribution synthetic polymer. Other molecular weight averages, like M_z and M_{z+1}, are less commonly used. Each of these attaches one higher order of significance to molecular weight than M_w. Full details are available in standard texts [1].

Before we leave molecular weight discussion mention should be made of the heterogeneity index (HI), or polydispersity index. This is defined as the weight average molecular weight divided by the number average molecular weight:

$$HI = \frac{M_w}{M_n} \qquad 20.9$$

The result of this calculation gives us a coarse indication of the breadth of the molecular weight distribution in the polymer sample. A "normal" (statistical) polymer molecular weight distribution gives a value of about 2 for the HI. A monodisperse (single molecular weight) polymer would give the same weight and number average molecular weights and so would give a HI of 1.0. For the hypothetical example we have been considering,

$$HI = \frac{5500}{1818} = 3.03$$

This is evidently not a very "normal" distribution. Very wide bimodal or very broad molecular weight distributions can give heterogeneity indices of 5 or higher.

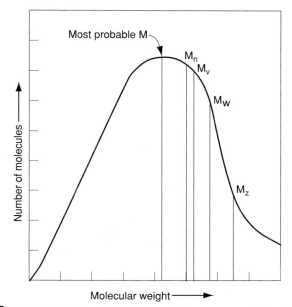

FIGURE 20.2 Relationships of the most probable (M), number average (M_n), viscosity average (M_v), weight average (M_w), and Z average (M_z) molecular weights used to define synthetic polymers.

20.5. POLYCONDENSATION POLYMERS

A characteristic common to the monomers of all condensation, or step-growth, polymers is that they possess two (or more than two) more-or-less reactive functional groups. The functional groups of a single monomer may be different, in which case the polymer may be produced by reaction of the single monomer with itself (Table 20.1, first two examples); or the functional groups required for condensation reactions may be on different monomers, in which case two monomers are required (Table 20.1, last three examples). Examples 1, 2, and 4 from this table are also typical of polycondensations in which a small molecule is split out, water in the first two cases and alcohol in

TABLE 20.1   Generalized Equations for the Formation of Some Common Types of Condensation Polymers

Polymer type	Condensation process
Polyesters	$n\ HO(CH_2)_xCO_2H \rightarrow \left[(CH_2)_xCO\text{-}O\right]_n + (n-1)\ H_2O$
Polyamides	$n\ H_2N(CH_2)_xCO_2H \rightarrow \left[NH(CH_2)_xCO\right]_n + (n-1)\ H_2O$
Polyurethanes	$n\ HO\text{-}R\text{-}OH + n\ OCNR'NCO \rightarrow \left[OROCO\text{-}NHR'NHCO\right]_n$
Polyacetals	$n\ HO\text{-}R\text{-}OH + n\ CH_2(OR')_2 \rightarrow \left[OROCH_2\right]_n + 2n\ R'OH$
Phenol–formaldehyde	$PhOH + HCHO \rightarrow$ resins of complex structure

Example 4. Polyurethane formation, in which the alcohol adds across the carbon nitrogen double bond of the isocyanate function ($O=C=N$-R) to form the urethane (-O-CO-NH-) link without loss of a small molecule, represents an exception to this general recognition feature of polycondensations. Bakelite is one of several possible products from phenol–formaldehyde resins, and these may or may not involve loss of a small molecule. This more complicated process will be discussed in further detail later.

20.6. POLYCONDENSATION MECHANISMS

For efficient polycondensation to high molecular weight polymers, one requires monomers with a functionality of 2 (or higher), high monomer purity, stoichiometric (equimolar) presence of the monomer functional groups, relatively long reaction times, at least moderate temperatures, and sometimes a suitable catalyst. Wallace Carothers (assignor of the original nylon patent U.S. 2,071,250 of Feb. 16, 1937 to the Du Pont Company) was the chemist who first established the importance of exact stoichiometry to the preparation of high molecular weight condensation polymers. Given the right conditions, reactions between the monomer functional groups will occur along the following lines:

1. H_2N-R-NH_2 + HOOC-R'-COOH $\rightarrow$

$$H_2N\text{-R-NH-CO-R'-COOH} + H_2O$$
$$\text{dimer}$$

2. dimer + monomer $\rightarrow$ trimer + H_2O
3. trimer + monomer $\rightarrow$ tetramer + H_2O
4. trimer + dimer $\rightarrow$ pentamer + H_2O
5. trimer + trimer $\rightarrow$ hexamer + H_2O
6. trimer + tetramer $\rightarrow$ heptamer + H_2O

All kinds of possibilities exist, and in practice all will occur (i.e., any two molecular species present in a polycondensation system can react). Detailed observation of many polycondensations has established that the monomer is completely consumed, or nearly so, very early in the reaction.

To be able to describe polycondensations kinetically, and to be able to predict rates, we have to consider a large number of different component reactions. How can we do this? It was Paul Flory (Nobel Prize for polymer chemistry, 1974) who first suggested that one make an overriding assumption that the reactivity of the functional groups of all the oligomers was the same. This assumption has been shown to be a reasonable one, by experiments, which compare the rates of condensation reactions of a homologous series of monomers. At trimer, and for higher oligomers, the rate constants remain unchanged regardless of molecular size. Verification was also obtained from the good agreement between experimental results and the kinetic theory derived from the use of this assumption.

To develop the kinetics, let us take an A-B type of condensation monomer, such as p-hydroxymethylbenzoic acid. This monomer, if pure,

$$HOCH_2\text{-}C_6H_4\text{-}COOH$$
A-B Type Monomer

makes the required 1:1 stoichiometry automatic. Let N_0 be the original number of functional groups present in the system, and N the total number of unreacted functional groups left after any given reaction time, t. Then the extent of reaction,

$$p = \frac{N_0 - N}{N_0}.$$

That is, p is the fraction of the functional groups left unreacted. This expression can be rearranged to the useful form:

$$N = N_0(1 - p) \qquad 20.10$$

which is referred to as the *Carothers equation*.

The number average degree of polymerization, X_n or $\overline{X}_n$, is given by

$$\overline{X}_n = \frac{N_0}{N}$$

Substituting for N from Eq. 20.10 into this expression gives

$$\overline{X}_n = \frac{1}{(1 - p)} \qquad 20.11$$

Equation 20.11 relates the number average degree of polymerization, and ultimately the molecular weight, to the extent of reaction, p.

As an example, suppose we wish to make a polymer with 50 monomer units in the average chain. This would not give a particularly high molecular weight. How far must one carry the polycondensation reaction to obtain this? Substituting, and solving for p:

$$50 = \frac{1}{(1 - p)},$$

$$p = 0.98, \text{ or } 98\% \text{ of complete reaction would be required.}$$

When one considers the usual extent of reaction for organic preparations in relation to this requirement it is evident that this is not going to be easy. In fact, quite forceful conditions are required. For commercially useful polymer, 99.9% reaction or better is required (Figs. 20.3 and 20.4). At least by using an A-B type of monomer, as in this example, good stoichiometry is assured. This removes one factor, which might detract from obtaining a high extent of reaction. However, if the condensation polymerization is between two different monomers (i.e., a system of the A-A, B-B type), then the importance of careful mass measurement at the start to ensure good stoichiometry becomes more evident.

20.7. POLYCONDENSATION KINETICS

Let us set up a general series of steps that describe the mechanism of a polyesterification, and use this to illustrate the progress of a step polymerization.

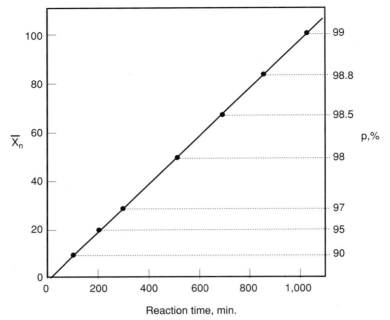

FIGURE 20.3 Relationship between $\overline{X}_n$ and extent of reaction, p, for various reaction times during a typical polycondensation.

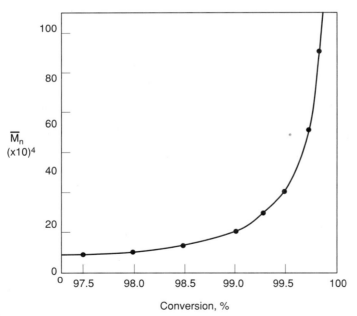

FIGURE 20.4 Asymptotic relationship between the percent conversion, as a step-growth polymerization proceeds, and the polymer molecular weight.

For a self-catalyzed polyesterification the initial step will be protonation of the carbonyl of one carboxyl by a proton from another, followed by the series of steps illustrated (Eqs. 20.12–20.14).

$$20.12$$

$$20.13$$

$$20.14$$

For this process the rate will be given by the following:

$$\text{rate} = \frac{-d[COOH]}{dt} = k[COOH]^2[OH] \qquad 20.15$$

To simplify this expression, and in keeping with the stoichiometry required for efficient polycondensation, we can assume the concentrations of the functional groups are the same, to give:

$$\frac{-dc}{dt} = kc^3 \qquad 20.16$$

where c is the initial concentration of each functional group. Integrating gives:

$$2\,kt = \frac{1}{c^2} - \frac{1}{c_0^2}, \text{ as long as } c = c_0 \text{ at zero time.}$$

Water is removed from the system, and may be neglected.

To be kinetically useful we need to put this equation entirely in terms of c_0. Replacing the N's in the Carothers equation by c's gives $c = c_0(1 - p)$. Substituting this value for c in the integrated expression:

$$2\,kt = \frac{1}{c_0^2(1 - p)^2} - \frac{1}{c_0^2} \text{ or } 2\,c_0^2 kt = \frac{1}{(1 - p)^2} - 1 \qquad 20.17$$

Equation 20.17 puts the kinetic expression entirely in terms of the initial concentration of monomer, c_o, and is in the form useful for following the rates of linear (only bifunctional monomers) self-catalyzed polycondensations. If $1/p^2$ is plotted against t one gets a straight line. Under self-catalyzed conditions, the reaction is third order, and the degree of polymerization (DP, or $\overline{X}_n$) is approximately proportional to $t^{1/2}$. This third-order dependency of the rate of self-catalyzed polycondensations also means that the rate of increase in the molecular weight (degree of polymerization) slows down quite rapidly as the reaction proceeds.

The rate reduction observed for self-catalyzed polycondensations may be remedied by using a strong acid catalyst. This also changes the kinetic treatment. The rate equation (Eq. 20.15) becomes

$$\frac{-d[COOH]}{dt} = k[COOH][OH][cat.] = k[COOH][OH][H^+] \qquad 20.18$$

Since the $[H^+]$ contributed by the mineral acid will remain constant, we can simplify this rate expression by substituting k', a new constant, for $k[H^+]$, which gives

$$\frac{-d[COOH]}{dt} = k'[COOH][OH]$$

The concentration of carboxyls in this system equals the concentration of hydroxyls so that we can substitute c for each of these terms, which gives

$$\frac{-d[COOH]}{dt} = k'c^2$$

which is easier to integrate. Integrating this expression then gives

$$c_0 k't = \frac{1}{(1-p)} - 1 \qquad 20.19$$

From this equation we can see that a plot of $1/p$ versus t for a strong acid-catalyzed polycondensation would be expected to give a straight line. The second-order kinetics obtained gives more rapid progression toward high molecular weight material under these conditions than possible under self-catalyzed conditions, particularly toward the end of the polymerization. This is because the hydrogen ion concentration in a mineral acid catalyzed polycondensation does not decrease as the polymerization proceeds, unlike in self-catalyzed polycondensations. These second-order kinetics have been found to hold to a DP of at least 90, which is equivalent to an M_n of about 10,000 (i.e., high polymer).

20.8. LINEAR POLYMERIZATIONS

Many useful condensation polymers employ two bifunctional monomers, one having two functional groups of one type (e.g., a di-alcohol or diol), and the other having two complementary functional groups (e.g., a diacid). This type of polycondensation is often referred to as an A-A, B-B type, and gives only linear polymers.

20.8.1. Quantitative Molecular Weight Control

To develop the theory that relates to molecular weight control we need to define a few new terms. Let

N_A = moles of A functional groups present
N_B = moles of B functional groups.

Then the term "imbalance ratio,"

$$r = \frac{N_A}{N_B}.$$

If $r = 1.000$, we have perfect balance. The imbalance ratio is always expressed as unity, or less than unity (i.e., $N_B \geq N_A$). Since each monomer molecule in a linear polycondensation possesses two functional groups, the total number of monomer molecules present,

$$N = \frac{N_A + N_B}{2}$$

Since $N_B = N_A/r$, the total number of monomer molecules present expressed in terms of just N_A, is

$$N = \frac{N_A\left(1 + \dfrac{1}{r}\right)}{2}$$

For the purposes required here, the extent of reaction, p, is defined as the fraction of A groups reacted in a given time.

The fraction of B groups reacted is therefore $= rp$.
The fraction of unreacted A groups $= (1 - p)$.
The fraction of unreacted B groups $= (1 - rp)$. (And, if $r = 1.000$, the fraction of unreacted B groups would be the same as the unreacted A groups.)
The total *number* of unreacted A groups $= N_A(1 - p)$, and the total number of unreacted B groups $= N_B(1 - rp)$.

To proceed from functional group considerations to polymer chains, the total number of polymer chain ends will be equal to the sum of the unreacted groups. Since each polymer chain has two end groups, the number of polymer chains becomes equal to

$$\frac{N_A(1 - p) + N_B(1 - rp)}{2}$$

The number average degree of polymerization is given by

$$\overline{X}_n = \frac{\text{(total no. of A-A and B-B molecules initially present)}}{\text{(total number of polymer molecules)}}$$

$$= \frac{\dfrac{N_A\left(1 + \dfrac{1}{r}\right)}{2}}{\dfrac{N_A(1 - p) + N_B(1 - rp)}{2}}$$

This complex fraction can be simplified to the form

$$\overline{X}_n = \frac{1 + r}{1 + r - 2rp} \qquad\qquad 20.20$$

Equation 20.20 is the generalization of the Carothers equation, which places the number average degree of polymerization in terms of the extent of reaction, p, and the imbalance ratio, r.

To obtain an idea of how Eq. 20.20 functions, let us consider two limiting cases. If the two bifunctional monomers are present in exactly stoichiometric amounts, $r = 1.000$, and Eq. 20.20 simplifies to

$$\overline{X}_n = \frac{2}{2 - 2p} = \frac{1}{(1 - p)}$$

That is, the number average degree of polymerization becomes dependent only on the extent of reaction, p, as in the first version of the Carothers' equation discussed.

If we consider the hypothetical situation of complete polymerization (hypothetical because in practice it could never be 100.00%), then p = 1.0000 and the generalized Carothers' equation becomes

$$\overline{X}_n = \frac{1 + r}{1 + r - 2r} = \frac{(1 + r)}{(1 - r)}$$

$$\text{For } r = 0.98, \overline{X}_n = \frac{1 + 0.98}{1 - 0.98} = 99$$

$$\text{For } r = 0.99, \overline{X}_n = \frac{1 + 0.99}{1 - 0.99} = 199$$

$$\text{For } r = 1.00, \overline{X}_n = \frac{1 + 1}{1 - 1} = \frac{2}{0} \text{ (i.e., an invalid solution)}$$

The invalidity of the last scenario occurs because in practice it is never possible to have 100.00% complete polymerization. However, the extent of reaction can approach very close to 1, at which time Eq. 20.20 gives a good approximation of the number average degree of polymerization to be expected (cf. Fig. 20.4).

20.8.2. Use of the Imbalance Ratio in Practice

The imbalance ratio, r, in conjunction with the generalized Carothers equation, Eq. 20.20, may be used to control either the chain length of a linear polycondensation, or the functional groups left on the ends of the polymer chains, or both product characteristics. The initial ratio of the two bifunctional monomers present could be adjusted to be nonstoichiometric to obtain the desired molecular weight. However, it is usually more convenient, and easier to obtain high accuracy in adjusting the ratio of the functional groups present at the start of the polymerization by adding a third, monofunctional component with the same functionality as one of the two bifunctional components, e.g.,

A-A + B-B + B' (added, with single functional group)

How can this be accommodated by the generalized Carothers' equation? First we need to redefine the imbalance ratio to take this change into account.

$$\text{imbalance ratio, } r = \frac{N_A}{(N_B + 2N_{B'})} \qquad\qquad 20.21$$

This redefinition can be explained by recognizing that one B' molecule will have the same quantitative effect on limiting the growth of a polymer chain as one excess B-B molecule. Either option will serve to stop chain growth. Only a small proportion of added monofunctional component is sufficient to have a significant effect on the DP, and ultimately on the molecular weight to be

TABLE 20.2 Predicted Number Average Degree of Polymerization (DP) for Different Stoichiometric Ratios, r, for an A-A, B-B Type Polycondensation[a]

Mole % excess		Stoichiometric ratio, r	DP at p = 0.980	DP at p = 0.990	DP at p = 1.000[c]
B-B	**B[,b]**				
0.20	0.10	0.9990	49	95	1999
0.50	0.25	0.9975	47	89	799
1.0	0.50	0.9950	45	80	399
2.0	1.0	0.9901	40	67	201

[a]Calculated using the generalized Carothers equation (Eqs. 20.20 and 20.21).
[b]For moles of A-A = moles of B-B, in the system polymerized.
[c]Limiting case.

expected for the product. The range of results given in Table 20.2 for extents of reaction 0.98, 0.99, and the hypothetical 1.00 indicate that to obtain commercially useful DPs in the 50–100 range, an extent of reaction of at least 0.990 is routinely required.

20.8.3. Estimation of Molecular Weight Control

Suppose we wish to prepare a nylon with an M_n of 16,000 using adipic acid $(HO_2C(CH_2)_4CO_2H)$ and hexamethylene diamine $(H_2N(CH_2)_6NH_2)$, which represents an A-A, B-B type of polycondensation. The common, or trade name of the product, poly(hexamethylene adipamide), $HO[OC(CH_2)_4\ CONH(CH_2)_6NH]_nH$, is nylon 6,6, or nylon 66. The first 6 refers to the number of carbon atoms between amide nitrogens for the amine component, and the second 6 refers to the number of carbon atoms separating nitrogens for the acid component. What monomer feed ratio would be required to achieve this M_n?

Polymerization in this system will stop when all the remaining unreacted groups on polymer chain ends are the same. The number average degree of polymerization is given by

$$\overline{X}_n = \frac{\text{molecular weight of polymer}}{\text{molecular weight of monomer unit}}$$

For A-A, B-B types of condensation polymers, the molecular weight of the monomer unit is defined as being one-half of the molecular weight of the repeating unit (the portion of the formula within square brackets).

Formula of repeating unit: $C_{12}H_{22}N_2O_2$
Molecular wt. of the repeating unit = 226.32
$\therefore M_0$ (mol. wt. of monomer unit) $= \dfrac{226.32}{2}$, or 113.16, and
$\overline{X}_n = \dfrac{16000}{113.16} = 141.39$

We need to determine the imbalance ratio, r, required to obtain an $\overline{X}_n$ of 141.39 after a reasonably attainable extent of reaction. Let's say, for this example, that p = 0.995. Then, substituting this information into the generalized Carothers equation (Eq. 20.20), we obtain

$$\overline{X}_n = 141.39 = \frac{1+r}{1+r-2r(0.995)}$$

Solving for r gives r = 0.99584, or rounded becomes 0.9960.

That is, we require that [COOH]/NH₂ = 0.9960, which would be met by a 4.0×10^{-3} mol deficiency of diacid per mole of diamine. Or the same value for r can be obtained by inverting the functional group ratio to: [NH₂]/[COOH] = 0.9960, which would be met by 4.0×10^{-3} mol deficiency of diamine per mole of diacid. Or, this value for r could be calculated from the monofunctional additive approach using: $r = N_A/(N_B + 2\,N_{B'})$ in which case it would require a 2.0×10^{-3} mol deficiency of monoamine per mole of diacid to obtain r = 0.9960. Or, the same value for r can be obtained by a 2.0×10^{-3} mol deficiency of mono-acid, e.g., benzoic acid, per mole of diamine in the system.

If we wish to prepare a single monomer polyester such as p-hydroxymethyl-benzoate (HOCH₂-C₆H₄-COₙ-OH), then both required functionalities are in a single molecule, and as long as the monomer is pure it assures correct stoichiometry. Molecular weight control may still be exercised by addition of a component with a single functionality. However, in the computation the molecular weight of the repeating unit of the polymer in square brackets is the same as the monomer used (less portions lost as the "small molecule"), so that now one does not divide the molecular weight of the repeating unit by two. Thus the formula of the repeating unit is $C_8H_6O_2$. Molecular weight of the repeating unit = 134.15 and M_0 (molar wt. of the monomer unit) = 134.15 (i.e., the same). The rest of the calculation method is the same as given for A-A, B-B type polymerizations.

20.9. NONLINEAR POLYCONDENSATIONS

The discussion concerning polycondensation polymers so far has been restricted to systems using only bifunctional monomers. By doing this we have seen that high purity, very close to equimolar (stoichiometric) proportions of functional groups, and nearly complete reactions are needed to obtain useful high molecular weight material. The linear products from these types of polymerizations may be more or less readily melted, or dissolved in a solvent for production of useful fibers, shapes, etc., from the crude polymer resins.

If one introduces a component into a polycondensation with a functionality of 3 or more the situation dramatically changes. Branches, branched branches, and crosslinks between polymer chains will cause a more rapid increase in molecular weight as the polymerization proceeds, and ultimately to higher final values than possible from linear polycondensations (see Section 20.2). The extent to which branches and crosslinks will occur will be proportional to the ratio of the polyfunctional to the bifunctional components present, and to the number of functionalities on the polyfunctional component. Usually 3, but up to 8, functionalities per monomer may be used for this component of polyfunctional polycondensations.

The theory for estimation of molecular weight and the "gel point," or the onset of significant crosslinking and setting or gelation during the course of a polyfunctional polycondensation, is well established [2], but is beyond the scope of this outline. The need for high-purity, stoichiometric proportions of reacting functional groups, and nearly complete reaction for useful linear polycondensation products also changes when a 3 or higher functionality component is introduced into the system. All these factors still have an effect on polyfunctional polycondensations, but the polyfunctionality introduces more resilience into the system so that these requirements are not nearly as stringent for useful products as they are for strictly linear polycondensations.

REVIEW QUESTIONS

1. (a) Draw the structure of the repeating unit obtained when the following monomers are polymerized:

$$HOOC\text{-}(CH_2)_4\text{-}COOH + H_2N\text{-}(CH_2)_6\text{-}NH_2$$

$$HOOC\text{-}C_6H_4\text{-}COOH + HO\text{-}CH_2\text{-}CH_2\text{-}OH$$

(b) Name the polymers using a systematic name, and a commercial name.

2. (a) Calculate the average degree of polymerization expected for an equimolar mixture of adipic acid and hexamethylene tetramine for each of the following extents of reaction: 0.500, 0.800, 0.950, 0.995, 0.998.

(b) What molar ratio of benzoic acid has to be added to either one of the monomers of the system of part (a) in order to limit the molecular weight of the product to 5,600, at 99.5% conversion?

(c) How else could the molecular weight of the product be controlled to 5,600 while still working at 99.5% conversion? Quantitatively, how should the polymerization be adjusted to achieve this?

3. (a) Calculate the extent of reaction necessary to obtain polymer of number average molecular weight 20,000, from stoichiometric amounts of hexamethylene diamine and sebacic acid (diacid).

(b) If a portion of the product from 3.(a) were hydrolyzed, what *weight* percentages of the two monomers would be obtained?

(c) What molar proportion of benzoic acid would have to be added to the polymerization of 3.(a) for extents of reaction of 0.998, *and* for extent of reaction 0.999, to limit the molecular weight ($\overline{M}_n$) in each case to 10,000?

4. (a) A polymer blend has been prepared from 5,000 and 200,000 molecular weight material in a weight ratio of 1:4. What would be the $\overline{M}_n$, $\overline{M}_w$, and the heterogeneity index of this blend?

(b) Answer the same questions as in part (a) for a 1:4 mol ratio of 5,000 and 200,000 molecular weight material.

(c) Comment on the normalcy of the calculated heterogeneity indexes (the polydispersities).

5. (a) A polymer blend from 40,000, 80,000, and 120,000 molecular weight material is prepared on a 1:2:1 weight ratio. What would be the $\overline{M}_n$, $\overline{M}_w$, and polydispersity of this material?
(b) How "normal" is the calculated polydispersity?
(c) Explain qualitatively the reasons for the differences from a normal polydispersity calculated for the systems of Review Question 4, as compared to this system.

FURTHER READING

F.W. Billmeyer, Jr., "Textbook of Polymer Science," 3rd ed. Wiley, New York, 1984.

C.E. Carraher and G. Swift, "Functional Condensation Polymers," Kluwer Academic/Plenum Publishers, New York, 2002.

Y. Furukawa, "Inventing Polymer Science: Staudinger, Carothers, and the Emergence of Macromolecular Chemistry," University of Pennsylvania Press, Philadelphia, 1998.

H.F. Mark and J. Kroschwitz, "Encyclopedia of Polymer Science and Engineering," Wiley, New York, 1985.

M.P. Stevens, "Polymer Chemistry: An Introduction," Oxford University Press, New York, 1999.

REFERENCES

1. R.B. Seymour and C.E. Carraher, Jr., "Polymer Chemistry: An Introduction," 3rd ed., Chap. 4. Dekker, New York, 1992.

2. G. Odian, "Principles of Polymerization," 3rd ed., pp.108–111. Wiley, New York, 1991.

21

COMMERCIAL POLYCONDENSATION (STEP-GROWTH) POLYMERS

Dragline silk, used by spiders... exhibits a combi-nation of strength and toughness unmatched by other high-performance synthetic fibers. [Kevlar]
—David A. Tirrell, 1996

A... fiber twice as strong as nylon, four times as strong as rayon, eight times as strong as cotton, and pound for pound five times as strong as steel... Kevlar...
—Eugene Magat, 1997

21.1. EARLY DEVELOPMENT

From almost zero volume in 1910, synthetic condensation polymers are currently produced on a very large scale. Some of the products are used for thermosetting applications, when the final shape of the article is decided at the time of forming the cross-links. Some are thermoplastic products from which articles are shaped by melt-forming, and fibers are formed by melt or solution spinning. Smaller producers are often specialists, producing a large volume of a single polymer type or a single class of polymer. Larger companies may have facilities to produce several polycondensation polymers while also processing vinyl-type monomers to high-polymer products (cf. Chaps. 22 and 23).

Table 21.1 gives production data for some of the commercially important condensation polymers [5, 6]. It is evident that the phenolics and the saturated polyesters and nylons destined for fiber production are all major players with American production levels for each of these of well over a million metric tonnes per year by 1994. Also evident in the American context is the strong rate of growth in the listed polymers.

■■   **TABLE 21.1   Production of Some Commercial Condensation Polymers in Thousand Metric Tonnes**

	W. Europe[a]	U.S.A.[b]			World[c]
	2003	1984	1994	2003	2000
Thermosetting resins					
Alkyds	370	–	–	–	–
Amino	2,630	–	–	–	–
Epoxy	398	184	271	262	–
Melamine	–		95	129	–
Phenolics	980	1,135	1,465	2,015	–
Polyesters (unsat'd)	490	559	663	1,954	–
Polyurethanes	2,672	–	–	–	–
Urea formaldehyde	–		543	1,005[d]	1,437[d]
Others	3,100	–	–	–	–
Total	10,640	2,517	3,404	5,668	
Thermoplastics					
Polyamides	1,328	–	1,671[e]	1,954[e]	1,700[f,g]
Polycarbonates	471	–	–	–	1,600[g]
Polyesters (PET)	3,802	–	1,750	1,374	8,390[g]
Other	–		–	–	1,050
Total	5,601	–	–	–	12,740[g]
(Polyethylenes)[g]	(13,492)[g]		(9,398)[g]	(15,717)[g]	

[a]Selected from *Plastics Europe* [1]. Given as consumption, rather than production.
[b]Selected from *Chemical and Engineering News* [2].
[c]Selected from Tullo [3] and Reisch [4].
[d]Includes melamine.
[e]Includes 1,243 thousand tonnes in nylon fiber production, for 1994 and 2003, respectively.
[f]Does not include products used in fiber production.
[g]Estimates.

Among the fibers, all except acrylics and rayon have shown significant growth in production in the 1994–2003 period (Table 21.2). Rayon production in the U.S.A. has declined about 70% since 1984. The American proportion of the world output of modified cellulosic and noncellulosic fibers production is about 3 and 13%, respectively. The preparative details of several of the more important condensation polymers are discussed in the following sections.

21.2. POLYESTER RESINS

21.2.1. Polyesters as Fibers

Polyesters in general have less intermolecular cohesion (less interchain non-bonding interactions) than polyamides so that poly(ethylene terephthalate) (PET) is the only polyester, which is commercially useful as a fiber. This polymer can be prepared by direct polyesterification of terephthalic acid with 1,2-ethane diol (ethylene glycol), usually with the help of a strong acid catalyst

■ **Table 21.2 Production Trends of Some Noncellulosic, Modified Cellulosic, and Cellulose Fibers in Thousands of Metric Tonnes**[a]

	U.S.A.			World			
	1984	1994	2003	1970	1993	2000	% Change[b]
Noncellulosic Fibers							
Acrylic		200	100	1,008	–	2,700	128
Nylon		1,243	1,115	2,016	–	3,600	78
Olefin		1,083	1,348	c	–	c	–
Polyester		1,750	1,374	1,680	–	18,600	1,007
Other[d]		–	–	84	–	2,700	3,100
Totals	3,389	4,287	3,938	4,788	17,100	31,200	
Modified cellulosic Fibers							
Rayon and acetate	267	225	75	–	2,630	–	
Cellulose, as cotton	2,830	4,290[e]	4,420[e]	3,612	–	2,400	<34>[b]

[a]Derived from *Chemical and Engineering News* [2], *Statistical Abstract of the United States* [7], and Short [8].
[b]From 1970 to 2000, For cellulose there was a 34% decrease
[c]Included in Other category.
[d]Includes aramid, elastane, polyethylene, poly(vinyl alcohol), poly(vinyl chloride), and poly(vinylidene chloride).
[e]Data given for 1994 and 2003 are from 1995 and 2002.

since this step is less favorable than direct polyamidation (Eqs. 19.3 and 19.4). However, thermally induced side reactions of monomer or oligomer can interfere with the direct polyesterification process under the high-temperature forcing conditions required. For this reason, and because it is a faster reaction, ester interchange is usually the method of choice (Eqs. 21.1 and 21.2).

First step:

$$CH_3O_2C\text{-}C_6H_4\text{-}CO_2CH_3 + 2\ HOCH_2CH_2OH \xrightarrow{105-200°C} \quad 21.1$$
$$\text{dimethyl terephthalate} \qquad \text{1,2-ethanediol}$$

$$HOCH_2CH_2O_2C\text{-}C_6H_4\text{-}CO_2CH_2CH_2OH + 2CH_3OH$$
$$\text{bis(2-hydroxyethyl)terephthalate}$$

Second step:

$$n\ HOCH_2CH_2O_2C\text{-}C_6H_4\text{-}CO_2CH_2CH_2OH \xrightarrow{ca.\ 260°C} \quad 21.2$$

$$(n-1)\ HOCH_2CH_2OH + HOCH_2CH_2O\text{+}CO\text{-}C_6H_4\text{-}CO_2CH_2CH_2\text{-}O\text{+}_n$$
$$\text{poly(ethylene terephthalate), or PET}$$

Also, when using the methyl ester the methanol formed is easier to remove than water, which facilitates the first step. Optimization of the prepolymerization step has been examined [9]. Removal of ethylene glycol (b.p. 198°C) to help drive the second stage of the polymerization is assisted by using reduced pressure or by sparging the hot mixture with an inert gas. There is no problem with stoichiometric imbalance since this is established in the second stage. The excess ethylene glycol is distilled from the polymerizing mixture. As a result the PET product will be obtained with entirely alcohol end groups.

Poly(ethylene terephthalate) may also be made by Schotten–Baumann condensation of terephthaloyl chloride with 1,2-ethanediol (Eq. 21.3).

$$n \; ClCO\text{-}C_6H_4\text{-}COCl \; + \; n \; HOCH_2CH_2OH \; \rightarrow \; PET \; + \; 2n \; HCl \qquad 21.3$$

Using the much more reactive acid chloride monomer allows polyester production at much lower temperatures than when using the ester, or free acid. Even under ambient conditions it is so vigorous that interfacial polymerization of diluted solutions of the monomers in an appropriate, immiscible solvent pair may be needed to control the rate. However, the high cost of the acid chloride as compared to terephthalic acid or the ester makes this route commercially unattractive. Also with this monomer pair the kinetic treatment and molecular weight distributions will differ from the general cases outlined in Chapter 20.

Commercial PET has a melting point of about 270°C. Thus, a fiber is produced from the bulk resin by melt spinning into a dry inert gas using a melt temperature of 280–300°C (Fig. 21.1). Fabrics woven from PET resin fibers include the well known Dacron, Fortrel, Crimplene, and Trevira trade names. Although fibers dominate PET processing, a substantial fraction is also blow

FIGURE 21.1   (a) Melt spinning of a fiber from a thermoplastic polymer. (b) The drawing process increases the degree of alignment of polymer chains into the long dimension of the fiber.

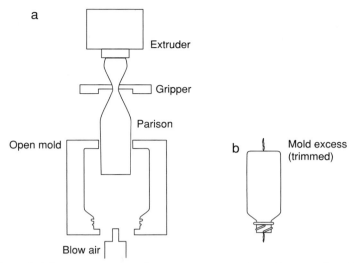

Extruder

Gripper

Parison

Open mold

b   Mold excess
(trimmed)

Blow air

FIGURE 21.2 Outline of the machinery used for blow molding. (a) A tube of molten polymer (the parison) is clamped into the mold. Then a puff of air expands the parison to the shape of the mold, the mold is briefly chilled, and (b) the blow-molded article is ejected.

molded into the familiar clear plastic soft drink bottles (Fig. 21.2) and made into Mylar film.

21.2.2. Polyester in Fiberglassing Resins

Polyester-based resins are also used in bulk plastics in fiberglass reinforced fabrications such as boats, auto body repairs, and for very large structures such as industrial vessels and pipes, and ships for mine sweeping [10]. This technology uses a combination of polycondensation and vinyl-type (chain reaction) technologies to obtain the final composite product. The viscous, still fluid resin used for layup is a linear unsaturated polyester of relatively low-molecular weight. Unsaturation is introduced into the backbone of the polymer by using maleic anhydride to replace a part of the phthalic anhydride (e.g., Eq. 21.4).

$$+ \quad + \quad 2\ HOCH_2CH_2OCH_2CH_2OH \quad \longrightarrow \qquad\qquad 21.4$$

$$+ \quad 2\ H_2O$$

The proportion of maleic anhydride used is kept low to avoid introducing too many cross-links, which would make the product brittle. This polycondensation resin is then dissolved in styrene monomer to form the so-called "fiberglassing resin." For use, a small proportion of peroxide initiator, (the "catalyst") is added to initiate vinyl-type chain reaction polymerization between the styrene monomer present and the residual unsaturation from the maleic anhydride in the polyester resin (Eq. 21.5).

21.5

This process very rapidly raises the molecular weight and introduces cross-links to set the resin to a rigid, three-dimensional network polymer incorporating the glass fiber reinforcing. Details of the mechanism of the cross-linking process are discussed more fully in Chap. 22.

21.2.3. Polyesters in Oil-modified Alkyds

Polyesters may be used to prepare durable paints and other surface coatings, again by employing a composite of polymer technologies as used in the formation of fiberglassing resins. In the initial step, a polyol of a 3 or higher functionality alcohol, such as glycerol, or pentaerythritol, is transesterified with one or more natural drying oils (e.g., linseed, tung, or cottonseed oils, or dehydrated castor oil) (Eq. 21.6).

21.6

The feature, which gives the natural drying oils their name is that there has to be at least some unsaturation of the carboxylic acid groups of the glyceryl triesters present in these oils (Table 21.3). The unsaturation is what provides the opportunity for cross-linking when the final product is exposed to air as a film. The transesterification process mixes the acids of the drying oil to form a mixture of esters among the glycerol components present and leaves some unreacted hydroxyls in the mixture.

The next stage is to take the ester mixture and react the residual alcohol functions with a di- or higher functionality carboxylic acid such as phthalic

■■■■ **TABLE 21.3** **Typical Glyceride Content of Some Representative Drying Oils**[a]

	Saturated[b]	Oleate[c]	Linoleate[c]	Linolenate[c]	Eleostearate[c]
Double bonds:	none	9,10	9,10; 12,13	9,10; 12,13; 15,16	9,10; 11,12 13,14
Geometry:	–	cis	both cis	all cis	trans, trans, cis
Cottonseed oil	25	40	35	–	–
Corn oil	11	46	43	–	–
Dehydrated castor oil	5	10	85	–	–
Linseed oil	10	18	17	55	–
Soybean oil	14	26	52	8	–
Tung oil	5	7	3	–	85

[a]Compiled from Fieser and Fieser [11] and Lange [12].

[b]Consists of varying proportions of the glycerides of the saturated n-carboxylic acids, palmitic ($C_{16}H_{32}O_2$), and stearic ($C_{18}H_{36}O_2$), which are of less importance in the film-forming functions of the oil-modified alkyds than the unsaturated acids. They serve to control the degree of cross-linking.

[c]These are all linear, 18-carbon unsaturated acids.

acid (or citric, or trimellitic acids; or pyromellitic dianhydride) to form a complex, already partially crossl-inked polymer (Eq. 21.7).

| phthalic acid | citric acid | trimellitic acid | pyromellitic dianhydride | 21.7 |

This process involves both direct esterification and *trans*-esterification reactions, and is carried to a sufficient extent of reaction to give a viscous liquid product, an oil-modified alkyd.

To formulate a paint or varnish, the oil-modified alkyd is dissolved in a solvent together with driers such as cobalt or manganese naphthenate. When applied as a coating, the driers in this solution of resin are exposed to air and catalyze the air oxidation of the residual unsaturation in the alkyd film. This process generates radicals, which serve to cross-link the polymer further as the solvent dries, to generate a tough, strongly adhering film. A pigment or a mixture of pigments is blended into the formulation to produce opaque, colored paint coatings. If a semitransparent resin is built into the oil-modified alkyd formulation instead of pigments, one obtains a varnish, an almost transparent coating.

21.3. THE POLYAMIDES

Synthetic polyamides are almost as important as polyesters for synthetic fiber production (Table 21.1). The structure of synthetic polyamides mimics the structure of proteins in some respects, in that they, too, contain recurring amide groups. The similarity ends there, however, since proteins possess a far more complicated molecular architecture with various sequences of about 15 different amino acids, as compared to the one or two monomer units typical of a synthetic polyamide.

The term "nylon" is said to have been coined from the simultaneous announcement of this new fiber-forming material in New York and London and is a generic term used for synthetic polyamides. This name was selected as a common name for polyamides by the Du Pont Company, but apparently was never registered as a trade name. The commercialization of nylon resin and fiber formation by melt spinning grew out of the fundamental research begun by Wallace Carothers of the Du Pont Company in 1929. An early patent covered polyester, polyamide, polyanhydride, and polyacetal synthesis, and was the forerunner to a whole series of patents covering this technology [13]. Nylon 6,6 was first synthesized in 1935 and commercial production started by 1938. Initial resin production was directed entirely toward fiber production, and nylon stockings were an instant success. The high strength and long wearing qualities appealed in this application and for ropes, carpeting, tire cord, and fabrics. By the early 1940s moldings and engineering parts were also being produced.

The excellent properties of fiber melt spun from nylon 6,6 led to research by would-be competitive companies to produce similar polymers that did not infringe on the Du Pont patent. Nylon 6 was such a product developed by I.G. Farbenindustrie, a German company, who started production of this resin in 1940. Nylon 6 is a single monomer polyamide of the AB type made by the ring-opening cyclization of caprolactam (Eq. 21.8). Nylon 6,6 and nylon 6 were initially, and remain, the dominant commercial polyamides.

$$n \quad \text{caprolactam} \quad \rightarrow \quad \text{—}(CH_2)_5CONH\text{—}_n \qquad \qquad 21.8$$

caprolactam   nylon 6

21.3.1. Poly(hexamethylene adipamide), Nylon 6,6

For the preparation of nylon 6,6 resin, exact stoichiometry is assured by the initial preparation of the 1:1 salt, and purification of this before polymerization (Eq. 21.9).

$$\begin{array}{cc} n \ H_2N(CH_2)_6NH_2 & + \ n \ HO_2C(CH_2)_4CO_2H \rightarrow \\ \text{hexamethylenediamine} & \text{adipic acid} \end{array} \qquad 21.9$$

$$n \left(\begin{array}{c} {}^-O_2C(CH_2)_4CO_2^- \\ H_3\overset{+}{N}-(CH_2)_6-\overset{+}{N}H_3 \end{array} \right)$$

1:1 nylon salt

The polyamidation is sufficiently favorable under the conditions used that no catalyst is required. Initially a slurry of 60–80% of the 1:1 nylon salt in water is heated to about 80–90% reaction, which is achieved at 200°C and 15 atm pressure (Eq. 21.10).

$$1:1 \text{ nylon salt} \rightarrow H\{NH(CH_2)_6NHCO(CH_2)_4CO\}_nOH + (2n-1)H_2O \quad 21.10$$
$$\text{nylon 6,6}$$

The elevated pressure used at this stage helps to retain the water, which aids in heat transfer and mixing. To complete the poly merization the temperature is slowly raised to 270–300°C while steam is gradually released to allow the pressure to drop to normal atmospheric. This last step is referred to as melt polymerization because it is carried out at temperatures above the 250°C melting point of nylon 6,6 polymer to help drive the polyamidation to completion. The molten condition also assures continuance of good heat transfer and mixing conditions during this period. Batch-type polymerization in a vessel with a heating jacket, and continuous polymerization in a long tube with zones controlled to the appropriate conditions are used for polyamidation of the 1:1 salt. Production of particularly high-molecular weight material is aided by evacuating the reaction vessel to assist in water removal.

21.3.2. Polycaproamide, Nylon 6

Batch and continuous processes are also used for the preparation of nylon 6 resin. To initiate the process, caprolactam and water are heated together at about 250°C for a period of 10–12 hr in an inert atmosphere. The ring-opened caprolactam product, 6-aminohexanoic acid (Eq. 21.11),

$$n \; [\text{caprolactam}] \; N-H + H_2O \;\rightarrow\; H_2N(CH_2)_5CO_2H \qquad 21.11$$

reacts rapidly with unchanged caprolactam to first form oligomers and then true polymers from this AB-type single monomer (Eq. 21.12).

$$n \; [\text{caprolactam}] \; N-H + H_2N(CH_2)_5CO_2H \;\rightarrow\; H\{NH(CH_2)_5CO\}_{n+1}OH \quad 21.12$$

There is some tendency for the 6-aminohexanoic acid to recyclize, which contributes to a crude product containing about 90% useful high-molecular weight resin and 10% caprolactam and oligomers. To obtain optimum properties the crude product may be heated at 180–200°C in a partial vacuum to form useful polymer from the residual oligomers. Or the mixture may be leached with water at 80–85°C to wash out the somewhat soluble caprolactam and oligomers from the product. Finished nylon 6 resin, poly (6-aminohexanoamide), has a melting point of 223°C, somewhat lower than nylon 6,6, but still high enough to be useful. The lower melting point is probably because a higher degree of order is required from nylon 6 polymer chains to obtain equivalent interchain nonbonded interactions as those obtained with nylon 6,6 (try sketching the respective H bonds).

21.3.3. Other Aliphatic Polyamides

The three other commercially important polyamides are nylon 6,10, nylon 11, and nylon 12. Nylon 6,10 is prepared from the purified 1:1 salt, hexamethylenediammonium sebacate, in a manner parallel to that described for nylon 6,6 (Eq. 21.13).

$$\text{n } H_2N(CH_2)_6NH_2 \; + \; \text{n } HO_2C(CH_2)_8CO_2H \; \rightarrow \qquad\qquad 21.13$$
$$\quad \text{hexamethylene} \qquad\qquad\quad \text{sebacic acid}$$
$$\quad\quad \text{diamine}$$

$$H\text{+}NH(CH_2)_6NHCO(CH_2)_8CO\text{+}_nOH$$
$$\text{nylon 6, 10}$$

The irregular intervals between linking amide units, 6 carbons between amide nitrogens staggered with the 10-carbon diacid units, confers a somewhat lower interchain cohesion to the product. As a result nylon 6,10 has a slightly lower melting point of 215°C.

Nylon 11, another AB-type polymer, is made by the ring-opening cyclization of 11-aminoundecanoic acid at 200–220°C in a manner similar to that described for nylon 6 (Eq. 21.14).

$$\text{cyclic lactam } + \; H_2N(CH_2)_{11}CO_2H \; \rightarrow \; H\text{+}NH(CH_2)_{10}CO\text{+}_nOH \qquad 21.14$$
$$\text{nylon 11}$$

Removal of water in the later stages is facilitated by reduced pressure. Since the ring-chain equilibrium for the 12-membered ring lactam is much less ring favorable than it is for the seven-membered lactam used to prepare nylon 6, there is only about 0.5% of the residual lactam left in crude nylon 11 product. This percentage of monomer has so little effect on the properties that it is normally left in the polymer.

Nylon 12 is also produced by ring-opening polymerization of the 13-membered ring dodecyl lactam catalyzed by phosphoric acid (Eq. 21.15).

$$\text{cyclic lactam } + \; H_2N(CH_2)_{12}CO_2H \; \rightarrow \; HNH(CH_2)_{11}CO_nOH \qquad 21.15$$
$$\text{nylon 12}$$

It requires finishing temperatures of up to 300°C to complete the polymerization. Yields are almost quantitative so the finished polymer may be used directly. Nylon 12, too, has a relatively low-melting point and lower tensile strength in fiber or engineering applications than nylon 6,6 or nylon 6, like the other aliphatic polyamides described in this section (Table 21.4). It is interesting to note that nylon 6,6, the first of the aliphatic polyamides to be commercialized, also has the best properties for many purposes.

21.3.4. Recent Polyamide Developments

Some recent introductions have been made to the commercial polyamide market. Nylon 4,6 is produced by replacing the hexamethylenediamine com-

TABLE 21.4. **Typical Properties of Commercial Grades of Polyamides (Nylons)a,b**

Nylon type	Specific gravity	Melting point (°C)	Tensile strength (MPa)[c]	Elongation at break (%)
4,6	1.18	290	89	–
6,6	1.14	255	77–80	300
6	1.12–1.15	223	70–80	300
6,10	1.09	215	58–60	100–150
6,12	1.06	212	61	~300
11	1.04	185	54	330
12	1.01	175	52	250
6,6–6,10[d]	1.08	195	38	200

[a]Compiled from Brandrup and Immergut [14], Saunders [15], and Zimmerman [16].

[b]Special postspinning heat treatments and orienting by drawing can significantly alter properties.

[c] 1 MPa $\equiv$ 1 N/mm^2. To convert to pounds per square inch, multiply by 145.

[d]A 35:65 by mass copolymer.

ponent of nylon 6,6 by 1,4-diaminobutane. This yields a regular repetition of four methylene groups between every pair of amide links, in contrast to the 4,6,4,6 alternation of amide link spacing in nylon 6,6 chains (Eq. 21.16).

$$n \ H_2N\text{-}(CH_2)_4\text{-}NH_2 \ + \ n \ HO_2C\text{-}(CH_2)_4\text{-}CO_2H \ \rightarrow \quad\quad 21.16$$

$$H \left[NH\text{-}(CH_2)_4\text{-}NH\text{-}CO\text{-}(CH_2)_4\text{-}CO \right]_n OH \ + \ (n-1) \ H_2O$$

The more regular repetition of amide functionalities together with the greater concentration of amide groups in the polymer gives a greater interchain cohesion for nylon 4,6. This raises the melting point to almost 300°C and enables a rapid maximum crystallization rate of up to 8 per sec [17] (cf. Table 21.4). Nylon 6,12, nylon 13, and blends of these resins with polyolefins, with and without fiberglass reinforcing were also introduced in the mid 1980s [18].

Copolymers of the homopolymer polyamides described may be obtained by heating a blend of the homopolymers above its melting point. A short heating period produces mostly block copolymer. After 1–2 hr of heating, amide interchange proceeds further and random copolymer results. The same copolymers may be obtained by heating the appropriate constituent monomers. The properties of commercial 6,6–6,10 copolymer, and 6,6–6, 10–6 terpolymer lie at values intermediate to those of the constituent homopolymers [19]. This is an important factor to consider in the recycling of misidentified polyamide, which may result in the accidental reprocessing of a mixture of different nylon resins. This occurrence would not waste a batch of nylon 6,6, or nylon 6, although it would cause some lowering of the melting point and the tensile strength below those of the values of the constituent nylon homopolymers. More background is given in Section 21.7.

21.3.5. The Aromatic Polyamides or Aramids, Nomex, and Kevlar

Also a relatively recent development is the use of aromatic diamines and diacids to form polyamide chains of inherently increased stiffness and order

[20]. These contributions to chain geometry and order confer both a much higher melting point to the resin and much higher tensile strength to spun fiber than is possible with the aliphatic nylons.

Initial research in this area by Stephanie Kwolek working at Du Pont in 1965 discovered that p-aminobenzoic acid could be polymerized, and that the product could be solution spun to produce superior fibers [21]. Experience clearly established the commercial value of the highly ordered fibers produced [22]. Unfortunately the high cost of p-aminobenzoic acid discouraged pursuit of this approach.

Nomex aramid, made from m-phenylenediamine and isophthaloyl chloride, produces poly(m-phenylene isophthalamide) with a melting point of 380–390°C, and substantially avoided the problems of high-monomer cost (Eq. 21.17) [23].

$$+ (n-1)HCl$$

21.17

Nomex fiber chars rather than burns at its melting point. This resistance to melting or burning has led to extensive use of Nomex fiber for the production of protective clothing for fire fighters, racing car drivers, petroleum industry workers, and any others whose occupation expose them to high fire risks [24]. It is also extensively used in a premium insulating paper for electrical equipment.

Kevlar (Du Pont) and Twaron (Enka/Akzo) are trade names for the all *para* aramid, poly(p-phenylene terephthalamide), produced by these companies (Eq. 21.18).

$$n\ H_2N\text{-}C_6H_4\text{-}NH_2\ +\ n\ ClCO\text{-}C_6H_4\text{-}COCl\ \xrightarrow[\text{N-methylpyrrolidone}]{\text{CaCl}_2\ \text{in solvent}}$$

21.18

$$H\{\,NH\text{-}C_6H_4\text{-}NH\text{-}CO\text{-}C_6H_4\text{-}CO_n\,\}\,OH\ +\ (n-1)\ HCl$$

The melting point of the product is so high, like Nomex, so that it cannot be melt spun. However, 20% Kevlar solution in hot 100% sulfuric acid may be solution spun into water [20, 25]. The extreme conditions required for resin production and fiber spinning contribute to the high cost of the fiber. However, the ability of Kevlar cable to match steel cable, strength for strength in the same diameter and with 20% of the weight, confers the wide utility of Kevlar fiber. It has a tensile strength of 3,800 MPa, nearly 500 times that of the aliphatic nylons, enabling its use in shipping applications such as cable for the anchoring of offshore drilling platforms and other situations where its light weight and corrosion resistance make for easier handling. The fiber is also used for premium tire cord, as a reinforcing fiber for production of sporting goods, bulletproof vests, aircraft components, and in chopped form as a replacement for asbestos [26]. The world *para*-aramid fiber market is estimated to be 20,000 tonne/year, split about two-third to Kevlar and one-third to Twaron [27].

21.4. POLYURETHANES

Initial work in the development of polyurethane technology, like nylon 6 development, was also stimulated by an interest in producing a nylon 6,6 fiber type by a polymer chemistry that did not infringe on the Du Pont patents. Polyurethane fiber from hexamethylene diisocyanate and 1,4-butanediol chemistry did have properties similar to nylon 6,6 fiber, but high-diisocyanate cost made this commercially unattractive (Eq. 21.19).

$$n \text{ OCN(CH}_2)_6\text{NCO } + \text{ n HO(CH}_2)_4\text{OH } \rightarrow \quad 21.19$$

$$H \text{-} OCONH(CH_2)_6NHCO\text{-}O(CH_2)_4O \text{-}_n H$$

However, during these exploratory experiments polyurethanes were discovered with novel properties, which were commercially attractive. Fiber, coating, molding, cushioning, and insulating applications for some of these materials are now exploited on a large scale [28].

The initial step in the production of cast elastomers is to terminate a low-molecular weight difunctional polyether or polyester by a mixture of toluene 2,4- and 2,6-diisocyanate using an excess of the diisocyanate (Eq. 21.20).

21.20

Mixtures of isocyanates are commonly used for convenience in commercial production of the diisocyanate, since the pure toluene 2,4-diisocyanate is more expensive to produce. The resulting prepolymer is then mixed with either a glycol, such as 1,6-hexanediol, or a deactivated (sterically hindered) diamine plus pigment if required, and then promptly poured into a preheated mold of the desired shape. In about half an hour the mixture sets to a pliable shape with stiffness and elasticity controlled by the components and processing details used [29]. Similar procedures produce high-strength polyurethane fiber (e.g., Perlon U) or elastomeric fibers (e.g., Spandex and Lycra).

Rigid polyurethanes employ a 3 or higher functionality polyglycol of 2,000–3,000 molecular weight (e.g., made from glycerol and ethylene oxide) (Eq. 21.21).

21.21

The oligomeric branches will be approximately the same length. Reacting this polyglycol with toluene diisocyanate rapidly yields a rigid, branched product. As would be expected from background theory this approach should yield high-molecular weight material more rapidly than when only bifunctional components are used. The degree of cross-linking is controlled by the functionality and molecular weight of the polyglycol component selected. Choice of glycerol as the substrate gives a trifunctional component, as in this example. Pentaerythritol contributes tetrafunctionality, and use of sucrose to form the polyglycol would give a functionality of 8 to this component.

21.4.1. Urethane Coatings

A mixture of intermediate molecular weight polyesters and natural drying oils (Table 21.3) is reacted with toluene diisocyanate to produce a viscous resin. A solution of this resin in solvent together with driers, and pigments if desired, comprises the basic clear or colored polyurethane coating. The transition metal complex driers are added to the coating formulation to catalyze the cross-linking reactions of the natural drying oil component on exposure to air in a manner similar to that described for the oil-modified alkyds.

21.4.2. Polyurethane Foams

In the polyurethane applications described so far, care has to be taken to exclude moisture. Any water, which is present, will react with isocyanate groups to give carbon dioxide gas, which can form bubbles in the product (Eq. 21.22).

$$R \sim NCO + H_2O \rightarrow R \sim NHCOOH \rightarrow R \sim NH_2 + CO_2 \uparrow \qquad 21.22$$
$$\underset{\text{isocyanate}}{} \qquad \underset{\text{a carbamic}}{} \qquad \underset{\text{primary}}{}$$
$$\underset{\text{acid (unstable)}}{} \qquad \underset{\text{amine}}{}$$

This occurrence significantly affects the stoichiometry, not only from consumption of the isocyanate functionality in the process, but also by production of an amine product with two active hydrogens, which adds reactive functional groups to the "glycol" component of the polyurethane reaction. For these reasons care has to be taken to exclude moisture for the polyurethane applications described so far. However, for the production of flexible and rigid polyurethane foams the reaction of isocyanate with water is exploited to produce bubbles.

The "one-shot" methods used to produce flexible polyurethane foams employ quick mixing of a (usually) triol-based polyether of fairly high-molecular weight with toluene diisocyanate, catalyst, and water for gas production (Eq. 21.19). The reaction of water with the diisocyanate rapidly raises the average functionality in the polymerizing system by forming urea, as well as urethane links (Eq. 21.23).

$$OCN \sim R \sim NH_2 + OCN \sim R' \sim NCO \rightarrow \qquad 21.23$$
partially hydrolyzed
diisocyanate

$$OCN \sim R \sim NHCONH \sim R' \sim NCO$$
a urea link

As a result a viscous polymer is formed, which traps the carbon dioxide as it is generated. A continuing increase in molecular weight, together with a greater degree of cross-linking as the reaction proceeds, rapidly brings about a flexible set to the foam. By varying the details of the formulation and temperatures used, flexible foams may be obtained with densities of $15-70 \, kg/m^3$, small or large, and open or closed (low-water uptake or low-gas exchange) cells, and in a variety of stiffnesses and tensile strengths [30]. Applications include furniture and crash padding, bedding and blankets, and car seat cushioning.

Rigid polyurethane foams are produced by analogous technology to the flexible foams except for two aspects. Higher functionality, or a somewhat lower molecular weight, will be used in the polyether or polyester component. Either change increases the frequency of cross-linking in the product. Also, lower cost bubble production is achieved via introduction of a low-boiling solvent such as Freon 11 (Cl_3CF, b.p. 24°C), or carbon dioxide gas directly, rather than by the isocyanate-water reaction. This conserves the more expensive isocyanate component for the formation of cross-linked polymer to trap the gas or vapor, as it is generated. Since 97% of the insulating capability of the rigid polyurethane is the consequence of the voids present, use of a low-thermal conductivity gas to fill the voids substantially improves the insulation capacity of the final foam. Freon-11 and carbon dioxide have about one-third and two-third of the thermal conductivity of air, respectively, which makes these such good candidates for foam generation. Their excellent insulating capacity and structural contribution lead to their use in modern, thin-walled refrigerators and freezers, foamed-in-place insulation for freezer-truck bodies and residential housing, and for flotation billets used in floating docks, etc. A representative breakdown of polyurethane usage according to type is given in Table 21.5.

21.5. EPOXY RESINS

This class of condensation polymers is characteristically recognized by cross-linking from the reactions of epoxide groups. The products are useful as one of the best adhesives for cementing of rigid materials, and as potting compounds for physical protection or encapsulation of electronic devices.

Two components are required, which are mixed just prior to use. The prepolymer, in which the cross-linkable epoxide functions reside, is made from epichlorohydrin and *bis*-phenol A. Epichlorohydrin in turn is derived from allyl chloride (Eq. 21.24).

$$CH_2 = CHCH_2Cl \xrightarrow{\text{HOCl}} HOCH_2CH(Cl)CH_2Cl \rightarrow \qquad 21.24$$

$$\underset{\substack{\diagdown\diagup \\ O}}{CH_2-CH-CH_2Cl}$$

epichlorohydrin

bis-Phenol A is produced from the acid-catalyzed reaction of phenol with acetone (Eq. 21.25).

TABLE 21.5   Breakdown of Polyurethane Demand in Selected Areas, in Thousands of Metric Tonnes[a]

	Canada 1981	Latin America 1981	United States[b]		
			1981	1995[b]	1996
Flexible foam	55.3	147.8	553	844	927[c]
Rigid foam	22.2	21.4	263	449	575
Elastomers	–	–	77	–	92[d]
Surface coatings	10.0	26	40	–	140
Adhesives, sealants	–	–	59	–	238[e]
Other[f]	–	–		–	118
Total	87.5	195.2	992	–	2090

[a]Data from Upjohn Polymer Chemicals [31] and Stinson [32]. The total European polyurethane market was estimated to be 1.41 million metric tonnes in 1988 with the largest use sectors being transportation, 27.8%; furniture, 21.4%; buildings, 16.6%; appliances, 5.4%; and footware, 5.2% Chem.Brit [33].
[b]Projections for 1995 given in parentheses, Kirk-Othmer [34]. In the same year it was projected that a total of 669,000 tonnes of polystyrene foam products would be produced in the United States.
[c]Includes slab and molded foam products.
[d]Includes cast elastomers and fibers (Spandex).
[e]Includes binders and fillers.
[f]Includes molded thermoplastics, and automotive and nonautomotive reaction injection molded products.

$$2\ C_6H_5OH\ +\ CH_3COCH_3\ \xrightarrow{H^+}\ HO-\bigcirc-\underset{CH_3}{\overset{CH_3}{C}}-\bigcirc-OH \quad 21.25$$

bis-phenol A, m.p. 158°

Combining an excess of epichlorohydrin with the *bis*-phenol A (4, 4′-isopropylidenediphenol) gives the required prepolymer component of epoxy resin systems (Eq. 21.26).

$$4n\ CH_2\text{-}CH\text{-}CH_2Cl\ +\ n\ HO-\bigcirc-\underset{CH_3}{\overset{CH_3}{C}}-\bigcirc-OH\ \longrightarrow \quad 21.26$$

epoxide prepolymer

glycidyl ether end group

Epoxide end groups (glycidyl ether) in the polymer are assured by using an excess of epichlorohydrin at this stage.

There are two options for the other component of an epoxy resin system. Use of mono- or di-anhydrides as curing agents, usually catalyzed by a tertiary amine, causes reactions with the residual secondary hydroxyls in the repeating unit of the prepolymer forming esters and free carboxylic acids. The carboxylic acids formed also react with the epoxide end groups forming cross-links and further free secondary hydroxyl groups. Maleic anhydride, phthalic anhydride, or pyromellitic dianhydride are suitable for this process (Eq. 21.27).

$$21.27$$

However, the reactions between these anhydrides and the prepolymer are sluggish at room temperature so that heating is required to enable the cross-linking reactions to proceed at a reasonable rate. Stationary shelving or a moving belt inside a thermostatted oven can be used to heat small-scale epoxy resin castings and encapsulated electronic components (the "potting compounds") to accelerate polymerization. Castings and encapsulation are the primary end uses of this class of epoxy resins.

For adhesive applications, it is more convenient to have the two components of the epoxy resin system react at a reasonable rate at room temperature. Replacing the anhydride component by a suitable, more reactive polyfunctional amine enables this to occur at room temperature. Several aliphatic and aromatic polyfunctional amines are used for this purpose (Eq. 21.28).

$H_2N\text{-}(CH_2)_2\text{-}NH\text{-}(CH_2)_2\text{-}NH_2$
diethylene triamine

$$21.28$$

4,4'-diaminodiphenylmethane   m-phenylenediamine

The so-called "five minute epoxy" employs a highly active amine that enables setting in 5 min and substantial strength in an hour under ambient conditions. Recent production of epoxy resins in the U.S.A. has totaled 134,000 metric tonnes per year [35].

21.6. PHENOL–FORMALDEHYDE RESINS

The constituents of phenol-formaldehyde resins owe their polymer-forming capability to the potential trifunctionality of phenol, and the bifunctional capability of formaldehyde. Delocalization of the electron pairs on the oxygen of phenol into the aromatic ring produces electron-rich sites at the *ortho* and *para* positions, which activates these sites, producing the potential trifunctionality of the phenol. The phenol–formaldehyde reaction is usually carried out in aqueous solution. Under these conditions, formaldehyde adds water in an equilibrium reaction to form formaldehyde hydrate (Eq. 21.29).

$$H_2C = O + H_2O \rightleftharpoons HOCH_2OH \qquad 21.29$$
formaldehyde   formaldehyde hydrate

Despite the favoring of formaldehyde hydrate by this equilibrium, there is sufficient free formaldehyde available for an *ortho* or *para* position of phenol to add to the highly electrophilic carbon of formaldehyde. As formaldehyde is consumed by this process, the equilibrium is displaced to the left providing further formaldehyde for reaction until all the phenol potential functionalities are taken up or all the formaldehyde is consumed. The structures of the phenol-formaldehyde polymers produced are difficult to study because the final product is infusible and insoluble. However, current thinking is that all possible monomer links can occur in a typical Bakelite sample (Eq. 21.30).

21.30

+ n H$_2$O

A representative segment of phenol-formaldehyde polymer

The phenol-formaldehyde resin system has such a high potential for crosslinking that in practice these monomers could rapidly yield an intractable mass. Strategies such as reagent imbalance may be used to effect control. This strategy is normally used for polymerizations carried out under acidic conditions to initially produce an oligomeric linear prepolymer. Polymerization is completed at a later stage by heating the prepolymer with a component to contribute formaldehyde to the system. Or, alternatively, the reaction in bulk may be stopped by dehydration well before completion, and then completed in molds of the desired final fabrication shape.

Polymerization under acidic conditions, the first control scenario mentioned, employs an initial formaldehyde to phenol ratio of 0.75–0.85 to 1. Stoichiometric ratio would be 1.5 to 1. Catalysis is achieved by using a strong acid; 0.1–0.3%, hydrochloric acid:phenol ratio, or 0.5–2% oxalic acid to phenol ratio. The nonstoichiometry restricts polymerization to the oligomer stage and avoids cross-linking (Eq. 21.31).

$$H_2CO + H_3O^+ \rightleftharpoons H_2\overset{+}{C}OH \overset{PhOH}{\longrightarrow}$$

$$21.31$$

These prepolymers, called novolacs, are dehydrated, reduced to a powder and then blended with about 10% by weight hexamethylenetetramine ("hexa," Eq. 21.32). Hexa is made by reacting

$$21.32$$

4 mole of ammonia with 6 mole of formaldehyde. When the powder mixture is placed in a mold of the desired shape and heated, rapid polymerization and cross-linking of the novolac occurs by the formation of methylene (mostly) with a few benzylamine bridges.

Polymerization under alkaline conditions, using about 1% sodium hydroxide catalyst based on the weight of the phenol, proceeds via a somewhat different mechanism. Methylol phenols, and oligomers consisting of 5 or 6 phenol units connected by methylene bridges, result in the so-called resole resin prepolymer (Eq. 21.33).

methylol phenols   representation of resole prepolymer

$$21.33$$

Resole resins may be converted into high-molecular weight, cross-linked materials simply by heating. For this reason, the resol process is sometimes referred to as the "one-stage" process to phenol-formaldehyde resins. With this process the potential for the formation of high-molecular weight, cross-linked material exists right from the start. So care has to be exercised to avoid gelation (solidification via cross-linking) of the product in the primary reactor. Mechanistic details have been presented, and recent detailed characterizations of resol resins have been published [36, 37].

The single largest use for the phenol-formaldehyde resins is in adhesive applications for the production of plywood, chipboard, and particle board. The resin can comprise as much as one-third of the weight of the board, particularly of particle boards, which contributes to a total demand for phenolics in the U.S.A. of over half a million metric tonnes per year. They are also used as the matrix adhesives for the production of several types of grindstones. In combination with paper, woven cotton, glass fiber, etc., components, phenolics contribute to the production of engineering and decorative laminates in the form of rods, tubes, and sheets. The sheet products Arborite and Formica are familiar as the finished surfaces of furniture, bathroom, and kitchen counter tops and other areas where attractive patterns and water resistance are important characteristics. Molded products from phenolics are also important where heat or electrical resistance is required, such as saucepan and toaster handles, switches, and the printed circuit boards used in computers. Recent phenolics production in the U.S.A. has totaled over 500,000 metric tonnes per year, not including fillers [38].

21.7. ENVIRONMENTAL AND RECYCLE ASPECTS

Not long after the large-scale production of plastics, interest and motivation for reuse and recycling of these materials developed, the details of which continue to be promoted and refined [39, 40]. These cover aspects relating to in-plant recycling of all types of polymer scrap (producer recycling) as well as all aspects of postconsumer recycling. They extend from applications such as drainage tile production from polyethylene milk bottles, to the conversion of used PET soft drink bottles into PET fiber for carpet production, sleeping bag insulation, and insulated clothing [41].

More specifically, a case study, which outlines a vapor combustion approach to the handling of an odor problem during polyester primary production could also be usefully applied to other polymer producing facilities [42]. Recently it has been established that recycled PET resin may be reused without substantial loss of grade under the appropriate conditions, even for applications involving food contact [43, 44].

Operating details for more benign batch reactors for nylon production have been studied [45]. Perhaps of greater concern, however, is the significant contribution to atmospheric nitrous oxide from the large-scale use of nitric acid to produce nylon monomers (Eq. 19.56) [46]. Nylon resins may be recycled in the form in which they are received, generally with a lowering of the grade. Thus, nylon engineering parts may be recycled into less demanding carpet fiber applications or employed in multilayer film production [47, 48]. Or the resins, segregated as to type as far as possible, may be subjected to hydrolysis back to the monomers, the monomers purified, and then repolymerized to a resin, which is indistinguishable from virgin material. Although hydrolysis-based recycling is possible for both nylon 6 and nylon 6,6, the two largest scale commercial polyamides, it is much more straightforward for nylon 6 because no posthydrolysis monomer separation is required since only a single monomer results.

An interest in reduced use of chlorofluorocarbon blowing agents (e.g., Freon-11 and CCl_3F) for production of rigid polyurethane foams has led to the study of suitable alternatives such as carbon dioxide [49, 50]. Even though the rigid polyurethane foams are mainly used in closed systems, eventual chlorofluorocarbon loss to the atmosphere is still almost certain in this application when the appliance is discarded. Usage precautions have been compiled [51]. Polyurethane scrap can be reusable, despite the cross-linked nature of the resin, by a variety of processes [52, 53]. The polyol component may be recovered, and the diisocyanate(s) in the form of the corresponding 2,4- and 2,6-toluenediamines using hydrolytic techniques at temperatures of 190–250°C [54].

Destructive distillation or anaerobic pyrolysis may be used, as a last expedient, to derive fuel or feedstock values from low grades of recovered plastics. This option is more often used for vinyl-type polymers, some of which can yield fairly uniform product by this means. This option is discussed more fully in Chapter 23, which covers the details of these types of polymers.

REVIEW QUESTIONS

1. (a) Draw the repeating unit obtained when the following monomers are polymerized: adipic acid and hexamethylene diamine; p-carboxybenzyl alcohol; terephthalic acid and 1,2-ethanediol; caprolactam.
(b) Name each of the polymers in part (a) using a systematic and a commercial name.
2. Titration of a solution of a 38.6 g sample of nylon 6,6 with base indicates that the sample contains 3.11×10^{-3} mole of carboxyl groups. Calculation using this information gives an $\overline{M}_n$ of 12,400 for the sample.
(a) What is the assumption required to obtain this result?
(b) What further experimental information is necessary in order to verify or correct this result? Explain the need and describe a feasible method.
3. (a) What would be an experimentally feasible approach to the synthesis of the following random copolymers?

$$\{O\text{-}(CH_2)_2\text{-}O\text{-}(CH_2)_2O(CH_2)_2\text{-}O\,\}_x(CO\text{-}(pC_6H_4)\text{-}CO\}_y$$

$$\{NH\text{-}(pC_6H_4)\text{-}CO\,\}_x\text{-}\{NH\text{-}(CH_2)_{10}\text{-}CO\,\}_y$$

(b) How should the procedure be altered to produce the equivalent block copolymers of any particular designated block length for the structures of part (a)?
4. Drastic hydrolysis of a sample of a nylon of number average molecular weight 20,575 yields 24.07 g of 1,4-diaminobutane, 39.51 g of adipic acid, and 0.32 g of benzoic acid.
(a) Draw the repeating unit of this polymer.
(b) What would be the systematic and the commercial names for this polymer?

(c) What would be the number average degree of polymerization and the extent of reaction used to prepare this polymer?

(d) What number average molecular weight product would be expected for the same extent of reaction if the proportion of benzoic acid used was halved?

5. What are the incentives to a nylon reprocessor to keep the various types of nylon segregated in their collection and recovery system? Explain.

6. What is a probable outcome of the thermal reprocessing of a recovered mixture of nylon resins? Discuss options.

7. (a) Discuss the factors that might dictate the need to hydrolyze a polycondensation polymer to constituent monomer(s), to permit useful recycling.

(b) Why might this be a less attractive option than cleaning and reextrusion of the resin itself?

8. Why might one monomer nylon and one monomer polyester polymers be easier to reprocess to the original monomer than those condensation polymers which contain two or more monomers?

FURTHER READING

S. Fakirov, "Transreactions in Condensation Polymers." Wiley-VCH, Weinheim & New York, 1999.

J.L. Hedrick and J.W. Labadie, eds., "Step-Growth Polymers for High Performance Materials: New Synthetic Methods," ACS Symp. Ser., No. 624. American Chemical Society, Washington, DC, 1996.

K.-H. Reichert and W. Geiseler, "Polymer Reaction Engineering: Emulsion Polymerization, High Conversion Polymerization Polycondensation." Hüthig & Wepf, New York, 1986.

M.E. Rogers, E. Martin, and T.E. Long, "Synthetic Methods in Step-growth Polymers." Wiley-Interscience, Hoboken, NJ, 2003.

REFERENCES

1. Plastics Europe, Plastics in Europe 2002 & 2003, Assoc. of Plastics Manufacturers in Europe, Brussels, Belgium, 2004. Available at www.plasticseurope.org
2. Facts & Figures, *Chem. Eng. News*, **82**(27), 49–60, July 5 (2004), and earlier listings.
3. A.H. Tullo, Polymers Heat Up, *Chem. Eng. News*, **79**(39), 24–28, Sept. 24 (2001).
4. M. Reisch, Polyester Resin Producers, *Chem. Eng. News*, **75**(21), 17, May 26 (1997).
5. Polymer Production Jumped 7.2% in 1994, *Chem. Eng. News*, **73**(15), 20, Apr. 10 (1995).
6. Facts and Figures for the Chemical Industry, *Chem. Eng. News*, **73**(26), 36, Jun. 26 (1995).
7. "Statistical Abstract of the United States," U.S. Dept. of Commerce, 2003 and earlier editions.
8. P. Short, Growth in Polyester Masks Stagnation, *Chem. Eng. News*, **78**(20), 25, May 15 (2000).
9. J. Shin, Y. Lee, and S. Park, Optimization of the Prepolymerization Step of Polyethylene Terephthalate, *Chem. Eng. J.* **75**(1), 47–55, Aug. (1999).
10. Plastic Warship Launched, *Chem. Br.* **18**(5), 334 (1982).
11. L.F. Fieser and M. Fieser, "Advanced Organic Chemistry," p. 990. Reinhold, New York, 1961.
12. N.A. Lange, ed., "Handbook of Chemistry," 10th ed., p.792. McGraw-Hill, New York, 1969.

13. W.H. Carothers, Linear Condensation Polymers, U.S. Patent. 2,071,250 (to the Du Pont Company) (1937).

14. J. Brandrup and E.H. Immergut, eds., "Polymer Handbook," 3rd ed., pp. V/113–V/114. Wiley, New York, 1989.

15. K.J. Saunders, "Organic Polymer Chemistry," p. 186. Chapman & Hall, London, 1973.

16. J. Zimmerman, Polyamides.in H.F. Mark *et al*., eds., "Encyclopedia of Polymer Science and Engineering," 2nd ed., Vol. 11, pp. 315–381. Wiley, New York, 1985.

17. D.O'Sullivan, Conventional Nylons Encounter Strong New Competitor in Nylon 4,6, *Chem. Eng. News*, **62**(21), 33, May 21 (1984).

18. S. Stinson, Nylon Impact Resistance Tailored to User Need, *Chem. Eng. News*, **64**(38), 39, Sept. 22 (1986).

19. C. Puglisi, F. Samperi, S. Di Giorgi *et al*., Exchange...Copolymers from Nylon 6,6 and Nylon 6,10, *Macromol.* **36**(4), 1098–1107, Feb. 25 (2003).

20. D. Tanner, J.A. Fitzgerald, and B.R. Phillips, The Kevlar Story—An Advanced Materials Case Study, *Angew. Chem., Int. Ed. Engl. Adv. Mater.* **28**(5), 649 (1989).

21. S.L. Kwolek (Du Pont), Br. Pat. 1,283,064 (1972); U.S. Patent. Reissue 30,352 (1980).

22. K.A. Hodd and D.C. Turley, High Modulus Fibers, *Chem. Br.* **14**(11), 545 (1978).

23. H.J. Sanders, Flame Retardants. *Chem. Eng. News*, **56**(17), 22, Apr. 24 (1978).

24. A. Eldad, A.Y. Salmon, S.Breiterman *et al*., Flame...: Assessment of a New, Air-cooled Fireproof Garment, *Milit. Med.* **168**(8), 595–599, Aug. (2003).

25. H. Short, A First Look at Aramid Processing, *Chem. Eng.* **96**, 37, Apr. (1989).

26. P.L. Layman and M.S. Reisch, DuPont, Akzo Resolve Long Hassle over Aramid Fiber Patents, *Chem. Eng. News*, **34**(9), 10, Aug. 22 (1988).

27. Akzo Nobel's Bulletproof Strategy, *Chem. Ind. (London)*. p. 744, Oct. 3 (1994).

28. D. Birkett, Urethanes Rise to the Challenges, *Chem. Brit.* **33**(6), 30–32, June (1997).

29. G. Woods, "The ICI Polyurethanes Book, ICI Polyurethanes," Wiley, New York, 1990.

30. G. Woods, "Flexible Polyurethane Foams: Chemistry and Technology," Appl. Sci. Publ., London, 1982.

31. Upjohn Polymer Chemicals, Polyurethane 81, *Chem. Eng. News*, **59**(23), 55, June 8 (1981).

32. S. Stinson, Polyurethane Use Continues to Grow, *Chem. Eng. News*, **75**(31), 22, Aug. 4 (1997).

33. Polyurethane, *Chem. Brit.* **26**(4), 315 (1990).

34. "Kirk-Othmer Encyclopedia of Chemical Technology," 4th ed., Vol. 11, p. 759. Wiley, New York, 1994.

35. Key Polymers, Epoxies, *Chem. Eng. News*, **59**(39), 18, Sept. 28 (1981).

36. K.J. Saunders, "Organic Polymer Chemistry," pp. 282–297. Chapman & Hall, 1973.

37. R.A. Haupt and T. Sellers, Jr., Characterization of Phenol–Formaldehyde Resol Resins. *Ind. Eng. Chem. Res.* **33**, 693 (1994).

38. Key Polymers, Phenolics, *Chem. Eng. News*, **59**(39), 16, Sept. 28 (1981).

39. R.J. Ehrig, ed., "Plastics Recycling: Products and Processes." Hanser Publishers, New York, 1992.

40. L. Avadanei, N.M. Szilagyi, and V. Avadanei, Recycling Possibilities of PET Wastes. *Mater. Plast.* **39**(4), 234–239 (2002).

41. D.O'Sullivan, PET Bottles Recycled into Insulation Foam. *Chem Eng. News*, **68**, 25, Feb. 12 (1990).

42. P. Mattli, Control of Odor Nuisance from the Chemical Industry. *Chem. Ind. (London)*. p. 338, May 7 (1984).

43. J.H. Schut and R-PET Cleared for Food Contact. *Plast. World*, **52**, 11, Oct. (1994).

44. B. Miller, Coinjection Opens Door for Recycled PET for Foods, *Plast. World*, **52**, 14, Apr. (1994).

45. R.M. Wajge and S.K. Gupta, Multiobjective Dynamic Optimization of a Nonvaporizing Nylon 6 Batch Reactor, *Polym. Eng. Sci.* **34**, 1161 (1994).

46. M.H. Mark *et al*., Nylon Production: An Unknown Source of Atmospheric Nitrous Oxides. *Science*, **251**, 932 (1991).

47. R. Welgos, Nylon 6 and 6,6 Aren't Always the Same, *Mach. Des.* **66**, 55, Nov. 21 (1994).

48. M.M. Nir, A. Ram, and J. Miltz, Performance of Reprocessed Multilayer LDPE/Nylon-6 Film, *Polym. Eng. Sci.* **35**, 1878 (1995).

49. L.M. Sherman, Third Generation Blowing Agents Starred at Polyurethanes' 95, *Plast. Technol.* **41**, 25, Oct. (1995).

50. B. Miller, Urethanes Technology Is Adapting to Environmental Concerns, *Plast. World* **51**, 12, Oct. (1993).

51. Upjohn Company, "Precautions for the Proper Usage of Polyurethanes, Polyisocyanurates, and Related Materials," Tech. Bull. 107. Chem. Div., The Upjohn Company, Kalamazoo, MI, 1980.

52. E. Weigand, W. Rasshofer, and M. Herrman, Polyurethane Recycling Processes Are Reaching Commercial Maturity, *Mod. Plast.* **70**, 71, Nov. (1993).

53. P. Mapelston, Bayer Process Makes Urethane Scrap Reusable, *Mod. Plast.* **68**, 20, Feb. (1991).

54. J. Braslaw and J.L. Gerlock, Polyurethane Waste Recycling. Polyol Recovery and Purification, *Ind. Eng. Chem. Process. Des. Dev.* **23**, 552 (1984). (series of articles).

22
ADDITION (CHAIN REACTION) POLYMER THEORY

Our time has seen the synthesis of Polyisoprene
And many cross-linked Helices unknown
To Robert Hooke;...
                    — John Updike, 1968

22.1. ADDITION (VINYL-TYPE) POLYMERS

Synthetic polymers may be broadly classified into three major applications on the basis of the suitability of their properties as plastics, fibers, or elastomers. The main differentiating feature of these divisions is their modulus or stiffness; fibers have the highest modulus and elastomers the lowest. A subdivision of the plastics is made on whether they are melt formable, or must have final polymerization take place in a mold of the final shape desired. The melt-formable polymers are referred to as thermoplastics, and the mold-formed or infusible polymers as thermosetting plastics. Thermosetting plastics, such as the phenol-formaldehyde resins discussed in Chap. 20, are structurally distinct from the thermoplastics in having substantial cross-linking to connect adjacent polymer chains into an interlocking network. This is the feature that makes these materials nonmeltable, although they will still char or burn at sufficiently high temperatures. Thermosets comprise about 10% of the current mass of production of all types of plastics.

Thermoplastics may be further subdivided into two broad categories on the basis of their cost and suitable end uses. "Commodity" plastics are typified by high volume production, good properties, and low resin cost. The four major commodity plastics are polyethylene, polypropylene, poly(vinyl chloride), and polystyrene. Their adequate properties and low cost have led to the extensive use of these plastics in packaging applications where they are very competitive with paper, steel, and glass. They are also used for some less demanding applications as components of durable goods (Table 22.1).

███████ **TABLE 22.1 Some Representative Thermoplastic Polymers**

Name	Repeating unit	Typical uses
Commodity plastics (vinylic polymers):		
Low-density polyethylene (LDPE or PE)	$\{CH_2 - CH_2\}_n{}^a$	Packaging/construction film, house wares, cable insulation, flexible bottles, toys
High-density polyethylene (HDPE)	$\{CH_2 - CH_2\}_n$	Drums, pipe, bottles, sheet, packaging film, cable insulation
Polypropylene (PP)	$\{CH_2 - CH(CH_3)\}_n$	Appliance and auto parts, carpeting, rope, film
Poly(vinyl chloride) (PVC)	$\{CH_2 - CHCl\}_n$	Housing, pipe, flooring, film, sheet, wire insulation
Polystyrene (PS)	$\{CH_2 - CHPh\}_n$	Foam and film packaging, appliance parts, housewares, foam insulation, disposable ware
Engineering plastics (condensation polymers):		
Polyamides, (Nylons) e.g., Poly(hexamethylene adipamide)	Various (Chap. 21), e.g., $\{NH(CH_2)_6NHCO(CH_2)_4CO\}_n$	Gears, bearings, latches, high tensile cordage, rope, aircraft parts
Polycarbonate (PC, Lexan)	$\{OC_6H_4C(CH_3)_2 - C_6H_4OCO\}_n$	Safety helmets, bullet-proof windows, gears, auto parts, machine casings
Polyesters, e.g., PET Poly(ethylene terephthalate)	$\{OCH_2CH_2OCOC_6H_4CO\}_n$	High-strength moldings Electrical insulation
PBT Poly(butylene terephthalate)	$\{O(CH_2)_4OCOC_6H_4CO\}_n$	Engineering products, Blends with PC in high-impact applications

^{a}With some branching; see Section 23.1

"Engineering" plastics are so called because of their significantly better mechanical properties than the commodity plastics. They are also more expensive to produce. As a class their production volume is about one-tenth of that of the commodity plastics (Table 22.2).

The repeating units of the engineering plastics also identify these as being predominantly condensation polymers, the class discussed in Chaps. 20 and 21. The commodity plastics are prime examples of the utility of the vinyl-type or chain reaction, polymers to be discussed here.

22.2. COMPARISON OF CONDENSATION AND CHAIN REACTION POLYMERIZATION

As outlined in Chap. 20, condensation polymerization can involve the reaction of any two species present in a condensation polymerization system. The result is a generally slow progression toward moderate molecular weights if conditions are chosen to help drive the reaction to very near completion. These requirements and the results are listed for convenience of comparison with chain reaction (vinyl-type) polymerization discussed in detail here (Table 22.3).

Chain reaction polymerization involves use of a monomer with at least one carbon-carbon or carbon-heteroatom double bond. The double bond

■■■ TABLE 22.2 Utility, American Volume, and Cost Profiles of Some Large Scale Synthetic Polymers[a]

Type	Thousand metric tonnes	Average prices (US $/kg)	
	2001	1995	2001
Commodity plastics (vinylic addition)			
Polyethylenes	15,083	0.70	1.04
Polypropylene	7,228	0.73	0.79
Polystyrene	2,773	0.90	1.12
Poly(vinyl chloride)[b]	6,467	0.77	0.79
Engineering plastics (condensation)			
Polycarbonate	—	—	3.52
Polyester, thermoplastic	1,464	1.25[c]	1.54
Synthetic elastomers (vinylic addition copolymers)			
Acrylonitrile-butadiene-styrene (ABS)	—	1.80	1.81
Styrene-acrylonitrile (SAN)	—	—	2.09
Polybutadiene	583	1.25[d]	—

[a]Data compiled from *Chem. Eng. News* [1, 3, 4], Kirschner [2], and *Chemical Market Reporter* [5, 6].
[b]Includes copolymers.
[c]U.S. demand, for 2002.
[d]Estimated.

represents a potential functionality of two, which can form new single bonds to link two or more monomer units assisted by an initiator. Initiation can be via a radical species or a charged molecule or atom, and represents the first of the three distinct kinetic divisions recognized for chain reaction polymerization: initiation, propagation, and termination.

For a free radical-initiated process with a monomer, such as ethylene, the three stages can be represented as follows (Eqs. 22.1–22.6). Species suitable for initiation of a free radical-promoted process are normally generated by the homolytic dissociation of weak bonds of a precursor by heat or light (Eq. 22.1).

Initiation:

$$\underset{\text{initiator}}{I} \xrightarrow[\text{light}]{\text{heat or}} 2\,R^{\cdot} \qquad\qquad 22.1$$

$$R^{\cdot} + CH_2{=}CH_2 \rightarrow RCH_2{-}\dot{C}H_2 \qquad\qquad 22.2$$

Propagation:

$$RCH_2\dot{C}H_2 + CH_2{=}CH_2 \rightarrow RCH_2CH_2CH_2\dot{C}H_2 \qquad\qquad 22.3$$

$$RCH_2CH_2CH_2\dot{C}H_2 + n\,CH_2{=}CH_2 \rightarrow R(CH_2CH_2)_{n+1}CH_2\dot{C}H_2 \qquad\qquad 22.4$$

Termination:

By combination:

$$2\,R(CH_2CH_2)_n CH_2\dot{C}H_2 \rightarrow \qquad\qquad 22.5$$
$$R(CH_2CH_2)_n CH_2CH_2CH_2CH_2(CH_2CH_2)_n R$$

Characteristic	Condensation (step-reaction) polymerization	Chain reaction (vinyl-type) polymerization
Chain growth	By reactions between monomers, oligomers, and polymers	Only by monomer addition to initiating radicals
	Usual loss of a small molecule	Whole monomer incorporated
Processes involved	Same reaction involved with all reacting species present	Initiation, propagation, and termination mechanisms different
	No initiator required (catalyst sometimes)	Suitable radical, or charged (+, or −) initiator required
	Monomer consumed early in polymerization; molecular wt. increases slowly	Only monomer and high polymer present; each polymer molecule forms very rapidly
Rate	Decreases steadily as polymerization proceeds	Increases initially, then stays nearly constant until monomer consumed
Product $\overline{DP}^a$	Low, to moderate for linear product	High, to very high
Heterogeneity index, $\dfrac{\overline{M}_w}{\overline{M}_n}$	ca. 2	ca. 2^b

^{a}Degree of polymerization, or $\overline{X}_n$.
^{b}Except for anionic "living" polymerizations where the value is about 1; see Section 22.5.

By disproportionation:

$$2\ R(CH_2CH_2)_n CH_2\overset{\cdot}{C}H_2\ \rightarrow \hspace{3cm} 22.6$$
$$R(CH_2CH_2)_n CH{=}CH_2\ +\ R(CH_2CH_2)_n CH_2CH_3$$

The initiating radical then reacts with monomer to form a propagating radical in the second phase of the initiation process. The propagating radical can then add many more monomer units in a chain reaction before intervention by a termination step stops the process to produce a finished (dead) polymer chain. The propagation steps consume by far the largest proportion of monomer. For a free radical-initiated process, termination can occur by combination (coupling) or by disproportionation. Coupling forms a saturated polymer product with a molecular weight equal to the sum of the molecular weights of the coupling radical species (Eq. 22.5). Termination by disproportionation occurs by abstraction of a hydrogen atom from one of the two radical species by the other, to form two stable polymer molecules of molecular weights equivalent to that of each of the initial species (Eq. 22.6). One of these is saturated and the other will have a terminal olefin. In theory, the olefin-terminated polymer could react with a new radical species and so increase in molecular weight. But in practice the olefinic terminus is so sterically hindered by the bulk of the polymer chain that this molecule too is essentially "dead" or finished polymer.

Representative features of chain reaction polymerization are listed in Table 22.3 for comparison with the characteristics of typical condensation polymerizations. In both cases, the degree of polymerization is determined by dividing the molecular weight of the polymer by the molecular weight of the

FIGURE 22.1   Relationship of the product number average molecular weight, $\overline{M}_n$, to the percent monomer conversion for (a) a step-growth, polycondensation versus (b) a chain growth (vinyl-type) polymerization.

monomer unit (Section 20.8.2). However, the relationship of the number average molecular weights to the degree of conversion is quite different in the two processes (Fig. 22.1). High molecular weight material is only obtained quite near the end of the polymerization for a polycondensation, whereas the molecular weight of the product is normally independent of the degree of polymerization in chain (vinyl-type) polymerizations.

22.3. RADICAL CHAIN POLYMERIZATION METHODS

22.3.1. Bulk or Mass Polymerization

Free radical polymerization of neat monomer in the absence of solvent and with only initiator present is called bulk or mass polymerization. Monomer in the liquid or vapor state is well mixed with initiator in a heated or cooled reactor as appropriate. The advantages of this method are that it is simple, and because of the few interacting components present, there is less possibility for contamination. However, vinyl-type polymerizations are highly exothermic so that control of the temperature of bulk polymerization may be difficult. Also, in the absence of a solvent viscosities may become very high toward the end of a polymerization, which could make stirring difficult, and add to the difficulty of heat removal from the system. The advantages of this system, however, are sufficiently attractive for this to be used commercially for the free radical polymerization of styrene, methyl methacrylate, vinyl chloride, and also for some of the polymerization processes of ethylene [7].

Incidentally, bulk polymerization is also the chief method used for commercial polycondensations. Polycondensations are not as exothermic as free radical catalyzed vinyl-type polymerizations, so thermal control is less of a problem. Bulk polycondensation also favors formation of linear polymer rather than the cyclic products that are favored by dilute solution polymerization, particularly if AB-type monomers are being used. Finally, since a high degree of polymerization (i.e., high DP, or high molecular weight) is only obtained at very near complete reaction of the functional groups, the viscosity stays low throughout most of the reaction. Hence, stirring remains easy until very near the end of the polymerization. Since the high temperatures that are used near the end of the process to help drive a typical polycondensation to completion are often above the melting point of the polymer; high viscosity is seldom a problem with these systems.

22.3.2. Solution Polymerization

Free radical polymerization of a monomer in solution is often more versatile and more amenable to temperature control than bulk polymerization. The presence of the solvent avoids any potential viscosity or stirring problems. Also the exotherm of the polymerization is moderated by the lower monomer concentrations under these conditions, which slows down rates, and by the thermal mass contributed by the solvent. Sometimes the polymerization temperature will be conveniently controlled by the reflux temperature of the solvent used.

To optimize these advantages of solution polymerization, care has to be exercised in certain production parameters when monomer and solvent are present. Appropriate solvents for free radical polymerization need to be selected carefully, since radical chain transfer from the propagating polymer chain end to solvent molecules may occur. Hydrogen abstraction from the solvent by the propagating radical will tend to shorten the propagating chain, which will tend to decrease the average molecular weight of the polymer product obtained. The extent of transfer will depend on the choice of solvent and the conditions (concentrations, temperature, etc.) used to carry out the polymerization. Also the purity of the polymers produced by solution polymerization may be affected, either by impurities present in the solvent, or by residual traces of the solvent itself if this is difficult to remove. Nevertheless the advantages of solution polymerization have led to it being commonly used for commercial production of polymers of the acrylic esters, ethylene, vinyl acetate, and acrylonitrile. Solution polymerization of acrylonitrile in water is a particularly convenient example since the monomer is soluble in water, but the polymer is not. On completion of the polymerization the finished polymer may be recovered from the solvent and any residual monomer by simple filtration.

22.3.3. Suspension Polymerization

Water-immiscible monomers, or not more than slightly water-soluble monomers, may be polymerized as a suspension of large droplets in water. The droplets are kept in suspension by agitation and by the use of stabilizers, such as gelatin, talc, or bentonite clay. The free radical initiator used must be soluble in the monomer. Droplet size is 0.01–0.5 cm in diameter in typical operating modes. Polymerization in this way can be pictured as the simultaneous operation of many droplet-sized reactors, which on completion give beads, or "pearls," of polymer. In fact these are the names, which are sometimes applied to this method of polymerization and to the product obtained.

The polymer product is recovered by filtration, washing with water to remove any adhering stabilizer(s), and then dried with warm air.

The kinetic behavior and the methods of predicting molecular weight for suspension polymerizations are virtually the same as for bulk polymerizations. Operating in suspension mode eliminates any temperature control problems since the water phase has both a large heat capacity and a high heat transfer capability. Also there are no difficulties with high viscosities or stirring because of the large proportion of water present. The main negative factors experienced with suspension polymerization relate to the occasional difficulty with complete removal of residual suspending agents that may get incorporated into the finished polymer. There may also be a greater difficulty in controlling monomer vapor loss during the polymerization than for the corresponding bulk polymerization system. Suspension polymerization is commercially appropriate for monomers such as methyl methacrylate, styrene, and vinyl chloride.

22.3.4. Emulsion Polymerization

Polymerization of a monomer in an emulsion system is similar to suspension polymerization in that the bulk medium in both cases is water. But there the similarity ends [8–10]. Monomer and initiator are still present, but this time the initiator must be a water-soluble, monomer-insoluble type. Suitable for this purpose is the persulfate-ferrous ion or peroxide-ferrous ion systems (Eqs. 22.7 and 22.8).

$$S_2O_8^{2-} + Fe^{2+} \rightarrow Fe^{3+} + SO_4^{2-} + SO_4^{-} \qquad 22.7$$

$$PhC(CH_3)_2OOH + Fe^{2+} \rightarrow Fe^{3+} + HO^{-} + PhC(CH_3)_2O^{\cdot} \qquad 22.8$$

Surfactants are used in combination with stirring to produce a suspension of very small aggregates of monomer known as micelles. A small proportion of the total monomer will be present in these micelles of 100–300 nm (0.1–0.3 µm) diameter, and stabilized by a high concentration of surfactant molecules on the surface (Fig. 22.2). Initially, the bulk of the monomer will be present in droplets of 1,500 nm in diameter and larger with no more than a few surfactant molecules present on the surface of each. Agitation performs the primary stabilizing function for the droplets. To visualize a typical emulsion polymerization, there will be up to 10^{18} micelles/cm^3 present, and up to 10^{11} droplets/cm^3. Thus, the micelles present a much larger surface area exposed to the water-soluble initiator, relative to the mass of monomer, than the droplets. This factor, plus the orders of magnitude higher concentration of surfactant molecules on the surface of the micelles, focuses most of the progress of the polymerization in the micelles. As the polymerization proceeds there is a gradual transfer of monomer from the droplets, which slowly decrease in size and number while the micelles gradually increase in size.

The very small size of the micelles means that even the presence of one propagating radical in the micelle represents a very high radical concentration. For this reason there is usually only one propagating radical present per micelle. When a second radical penetrates the polar surfactant

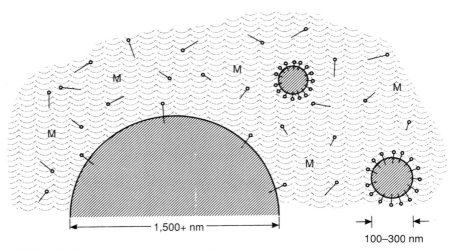

FIGURE 22.2 Pictorial representation of monomer distribution in emulsion polymerization. Dissolved monomer is represented by letters M, molecules of surfactant with hydrophilic ends indicated by open circles, connected to their hydrophobic alkyl tails represented by a solid line.

water-monomer interface, it generally terminates the process. The next radical entering initiates propagation again. Thus, the monomer in one-half of the micelles is actively being polymerized at any point in time, and the monomer in the other half is dormant. Emulsion polymerization, because of this dominant mode of progression, operates via kinetics, which differs significantly from those of bulk polymerization. Molecular weight control, rate, etc., are all affected by different variables. Details of the methods used to evaluate these parameters have been developed and are outlined in standard texts.

The main advantage of emulsion polymerization is that higher degrees of polymerization (DPs) and molecular weights are possible from some monomers by this method than by use of any of the other methods discussed. In some cases, molecular weights obtained may be so high that subsequent processing could be made difficult with the unmodified process. Fortunately, the DPs obtained may be moderated if desired by addition of a chain transfer agent. This additive can decrease the number of monomer molecules, which can add to the propagating radical before termination occurs by radical transfer to the additive. Thus, the concentration of chain transfer agent used can be used to control the molecular weight of the product to the desired range.

Emulsion polymerization presents similar processing difficulties to those of suspension polymerization. The product has to be recovered, in this case usually by coagulation, and then washed and dried. Again it may be difficult to remove all traces of the surfactant, etc., used to stabilize the emulsion, as with the product from suspension polymerization. However, for some applications, such as for latex (water-based) paints and carpet adhesives, the aqueous product from emulsion polymerization may be used directly. For applications, such as these, it is possible to produce lattices (latexes) containing as high as 50% solids.

22.4. IONIC INITIATORS FOR VINYL MONOMERS

Vinylic monomers that are capable of forming a propagating radical on introduction of a radical species can generally be polymerized using a free radical initiator, as just discussed. However, if the double bond of the vinylic monomer has a high electron density contributed by electron-donating substituents on the double bond, such as shown by isobutylene (CH_2=$C(CH_3)_2$) or vinyl ethers, then radical initiation fails to produce polymer. These types of monomers, however, are amenable to polymerization by cationic catalysts because the high electron density on the double bond can stabilize a propagating carbocation (formerly called carbonium ion; proposed new name, carbenium ion), as it forms.

For analogous reasons to the monomer requirements that favor cationic initiators, vinylic monomers with electron-withdrawing substituents on the carbon-carbon double bond are amenable to polymerization by anionic catalysts, since under these conditions the electron-withdrawing substituent assists in stabilization of the propagating carbon ion as it forms. However, this class of monomer is usually still sensitive to free radical-initiated polymerization because of electronic back donation from the electron-withdrawing group to the carbon-carbon double bond (Table 22.4 [11, 12]).

Both cationic and anionic polymerizations proceed via chain mechanisms, like free radical polymerizations. In free radical processes the initiator is incorporated into the polymer chains, as they are formed. But in ionic polymerizations the "termination" processes that produce dead polymer chains

TABLE 22.4 Suitable Initiating Systems for the Chain Polymerization of Representative Vinylic Monomers[a]

Monomer types	Radical	Cationic	Anionic	Coordination
		Suitable initiating systems[b]		
Ethylene	+	+	−	+
Propylene, monosubstituted α-olefins	−	−	−	+
Isobutylene, dialkyl olefins	−	+	−	−
Dienes	+	−	+	+
Styrene	+	+	+	+
Vinyl chloride, tetrafluoroethylene[c]	+	−	−	+
Vinylidene chloride, CH_2=CCl_2[d]	+	−	+	−
Vinyl ethers, CH_2=$CHOR$	−	+	−	+
Vinyl esters, CH_2=$CHOCOR$	+	−	−	−
Acrylic, methacrylic esters, CH_2=$CHCO_2R$, CH_2=$C(CH_3)CO_2R$	+	−	+	+
Acrylonitrile	+	−	+	+
Aldehydes, ketones	−	+	+	n/a

[a]Compiled from Lenz [11] and Billmeyer [12].

[b]+, forms high polymers; −, no reaction or forms only oligomers; n/a information not available.

[c]Product trade names: Teflon, Goretex.

[d]Product trade names: Saran wrap, Saran fiber.

very often also release the active initiating species again. Thus, in ionic polymerizations these are often referred to as catalysts, rather than initiators, to make this distinction.

22.4.1. Cationic Initiating Systems

Suitable catalysts for cationic polymerization may be conveniently classified into three groups. The classic protonic mineral acids, such as H_2SO_4, $HClO_4$, and H_3PO_4, are often useful. Effective catalysis is obtained if the acid selected has high acid strength, and an anion of low nucleophilicity (Eq. 22.9).

$$H^+ + XO_4^- + CH_2{=}CR_2 \rightleftharpoons CH_3{-}\overset{+}{C}R_2 + XO_4^- \qquad 22.9$$

It is this latter requirement that makes the halogen acids a poor choice despite their high acid strength, since the halide ions are good nucleophiles and will also add to the double bond at the same time as hydrogen ion does. Some halogen acids can serve as efficient co-catalysts, however, in combination with a Lewis acid.

Friedel-Crafts reagents (Lewis acids) represent the most widely used class of cationic polymerization catalysts.

General formula, MX_n: $AlCl_3$, $TiCl_4$, BF_3, $SnCl_4$, $ZnCl_2$, etc.

They usually require a co-catalyst, such as water, an organic acid, a halogen acid, or an organic halide, which serves as the proton or cation donor as it associates with the Lewis acid (Eq. 22.10–22.12).

$$AlCl_3 + RCl \rightleftharpoons R^+ + AlCl_4^- \qquad 22.10$$

$$TiCl_4 + H_2O \rightleftharpoons H^+ + TiCl_4OH^- \qquad 22.11$$

$$TiCl_4 + HCl \rightleftharpoons H^+ + TiCl_5^- \qquad 22.12$$

The Lewis acid can function in the absence of a co-catalyst only in special cases, when autoionization is possible (Eq. 22.13).

$$2\,AlBr_3 \rightleftharpoons AlBr_2^+\,AlBr_4^- \qquad 22.13$$

A number of other materials can function as catalysts for cationic polymerization, among them iodine, tropylium salts, triphenylmethyl halides, and t-butyl halides (Eqs. 22.14–22.16).

$$2\,I_2 \rightleftharpoons I^+I_3^- \qquad 22.14$$

$$I_2 + CH_2{=}C(CH_3)_2 \rightleftharpoons ICH_2{-}CI(CH_3)_2 \rightleftharpoons \qquad 22.15$$

$$ICH{=}C(CH_3)_2 + H^+I^-$$

$$\text{or, } ICH_2{-}CI(CH_3)_2 \rightleftharpoons ICH_2^+{-}C(CH_3)_2I^-$$

$$Ph_3CCl \rightleftharpoons Ph_3C^+Cl^- \qquad 22.16$$

Ionizing radiation of energies sufficient to cause heterolytic cleavage may also be used (Eq.22.17).

$$CH_2{=}CHR \xrightarrow[\gamma\text{-rays}]{} CH_2{=}CR^- + H^+ \qquad 22.17$$

The type and activity of a suitable catalyst is often quite closely related to the activity of the monomer to be polymerized. The relatively stable carbocation triphenylmethyl, for example, is only effective with very reactive monomers, such as methyl vinyl ether. Zinc chloride plus co-catalyst will polymerize an appropriate monomer more slowly than a catalyst combination, which uses one of the more active Lewis acids. Also the classic proton acids only rarely give high molecular weight polymers.

22.4.2. Example of Cationic Polymerization

While the detailed mechanistic understanding of many cationic polymerization systems has not as yet been clearly established, there are some that are reasonably well understood. Let us consider the detailed progression of the cationic polymerization of isobutylene using a $TiCl_4/H_2O$ catalyst system as an example. This system cannot be used as a general model for all cationic polymerizations as is possible for free radical-initiated systems, since there is a wider variety of mechanisms seen for cationic, than for free radical, systems.

Initiation involves the reversible reaction of water with $TiCl_4$ to produce the active catalyst, which is then in a form capable of protonating the methylene of isobutylene (Eqs. 22.18 and 22.19). The active propagating species formed is a tightly bound ion pair.

Initiation:

$$TiCl_4 + \underset{\text{co-catalyst}}{H_2O} \rightleftharpoons \underset{\text{active catalyst}}{H^+(TiCl_4OH)^-} \qquad 22.18$$

$$H^+(TiCl_4OH)^- + \underset{\text{isobutylene}}{(CH_3)_2C{=}CH_2} \rightarrow \underset{\text{ion pair}}{(CH_3)_3C^+(TiCl_4OH)^-} \qquad 22.19$$

In a nonpolar, hydrocarbon-type solvent, polymerization may proceed rather slowly because of the steric hindrance of the tightly bound gegen ion, or counter ion, to the active carbocationic center. The rate of polymerization can be increased by use of a more polar solvent, which will increase the spacing between the gegen ion and carbocation, which in turn will increase the rate.

Propagation involves monomer insertion into the carbocationic center, and sigma bond formation via the pi electrons of the entering isobutylene (Eq. 22.20).

Propagation:

$$\underset{+}{\overset{\text{gegen ion}}{(CH_3)_3C^+(TiCl_4OH)^-}} \underset{\underset{\text{single step}}{\rightarrow}}{\overset{CH_2{=}C(CH_3)_2}{}} (CH_3)_3CCH_2\overset{+}{C}(CH_3)_2 \, (TiCl_4OH)^-$$

$$\underset{\text{many steps}}{\overset{(n-1)CH_2{=}C(CH_3)_2}{\rightarrow \rightarrow \rightarrow}} H{\Big[}CH_2C(CH_3)_2{\Big]}_n CH_2\overset{+}{C}(CH_3)_2 \, (TiCl_4OH)^-$$

$$22.20$$

This polymerization is normally carried out at about 170 K ($-100°C$) and produces extremely rapid polymerization, comparable or even faster than the rates observed for many free radical-initiated polymerizations. High molecular weight product begins to appear within a few seconds of initiation.

If the cationic polymerization is carried out at room temperature or higher, both the rate and the molecular weight of the product obtained are lower. This is probably the result of an increased rate of transfer to monomer competing more effectively with the propagation process at the higher temperatures (Eq. 20.21).

Transfer:

$$H[CH_2C(CH_3)_2]_n \ CH_2\overset{+}{C}(CH_3)_2 \ (TiCl_4OH)^- \ + \ CH_2{=}C(CH_3)_2 \ \rightarrow \ \ \ \ 22.21$$

$$\underset{\text{dead polymer}}{H[CH_2C(CH_3)_2]_n \ CH_2\text{-}C(CH_3){=}CH_2} \ + \ \underset{\text{initiating species}}{(CH_3)_3C^+ \ (TiCl_4OH)^-}$$

Hydrogen transfer from the propagating species to monomer produces "dead" polymer, which has a terminal olefin, and regenerates the close ion pair product as given in the second step of the initiation process (Eq. 20.19). This is available to start a new polymer chain all over again. Since the activation energy for this transfer process is higher than the activation energy required for propagation, the rate of transfer competes more effectively with the rate of propagation at higher temperatures than at low temperatures.

Cationic polymerizations generally have no formal termination process. Dead chains are produced by what amounts to an ionic rearrangement of the propagating species, which produces an olefin-terminated chain and regenerates the first active catalyst species (Eq. 22.22).

"Termination":

$$H[CH_2C(CH_3)_2]_n \ CH_2\overset{+}{C}(CH_3)_2 \ (TiCl_4OH)^- \ \rightarrow \ \ \ \ \ \ \ \ \ \ \ \ \ \ \ \ 22.22$$

$$\underset{\text{dead polymer}}{H[CH_2C(CH_3)_2]_n \ CH_2\text{-}C(CH_3){=}CH_2} \ + \ \underset{\text{still active}}{H^+(TiCl_4OH)^-}$$

Thus, like the transfer process just discussed, this does produce dead polymer but also regenerates an active species, which can continue to propagate, unlike a true termination step. Also, as with the intermolecular transfer event, this intramolecular transfer process generally requires larger activation energy than propagation. So this type of transfer also begins to compete more effectively with propagation at higher temperatures. More details of the kinetics for at least some specific cases of cationic polymerization are accessible from standard texts (see Further Reading Section at the end of this chapter).

22.4.3. Suitable Catalysts for Anionic Polymerization

Progress of anionic polymerizations is most favorable with monomers, which have one or more electronegative (electron-withdrawing) groups, such as nitrile or halogen, directly bonded to the vinyl group. Suitable catalysts for initiation of anionic polymerizations fall into one of two classes, according to their mode of action.

The basic class of initiators functions by the addition of the basic anion of a salt to the double bond of a suitable monomer in order to generate the propagating species (Eq. 21.23).

$$\text{MX} + \text{CH}_2\!=\!\text{CHZ} \rightarrow \text{X}\!-\!\text{CH}_2\!-\!\overset{..}{\text{C}}\text{HZ}^-\text{M}^+ \qquad \text{21.23}$$

e.g., where Z $= -\text{CN}, -\text{COR}$

Relatively weakly basic catalysts, such as hydroxide ion or cyanide ion, may be used to initiate polymerization of monomers, which have strongly electron-withdrawing substituents on the vinyl group, such as a nitrile or carbonyl. Thus, suitable monomers would be acrylonitrile or methylmethacrylate.

At the other extreme, strongly basic anionic initiators are required for polarizable monomers or for monomers with relatively poor electron-withdrawing substituents, such as styrene, isoprene, and 1,3-butadiene. With these monomers that have no electrophilic (or hetero) atom to decrease the electron density of the vinyl group, strongly basic anions, such as amide ion or alkyl anion, are required for effective initiation. Amides of sodium or potassium and n-butyllithium are appropriate examples.

Of intermediate utility between these two extremes for anionic polymerization is alkoxide ion, which has adequate base strength for initiation of the ring-opening polymerization of epoxides (Eqs. 22.24–22.26).

$$\text{NaOCH}_3 \rightleftharpoons \text{Na}^+ + \text{CH}_3\text{O}^- \qquad \text{22.24}$$

$$\text{CH}_3\text{O}^- + \underset{\diagdown \; \underset{\text{O}}{} \; \diagup}{\text{CH}_2\!-\!\text{CH}_2} \rightarrow \text{CH}_3\text{O}\!-\!\text{CH}_2\!-\!\text{CH}_2\text{O}^- \qquad \text{22.25}$$

$$\text{CH}_3\text{O}\!-\!\text{CH}_2\!-\!\text{CH}_2\text{O}^- + \text{n}\,\underset{\diagdown \; \underset{\text{O}}{} \; \diagup}{\text{CH}_2\!-\!\text{CH}_2} \rightarrow \qquad \text{22.26}$$

$$\text{CH}_3\text{O}\{\text{CH}_2\!-\!\text{CH}_2\!-\!\text{O}\}_\text{n}\text{CH}_2\!-\!\text{CH}_2\text{O}^-$$

Thus, the relative strength of the base used is important to the effectiveness of the catalysis. A weakly basic anion will not function for a mildly activated monomer, and a strongly basic anion may not function well for a strongly activated monomer.

Donor-acceptor catalysts represent the other type of initiator useful in anionic polymerizations. In this class, the donor entity may be an electronically neutral species, which gives up an electron to the monomer to form the cationic propagating species with monomer (Eq. 22.27).

$$\text{D} + \text{M} \rightarrow \text{D}^+ + \text{M}^- \qquad \text{22.27}$$

Use of sodium metal to initiate the polymerization of methacrylonitrile is an example of this type of process (Eq. 22.28).

$$\text{Na} + \text{CH}_2\!=\!\text{C(CH}_3)\!-\!\text{CN} \rightarrow \text{Na}^+(\!^{.}\text{CH}_2\!-\!\overset{..}{\text{C}}\text{(CH}_3)\text{CN})^- \qquad \text{22.28}$$
$$\text{propagating species}$$
$$\text{(may dimerize)}$$

Another donor-acceptor approach is to use a donor entity that is already charged. In the initiation process this gives up its charge to monomer to form a neutral entity and the propagating species (Eq. 22.29).

$$D^- + M \rightarrow D + M^-$$ 22.29

22.4.4. Anionic Polymerization of Styrene Using Amide Ion

Polymerization of styrene in liquid ammonia at low temperatures catalyzed by potassium metal represents a good example of a base-initiated anionic polymerization. Styrene, being relatively nonpolar, requires the strongly basic amide ion for effective anionic polymerization. The catalyst is made, *in situ*, usually before the styrene is added, by the addition of small pieces of potassium metal to liquid ammonia kept at dry ice (solid carbon dioxide) temperatures (Eq. 22.30).

$$2 \, K + 2 \, NH_3 \rightarrow 2 \, KNH_2 + H_2$$ 22.30

Initiation involves two steps, ionization of the potassium amide in the extremely polar liquid ammonia, then addition of amide anion to the styrene, which is added to the system at this stage (Eqs. 22.31 and 22.32).

Initiation:

$$KNH_2 \rightleftharpoons K^+ + H_2N:^-$$ 22.31

$$H_2N:^- + CH_2{=}CHPh \rightarrow H_2N{-}CH_2{-}(Ph)\ddot{C}H^-$$ 22.32
$$\text{propagating species}$$

Potassium ions are also present in this solution, but because of the high dielectric constant and high solvating capability of liquid ammonia the propagating species formed here is truly a free ion, not more than very loosely associated with the cation. The less polar the solvent used, the greater would be the degree of association between the propagating anion and the cation present. Thus, the polarity of the solvent chosen can affect the rate of formation and sometimes also the stereochemistry of the anionic polymerization product.

Propagation involves addition of the propagating species to styrene monomer (Eqs. 22.33 and 22.34). Head-to-tail addition, the normal mode, retains the negative charge at a benzylic, resonance-stabilized position.

Propagation:

$$H_2N{-}CH_2{-}(Ph)\ddot{C}H^- + CH_2{=}CHPh \rightarrow$$ 22.33
$$H_2N{-}CH_2{-}(Ph)CH{-}CH_2{-}(Ph)\ddot{C}H^-$$

$$H_2N{-}M_n^- + M \xrightarrow{k_p} (H_2N{-}M_{n+1})^-$$ 22.34

As seen with cationic polymerization, there is also no formal termination step with anionic-initiated systems. Dead chains are produced by transfer of a proton (or other positive fragment), usually from the solvent to the anion. When this happens with this particular example, it regenerates amide ion again, the primary initiating species (Eq. 22.35).

Termination:

$$H_2N \{ CH_2-(Ph)CH \} CH_2-(Ph)\ddot{C}H^- + NH_3 \xrightarrow{k_{tr}} \qquad 22.35$$
$$H_2N \{ CH_2-(Ph)CH \}_{n+1}H + H_2N:^-$$

This is now available to generate a new propagating species and start the process all over again. The following experimental evidence supports this mechanistic interpretation. The rate is found to increase with increased amide ion concentration or with an increased concentration of styrene. Rate decreases with added potassium ion, presumably by increasing the extent to which the propagating species was associated with it. Also each polymer chain in the product contains only one nitrogen atom and has no residual unsaturation.

An anionic polymerization carried out under an inert atmosphere, since it has no formal termination step and uses relatively stable propagating species, is sometimes referred to as a living polymerization, and the product as a living polymer. As such, the DP of these systems becomes simply the moles of monomer present divided by the moles of initiator used (Eq. 22.36).

$$\overline{X}_n = \frac{\text{moles of monomer}}{\text{moles of initiator}} \qquad 22.36$$

The other unique features of living polymerizations are that $\overline{M}_n$ varies directly with conversion (Fig. 22.3, cf. Fig. 22.1) and the heterogeneity index (polydispersity) is 1, or very close to it.

22.4.5. Sodium–Naphthalene Anionic Initiation

An example of the donor–acceptor initiation process is demonstrated by the sodium–naphthalene system devised by Szwarc [13]. The initiation sequence begins with the transfer of an electron from sodium metal to naphthalene. In an ether-type solvent, such as tetrahydrofuran (THF) sodium naphthalide,

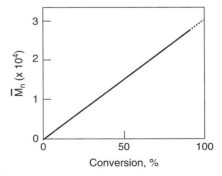

FIGURE 22.3 Relationship of number average molecular weight, $\overline{M}_n$, to percent conversion for a living polymerization.

$$\text{Na} \quad + \quad \text{(naphthalene)} \quad \xrightarrow{\text{THF}} \quad \text{Na}^+ \left[\text{(naphthalene radical anion)} \right]^{-} \qquad 22.37$$

green solution

in which the naphthalene exists as an anion radical, is formed quantitatively (Eq. 22.37).

The solution becomes an intense green, and an electron spin resonance (ESR) measurement at this point confirms the presence of the naphthalene radical anion. Addition of styrene at this stage causes the color of the solution to change to red and the ESR signal disappears. The styryl radical anion formed initially is correctly viewed as a resonance hybrid (Eqs. 22.38 and 22.39).

$$\text{Na}^+ \left[\text{(naphthalene radical anion)} \right]^{-} \quad + \quad PhCH=CH_2 \quad \rightarrow \qquad 22.38$$

green solution

$$(Ph\dot{C}H \overset{\cdots}{=} CH_2)^- Na^+ \quad + \quad \text{(naphthalene)} $$

$$(Ph-\dot{C}H-\ddot{C}H_2 \quad \leftrightarrow \quad Ph\ddot{C}H-\dot{C}H_2)^- \qquad 22.39$$

The styryl radical anion species is much more reactive than the naphthalene radical anion, and rapidly couples to form a dimeric dianion, the source of the red color and the reason for the disappearance of the ESR signal (Eq. 22.40). The dimeric dianion is a double-ended anionic propagating species useful for the initiation of a number of valuable homopolymerizations (Eq. 22.41).

$$2 (Ph\dot{C}H \overset{\cdots}{=} \ddot{C}H_2)^- Na^+ \quad \rightarrow \quad Na^+ \overline{C}H(Ph)-CH_2CH_2-\overline{C}H(Ph)Na^+ \qquad 22.40$$
$$\text{1, gives a red solution}$$

$$1 \; + \; nPhCH=CH_2 \rightarrow \qquad 22.41$$
$$+PhCH-CH_2)_{n/2}CH(Ph)-CH_2CH_2-CH(Ph)(CH_2-CHPh+_{n/2}$$

The same dianion is also an invaluable initiator for the block copolymerization of compatible monomers. Its ability to add monomer to both ends simultaneously doubles the rate of accumulation of blocks as compared to the number of steps required with a single ended propagating species (Eq. 22.42).

$$1(=\sim R \sim) \xrightarrow[\text{monomer A}]{6A} AAA \sim R \sim AAA \xrightarrow[\text{monomer B}]{8B} \qquad 22.42$$
$$BBBBAAA \sim R \sim AAABBBB$$

22.5. BASIC POLYMER STEREOCHEMISTRY

Polymerization of ethylene itself gives essentially a high molecular weight alkane product with which there is no opportunity for differing stereostructures. However, many important vinyl monomers are monosubstituted, or 1,1-(gem) disubstituted. Most polymerizations progress via a head-to-tail addition of the propagating center to the monomer. For this reason when a substituted ethylene is polymerized the resulting product can have various stereochemical possibilities.

A monosubstituted monomer that polymerizes in a head-to-tail fashion can give two possible stereoregular polymer structures. One of these will have all chiral centers of the chain in the same configuration, and the other will have every second chiral center of the same configuration and the alternating chiral center of the opposite configuration. The terms for these configurations, originally proposed by Guilio Natta, are tactic for the first type and syndiotactic for the second (Fig. 22.4). The nonstereoregular structure that results from random configurations along the chain is referred to as atactic, literally "without arrangement."

A stereoregular polymer is identified by adding a prefix to the name, which signifies the type of tacticity (e.g., isotactic polypropylene). If the polymer is atactic then the generic term for the polymer is used (e.g., use of polypropylene alone means atactic polypropylene). One has to be careful, however, because common usage sometimes ignores or overlooks these naming conventions of polymers of monosubstituted ethylenes. Introduction of two, or very occasionally three, substituents on the ethylene or incorporation of a heteroatom into the backbone of a synthetic polymer can increase the number and complexity of possible stereoregular forms. Many of these more complex possibilities are discussed in detail in the text by Odian [14].

Knowledge of the existence of the various stereoregular forms and a comprehension of their respective structures adds both to the scientific interest and an appreciation of the physical properties and the potential utility of these materials. In general terms, atactic (heterotactic) polymers tend to be amorphous (noncrystalline) in the solid state. They will have a lower melting

Fischer Projections   Newman Projections

Atactic   Isotactic   Syndiotactic   Atactic   Isotactic   Syndiotactic

FIGURE 22.4   Different representations of three types of polymer stereostructures.

point than their stereoregular equivalents, and lower tensile strength. Some will be soft and waxy in nature. In contrast to this, the stereoregular polymers of either the isotactic or the syndiotactic form possess a stereochemistry that lends itself to more rapid crystallization and usually improved physical properties. The regular array of substituents in stereoregular polymer chains allows closer packing. For this reason stereoregular polymers often spontaneously crystallize to a significant extent, or may be processed in ways to promote crystallization during the fabrication or spinning of these materials. The highly ordered structures obtained in crystalline polymers give these resins higher melting points, higher tensile strengths, greater solvent and chemical resistance, and lowered gas permeability than that obtained for the corresponding atactic materials. So scientifically and commercially the stereochemistry of polymers is of great interest and significance. Variations in the method used to prepare synthetic polymers, such as solvent or catalyst choice, or polymerization temperature, can significantly influence the stereochemistry of the polymer product obtained.

22.5.1. Stereochemistry of Free Radical-Initiated Polymerization

The configuration of an asymmetric carbon in a polymer is determined at the time of monomer addition to the propagating center. To visualize the situation for the free radical-initiated polymerization of vinyl chloride, approaching monomer may produce either the same or the opposite configuration for the chlorine-substituted carbon in the adding monomer, as already present in the adjacent unit of the propagating center (Eqs. 22.43 and 22.44).

22.43

sp2 type hybrid,
nearly planar

isotactic

22.44

syndiotactic

Addition with the same configuration is favored if the substituent of the incoming monomer is on the same side as the substituent on the active site of the propagating center and produces an isotactic link or unit. Addition of monomer with the opposite configuration is favored if the substituent of the incoming monomer is on the opposite side to that of the propagating center and produces a syndiotactic link or unit in the polymer. Since there is generally less steric interaction at the time of bond formation for syndiotactic addition under these circumstances, syndiotactic polymer is generally the favored configuration for free radical-initiated polymerizations.

The foregoing comments are not meant to imply that the configuration of substituents in polymers produced under free radical conditions will all be syndiotactic, since the extent of this influence is strongly affected by the size and polarity of the substituent group in the monomer, as well as other factors. Also the configuration of the monomer unit just added is not finally determined until the next monomer bonds to the propagating center. The higher temperatures often used to activate typical free radical initiators may also provide sufficient collision energy for some isotactic addition to occur. Therefore, normally significant blocks of syndiotactic polymer would be observed with the occasional isotactic link. Thus, for steric reasons free radical-initiated polymerizations generally favor syndiotactic addition over isotactic addition. The preference for syndiotacticity may be increased with a bulky or a highly polar substituent and may be enhanced slightly by carrying out the polymerization at lower temperatures. This is feasible by changing from a thermally activated to a photolytically activated initiator, such as AIBN (azo-*bis*-isobutyronitrile) or a disulfide. In these ways it is possible to influence the stereoregularity of free radical-initiated polymerizations, but not to control it completely.

22.5.2. Stereocontrol in Ionic Polymerizations

In the cationic-initiated polymerization of alkyl vinyl ethers it is possible to exercise fairly rigorous control of the configuration of the product by appropriate choice of the monomer and conditions. For example, isobutyl vinyl ether polymerized by BF_3 etherate at 195 K in toluene can give isotactic polymer [15]. In this low polarity solvent, close association of the gegen ion with the cationic propagating center helps to block one mode of entry of fresh monomer (Eq. 22.45).

22.45

$R=-CH_2CH(CH_3)_2$

4-centered transition state

isotactic polyether product

Stereoregulation is not obtained if the isobutyl vinyl ether is replaced by ethyl vinyl ether or isopropyl vinyl ether. This makes it clear that the bulk and placement of the isobutyl group also plays a role in the stereoregulation. It is thought that a loosely associated six-membered ring helps to both stabilize the propagating carbocation and, with the help of the closely associated gegen ion, guide the approach of monomer to it in a single, reproducible orientation. Actual bond formation could occur via a four-centered cyclic transition state to yield the observed isotactic product. Thus, it is possible to exercise stereocontrol of cationic polymerizations but the examples available are generally special cases.

It is also possible to exercise control of the stereochemistry of vinyl polymerizations, which use anionic initiators. As with the cationic systems these are often special cases. For example, anionic polymerization of methyl methacrylate in a polar solvent, such as 1,2-dimethoxyethane, will give a highly syndiotactic product (Table 22.5). The polar solvent provides good gegen ion separation from the propagating anion so that it behaves almost like a free ion as the polymer forms. Under these conditions, the minimum steric interaction between the propagating anion and the incoming gem-disubstituted monomer is obtained when there is an alternation of orientation of the incoming monomer molecules at the time of bond formation. This results in syndiotactic product.

For the anionic polymerization of methyl methacrylate in toluene, a much less polar solvent, the gegen ion remains closely associated with the propagating center (Eq. 22.46).

$$ \text{(22.46)} $$

TABLE 22.5   Influence of Solvent Polarity on the Polymerization of Methyl Methacrylate[a]

Catalyst, conditions	Solvent	Polymer tacticity
n-Butyllithium		
0°C, −78°C	Toluene	Isotactic
−78°C	Tetrahydrofuran	Syndiotactic
9-Fluorenyl lithium		
−68°C	Liquid ammonia	Syndiotactic
−60°C	1,2-dimethoxyethane	Highly syndiotactic
−60°C	Toluene	Isotactic
−60°C	Mixed 1,2-dimethoxy-ethane/toluene	Mixed block copolymer, "blocky" atactic
Phenyl magnesium bromide		
0°C, 30°C	Toluene	100% isotactic
γ-Rays, −78°C	No solvent, i.e., bulk polymerization	95% syndiotactic

[a]Data selected from Odian [14, 16]. All conducted under an inert atmosphere.

Approach of methyl methacrylate monomer is possible in only one orientation because of the steric hindrance of the gegen ion. As a result, at the critical bond-forming step the substituents of the incoming monomer are always oriented in the same way, which results in isotactic product. The outcomes for the other examples tabulated can be rationalized in similar ways.

22.6. COORDINATION POLYMERIZATION

Until 1953 there were only three mainline methods in common use for the initiation or catalysis of the polymerization of vinylic monomers. These were based on free radical, cationic, and anionic techniques. However, in that year Karl Ziegler, working in Germany, announced his discovery that ethylene could be polymerized to a high molecular weight, easily crystallized product under relatively mild conditions. The key to this success lay in the preparation of a heterogeneous catalyst from titanium tetrachloride or titanium trichloride and an aluminum alkyl. Polyethylene prepared using this catalyst had significantly different properties than the first commercial product and so was referred to as high density polyethylene (HDPE) (Table 23.1). The first commercial polyethylene produced by free radical initiation was called low density polyethylene (LDPE) to distinguish it. The HDPE resin was higher melting, more dense, and tougher than LDPE because it was substantially free of the spurious branches that are present in LDPE. Despite predictions that the new HDPE would supersede and replace the slightly older product, each has properties, which are more suitable for particular applications so that both continue in large-scale production today (Table 22.2). Details of the structural differences are discussed in Section 23.1.

Within 2 or 3 years of Ziegler's announcement Guilio Natta, working for Montecatini in Italy, tried Ziegler's catalyst systems for propylene. This could not be predicted to work without experiment, since the free radical systems that could be used to polymerize ethylene only yielded oligomeric, low molecular weight oils with propylene. He had some initial success, and soon realized that a stereoregular polypropylene would be required for the product to be commercially useful. Changes in the catalyst, in particular a change from $TiCl_4$ to $TiCl_3$ in the transition metal component of this, gave the desired change in activity and yielded substantially isotactic polypropylene. Commercially useful isotactic polypropylene, like high-density polyethylene, only became feasible with the coordination polymerization discoveries of Ziegler and Natta. American production of these two polymers alone now totals of the order of 5 million metric tonnes per year (Table 22.2). The Nobel Prize in chemistry was awarded jointly to Otto Ziegler and Guilio Natta in 1953 in recognition of their discoveries.

22.6.1. Coordination (Ziegler–Natta) Catalysts and Mechanisms

There is a very wide range of catalyst possibilities with which it is feasible to obtain coordination polymerization, depending on the monomer to be polymerized. In the early forms they consisted of a poorly defined product obtained from the mixture of an alkyl or aryl compound of an element from Groups IA to IVA of the periodic table, with a halide or ester of a transition element from Groups IVB to VIIIB. Each catalyst produced was—and still is—highly specific for a particular monomer or monomer mixture (to produce copolymer), and there is still little theory available to enable prediction of which particular catalyst components should be combined for a new monomer substrate, or to yield a different polymer tacticity from an existing monomer. The majority of Ziegler–Natta catalysts are heterogeneous, and the stereoregularity induced in the polymer product is thought to be the consequence of this. Support for this suggestion is obtained from the discovery that vigorous mixing of the first-generation, sludgy catalyst under high shear conditions sufficient to homogenize it simultaneously decreases catalytic activity and yields only an atactic product. Homogeneous Ziegler–Natta catalysts are known, but in general these yield atactic polymers and hence are not of commercial importance.

While the exact nature of the active material present is still poorly understood, it is thought that the catalytic material produced from aluminum triethyl and titanium trichloride has a bridged structure as shown (first structure, Fig. 22.5). A bimetallic mechanism, originally proposed by Natta, illustrates a probable mode of progression of a substituted vinylic monomer to stereoregular polymer.

From the utility of coordination polymerization to produce a useful, stereoregular (easily crystallized) product from propylene in the late 1950s this method has now been expanded to be useful for the polymerization of a variety of monosubstituted α-olefins. This polymerization method

FIGURE 22.5 A bimetallic mechanism for coordination (Ziegler–Natta) polymerization.

provides the only avenue to useful products from these particular monomers (Table 22.4). Coordination polymerization also provides a useful alternative to other methods of polymerization of monomers such as ethylene, the dienes, styrene, haloalkenes, etc. Details of some of these processes are given in Chap. 23.

REVIEW QUESTIONS

1. Outline methods that may be used to experimentally determine whether a polymerization is proceeding via a step condensation or a chain reaction mechanism. Explain your suggestions.
2. (a) Describe and explain the monomer characteristics that best lend themselves to each of the following initiating systems: free radical, anionic, and cationic.
   (b) Which of the initiating systems of part (a) has the widest applicability to monomer polymerization and why?
3. (a) Explain the stereochemistry of vinyl-type addition polymerizations using each of the four available types of initiating systems.
   (b) Describe for each system of part (a) the effect of solvent (or excess monomer) polarity on the degree of stereocontrol which may be possible.
4. (a) Sketch 3 or 4 repeating units of Fischer projections of the following polymers: polyacrylonitrile, poly(methyl methacrylate), polypropylene, polystyrene, poly(vinyl acetate), poly(vinyl chloride), poly(vinylidene chloride), and Teflon.
   (b) Which of these polymers could exist as isotactic and syndiotactic stereoisomers?
5. (a) List and briefly describe variables which could be altered, which would tend to decrease the molecular weight of the product of a free radical chain type polymerization.
   (b) Outline all possible measures that could be taken and the direction in which they should be altered, to increase the molecular weight of a free radical addition polymer product.
6. Name the process and describe the type of polymer that would be expected from a radical chain polymerization if some chain transfer to polymer occurred during the process.

FURTHER READING

J. Brandrup and E.H. Immergut, "Polymer Handbook," Wiley, New York, 1999.
R.G. Gilbert, "Emulsion Polymerization: A Mechanistic Approach," Academic Press, San Diego, California, 1995.
T. Meyer and J. Keurentjes, "Handbook of Polymer Reaction Engineering," Wiley, New York, 2005.
E.P. Moore, Jr., "Polypropylene Handbook," Hanser, Munich, 1996.
R.B. Seymour and C.E. Carraher, Jr., "Polymer Chemistry: An Introduction," 4th ed., Dekker, New York, 1996.

REFERENCES

1. Facts and Figures for the Chemical Industry, *Chem. Eng. News*, **73**(26), 36–79, June 26 (1995).
2. E.M. Kirschner, Growth of Top 50 Chemicals Slowed. *Chem. Eng. News*, **74**(15), 16–22 Apr. 8 (1996).
3. Production, Some Improvement, *Chem. Eng. News*, **82**(27), 49–60, July 5 (2004).
4. A.H. Tullo, Synthetic Rubber, *Chem. Eng. News*, **81**(15), 23–26, Apr. 4 (2003).
5. Chemical Prices, *Chemical Market Reporter (New York)*. **246**(26), 26–33, June (1995).
6. Prices and People, *Chemical Market Reporter (New York)*. **259**(25), 31–35, June 25 (2001).
7. M.F. Cunningham, K.F. O'Driscoll and H.K. Mahabadi, Bulk Polymerization in Tubular Reactors, I. Experimental Observations on Fouling. *Can. J. Chem. Eng.* **69**, 630–638 (1991).
8. M. Morton, Mechanisms of Emulsion Polymerization. *Elastomerics*, **120**, 2–23, Mar. (1988).
9. D.-Y. Lee, J.-F. Kuo, and J.-H. Wang, Study on the Continuous Tubular Reactor for Emulsion Polymerization of Styrene. *Polym. Eng. Sci.* **30**, 187–192 (1990).
10. D.A. Paquet and W.H. Ray, Tubular Reactors for Emulsion Polymerization (I. Experiments; II. Model Comparisons). *AIChE J.* **40**, 73–87, 88–96, Jan. (1994).
11. R.W. Lenz, "Organic Chemistry of Synthetic High Polymers," Wiley-Interscience, New York, 1967.
12. F.W. Billmeyer, Jr., "Textbook of Polymer Science," 3rd ed. Wiley-Interscience, New York, 1984.
13. M. Szwarc, "Carbanions, Living Polymers and Electron Transfer Processes," Interscience, New York, 1968.
14. G. Odian, "Principles of Polymerization," 3rd ed., pp. 604–690. Wiley, New York, 1991.
15. C.E.H. Bawn and A. Ledwith, Stereoregular Addition Polymerization. *Q. Rev. Chem. Soc.* **16**, 361–434 (1962).
16. G. Odian, "Principles of Polymerization," 2nd ed., p. 588. Wiley, New York, 1991.

23

COMMERCIAL ADDITION (VINYL-TYPE) POLYMERS

The Federal Liberals are the Teflon Party. Nothing sticks to them.
—Dave Barrett (NDP Party), 1996

23.1. SYNTHETIC VINYL ADDITION POLYMERS

As with condensation polymers many examples of biochemically formed vinyl addition polymers, such as the poly-*cis*-isoprene found in the sap of rubber trees, were known long before we were able to replicate these materials even in the laboratory. Our ability to initiate and control the preparation of vinylic polymers on a laboratory scale came in the early 1930s, substantially later than the commercialization of phenol-formaldehyde condensation polymers. Since then, however, starting with the synthesis of polyethylene, then poly(vinyl chloride) (PVC), synthetic rubbers and polystyrene, the scale of production of this class of polymer has outstripped the polycondensation class by more than an order of magnitude. Table 23.1 displays some representative production figures to illustrate this.

Most of these products now supply large scale applications, which were not even anticipated at the time of their process development. The preparative details and properties of several of the more important processes used to produce commercially important vinylic addition polymers are discussed in the following sections.

23.2. POLYETHYLENE (PE)

Polyethylene was the first of the vinylic polymers to be produced commercially and the total of its various forms are still produced on the largest scale of any commercial plastic (Table 23.1). It was first synthesized unintentionally at the facilities of Imperial Chemical Industries (now ICI, U.K.) in 1933, when a

Table 23.1   Production of Some Commercial Vinyl Addition Polymers[a]

	Thousand metric tonnes		
	Western Europe [1]	**U.S.A. [2]**	**World [3]**
	2003	**1994**	**2003**	**1996**
Polyethylenes, total	13,492	9,397	16,605	40,100
Low density	8,062[b]	2,077	3,540	–
Linear low density	–	2,278	5,052	–
High density	5,430	5,042	7,125	–
Polypropylene	7,879	4,327	8,013	20,500
Polystyrene and copolymers:				
Polystyrene	3,136	2,653	2,900	10,700
ABS copolymer, and others	803	1,465	1,296	3,910
(Acrylonitrile-Butadiene-Styrene)				
Styrene-butadiene rubber	–	851	781	–
Poly(vinyl chloride) & copolymers	5,832	5,312	6,669	22,500

[a]Data from PlasticsEurope[1], *Chemical & Engineering News* [2] and Tullo [3]
[b]Includes linear low density polyethylene.

high-pressure reaction of benzaldehyde with ethylene produced a waxy, white solid product that did not contain oxygen. Closer examination revealed this unexpected product to be polyethylene, and subsequent development work led to the construction of a commercial plant that started production in 1939.

The pressures used in the polymerization vessels of the early ICI polyethylene plants were among the highest used in the chemical industry, 50–300 MPa (500–3,500 atm). Until 1955 all commercial polyethylene was produced by the high-pressure process, or minor variants of this, at pressures of up to 350 MPa. In that year the first of the polyethylene plants that used the Ziegler-devised low-pressure polymerization process began production, and 2 years later an analogous process developed by Phillips Petroleum also came on stream. These production units used much more moderate conditions and were followed by other facilities, which also used low-pressure techniques. The new low-pressure polyethylene product of these plants was immediately distinguishable from the original product from its higher density, $0.94–0.96 \, \text{g/cm}^3$, compared to the $0.91–0.93 \, \text{g/cm}^3$ of the high-pressure product. It also differed in other important respects (Table 23.2). The initial high pressure product became known as low-density polyethylene (LDPE) to distinguish it from the newer, high-density polyethylene (HDPE) product.

23.2.1. Low Density Polyethylene (LDPE)

High-pressure processes of 50–350 MPa (500–3,500 atm) catalyzed using traces of oxygen or peroxides are used at temperatures of 80–300°C

■ **TABLE 23.2   Properties and Product Breakdown of Commercial Polyethylenes**

	High-pressure	Low-pressure	
Commercial symbol	LDPE	HDPE	LLDPE[a]
Density, g/cm^3	0.91–0.93	0.94–0.96	0.918–0.94
Crystallinity	50–70%	80–90%	–
Crystalline m.p., T_m, °C	110	135	123
Glass transition temperature, T_g, °C	−120	−120	–
Tensile strength, MPa	10	10	30
Elongation at break, %	450	700	500
Proportion of total, 1988:			
U.S.A., %	35	46	19
EEC,[b] %	57	34	9

[a]Refers to linear low-density polyethylene, also produced by coordination polymerization at low pressures. It is not a homopolymer of ethylene; see text for details.

[b]European Economic Community.

to convert high purity ethylene into the low-density polymer product (Eq. 23.1).

$$n\ CH_2 = CH_2 \xrightarrow{\ R^{\cdot}\ } \ \dashv CH_2 - CH_2 \dashv n \qquad 23.1$$

$$\text{b.p.} -104°C\ (169\ K) \qquad T_m(\text{m.p.})137°C$$

High pressures are necessary because the propagating radicals are short lived. High monomer concentrations produced by the high pressures help to trap the short-lived radicals involved with a sufficient number of molecules of ethylene before termination to produce a high molecular weight product. When oxygen (0.03–0.1%) is used as the catalyst, conditions of 150 MPa and 190–200°C are sufficient to generate the initiating radical species *in situ*. Tubular reactors or high-pressure autoclaves are operated in continuous mode to ethylene conversions of 10–30%, depending on the process. Good heat dissipation and mixing are essential as the reaction is highly exothermic. Without these precautions a hot spot could develop to give extremely rapid localized formation of carbon, hydrogen, and methane from the 99+% ethylene initially contained in the reactor.

Once the desired ethylene conversion is obtained, the reaction stream passes to a high-pressure separator where the bulk of the unreacted ethylene is removed from the product and recycled (Fig. 23.1 [4]). High-pressure separation of most of the unreacted ethylene conserves compression costs in much the same way as high-pressure hydrogen recovery does during ammonia production. After high-pressure ethylene separation the hot product moves to a low-pressure separator where the rest of the unreacted ethylene is removed. The polyethylene product is then extruded into cold water and pelletized, dried, and then either bagged or shipped in bulk containers for delivery to fabricators.

The essential reaction to polymer, which occurs in the high-pressure process to polyethylene, is correctly represented by Eq. 23.1. However, careful examination of the high-pressure product by infrared spectroscopy shows that it has 20–30 methyl groups per 1,000 carbon atoms of chain, i.e., far more than can be accounted for by the 1 or 2 chain end methyl groups expected

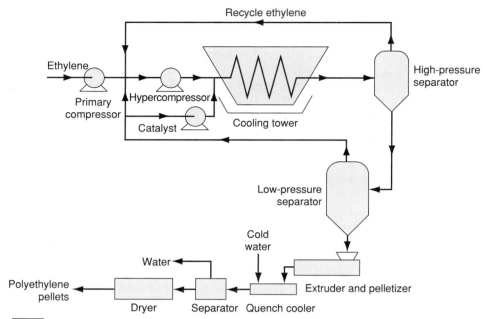

FIGURE 23.1   The high-pressure process to low-density polyethylene. (From Austin [4] with permission.)

per 1,000 carbon atoms. This is explained by the existence of branches on the polyethylene produced by the high-pressure process. Short branches are produced by a process called "back-biting," involving a six-membered cyclic transition state (Eq. 23.2).

23.2

Depending on the mode of back-biting and the extent of subsequent propagation before termination, this can give ethyl, butyl, and occasionally longer branches. These outcomes are at least partly the consequence of the high pressure used, which tends to cause the propagating chains to fold back onto themselves. In practice, the ratio of butyl to ethyl branches in high-pressure polyethylene is about 1:2, with the occasional longer branch and branched branch. The existence of branching in the product from this process is what confers the less dense packing of polymer chains in the solid, which leads to lower crystallinities and the lower density for this product.

23.2.2. High Density Polyethylene (HDPE)

Several commercial processes are used to produce high-density polyethylene. All employ more moderate pressures and most also use lower temperatures than the low-density polyethylene processes. The Ziegler-developed process uses the mildest conditions, 200–400 kPa (2–4 atm) and 50–75°C, to polymerize a solution of ethylene in a hydrocarbon solvent using a titanium tetrachloride/aluminum alkyl-based coordination catalyst. After quenching the polymerized mixture with a simple alcohol, the catalyst residues may be removed by extraction with dilute hydrochloric acid or may be rendered inert by a proprietary additive. The product is almost insoluble in the hydrocarbon solvent, so is recovered by centrifuging and drying. The final product is extruded into uniform pellets and cooled for shipping to fabricators.

The other processes all use catalysts placed on the surface of a solid support to effect the polymerization of ethylene. This may be conducted in solution in a hydrocarbon solvent, such as in the Phillips and Standard oil processes, or in the gas phase in a fluidized bed reactor, as used in the Union Carbide process [5] (Fig. 23.2). The pressures and temperatures used, 1–10 MPa and 90–300°C depending on the process, are higher in both cases than for the Ziegler process but much more moderate than those used in the high-pressure process. The product is recovered from the solution processes in a manner similar to that used for the Ziegler process. In the Union Carbide fluidized-bed process polyethylene granules (with traces of catalyst) form the bed material, and grow in size with time. Once the granules reach about 1-mm average diameter they are removed from the bottom of the bed, where they tend to accumulate by gravity, via a gas lock chamber. The granules are then flushed with nitrogen after which they are ready for storage or shipment without requiring further processing. The concentration of catalyst residues

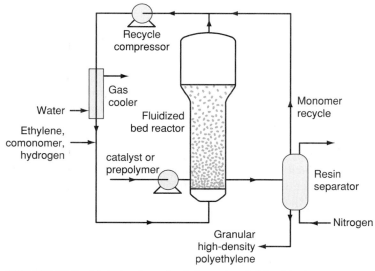

FIGURE 23.2 High-density polyethylene via a fluidized bed process.

is so low that removal is unnecessary. Stabilizing additives may be used in some cases.

High-density polyethylene has less spurious branching than the low-density material, about 0.3–5 ethyl branches per 1,000 carbon atoms. The much lower extent of branching in HDPE enables closer packing of the polymer chains in the solid state. Closer packing explains both the higher density and the higher crystallinities observed for this material as compared to those of low-density polyethylene.

23.2.3. Linear Low-Density Polyethylene (LLDPE)

The LLDPE product represents the outcome of a method developed to produce a low-density polyethylene but by using the more moderate, and therefore less costly, conditions employed by the processes used to produce high-density polyethylene. It is not, strictly speaking, a "polyethylene" since it is not a homo-polymer of ethylene. An LLDPE is actually a copolymer of ethylene, which includes traces of 1-octene (Dow and Du Pont), 1-hexene (Phillips), or 1-butene (Union Carbide). This results in a polymer that has entirely short branches, and these are more uniformly spaced along the backbone than in LDPE. The spacing of the branches obtained in these cases can be closely controlled by the propor-tion of α-olefin to ethylene used in the feed, and the lengths by the choice of the α-olefin comonomer. Properties intermediate to those of low- and high-density polyethylene are obtained for the product (Table 23.2).

23.2.4. New Polyethylenes

It is appropriate to describe here the details of a high-tensile-strength, ultrahigh-molecular-weight polyethylene fiber recently developed by Allied and Honeywell [6, 7]. This material, produced by methods, which results in a very high degree of chain orientation along the axis of the fiber, is said to be 30–40% stronger on a weight for weight basis than the aramids, its nearest competitor. An interesting application is as a thin, lightweight knitted glove to be worn under latex surgical gloves. These confer about 30 times the cut resistance of latex gloves alone, and 15 times the resistance of a medium-weight leather work glove, while retaining "feel" for the surgeon.

In 2002 Dow Chemical announced their initial market offering of an elastomeric fiber based on monofilament polyethylene to be called Lastol. This is resistant to chlorine and is to be priced to compete directly with Spandex fiber derived from polyurethane technology [8].

Also worthy of mention are the new very-low-density polyethylenes (VLDPEs), which are produced using metallocene-based catalysts [9]. These materials, which remain largely experimental at the moment, have a density of less than $0.9\,\text{g/cm}^3$.

23.2.5. Uses of Polyethylene

Applications of polyethylene depend on the resin type. An interesting early application for LDPE was as a low power factor electrical insulation for the wiring of prototype radar installations, which were developed in the U.K. during World War II. The existing plasticized PVC was poor in this respect because of the very polar carbon-chlorine bonds. About two-thirds of the LDPE is used to make film, mostly for packaging. A further 10% is used as a paper coating and barrier layer for paper products used in food packaging, such as milk cartons. Only about 5% of the total is used to make injection-molded products. These are formed by forcing the molten resin into a closed, stainless steel mold cavity of the desired shape followed by rapid cooling of the mold to solidify the shaped resin. The sectional mold is then opened and the article ejected from the cavity, the whole process being repeated every few seconds.

In contrast with the applications for LDPE, about two-thirds of high-density polyethylene is used in molding applications, split about 40% into blow molding of bottles and containers, and 25% into injection-molded products. A further 10% of the total HDPE resin output is consumed in each of the pipe and electrical conduit fabrication areas, and for film and sheet products.

Linear low-density polyethylene (LLDPE) is used for substantially the same applications as LDPE. As the markets for LDPE have expanded, LLDPE plants have been built to supplement the supply of resin to these markets. Very few, if any, new LDPE plants have been built in recent years; in fact, there are indications that LLDPE plants are being built to replace some aging LDPE units. Fortunately this is usually long after the original capital cost of the LDPE plant has been recovered from marketed product.

23.3. POLYPROPYLENE (PP)

Early attempts to obtain polypropylene by vinylic polymerization methods yielded atactic, low molecular weight material of little practical utility. Giulio Natta first experimentally produced isotactic polypropylene of useful molecular weight in 1954 (Eq. 23.3).

$$\text{n } CH_2 = CHCH_3 \quad \rightarrow \quad \text{isotactic } \{CH_2 - CHCH_3\}_n \qquad 23.3$$
$$\text{b.p. } -48°C \text{ (225K)} \qquad \qquad T_m \text{ (m.p.) } 175°C$$

The Montecatini Company began to produce the isotactic resin on a commercial scale in 1957. It is now produced by several slurry and gas phase processes similar to those used for high-density polyethylene. The first-generation catalysts for slurry-based commercial processes yielded about 1 kg of polypropylene (PP) per gram of titanium, with 90–95% isotacticity. Any atactic, coproduced product was soluble in the suspending solvent and could be removed with this during product recovery. Better understood second-generation commercial catalysts gave 30 kg of PP per gram of titanium in the catalyst, and a sufficiently high degree of isotacticity that atactic material did not have to be removed. Modern third-generation catalysts yield 2,000 kg of essentially isotactic PP per gram of titanium by a gas phase (Unipol) process in which the residual catalyst concentration is so low that it may be left in the final polymer [10].

Isotactic polypropylene has the lowest density of the commercially dominant thermoplastics, $0.90 \, \text{g/cm}^3$, which contributes to its high strength to weight ratio. It has a crystalline melting point (T_m) of 175°C, which gives the resin a useful upper applications temperature limit of about 120°C, and a tensile strength of 28 MPa (4,000 lb/in.²). Since it is priced about the same as high-density polyethylene ($T_m = 135°C$), it fills the application niches where its significantly higher melting point is important. Injection-molding products for the auto industry, appliances, toys and housewares, and fiber uses in carpeting, decorator fabrics, specialized outdoor wear, and lighter-than-water ropes dominate the market. About 10% of the resin is used for film, mostly for packaging applications [11].

23.4. POLY(VINYL CHLORIDE) (PVC)

Laboratory polymerization of vinyl chloride was accomplished as early as the 1870s, but the high-melting point polymer product was not adopted commercially for over 50 years. The practical factor that caused the delay in commercial development of this material was the discovery that temperatures high enough to permit melt processing also caused significant thermal decomposition of the polymer. Initial measures to lower the processing temperatures involved copolymerization of vinyl chloride with vinyl acetate. But in 1930, before development of this remedy for the difficult processing got very far, it was discovered that intimate mixing of poly(vinyl chloride) (PVC) with certain high boiling point liquids called plasticizers produced a material much more amenable to processing and melt-forming techniques. This process is referred to as plasticization. Production of articles from plasticized PVC began in both Germany and the U.S. in 1933, and today PVC is produced on the third-largest scale of all the plastic resins. It has reached this position for reasons of its ease of processing in compounded form, its excellent physical properties, and its competitive cost relative to alternative materials.

Most commercial PVC is produced by suspension polymerization, although bulk, emulsion, and solution processes are also used. Vigorous stirring plus a suspending agent are used to form a suspension of small droplets of liquid vinyl chloride (boiling point $-14°C$) in water, as it is forced into the reactor under pressure using oxygen-free nitrogen. Trichloroethylene may be added as a chain transfer agent, to help control molecular weight. The contents are then heated, usually to about 50°C, to promote radical formation from the monomer-soluble initiator and start the polymerization. This also raises the pressure to about 0.5 MPa (5 atm) (Eq. 23.4).

$$\text{n CH}_2 = \text{CHCl} \quad \xrightarrow{\text{R}^{\cdot}} \quad \{\text{CH}_2 - \text{CHCl}\}_n \qquad \text{23.4}$$
$$\text{b.p.} -15°C \, (258\text{K}) \qquad\qquad T_m \, (\text{m.p.}) \, 273°C$$

As the polymerization proceeds vinyl chloride monomer is consumed and the pressure drops. At about 0.05 MPa gauge pressure (pressure above normal atmospheric) sufficient monomer has been converted to polymer. Unreacted monomer is vented and the residual vinyl chloride is removed by blowing steam through the system. Both vapor streams are led to a vinyl chloride recovery system. These steps are taken to recover the unreacted vinyl chloride and to minimize the concentration of residual monomer left in the product. The suspension of polymer beads in water is then cooled, and the polymer separated and dried. PVC produced in this way will have a number average molecular weight in the 30,000–80,000 range (or a DP or X_n of = 480–1,280 vinyl chloride units).

The monomer chain transfer constant for vinyl chloride at 60°C is in the range $10.8–16 \times 10^{-4}$, one of the highest of the common vinyl monomers. This normally restricts the reasonably achievable number average molecular weight to the 50,000–100,000 range. Fortunately, number average molecular weights for PVC of around 50,000 are high enough to be commercially useful.

Poly(vinyl chloride) always requires one or more additives before it is processed to final products because of its susceptibility to thermal decomposition. It may be stabilized for processing by the addition of one of barium, cadmium, calcium, or zinc stearates, or organic derivatives of lead or tin. In addition, a processing lubricant such as stearic acid or a metal stearate may be added. Rigid PVC products, such as pipe, sheet, and trim profiles, consume about half of the commercial resin production. For rigid pipe, a specially lubricated melt of the thermally stabilized resin is extruded through an annular die with the bore-forming disk supported on slender stays extended upstream of the annulus. PVC for use in higher temperature pipe applications may be readily chlorinated to increase its stiffness under these conditions, before it is formed into pipe. This raises the usable service temperature to about 100°C as compared to the 65°C rating for normal chlorine content rigid PVC pipe. The rigid product is also formed into siding, gutters, window frames, and other durable building products by similar extrusion methods.

The other half of commercial products from PVC is made from a form of the resin, which is made flexible (plasticized) by vigorous mixing of the powdered resin at high temperatures with a suitable plasticizer. The plasticizers used must be resistant to extraction by water, nonflammable and stable, low in toxicity, and economical. Tricresyl phosphate was the first plasticizer used for PVC, in 1933, but this was rapidly replaced by less toxic materials (such as di-octyl or by di-2-ethylhexyl) phthalate or the esters of adipic acid or fatty acids. Phthalate esters now fill about three-quarters of the PVC plasticizer market. Addition of 30–80% by weight plasticizer, together with varying proportions of a functional filler, lubricant, and pigment, yields products, which are flexible, elastic to the extent required by the final application, and attractively colored. Most pigments and some fillers also provide a degree of additional UV protection for the product, over and above that usually provided by the plasticizer.

Products from plasticized PVC include flexible sheeting ("vinyl") and tubing, floor tiles, etc. These raw products, in turn, are used to produce auto upholstery, components of medical appliances, shoe soles, and in the construction and renovation industry. User evidence of the presence of the plasticizer in "vinyl" products is obtained from the distillation of a small fraction of it from auto upholstery or lining materials onto the inner surfaces of car windows, in hot weather. Also it has recently been noticed that traces of plasticizer were extracted into blood and blood products from the plasticized PVC used in the medical tubing and bags used to handle these materials, which has led to some controversy [12, 13].

Poly(vinyl chloride) may also be internally plasticized by copolymerization of vinyl chloride with vinyl acetate, as mentioned earlier. It is not possible to obtain the same result from a mixture of the same proportions of PVC and poly(vinyl acetate) homopolymers. Incorporation of vinyl acetate into the polymer chains produces the intimate association of components needed to lower the temperature required for processing in much the same way as an added plasticizer does. Important end uses for poly(vinyl chloride-co-vinyl acetate) at low levels of vinyl acetate incorporation are phonograph records and floor tiles, and also for sheeting, when added filler will also be used.

23.5. POLYSTYRENE (PS)

Styrene is one of the few monomers that may be polymerized by free-radical, anionic, cationic, or coordination (Ziegler-Natta) methods. This property, common to styrene and most of its derivatives, is the consequence of the availability of a benzylic position in these monomers, which is capable of stabilizing a radical, carbanionic, or carbocationic center, as well as possessing a polarizability amenable to the charge distributions required by coordination methods of polymerization.

Styrene (and derivatives) also possesses the rare monomer quality that the neat material, without initiator, may be spontaneously polymerized by simply heating to 80–100°C for 24–48 hr. It is thought that this occurs via the initial Diels-Alder dimerization of styrene to the two diastereomers A and B [14]. The two diastereomers appear to have an extremely labile hydrogen, which is both doubly allylic and tertiary. However, only dimer A has the correct stereochemistry (an axial phenyl), which enables the excess styrene to abstract a hydrogen atom from it, producing two radical species (Eq. 23.5).

23.5

Each of the radical species produced is capable of initiating radical chain polymerization of styrene to yield useful polymer. The other diastereomer is probably consumed along with monomer during the process.

Styrene, as a monomer, also has an uncommon versatility in that bulk, solution, suspension, and emulsion techniques may be used to produce high polymer. All four methods are used commercially, though solution polymerization is the most common because it reduces the viscous material transfer problems of bulk polymerization and eliminates the higher risk of contaminated product and the need for drying required by suspension and emulsion methods. Solution polymerization also lends itself to continuous processing, unlike the batch methods usually used for bulk polymerization.

Solution polymerization of styrene employs a reactor train of polymerization vessels, usually three, each progressively hotter than the first. A stream of about 10% styrene dissolved in ethylbenzene plus a suitable thermally activated initiator, is fed into the first reactor under an inert atmosphere and kept at 110–130°C. The solution moves through each reactor in turn until it leaves the last reactor at 150–170°C, the polymerization at this point being completed to the extent desired (Eq. 23.6).

$$n\ PhCH = CH_2 \xrightarrow{R^{\cdot}} \{PhCH - CH_2\}_n \qquad 23.6$$
$$\text{b.p. } 145°C \qquad\qquad T_m \text{ (m.p.) } 250°C$$

To remove the solvent and unreacted monomer the solution is then extruded as fine strands into a devolatilization unit kept at >200°C. Solvent and monomer are recovered leaving strands of polymer containing only traces of ethylbenzene and monomer. Extruding into coarse filaments, chopping, and cooling complete the process to produce polystyrene resin pellets suitable for sale.

Commercial polystyrene has a weight average molecular weight in the $2-3 \times 10^5$ g/mol range representing 2000–3000 monomer units. It is an atactic material, with significant syndiotacticity, and has a low degree of crystallinity, a density of 1.04–1.06 g/cm^3, and melting point (T_m) 240–250°C. It is possible to make isotactic polystyrene via Ziegler-Natta catalysis, but this still has a low degree of crystallinity, and is less transparent and more brittle so is of little commercial interest. Preparation of syndiotactic polystyrene is attracting research interest, but is not a commercial product as yet.

Pure polystyrene resin is a highly transparent brittle material, and relatively inexpensive to produce. About half the total production is consumed in packaging, as film and in expanded form, and in disposables (drinking cups, etc.). Another 15% is used to produce electrical goods, such as switch cover plates, and as components of appliances such as refrigerator and freezer liners. In the latter application high-impact grades of polystyrene, described shortly, are favored for their greater toughness. Furniture components and the construction industry consume a further 10%. The chief application in construction is in expanded form, for excellent thermal insulation. Housewares, toys, and recreation and personal products consume the bulk of the remainder.

23.5.1. Polystyrene Foam Products

Expanded polystyrene is of such importance that it is of interest to briefly outline the methods used in its production. The easiest technology for a fabricator to use employs expandable polystyrene beads to produce foam products. To make the beads, styrene is suspension polymerized in the presence of n-pentane (b.p. 36°C), or other hydrocarbon with a low boiling point. The polystyrene beads, as they form, contain 5–6% of the n-pentane, or more recently water [15], incorporated into the bead, the dense precursor form in which this material is shipped. At the fabricator the expandable beads are heated by steam in a chamber, which allows free expansion and which causes the polystyrene to soften and the pentane or water to vaporize to "puff up" the beads like popcorn, to about 40 times their original size. These once-expanded beads are then cooled in the presence of air for a few hours so that air may diffuse into the expanded beads to equalize the pressure. Then, in a mold of the desired shape for the final application such as flotation billets, television/computer packing, or disposable cups, the requisite quantity of once-foamed beads is placed in the mold and is again heated by steam [16,17]. The secondary expansion plus slight softening of the bead surface in a limited space fuses the mass into a low-density product the shape of the mold. Cooling and then opening of the mold releases the item. Construction formats are referred to as "bead board."

Expanded polystyrene is also produced by extrusion of the melt containing a volatile solvent, or more recently carbon dioxide, through a slit. The pressure drop caused by the emergence of the molten polymer from the slit of the extruder allows the hot solvent to rapidly vaporize, and in the process expands the extruded shape to an extent related to the temperature and proportion of solvent used. When carbon dioxide is used as the blowing agent, its expansion on emergence of the melt from the slit can perform the same function. This product is referred to as "extruded polystyrene" foam. Some manufacturers incorporate a flame retardant into grades intended for construction.

23.5.2. Polystyrene Copolymers

High-impact polystyrenes, in which the brittle nature of the pure polystyrene resin is decreased by incorporation of a rubber, give increased durability and resistance to fracture to the resin when shock loads are expected. Simple mixing of finely divided polystyrene with *cis*-1,4-polybutadiene (a synthetic rubber) gives only slight improvement in impact strength. Optimum results are obtained by copolymerization of the rubber into the polystyrene chains. This is accomplished by solution polymerization of styrene in a solvent in which a rubber is already dissolved. Chain transfer processes result in copolymerization of the styrene with the rubber. The product has much improved impact resistance, but at the same time has lower tensile strength and is less transparent than the pure polystyrene product. Since pigments are normally added to color the high-impact resins before forming into product shapes, reduced clarity of the product in commercial products is not a problem.

23.6. POLYTETRAFLUOROETHYLENE (PTFE)

Discussion of polytetrafluoroethylene is included here as one example of a high value, relatively low volume specialty polymer to illustrate the diversity of monomers, which are amenable to vinyl-type polymerization. Very special polymer properties may be achieved by these specialty materials. PTFE technology, like the development of isotactic polypropylene, is also one of the more recently commercialized vinyl polymers. PTFE was first prepared experimentally in 1938 by Plunkett, in the United States and its novel properties were immediately recognized. A pilot plant for this product came on stream in 1943 followed by a full-scale production unit in 1950, both built by the Du Pont company. Du Pont coined the trade name Teflon for this material.

Pure tetrafluoroethylene monomer under ambient conditions is an odorless, colorless, tasteless gas with low toxicity. It may be polymerized by either suspension or emulsion techniques. Both procedures require use of high pressures in an autoclave in order to maintain the monomer in liquid form. These techniques produce chemically identical product, the first a granular resin, and the second a fine powder (Eq. 23.7).

$$\text{n CF}_2 = \text{CF}_2 \quad \xrightarrow{\text{R}^{\cdot}} \quad +\!\text{CF}_2 - \text{CF}_2\!+_n \qquad 23.7$$
$$\text{b.p. } -76°\text{C (197K)} \qquad \qquad T_m \text{ (m.p.) } 327°\text{C}$$

If the emulsion polymerization product is retained in the water phase for use it is referred to as a PTFE dispersion. The very strong C–F bonds, plus the strengthening effect of these on the C–C bonds confers extraordinarily stable properties to this polymer. The chemical inertness, excellent electrical resistance, high-heat resistance, "nonstick" properties, and low coefficient of friction combine to make PTFE one of the highest performance commercial vinyl plastics produced (Table 23.3).

A hint of the low intermediate forces to be expected in the polymer could be gained from the same feature in the monomer, which has about the same molecular weight as styrene yet a boiling point more than 222°C lower.

Unprocessed PTFE resin is a white solid with a waxy appearance and feel. It has only a moderate tensile strength, much less than the nylons and polyesters. It also has a tendency to creep (or sag) under compression, but this can be accommodated in bearing applications by proper engineering

TABLE 23.3 Properties of Typical Polytetrafluoroethylene (Teflon) Resin

Molecular weight, M_w, g/mol	$10^6 – 10^7$
Density, g/cm^3	2.18
Crystalline melting point, T_m, °C	327[a]
Useful temperature range, °C	−100–260
K	170–530
Static coefficient of friction (against polished steel)	0.05–0.08
Tensile strength, MPa	20–28
lb/in^2	2,900–4,000
Solvent resistance	Insoluble, ordinary conditions[b]
Chemical resistance	Inert, ordinary conditions[c]

[a]Holds shape of melt. Viscosity of the melt at 380°C is 10 GPa (10^{11} Pa).

[b]But is swollen by fluorocarbon oils.

[c]Under severe conditions reacts with alkali metals, fluorine, strong fluorinating agents, and sodium hydroxide at temperatures above 300°C.

design of the bearing area and housing shape. The very high softening point (it doesn't form a true melt), high viscosity at the softening point, and low solubility in solvents make PTFE resins difficult to fabricate by conventional means. These difficulties slowed the early development of applications for this material. Powder sintering techniques related to procedures used in powder metallurgy are used to form films and coatings. Care is required in these techniques and in the cold extrusion procedures used to form fabricated shapes in order to minimize shear since these very high molecular weight resins are susceptible to shear-induced chain shortening. Some of the applications of this specialty polymer are gaskets and seals, particularly for chemically aggressive environments, low friction (bearings) and nonstick applications, high-performance tubing and pipe, and insulating wire wrappings.

Gore Tex (Gore, Inc.) membranes, with a pore size large enough to permit the passage of water vapor but small enough to exclude water droplets, are extruded from a Teflon melt under stress. The stress is applied to the extrusion by taking up the extruded membrane onto a roller, which is operated at a speed substantially higher than the extruder. This stretches the membrane, as it forms, and produces the close distribution of micropores that is required for this application. Final adjustment of pore size, etc., may be made by stretching the membrane again at temperatures below the melting point, sometimes in the presence of an aromatic hydrocarbon.

23.6.1. Polytetrafluoroethylene (Teflon) Copolymer

As with the higher impact resistance achievable with polystyrene when styrene is copolymerized with a rubber, the melt-forming temperatures of PTFE resin may be conveniently lowered by copolymerization of tetrafluoroethylene with low ratios of perfluoropropyl vinyl ether (boiling point, 36°C). The polymerization may be conducted in either an aqueous medium or in an organic solvent (Eq. 23.8).

$$\text{n } CF_2 = CF_2 \; + \; \text{x } CF_3CF_2CF_2OCF = CF_2 \; \xrightarrow{R^\cdot} \qquad \qquad 23.8$$
$$+CF_2 - CF_2 +_n +FC(OC_3F_7) - CF_2 +_x$$

The product, commonly referred to as Teflon PFA (for perfluoro alkoxy), has a melting point (melt transition temperature, T_m) of about 305°C. It retains many of the desirable properties of PTFE (Teflon) homopolymer, and is much more readily melt-formable into products than PTFE itself.

Polakoff *et al.* recorded the results of a study of urinary fluoride levels in PTFE fabricators on the line [18]. There were no markedly elevated fluoride levels found, though fluoride levels did correlate with exposures.

23.7. ENVIRONMENTAL ASPECTS OF VINYLIC POLYMERS (PLASTICS)

23.7.1. Biodegradable Polymers

Medical applications originally stimulated research in the area of biodegradable polymers, but more recently an interest in accelerated litter decomposition and landfill degradation interests have provided additional incentives. These objectives have been met by a variety of strategies [19]. Among these incorporation of a photosensitive species, such as a carbonyl group or certain metal complexes, has been used to promote accelerated photodegradation with some success. Carbonyl incorporation can be accomplished readily, for instance, by copolymerization of carbon monoxide with ethylene, which is feasible with the low density polyethylene processes. Photosensitization may also be achieved after polymerization by blending certain proprietary additives into a melt of the finished polymer [19]. It should be mentioned here that far more attention is usually devoted to the photo*stabilization* of synthetic polymers in order to increase their useful life in outdoor applications. Methods of assessment of environmental degradation rates have been reviewed [20].

Starch addition at 6–15%, or gelatinized starch addition at 40–60% by weight blended into the melt, provides an alternative strategy, e.g., [21]. Physical incorporation of a degradable constituent into the polymer in this manner facilitates breakdown by a combination of oxidative, biological, and physical processes.

Special types of purpose-built polymers can also be made with the intention of enabling more rapid direct microbiological attack and decomposition of these materials. Classes of both condensation and addition polymers among the polyesters, vinylic polymers, and polyhydroxyalkanoates typify current candidate materials being tested in these applications (e.g., Eq. 23.9), but as yet high costs have discouraged large scale exploitation.

$$
\underset{\substack{|\\ CH_3 \\ poly(\text{L-lactic acid})}}{\left.\!\!+\!OCH\!-\!\overset{\displaystyle O}{\overset{\displaystyle \|}{C}}\!\!\right\}_n} \qquad \underset{\substack{|\\ OH \\ poly(\text{vinyl alcohol})}}{\left.\!\!+\!CH_2\!-\!CH\!\right\}_n} \qquad \underset{\substack{|\qquad\quad \| \\ CH_3 \quad\; O \\ poly(\text{3-hydroxybutyrate})}}{\left.\!\!+\!CH\!-\!CH_2\!-\!\overset{O}{\overset{\|}{C}}\!-\!O\!\right\}_n}
$$

23.9

If further development of biodegradable packaging makes these options economically feasible then care is going to be required to enable differentiation and separation of these polymers from very similar stabilized polymers so that they do not cause a loss of properties of the products of recycle streams from misidentification.

23.7.2. Recycling of Vinylic Polymers (Plastics)

Since the plastics are produced from petrochemicals derived from hydrocarbons, the motivation to reuse, recycle, or reprocess for energy recovery is primarily driven by an interest in conservation of petroleum resources. Economic factors are also important, but the potential saving of landfill space is more a perception rather than a reality [9]. Most of the categories of vinylic polymers discussed in this chapter are melt-formable, that is, they are "thermoplastic" materials, rather than nonmelting or "thermosetting" as are several of the condensation polymers discussed in Chapters 20 and 21. Thus, from a technical point of view, all of the plastics discussed here are relatively easily reprocessed and reshaped into "new" articles merely by cleaning and remelting the polymer. One notable exception to this generalization is vulcanized rubber. Although synthetic rubber is produced by a vinyl-type polymerization, which in theory should be melt-formable, the final shape of rubber products such as tires and footwear is fixed by vulcanization. This process introduces disulfide crosslinks into the rubber, which thereby stabilizes the shape of the article for its intended purpose. The presence of these crosslinks alters the reprocessing methods required for vulcanized rubber as compared to any of the other plastic materials.

23.7.3. Industrial Scrap

Fabricator scrap produced from the trimming of the sprues, and gates used to direct molten polymer into the mold cavity, or from molded products which are off-specification, are usually recycled on site. The material for recovery is reduced to a uniform size and additional processing aids or stabilizers are added, if required. This suitably pretreated scrap may be blended directly with virgin material for production of the originally produced item. If blending of scrap with virgin material is unable to meet the specifications for the original product, the reprocessed producer scrap will have to be downgraded and used to form articles with less stringent specifications. This still represents an efficient form of producer recycling.

23.7.4. Postconsumer Plastic Waste Recovery

A large proportion of the vinylic plastics discussed here is used in packaging, mostly in small units. This wide dispersal in small units is what makes collection and recovery of these plastic materials difficult and costly in terms of sorting labor and shipping energy. Mechanical means are available to separate the major constituents of municipal waste after household collection. But the rejection capability of these processes for foreign material is imperfect, causing some nonplastic materials to enter plastic recycle streams, and some misidentified plastics to enter the wrong segregated plastics stream. Also, inevitably, some plastics are inappropriately missorted into a compostable stream. The first outcome leads to a downgraded, and therefore much lower value recycle resin, and the second to a loss of one of the more valuable constituents of the municipal waste stream as well as contamination of the compost component stream. Most plastics are highly resistant to decomposition by composting but are not harmful to the process, although they may interfere with the end uses of the compost.

Much better differentiation of plastic waste from general refuse and segregation of the recovered plastic materials according to resin type is obtained by presorting of the waste at the householder level. As already discussed in connection with the properties of mixtures of PVC and poly(vinyl acetate), or polystyrene with rubber, crude mixtures of two or more polymers usually result in degraded properties relative to those achievable from any of the more rigorously segregated component materials. For this reason, for

TABLE 23.4 Symbolism, Uses, and Properties of Several Important Packaging Plastics

Symbol for recycling	Name, and (industry acronym)	Typical uses	Density (g/cm^3)	Melting point, T_m, (°C)
1 PETE	Poly(ethylene terephthalate) (PET)	soft drink and water bottles	1.40	240
2 HDPE	High-density polyethylene (HDPE)	milk, bleach bottles	0.94–0.96	130
3 PVC	Poly(vinyl chloride) (PVC)	water, glass cleaner bottles	1.40^a	ca. 80
4 LDPE	Low-density polyethylene (LDPE)	plastic bags; bread, food wrap	0.91–0.93	108
5 PP	Polypropylene (PP)	yogurt, margarine containers; shampoo, syrup bottles	0.90	176
6 PS	Polystyrene (PS), often foamed	disposable picnic ware, fast food clamshells, egg cartons, meat trays	1.05^b	ca. 100
7 OTHER	Other plastics, composites, and laminates	beverage and juice boxes	various	varies

^{a}Unplasticized. Decreases to 1.30–1.32, depending on the formulation used when plasticized.
^{b}General purpose, not expanded.

optimum properties from a recycle plastics stream the component polymers must be segregated as to resin type. A decade ago this would have been very difficult by mechanical means, because the densities of several of these plastics are quite similar. Separation by hand would also have been poorly feasible because of the lack of identifying marks on packaging items. Today, however, numerical codes are embossed on most types of plastic packaging, which identify the type of plastic used in the package (Table 23.4).

These codes, usually given together with the letter short forms commercially used for the polymer, now make handsorting relatively unambiguous, though still labor intensive. It is possible that machine-readable bar codes could supplement this information to make for rapid, relatively error-free machine sorting as a cost-effective means of providing segregated polymer streams ideal for recycling. All that is needed to return these streams into production is shredding, washing, drying, and then reextrusion to pellets suitable for all types of products. In some cases additive levels may need to be adjusted or restored. The resulting resins may either be used on their own or may be blended with virgin material, depending on the specifications required for the product.

If downgraded specifications are acceptable it is possible to reprocess a mixture of recycled plastics to form usable articles more tolerant of minor variations in specifications such as field drainage tile and building products such as base layer flooring. One option to enable this is to thoroughly melt-blend the mixed plastic stream with a compatibilizer, such as a chlorinated polyethylene [22]. The compatibilizer aids in producing a mutual solution of all the components present in the melt so that articles formed directly from the blended resin have acceptable properties. Alternatively, hot mix machinery, such as the Mitsubishi Reverzer, can produce an intimate, uniform mixture of plastics from a feed raised to a somewhat higher than normal processing temperature to obtain a lowered viscosity [23]. Alternatively, reactive melt compatibilization may be achieved using an additive such as maleic anhydride [24]. To avoid an extra processing step the product desired is preferably formed directly from the well-mixed melt using the extruder of this machine to force the product into molds of the desired shape.

23.7.5. Anaerobic Pyrolysis

Some segregated vinyl polymers will very readily revert to the monomer on heating the polymer out of contact with air. This process is sometimes referred to as tertiary recycling. Poly(methyl methacrylate) and poly(α-methylstyrene) can yield practically 100% monomer in this way, and polystyrene can yield 80–90% monomer [9, 25, 26]. Polytetrafluoroethylene, too, can yield about 95% monomer on heating to 500°C. However, polyethylene, polypropylene, and PVC are more indiscriminate in their decomposition pathways and yield a mixture of products [27, 28]. Pyrolysis of a mixture of polymers, such as may be recovered from automobile shredder residues, also gives more complex fragments in the recovery system, but is nevertheless a recommended option for retrieval of usable feedstock from a mixed plastics waste stream [29, 30]. Of course, mixed plastics with an energy content of about twice that of paper may be burned directly for energy recovery with appropriate emission controls. This may be managed as a segregated stream or as a mixture with other combustible waste.

23.7.6. Laminates, Composites

Some of the procedures of value for processing of mixed plastic wastes may also be useful for the recovery of value from the laminates used for efficient packaging of food and beverages (e.g., milk cartons, Tetrapaks). A combination of heat and pressure may be employed on a recycle stream of this type of material to fuse the mixed fibrous and resin components into a dense board-like product [31, 32]. This highly weather and insect resistant product, "Eco Superwood Plastic Lumber," may be sawn, nailed, drilled, etc., like wood, and it resists rotting in direct contact with the soil without the need for preservatives.

Other types of composite packaging in which a plastic and another material, or two different plastics are used, such as for a cap and a bottle, are increasingly being recognized as posing a problem for the higher value segregated plastics recycling programs. Package designers are increasingly avoiding these problems by making a packaging unit entirely from a single resin. A further development in package design for commonplace liquid products involves initial sale of the liquid detergent, syrup, etc., in a convenient, sturdy reusable dispenser. This may or may not be made from a single plastic, a factor which has less significance in this role. Refills are offered in lightweight plastic film pouches saving the weight difference in packaging on each use of a pouch refill.

23.7.7. Vulcanized Rubber

The cross-linked rubber remaining in used tires may not be reclaimed directly by remelting. However, the rubber may be recovered by chilling the used tire carcasses and milling to produce a crumb rubber product. The fibrous plies and any steel belts are separated. Crumb rubber may be blended with an appropriate flexible polyurethane mix to bind the granules into the durable and commercially tested "Rubberloc Pavers" patio bricks [33]. Crumb rubber incorporation into asphalt has also been found to confer greater flexibility and elasticity to built-up roofing.

Natural and butyl rubber from inner tubes and other applications, which are fiber-free may be reclaimed by the addition of reclaiming oils and other additives and then subjecting the mix to temperatures of 175–205°C and high shear in a Reclaimator. These conditions are sufficiently severe to break the relatively weak bonds of the disulfide cross-links in the feed, effectively devulcanizing the rubber in a period of 1–3 min under these conditions.

With waste rubber goods, as with nonsegregated plastic wastes, it is possible to simply burn the rubber and regain the considerable fuel value. Whole scrap tires have been burned in cement kilns, for example, to obtain a proportionate saving in the usual oil or gas fuel supply used in these operations. At the extremely high kiln temperatures the combustion is complete and any steel belts or beading are incorporated into the cement klinker without problems.

Recently developed thermoplastic rubbers (e.g., Santoprene developed by Monsanto) are taking over some of the market niches, which were formerly reserved for vulcanized (cross-linked) rubbers [34]. Not only does this material lend itself to convenient processing like a thermoplastic, but once an item made from a thermoplastic rubber has outlived its usefulness it can be much more readily reprocessed. However, thermoplastic elastomers generally have lower strength, less abrasion resistance, and lower tolerance to high temperatures than vulcanized rubber so these materials will only encroach on conventional rubber applications in areas where these properties are less important.

REVIEW QUESTIONS

1. Why might isotactic polypropylene be absolutely essential for commercial utility whereas isotactic polystyrene, while possible to prepare, is not sought for commerce?

2. When PVC is chlorinated to raise its service temperature for rigid hot water piping, chloroethyl repeating units are converted to 1,2-dichloroethyl repeat units.

(a) What chlorine content, in percent by weight, would be expected for neat PVC, as first polymerized?

(b) What ratio of 1,2-dichloroethyl to monochloroethyl units would be present in chlorinated PVC that gave a 68% by weight chlorine content on analysis?

(c) Would the original PVC or the chlorinated PVC be more resistant to combustion and why?

3. (a) What were the motivating factors for development of linear low-density polyethylene (LLDPE) production technology?

(b) Why might LLDPE producing facilities be favored over LDPE technology for replacement of low-density polyethylene production facilities?

4. Compare and contrast the structures and properties of low-density polyethylene, high-density polyethylene, and linear low-density polyethylene.

5. (a) List and briefly describe the polymerization variables, which could be altered and which would tend to decrease the molecular weight of the product of a free radical chain type polymerization.

(b) Outline all possible measures, which could be taken, and the direction in which they should be altered, to increase the molecular weight of a free radical addition polymer product.

6. (a) Describe the precautions necessary in a plastics recycling operation to obtain optimum properties for the reprocessed plastics.

(b) If the procedures required for efficient recycling have not been taken, give two reasonable options for remedies that may be taken with the recycled plastics stream to conserve either material or energy values from these.

7. Efficient incineration may be practiced for energy recovery from a severely degraded waste plastic stream. What precaution(s) is necessary for safe disposal of waste PVC by this method, as compared to the procedure used for most other plastics?

8. One thousand kilograms of 99.6% pure ethylene is polymerized in a heterogeneous process to 850 kg of polyethylene (HDPE) of 80,000 number average molecular weight. Ethylene (99%) totaling 148 kg is recovered.

(a) What is the degree of polymerization of the product?

(b) What is the percent industrial yield of polyethylene in this instance?

FURTHER READING

California Integrated Waste Management Board, "State of California, Plastics: Waste Management Alternatives Use, Recyclability, and Disposal." California Integrated Waste Management Board, Sacramento, 1992.

R.J. Ehrig, ed., "Plastics Recycling: Products and Processes." Hanser Publishers, New York, 1992.

J.R. Fried, "Polymer Science and Technology." Prentice-Hall PTR, Upper Saddle River, NJ, 1995.

G.J.L. Griffin, ed., "Chemistry and Technology of Biodegradable Polymers." Chapman & Hall, Glasgow, 1994.

E.P. La Mantia, ed., "Recycling of PVC and Mixed Waste." ChemTec Publishing, Toronto, 1995.

M. Szwarc and M. van Beylen, "Ionic Polymerization and Living Polymers." Chapman & Hall, New York, 1993.

REFERENCES

1. PlasticsEurope, "Plastics in Europe," pp. 1–18. Assoc. of Plastics Manufacturers, Brussels, Summer, 2004.

2. Facts and Figures, Production, *Chem. Eng. News*, **82**(27), 49–60, July 4 (2004).

3. A.H. Tullo, Polymers Heat Up, *Chem. Eng. News*, **79**(39), 24–28, Sept. 24 (2001).

4. G.T. Austin, "Shreve's Chemical Process Industries," 5th ed. p. 657. McGraw-Hill, New York, 1984.

5. T. Xie, K.B. McAuley, J.C.C. Hsu *et al.*, Gas Phase Ethylene Polymerization: Production Processes, Polymer Properties, and Reactor Modeling. *Ind. Eng. Chem. Res.* **33**, 449–479 (1994).

6. S. Borman, New High-strength Fiber Finds Innovative Use in Protective Clothing, *Chem. Eng. News*. **67**(41), 23–24, Oct. 9 (1989).

7. M.S. Reich, Getting Beyond [Spectra Fiber]..., *Chem. Eng. News*, **80**(24), 16–17, June 17 (2002).

8. Dow Debuts Stretch Fiber, *Chem. Eng. News*, **80**(40), 11, Oct. 7 (2002).

9. V. Gibson and D. Wass, Iron Age Polymers, *Chem. Brit.* **35**(7), 20–23, July (1999).

10. J. Haggin, Unipol Polypropylene Process Garners Licensees, *Chem. Eng. News*, **64**(13) 15–45, Mar. 15 (1986).

11. A.M. Thayer, U.S. Polypropylene Producers Confront Troubling Supply-and-demand Scenario, *Chem. Eng. News*, **72**(38) 17–18, Sept. 19 (1996).

12. B. Hileman, One Plasticizer Okay, Another Still Suspect...*Chem. Eng. News*, **78**(1), 8–9, Jan. 3 (2000).

13. L. Ember, In the Name of Prudence, Switch, *Chem. Eng. News*, **77**(11), 41, Mar. 15 (1999).

14. W.D. Graham, J.G. Green, and W.A. Pryor, Chemistry of Methylenecyclohexadiene and the Thermal Polymerization of Styrene, *J. Org. Chem.* **44**(6), 907–914 (1979).

15. A. Tullo, Nova Revamps Styrenics, *Chem. Eng. News*, **80**(27), 8, July 8 (2002).

16. M.B. Hocking, Paper Versus Polystyrene: A Complex Choice. *Science*, **251**, 504–505 (1991).

17. M.B. Hocking, Relative Merits of Polystyrene Foam and Paper in Hot Drink Cups, *Environ. Manag.* **15**(6), 731–747 (1991).

18. P.L. Polakoff, K.A. Busch, and M.T. Okawa, Urinary Fluoride Levels in Polytetrafluoroethylene Fabricators, *J. Am. Ind. Hyg. Assoc.* **35**, 99–106 (1974).

19. P.J. Hocking, The Classification, Preparation, and Utility of Degradable Polymers, *J.M.S.-Rev. Macromol. Chem. Phys.* **C32**(1), 35–54 (1992).

20. A.L. Andrady, Assessment of Environmental Biodegradation of Synthetic Polymers, *J.M.S.-Rev. Macromol. Chem. Phys.* **C34**(1), 25–76 (1994).

21. S.J. Wang, J.G. Yu, and J.L. Yu,...Compatible Thermoplastic Starch/Polyethylene Blends, *Polym. Degrad. Stability*, **87**(3), 395–401, Mar. (2005).

22. J.N. Schramm and R.R. Blanchard, "The use of CPE [chlorinated polyethylene] as a compatibilizer for reclamation of waste plastic materials," presented at Palisades Section, SPE RETEC, Plastics and Ecology, Cherry Hill Inn, Cherry Hill, NJ, 1970.

23. J. Paul, Recycling plastics. In "Kirk-Othmer Concise Encyclopedia of Chemical Technology," p. 1007. Wiley, New York, 1985.

24. M. Pracella, D. Chionna, R. Ishak *et al.*, Recycling of PET and Polyolefin-based Packaging...by Reactive Blending, *Polym.-Plastics Technol. Eng.* **43**(6), 1711–1722 (2004).

25. N. Grassie, Depolymerization Reactions in Vinyl Polymers. *Chem. Ind. (London)*. p. 622, June 27 (1953).

26. C. Bouster, P. Vermande, and J. Vernon, Evolution of the Product Yield with Temperature and Molecular Weight in the Pyrolysis of Polystyrene, *J. Anal. Appl. Pyrolysis*. **15**, 249 (1989).

27. J. Leidner, "Plastics Waste, Recovery of Economic Value," Dekker, New York, 1981.

28. M. Day, J.D. Cooney, C. Klein *et al.*,...Thermal Processing Effects on Polypropylene, *J. Therm. Anal.* **41**, 225 (1994).

29. H. Kastner and W. Kaminsky, Recycle Plastics into Feedstocks, *Hydrocarbon Process.* **74**, 109, May (1995).

30. I.H. Metecan, A.R.Ozkan, R. Isler *et al.*, Naphtha Derived from Polyolefins, *Fuel*, **84**(5), 619–628, Mar. (2005).

31. B.A. Hegberg, "Technologies for Recycling Postconsumer Mixed Plastics: Emerging Plastic Lumber Production," Office of the Univ. of Illinois Center for Solid Waste Management and Research, Chicago, 1991.

32. W.C. Fergusson, Recovery of Resources from Plastics Industrial Waste, *Chem. Ind.* (*London*). p. 725, Sept. 4 (1976).

33. G. Rodden, From Window Scrapers to Patio "Stones," Old Tires Continue to Roll Along, *Can. Chem. News*, **46**(4), 24 (1994).

34. M.S. Reisch, Thermoplastic Elastomers Target Rubber and Plastics Markets, *Chem. Eng. News*, **74**(32), 10–14, Aug. 5 (1996).

INFORMATION RELATED TO SOIL POLLUTION TOPICS

BACKGROUND BOOKS

Ibrahim Mirsal, "Soil Pollution, Origin, Monitoring, & Remediation," Springer-Verlag, 2004.

R.E. Hester and R.M. Harrison, eds., "Assessment and Reclamation of Contaminated Land," *Royal Society of Chemistry*, Cambridge, 2001.

D.L. Wise *et al.*, eds., "Remediation Engineering of Contaminated Soils," M. Dekker, New York, 2000.

D.M. McNicoll and A.S. Baweja, "Bioremediation of Petroleum-contaminated Soils: An Innovative, Environmentally Friendly Technology," *Environment Canada*, Ottawa, 1995.

P.B. Meyer, R.H. Williams, and K.R. Yount, "Contaminated Land: Reclamation, Redevelopment, and Reuse in the United States and the European Union," Brookfield, VT, E. Elgar, Aldershot, U.K., 1995.

Committee on Long-Range Soil and Water Conservation, "Soil and Water Quality: An Agenda for Agriculture," National Academy Press, Washington, DC, 1993.

"Subsurface Assessment Handbook for Contaminated Sites/Waterloo Centre for Groundwater Research, University of Waterloo, Canadian Council of Ministers of the Environment for the Hazardous Waste Management Branch," *Environ. Prot. Service, Environ. Canada*, Ottawa, The Branch, 1994.

T. Cairney, ed., "Reclaiming Contaminated Land," Blackie, Glasgow, 1987.

SOIL POLLUTION RELEVANT REFERENCES BY SUBJECT

Analysis, Evaluation

T.D. Hettipathirana, Simultaneous Determination of Parts-per-million level Cr, As, Cd, and Pb, and Major Elements in Low-level Contaminated Soils Using Borate Fusion and Energy Dispersive X-ray Fluorescence Spectrometry with Polarized Excitation, *Spectrochim. Acta Part B-Atomic Spectros.* **59**(2), 223–229, **Feb. 27 (2004)**.

A.M. Ahmed and W.N. Sulaiman, Evaluation of Groundwater and Soil Pollution in a Landfill Area Using Electrical Resistivity Imaging Survey, *Environ. Manag.* **28**(5), 655–663, Nov. (2001).

J.D. Mckinley, H.R. Thomas, K.P. Williams *et al.*, Chemical Analysis of Contaminated Soil Strengthened by the Addition of Lime, *Engin. Geol.* **60**(1–4), 181–192, June (2001).

K.H. Karstensen, O. Ringstad, I. Rustad *et al.*, Methods for Chemical Analysis of Contaminated Soil Samples—Tests of Their Reproducibility Between Nordic Laboratories, *Talanta.* **46**(3), 423–437, July (1998).

D.M. White, H. Luong, and R.L. Irvine, Pyrolysis-GC/MS Analysis of Contaminated Soils in Alaska, *J. Cold Regions Engin.* **12**(1), 1–10, Mar. (1998).

A. Argyraki, M.H. Ramsey, and P.J. Potts, Evaluation of Portable X-ray Fluorescence Instrumentation for *In situ* Measurements of Lead on Contaminated Land, *Analyst*, **122**(8), 743–749, Aug. (1997).

D.K. Banerjee and M.R. Murray, Analysis of Hydrocarbon-contaminated Soil by Thermal Extraction—Gas Chromatography, *Environ. Sci. Technol.* **31**(3), 646–650, Mar. (1997).

W. Rasemann, U. Seltmann, and M. Hempel, Statistical Evaluation of the Influence of Several Sample Pretreatment Methods on the Mercury Content Detectable by Chemical-analysis of Contaminated Soil Samples Under Practical Conditions, *Fresenius J. Anal. Chem.* **351**(7), 632–641, Apr. (1995).

Arsenic, Anthropogenic

C. Patinha, E.F. da Silva, and E.C. Fonseca, Mobilization of Arsenic at the Talhadas Old Mining Area—Central Portugal, *J. Geochem. Explor.* **84**(3), 167–80, Nov. (2004).

N.K. Blute, D.J. Brabander, H.F. Hemond *et al.*, Arsenic Sequestration by Ferric Iron Plaque on Cattail Roots, *Environ. Sci. Technol.* **38**(22), 6074–6077, Nov. 15 (2004).

P. Richter, R. Seguel, I. Ahumada *et al.*, Arsenic Speciation in Environmental Samples of a Mining Impacted Sector of Central Chile, *J. Chilean Chem. Soc.* **49**(4), 333–339, Dec. (2004).

H.R. Pfeifer, A. Gueye-Girardet, D. Reymond *et al.*, Dispersion of Natural Arsenic in the Malcantone Watershed, Southern Switzerland: Field Evidence for Repeated Sorption–Desorption and Oxidation–Reduction Processes, *Geoderma.* **122**(2–4), 205–234, Oct. (2004).

V. Chatain, K. Hanna, C. de Brauer, R. Bayard *et al.*, Enhanced Solubilization of Arsenic and 2,3,4,6-tetrachlorophenol from Soils by a Cyclodextrin Derivative, *Chemosphere*, 57(3), 197–206, Oct. (2004).

P. Madejon, T. Maranon, J.M. Murillo *et al.*, White Poplar (Populus Alba) As a Biomonitor of Trace Elements in Contaminated Riparian Forests, *Environ. Pollut.* 132(1), 145–155, Nov. (2004).

W. Hartley, R. Edwards, and N.W. Lepp, Arsenic and Heavy Metal Mobility in Iron Oxide-amended Contaminated Soils As Evaluated by Short- and Long-term Leaching Tests, *Environ. Pollut.* 131(3), 495–504, Oct. (2004).

J. Whyte, Surface Appearances Can Deceive: Groundwater Conditions a Major Factor in Responsible Mine Closure, *Hazardous Mater. Manag.* 15(2), 17–18, Apr./May (2003).

M. Farago, Arsenic Contamination in South West England—A Legacy of Mining, *Environ. Chem. Group Newsletter, Roy. Soc. Chem.* 10, 5–6, July (1999).

D. Hocking, P. Kuchar, J.A. Plambeck *et al.*, The Impact of Gold Smelter Emissions on Vegetation and Soils of a sub-Arctic Forest-tundra Transition Ecosystem, *APCA J.* 28(2), 133–137, Feb. (1978).

Arsenic, Geological

M. Sharpe, Deadly Waters Run Deep: The Global Arsenic Crisis, *J. Environ. Monit.* 5(5), 81–85N (2003).

B. Clark and D. Laughlin, Just "Do it": *In situ* Bioremediation for TPH and MTBE Contaminants, *Hazardous Mater. Manag.* 12(2), 17–19, Apr./May (2000).

D. Kinniburgh and P. Smedley, Trace Element Contamination of Ground-waters and Soils, *Environ. Chem. Group Newsletter, Roy. Soc. Chem.* 10, 2–4, July (1999).

M. Brooks, Drinking Water Contaminated by Arsenic in Bangladesh—An Update, *Environ. Chem. Group Newsletter, Roy. Soc. Chem.* 9, 2–4, Jan. (1999).

Arsenic Removal, Remediation

V. Campos, M. Aparecida, and F. Pires, Phytoremoval of Arsenic from Soil, *Commun. Soil Sci. Plant Anal.* 35(15–16), 2137–2146, Sept. (2004).

S. Dawadi, M.R. Hansen, and B.W. Berdanier, Encapsulation of Contaminated Soil in Concrete Mortar, *ACI Mater. J.* 101(5), 347–352, Sept.–Oct. (2004).

M. Freemantle, Arsenic Rooted from Water, *Chem. Eng. News*, 83(14), 12, Apr. 4 (2005).

C. Hogue, A Sip of Arsenic? Debate over Allowable Level in [U.S.] Drinking Water Focuses on Health Effects, Removal Cost, *Chem. Eng. News*, 79(21), 51–58, May 21 (2001).

Groundwater Contamination

M. Paul, L. W. Fund *et al.*, Microbiological Condition of Urban Groundwater in the Vicinity of Leaky Sewer Systems, *Acta Hydrochim. et Hydrobiol.* **32**(4–5), 351–360, Nov. (2004).

A. Chofqi, A. Younsi, Lhadi *et al.*, Environmental Impact of an Urban Landfill on a Coastal Aquifer (El Jadida, Morocco), *J. African Earth Sci.* **39**(3–5), 509–516, June–Aug. (2004).

I. Melenova and K. Demnerov, New Possibilities of Pollutant Removal from the Environment and Use of a Multifunctional Permeable Barrier System—Multibarrier, *Chemicke Listy.* **98**(10), 908–915 (2004).

K.Y. Lee, Modeling Long-term Transport of Contaminants Resulting from Dissolution of a Coal Tar Pool in Saturated Porous Media, *J. Environ. Engin.-ASCE,* **130**(12), 1507–1513, Dec. (2004).

J. Zhu, J, Luo, A, Zhou, Y *et al.*, The Age Effect of Dairy Feedlots on Manure Nutrient Seepage in Loam Soils, *Biosyst. Engin.* **89**(2), 223–229, Oct. (2004).

B.F. Alemaw, E.M. Shemang, and T.R. Chaoka, Assessment of Groundwater Pollution Vulnerability and Modelling of the Kanye Wellfield in SE Botswana—A GIS Approach, *Phys. Chem. Earth*, **29**(15–18), 1125–1128 (2004).

S. Ravengai, R. Owen, and D. Love, Evaluation of Seepage and Acid Generation Potential from Evaporation Ponds, Iron Duke Pyrite Mine, Mazowe Valley, Zimbabwe, *Phys. Chem. Earth*, **29**(15–18), 1129–1134 (2004).

R.M. Singh, B. Datta, and A. Jain, Identification of Unknown Groundwater Pollution Sources Using Artificial Neural Networks, *J. Water Res. Planning Manag.-ASCE*, **130**(6), 506–514, Nov.–Dec. (2004).

V.F. Uricchio, R. Giordano, and N. Lopez, A Fuzzy Knowledge-based Decision Support System for Groundwater Pollution Risk Evaluation, *J. Environ. Manag.* **73**(3), 189–197, Nov. (2004).

F. Celico, I. Musilli, and G. Naclerio, The Impacts of Pasture- and Manure-spreading on Microbial Groundwater Quality in Carbonate Aquifers, *Environ. Geol.* **46**(2), 233–236, Aug. (2004).

R.M. Kalin, Engineered Passive Bioreactive Barriers: Risk-managing the Legacy of Industrial Soil and Groundwater Pollution, *Curr. Opinion Microbiol.* **7**(3), 227–238, June (2004).

C.J. von der Heyden and M.G. New, Groundwater Pollution on the Zambian Copperbelt: Deciphering the Source and the Risk, *Sci. Total Environ.* **327**(1–3), 17–30, July 5 (2004).

S.S.D. Foster and P.J. Chilton, Downstream of Downtown: Urban Wastewater as Groundwater Recharge, *Hydrogeol. J.* **12**(1), 115–120, Feb. (2004).

J.K. Böhlke, K. Révész, E. Busenberg *et al.*, Groundwater Record of Halocarbon Transport by the Danube River, *Environ. Sci. Technol.* **31**, 3293–3299 (1997).

Hydrocarbon Contaminated

J.G. Bundy, G.L. Paton, and C.D. Campbell, Combined Microbial Community Level and Single Species Biosensor Responses to Monitor Recovery of Oil Polluted Soil, *Soil Biol. Biochem.* **36**(7), 1149–1159, July (2004).

P. Juteau, J.G. Bisaillon, F. Lepine *et al.*, Improving the Biotreatment of Hydrocarbons-contaminated Soils by Addition of Activated Sludge Taken from the Wastewater Treatment Facilities of an Oil Refinery, *Biodegradation*, 14(1), 31–40 (2003).

K. Unlu and E. Demirekler, Modeling Water Quality Impacts of Petroleum Contaminated Soils in a Reservoir Catchment, *Water Air Soil Pollut.* 120(1–2), 169–193, May (2000).

A. Saterbak, R.J. Toy, D.C.L. Wong *et al.*, Ecotoxicological and Analytical Assessment of Hydrocarbon-contaminated Soils and Application to Ecological Risk Assessment, *Environ. Toxicol. Chem.* 18(7), 1591–1607, July (1999).

M.H. Huesemann and M.J. Truex, The Role of Oxygen Diffusion in Passive Bioremediation of Petroleum Contaminated Soils, *J. Hazard. Mater.* 51(1–3), 93–113, Nov. (1996).

K. Karan, A. Chakma, and A.K. Mehrotra, Air Stripping of Hydrocarbon-contaminated Soils—Investigation of Mass-transfer Effects, *Can. J. Chem. Eng.* 73(2), 196–203, Apr. (1995).

S.J.T. Pollard, S.E. Hrudey, and P.M. Fedorak, Bioremediation of Petroleum—And Creosote-contaminated Soils—A Review of Constraints, *Waste Manag. Res.* 12(2), 173–194, Apr. (1994).

Landfills

S.I. Darmawan Wada, Effect of Clay Mineralogy on the Feasibility of Electrokinetic Soil Decontamination Technology, *Appl. Clay Sci.* 20(6), 283–293, Feb. (2002).

A.M. Ahmed and W.N. Sulaiman, Evaluation of Groundwater and Soil Pollution in a Landfill Area Using Electrical Resistivity Imaging Survey, *Environ. Manag.* 28(5), 655–663, Nov. (2001).

F. Dijkstra, Sanitation of Polluted Soil Areas and Hazardous-waste Management at DSM, *Water Sci. Technol.* 24(12), 113–122 (1991).

Landfill Gas(es)

B. Eklund, E.P. Anderson, B.L. Walker *et al.*, Characterization of Landfill Gas Composition at the Fresh Kills Municipal Solid-waste Landfill, *Environ. Sci. Technol.* 32(15), 2233–2237, Aug. 1 (1998).

M.R. Allen, A. Braithwaite, and C.C. Hills, Trace Organic Compounds in Landfill Gas at Seven UK Waste Disposal Sites, *Environ. Sci. Technol.* 31(4), 1054–1061, Apr. (1997).

C.T. He, D.J. Herman, R.G. Minet *et al.*, A Catalytic/Sorption Hybrid Process for Landfill Gas Cleanup, *Ind. Eng. Chem. Res.* 36(10), 4098–4105, Oct. (1997).

R.J. Spiegel, J.C. Trocciola, and J.L. Preston, Test Results for Fuel-cell Operation on Landfill Gas, *Energy*, 22(8), 777–786, Aug. (1997).

G.J. Sandelli, J.C. Trocciola, and R.J. Spiegel, Landfill Gas Pretreatment for Fuel-cell Applications, *J. Power Sources*, **49**(1–3), 143–149, Apr. (1994).

G.P. Mueller, Landfill Gas Application Development of the Caterpillar G3600 Spark-ignited Gas Engine, *J. Eng. Gas Turb. Power*, **117**(4), 820–825, Oct. (1995).

Mercury, Anthropogenic

M.H.D. Pestana and M.L.L. Formoso, Mercury Contamination in Lavras do Sul, South Brazil: a Legacy from Past and Recent Gold Mining, *Sci. Total Environ.* **307**(1–3), 125–140, May 20 (2003).

O. Lindqvist, K. Johansson, and M. Aastrup, Mercury in the Swedish Environment—Recent Research on Causes, Consequences and Corrective Methods, *Water Air Soil Pollut.* **55**(1–2), R11, Jan. (1991).

Mercury, Geological

L. Liang, M. Horvat, X.B. Feng *et al.*, Reevaluation of Distillation and Comparison with HNO_3 Leaching/Solvent Extraction for Isolation of Methylmercury Compounds from Sediment/Soil Samples, *Appl. Organometal. Chem.* **18**(6), 264–270, June (2004).

M.S. Gustin, Are Mercury Emissions from Geologic Sources Significant? A Status Report, *Sci. Total Environ.* **304**(1–3), 153–167, Mar. 20 (2003).

T. Tomiyasu, M. Okada, R. Imura, and H. Sakamoto, Vertical Variations in the Concentration of Mercury in Soils Around Sakurajima Volcano, Southern Kyushu, Japan, *Sci. Total Environ.* **304**(1–3), 221–230, Mar. 20 (2003).

M.S. Gustin, M.F. Coolbaugh, M.A. Engle *et al.*, Atmospheric Mercury Emissions from Mine Wastes and Surrounding Geologically Enriched Terrains, *Environ. Geol.* **43**(3), 339–351, Jan. (2003).

M.F. Coolbaugh, M.S. Gustin, and J.J. Rytuba, Annual Emissions of Mercury to the Atmosphere from Natural Sources in Nevada and California, *Environ. Geol.* **42**(4), 338–349, July (2002).

P.V. Koval, G.V. Kalmychkov, V.F. Gelety *et al.*, Correlation of Natural and Technogenic Mercury Sources in the Baikal Polygon, Russia, *J. Geochem. Explor.* **66**(1–2), 277–289, July (1999).

W. Gu, C.Y. Zhou, M.K. Wong *et al.*, Orthogonal Array Design (OAD) for the Optimization of Mercury Extraction from Soils by Dilute Acid with Microwave Heating, *Talanta.* **46**(5), 1019–1029, Aug. (1998).

R. Ferrara, B. Mazzolai, H. Edner *et al.*, Atmospheric Mercury Sources in the Mt. Amiata Area, Italy, *Sci. Total Environ.* **213**(1–3), 13–23, June 10 (1998).

Reclamation and Remediation

R.R. Suleimanov and R.N. Sitdikov, Reclamation of Oil-contaminated Soils and Grounds Using Industrial Detergent, *Neftyanoe Khozyaistvo.* **12**, 115–117, Dec. (2003).

A.K. Wanekaya and A.S. Omowunmi, Pressurized Extractions for the Remediation of Heavy Metal Contaminants in Soil and Sediments, *Amer. Laboratory*, **35**(17), 10–16, Aug. (2003).

G.P. Warren, B.J. Alloway, N.W. Lepp *et al.*, Field Trials to Assess the Uptake of Arsenic by Vegetables from Contaminated Soils and Soil Remediation with Iron Oxides, *Sci. Total Environ.* **311**(1–3), 19–33, July 20 (2003).

L. Di Palma, P. Ferrantelli, and E. Petrucci, Experimental Study of the Remediation of Atrazine Contaminated Soils Through Soil Extraction and Subsequent Peroxidation, *J. Hazard. Mater.* **99**(3), 265–276, May 30 (2003).

L. Di Palma, Experimental Assessment of a Process for the Remediation of Organophosphorous Pesticides Contaminated Soils Through In situ Soil Flushing and Hydrolysis, *Water Air Soil Pollut.* **143**(1–4), 301–314, Feb. (2003).

B.W. Bogan, V. Trbovic, and J.R. Paterek, Inclusion of Vegetable Oils in Fenton's Chemistry for Remediation of PAH-contaminated Soils, *Chemosphere*, **50**(1), 15–21, Jan. (2003).

J. Rohrs, G. Ludwig, and D. Rahner, Electrochemically Induced Reactions in Soils—A New Approach to the In-situ Remediation of Contaminated Soils? Part 2: Remediation Experiments with a Natural Soil Containing Highly Chlorinated Hydrocarbons, *Electrochim. Acta* **47**(9), 1405–1414, Feb. 15 (2002).

C.W. Lin, Mapping Soil Lead and Remediation Needs in Contaminated Soils, *Environ. Geochem. Health*, **24**(1), 23–33, Mar. (2002).

B. Alpaslan and M.A. Yukselen, Remediation of Lead Contaminated Soils by Stabilization/Solidification, *Water Air Soil Pollut.* **133**(1–4), 253–263, Jan. (2002).

C.L. Zhang, R.C. Daprato, S.F. Nishino *et al.*, Remediation of Dinitrotoluene Contaminated Soils from Former Ammunition Plants: Soil Washing Efficiency and Effective Process Monitoring in Bioslurry Reactors, *J. Hazard. Mater.* **87**(1–3), 139–154, Oct. 12 (2001).

G.C.C. Yang and C.Y. Liu, Remediation of TCE Contaminated Soils by In situ EK–Fenton Process, *J. Hazard. Mater.* **85**(3), 317–331, Aug. 17 (2001).

R.G. Burns, Degradation of Xenobiotics: Soil Surfaces and Biofilms and Their Relevance to Bioavailability and Bioremediation, *Environ. Chem. Group Newsletter, Roy. Soc. Chem.* **14**, 3–8, July (2001).

C.N. Mulligan, R.N. Yong, and B.F. Gibbs, Remediation Technologies for Metal-contaminated Soils and Groundwater: An Evaluation, *Engin. Geol.* **60**(1–4), 193–207, June (2001).

J. Lynn, Bioremediation of Oil Spills, *Hazard. Mater. Manag.* **13**(1), 24–28, Feb./Mar. (2001).

L.J. Pinto and M.M. Moore, Release of Polycyclic Aromatic Hydrocarbons from Contaminated Soils by Surfactant and Remediation of this Effluent by Penicillium spp. *Environ. Toxicol. Chem.* **19**(7), 1741–1748, June (2000).

J. Roed, K.G. Andersson, C.L. Fogh *et al.*, Triple Digging—A Simple Method for Restoration of Radioactively Contaminated Urban Soil Areas, *J. Environ. Radioactivity*, **45**(2), 173–183 (1999).

P. Zurer, Lasagna: Field-tested Recipe for Soil Cleanup, *Chem. Eng. News*, **77**(15), 14, Apr. (1999).

A. Davis, S. Kamp, G. Fennemore *et al.*, A Risk-based Approach to Soil Remediation Modeling, *Envir. Policy Anal.* **31**(11), 521A–525A (1997).

A.M. Thayer, Bioremediation: Innovative Technology for Cleaning up Hazardous Waste, *Chem. Eng. News*, **69**(34), 23–44, Aug. 26 (1991).

Sampling Technique

L. Malherbe, Designing a Strategy for Contaminated Soil Sampling Human Health Risk Assessment, *Accredit. Qual. Assur.* **7**(5), 189–194, May (2002).

X.J. Wang and F. Qi, The Effects of Sampling Design on Spatial Structure Analysis of Contaminated Soil, *Sci. Total Envir.* **224**(1–3), 29–41, Dec. 11 (1998).

F.P.J. Lame and P.R. Defize, Sampling of Contaminated Soil—Sampling Error in Relation to Sample Size and Segregation, *Environ. Sci. Technol.* **27**(10), 2035–2044, Oct. (1993).

Salinity, Salinization, and Fertility

J. Akhter, R. Murray, K. Mahmood *et al.*, Improvement of Degraded Physical Properties of a Saline–Sodic Soil by Reclamation with Kallar Grass (Leptochloa fusca), *Plant Soil*, **258**(1), 207–216, Jan. (2004).

M.J. Mohammad and N. Mazahreh, Changes in Soil Fertility Parameters in Response to Irrigation of Forage Crops with Secondary Treated Waste-water, *Communic. Soil Sci. Plant Anal.* **34**(9–10), 1281–1294 (2003).

D.P. Patil, M.V. Kulkarni, V.L. Maheshwari *et al.*, A Sustainable Agro–Biotechnology for Bioremediation of Saline Soil, *J. Sci. Ind. Res.* **61**(7), 517–528, July (2002).

B. Zenjari and A. Nejmeddine, Impact of Spreading Olive Mill Wastewater on Soil Characteristics: Laboratory Experiments, *Agronomie.* **21**(8), 749–755, Dec. (2001).

R.P. Courtemanche, M.W. Hester, and I.A. Mendelssohn, Recovery of a Louisiana Barrier Island Marsh Plant Community Following Extensive Hurricane-induced Overwash, *J. Coastal Res.* **15**(4), 872–883, Fall (1999).

G. Singh, H. Singh, and P.P. Bhojvaid, Amelioration of Sodic Soils by Trees for Wheat and Oat Production, *Land Degrad. Devel.* **9**(5), 453–462, Sept.–Oct. (1998).

J. Albaladejo, M. Stocking, E. Diaz, *et al.*, Land rehabilitation by urban refuse amendments in a semiarid environment - effect on soil chemical-properties, *Soil Technol.* **7**(3), 249–260, Oct. (1994).

K.E. Elsiddig and P. Ludders, Interactive Effects of Nitrogen Nutrition and Salinity on Reproductive Growth of Apple Trees, *Gartenbauwissenschaft.* **59**(3), 127–131, May–June (1994).

D.W. James and J.J. Jurinak, Irrigation Resources in Utah—Water-quality Versus Soil–Salinity, Soil Fertility, and the Environment, *Utah Agric. Exper. Station Res. Bull.* **514**, 1–41, Dec. (1986).

T.G. Selassie and R.J. Wagenet, Interactive Effects of Soil–Salinity, Fertility, and Irrigation on Field Corn, *Irrig. Sci.* **2**(2), 67–78 (1981).

2
RELEVANT TECHNICAL WEBSITES BY TOPIC

Aircraft Builders

Airbus Industries, Toulouse/Hamburg: http://www.airbus.com/dynamic/about/index_h.asp

Freq. Asked Questions: http://www.airbus.com/about/faq.asp

Boeing Aircraft: http://www.boeing.com/flash.html

Bombardier Aerospace http://www.aerospace.bombardier.com/index

General Builder Listing: http://www.landings.com/_landings/pages/aircraft-manuf.html

Air Quality Related

AMA Council on Scientific Affairs: http://www.asma.org/Publication/tips.html

Aerospace Medical Association: http://www.asma.org/Publication/tips.html

Publications: http://www.epa.gov/iaq/pubs/airclean.html

Respiratory Reviews: http://www.respiratoryreviews.com/archives.html

Applied Research Agencies

Alberta Research Council: http://www.arc.ab.ca/

B.C. Research Council: http://bcresearch.com/default.htm

Arctic Information, Studies

http://www.uarctic.org/

Brewing

101 Beer Links: http://www.iwantadrink.com/directory/bartools/101-beer-links.htm

Master Brewers Association of America: http://www.mbaa.com/

Chemicals

Structure, nomenclature, toxicology (National Library of Medicine): http://sis.nlm.nih.gov/

(University of Hamburg): http://lb.chemie.uni-hamburg.de/static/data2/42_12xrc3np.html

Chemical Company Information

General: http://finance.yahoo.com/q?s

http://finance.yahoo.com/

http://www.hoovers.com/

http://webbolt.ecnext.com/

Oil and Gas Companies: http://www.infield.com/links.htm

Specific Companies:

BASF: http://www.corporate.basf.com/en/?id = V00-1nt- 168tjbcp*1R

Bayer: http://www.bayer.com/

Dow: http://www.dow.com/Homepage/index.html

Du Pont: http://www1.dupont.com/NASApp/dupontglobal/corp/index.jsp

Hoechst: http://www.hoechst.com/homepage/homepage.htm

ICI: http://www.ici.com/ICIPLC/home/index.jsp

Nova Chemicals: http://www.novachemicals.com

Total: http://www.total.com/fr/home_page

Chemical Structures:

http://chemfinder.cambridgesoft.com/

Chemical Technology and Related

Books: http://www.southalabama.edu/univlib/guides/chemistry.htm

Chemistry Resources

Latest Fundamental Physical Constants: http://physics.nist.gov/cuu/Constants/index.html

NIST Databases (see chemistry Web book): http://www.nist.gov/srd/online.htm

Virtual Kinetics Laboratory: http://vklab.hec.utah.edu/

Cleaner Production Platforms:

Switzerland (opened Nov 2002): http://www.cpplatform.ch

Company Profiles

General: http://gale.ecnext.com/

http://gale.ecnext.com/coms2/dbcomp_0000551188

Currency Conversion, Historical and Present

OANDA.com: http://www.oanda.com/convert/fxhistory

Dictionaries, Language

English: http://dictionary.cambridge.org/

French (& other) < > English: http://www.freedict.com/onldict/fre.html

German < > English, Chemical: http://www-sul.stanford.edu/depts/swain/beilstein/bedict1.html

General: http://www.iee.et.tu-dresden.de/cgi-bin/cgiwrap/wernerr/search.sh

IUPAC Compendium of Chemical Terminology: http://www.chemsoc.org/chembytes/goldbook/index.htm

Dictionaries, Technical

Chemical dictionary: http://www.chemnet.com/dict/

Chemical engineering dictionary, encyclopedia listing: http://www.library.
gatech.edu/research_help/subject/chemeng_dictionaries.htm

Chem-cyclopedia: http://www.mediabrains.com/client/chemcyclop/BG1/
search.asp

Chemical technology references: http://www.southalabama.edu/univlib/
guides/chemistry.htm

Chemindustry.com Website directory: http://www.chemindustry.com/
chem2ask.asp?cmd = search&id = 0

IUPAC Compendium of Chemical Terminology:
http://www.chemsoc.org/chembytes/goldbook/index.htm

Electrochemistry Resources

Electrochemical Science and Technology Information Resource (ESTIR):
http://electrochem.cwru.edu/estir/

Impedance-Research Solutions and Resources: http://www.consultrsr.com/

Fuel Cell Resources

Hydrogen and Fuel Cells in Canada: http://www.port-h2.com/

Electronic Journal Access

Science Articles: http://www.scirus.com/

Element Cycles:

The "matter cycles" and their pollution problems due to human interference: http://www.lenntech.com/matter-cycles.htm
http://www.lenntech.com/matter-cycles-pollution.htm

Environmental Law

A gateway into, operated jointly by FAO, IUCN, and UNEP: http://
www.ecolex.org/ecolex/index.php

Environmental Regulations

Canada: http://www.qc.ec.gc.ca/dpe/Anglais/dpe_liens_en.asp?dpe_related_
links

Environmentally Related Sites

Springer-Verlag: The Handbook of Environmental Chemistry
http://link.springer.de/link/service/series/0698/index.htm

U.S. Environmental Protection Agency Public Archive: http://www.epa.
gov/eims/

Worldwatch site: http://www.worldwatch.org/pubs/signposts/2002/

Facts and Figures

U.S.A. (plus some other countries), chemical industry: http://pubs.acs.org/
cen/index.html

World: http://www.worldfactsandfigures.com/

Fertilizers

International Fertilizer Industry Association: http://www.fertilizer.org/ifa/

Filters

Airtesters filters: http://www.airtesters.com/HEPA_filters.cfm

Global Warming

Intergovernmental Panel on Climate Change (IPCC), 2001: http://www.
climatehotmap.org/

"An increasing body of observations gives a collective picture of a warming world and other changes in the climate system."

EPA Website: http://yosemite.epa.gov/oar/globalwarming.nsf/content/Climate.html

Greenhouse Gas Online: http://www.ghgonline.org

Handbooks

Knovel Critical Tables: http://www.knovel.com/knovel2/crc_handbook.jsp

International Critical Tables of Numerical Data, Physics, Chemistry and Technology: http://www.knovel.com/knovel2/Toc.jsp?BookID = 735

Handbook of Environmental Chemistry (Springer-Verlag): http://194.94.42.12/licensed_materials/0698/tocs.htm

http://www.springerlink.com/app/home/journal.asp?wasp = 570cd1wyl-g2yyl4lmv56& referrer = parent&backto = searchpublicationsresults,1,2;

Health Related

Canada Communicable Disease Report: http://www.hc-sc.gc.ca/pphb-dgspsp/publicat/ccdr-rmtc/01vol27/27s2/index.html

Health Canada: http://www.hc-sc.gc.ca/pphb-dgspsp/publicat/info/infflu_e.html

Journal of the American Medical Assoc'n: http://jama.ama-assn.org/

Respiratory Reviews: http://www.respiratoryreviews.com/main.html
http://www.respiratoryreviews.com/archives.html

Indoor Air Quality

US EPA: http://www.epa.gov/iaq/pubs/airclean.html

Industrial Hygiene

WHMIS: http://ohs.uvic.ca/training/whmis/whmiscourse.pdf

United Nations Department of Economic and Social Affairs

United Nations Division for Sustainable Development: http://www.un.org/esa/sustdev/

Johannesburg Plan of Implementation:
http://www.un.org/esa/sustdev/documents/WSSD_POI_PD/English/POIToc.htm

Journals, Relevant

American Chemical Society Journals: http://pubs.acs.org/

Chemical and Engineering News: http://pubs.acs.org/cen/

AIChE (American Inst. of Chemical Engineers) Journal: http://www.aiche.org/aichejournal/

Archives of Internal Medicine: http://archinte.ama-assn.org/

Blackwell Synergy: http://www.blackwell-synergy.com/

Canadian Chemical News: http://www.accn.ca/

Environmental Journals

Chemical and Engineering News: http://pubs.acs.org/cen/index.html

Elsevier, Environmental Journals: http://www.elsevier.com/wps/find/homepage.cws_home

Environmental Health Journal: http://www.ehj-online.com/currentissue.html

Environmental Science and Technology: http://pubs.acs.org/journals/esthag/index.html

Indoor Air, International Journal of Indoor Air Quality and Climate:
http://www.munksgaard.dk/indoorair
Journal of Environmental Health Research:
http://www.jehr-online.org/volume2/issue1/index.asp
Reviews on Environmental Health: http://www.angelfire.com/il/freund/
Royal Society of Chemistry Journals: library@rsc.org or www.rsc.org/
library
Springer-Verlag: The Handbook of Environmental Chemistry
http://link.springer.de/link/service/series/0698/index.htm

Libraries, Document Access
CISTI: http://doc.cisti-icist.nrc-cnrc.gc.ca/CISTI/
http://cisti-icist.nrc-cnrc.gc.ca/cisti_e.shtml
Royal Society of Chemistry: library@rsc.org or, www.rsc.org/library
IngentaConnect: Academic and Professional Publications www.ingenta.
com (only subscribers?)

Lifecycle Analysis
Institute for Lifecycle Energy Analysis (Seattle): www.ilea.org/lcas/hocking
1994.html

Material Safety Information
Workplace Hazardous Materials Information System (WHMIS) Website:
http://www.hc-sc.gc.ca/hecs-sesc/whmis/index.htm
Canadian Centre for Occupational Health and Safety, MSDS:
http://ccinfoweb.ccohs.ca/msds/search.html (payment req'd ?)

Materials Information
MatWeb material property data: http://www.matweb.com/index.asp?
ckck = 1

Meetings, Scientific: http://www.science-events.com/

Mercury Management
Environmental Impact Statement (etc.): http://www.mercuryeis.com

Newspapers
Canada.com: tp://www.canada.com/national/index.html
Canadian Newspaper Association: http://www.cna-acj.ca/client/cna/
ult.nsf/CnaSearch
Internet Public Library: http://www.ipl.org/div/news/browse/CA/

Organic Chemistry
Outlined: http://www.cem.msu.edu/~reusch/VirtualText/intro1.htm#
contnt

Patent Databases
Generic: http://www.orbit.com/index.htm
Canadian Patents: http://patents1.ic.gc.ca/intro-e.html
European Patents: http://ep.espacenet.com/
UK Patents (and European, EP prefix): ttp://www.patent.gov.uk/
US Patents: http://www.uspto.gov/patft/-download TIFF viewer from
http://www.alternatiff.com/
World Patents:
via Chemical Abstracts Service: http://www.cas.org/
via Derwent: http://www.derwent.com

Periodic table

Interactive: http://www.lenntech.com/periodic-chart.htm

With this periodic chart you can click on the respective elements and it will show physical and chemical data of each element, plus health and environmental data.

Pharmacological Information

Elsevier/MDL site: http://www.mdl.com

Plastics Production

The Society of the Plastics Industry (U.S.A.): http://www.socplas.org/

Europe, Association of Plastics Manufacturers in (APME): http://www.apme.org/

Polymer Background

http://www.cem.msu.edu/~reusch/VirtualText/polymers.htm#polmr3

Glossary, Polymer Doctor: http://www.polymerdoctor.com/fine/plsql/terminology_main_new

Production Data

OECD Countries: http://www.sourceoecd.org/content/html/index.htm

Petroleum: http://www.hubbertpeak.com/

http://www.oilempire.us/peakoil.html

Publishers, Technical

Academic Press, via Elsevier's Website at: http://www.elsevier.com/wps/find/homepage.cws_home

Beuth Verlag GmbH: http://www.beuth.de/index_en.php

Springer-Verlag GmbH: http://www.springeronline.com/sgw/cda/frontpage/

Pulp and Paper

Australian: http://www.apic.asn.au/statistics/archive/stat9394.htm

Canadian PP Association: http://www.cppa.org/english/

U.K.: http://www.ppic.org.uk/

Radiation

National Radiation Protection Board, U.K.: http://www.nrpb.org/

Recycling

Global Recycling Network: http://grn.com/grn/

Used oil recycling in Canada: http://www.usedoilrecycling.com/index.cfm

U.S.A.: http://www.42explore.com/recycle.htm

Remote Sensing

Canada Centre for Remote Sensing: www.ccrs.nrcan.gc.ca

Francais: www.cct.rncan.gc.ca

Safety, Chemicals

Chemical and Engineering News: www.pubs.acs.org/cen/safety/index.html

Salt, The Salt Institute

http://www.saltinstitute.org/

Science, General

Science Treasure Troves (Weisstein's): http://www.treasure-troves.com

Scientific Societies

American Chemical Society: http://www.chemcenter.org/portal/Chemistry

http://chemistry.org/education/DGRweb

American Institute of Physics: http://www.aip.org/history/
American Vacuum Society: http://www.avs.org/
The Electrochemical Society: http://www.electrochem.org/
International Society of Electrochemistry: http://www.ise-online.org/
Canadian Association of Physicists: http://www.cap.ca/
Canadian Institute of Chemistry: http://www.cheminst.ca/
  http://www.chemistry.ca
Royal Society of Chemistry: http://www.rsc.org/
  Student Contact: http://www.rsc.org/lap/educatio/studentzone.htm

Space Information
NASA: http://search.nasa.gov/nasasearch/search/search.jsp
  http://www.nasa.gov/home/index.html?skipIntro = 1
Man-Systems Integration Standards Information: http://msis.jsc.nasa.gov/
  sections/section05.asp

Statistics, Relevant
Statistics Canada: www.statcan.ca

Surface Science Resources
Royston Paynter's list of Surface Science sites: http://goliath.inrs-ener.
  uquebec.ca/surfsci/links.html (A surface science tutorial)
Surface Science Division of CIC/CAP: http://goliath.inrs-ener.uquebec.ca/
  surfsci/

Tradenames
Plastics http://www.plastics.com/tradenames.php?l = d

Toxicology, Toxicity information
Distributed Structure Searchable Toxicity (DSSTox): http://www.epa.gov/
  nheerl/dsstox/
National Library of Medicine Website: http://toxtown.nlm.nih.gov/#
RTECHS Data Base: http://ccinfoweb.ccohs.ca/rtecs/search.html
(Canadian Centre for Occupational Health and Safety)

Unit Conversions
Via Knoevel Website: http://www.knovel.com/knovel2/crc_handbook.jsp

Universities, Colleges:
http://gateway.uvic.ca/erf/education.html

Water Frequently Asked Questions (FAQ), and a water glossary:
http://www.lenntech.com/Water-Glossary.htm
http://www.lenntech.com/Water-FAQ.htm

Water Quality
International Joint Commission, Canadian Office, Home Page: http://
  www.ijc.org

Water Treatment Systems
American Water Works Association: http://www.awwa.org
Aquasmart Technologies Inc., Burnaby, BC:
  http://www.aquasmart.com/products.html
Atlantic Ultraviolet Corporation: http://www.ultraviolet.com/water/
  water01.htm
Victoria Reservoir, treatment data: http://www.crd.bc.ca/water/

Xinix Disinfection Technologies (ClO_2): http://www.xinix.com/

Zenon (membrane technologies): http://www.zenon.com/markets/markets.shtml

Winemaking and Related

Selection of hundreds of relevant Websites: http://www.ryerson.ca/~dtudor/wine.htm

California Wine Institute: http://www.wineinstitute.org/

Statistics: http://www.wineinstitute.org/communications/statistics/index.htm

World wine production by country: http://www.wineinstitute.org/communications/statistics/keyfacts_worldwineproduction02.htm

World per capita wine consumption by country: http://www.wineinstitute.org/communications/statistics/keyfacts_worldpercapitaconsumption02.htm

World Facts and Figures:

http://www.worldfactsandfigures.com/

3
CONSTANTS, SI UNITS, AND MULTIPLES AND FORMULAS

CONSTANTS

Speed of light in a vacuum, c	2.99793×10^{10} cm/sec	Ideal gas constant, R:	8.2056×10^{-2} L atm/K mole
Faraday's constant, F	9.6487×10^{4} C/g equiv wt.		8.3143 J/K mole
Avogadro's number	6.02252×10^{23} molecules/mole		1.9872 cal/K mole
Ice pont	273.15 K, 0°C	Molar gas volume	22.4136 L/mole (at 0°C, 1 atm)

Seawater is about 3.5 wt. % dissolved solids.

Potable water contains <500 ppm dissolved solids.

One mole = Avogadro's number of atoms, molecules, ions, electrons, etc.

INTERNATIONAL SYSTEM OF UNITS AND MULTIPLES (SYSTÈME INTERNATIONAL D'UNITES, SI)

Fundamental Units:

Quantity	Unit	Abbreviation
length	meter	m
mass	kilogram	kg
time	second	s or sec
electric current	ampere	A or amp
temperature	Kelvin	K
luminous intensity	candela	cd

Prefix Factors:

Multiple or fraction	Prefix	Symbol
10^{12}	tera	T
10^9	giga	G
10^6	mega	M
10^3	kilo	K
10^2	hecto[a]	h
10	deka[a]	da
10^{-1}	deci[a]	d
10^{-2}	centi[a]	c
10^{-3}	milli	m
10^{-6}	micro	μ
10^{-9}	nano	n
10^{-12}	pico	p
10^{-15}	femto	f
10^{-18}	atto	a

[a]These are normally restricted to steps of 10^3 or 10^{-3}, which exclude hecto, deci, and centi. However, these are in common use, so are included here for reference.

FORMULAS

Area		Volume	
Triangle	1/2 bh	cylinder	$\pi r^2 h$
Regular polygon[b]	$nl^2/(4\tan\pi/n)$	Sphere	$4/3\,\pi r^3$
Circle	πr^2		
Sphere	$4\pi r^2$		

Gas Laws	Electrochemical
$pv = nRT$	watts, W = amperes × volts
$\dfrac{p_1 v_1}{T_1} = \dfrac{p_2 v_2}{T_2}$	Coulombs, C = amperes × seconds
	1 g equiv. wt. = formula weight ÷ charge
	1 Faraday, F, deposits 1 g equiv. wt.

[b] n = number of sides, l = length of side.

4
CONVERSION FACTORS, VISCOSITY DATA

From	To	Factor
Length		
miles	m	1.609×10^3
foot	m	0.3048
inches	m	2.54×10^{-2} (exact)
angstroms	m	10^{-10} (exact)
Mass		
ton, long (2240 lb)	kg	1.015×10^3
tonne, metric (1000 kg; 1 Mg)	lb	2.204×10^3
ton, short (2000 lb)	kg	9.07×10^2
pound (avdp)	kg	0.4536
ounce (avdp)	g	28.350
ounce (troy)	g	31.103
Area		
square mile	acres	640
acres	ha (hectares)	0.40469
hectare	m^2	10^4
square foot	m^2	0.0929
Volume		
barrel (bbl)	m^3	0.159
	Imp. gal	34.971
	U.S. gal (liq.)	42 (exact)
cubic foot	m^3	2.832×10^{-2}

(continued)

Conversion Factors, Viscosity Data (continued)

From	To	Factor
Imperial gallon	L	4.5460
U.S. gallon (liq.)	L	3.7854
Imperial gallon	U.S. gal (liq.)	1.20095 (i.e., 6/5)
Pressure		
bar	atmosphere, atm	0.98692
	MPa	0.1
	Pa(Nm^{-2})	10^5
atmosphere	bar	1.01325
	Pa (pascals)	1.013×10^5
	lb/in.^{2}	14.696
pound/in.^{2}	kPa	6.89
Energy, Power, work		
therm	MJ	106
British thermal unit (BTU)	kJ	1.06
calorie	J	4.1840
watt	J/sec	1
watt hour	J	3600
horsepower	W	746
horsepower-hours	kWh	0.746

Some Viscosity Units

	To	Factor
Poise (c.g.s. unit of absolute viscosity),	P = (g/cm $\times$ sec):	
Poise	cP	100
Poise	μP	10^6
Stokes (c.g.s. unit of kinematic viscosity),	S = g/(cm $\times$ sec $\times$ density):	
Stokes	cS	100
Stokes	μS	10^6

Viscosities of Some Common Gases and Liquids

Gases	Micropoises	Liquids	Centipoises
Air at 0°C	170.8	Acetone at 0°C	0.399
Hydrogen at 0°C	83.5	Ethanol at 0°C	1.773
Methane at 0°C	102.6	Hexane at 0°C	0.401
Oxygen at 0°C	189	Octane at 0°C	0.706
Water vapor at 100°C	125.5	Water at 0°C	1.687
at 150°C	144.5	at 20°C	1.002
at 200°C	163.5	at 100°C	0.282

*The very convenient 1.002 cP viscosity of water at 20°C is recommended for use for calibration of viscometers.

■ INDEX

Common Process Flowsheet Symbols

Ball mill

Blower

Classifier,
rake

Crusher,
gyratory

Cyclone
separator

Elevator,
bucket

Feeder,
rotary

Filter press

Filter,
rotary

Flaker

Gas holder

Gauge,
liquid level

Heat
exchanger

Kiln, rotary

Rotameter

Screener

Tower,
packed

Tower,
sieve tray

Turbine

Vessel,
internal coils

PERIODIC TABLE OF THE ELEMENTS

GROUP

IA										
1 1.0079 **H** 1s										

IIA

3 6.941 **Li** 180.5 (He) 2s	4 9.01218 **Be** 1278 (He) 2s^2
11 22.93977 **Na** 97.8 (He) 3s	12 24.305 **Mg** 650 (He) 3s^2

IIIB · **IVB** · **VB** · **VIB** · **VIIB** · **VIII**

19 39.0983 **K** 63.6 (Ar) 4s	20 40.08 **Ca** 848 (Ar) 4s^2	21 44.9559 **Sc** 1541 (Ar) 3d4s^2	22 47.90 **Ti** 1660 (Ar) 3d^{2}4s^2	23 50.9415 **V** 1890 (Ar) 3d^{3}4s^2	24 51.996 **Cr** 1857 (Ar) 3d^{5}4s	25 54.9380 **Mn** 1244 (Ar) 3d^{5}4s^2	26 55.847 **Fe** 1535 (Ar) 3d^{6}4s^2	27 58.9332 **Co** 1495 (Ar) 3d^{7}4s^2	28 58.70 **Ni** 1453 (Ar) 3d^{8}4s^2
37 85.4678 **Rb** 38.9 (Kr) 5s	38 87.62 **Sr** 769 (Kr) 5s^2	39 88.9059 **Y** 1522 (Kr) 4d5s^2	40 91.22 **Zr** 1852 (Kr) 4d^{2}5s^2	41 92.9064 **Nb** 2468 (Kr) 4d^{4}5s	42 95.94 **Mo** 2617 (Kr) 4d^{5}5s	43 (98) **Tc** 2310 (Kr) 4d^{6}5s	44 101.07 **Ru** 1966 (Kr) 4d^{7}5s	45 102.9055 **Rh** 1552 (Kr) 4d^{8}5s	46 106.4 **Pd** (Kr) 4d^{10}
55 132.9054 **Cs** 28.4 (Xe) 6s	56 137.33 **Ba** 725 (Xe) 6s^2	57 138.9055 **La** 921 (Xe) 5d6s^2	72 178.49 **Hf** 2227 (Xe)4f^{14}5d^{2}6s^2	73 180.9479 **Ta** 2996 (Xe)4f^{14}5d^{3}6s^2	74 183.85 **W** 3440 (Xe)4f^{14}5d^{4}6s^2	75 186.207 **Re** 3180 (Xe)4f^{14}5d^{5}6s^2	76 190.2 **Os** 3045 (Xe)4f^{14}5d^{6}6s^2	77 192.22 **Ir** 2410 (Xe)4f^{14}5d^9	78 195.09 **Pt** 1772 (Xe)4f^{14}5d^{9}6s
87 (223) **Fr** 700 (Rn) 7s	88 266.0254 **Ra** 1050 (Rn) 7s^2	89 227.0278 **Ac** (Rn) 6d7s^2							

58 140.12 **Ce** 799 (Xe)4f^{2}6s^2	59 140.9077 **Pr** 931 (Xe) 4f^{3}6s^2	60 144.24 **Nd** 840 (Xe) 4f^{4}6s^2	61 (145) **Pm** 1168 (Xe) 4f^{5}6s^2	62 150.4 **Sm** 1077 (Xe) 4f^{6}6s^2	63 151.96 **Eu** 822 (Xe) 4f^{7}6s^2	64 157.25 **Gd** 1313 (Xe)4f^{7}5d6s^2
90 232.0381 **Th** 1730 (Rn) 6d^{2}7s^2	91 231.0359 **Pa** 1132 (Rn)5f^{2}6d7s^2	92 238.029 **U** 640 (Rn)5f^{3}6d7s^2	93 237.0482 **Np** (Rn)5f^{4}6d7s^2	94 (244) **Pu** 641 (Rn) 5f^{6}7s^2	95 (243) **Am** (Rn) 5f^{7}7s^2	96 (247) **Cm** (Rn)5f^{7}6d7s^2

Atomic number → 26   Atomic weight* → 55.847

Melting point (°C) → 1535   **Fe** ← Element

(Ar) 3d^{6}4s^2 ← Electron configuration